SOME PHYSICAL PROPERTIES

Air (dry, at 20° C and 1 atm)
    Density 1.29 kg/m³
    Specific heat at constant pressure 1.00×10^3 J/kg·K
                                           0.240 cal/gm·K

    Ratio of specific heats (γ) 1.40
    Speed of sound 331 m/s
                                           1090 ft/s

Water (20° C and 1 atm)
    Density 1.00×10^3 kg/m³
                                           1.00 gm/cm³

    Speed of sound 1460 m/s
                                           4790 ft/s

    Index of refraction ($\lambda = 5890\text{Å}$) 1.33
    Specific heat at constant pressure 4180 J/kg·K
                                           1.00 cal/gm·K

    Heat of fusion (0° C) 3.33×10^5 J/kg
                                           79.7 cal/gm

    Heat of vaporization (100° C) 2.26×10^6 J/kg
                                           539 cal/gm

The Earth
    Mass 5.98×10^{24} kg
    Mean radius 6.37×10^6 m
                                           3960 mi

    Mean earth-sun distance 1.49×10^8 km
                                           9.29×10^7 mi

    Mean earth-moon distance 3.80×10^5 km
                                           2.39×10^5 mi

    Standard gravity 9.81 m/s²
                                           32.2 ft/s²

    Standard atmosphere 1.01×10^5 Pa
                                           14.7 lb/in²
                                           760 mm-Hg
                                           29.9 in-Hg

physics

PART I

BOOKS BY HALLIDAY (D.)
Introductory Nuclear Physics. Second Edition

BOOKS BY HALLIDAY (D.) AND RESNICK (R.)
Physics, Parts I and II Combined, Third Edition
Physics, Part II, Third Edition
Fundamentals of Physics

BOOKS BY RESNICK (R.) AND HALLIDAY (D.)
Physics, Part I, Third Edition

BOOKS BY R. RESNICK
Introduction to Special Relativity
Available in Paper Edition
Basic Concepts in Relativity and
Early Quantum Theory
Available in Cloth and Paper Editions

BY ROBERT EISBERG AND ROBERT RESNICK
Quantum Physics of Atoms, Molecules,
Solids, Nuclei, and Particles

physics

PART ONE
THIRD EDITION

PROFESSOR OF PHYSICS **ROBERT RESNICK**
RENSSELAER POLYTECHNIC INSTITUTE

PROFESSOR OF PHYSICS **DAVID HALLIDAY**
UNIVERSITY OF PITTSBURGH

JOHN WILEY & SONS,
NEW YORK   SANTA BARBARA   LONDON   SYDNEY   TORONTO

SUPPLEMENTAL MATERIAL
Student Study Aid

Available for student use with *Physics* as well as with *Fundamentals of Physics* is the *Student Study Guide* by Williams, Brownstein, and Gray. ISBN 94801-2

This study guide is available from John Wiley and Sons.

Cover and text design by Suzanne G. Bennett.
Cover photograph by Dan Lenore.

Library of Congress Cataloging in Publication Data:

Resnick, Robert, 1923–
   Physics.

   Published in 1960 and 1962 under title: Physics for students of science and engineering.
   Includes bibliographical references and index.
   1. Physics.   I. Halliday, David, joint author.
II. Title.
QC21.2.R47 1977   530   77-1295
ISBN 0-471-71716-9

Printed in the United States of America

10 9 8 7 6 5

preface to the third edition of part one

Physics is available in a single volume or in two separate parts; Part I includes mechanics, sound and heat, and Part II includes electromagnetism, optics and quantum physics. The first edition was published in 1960 (*Physics for Students of Science and Engineering*) and the second in 1966 (*Physics*).

The text is intended for students studying calculus concurrently, such as students of science and engineering. The emphasis is on building a strong foundation in the principles of classical physics and on solving problems. Attention is given, however, to practical application, to the most modern theories, and to historical and philosophic issues throughout the book. This is accomplished by inclusion of special sections and thought questions, and by the entire manner of presentation of the material. There is a large set of worked-out examples, interspersed throughout the book, and an extensive collection of problems at the end of each chapter. Much care has been given to pedagogic devices that have proved effective for learning.

It has been eleven years since the publication of the second edition of *Physics*. During that time the book has continued to be well received throughout the world. We have had abundant correspondence with users over those years and concluded that a new edition is now appropriate.

In accordance with the increasing use of metric units in the United States and their general use throughout the world, we have greatly increased the emphasis on the metric system, using the Système Internationale (SI) units and nomenclature throughout. Where it seems to be sensible, in this transition period for the United States, we retain some features of the British Engineering system. To help the student making the transition to the SI to get a physical feeling for its units, we have stressed equivalencies between the two systems, especially in problems and worked-out examples, by frequently presenting the same data in both systems.

The entire book was carefully reviewed for pedagogic improvement, based chiefly on the experience of users and on the most recent scientific literature. As a result, we have rewritten selected areas significantly for improvements in presentation, accuracy, or physics. We have included new worked-out examples for topics or areas needing them. We have modernized all references, added new ones, and have improved many figures for greater clarity. The tables and the appendices have been expanded and updated to give newer data and more information than before. And we have added a supplementary topic on special relativity.

Major improvements have been made in the questions and problems. Overall in Part I there has been a net increase over the second edition of 35% in their number, with 430 out of the total of 1567 being new. The set of questions, now numbering 611 compared with 413 before, covers a wider range of ideas, puts somewhat more stress on current and applied topics, and contains a large increase in up-to-date useful references to the popular scientific literature. We encourage students and teachers to make use of them. As with the questions, most of the previous problems have been retained, though some have been revised for greater clarity. But 225 new tested problems have been added to Part I to improve the coverage of the material and the spread of level for the student and to give the teacher a fresher choice.

To assist students and teachers in organizing and evaluating this large number of problems, 956 now compared with 746 before, we have done several things. First, we have grouped problems within each chapter by section number; namely the first section needed to be covered in order to be able to work out the problem. Then, within each set of section problems, we have arranged the problems in the approximate order of increasing difficulty. Naturally, neither the assignment by section nor by difficulty is absolute, given different ways of solving some problems and different pedagogic values and tastes. Finally, we have coded the illustrations to the problems and have put the answers to the odd-numbered problems right at the end of these problems rather than at the end of the book.

Lastly, we have restyled the physical layout of the book to give it a less crowded appearance than formerly, making it easier now for the student to read the material, to make notations and to differentiate between the various components of each chapter (text, figures, examples, tables, quotes, references, questions, problems, and so forth).

We are grateful to John Wiley and Sons and to Donald Deneck, physics editor, for outstanding cooperation. We acknowledge the valuable assistance of Dr. Edward Derringh with the problem sets and of Mrs. Carolyn Clemente with the wide range of secretarial services required.

We hope that this third edition of *Physics* will contribute to the improvement of physics education.

January 1977
Troy, New York 12181

ROBERT RESNICK
Department of Physics
Rensselaer Polytechnic Institute

Hanover, New Hampshire 03755

DAVID HALLIDAY
3 Clement Road

contents

physics

1
measurement

The building blocks of physics are the physical quantities that we use to express the laws of physics. Among these are length, mass, time, force, velocity, density, resistivity, temperature, luminous intensity, magnetic field strength, and many more. Many of these words, such as length and force, are part of our everyday vocabulary. You might say for example: "I will go to any *length* to satisfy you as long as you do not *force* me to do so." In physics, however, we must define words that we associate with physical quantities, such as force and length, clearly and precisely and we must not confuse them with their everyday meanings. In this example the precise scientific definitions of length and force have no connection at all with the uses of these words in the quoted sentence.

We say that we have defined a physical quantity such as mass, for example, when we have laid down a set of procedures, a recipe if you will, for measuring that quantity and assigning a unit, such as the kilogram, to it. That is, we set up a standard. The procedures are quite arbitrary. We can define the kilogram in any way we want. The important thing is to define it in a useful and practical way, and to obtain international acceptance of the definition.

There are so many physical quantities that it becomes a problem as to how to organize them. They are not independent of each other. For a simple example, a speed is the ratio of a length to a time. What we do is select from all possible physical quantities a certain small number that we choose to call basic, all others being derived from them. We then assign standards to each of these basic quantities and to no others. If, for example, we select length as a basic quantity, we choose a standard called the meter (see Section 1-3) and we define it in terms of precise laboratory operations.

Several questions arise: (a) How many basic quantities should be selected? (b) Which quantities should they be? (c) Who is going to do the selecting?

The answers to the first two questions are that we select the smallest number of physical quantities that will lead to a complete description of physics in the simplest terms. Many choices are possible. In one system, for example, force is a basic quantity. In the system that we select (see Section 1-2) it is a derived quantity.

The answer to the third question depends on international agreement. The International Bureau of Weights and Measures, located near Paris and established in 1875, is the fountainhead for these matters. It maintains contact with standardizing laboratories throughout the world, our own National Bureau of Standards being an example.* Periodically the General Conference on Weights and Measures, an international body, meets and makes its resolutions and recommendations. Its first meeting was in 1889 and its fourteenth meeting was in 1971.

Once you have set up a basic standard, for length say, you must also set up procedures that allow you to measure the length of any object by comparing it with the standard. This means that the standard must be accessible. Also you want within acceptable limits to get the same answer every time you compare the standard with a given object. This means that the standard must be invariable. These two requirements are often incompatible. If you choose length as a basic quantity, define its standard as the distance between a person's nose and the fingertips of the outstretched arm, and assign the yard as the unit, you have a standard that is certainly accessible but it is not invariable. The demands of science and technology steer us just the other way. We achieve accessibility by creating more readily available secondary, tertiary, etc., standards and we strongly stress invariability.

We often make comparisons with a basic standard in very indirect ways. In the case of length consider the problems of measuring: (a) the distance to the Great Nebula in Andromeda, (b) your height, and (c) the distances between nuclei in the molecule NH_3. It is clear that the comparison techniques vary greatly. For example, you cannot use a ruler directly in either the first or third problem.

1-2 THE INTERNATIONAL SYSTEM OF UNITS**

The 14th General Conference on Weights and Measures (1971), building on the work of preceding conferences and international committees, selected as base units the seven quantities displayed in Table 1-1. This is the basis of the International System of Units, abbreviated SI from the French "Le Système International d'Unites."

Throughout the book we will give many examples of SI derived units, such as velocity, force, electric resistance, and so on that follow from Table 1-1. For example, the SI unit of force, called the *newton* (abbreviation N), is defined in terms of the SI base units as

$$1 \ N = 1 \ m \cdot kg/s^2$$

as we will make clear in Chapter 5.

---

* See "Standards of Measurement," Allen V. Astin, *Scientific American*, June 1968, and "A Short History of Measurement Standards at the National Physical Laboratory," H. Barrell, *Contemporary Physics 9*, 205 (1968).

** See NBS Special Publication 330, 1972, "The International System of Units (SI)." Write to the U.S. Government Printing Office, Washington, D.C. 20402.

Table 1-1
SI base units

Quantity	Name	Symbol
Length	meter[a]	m
Mass	kilogram	kg
Time	second	s
Electric current	ampere	A
Thermodynamic temperature	kelvin	K
Amount of substance	mole	mol
Luminous intensity	candela	cd

[a] The officially recommended spelling is "metre." However, many SI supporters in this country prefer "meter," which we adopt. We will also use "liter" in preference to the recommended "litre."

Often if we express physical properties such as the radius of the earth or the time interval between two nuclear events in SI units (base or derived), we end up with very large or very small numbers. For convenience, the 14th General Conference on Weights and Measures, again building on previous work, recommended the prefixes shown in Table 1-2. Thus we can write the mean radius of the earth ($=6.37 \times 10^6$ m) as 6.38 Mm and a time interval of the size often encountered in nuclear physics, 2.35×10^{-9} s say, as 2.35 ns. Prefixes for factors greater than unity have Greek roots; those for factors less than unity have Latin roots (except that femto and atto, recently added, have Danish roots).

Table 1-2
SI prefixes

Factor	Prefix	Symbol	Factor	Prefix	Symbol
10^1	deka	da	10^{-1}	deci	d
10^2	hecto	h	10^{-2}	centi	c
10^3	kilo	k	10^{-3}	milli	m
10^6	mega	M	10^{-6}	micro	μ
10^9	giga	G	10^{-9}	nano	n
10^{12}	tera	T	10^{-12}	pico	p
10^{15}	peta	P	10^{-15}	femto	f
10^{18}	exa	E	10^{-18}	atto	a

To fortify Table 1-1 we need seven sets of operational procedures that tell us how to produce in the laboratory the seven SI base units. We will explore those for length, mass, and time in the next three sections.

Two other major systems of units compete with the International System (SI). One is the Gaussian system, in terms of which much of the literature of physics is expressed. We will not use this system in this book. Appendix G gives conversion factors to SI units.

The second is the British system, still in daily use in this country, Britain, and elsewhere. The basic units, in mechanics, are length (the foot), force (the pound), and time (the second). Again Appendix G gives conversion factors to SI units. We will use SI units in this book except that in mechanics we will sometimes use the British system, especially in the early chapters. The British system is being phased out in Britain in favor of the officially adopted International System. In fact, as of 1970, the countries Ceylon (later renamed Sri Lanka), Gambia, Guyana, Jamaica, Liberia, Malawi, Nigeria, Sierra Leone, and the United States had in common the fact that they had not by that date adopted the metric system (which later emerged as SI), or officially indicated that they intended to do so.*

The first international standard of length was a bar of a platinum-iridium alloy called the standard meter, and was kept at the International Bureau of Weights and Measures. The distance between two fine lines engraved on gold plugs near the ends of the bar, when the bar was held at a temperature of 0°C and supported mechanically in a prescribed way, was defined to be one meter. Historically, the meter was intended to be one ten-millionth of the distance from the north pole to the equator along the meridian line through Paris. However, accurate measurements taken after the standard meter bar was constructed showed that it differs slightly (about 0.023%) from its intended value.

Because the standard meter is not very accessible, accurate master copies of it were made and sent to standardizing laboratories throughout the world. These secondary standards were used to calibrate other, still more accessible, measuring rods. Thus until recently every measuring rod or device derived its authority from the standard meter through a complicated chain of comparisons using microscopes and dividing engines. Since 1959 this statement had also been true for the yard, whose legal definition in this country was adopted in that year to be

$$1 \text{ yard} = 0.9144 \text{ meter (exactly)}$$

which is equivalent to

$$1 \text{ in.} = 2.54 \text{ cm (exactly)}$$

There are several objections to the meter bar as the primary standard of length: It is potentially destructible, by fire or war for example, and it is not very accessible. These are not idle threats. When the British Houses of Parliament burned in 1834 the British standard yard and standard pound were destroyed. The International Bureau of Weights and Measures was established by France as a neutral international zone and was, fortunately, so respected by the Nazis during World War II.

Most important, the accuracy with which the necessary intercomparisons of length can be made by the technique of comparing fine scratches using a microscope is no longer satisfactory for modern science and technology. Evidence of this is suggested by the trifling mid-course corrections required on space missions. If, among other things, we did not know the distance to the moon in meters as a function of time with some precision, these missions would be much more difficult.

The suggestion that the length of a light wave be used as a length standard was first made in 1828 by J. Babinet. The later development of the interferometer (see Chapter 45) provided scientists with a precision optical device in which a light wave can be used as a length comparison probe. Visible light has a wavelength of about 0.5 μm (see Table 1-2) and length measurements of bars of even many centimeters long can be made to a small fraction of a wavelength. An accuracy of 1 part in 10^9 in the intercomparison of lengths using light waves is possible.

---

* See "Conversion to the Metric System," Lord Ritchie-Calder, *Scientific American*, July 1970. The journal *Metric News* (Swani Publishing Company, P.O. Box 248, Roscoe, Illinois 61073) gives up-to-date information about "metrification" problems, as does the *Metric System Guide*—Bulletin (J. J. Keller Associates, 145 W. Wisconsin Avenue, Neenah, Wisconsin 54956).

** See "The Metre," H. Barrell, *Contemporary Physics* 3, 415 (1962).

In 1960 the 11th General Conference on Weights and Measures adopted an atomic standard for the meter. The wavelength in vacuum of a particular orange-red radiation, identified by the spectroscopic notation $2p_{10} - 5d_5$ and emitted by atoms of a particular isotope of krypton, Kr^{86}, in electrical discharge was chosen (see Fig. 1-1). Specifically, one meter is now defined to be 1,650,763.73 wavelengths of this light. This number of wavelengths was arrived at by carefully measuring the length of the standard meter bar in terms of these light waves. This comparison was done so that the new standard, based on the wavelength of light, would be as consistent as possible with the old standard based on the meter bar. The new standard permits length comparisons to a factor of ten better than is possible with the meter bar.

figure 1-1
A Kr^{86} light source shown removed from the container in which it is housed. In operation the lamp is cooled with liquid nitrogen. (Courtesy the National Physical Laboratories, Teddington, England. Crown copyright reserved.)

The choice of an atomic standard offers advantages other than increased precision in length measurements. The Kr^{86} atoms are available everywhere, are identical, and emit light of the same wavelength. The particular wavelength chosen is uniquely characteristic of Kr^{86} and is sharply defined. The isotope can readily be obtained in pure form.

Given the atomic length standard as basic we still need convenient secondary standards calibrated against it for practical use. Often, as in measuring intramolecular or interstellar distances, we cannot make a direct comparison to a standard. We must use indirect methods to relate the distance in question to the primary standard of length. For example, we know the distances to nearby stars because their positions against the background of much more distant stars shift as the earth moves around its orbit. If we measure this angular shift (parallax), and if we know the diameter of the earth's orbit in meters, we can calculate the distance to the nearby star.

Table 1-3 shows some measured lengths. Note that they vary by a factor of about 10^{37}.

Table 1-3
Some measured lengths

Length	Meters
Distance to the nearest galaxy (in Andromeda)	2×10^{22}
Radius of our galaxy	6×10^{19}
Distance to the nearest star (Alpha Centauri)	4.3×10^{16}
Mean orbit radius for our most distant planet (Pluto)	5.9×10^{12}
Radius of the sun	6.9×10^{8}
Radius of the earth	6.4×10^{6}
Height of Mt. Everest	8.9×10^{3}
Height of a typical person	1.8×10^{0}
Thickness of a page in this book	1×10^{-4}
Size of a poliomyelitis virus	1.2×10^{-8}
Radius of a hydrogen atom	5.0×10^{-11}
Effective radius of a proton	1.2×10^{-15}

1-4
THE STANDARD OF MASS

The SI standard of mass is a platinum-iridium cylinder kept at the International Bureau of Weights and Measures and assigned, by international agreement, a mass of one kilogram. Secondary standards are sent to standardizing laboratories in other countries and the masses of other bodies can be found by an equal-arm balance technique to a precision of two parts in 10^{8}.

The U.S. copy of the international standard of mass, known as Prototype Kilogram No. 20, is housed in a vault at the National Bureau of Standards (see Fig. 1-2). It is removed no more than once a year for checking the values of tertiary standards. Since 1889 Prototype No. 20 has been taken to France twice for recomparison with the master kilogram. When it is removed from the vault two people are always present, one to carry the kilogram in a pair of forceps, the second to catch the kilogram if the first person should fall.

Table 1-4 shows some measured masses. Note that they vary by a factor of about 10^{70}. Most masses have been measured in terms of the standard kilogram by indirect methods. For example, we can measure the mass of the earth (see Section 16.3) by measuring in the laboratory the gravitational force of attraction between two lead spheres. Their masses must be known by direct comparison with the standard kilogram, using, say, an equal-arm balance.

On the atomic scale we have a second standard of mass, not an SI unit. It is the mass of the C^{12} atom which, by international agreement, has been assigned an atomic mass of 12 unified atomic mass units (abbreviation u), exactly and by definition. We can find the masses of other atoms to considerable accuracy by using a mass spectrometer. Table 1-5 shows some selected atomic masses, including the probable errors of measurement. We need a second standard of mass because present laboratory techniques permit us to compare atomic masses to each other with greater precision than we can compare them to the standard kilogram. The relationship is approximately

$$1 \text{ u} = 1.661 \times 10^{-27} \text{ kg.}$$

figure 1-2
This is national standard kilogram No. 20 which is kept at the United States National Bureau of Standards. It is an accurate copy of the International standard kept at the International Bureau of Weights and Measures near Paris. The standard kilogram is the platinum cylinder housed under the double bell-jar.

Table 1-4
Some measured masses

Object	Kilograms
Our galaxy	2.2×10^{41}
The sun	2.0×10^{30}
The earth	6.0×10^{24}
The moon	7.4×10^{22}
The waters of the oceans	1.4×10^{21}
An ocean liner	7.2×10^{7}
An elephant	4.5×10^{3}
A person	5.9×10^{1}
A grape	3.0×10^{-3}
A speck of dust	6.7×10^{-10}
A tobacco mosaic virus	2.3×10^{-13}
A penicillin molecule	5.0×10^{-17}
A uranium atom	4.0×10^{-26}
A proton	1.7×10^{-27}
An electron	9.1×10^{-31}

Table 1-5
Some measured atomic masses

Isotope	Mass in Atomic mass units	
H^1	1.00782522	± 0.00000002
C^{12}	12.00000000	(exactly)
Cu^{64}	63.9297568	± 0.0000035
Ag^{102}	101.911576	± 0.000024
Cs^{137}	136.907074	± 0.000005
Pt^{190}	189.959965	± 0.000026
Pu^{238}	238.049582	± 0.000011

1-5
*STANDARD OF TIME**

The measurement of time has two aspects. For civil and for some scientific purposes we want to know the time of day so that we can order events in sequence. In most scientific work we want to know how long an event lasts (the time interval). Thus any time standard must be able to answer the questions "At what time does it occur?" and "How long does it last?" Table 1-6 shows the range of time intervals that can be measured. They vary by a factor of about 10^{40}.

We can use any phenomenon that repeats itself as a measure of time. The measurement consists of counting the repetitions. We could use an oscillating pendulum, a mass spring system, or a quartz crystal, for example. Of the many repetitive phenomena in nature the rotation of

Table 1-6
Some measured time intervals

Time Interval	Seconds
Age of the earth	1.3×10^{17}
Age of the pyramid of Cheops	1.2×10^{11}
Human life expectancy (USA)	2×10^{9}
Time of earth's orbit around the sun (1 year)	3.1×10^{7}
Time of earth's rotation about its axis (1 day)	8.6×10^{4}
Period of a typical earth satellite	5.1×10^{3}
Half-life of the free neutron	7.0×10^{2}
Time between normal heartbeats	8.0×10^{-1}
Period of concert-A tuning fork	2.3×10^{-3}
Half-life of the muon	2.2×10^{-6}
Period of oscillation of 3-cm microwaves	1.0×10^{-10}
Typical period of rotation of a molecule	1×10^{-12}
Half-life of the neutral pion	2.2×10^{-16}
Period of oscillation of a 1-MeV gamma ray (calculated)	4×10^{-21}
Time for a fast elementary particle to pass through a medium-sized nucleus (calculated)	2×10^{-23}

* See "Accurate Measurement of Time," Louis Essen, *Physics Today*, 1960.

the earth on its axis, which determines the length of the day, has been used as a time standard for centuries. It is still the basis of our civil time standard, one (mean solar) second being defined to be 1/86,400 of a (mean solar) day. Time defined in terms of the rotation of the earth is called universal time (UT).

Universal time must be measured by astronomical observations extended over several weeks. Thus we need a good terrestrial clock, calibrated by the astronomical observations. Quartz crystal clocks based on the electrically sustained periodic vibrations of a quartz crystal serve well as secondary time standards. The best of these have kept time for a year with a maximum error of 0.02 s.

One of the most common uses of a time standard is to measure frequencies. In the radio range frequency comparisons to a quartz clock can be made electronically to a precision of at least 1 part in 10^{10} and, indeed, we often need such precision. However this precision is about 100 times greater than that with which a quartz clock itself can be calibrated by astronomical observations. To meet the need for a better time standard, atomic clocks have been developed in several countries, using periodic atomic vibrations as a standard.

A particular type of atomic clock, based on a characteristic frequency associated with the Cs^{133} isotope, has been in continuous operation at the National Physical Laboratory in England since 1955. Figure 1-3 shows a similar clock at the National Bureau of Standards in this country.

figure 1-3
Atomic cesium beam frequency standard at the Boulder laboratories of the National Bureau of Standards.

In 1967 the second based on the cesium clock was adopted as an international standard by the Thirteenth General Conference on Weights and Measures. The second was defined as 9,192,631,770 periods of the particular Cs^{133} transition selected. This action increased the accuracy of time measurements to 1 part in 10^{12}, an improvement over the accuracy associated with astronomical methods of about 10^3. If two cesium clocks are operated at this precision, and if there are no other sources of error, the clocks will differ by no more than one second after running for 6000 years. Even better potential atomic clocks are being studied.

Figure 1-4 shows, by comparison with the cesium clock, variations in the rate of rotation of the earth over nearly a three-year period. Note that the earth's rotation rate is high in summer and low in winter (northern hemisphere) and decreases steadily from year to year. You may ask how we can be sure that the rotating earth and not the cesium clock is at fault. There are two answers. (1) The relative simplicity of the atom compared to the earth leads us to account for any difference between these two timekeepers to the earth. Tidal friction between the water and the land, for example, causes a slowing down of the earth's rotation. Also the seasonal motion of the winds introduces a seasonal variation in the rotation. Other variations may be associated with the melting and refreezing of polar icecaps. (2) The solar system contains other timekeepers such as the orbiting planets and the orbiting moons of the planets. The rotation of the earth shows variations with respect to these, too, which are similar to but less accurately observable than the variations shown in Fig. 1-4.

The time standard can be made available at remote locations by radio transmission.* WWV in Colorado and WWVH in Hawaii, operated by the National Bureau of Standards, are examples of such stations. They broadcast on frequencies of 2.5, 5, 10, 15, 20, and 25×10^6 Hz stabilized to 1 part in 10^{11} by comparison with a cesium clock. One hertz (abbreviation Hz) is 1 cycle/second. At 5-min intervals WWV alternately broadcasts an accurate 440 Hz tone (concert A) and a 600 Hz tone. Ten times per hour it broadcasts time signals using a binary digit coding system.

figure 1-4
Variation in the rate of rotation of the earth as revealed by comparison with a cesium clock. (Adapted from L. Essen, *Physics Today*, July 1960.)

questions

1. How would you criticize this statement: "Once you have picked a standard by the very meaning of 'standard' it is invariable"?

2. Many capable investigators, on the evidence, believe in the reality of extrasensory perception. Assuming that ESP is indeed a fact of nature, what physical quantity or quantities would you seek to define to describe this phenomenon quantitatively?

3. According to a point of view adopted by some physicists and philosophers, if we cannot describe procedures for determining a physical quantity, we say that the quantity is undetectable and should be given up as having no physical reality. Not all scientists accept this view. What in your opinion are the merits and drawbacks of this point of view?

4. Do you think that a definition of a physical quantity for which no method of measurement is given has meaning?

5. List characteristics other than accessibility and invariability that you would consider desirable for a physical standard.

6. Can you imagine a system of base units (Table 1-1) in which time was not included?

7. Of the seven base units listed in Table 1-1 only one, the kilogram, has a prefix (see Table 1-2). Would it be wise to redefine the mass of that platinum-iridium cylinder at the International Bureau of Weights and Measures as one gram, rather than one kilogram?

* See "NBS Time and Frequency Dissemination Services; Special Publication 432," National Bureau of Standards, January, 1976 (write to the U.S. Government Printing Office, Washington, D.C. 20402).

8. Can we define temperature as a derived quantity, in terms of length, mass, and time? Think of a pendulum.

9. The meter was originally intended to be one ten-millionth of the meridian line from the north pole to the equator, passing through Paris. In Section 1-3 we learned that this definition was in disagreement with the standard meter bar by 0.023%. Does this mean that the standard meter bar is inaccurate to this extent?

10. In defining the meter bar as the standard of length why specify its temperature? Can length be called a fundamental quantity if another physical quantity, such as temperature, must be specified in choosing a standard?

11. If someone told you that every dimension of every object had shrunk to half its former value overnight, how could you refute this statement?

12. Can length be measured along a curved line? If so, how?

13. Can you suggest a way to measure (a) the radius of the earth, (b) the distance between the sun and the earth, (c) the radius of the sun?

14. Can you suggest a way to measure (a) the thickness of a sheet of paper, (b) the thickness of a soap bubble film, (c) the diameter of an atom?

15. Why do we find it useful to have two standards of mass, the kilogram and the C^{12} atom?

16. How does one obtain the relation between the mass of the standard kilogram and the mass of the C^{12} atom?

17. Is the current standard kilogram of mass accessible, invariable, reproducible, indestructible? Does it have simplicity for comparison purposes? Would an atomic standard be better in any respect? Why don't we use an atomic standard, as we do for length and time?

18. Suggest practical ways by which one could determine the mass of the various objects listed in Table 1-4.

19. Suggest objects whose mass would fall in the wide range in Table 1-4 between that of an ocean liner and all the water in the oceans and estimate their mass.

20. Name several repetitive phenomena occurring in nature which could serve as reasonable time standards.

21. You could define "one second" to be 1.20 pulse beats of the current president of the American Physical Society. Galileo used a similar definition in some of his work. Putting aside considerations of invariability, why is a definition based on the atomic clock better?

22. What criteria should a good clock satisfy?

23. The time it takes the moon to return to a given position as seen against the background of the fixed stars is called a sidereal month. The time interval between identical phases of the moon is called a lunar month. The lunar month is longer than a sidereal month. Why?

24. From what you know about pendula, cite the drawbacks to using the period of a pendulum as a time standard.

25. Can you think of a way to define a length standard in terms of a time standard or vice versa? Think about a pendulum clock. If so, can length and time both be considered as basic quantities?

26. Critics of the metric system often cloud the issue by saying things such as "Instead of buying one pound of butter you will have to ask for 0.452 kg of butter." The implication is that life would be more complicated. How would you refute this?

SECTION 1-2

1. Use the prefixes in Table 1-2 and express (a) 10^6 phones; (b) 10^{-6} phones; (c) 10^1 cards; (d) 10^9 los; (e) 10^{12} bulls; (f) 10^{-1} mates; (g) 10^{-2} pedes; (h) 10^{-9} Nannettes; (i) 10^{-12} boos; (j) 10^{-18} boys; (k) 2×10^2 withits; (l) 2×10^3 mockingbirds. Now that you have the idea, invent a few more similar expres-

problems

sions. (See, in this connection, p. 61 of *A Random Walk in Science*, edited by R. L. Weber, Crane, Russak, and Co., Inc., New York, 1974).

SECTION 1-3

2. What is your height in meters?

3. Calculate the number of kilometers in 20 miles using only the following conversion factors: 1 mile = 5280 ft, 1 ft = 12 in., 1 in. = 2.54 cm, 1 meter = 100 cm, and 1 km = 1000 meters. *Answer:* 32.2 km.

4. A rocket attained a height of 300 km. What is this distance in miles?

5. (*a*) In track meets both 100 yards and 100 meters are used as distances for dashes. Which is longer? By how many meters is it longer? By how many feet? (*b*) Track and field records are kept for the mile and the so-called metric mile (1500 meters). Compare these distances.
   Answer: (*a*) 100 meters exceeds 100 yards by 8.56 meters or 28.1 feet. (*b*) One mile exceeds one metric mile by 109 m or 358 ft.

6. Astronomical distances are so large compared to terrestrial ones that much larger units of length are used for easy comprehension of the relative distances of astronomical objects. An *astronomical unit* (AU) is equal to the average distance from the earth to the sun, about 92.9×10^6 miles. A *parsec* is the distance at which one astronomical unit would subtend an angle of 1″. A *light-year* is the distance that light, traveling through a vacuum with a speed of 186,000 miles/s, would cover in one year. (*a*) Express the distance from earth to sun in parsecs and in light years. (*b*) Express a light year and a parsec in miles.

7. Master machinists would like to have master gauges (1 in. long, for example) good to 0.0000001 in. Show that the platinum-iridium meter is not measurable to this accuracy but that the Kr^{86} meter is. Use data given in this chapter.
   Answer: Pt-Ir meter bar good to 10^{-7} meter; Kr^{86} standard good to 10^{-9} meter; 10^{-7} in. = 2.5×10^{-9} meter; 10^{-7} meter > 10^{-9} meter.

8. Give the relation between (*a*) a square inch and a square centimeter; (*b*) a square mile and a square kilometer; (*c*) a cubic meter and a cubic centimeter; (*d*) a square foot and a square yard.

9. Assume that the average distance of the sun from the earth is 400 times the average distance of the moon from the earth. Now consider a total eclipse of the sun and state conclusions that can be drawn about (*a*) the relation between the sun's diameter and the moon's diameter; (*b*) the relative volumes of the sun and the moon. (*c*) Find the angle intercepted at the eye by a dime that just eclipses the full moon and from this experimental result and the given distance between the moon and the earth (= 3.80×10^5 km) estimate the diameter of the moon.
   Answer: (*a*) $d_{sun}/d_{moon} = 400$. (*b*) $V_{sun}/V_{moon} = 6.4 \times 10^7$. (*c*) 3.5×10^3 km.

SECTION 1-4

10. Using appropriate conversions and data in the chapter, determine the number of hydrogen atoms (isotope number 1) required to obtain one kilogram of mass.

11. If you remember Avogadro's number, you can think of the mass of the earth as being 10 moles of kilograms. What does this statement mean, and how accurate is it? The actual mass of the earth is 5.98×10^{24} kg.
    Answer: Error = 0.67%.

12. (*a*) Assuming that the density (mass/volume) of water is exactly one gram per cubic centimeter, express the density of water in kilograms per liter. (*b*) Suppose that it takes exactly 10 hours to drain a container of 1.00 liter of water. What is the average mass flow rate, in kilograms per second, of water from the container?

13. A convenient substitution for the number of seconds in a year is $\pi \times 10^7$. To within what percentage error is this correct?   *Answer: −0.44%.*

14. (a) A unit of time sometimes used in microscopic physics is the *shake*. One shake equals 10^{-8} s. Are there more shakes in a second than there are seconds in a year? (b) Mankind has existed for about 10^6 years, whereas the universe is about 10^{10} years old. If the age of the universe is taken to be one day, for how many seconds has mankind existed?

15. The maximum speeds of various animals are given roughly as follows in miles per hour: (a) snail, 3×10^{-2}; (b) spider, 1.2; (c) squirrel, 12; (d) man, 28; (e) rabbit, 35; (f) fox, 42; (g) lion, 50; and (h) cheetah, 70. Convert these data to meters per second.
Answer: (a) 0.013. (b) 0.54. (c) 5.4. (d) 13. (e) 16. (f) 19. (g) 22. (h) 31 m/s.

16. From Fig. 1-2 calculate by what length of time the earth's rotation period in midsummer differs from that in the following spring.

17. Five clocks are being tested in a laboratory. Exactly at noon, as determined by the WWV time signal, on the successive days of a week the clocks read as follows:

Clock	Sun.	Mon.	Tues.	Wed.	Thurs.	Fri.	Sat.
A	12:36:40	12:36:56	12:37:12	12:37:27	12:37:44	12:37:59	12:38:14
B	11:59:59	12:00:02	11:59:57	12:00:07	12:00:02	11:59:56	12:00:03
C	15:50:45	15:51:43	15:52:41	15:53:39	15:54:37	15:55:35	15:56:33
D	12:03:59	12:02:52	12:01:45	12:00:38	11:59:31	11:58:24	11:57:17
E	12:03:59	12:02:49	12:01:54	12:01:52	12:01:32	12:01:22	12:01:12

How would you arrange these five clocks in the order of their relative value as good timekeepers? Justify your choice.
Answer: C, D, A, B, E (best to worst). The important criterion is the constancy of the daily variation, not its magnitude.

18. Assuming that the length of the day uniformly increases by 0.001 s in a century, calculate the cumulative effect on the measure of time over twenty centuries. Such a slowing down of the earth's rotation is indicated by observations of the occurrences of solar eclipses during this period.

19. Express the speed of light, 3×10^8 m/s, in (a) feet/nanosecond and (b) in millimeters/picosecond.   *Answer: (a) 0.98 ft/ns. (b) 0.3 mm/ps.*

20. An astronomical unit (AU) is the average distance of the earth from the sun, approximately 149,000,000 km. The speed of light is about 3.0×10^8 m/s. Express the speed of light in terms of astronomical units per minute.

21. A certain spaceship has a speed of 18,600 mi/h. What is its speed in light-years per century? A light-year is the distance light travels in one year with a speed of 186,000 mi/s.   *Answer: 2.8×10^{-3} light-years/century.*

22. (a) The radius of the proton is about 10^{-15} m; the radius of the observable universe is about 10^{28} cm. Identify a physically meaningful distance which is approximately halfway between these two extremes on a logarithmic scale. (b) The mean life of a neutral pion (an elementary particle) is about 2×10^{-16} s. The age of the universe is about 4×10^9 years. Identify a physically meaningful time interval that is approximately halfway between these two extremes on a logarithmic scale.

2
vectors

A change of position of a particle is called a *displacement*. If a particle moves from position A to position B (Fig. 2-1a), we can represent its displacement by drawing a line from A to B; the direction of displacement can be shown by putting an arrowhead at B indicating that the displacement was *from A to B*. The path of the particle need not necessarily be a straight line from A to B; the arrow represents only the net effect of the motion, not the actual motion.

In Fig. 2-1b, for example, we plot an actual path followed by a particle from A to B. The path is not the same as the displacement AB. If we were to take snapshots of the particle when it was at A and, later, when

2-1
VECTORS AND SCALARS

figure 2-1
Displacement vectors. *(a)* Vectors AB and $A'B'$ are identical since they have the same length and point in the same direction. *(b)* The actual *path* of the particle in moving from A to B may be the curve shown; the *displacement* remains the vector AB. At some intermediate point P the displacement from A is the vector AP. *(c)* After displacement AB the particle undergoes another displacement BC. The net effect of the two displacements is represented by the vector AC.

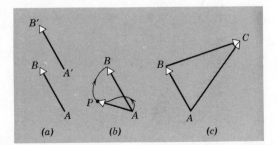

it was at some intermediate position P, we could obtain the displacement vector AP, representing the net effect of the motion during this interval, even though we would not know the actual path taken between these points. Furthermore, a displacement such as $A'B'$ (Fig. 2-1a), which is parallel to AB, similarly directed, and equal in length to AB, represents the same *change* in position as AB. We make no distinction between these two displacements. A displacement is therefore characterized by a *length* and a *direction*.

In a similar way, we can represent a subsequent displacement from B to C (Fig. 2-1c). The net effect of the two displacements will be the same as a displacement from A to C. We speak then of AC as the *sum* or *resultant* of the displacements AB and BC. Notice that this sum is not an algebraic sum and that a number alone cannot uniquely specify it.

Quantities that behave like displacements are called *vectors*.* Vectors, then, are quantities that have both magnitude and direction and combine according to certain rules of addition. These rules are stated below. The displacement vector is a convenient prototype. Some other physical quantities which are vectors are force, velocity, acceleration, the electric field, and the magnetic field. Many of the laws of physics can be expressed in compact form using vectors; derivations involving these laws are often greatly simplified if we do this.

Quantities that can be completely specified by a number and unit and that therefore have magnitude only are called *scalars*. Some physical quantities which are scalars are mass, length, time, density, energy, and temperature. Scalars can be manipulated by the rules of ordinary algebra.

To represent a vector on a diagram we draw an arrow. We choose the length of the arrow proportional to the magnitude of the vector (that is, we choose a scale), and we choose the direction of the arrow to be the direction of the vector, with the arrowhead giving the sense of the direction. For example, a displacement of 40 ft north of east on a scale of 1.0 in. per 10 ft would be represented by an arrow 4.0 in. long, drawn at an angle of 45° above a line pointing east with the arrowhead at the top right extreme. A vector such as this is represented conveniently in printing by a boldface symbol such as **d.** In handwriting it is convenient to put an arrow above the symbol to denote a vector quantity, such as $\vec{d}$.

Often we shall be interested only in the magnitude of the vector and not in its direction. The magnitude of **d** may be written as $|\mathbf{d}|$, called the absolute value of **d;** more frequently we represent the magnitude alone by the italic letter symbol d. The boldface symbol is meant to signify both properties of the vector, magnitude and direction.

Consider now Fig. 2-2 in which we have redrawn and relabeled the vectors of Fig. 2-1c. The relation among these displacements (vectors) can be written as

$$\mathbf{a} + \mathbf{b} = \mathbf{r}. \tag{2-1}$$

The rules to be followed in performing this (vector) addition geometrically are these: On a diagram drawn to scale lay out the displacement

2-2
ADDITION OF VECTORS, GEOMETRICAL METHOD

figure 2-2
The vector sum $\mathbf{a} + \mathbf{b} = \mathbf{r}$. Compare with Fig. 2-1c.

---

* The word *vector* means *carrier* in Latin, which suggests a displacement. You might want to review what your analytic geometry and calculus text says about vectors. A good reference that explores the matter in depth is *About Vectors*, by Banesh Hoffman, Prentice-Hall, Englewood Cliffs, N.J., 1966.

vector **a**; then draw **b** with its tail at the head of **a**, and draw a line from the tail of **a** to the head of **b** to construct the vector sum **r**. This is a displacement equivalent in length and direction to the successive displacements **a** and **b**. This procedure can be generalized to obtain the sum of any number of successive displacements.

Since vectors are new quantities, we must expect new rules for their manipulation. The symbol "+" in Eq. 2-1 simply has a different meaning from its meaning in arithmetic or scalar algebra. It tells us to carry out a different set of operations.

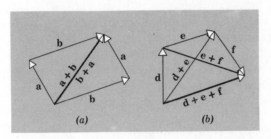

figure 2-3
(a) The commutative law for vector sums, which states that **a** + **b** = **b** + **a**. *(b)* The associative law, which states that **d** + (**e** + **f**) = (**d** + **e**) + **f**.

Using Fig. 2-3 we can prove two important properties of vector addition:

$$\mathbf{a} + \mathbf{b} = \mathbf{b} + \mathbf{a}, \qquad \text{(commutative law)} \qquad (2\text{-}2)$$

and

$$\mathbf{d} + (\mathbf{e} + \mathbf{f}) = (\mathbf{d} + \mathbf{e}) + \mathbf{f}. \qquad \text{(associative law)} \qquad (2\text{-}3)$$

These laws assert that it makes no difference in what order or in what grouping we add vectors; the sum is the same. In this respect, vector addition and scalar addition follow the same rules.

The operation of subtraction can be included in our vector algebra by defining the negative of a vector to be another vector of equal magnitude but opposite direction. Then

$$\mathbf{a} - \mathbf{b} = \mathbf{a} + (-\mathbf{b}) \qquad (2\text{-}4)$$

as shown in Fig. 2-4.

Remember that, although we have used displacements to illustrate these operations, the rules apply to *all* vector quantities.

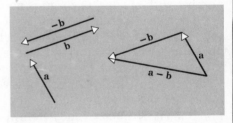

figure 2-4
The vector difference **a** − **b** = **a** + (−**b**).

The geometrical method of adding vectors is not very useful for vectors in three dimensions; often it is even inconvenient for the two-dimensional case. Another way of adding vectors is the analytical method, involving the resolution of a vector into components with respect to a particular coordinate system.

Figure 2-5*a* shows a vector **a** whose tail has been placed at the origin of a rectangular coordinate system. If we drop perpendicular lines from the head of **a** to the axes, the quantities a_x and a_y so formed are called the *components* of the vector **a**. The process is called resolving a vector into its components. Figure 2-5 shows a two-dimensional case for convenience; the extension of our conclusions to three dimensions will be clear.

A vector may have many sets of components. For example, if we rotate the *x*-axis and *y*-axis in Fig. 2-5*a* by 10° counterclockwise, the components of **a** would be different. Furthermore, we may use a nonrec-

2-3
RESOLUTION AND ADDITION OF VECTORS, ANALYTIC METHOD

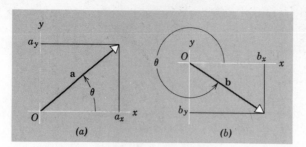

figure 2-5
Two examples of the resolution of a vector into its scalar components in a particular coordinate system.

tangular coordinate system, that is, the angle between the two axes need not be 90°. Thus the components of a vector are only uniquely specified if we specify the particular coordinate system being used. The vector need not be drawn with its tail at the origin of the coordinate system to find its components—although we have done so for convenience; the vector may be moved anywhere in the coordinate space and, as long as its angles with the coordinate directions are maintained, its components will be unchanged.

The components a_x and a_y in Fig. 2-5a are readily found from

$$a_x = a \cos \theta \quad \text{and} \quad a_y = a \sin \theta, \quad (2\text{-}5)$$

where θ is the angle that the vector **a** makes with the positive x-axis, measured counterclockwise from this axis. Note that, depending on the angle θ, a_x and a_y can be positive or negative. For example, in Fig. 2-5b, b_y is negative and b_x is positive. The components of a vector behave like scalar quantities because, in any particular coordinate system, only a number with an algebraic sign is needed to specify them.

Once a vector is resolved into its components, the components themselves can be used to specify the vector. Instead of the two numbers a (magnitude of the vector) and θ (direction of the vector relative to the x-axis), we now have the two numbers a_x and a_y. We can pass back and forth between the description of a vector in terms of its components a_x, a_y and the equivalent description in terms of magnitude and direction a and θ. To obtain a and θ from a_x and a_y, we note from Fig. 2-5a that

$$a = \sqrt{a_x{}^2 + a_y{}^2} \quad (2\text{-}6a)$$

and

$$\tan \theta = a_y/a_x. \quad (2\text{-}6b)$$

The quadrant in which θ lies is determined from the signs of a_x and a_y.

When resolving a vector into components it is sometimes useful to introduce a vector of unit length in a given direction. Thus vector **a** in Fig. 2-6a may be written, for example, as

$$\mathbf{a} = \mathbf{u}_a a, \quad (2\text{-}7)$$

where $\mathbf{u}_a$ is a *unit vector* in the direction of **a**. Often it is convenient to

figure 2-6
(a) The vector **a** may be written as $\mathbf{u}_a a$ in which $\mathbf{u}_a$ is a unit vector in the direction of **a**. (b) The unit vector **i**, **j**, and **k**, used to specify the positive x-, y-, and z-directions respectively.

draw unit vectors along the particular coordinate axes chosen. In the rectangular coordinate system the special symbols **i**, **j**, and **k** are usually used for unit vectors in the positive x-, y-, and z-directions, respectively; see Fig. 2-6b. Note that **i**, **j**, and **k** need not be located at the origin. Like all vectors, they can be translated anywhere in the coordinate space as long as their directions with respect to the coordinate axes are not changed.

The vectors **a** and **b** of Fig. 2-5 may be written in terms of their components and the unit vectors as

$$\mathbf{a} = \mathbf{i}a_x + \mathbf{j}a_y \qquad (2\text{-}8a)$$

and

$$\mathbf{b} = \mathbf{i}b_x + \mathbf{j}b_y; \qquad (2\text{-}8b)$$

see Fig. 2-7. The vector relation Eq. 2-8a is equivalent to the scalar relations of Eq. 2-6; each equation relates the vector (**a**, or a and θ) to its components (a_x and a_y). Sometimes we will call quantities such as $\mathbf{i}a_x$ and $\mathbf{j}a_y$ in Eq. 2-8a the *vector components* of **a**; they are drawn as vectors in Fig. 2-7a. The word *component* alone will continue to refer to the scalar quantities a_x and a_y.

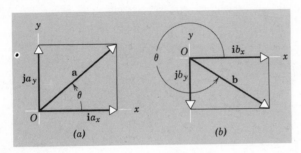

(a)   (b)

figure 2-7
Two examples of the resolution of a vector into its vector components in a particular coordinate system; compare with Figure 2-5.

We now consider the addition of vectors by the analytical method. Let **r** be the sum of the two vectors **a** and **b** lying in the x-y plane, so that

$$\mathbf{r} = \mathbf{a} + \mathbf{b}. \qquad (2\text{-}9)$$

In a given coordinate system, two vectors such as **r** and **a** + **b** can only be equal if their corresponding components are equal, or

$$r_x = a_x + b_x \qquad (2\text{-}10a)$$

and

$$r_y = a_y + b_y. \qquad (2\text{-}10b)$$

These two algebraic equations, taken together, are equivalent to the single vector relation Eq. 2-9. From Eqs. 2-6 we may find r and the angle θ that **r** makes with the x-axis; that is,

$$r = \sqrt{r_x{}^2 + r_y{}^2}$$

and

$$\tan\theta = r_y/r_x.$$

Thus we have the following analytic rule for adding vectors: Resolve each vector into its components in a given coordinate system; the algebraic sum of the individual components along a particular axis is the component of the sum vector along that same axis; the sum vector can be reconstructed once its components are known. This method for adding vectors may be generalized to many vectors and to three dimensions (see Problems 13 and 18).

The advantage of the method of breaking up vectors into components, rather than adding directly with the use of suitable trigonometric relations, is that we always deal with right triangles and thus simplify the calculations.

In adding vectors by the analytical method, the choice of coordinate axes determines how simple the process will be. Sometimes the components of the vectors with respect to a particular set of axes are known to begin with, so that the choice of axes is obvious. Other times a judicious choice of axes can greatly simplify the job of resolution of the vectors into components. For example, the axes can be oriented so that at least one of the vectors lies parallel to an axis.

EXAMPLE 1

An airplane travels 130 miles (= 209 km) on a straight course making an angle of 22.5° east of due north. How far north and how far east did the plane travel from its starting point?

We choose the positive x-direction to be east and the positive y-direction to be north. Next (Fig. 2-8) we draw a displacement vector from the origin (starting point), making an angle of 22.5° with the y-axis (north) inclined along the positive x-direction (east). The length of the vector is chosen to represent a magnitude of 130 miles. If we call this vector **d,** then d_x gives the distance traveled east of the starting point and d_y gives the distance traveled north of the starting point. We have

$$\theta = 90.0° - 22.5° = 67.5°,$$

so that (see Eqs. 2-5)

$$d_x = d \cos \theta = (130 \text{ miles}) \cos 67.5° = 50.0 \text{ miles } (= 80.5 \text{ km}),$$

and

$$d_y = d \sin \theta = (130 \text{ miles}) \sin 67.5° = 120 \text{ miles } (= 193 \text{ km}).$$

figure 2-8
Example 1

EXAMPLE 2

An automobile travels due east on a level road for 30 km. It then turns due north at an intersection and travels 40 km before stopping. Find the resultant displacement of the car.

We choose a coordinate system fixed with respect to the earth, with the positive x-direction pointing east and the positive y-direction pointing north. The two successive displacements, **a** and **b,** are then drawn as shown in Fig. 2-9. The resultant displacement **r** is obtained from **r** = **a** + **b.** Since **b** has no x-component and **a** has no y-component, we obtain (see Eqs. 2-10)

$$r_x = a_x + b_x = 30 \text{ km} + 0 = 30 \text{ km},$$
$$r_y = a_y + b_y = 0 + 40 \text{ km} = 40 \text{ km}.$$

The magnitude and direction of **r** are then (see Eqs. 2-6)

$$r = \sqrt{r_x^2 + r_y^2} = \sqrt{(30 \text{ km})^2 + (40 \text{ km})^2} = 50 \text{ km},$$

$$\tan \theta = r_y/r_x = \frac{40 \text{ km}}{30 \text{ km}} = 1.33, \qquad \theta = \tan^{-1} (1.33) = 53°.$$

The resultant vector displacement **r** has a magnitude of 50 km and makes an angle of 53° north of east.

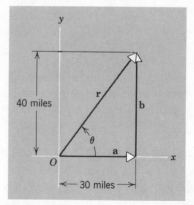

figure 2-9
Example 2

EXAMPLE 3

Three coplanar vectors are expressed, with respect to a certain rectangular coordinate system, as

$$\mathbf{a} = 4\mathbf{i} - \mathbf{j},$$

$$\mathbf{b} = -3\mathbf{i} + 2\mathbf{j},$$

and

$$\mathbf{c} = -3\mathbf{j},$$

in which the components are given in arbitrary units. Find the vector $\mathbf{r}$ which is the sum of these vectors.

From Eqs. 2-10 we have

$$r_x = a_x + b_x + c_x = 4 - 3 + 0 = 1,$$

and

$$r_y = a_y + b_y + c_y = -1 + 2 - 3 = -2.$$

Thus

$$\mathbf{r} = \mathbf{i}r_x + \mathbf{j}r_y$$

$$= \mathbf{i} - 2\mathbf{j}.$$

Figure 2-10 shows the four vectors. From Eqs. 2-6 we can calculate that the magnitude of $\mathbf{r}$ is $\sqrt{5}$ and that the angle that $\mathbf{r}$ makes with the positive x-axis, measured counterclockwise from that axis, is

$$\tan^{-1}(-2/1) = 297°.$$

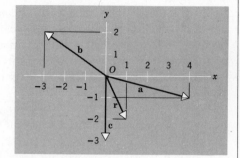

figure 2-10
Three vectors, $\mathbf{a}$, $\mathbf{b}$, and $\mathbf{c}$, and their vector *sum* $\mathbf{r}$.

2-4
*MULTIPLICATION OF VECTORS**

We have assumed in the previous discussion that the vectors being added together are of like kind; that is, displacement vectors are added to displacement vectors, or velocity vectors are added to velocity vectors. Just as it would be meaningless to add together scalar quantities of different kinds, such as mass and temperature, so it would be meaningless to add together vector quantities of different kinds, such as displacement and electric field.

However, like scalars, vectors of different kinds can be multiplied by one another to generate quantities of new physical dimensions. Because vectors have direction as well as magnitude, vector multiplication cannot follow exactly the same rules as the algebraic rules of scalar multiplication. We must establish new rules of multiplication for vectors.

We find it useful to define three kinds of multiplication operations for vectors: (1) multiplication of a vector by a scalar, (2) multiplication of two vectors in such a way as to yield a scalar, and (3) multiplication of two vectors in such a way as to yield another vector. There are still other possibilities, but we shall not consider them here.

The multiplication of a vector by a scalar has a simple meaning: The product of a scalar k and a vector $\mathbf{a}$, written $k\mathbf{a}$, is defined to be a new vector whose magnitude is k times the magnitude of $\mathbf{a}$. The new vector has the same direction as $\mathbf{a}$ if k is positive and the opposite direction if k is negative. To divide a vector by a scalar we simply multiply the vector by the reciprocal of the scalar.

When we multiply a vector quantity by another vector quantity, we must distinguish between the *scalar* (or *dot*) *product* and the *vector*

---

* The material of this section will be used later in the text. The scalar product is used first in Chapter 7 and the vector product in Chapter 11. The instructor who wishes to postpone this section can do so. Its presentation here gives a unified treatment of vector algebra and serves as a convenient reference for later work.

(or *cross*) *product.* The *scalar product* of two vectors **a** and **b**, written as **a · b,** is defined to be

$$\mathbf{a} \cdot \mathbf{b} = ab \cos \phi, \qquad (2\text{-}11)$$

where a is the magnitude of vector **a**, b is the magnitude of vector **b**, and $\cos \phi$ is the cosine of the (smaller) angle ϕ between the two vectors* (see Fig. 2-11).

Since a and b are scalars and $\cos \phi$ is a pure number, the scalar product of two vectors is a scalar. The scalar product of two vectors can be regarded as the product of the magnitude of one vector and the component of the other vector in the direction of the first. Because of the notation **a · b** is also called the dot product of **a** and **b** and is spoken as "a dot **b.**"

We could have defined **a · b** to be any operation we want, for example, to be $a^{1/3}b^{1/4} \tan (\phi/2)$, but this would turn out to be of no use to us in physics. With our definition of the scalar product, a number of important physical quantities can be described as the scalar product of two vectors. Some of them are mechanical work, gravitational potential energy, electrical potential, electric power, and electromagnetic energy density. When such quantities are discussed later, their connection with the scalar product of vectors will be pointed out.

The *vector product* of two vectors **a** and **b** is written as **a × b** and is another vector **c,** where **c = a × b.** The *magnitude* of **c** is defined by

$$c = ab \sin \phi, \qquad (2\text{-}12)$$

where ϕ is the (smaller) angle* between **a** and **b.**

The *direction* of **c,** the vector product of **a** and **b,** is defined to be perpendicular to the plane formed by **a** and **b.** To specify the sense of the vector **c** we must refer to Fig. 2-12. Imagine rotating a right-handed screw whose axis is perpendicular to the plane formed by **a** and **b** so as to turn it *from* **a** *to* **b** through the angle ϕ between them. Then the direction of advance of the screw gives the direction of the vector product **a × b** (Fig. 2-12*a*). Another convenient way to obtain the direction of a vector product is the following. Imagine an axis perpendicular to the plane of **a** and **b** through their origin. Now wrap the fingers of the *right hand* around this axis and push the vector **a** into the vector **b** through the smaller angle between them with the fingertips, keeping the thumb erect; the direction of the erect thumb then gives the direction of the vector product **a × b** (Fig. 2-12*b*).† Because of the notation, **a × b** is also called the cross product of **a** and **b** and is spoken as "a cross **b.**"

Notice that **b × a** is not the same vector as **a × b,** so that the order of factors in a vector product is important. This is not true for scalars because the order of factors in algebra or arithmetic does not affect the resulting product. Actually, **a × b = −b × a** (Fig. 2-12*c*). This can be deduced from the fact that the magnitude $ab \sin \phi$ equals the magnitude

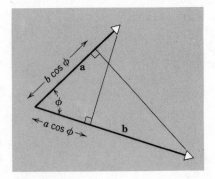

figure 2-11
The scalar product **a · b** $(= ab \cos \phi)$ is the product of the magnitude of either vector $(a$, say) by the component of the other vector in the direction of the first vector $(b \cos \phi$, say).

---

* There are two different angles between a pair of vectors, depending on the sense of rotation. We always choose the *smaller* of the two in vector multiplication. In Eq. 2-11 it does not matter because $\cos(2\pi - \phi) = \cos \phi$. But in Eq. 2-12 it does matter because $\sin(2\pi - \phi) = -\sin \phi$.

† The procedures described in Fig. 2-12 are a convention. Two vectors such as **a** and **b** form a plane and there are *two* directions that point away from any plane. We choose the right hand or right-handed screw convention; choosing the left hand or a left-handed screw would have led to the opposite choice for the direction of **a × b.**

COLLEGE SKI WEEKS

COME SKI WHAT YOU'VE BEEN MISSING!

Student Ski Association, the world's largest collegiate ski organization, invites you to celebrate our tenth anniversary by skiing **Stowe, Mt. Snow, Bromley/Stratton/Magic** and the **Mt. Washington Valley.** Join us for a week of great skiing, parties and lots of fun. You'll meet students from across the country, so come solo or with friends.

New this year! To celebrate our 10th Anniversary, we've added **two new destinations: Stowe,** and **Bromley/Stratton/Magic.** We've also added **two new trip dates:** a four day Thanksgiving trip and a 5 day pre-Christmas week, both at **Stowe, Vermont!** Like our famous January ski weeks, lift lines are short and slopes are uncrowded. Be a part of it all! Last year students from hundreds of schools were on hand.

ACTIVITIES: With attendance ranging from 100-500 per trip, SSA's **free** smorgasbord of activities will provide a great opportunity for you to mix and mingle, meet new people or enjoy old friends. Included with the events and activities listed on the front are disco dancing and live rock bands and ski movies. SSA's professional staff will assist you, but won't get in your way.

LODGING & OPTIONAL MEAL PLAN: SSA prides itself on selecting only **quality lodges** which are clean and comfortable. Unlike our competitors who use crowded bunk rooms, SSA chooses **1st class lodging.** Most average 4 students per room in double and single beds with private baths. Rooms for two or three available at additional supplement (see prices listed underneath trip prices/dates chart).
Some trips also offer luxury condominiums with fully-equipped kitchenettes, perfect for meals — a real money saver! The average condo comes with 2-3 bedrooms, living room with sleeper sofa and fireplace, 1-2 baths. and accommodates 6-10 people. Our standard package is based on quad occupancy in quality motel rooms. Drive to the restaurant of your choice or save money and time by taking advantage of our special discount meal plan. **Note: Cooking in motel rooms is not allowed.**
SSA offers an **unbeatable** optional meal plan on every trip: Only $7.75 a day (plus 15% tax and gratuities) for a full breakfast and delicious dinner, served in your own lodge. **Sample menu:** Breakfast of eggs, juice, toast and coffee. Dinners of soup, salad, rolls, beverage and dessert, complete with main course of roast beef, spaghetti or lasagne, turkey, chicken and meatloaf. Available on weekly basis only. Our most popular lodging plan must be prepaid, so **sign up early!**

STOWE, VERMONT. Exciting new trip — an SSA exclusive! Student Ski is pleased to welcome the "Aspen of the East" to it s College Ski Weeks program. Centered around a charming, nineteenth-century Vermont town, Stowe is the ski capital of the East. With a 2,050 foot vertical drop, it's Vermont's highest peak. Expert skiers will find an abundance of challenging terrain comparing favorably with Aspen Mountain, Sun Valley, Taos or other western areas. Intermediate and novice skiers will delight with Stowe's well-groomed slopes.
Our lodges will include the Alpine Lodge, Salzburg Inn (complete with indoor swimming facilities), Stowe Motel and the Snowdrift. You can sign-up for the standard package which includes 5 night's quality lodging, 5 day's lift tickets and full schedule of FREE activities, or opt for a luxury condominium/efficiency at Holm's or Snowdrift, complete with fully-equipped kitchenettes.
November Thanksgiving trip includes 4 night's lodging and 4 day's lift tickets plus activities. Prices start as low as $99 complete. The skiing is fabulous and the night life is the best in the east! Indoor tennis, ice skating and cross-country skiing are also available.

BROMLEY/STRATTON/MAGIC, VERMONT. SSA now brings together three great southern Vermont ski areas. **Stratton** offers something for everyone from steep mogul runs to well-groomed, wide-open slopes just perfect for carving long, gradual parallel turns. **Bromley,** one of Vermont's oldest areas, has extensive snowmaking facilities that ensure reliable snow cover. It's southern exposure enables skiers to enjoy warmer, sunnier days than most other eastern ski areas. With 41 years of experience, Bromley is considered the **"Ski Vacation Center of Vermont".** Two mile-long chair lifts whisk you to the top of **Magic Mountain** where you can choose from 24 trails: some tight and tough; some wide and gentle; all challenging and immensely satisfying. Well protected from the wind and immaculately groomed, Magic is one of **Vermont's best bargains.** The Valley nightlife is plentiful with disco dancing at famous Haig's and dancing to live rock music at Alfie's Roundhouse. SSA's superb new package includes 5 night's lodging, 5 day's lift tickets at Bromley or Magic with the option of converting to any number of days at Stratton ($7 supplement per day) plus admission to all SSA sponsored activities, **just $99 complete!** Lodging includes the Kandahar, Red Sled, Toll Road and Bromley Sun Lodge located right at the base of the mountain.

MT. WASHINGTON VALLEY, NEW HAMPSHIRE. With spectacular White Mountain scenery and forests of pine surrounding the unspoiled village of North Conway, it's easy to see why this destination area has become one of our most popular trips! **4 great mountains** on one interchangeable lift ticket, plenty of nightlife, cross country skiing, a wide range of activities and condominium lodging right on the ski slopes make this trip an experience you'll long remember.
Our lodges include the River Run Condominiums, Crestwood, Oxen Yoke and Beechwood. This years' complete package price includes motel room, 5 days' skiing at any of the 4 mountains (Wildcat, Cranmore, Attitash or Black), and all SSA sponsored activities for **just $99.**

MOUNT SNOW, VERMONT. Join us for our most popular trip to one of New England's largest resorts with 15 different lifts servicing 50 slopes and trails. Beginners and intermediates will enjoy the great variety of terrain and the highly-rated Mount Snow Ski School. Don't worry about conditions — Mount Snow has installed extensive snowmaking from top to bottom, including the expert North Face. Enjoy cross country skiing? There are miles of trails to explore. Low cost equipment rentals and lessons also available. At night, the Valley really lights up with an assortment of discos and live-entertainment watering spots. We'll be using only top-notch lodging including the Sitzmark, Nutmeg Inn, Mountaineer Lodge, Viking, Nordic Hills, Old Red Mill and other. Leave it to SSA to come up with this **unbeatable package** of 5 days' life tickets, 5 nights' lodging and a full schedule of free activities, all for **only $109, complete!**

DISCOUNT TRANSPORTATION. Roundtrip bus transportation and ski area shuttle service available from major cities, whenever student numbers warrant. Call our hotline at (203) 227-8764 for details and reservations. Or check appropriate box on coupon and we'll mail you the information. Approximate sample roundtrip bus fare and ski area shuttle service to any trip from NYC, $45.

LOW-COST OPTIONAL EQUIPMENT RENTALS AND LESSON PLAN. Whatever your skiing ability, we recommend lessons to improve skills, have more fun and meet new people. Five 1½ hour lessons just **$20** at Mt. Snow; **$25** at Mt. Washington Valley (any combination of areas). Daily lessons available for **$8** at Bromley, **$10** at Stowe. If you don't bring your own equipment, you can rent on-location, **excellent** skis, boots and poles. Just **$29** at Mt. Snow, **$35** at Stowe, Bromley and Mt. Washington.

TRIP	MT. WASHINGTON			BROMLEY/STRATTON/MAGIC		MT. SNOW			STOWE			
DATES	JAN 1-6	JAN 6-11	JAN 13-18	JAN 1-6	JAN 6-11	JAN 1-6	JAN 6-11	JAN 13-18	NOV 21-25	DEC 16-21	JAN 1-6	JAN 6-11
PRICE ☆		$99			$99		$109		$99	$109	$119	
CONDO SUPPLEMENT ★	$20			NOT AVAILABLE		NOT AVAILABLE			$23	$34		

All prices are complete including tax and gratuities.

☆ Complete prices per person/per week include daily lift tickets, 5 nights lodging (4 nights on Nov. 21 trip) and full schedule of free activities including Welcome Party with complimentary beer. Lodging based on average of 4 per room. Private rooms in motels or lodges available at additional supplement: $15 private triple, $35 private double; per person, per week. Couples are welcome.

★ Per person/per week based on average of 6-8 per condominium.

RESERVATIONS & EARLY BIRD SAVINGS. Advance reservations and $25 deposit are imperative. We have to know how many are coming. **Prices shown in this flyer reflect Earlybird Savings.** To qualify for prices on chart, mail deposit 60 days before trip and make final payment at least one month in advance of departure date. Trip total price is $5 higher for deposits mailed less than 60 days before trip. Final payments mailed within 30 days of trip subject to additional $10 surcharge. So, **plan ahead and save $15 on your SSA College Ski Week.**
Written confirmation sent immediately upon receipt of deposit. Everyone welcome including non-students, but our lowest prices are reserved for SSA members. If not already an SSA member, enclose $9 membership fee with deposit or final payment. Membership entitles students to ski equipment and lift ticket discounts at over 100 popular ski areas nationwide, plus a subscription to "Poor Howard's Discount Lodging Guide" and the "Student Skier" magazine. All high school, college and graduate students eligible to join SSA. Non-students pay $10 non-student fee. **Earlybirds get the best rooms,** so mail your reservations today.
To further ensure that your group will be lodged together, whenever possible, send entire groups' deposits in one envelope and list their names in the space provided on order coupon. You can always add other names later if desired. **NOTE: people wishing to together must all be on the same lodging plan: meals, no-meals, or condominium.** Rates based on **at least 4 people per room.** If less than 4 in your party, take advantage of our Roommate Matching Service, only **$6,** which **guarantees** you the **low quad rate,** even if we fail to find you roommates. Or pay additional supplement listed on chart for private double/triple room. **Important Note: all final payments must be received by Dec. 17th** or reservations subject to cancellation. Certified checks or money orders only, for final payments made within **3 weeks** of departure. Reservations and confirmations on a space available basis only, within 2 weeks of trip departure date.

CANCELLATION PROTECTION. All payments, other than $25 deposit are **fully refundable,** if cancellation in writing is received by Dec. 17th. Cancellations received between Dec. 17th and Dec. 24th subject to cancellation fee of $50, between Dec. 24th and Dec. 31st cancellation charge of $100 applies. No refunds after Dec. 31st. No refunds for no shows, late arrivals or early departures. If you must cancel, you may send a substitute without penalty. Join us for our tenth year anniversary. Every year our trips sell out early, so **reserve now.**

ba sin ϕ, but the direction of **a** × **b** is opposite to that of **b** × **a;** this is so because the right-handed screw advances in one direction when rotated from **a** to **b** through ϕ but advances in the opposite direction when rotated from **b** to **a,** through ϕ. You can obtain the same result by applying the right-hand rule.

If ϕ is 90°, **a**, **b**, and **c** (= **a** × **b**) are all at right angles to one another and give the directions of a three-dimensional right-handed coordinate system.

figure 2-12
The vector product. (*a*) In **c** = **a** × **b**, the direction of **c** is that in which a right-handed screw advances when turned from **a** to **b** through the smaller angle. (*b*) The direction of **c** can also be obtained from the "right-hand rule": If the right hand is held so that the curled fingers follow the rotation of **a** into **b**, the extended right thumb will point in the direction of **c**. (*c*) The vector product changes sign when the order of the factors is reversed: **a** × **b** = −**b** × **a.** Apply the right-hand rule or the rule for the advance of a right-handed screw to show that **c** and **c**′ have opposite directions.

The reason for defining the vector product in this way is that it proves to be useful in physics. We often encounter physical quantities that are vectors whose product, defined as above, is a vector quantity having important physical meaning. Some examples of physical quantities that are vector products are torque, angular momentum, the force on a moving charge in a magnetic field, and the flow of electromagnetic energy. When such quantities are discussed later, their connection with the vector product of two vectors will be pointed out.

The scalar product is the simplest product of two vectors. The order of multiplication does not affect the product. The vector product is the next simplest case. Here the order of multiplication does affect the product, but only by a factor of minus one, which implies a direction reversal. Other products of vectors are useful but more involved. For example, a tensor can be generated by multiplying each of the three components of one vector by the three components of another vector. Hence a tensor (of the second rank) has nine numbers associated with it, a vector three, and a scalar only one. Some physical quantities that can be represented by tensors are mechanical and electrical stress, moments and products of inertia, and strain. Still more complex physical quantities are possible. In this book, however, we are concerned only with scalars and vectors.

A certain vector **a** in the x-y plane is 250° counterclockwise from the positive x-axis and has a magnitude 7.4 units. Vector **b** has magnitude 5.0 units and is directed parallel to the z-axis. Calculate (a) the scalar product **a · b** and (b) the vector product **a × b**.

EXAMPLE 4

(a) Because **a** and **b** are perpendicular to one another, the angle ϕ between them is 90° and $\cos \phi = \cos 90° = 0$. Therefore, from Eq. 2-11, the scalar product is

$$\mathbf{a} \cdot \mathbf{b} = ab \cos \phi = ab \cos 90° = (7.4)(5.0)\,0 = 0,$$

consistent with the fact that neither vector has a component in the direction of the other.

(b) The *magnitude* of the vector product is, from Eq. 2-12,

$$|\mathbf{a} \times \mathbf{b}| = ab \sin \phi = (7.4)(5.0) \sin 90° = 37.$$

The *direction* of the vector product is perpendicular to the plane formed by **a** and **b**. Therefore, as shown in Fig. 2-13, it lies in the x-y plane (perpendicular to **b**) at an angle of 250° − 90° = 160° from the +x-axis (perpendicular to **a**) in accordance with the right-hand rule.

figure 2-13
Example 4

Vectors turn out to be very useful in physics. It will be helpful to look a little more deeply into why this is true. Suppose that we have three vectors **a**, **b**, and **r**, which have components a_x, a_y, a_z; b_x, b_y, b_z; and r_x, r_y, r_z, respectively, in a particular coordinate system xyz. Let us suppose further that the three vectors are related so that

2-5
*VECTORS AND THE
LAWS OF PHYSICS*

$$\mathbf{r} = \mathbf{a} + \mathbf{b}. \qquad (2\text{-}13)$$

By a simple extension of Eqs. 2-10 this means that

$$r_x = a_x + b_x; \qquad r_y = a_y + b_y; \qquad \text{and} \qquad r_z = a_z + b_z \qquad (2\text{-}14)$$

Now consider another coordinate system x'y'z' which has these properties: (1) its origin does not coincide with the origin of the first, or xyz, system and (2) its three axes are not parallel to the corresponding axes in the first system. In other words, the second set of coordinates has been both *translated* and *rotated* with respect to the first.

The components of the vectors **a**, **b**, and **r** in the new system would all prove, in general, to be different; we may represent them by $a_{x'}, a_{y'}, a_{z'}$; $b_{x'}, b_{y'}, b_{z'}$; and $r_{x'}, r_{y'}, r_{z'}$, respectively. These new components would be found, however, to be related (see Problem 39) in that

$$r_{x'} = a_{x'} + b_{x'}; \qquad r_{y'} = a_{y'} + b_{y'}; \qquad \text{and} \qquad r_{z'} = a_{z'} + b_{z'}. \qquad (2\text{-}15)$$

That is, in the new system we would find once again (see Eq. 2-13) that

$$\mathbf{r} = \mathbf{a} + \mathbf{b}.$$

In more formal language: Relations among vectors, of which Eq. 2-13 is only one example, are invariant (that is, are unchanged) with respect to translation or rotation of the coordinates. Now it is a fact of experience that the experiments on which the laws of physics are based and indeed the laws of physics themselves are similarly unchanged in form when we rotate or translate the coordinate system. Thus the language of vectors is an ideal one in which to express physical laws. If we can express a law in vector form, the invariance of the law for translation and rotation of the coordinate system is assured by this purely geometrical property of vectors.

It was thought until about 1956 that all laws of physics were invariant under another kind of transformation of coordinates, the substitution of a right-handed coordinate system for a left-handed one (see Fig. 2-14). In that year, however, some experiments involving the decay of certain elementary particles were studied in which the result of the experiment *did* turn out to depend on the "handedness" of the coordinate system used to express the results. In other words, the experiment and its image in a mirror would yield different results!* This surprising result led to a re-examination of the whole question of the symmetry of physical laws; these studies remain among the most challenging in modern physics.

figure 2-14
Showing (a) a left-handed and (b) a right-handed coordinate system. Notice that (a) and (b) are related in that each may be viewed as the image of the other in mirror *MM*. The "handedness" of a coordinate system cannot be changed by rotating it. Note that in (b), $\mathbf{i} \times \mathbf{j} = \mathbf{k}$, whereas in (a), $\mathbf{i} \times \mathbf{j} = -\mathbf{k}$.

questions

1. Three astronauts leave Cape Canaveral, go to the moon and back, and splash down in the Pacific Ocean. An Admiral bids them goodby at the Cape and then sails to the Pacific Ocean in an aircraft carrier where he picks them up. For their respective journeys do the astronauts or the Admiral have the larger displacement?

2. Can two vectors of different magnitude be combined to give a zero resultant? Can three vectors?

3. Can a vector have zero magnitude if one of its components is not zero?

4. Does it make any sense to call a quantity a vector when its magnitude is zero?

5. If three vectors add up to zero, they must all be in the same plane. Make this plausible.

6. Does a unit vector have units?

7. Name several scalar quantities. Is the value of a scalar quantity dependent on the coordinate system chosen?

8. We can order events in time. For example, event b may precede event c but follow event a, giving us a time order of events a, b, c. Hence there is a sense of time, distinguishing past, present, and future. Is time a vector therefore? If not, why not?

9. Do the commutative and associative laws apply to vector subtraction?

10. Can a scalar product be a negative quantity?

11. (a) If $\mathbf{a} \cdot \mathbf{b} = 0$, does it follow that $\mathbf{a}$ and $\mathbf{b}$ are perpendicular to one another? (b) If $\mathbf{a} \cdot \mathbf{b} = \mathbf{a} \cdot \mathbf{c}$, does it necessarily follow that $\mathbf{b}$ equals $\mathbf{c}$?

12. If $\mathbf{a} \times \mathbf{b} = 0$, must $\mathbf{a}$ and $\mathbf{b}$ be parallel to each other? Is the converse true?

13. (a) Show that if all of the components of a vector are reversed in direction, then the vector itself is reversed in direction. (b) Show that if the compo-

---

* C. N. Yang and T. D. Lee were awarded the Nobel prize in 1957 for their theoretical prediction that this would be the case. See "The Overthrow of Parity" by Phillip Morrison, *Scientific American*, April 1957, for a very readable review of this matter.

nents of a vector product are all reversed, then the vector product is not changed. (c) Is a vector product, then, a vector?

14. Thus far we have discussed addition, subtraction, and multiplication of vectors. Why do you suppose that we do not discuss the division of vectors? Is it possible to define such an operation?

15. Must you specify a coordinate system when you (a) add two vectors, (b) form their scalar product, (c) form their vector product, (d) find their components?

16. It is conventional to use the right hand in rules for vector algebra. What changes would be required if a left-hand convention were adopted instead?

problems

SECTION 2-2

1. Consider two displacements, one of magnitude 3 m and another of magnitude 4 m. Show how the displacement vectors may be combined to get a resultant displacement of magnitude (a) 7 m, (b) 1 m, and (c) 5 m.
   Answer: The displacements should be: (a) parallel, (b) antiparallel, (c) perpendicular.

2. What are the properties of two vectors $\mathbf{a}$ and $\mathbf{b}$ such that
   (a) $\mathbf{a} + \mathbf{b} = \mathbf{c}$ and $a + b = c$,

   (b) $\mathbf{a} + \mathbf{b} = \mathbf{a} - \mathbf{b}$,

   (c) $\mathbf{a} + \mathbf{b} = \mathbf{c}$ and $a^2 + b^2 = c^2$.

3. Two vectors $\mathbf{a}$ and $\mathbf{b}$ are added. Show that the magnitude of the resultant cannot be greater than $a + b$ or smaller than $|a - b|$, where the vertical bars signify absolute value.

4. A car is driven east for a distance of 50 km, then north for 30 km, and then in a direction 30° east of north for 25 km. Draw the vector diagram and determine the total displacement of the car from its starting point.

5. A golfer takes three putts to get his ball into the hole once he is on the green. The first putt displaces the ball 12 ft north, the second 6.0 ft southeast, and the third 3.0 ft southwest. What displacement was needed to get the ball into the hole on the first putt?   *Answer:* 6.0 ft, 20.5° E of N.

6. Vector $\mathbf{a}$ has a magnitude of 5.0 units and is directed east. Vector $\mathbf{b}$ is directed 45° west of north and has a magnitude of 4.0 units. Construct vector diagrams for calculating (a) $(\mathbf{a} + \mathbf{b})$ and (b) $(\mathbf{b} - \mathbf{a})$. Estimate the magnitudes and directions of $(\mathbf{a} + \mathbf{b})$ and $(\mathbf{b} - \mathbf{a})$ from your diagrams.

SECTION 2-3

7. Find the sum of the vector displacements $\mathbf{c}$ and $\mathbf{d}$ whose components in kilometers along three perpendicular directions are
   $$c_x = 5.0, \ c_y = 0, \ c_z = -2.0; \ d_x = -3.0, \ d_y = 4.0, \ d_z = 6.0.$$
   Answer: $r_x = 2.0$ km; $r_y = r_z = 4.0$ km.

8. (a) A man leaves his front door, walks 1000 ft east, 2000 ft north, and then takes a penny from his pocket and drops it from a cliff 500 ft high. Set up a coordinate system and write down an expression, using unit vectors, for the displacement of the penny. (b) The man then returns to his front door, following a different path on the return trip. What is his resultant displacement for the round trip?

9. Two vectors are given by $\mathbf{a} = 4\mathbf{i} - 3\mathbf{j} + \mathbf{k}$ and $\mathbf{b} = -\mathbf{i} + \mathbf{j} + 4\mathbf{k}$. Find (a) $\mathbf{a} + \mathbf{b}$, (b) $\mathbf{a} - \mathbf{b}$, and (c) a vector $\mathbf{c}$ such that $\mathbf{a} - \mathbf{b} + \mathbf{c} = 0$.
   Answer: (a) $3\mathbf{i} - 2\mathbf{j} + 5\mathbf{k}$. (b) $5\mathbf{i} - 4\mathbf{j} - 3\mathbf{k}$. (c) Negative of (b).

10. A room has the dimensions 10 ft × 12 ft × 14 ft. A fly starting at one corner ends up at a diametrically opposite corner. (a) What is the magnitude of its displacement? (b) Could the length of its path be less than this distance? Greater than this distance? Equal to this distance? (c) Choose a suitable

HALF PRICE LIFT TICKETS

On non-holiday weekdays save 50% off the regular weekend price at over 100 participating ski area
Save $1 to $3 on your weekend or holiday lift tickets. Also, save $1 to $5 on weekdays and/or wee
ends at over 30 additional areas. (See lists below.)

HALF PRICE LESSONS AND RENTALS

Save 50% on ski lessons and equipment rentals at over 75 ski areas on non-holiday weekdays.

BUY SKIS FOR HALF PRICE

SSA members can now purchase high quality Graves Skis and save 50%. The lightweight design ar
lively response are new. The fantastic control and legendary name are the same. Choose from 2 gre
models, both with fiberglass foam-cores and full manufacturer's guarantee. Equivalent skis sell f
$250 retail, you pay $115 or $135.

10% to 15% DISCOUNTS ON SKI EQUIPMENT

Over 50 ski shops offer SSA members discounts of 10% to 15% on the purchase of skis, boots, bin
ings and poles.

NATION'S ONLY DISCOUNT LODGING GUIDE

A complete guide to skiing, designed for the budget-minded college skier. Includes a list of over 20
inexpensive lodges close to major areas, many of whom offer substantial discounts to SSA membe
only.

THE STUDENT SKIER MAGAZINE

As an SSA member, you'll receive the nation's only magazine for college students who ski. It's fille
with timely and informative articles including "How to Ski on a Student's Budget".

LOW COST COLLEGE SKI VACATIONS

Every year SSA offers a variety of low cost all-college ski and party vacations. Lodging is always fir
class. This year's destinations include Steamboat, Stowe, Mt. Snow, Mt. Washington Valley ar
Bromley/Stratton.

"NO-WAY-TO-LOSE" GUARANTEE

If you are not completely satisfied with the program, return the unused Ski Card before January
1980. We'll refund the full purchase price immediately! If for any reason you are unable to use yo
Student Ski Card during the season, just return it to us over the summer and we'll mail you anoth
next fall for FREE!

SKI EQUIPMENT DISCOUNTS

Save Big! Typical ski shop prices include skis at $200, boots at $150, bindings at $90 and poles at $25. With an SSA card, save a total
$46 to $70. Listed below are many of the quality ski shops where SSA members receive 10% to 15% discounts. A complete list w
accompany your SSA membership card.

EASTERN: MASSACHUSETTS: Strand's (Worcester), Stuart's Sportmen's Center (Agawam & Indian Orchard), Mt. Tom Ski Shop (Ho
oke), Werner's Sporting Goods Ski Shop (Leominster), Matt's Ski Shop (Northampton & Westfield), Tonon's Ski Haus (Pittsfiel
CONNECTICUT: Mt. Tom Ski Shop (Enfield), Rinaldi's Sport & Ski (Meriden), Yale Co-op, (New Haven), Vality Ski Chalet (Gale's Ferr
Sport Mart (Vernon & Rockville), Bob's Chalet (Bristol). **NEW HAMPSHIRE:** Mittersill Sport Shop (Franconia), Gallagher's Sport Cen
(Portsmouth), Ski-N-Tennis (Manchester), Bob Skinner's Ski Shops (Mt. Sunapee). **VERMONT:** Jansen's Sport Shop (St. Albans), De
Way Ski Shop (Springfield), Clapp's Sporting Goods (Brattleboro). **MAINE:** Peter Webber Sports (Waterville). **NEW YORK:** Olympic S
Shops (Brooklyn, Manhasset, Rockville Centre, Huntington), McCoy's Sport Shops (Newburgh, Middletown), Ski & Sport Shop (Poug
keepsie), Tom Kat Sporting Goods (Mahopac), The Olde Soft Shoe (Mt. Kisco), BobCat at Catskill Ski Centre (Andes), Inside Edge Sk
Bike (Glen Falls), Utica Sporting Goods (Utica), Schuss Ski Shop (Rome), Berger's Ski Shop (Binghamton). **NEW JERSEY:** Werne
Ski & Sport Shop (West Orange), Bobby Knapp Ski Shop (Paramus, Ridgewood), Four Seasons Ski Shop (Hillsborough, Greenbroo
PENNSYLVANIA: D & E Sporting Goods (Bethlehem).
MIDWEST: ILLINOIS: Hubbard Woods Ski Chalet (Winnetka), Cook's Sportcraft Co. (Chicago). **MINNESOTA:** Hi Tempo (Mendc
Heights, White Bear Lake), Arnie's Ski & Garden Center (EauClaire). The Ski Den (Minneapolis), Joyne's Department Store (Gra
Marais). **WISCONSIN:** Miller's Ski Shop (Sturgeon Bay), Bertrand's Sporting Goods (Green Bay), Mel's Trading Post (Rhinelander).
WEST: COLORADO: Andel's Pro Ski Shop (Denver), Jerry Jolly's Ltd. (Denver, Glendale). **UTAH:** The Sportsman Mountain Pla
(Logan), Woodhaus Ski Shop (Park City). **IDAHO:** Sturtevant of Sun Valley (Ketchum). **MONTANA:** The Chalet (Bozeman). **OREGO**
North Slope Ski & Sports (Hood River).
Note: Discounts do not apply on items already on sale nor on special packages of skis, boots, poles and bindings together.

SKI AREA DISCOUNTS

Save big! Typical lift ticket price is $15 at many larger areas. With an SSA card you may pay as little as $7.50. Regular daily rental
skis, boots and poles cost $10, you can save $5. Regular lesson price is $8, you can save $4. Total possible daily savings: $16.50. List
below are many of the popular ski areas where SSA members enjoy ski savings. A complete list of ski areas will accompany your Si
membership card.

EASTERN: VERMONT: Stowe (Stowe), Jay Peak (Jay), Mt. Snow (West Dover), Big Bromley (Manchester), Okemo (Ludlow), Pico Pe
(Rutland), Mad River Glen (Waitsfield), Burke Mountain (E. Burke), Round Top (Plymouth Union), Magic Mountain (Londonderry), Mou
Ascutney (Brownsville), Haystack (Wilmington), Hogback (Marlboro). **NEW HAMPSHIRE:** Wildcat Mountain (Pinkham Notch), Mt. Cra
more Skimobile (N. Conway), Attitash (Bartlett), Bretton Woods (Bretton Woods), Black Mountain (Jackson), King Ridge (New Londo
Pat's Peak (Henniker), Crotched Mountain (Francestown), Bobcat (Bennington). **MAINE:** Saddleback (Rangeley), Pleasant Mounta
(Bridgton). **MASSACHUSETTS:** Mount Tom (Holyoke), Berkshire East (Charlemont), Jiminy Peak (Hancock), Catamount (S. Egremor
Wachusett Mountain (Princeton), Mt. Raimer (Williamstown). **CONNECTICUT:** Powder Ridge (Middlefield), Mt. Southington (Southir
ton), Mohawk Mountain (W. Cornwall), Woodbury Ski & Racquet (Woodbury). **EASTERN NEW YORK STATE:** Catamount (Hillsdal
Scotch Valley (Stamford), Ski Minnewaska (Lake Minnewaska), Big Vanilla at Davos (Woodridge), Holiday Mountain (Monticello), Mou
Cathalia (Ellenville), Mount Storm (Stormville), Silver Mine (Bear Mountain), Gunset Ski Bowl (Richfield Springs), Catskill Ski Cent
(Andes), Big Birch (Patterson), Highmount (Highmount), North Creek Ski Bowl (North Creek), Moon Valley (Malone), Adirondack S
Center (Corinth), Mount Peter (Greenwood Lake), Cortina Valley (Haines Falls), Mt. Whitney (Lake Placid). **WESTERN NEW YO**
STATE: Bluemont (Yorkshire), Mystic Mountain (New Woodstock), McCauley Mountain (Old Forge), Happy Valley Ski Center (Alfre
Snow Mountain (Ellicotville), Wing Hollow (Allegany), Fahneystock (Stormville). **NEW JERSEY:** Craigmeur (Newfoundland),
Eagle Rock (Hazelton). **PENNSYLVANIA:** Seven Springs (Champion), Big Boulder (Lake Harmony), Ja
Frost Mountain (White Haven), Blue Knob (Claysburg), Hidden Valley (Somerset), Tanglewood (Tafton), Mt. Pleasant (Cambrid
Springs), Pocono Manor (Pocono Manor), Ski Round Top (Dillsburg), Ski Liberty (Fairfield). **MARYLAND:** Wisp (Oakland). **VIRGINIA:** Bry
Mountain (Basye), Mountain Run (New Market), Massanutten (Harrisonburg). **NORTH CAROLINA:** Appalachian Ski Mountain (Blowi
Rock), Sapphire Valley (Sapphire). **QUEBEC:** Gray Rocks (St. Jovite).

WESTERN: COLORADO: Aspen Highlands (Aspen), Telluride (Telluride), Sunlight (Glenwood Springs). **WYOMING:** Jackson Hole (Jac
son Hole), Targhee (Alta). **MONTANA:** Big Sky (Big Sky), Red Lodge (Red Lodge). **CALIFORNIA:** Squaw Valley (Olympic Valley), Goldmi
Ski Area (Big Bear Lake).

MIDWEST: Plus over 50 major ski areas in the Midwest, including Boyne Mountain and Indianhead, Michigan; Mt. Telemark and Whi
cap, Wisconsin; and Lutsen, Minnesota.

coordinate system and find the components of the displacement vector in this frame. (d) If the fly walks rather than flies, what is the length of the shortest path it can take?

11. Given two vectors $\mathbf{a} = 4\mathbf{i} - 3\mathbf{j}$ and $\mathbf{b} = 6\mathbf{i} + 8\mathbf{j}$, find the magnitude and direction of $\mathbf{a}$, of $\mathbf{b}$, of $\mathbf{a} + \mathbf{b}$, of $\mathbf{b} - \mathbf{a}$, and of $\mathbf{a} - \mathbf{b}$.
    Answer: The magnitudes are 5, 10, 11, 11, and 11. The angles with the positive x-axis are 323°, 53°, 27°, 80° and 260°.

12. Two vectors of lengths a and b make an angle θ with each other when placed tail to tail. Prove, by taking components along two perpendicular axes, that the length of their sum is

    $$r = \sqrt{a^2 + b^2 + 2ab\, \cos \theta}.$$

13. Generalize the analytical method of resolution and addition to the case of *three or more vectors*, in two dimensions.

14. Two vectors $\mathbf{a}$ and $\mathbf{b}$ have equal magnitudes, say 10 units. They are oriented as shown in Fig. 2-15 and their vector sum is $\mathbf{r}$. Find (a) the x- and y-components of $\mathbf{r}$; (b) the magnitude of $\mathbf{r}$; and (c) the angle $\mathbf{r}$ makes with the x-axis.

figure 2-15
Problems 14 and 25

15. A particle undergoes three successive displacements in a plane, as follows: 4.0 m southwest, 5.0 m east, 6.0 m in a direction 60° north of east. Choose the y-axis pointing north and the x-axis pointing east and find (a) the components of each displacement, (b) the components of the resultant displacement, (c) the magnitude and direction of the resultant displacement, and (d) the displacement that would be required to bring the particle back to the starting point.
    Answer: (a) $a_x = -2.8$ m, $a_y = -2.8$ m;
    $\quad\quad\quad\quad b_x = +5.0$ m, $b_y = 0$;
    $\quad\quad\quad\quad c_x = +3.0$ m, $c_y = +5.2$ m.
    $\quad\quad\quad$ (b) $d_x = +5.2$ m, $d_y = +2.4$ m.
    $\quad\quad\quad$ (c) 5.7 m, 25° north of east.
    $\quad\quad\quad$ (d) 5.7 m, 25° south of west.

16. Use a scale of 2 m to the inch and add the displacements of Problem 15 graphically. Determine *from your graph* the magnitude and direction of the resultant.

17. A person flies from Washington to Manila. (a) Describe the displacement vector. (b) What is its magnitude if the latitude and longitude of the two cities are 39° N, 77° W, and 15° N, 121° E?   *Answer:* (b) 11,230 km.

18. Generalize the analytical method of resolving and adding two vectors to *three dimensions*.

19. Let N be an integer greater than one; then

    $$\cos 0 + \cos \frac{2\pi}{N} + \cos \frac{4\pi}{N} + \cdots + \cos (N-1) \frac{2\pi}{N} = 0;$$

    that is,

    $$\sum_{n=0}^{n=N-1} \cos \frac{2\pi n}{N} = 0.$$

    Also

    $$\sum_{n=0}^{n=N-1} \sin \frac{2\pi n}{N} = 0.$$

    Prove these two statements by considering the sum of N vectors of equal length, each vector making an angle of $2\pi/N$ with that preceding.

20. A vector **d** has a magnitude 2.5 m and points north. What are the magnitudes and directions of the vectors

$$(a) -\mathbf{d}, (b) \mathbf{d}/2.0, (c) -2.5\mathbf{d} \text{ and } (d) 4.0\mathbf{d}?$$

21. In the coordinate system of Fig. 2.6*b* show that

$$\mathbf{i} \cdot \mathbf{i} = \mathbf{j} \cdot \mathbf{j} = \mathbf{k} \cdot \mathbf{k} = 1$$

and

$$\mathbf{i} \cdot \mathbf{j} = \mathbf{j} \cdot \mathbf{k} = \mathbf{k} \cdot \mathbf{i} = 0.$$

22. In the right-handed coordinate system of Fig. 2-6*b* show that

$$\mathbf{i} \times \mathbf{i} = \mathbf{j} \times \mathbf{j} = \mathbf{k} \times \mathbf{k} = 0$$
$$\mathbf{i} \times \mathbf{j} = \mathbf{k}; \mathbf{k} \times \mathbf{i} = \mathbf{j}; \mathbf{j} \times \mathbf{k} = \mathbf{i}.$$

23. Show for any vector **a** that (*a*) $\mathbf{a} \cdot \mathbf{a} = a^2$ and that (*b*) $\mathbf{a} \times \mathbf{a} = 0$.

24. Use the standard (right-hand) *xyz* system of coordinates. Given vector **a** in the +*x*-direction, vector **b** in the +*y*-direction, and the scalar quantity *d*: (*a*) What is the direction of $\mathbf{a} \times \mathbf{b}$? (*b*) What is the direction of $\mathbf{b} \times \mathbf{a}$? (*c*) What is the direction of $\mathbf{b}/d$? (*d*) What is $\mathbf{a} \cdot \mathbf{b}$?

25. For the two vectors in Problem 14, find (*a*) $\mathbf{a} \cdot \mathbf{b}$, and (*b*) $\mathbf{a} \times \mathbf{b}$.
    Answer: (*a*) −26. (*b*) 97**k**.

26. A vector **a** of magnitude ten units and another vector **b** of magnitude six units point in directions differing by 60°. Find (*a*) the scalar product of the two vectors and (*b*) the vector product of the two vectors.

27. Show that the area of the triangle contained between the vectors **a** and **b** is $\frac{1}{2}|\mathbf{a} \times \mathbf{b}|$, where the vertical bars signify absolute value (see Fig. 2-16).

28. Show that the magnitude of a vector product gives numerically the area of the parallelogram formed with the two component vectors as sides (see Fig. 2-16). Does this suggest how an element of area oriented in space could be represented by a vector?

figure 2-16
Problems 27 and 28

29. Show that $\mathbf{a} \cdot (\mathbf{b} \times \mathbf{c})$ is equal in magnitude to the volume of the parallelepiped formed on the three vectors **a**, **b**, and **c**.

30. Prove that two vectors must have equal magnitudes if their sum is perpendicular to their difference.

31. *Scalar product in unit vector notation.* Let two vectors be represented in terms of their coordinates as

$$\mathbf{a} = \mathbf{i}a_x + \mathbf{j}a_y + \mathbf{k}a_z$$

and

$$\mathbf{b} = \mathbf{i}b_x + \mathbf{j}b_y + \mathbf{k}b_z$$

Show analytically that

$$\mathbf{a} \cdot \mathbf{b} = a_x b_x + a_y b_y + a_z b_z.$$

(*Hint:* See Problem 21.)

32. Use the definition of scalar product $\mathbf{a} \cdot \mathbf{b} = ab \cos \phi$ and the fact that $\mathbf{a} \cdot \mathbf{b} = a_x b_x + a_y b_y + a_z b_z$ (see Problem 31) to obtain the angle between the two vectors given by $\mathbf{a} = 3\mathbf{i} + 3\mathbf{j} - 3\mathbf{k}$ and $\mathbf{b} = 2\mathbf{i} + \mathbf{j} + 3\mathbf{k}$.

33. *Vector product in unit vector notation.* Show analytically that $\mathbf{a} \times \mathbf{b} = \mathbf{i}(a_y b_z - a_z b_y) + \mathbf{j}(a_z b_x - a_x b_z) + \mathbf{k}(a_x b_y - a_y b_x)$. (*Hint:* See Problem 22.)

34. Three vectors are given by $\mathbf{a} = 3\mathbf{i} + 3\mathbf{j} - 2\mathbf{k}$, $\mathbf{b} = -\mathbf{i} - 4\mathbf{j} + 2\mathbf{k}$, and $\mathbf{c} = 2\mathbf{i} + 2\mathbf{j} + \mathbf{k}$. Find (*a*) $\mathbf{a} \cdot (\mathbf{b} \times \mathbf{c})$, (*b*) $\mathbf{a} \cdot (\mathbf{b} + \mathbf{c})$ and (*c*) $\mathbf{a} \times (\mathbf{b} + \mathbf{c})$.

35. Let **b** and **c** be the intersecting face diagonals of a cube of edge *a*, as shown in Fig. 2-17. (*a*) Find the components of the vector **d**, where $\mathbf{d} = \mathbf{b} \times \mathbf{c}$. (*b*) Find the values of $\mathbf{b} \cdot \mathbf{c}$, of $\mathbf{d} \cdot \mathbf{c}$, and of $\mathbf{d} \cdot \mathbf{b}$. (*c*) Find the angle between the body diagonal **e**, as shown in Fig. 2-17, and the face diagonal **b**.
    Answer: (*a*) $d_x = d_z = a^2$, $d_y = -a^2$. (*b*) $\mathbf{b} \cdot \mathbf{c} = a^2$, $\mathbf{d} \cdot \mathbf{c} = \mathbf{d} \cdot \mathbf{b} = 0$. (*c*) 35°.

figure 2-17
Problem 35

36. Suppose **a**, **b**, and **c** are any three noncoplanar vectors. They are not necessarily mutually at right angles. (*a*) show that

$$\mathbf{a} \cdot (\mathbf{b} \times \mathbf{c}) = \mathbf{b} \cdot (\mathbf{c} \times \mathbf{a}) = \mathbf{c} \cdot (\mathbf{a} \times \mathbf{b}).$$

(b) Let

$$\mathbf{A} = \frac{\mathbf{b} \times \mathbf{c}}{v}, \, \mathbf{B} = \frac{\mathbf{c} \times \mathbf{a}}{v}, \, \mathbf{C} = \frac{\mathbf{a} \times \mathbf{b}}{v},$$

where $v = \mathbf{a} \cdot (\mathbf{b} \times \mathbf{c})$. Evaluate the dot product of each of $\mathbf{a}, \mathbf{b}, \mathbf{c}$ with each of $\mathbf{A}, \mathbf{B}, \mathbf{C}$. (c) If $\mathbf{a}, \mathbf{b}, \mathbf{c}$ have dimensions of length, what are the dimensions of $\mathbf{A}, \mathbf{B}, \mathbf{C}$?

37. Two vectors $\mathbf{a}$ and $\mathbf{b}$ have components, in arbitrary units $a_x = 3.2$, $a_y = 1.6$; $b_x = 0.50$, $b_y = 4.5$. (a) Find the angle between $\mathbf{a}$ and $\mathbf{b}$. (b) Find the components of a vector $\mathbf{c}$ which is perpendicular to $\mathbf{a}$, is in the x-y plane, and has a magnitude of 5.0 units.
    Answer: (a) 57°. (b) $c_x = \pm 2.2$ units; $c_y = \mp 4.5$ units.

38. (a) We have seen that the commutative law does *not* apply to vector products, that is, $\mathbf{a} \times \mathbf{b}$ does not equal $\mathbf{b} \times \mathbf{a}$. Show that the commutative law *does* apply to scalar products, that is, $\mathbf{a} \cdot \mathbf{b} = \mathbf{b} \cdot \mathbf{a}$. (b) Show that the distributive law applies to both scalar products and vector products, that is, show that

$$\mathbf{a} \cdot (\mathbf{b} + \mathbf{c}) = \mathbf{a} \cdot \mathbf{b} + \mathbf{a} \cdot \mathbf{c} \text{ and that } \mathbf{a} \times (\mathbf{b} + \mathbf{c}) = \mathbf{a} \times \mathbf{b} + \mathbf{a} \times \mathbf{c}.$$

(c) Does the associative law apply to vector products, that is, does $\mathbf{a} \times (\mathbf{b} \times \mathbf{c})$ equal $(\mathbf{a} \times \mathbf{b}) \times \mathbf{c}$? Does it make any sense to talk about an associative law for scalar products?

SECTION 2-5

39. *Invariance of vector addition under rotation of the coordinate system.* Figure 2-18 shows two vectors $\mathbf{a}$ and $\mathbf{b}$ and two systems of coordinates which differ in that the x and x' axes and the y and y' axes each make an angle ϕ with each other. Prove analytically that $\mathbf{a} + \mathbf{b}$ has the same magnitude and direction no matter which system is used to carry out the analysis.

figure 2-18
Problem 39

3
motion in one dimension

Mechanics, the oldest of the physical sciences, is the study of the motion of objects. The calculation of the path of a baseball or of a space probe sent from Earth to Mars is among its problems. So too is the analysis of tracks formed in bubble chambers, representing the collisions, decay, and interactions of elementary particles (see Fig. 10-11 and Appendix F).

When we describe motion we are dealing with that part of mechanics called *kinematics.* When we relate motion to the forces associated with it and to the properties of the moving objects, we are dealing with *dynamics.* In this chapter we shall define some kinematical quantities and study them in detail for the special case of motion in one dimension. In Chapter 4 we discuss some cases of two- and three-dimensional motion. Chapter 5 deals with the more general case of dynamics.

An object can rotate as it moves. For example, a baseball may be spinning while it is moving as a whole in some trajectory. Also, a body may vibrate as it moves, as, for example, a falling water droplet. These complications can be avoided by considering the motion of an idealized body called a *particle.* Mathematically, a particle is treated as a point, an object without extent, so that rotational and vibrational considerations are not involved.

Actually, there is no such thing in nature as an object without extent. The concept of "particle" is nevertheless very useful because real objects often behave to a very good approximation as though they were particles. A body need not be "small" in the usual sense of the word in order to be treated as a particle. For example, if we consider the distance

3-1
MECHANICS

3-2
PARTICLE KINEMATICS

from the earth to the sun, with respect to this distance the earth and the sun can usually be considered to be particles. We can find out a great deal about the motion of the sun and planets, without appreciable error, by treating these bodies as particles. Baseballs, molecules, protons, and electrons can often be treated as particles. Even if a body is too large to be considered a particle for a particular problem, it can always be thought of as made up of a number of particles, and the results of particle motion may be useful in analyzing the problem. As a simplification, therefore, we confine our present treatment to the motion of a particle.

Bodies that have only motion of translation behave like particles. An observer will call motion *translational* if the axes of a reference frame which is imagined rigidly attached to the object, say x', y', and z', always remain parallel to the axes of his own reference frame, say x, y, and z. In Fig. 3-1, for example, we show the translational motion of an object moving from positions A to B to C. Notice that the path taken is not necessarily a straight line. Notice too that throughout the motion every point of the body undergoes the same displacements as every other point. We can assume the body to be a particle because in describing the motion of one point on the body we have described the motion of the body as a whole.

figure 3-1
Translational motion of an object. Translation can occur in three dimensions, but only two are shown for simplicity.

3-3
AVERAGE VELOCITY

The displacement, the velocity, and the acceleration of a particle are vectors. Because this chapter deals with motion in one dimension only, we really do not need the full power of the vector method to deal with it. Nevertheless we find it useful to begin by considering motion in two dimensions (the extension to three is not difficult). From this vantage point we then specialize to the particular case of one-dimensional motion. This procedure allows us to keep in mind the essential vector character of all motion.

The velocity of a particle is the rate at which its position changes with time. The position of a particle in a particular reference frame is given by a position vector drawn from the origin of that frame to the particle. At time t_1, let a particle be at point A in Fig. 3-2a, its position in the x-y plane being described by position vector $\mathbf{r}_1$. At a later time t_2 let the particle be at point B, described by position vector $\mathbf{r}_2$. The displacement vector describing the *change* in position of the particle as it moves from A to B is $\Delta\mathbf{r}\ (=\mathbf{r}_2-\mathbf{r}_1)$ and the elapsed time for the motion between these points is $\Delta t\ (=t_2-t_1)$. The *average velocity* for the particle during this interval is defined by

$$\bar{\mathbf{v}} = \frac{\Delta \mathbf{r}}{\Delta t} = \frac{\text{displacement (a vector)}}{\text{elapsed time (a scalar)}} \qquad (3\text{-}1)$$

A bar above a symbol indicates an average value for the quantity in question.

The quantity $\bar{\mathbf{v}}$ is a vector, for it is obtained by dividing the vector $\Delta \mathbf{r}$ by the scalar Δt. Velocity, therefore, involves both direction and magnitude. Its direction is the direction of $\Delta \mathbf{r}$ and its magnitude is $|\Delta \mathbf{r}/\Delta t|$. The magnitude is expressed in distance units divided by time units, as, for example, meters per second or miles per hour.

The velocity defined by Eq. 3-1 is called an *average* velocity because the measurement of the net displacement and the elapsed time does not tell us anything at all about the motion between A and B. The path may have been curved or straight; the motion may have been steady or erratic. The average velocity involves simply the total displacement and the total elapsed time. For example, suppose a man leaves his house and goes on an automobile trip, returning to his house in a time Δt (five hours, say) after he left it. His average velocity for the trip is zero because his displacement for this particular time interval Δt is zero.

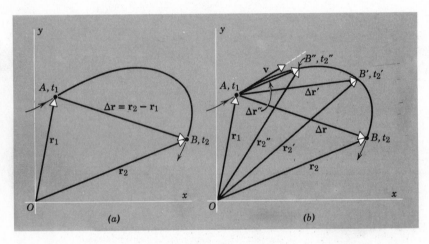

(a)   (b)

figure 3-2
(a) A particle moves from A to B in time Δt $(= t_2 - t_1)$ undergoing a displacement $\Delta \mathbf{r}$ $(= \mathbf{r}_2 - \mathbf{r}_1)$. *The average velocity $\bar{\mathbf{v}}$ between A and B is in the direction of $\Delta \mathbf{r}$. (b)* As B moves closer to A the average velocity approaches the *instantaneous velocity* $\mathbf{v}$ at A; $\mathbf{v}$ is tangent to the path at A.

If we were to measure the time of arrival of the particle at each of many points along the actual path between A and B in Fig. 3-2a, we could describe the motion in more detail. If the average velocity turned out to be the same (in magnitude and direction) between any two points along the path, we would conclude that the particle moved with *constant velocity*, that is, along a straight line (constant direction) at a uniform rate (constant magnitude).

Suppose that a particle is moving in such a way that its average velocity, measured for a number of different time intervals, does *not* turn out to be constant. This particle is said to move with variable velocity. Then we must seek to determine a velocity of the particle at any given instant of time, called the *instantaneous velocity.*

Velocity can vary by a change in magnitude, by a change in direction, or both. For the motion portrayed in Fig. 3-2a, the average velocity during the time interval $t_2 - t_1$ may differ both in magnitude and direction from the average velocity obtained during another time interval $t_2' - t_1$. In Fig. 3-2b we illustrate this by choosing the point B to be suc-

3-4
INSTANTANEOUS VELOCITY

cessively closer to point A. Points B' and B'' show two intermediate positions of the particle corresponding to the times t_2' and t_2'' and described by position vectors $\mathbf{r}_2'$ and $\mathbf{r}_2''$, respectively. The vector displacements $\Delta\mathbf{r}$, $\Delta\mathbf{r}'$, and $\Delta\mathbf{r}''$ differ in direction and become successively smaller. Likewise, the corresponding time intervals Δt $(= t_2 - t_1)$, $\Delta t'$ $(= t_2' - t_1)$, and $\Delta t''$ $(= t_2'' - t_1)$ become successively smaller.

As we continue this process, letting B approach A, we find that the ratio of displacement to elapsed time approaches a definite limiting value. Although the displacement in this process becomes extremely small, the time interval by which we divide it becomes small also and the ratio is not necessarily a small quantity. Similarly, while growing smaller, the displacement vector approaches a limiting direction, that of the tangent to the path of the particle at A. This limiting value of $\Delta\mathbf{r}/\Delta t$ is called the *instantaneous* velocity at the point A, or the velocity of the particle at the instant t_1.

If $\Delta\mathbf{r}$ is the displacement in a small interval of time Δt, following the time t, the velocity at the time t is the limiting value approached by $\Delta\mathbf{r}/\Delta t$ as both $\Delta\mathbf{r}$ and Δt approach zero. That is, if we let $\mathbf{v}$ represent the instantaneous velocity,

$$\mathbf{v} = \lim_{\Delta t \to 0} \frac{\Delta\mathbf{r}}{\Delta t}.$$

The direction of $\mathbf{v}$ is the limiting direction that $\Delta\mathbf{r}$ takes as B approaches A or as Δt approaches zero. As we have seen, this limiting direction is that of the tangent to the path of the particle at point A.

In the notation of the calculus, the limiting value of $\Delta\mathbf{r}/\Delta t$ as Δt approaches zero is written $d\mathbf{r}/dt$ and is called the *derivative* of $\mathbf{r}$ with respect to t. We have then

$$\mathbf{v} = \lim_{\Delta t \to 0} \frac{\Delta\mathbf{r}}{\Delta t} = \frac{d\mathbf{r}}{dt}. \tag{3-2}$$

The magnitude v of the instantaneous velocity is called the *speed* and is simply the absolute value of $\mathbf{v}$. That is,

$$v = |\mathbf{v}| = |d\mathbf{r}/dt|. \tag{3-3}$$

Speed, being the magnitude of a vector, is intrinsically positive.

Just as a particle is a physical concept making use of the mathematical concept of a point, so here velocity is a physical concept using the mathematical concept of differentiation. In fact, the calculus was invented in order to have a proper mathematical tool for treating fundamental mechanical problems.

In the next section we shall examine the concept of instantaneous velocity in detail for the special case of motion in one dimension, sometimes called rectilinear motion.

3-5 ONE-DIMENSIONAL MOTION—VARIABLE VELOCITY

Here again we approach one-dimensional motion by first considering two-dimensional motion and then considering the special case in which only one dimension is involved.

Figure 3-3 shows a particle moving along a path in the x-y plane. At time t its position with respect to the origin is described by position vector $\mathbf{r}$ (see Fig. 3-3a) and it has a velocity $\mathbf{v}$ (see Fig. 3-3b) tangent to its path as shown. We can write (see Eq. 2-8)

$$\mathbf{r} = \mathbf{i}x + \mathbf{j}y, \tag{3-4}$$

figure 3-3
A particle at time t has (a) a position described by $\mathbf{r}$, (b) an instantaneous velocity $\mathbf{v}$, and (c) an instantaneous acceleration $\mathbf{a}$. The vector components $\mathbf{i}x$ and $\mathbf{j}y$ of Eq. 3-4, $\mathbf{i}v_x$ and $\mathbf{j}v_y$ of Eq. 3-5, and $\mathbf{i}a_x$ and $\mathbf{j}a_y$ of Eq. 3-10 are also shown, as are the unit vectors $\mathbf{i}$ and $\mathbf{j}$.

where $\mathbf{i}$ and $\mathbf{j}$ are unit vectors in the positive x- and y-directions, respectively, and x and y are the (scalar) components of vector $\mathbf{r}$. Because $\mathbf{i}$ and $\mathbf{j}$ are constant vectors, we have, on combining Eqs. 3-2 and 3-4,

$$\mathbf{v} = \frac{d\mathbf{r}}{dt} = \mathbf{i}\frac{dx}{dt} + \mathbf{j}\frac{dy}{dt},$$

which we can express as

$$\mathbf{v} = \mathbf{i}v_x + \mathbf{j}v_y \qquad \text{(two-dimensional motion)}, \qquad (3\text{-}5)$$

where $v_x \, (= dx/dt)$ and $v_y \, (= dy/dt)$ are the (scalar) components of the vector $\mathbf{v}$.

We now consider motion in one dimension only, chosen for convenience to be the x-axis. We must then have $v_y = 0$ so that Eq. 3-5 reduces to

$$\mathbf{v} = \mathbf{i}v_x \qquad \text{(one-dimensional motion)}. \qquad (3\text{-}6)$$

Since $\mathbf{i}$ points in the positive x-direction, v_x will be positive (and equal to $+v$) when $\mathbf{v}$ points in that direction, and negative (and equal to $-v$) when it points in the other direction. Since, in one-dimensional motion, there are only two choices as to the direction of $\mathbf{v}$, the full power of the vector method is not needed, as we have pointed out; we can work with the (scalar) velocity component v_x alone.

The limiting process. As an illustration of the limiting process in one dimension, consider the following table of data taken for motion along the x-axis. The first four columns are experimental data. The symbols refer to Fig. 3-4 in which the particle is moving from left to right, that is, in the positive x-direction. The particle was at position x_1 (100 cm from the origin) at time t_1 (1.00 s). It was at position x_2 at time t_2. As we consider different values for x_2, and different corresponding times t_2, we find

EXAMPLE 1

x_1, cm	t_1, s	x_2, cm	t_2, s	$x_2 - x_1$ $= \Delta x$, cm	$t_2 - t_1$ $= \Delta t$, s	$\Delta x/\Delta t$, cm/s
100.0	1.00	200.0	11.00	100.0	10.00	+10.0
100.0	1.00	180.0	9.60	80.0	8.60	+9.3
100.0	1.00	160.0	7.90	60.0	6.90	+8.7
100.0	1.00	140.0	5.90	40.0	4.90	+8.2
100.0	1.00	120.0	3.56	20.0	2.56	+7.8
100.0	1.00	110.0	2.33	10.0	1.33	+7.5
100.0	1.00	105.0	1.69	5.0	0.69	+7.3
100.0	1.00	103.0	1.42	3.0	0.42	+7.1
100.0	1.00	101.0	1.14	1.0	0.14	+7.1

Equation 3-2, which holds for the general case of motion in three dimensions, is

$$\mathbf{v} = \lim_{\Delta t \to 0} \frac{\Delta \mathbf{r}}{\Delta t} = \frac{d\mathbf{r}}{dt}.$$

For one-dimensional motion along the x-axis we have a similar relation, scalar in character, in which each vector quantity is replaced by its corresponding component or

$$v_x = \lim_{\Delta t \to 0} \frac{\Delta x}{\Delta t} = \frac{dx}{dt}. \qquad (3\text{-}7)$$

It is clear from the table that as we select values of x_2 closer to x_1, Δt approaches zero and the ratio $\Delta x/\Delta t$ approaches the apparent limiting value $+7.1$ cm/s. At time t_1, therefore, $v_x = +7.1$ cm/s, as closely as we are able to determine from the data. Since v_x is positive, the velocity $\mathbf{v}$ $(= \mathbf{i}v_x$; see Eq. 3-6) points to the right in Fig. 3-4. This is tangent to the path in the direction of motion, as it must be.

figure 3-4
A particle is moving to the right along the x-axis.

EXAMPLE 2

Figure 3-5a shows six successive "snapshots" of a particle moving along the x-axis with variable velocity. At $t = 0$ it is at position $x = +1.00$ m to the right of the origin; at $t = 2.5$ s it has come to rest at $x = +5.00$ m; at $t = 4.0$ s it has returned to $x = +1.40$ m. Figure 3-5b is a plot of position x versus time t for this motion. The average velocity for the entire 4.0-s interval is the net displacement or change of position ($+0.40$ m) divided by the elapsed time (4.0 s) or $\overline{v_x} = +0.10$ m/s. (We call $\overline{v_x}$ average velocity and v_x velocity, for one-dimensional motion, even though velocity is a vector and not a scalar. This conforms to common usage and should cause no misunderstandings. These quantities are not speeds for they may be negative, whereas speed is intrinsically positive.) The average velocity vector $\overline{\mathbf{v}}$ points in the positive x-direction (that is, to the right in Fig. 3-5a) because the net displacement points in this direction. The quantity $\overline{v_x}$ can be obtained directly from the slope of the dashed line af in Fig. 3-5b, where by slope we mean the ratio of the net displacement gf to the elapsed time ga. (The slope is *not* the tangent of the angle fag measured on the graph with a protractor. This angle is arbitrary because it depends on the scales we choose for x and t.)

The velocity v_x at any instant is found from the slope of the curve of Fig. 3-5b at that instant. Equation 3-7 is in fact the relation by which the slope of the curve is defined in the calculus. In our example the slope at b, which is the value of v_x at b, is $+1.7$ m/s; the slope at d is zero and the slope at f is -6.2 m/s. When we determine the slope dx/dt at each instant t, we can make a plot of v_x versus t, as in Fig. 3-5c. Note that for the interval $0 < t < 2.5$ s, v_x is positive so that the velocity vector $\mathbf{v}$ points to the right in Fig. 3-5a; for the interval 2.5 s $< t < 4.0$ s v_x is negative so that $\mathbf{v}$ points to the left in Fig. 3-5a.

3-6
ACCELERATION

Often the velocity of a moving body changes either in magnitude, in direction, or both as the motion proceeds. The body is then said to have an acceleration. *The acceleration of a particle is the rate of change of its velocity with time.* Suppose that at the instant t_1 a particle, as in Fig. 3-6, is at point A and is moving in the x-y plane with the instantaneous velocity $\mathbf{v}_1$, and at a later instant t_2 it is at point B and moving

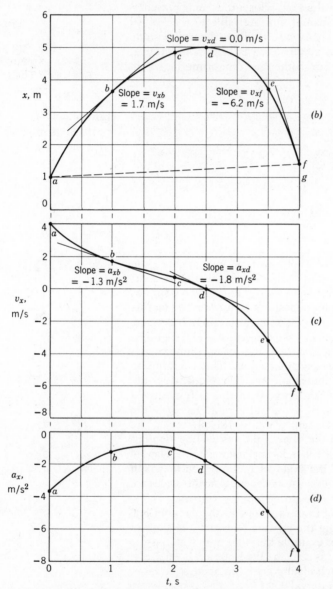

figure 3-5

(a) Six consecutive "snapshots" of a particle moving along the *x*-axis. The vector joined to the particle is its instantaneous velocity; that below the particle is its instantaneous acceleration.

(b) A plot of *x* versus *t* for the motion of the particle.

(c) A plot of v_x versus *t*.

(d) A plot of a_x versus *t*.

with the instantaneous velocity $\mathbf{v}_2$. The *average acceleration* $\bar{\mathbf{a}}$ during the motion from *A* to *B* is defined to be the *change of velocity* divided by the time interval, or

$$\bar{\mathbf{a}} = \frac{\mathbf{v}_2 - \mathbf{v}_1}{t_2 - t_1} = \frac{\Delta\mathbf{v}}{\Delta t}. \tag{3-8}$$

The quantity $\bar{\mathbf{a}}$ is a vector, for it is obtained by dividing a vector $\Delta\mathbf{v}$ by a scalar Δt. Acceleration is therefore characterized by magnitude and direction. Its direction is the direction of $\Delta\mathbf{v}$ and its magnitude is

$|\Delta\mathbf{v}/\Delta t|$. The magnitude of the acceleration is expressed in velocity units divided by time units, as for example meters per second per second (written m/s² and read "meters per second squared"), cm/sec², and ft/sec².

We call $\bar{\mathbf{a}}$ of Eq. 3-8 the *average* acceleration because nothing has been said about the time variation of velocity during the interval Δt. We know only the net change in velocity and the total elapsed time. If the change in velocity (a vector) divided by the corresponding elapsed time, $\Delta\mathbf{v}/\Delta t$, were to remain constant, regardless of the time intervals over which we measured the acceleration, we would have *constant* acceleration. Constant acceleration, therefore, implies that the *change* in velocity is uniform with time in direction and magnitude. If there is *no change* in velocity, that is, if the velocity were to remain constant both in magnitude and direction, then $\Delta\mathbf{v}$ would be zero for all time intervals and the acceleration would be zero.

If a particle is moving in such a way that its average acceleration, measured for a number of different time intervals, does not turn out to be constant, the particle is said to have a variable acceleration. The acceleration can vary in magnitude, or in direction, or both. In such cases we seek to determine the acceleration of the particle at any given time, called the instantaneous acceleration.

The *instantaneous acceleration* is defined by

$$\mathbf{a} = \lim_{\Delta t \to 0} \frac{\Delta\mathbf{v}}{\Delta t} = \frac{d\mathbf{v}}{dt}. \qquad (3\text{-}9)$$

That is, the acceleration of a particle at time t is the limiting value of $\Delta\mathbf{v}/\Delta t$ at time t as both $\Delta\mathbf{v}$ and Δt approach zero. The direction of the instantaneous acceleration $\mathbf{a}$ is the limiting direction of the vector change in velocity $\Delta\mathbf{v}$. The magnitude a of the instantaneous acceleration is simply $a = |\mathbf{a}| = |d\mathbf{v}/dt|$. When the acceleration is constant the instantaneous acceleration equals the average acceleration. You should note that the relation of $\mathbf{a}$ to $\mathbf{v}$, in Eq. 3-9, is the same as that of $\mathbf{v}$ to $\mathbf{r}$, in Eq. 3-2.

Two special cases illustrate that acceleration can arise from a change in either the magnitude or the direction of the velocity. In one case we have motion along a straight line with uniformly changing speed (as in Section 3-8). Here the velocity does not change in direction but its magnitude changes uniformly with time. This is a case of constant acceleration. In the second case we have motion in a circle at constant speed (Section 4-4). Here the velocity vector changes continuously in direction but its magnitude remains constant. This, too, is accelerated motion, though the direction of the acceleration vector is not constant. Later we will encounter other important cases of accelerated motion.

From Eqs. 3-5 and 3-9 we can write, for motion in two dimensions as in Fig. 3-3,

$$\mathbf{a} = \frac{d\mathbf{v}}{dt} = \mathbf{i}\frac{dv_x}{dt} + \mathbf{j}\frac{dv_y}{dt}$$

or

$$\mathbf{a} = \mathbf{i}a_x + \mathbf{j}a_y, \qquad (3\text{-}10)$$

where a_x $(= dv_x/dt)$ and a_y $(= dv_y/dt)$ are the (scalar) components of the acceleration vector $\mathbf{a}$ (see Fig. 3-3c).

figure 3-6
A particle has velocity $\mathbf{v}_1$ at point A and moves to point B, where its velocity is $\mathbf{v}_2$. The triangle shows the (vector) change in velocity $\Delta\mathbf{v}$ $(= \mathbf{v}_2 - \mathbf{v}_1)$ experienced by the particle as it moves from A to B.

3-7
ONE-DIMENSIONAL MOTION—VARIABLE ACCELERATION

We again restrict ourselves to motion in one dimension only, chosen for convenience to be the x-axis. Since v_y for such motion does not change with time (and is, in fact, zero), a_y, which is dv_y/dt, must also be zero so that

$$\mathbf{a} = \mathbf{i}a_x. \qquad (3\text{-}11)$$

Since $\mathbf{i}$ points in the positive x-direction, a_x will be positive when $\mathbf{a}$ points in this direction and negative when it points in the other direction.

The motion of Fig. 3-5a is one of variable acceleration along the x-axis. To find the acceleration* a_x at each instant we must determine dv_x/dt at each instant. This is simply the slope of the curve of v_x versus t at that instant. The slope of Fig. 3-5c at point b is -1.3 m/s² and that at point d is -1.8 m/s², as shown in the figure. The result of calculating the slope for all points is shown in Fig. 3-5d. Notice that a_x is negative at all instants, which means that the acceleration vector $\mathbf{a}$ points in the negative x-direction. This means that v_x is always decreasing with time, as is clearly seen from Fig. 3-5c. The motion is one in which the acceleration vector has a constant direction but varies in magnitude (see Fig. 3-5a).

EXAMPLE 3

Let us now further restrict our considerations to motion which not only occurs in one dimension (the x-axis) but for which $a_x = $ a constant. For such constant acceleration the average acceleration for any time interval is equal to the (constant) instantaneous acceleration a_x. Let $t_1 = 0$ and let t_2 be any arbitrary time t. Let v_{x0} be the value of v_x at $t = 0$ and let v_x be its value at the arbitrary time t. With this notation we find a_x (see Eq. 3-8) from

3-8
ONE-DIMENSIONAL MOTION – CONSTANT ACCELERATION

$$a_x = \frac{\Delta v}{\Delta t} = \frac{v_x - v_{x0}}{t - 0}$$

or

$$v_x = v_{x0} + a_x t. \qquad (3\text{-}12)$$

This equation states that the velocity v_x at time t is the sum of its value v_{x0} at time $t = 0$ plus the change in velocity during time t, which is $a_x t$.

Figure 3-7c shows a graph of v_x versus t for constant acceleration; it is a graph of Eq. 3-12. Notice that the slope of the velocity curve is constant, as it must be because the acceleration $a_x \ (= dv_x/dt)$ has been assumed to be constant, as Fig. 3-7d shows.

When the velocity v_x changes uniformly with time, its average value over any time interval equals one-half the sum of the values of v_x at the beginning and at the end of the interval. That is, the average velocity $\overline{v_x}$ between $t = 0$ and $t = t$ is

$$\overline{v_x} = \tfrac{1}{2}(v_{x0} + v_x). \qquad (3\text{-}13)$$

This relation would not be true if the acceleration were not constant, for then the curve of v_x versus t would not be a straight line.

---

* As for velocity, we commonly call a_x for one-dimensional motion the acceleration even though acceleration is a vector and a_x is correctly an acceleration component. For one-dimensional motion there is only one component if the axis is chosen along the line of the motion.

If the position of the particle at $t = 0$ is x_0, the position x at $t = t$ can be found from

$$x = x_0 + \overline{v_x}t$$

which can be combined with Eq. 3-13 to yield

$$x = x_0 + \tfrac{1}{2}(v_{x0} + v_x)t. \qquad (3\text{-}14)$$

The displacement due to the motion in time t is $x - x_0$. Often the origin is chosen so that $x_0 = 0$.

Notice that aside from initial conditions of the motion, that is, the values of x and v_x at $t = 0$ (taken here as $x = x_0$ and $v_x = v_{x0}$), there are four parameters of the motion. These are x, the displacement; v_x, the velocity; a_x, the acceleration; and t, the elapsed time. If we know only that the acceleration is constant, but not necessarily its value, from any two of these parameters we can obtain the other two. For example, if a_x and t are known, Eq. 3-12 gives v_x, and having obtained v_x, we find x from Eq. 3-14.

(a)

(b)

(c)

(d)

figure 3-7

(a) Five successive "snapshots" of rectilinear motion with constant acceleration. The arrows on the spheres represent **v**; those below represent **a**.

(b) The displacement increases quadratically according to $x = v_{x0}t + \tfrac{1}{2}a_x t^2$. Its slope increases uniformly and at each instant has the value v_x, the velocity.

(c) The velocity v_x increases uniformly according to $v_x = v_{x0} + a_x t$. Its slope is constant and at each instant has the value a_x, the acceleration.

(d) The acceleration a_x has a constant value; its slope is zero. Figure 3-5 shows similar plots for one-dimensional motion in which the acceleration is *not* constant.

In most problems in uniformly accelerated motion, two parameters are known and a third is sought. It is convenient, therefore, to obtain relations between any three of the four parameters. Equation 3-12 contains v_x, a_x, and t, but *not x*; Eq. 3-14 contains, x, v_x, and t but *not a_x*. To complete our system of equations we need two more relations, one containing x, a_x, and t but *not v_x* and another containing x, v_x, and a_x but *not t*. These are easily obtained by combining Eqs. 3-12 and 3-14.

Thus, if we substitute into Eq. 3-14 the value of v_x from Eq. 3-12, we thereby eliminate v_x and obtain

$$x = x_0 + v_{x0}t + \tfrac{1}{2}a_xt^2. \qquad (3\text{-}15)$$

When Eq. 3-12 is solved for t and this value for t is substituted into Eq. 3-14, we obtain

$$v_x{}^2 = v_{x0}{}^2 + 2a_x(x - x_0). \qquad (3\text{-}16)$$

Equations 3-12, 3-14, 3-15, and 3-16 (see Table 3-1) are the complete set of equations for motion along a straight line with constant acceleration.

Table 3-1
Kinematic equations for straight line motion with constant acceleration
(The position x_0 and the velocity v_{x0} at the initial instant $t = 0$ are the given initial conditions)

Equation Number	Equation	Contains x	v_x	a_x	t
3-12	$v_x = v_{x0} + a_xt$	×	✓	✓	✓
3-14	$x = x_0 + \tfrac{1}{2}(v_{x0} + v_x)t$	✓	✓	×	✓
3-15	$x = x_0 + v_{x0}t + \tfrac{1}{2}a_xt^2$	✓	×	✓	✓
3-16	$v_x{}^2 = v_{x0}{}^2 + 2a_x(x - x_0)$	✓	✓	✓	×

A special case of motion with constant acceleration is one in which the acceleration is zero, that is, $a_x = 0$. In this case the four equations in Table 3-1 reduce to the expected results $v_x = v_{x0}$ (the velocity does not change) and $x = x_0 + v_{x0}t$ (the displacement changes linearly with time).

EXAMPLE 4

The curve of Fig. 3-7b is a displacement-time graph for motion with constant acceleration; that is, it is a graph of Eq. 3-15 in which $x_0 = 0$. The slope of the tangent to the curve at time t equals the velocity v_x at that time. Notice that the slope increases continuously with time from v_{x0} at $t = 0$. The *rate of increase* of this slope should give the acceleration a_x, which is constant in this case. The curve of Fig. 3-7b is a parabola since Eq. 3-15 is the equation for a parabola having slope v_{x0} at $t = 0$. We obtain, on successive differentiation of Eq. 3-15,

$$x = x_0 + v_{x0}t + \tfrac{1}{2}a_xt^2$$

$$dx/dt = v_{x0} + a_xt \qquad \text{or} \qquad v_x = v_{x0} + a_xt,$$

which gives the velocity v_x at time t (compare Eq. 3-12), and

$$dv_x/dt = a_x,$$

the constant acceleration. The displacement-time graph for uniformly accelerated rectilinear motion will therefore always be parabolic.

3-9
CONSISTENCY OF UNITS AND DIMENSIONS

You should not feel compelled to memorize relations such as those of Table 3-1. The important thing is to be able to follow the line of reasoning used to obtain the results. These relations will be recalled automatically after you have used them repeatedly to solve problems, partly as a result of the familiarity acquired, but chiefly as a result of the better understanding obtained through application.

We can use any convenient *units* of time and distance in these equations. If we choose to express time in seconds and distance in meters, for self-consistency we must express velocity in m/s and acceleration in m/s². If we are given data in which the units of one quantity, as velocity, are not consistent with the units of another quantity, as acceleration, then before using the data in our equations we should transform both quantities to units that are consistent with one another. Having chosen the units of our fundamental quantities, we automatically determine the units of our derived quantities consistent with them. In carrying out any calculation, always remember to attach the proper units to the final result, for the result is meaningless without this label.

Suppose we wish to find the speed of a particle which has a uniform acceleration of 5.00 cm/s² for an interval of 0.50 h if the particle has a speed of 10.0 ft/s at the beginning of this interval. We decide to choose the foot as our length unit and the second as our time unit. Then

EXAMPLE 5

$$a_x = 5.00 \text{ cm/s}^2 = 5.00 \text{ cm/s}^2 \times \left(\frac{1 \text{ in.}}{2.54 \text{ cm}}\right) \times \left(\frac{1 \text{ ft}}{12 \text{ in.}}\right) = \frac{5.00}{30.5} \text{ ft/s}^2 = 0.164 \text{ ft/s}^2.$$

The time interval

$$\Delta t = t - t_0 = 0.50 \text{ h} \times \left(\frac{60 \text{ min}}{1 \text{ h}}\right) \times \left(\frac{60 \text{ s}}{1 \text{ min}}\right) = 1800 \text{ s}.$$

Note that the conversion factors in large parentheses are equal to unity. Taking the initial time $t_0 = 0$, as in Eq. 3-12, we then have

$$v_x = v_{x0} + a_x t = 10.0 \text{ ft/s} + (0.164 \text{ ft/s}^2)(1800 \text{ s}) = 305 \text{ ft/s}.$$

One way to spot an erroneous equation is to check the *dimensions* of all its terms. The dimensions of any physical quantity can always be expressed as some combination of the fundamental quantities, such as mass, length, and time, from which they are derived. The dimensions of velocity are length (L) divided by time (T); the dimensions of acceleration are length divided by time squared, etc. *In any legitimate physical equation the dimensions of all the terms must be the same.* This means, for example, that we cannot equate a term whose total dimension is a velocity to one whose total dimension is an acceleration. The dimensional labels attached to various quantities may be treated just like algebraic quantities and may be combined, canceled, and so on, just as if they were factors in the equation. For example, to check Eq. 3.15, $x = x_0 + v_{x0}t + \frac{1}{2}a_x t^2$, dimensionally, we note that x and x_0 have the dimension of a length. Therefore the two remaining terms must also have the dimension of a length. The dimension of the term $v_{x0}t$ is

$$\frac{\text{length}}{\text{time}} \times \text{time} = \text{length} \qquad \text{or} \qquad \frac{L}{T} \times T = L,$$

and that of $\frac{1}{2}a_x t^2$ is

$$\frac{\text{length}}{\text{time}^2} \times \text{time}^2 = \text{length} \qquad \text{or} \qquad \frac{L}{T^2} \times T^2 = L.$$

The equation is therefore dimensionally correct. You should check the dimensions of all the equations you use.

The speed of an automobile traveling due east is uniformly reduced from 45.0 miles per hour to 30.0 miles per hour in a distance of 264 ft.

EXAMPLE 6

(a) What is the magnitude and direction of the constant acceleration?

We choose, arbitrarily, the direction from west to east to be the positive x-direction. We are given x and v_x and we seek a_x. The time is not involved. Equation 3.16 is therefore appropriate (see Table 3-1). We have $v_x = +30.0$ mi/h, $v_{x0} = +45.0$ mi/h, $x - x_0 = +264$ ft $= 0.0500$ mi. From Eq. 3-16, $v_x^2 = v_{x0}^2 + 2a_x(x - x_0)$, we obtain

$$a_x = \frac{v_x^2 - v_{x0}^2}{2(x - x_0)},$$

or
$$a_x = \frac{(30.0 \text{ mi/h})^2 - (45.0 \text{ mi/h})^2}{2(0.0500 \text{ mi})} = -1.13 \times 10^4 \text{ mi/h}^2 = -4.58 \text{ ft/s}^2.$$

The direction of the acceleration **a** is due west, that is, in the negative x-direction because a_x is negative. The car is slowing down as it moves eastward, as it must do if it is being accelerated toward the west. When the speed of a body is decreasing, we often say that it is decelerating.

(b) How much time has elapsed during this deceleration?

If we use only the original data, Table 3-1 shows that Eq. 3-14 is appropriate. From Eq. 3-14, $x = x_0 + \frac{1}{2}(v_{x0} + v_x)t$, we obtain

$$t = \frac{2(x - x_0)}{v_{x0} + v_x},$$

or

$$t = \frac{(2)(0.0500 \text{ mi})}{(45.0 + 30.0) \text{ mi/h}} = \frac{1}{750} \text{ h} = 4.80 \text{ s}.$$

If we use the derived data of part (a), Eq. 3-12 is appropriate. This gives us a check. From Eq. 3-12, $v_x = v_{x0} + a_x t$, we have

$$t = \frac{v_x - v_{x0}}{a_x}$$

or
$$t = \frac{(30.0 - 45.0) \text{ mi/h}}{-1.13 \times 10^4 \text{ mi/h}^2} = 1.33 \times 10^{-3} \text{ h} = 4.80 \text{ s}.$$

(c) If one assumes that the car continues to decelerate at the same rate, how much time would elapse in bringing it to rest from 45.0 mi/h?

Equation 3-12 is useful here. We have $v_{x0} = 45.0$ mi/h, $a_x = -1.13 \times 10^4$ mi/h^2, and the final velocity $v_x = 0$. Then from Eq. 3-12, $v_x = v_{x0} + a_x t$, we obtain

$$t = \frac{v_x - v_{x0}}{a_x},$$

or
$$t = \frac{(0 - 45.0) \text{ mi/h}}{-1.13 \times 10^4 \text{ mi/h}^2} = 4.00 \times 10^{-3} \text{ h} = 14.4 \text{ s}.$$

(d) What total distance is required to bring the car to rest from 45.0 mi/h? Equation 3-15 is appropriate here. We have $v_{x0} = 45.0$ mi/h, $a_x = -1.13 \times 10^4$ mi/h^2, $t = 4.00 \times 10^{-3}$ h. From Eq. 3-15, $x = x_0 + v_{x0}t + \frac{1}{2}a_x t^2$, we obtain

$$x - x_0 = v_{x0}t + \frac{1}{2}a_x t^2$$

$$= (45.0 \text{ mi/h})(4.00 \times 10^{-3} \text{ h}) + \frac{1}{2}(-1.13 \times 10^4 \text{ m/h}^2)(4.00 \times 10^{-3} \text{ h})^2$$

$$= 0.0900 \text{ mi} = 475 \text{ ft}.$$

EXAMPLE 7

43

FREELY FALLING BODIES   SEC. 3-10

The nucleus of a helium atom (alpha-particle) travels along the inside of a straight hollow tube 2.0 m long which forms part of a particle accelerator. (a) If one assumes uniform acceleration, how long is the particle in the tube if it enters at a speed of 1.0×10^4 m/s and leaves at 5.0×10^6 m/s? (b) What is its acceleration during this interval?

(a) We choose an x-axis parallel to the tube, its positive direction being that in which the particle is moving and its origin at the tube entrance. We are given x and v_x and we seek t. The acceleration a_x is not involved. Hence we use Eq. 3-14, $x = x_0 + \frac{1}{2}(v_{x0} + v_x) t$ with $x_0 = 0$ or

$$t = \frac{2x}{v_{x0} + v_x},$$

$$t = \frac{(2)(2.0 \text{ m})}{(500 + 1) \times 10^4 \text{ m/s}} = 8.0 \times 10^{-7} \text{ s},$$

or 0.80 microseconds (= 0.80 μs).

(b) The acceleration follows from Eq. 3-12, $v_x = v_{x0} + a_x t$, or

$$a_x = \frac{v_x - v_{x0}}{t} = \frac{(500 - 1) \times 10^4 \text{ m/s}}{8.0 \times 10^{-7} \text{ s}} = +6.3 \times 10^{12} \text{ m/s}^2,$$

or 6 trillion meters per second per second! Although this acceleration is enormous by standards of the previous example, it occurs over an extremely short time. The acceleration **a** is in the positive x-direction, that is, in the direction in which the particle is moving, because a_x is positive.

3-10
FREELY FALLING BODIES

The most common example of motion with (nearly) constant acceleration is that of a body falling toward the earth. In the absence of air resistance we find that all bodies, regardless of their size, weight, or composition, fall with the same acceleration at the same point of the earth's surface, and if the distance covered is not too great, the acceleration remains constant throughout the fall. This ideal motion, in which air resistance and the small change in acceleration with altitude are neglected, is called "free fall."

The acceleration of a freely falling body is called the acceleration due to gravity and is denoted by the symbol **g**. Near the earth's surface its magnitude* is approximately 32 ft/s², 9.8 m/s², or 980 cm/s², and it is directed down toward the center of the earth. The variation of the exact value with latitude and altitude will be discussed later (Chapter 16).

The nature of the motion of a falling object was long ago a subject of interest in natural philosophy. Aristotle had asserted that "the downward movement . . . of any body endowed with weight is quicker in proportion to its size." It was not until many centuries later when Galileo Galilei (1564–1642) appealed to experiment to discover the truth, and then publicly proclaimed it, that Aristotle's authority on the matter was seriously challenged. In the later years of his life, Galileo wrote the treatise entitled *Dialogues Concerning Two New Sciences* in which he detailed his studies of motion.

Aristotle's belief that a heavier object will fall faster is a commonly held view. It appears to receive support from a well-known lecture demonstration in which a ball and a sheet of paper are dropped at the same instant, the ball reaching the floor much sooner than the paper. However, when the lecturer first crumples the paper tightly and then repeats the demonstration, both ball and

*See "Absolute value of g at the National Bureau of Standards" by D. R. Tate, *J. Res. NBS 70C*, April–June, 1966.

paper strike the floor at essentially the same time. In the former case, it is the effect of greater resistance of the air which makes the paper fall more slowly than the ball. In the latter case, the effect of air resistance on the paper is reduced and is about the same for both bodies, so that they fall at about the same rate. Of course, a direct test can be made by dropping bodies in vacuum. Even in easily obtainable partial vacuums we can show that a feather and a ball of lead thousands of times heavier drop at rates that are practically indistinguishable.

In Galileo's time, however, there was no effective way to obtain a partial vacuum, nor did equipment exist to time freely falling bodies with sufficient precision to obtain reliable numerical data. Nevertheless, Galileo proved his result by showing first that the character of the motion of a ball rolling down an incline was the same as that of a ball in free fall.* The incline merely served to reduce the effective acceleration of gravity and to slow the motion thereby. Time intervals measured, for example, by the volume of water discharged from a tank could then be used to test the speed and acceleration of this motion.** Galileo showed that if the acceleration along the incline is constant, the acceleration due to gravity must also be constant; for the acceleration along the incline is simply a component of the vertical acceleration of gravity, and along an incline of constant slope the ratio of the two accelerations remains fixed.

He found from his experiments that the distances covered in consecutive time intervals were proportional to the odd numbers 1, 3, 5, 7, . . ., etc. Total distances for consecutive intervals thus were proportional to $1 + 3$, $1 + 3 + 5$, $1 + 3 + 5 + 7$, and so on, that is, to the squares of the integers 1, 2, 3, 4, etc. But if the distance covered is proportional to the square of the elapsed time, velocity acquired is proportional to the elapsed time, a result which is true only if motion is uniformly accelerated. He found that the same results held regardless of the mass of the ball used.

3-11
EQUATIONS OF MOTION IN FREE FALL

We shall select a reference frame rigidly attached to the earth. The y-axis will be taken as positive vertically upward. Then the acceleration due to gravity g will be a vector pointing vertically down (toward the center of the earth) in the negative y-direction. (This choice is arbitrary. In other problems it may be convenient to choose down as positive.) Our equations for constant acceleration are applicable here. We simply replace x by y and set $y_0 = 0$ in Eqs. 3-12, 3-14, 3-15, and 3-16, obtaining

$$v_y = v_{y0} + a_y t,$$
$$y = \tfrac{1}{2}(v_{y0} + v_y)t,$$
$$y = v_{y0}t + \tfrac{1}{2}a_y t^2,$$
$$v_y^2 = v_{y0}^2 + 2a_y y,$$

(3-17)

and, for problems in free fall, we set $a_y = -g$. Notice that we have chosen the initial position as the origin, that is, we have chosen $y_0 = 0$ at $t = 0$. Note also that g is the magnitude of the acceleration due to gravity.

* See "Galileo's Discovery of the Law of Free Fall' by Stillman Drake, *Scientific American*, May, 1973.
** See "The Role of Music in Galileo's Experiments" by Stillman Drake, *Scientific American*, June, 1975.

A body is dropped from rest and falls freely. Determine the position and speed of the body after 1.0, 2.0, 3.0, and 4.0 s have elapsed.

EXAMPLE 8

We choose the starting point as the origin. We know the initial speed and the acceleration and we are given the time. To find the position we use

$$y = v_{y0}t - \tfrac{1}{2}gt^2.$$

Then, $v_{y0} = 0$ and $g = 32$ ft/s², and with $t = 1.0$ s we obtain

$$y = 0 - \tfrac{1}{2}(32 \text{ ft/s}^2)(1.0 \text{ s})^2 = -16 \text{ ft}.$$

To find the speed with $t = 1.0$ s, we use

$$v_y = v_{y0} - gt$$

and obtain $\qquad v_y = 0 - (32 \text{ ft/s}^2)(1.0 \text{ s}) = -32 \text{ ft/s}.$

After 1.0 s of falling from rest, the body is 16 ft (= 4.9 m) below its starting point and has a velocity directed downward whose magnitude is 32 ft/s (= 9.8 m/s); the negative signs for y and v_y show that the associated vectors each point in the negative y-direction, that is, downward.

Show that the values of y, v_y, and a_y obtained at times $t = 2.0$, 3.0, and 4.0 s are those shown in Fig. 3-8 and determine the metric equivalents.

figure 3-8
A body in free fall; showing y, v_y, and a_y at particular times t.

t s	y ft	v_y ft/s	a_y ft/s²
0	0	0	−32
1.0	−16	−32	−32
2.0	−64	−64	−32
3.0	−144	−96	−32
4.0	−256	−128	−32

A ball is thrown vertically upward from the ground with a speed of 80 ft/s (= 24.4 m/s).

EXAMPLE 9

(a) How long does it take to reach its highest point?

At its highest point, $v_y = 0$, and we have $v_{y0} = +80$ ft/s. To obtain the time t we use $v_y = v_{y0} - gt$, or

$$t = \frac{v_{y0} - v_y}{g}$$

$$t = \frac{(80 - 0) \text{ ft/s}}{32 \text{ ft/s}^2} = 2.5 \text{ s}.$$

(b) How high does the ball rise? Using only the original data, we choose the relation $v_y^2 = v_{y0}^2 - 2gy$, or

$$y = \frac{v_{y0}{}^2 - v_y{}^2}{2g},$$

$$= \frac{(80 \text{ ft/s})^2 - 0}{2 \times 32 \text{ ft/s}^2} = +100 \text{ ft } (= 30.5 \text{ m}).$$

(c) At what times will the ball be 96 ft (= 29 m) above the ground? Using $y = v_{y0}t - \frac{1}{2}gt^2$, we have

$$\tfrac{1}{2}gt^2 - v_{y0}t + y = 0,$$

$$\tfrac{1}{2}(32 \text{ ft/s}^2)t^2 - (80 \text{ ft/s})t + 96 \text{ ft} = 0,$$

or

$$t^2 - 5.0t + 6.0 = 0,$$

which yields $t = 2.0$ s and $t = 3.0$ s.

At $t = 2.0$ s, the ball is moving upward with a speed of 16 ft/s (= 4.9 m/s), for

$$v_y = v_{y0} - gt = 80 \text{ ft/s} - (32 \text{ ft/s}^2)(2.0 \text{ s}) = +16 \text{ ft/s}.$$

At $t = 3.0$ s, the ball is moving downward with the same speed, for

$$v_y = v_{y0} - gt = 80 \text{ ft/s} - (32 \text{ ft/s}^2)(3.0 \text{ s}) = -16 \text{ ft/s}.$$

Notice that in this 1.0-s interval the velocity changed by −32 ft/s (= −9.8 m/s), corresponding to an acceleration of −32 ft/s² (= −9.8 m/s²).

You should be able to convince yourself that in the absence of air resistance the ball will take as long to rise as to fall the same distance, and that it will have the same speed going down at each point as it had going up.

questions

1. Can you think of physical phenomena involving the earth in which the earth cannot be treated as a particle?

2. Each second a rabbit moves half the remaining distance from his nose to a head of lettuce. Does he ever get to the lettuce? What is the limiting value of his average velocity? Draw graphs showing his velocity and position as time increases.

3. Average speed can mean the magnitude of the average velocity vector. Another meaning given to it is that average speed is the total length of path traveled divided by the elapsed time. Are these meanings different? If so, give an example.

4. When the velocity is constant, does the average velocity over any time interval differ from the instantaneous velocity at any instant?

5. Is the average velocity of a particle moving along the x-axis $\frac{1}{2}(v_{x0} + v_x)$ when the acceleration is not uniform? Prove your answer with the use of graphs.

6. Does the speedometer on an automobile register speed as we defined it?

7. (a) Can a body have zero velocity and still be accelerating? (b) Can a body have a constant speed and still have a varying velocity? (c) Can a body have a constant velocity and still have a varying speed?

8. Can an object have an eastward velocity while experiencing a westward acceleration?

9. Can the direction of the velocity of a body change when its acceleration is constant?

10. Can a body be increasing in speed as its acceleration decreases? Explain.

11. Of the following situations, which one is impossible? (a) A body having velocity east and acceleration east; (b) a body having velocity east and acceleration west; (c) a body having zero velocity but acceleration not zero; (d) a body having constant acceleration and variable velocity; (e) a body having constant velocity and variable acceleration.

12. If a particle is released from rest ($v_{y0} = 0$) at $y_0 = 0$ at the time $t = 0$, Eq. 3-17 for constant acceleration says that it is at position y at two different times,

namely, $+\sqrt{2y/a_y}$ and $-\sqrt{2y/a_y}$. What is the meaning of the negative root of this quadratic equation?

13. What happens to our kinematic equations under the operation of time reversal, that is, replacing t by $-t$? Explain.

14. Consider a ball thrown vertically up. Taking air resistance into account, would you expect the time during which the ball rises to be longer or shorter than the time during which it falls?

15. (a) A body is thrown upwards with a certain speed on a world where the acceleration due to gravity is double that on earth. How high does it rise compared to the height it rises on earth? (b) If the initial speed were doubled, what change would that make?

16. Can there be motion in two dimensions with acceleration in only one dimension?

17. A person standing on the edge of a cliff at some height above the ground below throws one ball straight up with initial speed u and then throws another ball straight down with the same initial speed. Which ball, if either, has the larger speed when it hits the ground? Neglect air resistance.

18. A tube in the shape of a rectangle with rounded corners is placed in a vertical plane, as shown in Fig. 3-9. You introduce two ball bearings at the upper right-hand corner. One travels by path AB and the other by path CD. Which will arrive first at the lower left-hand corner?

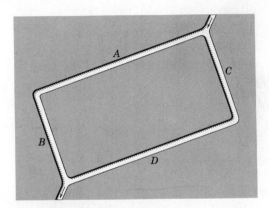

figure 3-9
Question 18.

19. We expect a truly general relation to be valid regardless of the choice of coordinate system. By demanding that general equations be dimensionally consistent we insure that the equations are valid regardless of the choice of units. Is there any need then for units or coordinate systems?

20. From what you know about angular measure, what *dimensions* would you assign to an angle? Can a quantity have units without having dimensions?

21. If m is a light stone and M is a heavy one, according to Aristotle M should fall faster than m. Galileo attempted to show that Aristotle's belief was logically inconsistent by the following argument. Tie m and M together to form a double stone. Then, in falling, m should retard M, because it tends to fall more slowly, and the combination would fall faster than m but more slowly than M; but according to Aristotle the double body $(M + m)$ is heavier than M and hence should fall faster than M.

   If you accept Galileo's reasoning as correct, can you conclude that M and m must fall at the same rate? What need is there for experiment in that case?

   If you believe Galileo's reasoning is incorrect, explain why.

SECTION 3-3

1. How far does a car, moving at 55 mi/h (88 km/h), travel forward during the one second of time that the driver takes to look at an accident on the side on the road?   *Answer:* 81 ft (24 m).

2. The legal speed limit on a thruway is changed from 65 mi/h (105 km/h) to

problems

55 mi/h (88.5 km/h) to conserve fuel. How much time is thereby added to the trip from the Buffalo entrance to the New York City exit of the New York Thruway for someone traveling at the legal speed limit over this 435-mile (700 km) stretch of highway?

3. Compare your average speed in the following two cases. (a) You walk 240 ft at a speed of 4.0 ft/s and then run 240 ft at a speed of 10 ft/s along a straight track. (b) You walk for 1.0 min at a speed of 4.0 ft/s and then run for 1.0 min at 10 ft/s along a straight track.   *Answer:* (a) 5.7 ft/s. (b) 7.0 ft/s.

4. A train moving at an essentially constant speed of 60 km/h moves east for 40 min, then in a direction 45° east of north for 20 min, and finally west for 50 min. What is the average velocity of the train during this run?

5. Two trains, each having a speed 40 km/h are headed for each other on the same straight track. A bird that can fly 60 km/h flies off one train when they are 80 km apart and heads directly for the other train. On reaching the other train it flies directly back to the first train, and so forth. (a) How many trips can the bird make from one train to the other before they crash? Explain. (b) What is the total distance the bird travels?
Answer: (a) an infinite number. (b) 60 km.

SECTION 3-6

6. A particle moving along the positive x-axis has the following positions at various times:

x(meters)	0.080	0.050	0.040	0.050	0.080	0.13	0.20
t(seconds)	0.0	1.0	2.0	3.0	4.0	5.0	6.0

(a) Plot displacement (not position) versus time. (b) Find the average velocity of the particle in the intervals 0.0 to 1.0 s, 0.0 to 2.0 s, 0.0 to 3.0 s, 0.0 to 4.0 s. (c) Find the slope of the curve drawn in part a at the points $t = 0.0$, 1.0, 2.0, 3.0, 4.0, and 5.0 s. (d) Plot the slope (units?) versus time. (e) From the curve of part (d) determine the acceleration of the particle at times $t = 2.0$, 3.0, and 4.0 s.

SECTION 3-7

7. The graph of x versus t (see Fig. 3-10a) is for a particle in straight line motion. (a) State for each interval whether the velocity v_x is +, −, or 0, and whether the acceleration a_x is +, −, or 0. The intervals are *OA, AB, BC,* and *CD.* (b) From the curve is there any *interval* over which the acceleration is obviously *not constant?* (Ignore the behavior at the end points of the intervals.)
Answer: (a)

	v_x	a_x	(b) No.
OA	+	0	
AB	+	−	
BC	0	0	
CD	−	+	

8. Answer the previous questions for the motion described by the graph of Fig. 3-10b.

figure 3-10a
Problem 7

figure 3-10b
Problem 8

9. An electron, starting from rest, has an acceleration that increases linearly with time, that is, $a = kt$, the change in acceleration being $k = (1.5 \text{ m/s}^2)/\text{s}$. (a) Plot a versus t during the first 10-s interval. (b) From the curve of part (a) plot the corresponding v versus t curve and estimate the electron's velocity 5.0 s after its motion starts. (c) From the v versus t curve of part (b) plot the corresponding x versus t curve and estimate how far the electron moved during the first 5.0 s of its motion.   *Answer:* (b) 19 m/s. (c) 31 m.

10. The position of a particle moving along the x-axis depends on the time according to the relation

$$x = \frac{v_{x0}}{k}(1 - e^{-kt})$$

in which v_{x0} and k are constants. (a) Plot a curve of x versus t. Notice that $x = 0$ at $t = 0$ and that $x = v_{x0}/k$ at $t = \infty$; that is, the total distance through which the particle moves is v_{x0}/k. (b) Show that the velocity v_x is given by

$$v_x = v_{x0}e^{-kt}$$

so that the velocity decreases exponentially with time from its initial value of v_{x0}, coming to rest only in infinite time. (c) Show that the acceleration a_x is given by

$$a_x = -kv_x$$

so that the acceleration is directed opposite to the velocity and has a magnitude proportional to the speed. (d) This particular motion is one with variable acceleration. Give a plausible physical argument explaining how it can take an infinite time to bring to rest a particle that travels a finite distance.

11. A particle moves along the x-axis with a displacement versus time as shown in Fig. 3-11. Sketch roughly curves of velocity versus time and acceleration versus time for this motion.

figure 3-11
Problem 11

SECTION 3-8

12. A jumbo jet needs to reach a speed of 225 mi/h (360 km/h) on the runway for takeoff. Assuming a constant acceleration and a runway 1.1 miles (1.8 km) long, what minimum acceleration from rest is required?

13. An automobile increases its speed uniformly from 25 to 55 km/h in one-half minute. A bicycle rider uniformly speeds up to 30 km/h from rest in one-half minute. Compare the accelerations.
Answer: Both accelerations are equal to 0.28 m/s².

14. A rocket-driven sled running on a straight level track is used to investigate the physiological effects of large accelerations on humans. One such sled can attain a speed of 1600 km/h in 1.8 s starting from rest. (a) Assume the acceleration is constant and compare it to g. (b) What is the distance traveled in this time?

15. A rocketship in free space moves with constant acceleration equal to 9.8 m/s². (a) If it starts from rest, how long will it take to acquire a speed one-tenth that of light? (b) How far will it travel in so doing?
Answer: (a) 36 days. (b) 4.6 × 10¹⁰ km.

16. An arrow while being shot from a bow was accelerated over a distance of 2.0 ft. If its speed at the moment it left the bow was 200 ft/s, what was the average acceleration imparted by the bow? Justify any assumptions you need to make.

17. A subway train accelerates from rest at one station at a rate of 1.20 m/s² for half of the distance to the next station, then decelerates at this same rate for the final half. If the stations are 1100 m apart, find (a) the time of travel between stations and (b) the maximum speed of the train.
Answer: (a) 60.6 s. (b) 36.4 m/s (= 81.4 mi/h).

18. Suppose that you were called upon to give some advice to a lawyer concerning the physics involved in one of her cases. The question is whether a driver was exceeding a 30 mi/h speed limit before he made an emergency stop, brakes locked and wheels sliding. The length of skid marks on the road was 19.2 ft. The police officer made the assumption that the maximum deceleration of the car would not exceed the acceleration of a freely falling body and arrested the driver for speeding. Was he speeding? Explain.

19. Two trains, one traveling at 60 mi/h and the other at 80 mi/h, are headed toward one another along a straight level track. When they are 2.0 miles apart, both engineers simultaneously see the other's train and apply their brakes. If the brakes decelerate each train at the rate of 3.0 ft/s², determine whether there is a collision. *Answer:* No.

20. A train started from rest and moved with constant acceleration. At one time it was traveling 30 ft/s and 160 ft farther on it was traveling 50 ft/s. Calculate (a) the acceleration, (b) the time required to travel the 160 ft mentioned, (c) the time required to attain the speed of 30 ft/s, (d) the distance moved from rest to the time the train had a speed of 30 ft/s.

21. An electron with initial velocity $v_{x0} = 1.0 \times 10^4$ m/s enters a region of width 1.0 cm where it is electrically accelerated (Fig. 3-12). It emerges with a velocity $v_x = 4.0 \times 10^6$ m/s. What was its acceleration, assumed constant? (Such a process occurs in the electron gun in a cathode-ray tube, used in television receivers and oscilloscopes.) *Answer:* 8.0×10^{14} m/s².

22. A meson is shot with speed 5.00×10^6 m/s into a region where an electric field produces an acceleration on the meson of magnitude 1.25×10^{14} m/s² directed opposite to the initial velocity. (a) How far does the meson travel before coming to rest? (b) How long does the meson remain at rest?

23. A car moving with constant acceleration covers the distance between two points 180 ft apart in 6.0 s. Its speed as it passes the second point is 45 ft/s. (a) What is its speed at the first point? (b) What is its acceleration? (c) At what prior distance from the first point was the car at rest?
Answer: (a) 15 ft/s. (b) 5.0 ft/s². (c) 23 ft.

24. The speed of an automobile traveling east is uniformly reduced from 45 mi/h to 30 mi/h in a distance of 264 ft. (a) What is the magnitude and direction of the constant acceleration? (b) How much time has elapsed during this deceleration? (c) If the car continues to decelerate at the same rate, how much time would elapse in bringing it to rest from 45 mi/h? (d) What distance is required to bring the car to rest from 45 mi/h? See Question 8.

25. At the instant the traffic light turns green, an automobile starts with a constant acceleration a_x of 6.0 ft/s². At the same instant a truck, traveling with a constant speed of 30 ft/s, overtakes and passes the automobile. (a) How far beyond the starting point will the automobile overtake the truck? (b) How fast will the car be traveling at that instant? (It is instructive to plot a qualitative graph of x versus t for each vehicle.)
Answer: (a) 300 ft. (b) 60 ft/s.

26. An automobile traveling 35 mi/h (56 km/h) is 110 ft (35 m) from a barrier when the driver slams on the brakes. Four seconds later the car hits the barrier. (a) What was the automobile's deceleration before impact? (b) How fast was the car traveling at impact?

figure 3-12
Problem 21

27. The engineer of a train moving at a speed v_1 sights a freight train a distance d ahead of him on the same track moving in the same direction with a slower speed v_2. He puts on the brakes and gives his train a constant deceleration a. Show that

$$\text{if } d > \frac{(v_1 - v_2)^2}{2a}, \text{ there will be no collision;}$$

$$\text{if } d < \frac{(v_1 - v_2)^2}{2a}, \text{ there will be a collision.}$$

(It is instructive to plot a qualitative graph of x versus t for each train.)

28. A driver's handbook states that an automobile with good brakes and going 50 mi/h can stop in a distance of 186 ft. The corresponding distance for 30 mi/h is 80 ft. Assume that the driver reaction time, during which the acceleration is zero, and the acceleration after he applies the brakes are both the same for the two speeds. Calculate (a) the driver reaction time and (b) the acceleration.

SECTION 3-9

29. The position of a particle along the x-axis depends on the time according to the equation

$$x = at^2 - bt^3,$$

where x is in meters and t in seconds. (a) What dimensions and units must a and b have? For the following, let their numerical values be 3.0 and 1.0, respectively. (b) At what time does the particle reach its maximum positive x-position? (c) What total length of path does the particle cover in the first 4.0 s? (d) What is its displacement during the first 4.0 s? (e) What is the particle's velocity at the end of each of the first four seconds? (f) What is the particle's acceleration at the end of each of the first four seconds? (g) What is the average velocity for the time interval $t = 2.0$ to $t = 4.0$ seconds?
Answer: (a) a: LT^{-2}, m/s²; b: LT^{-3}, m/s³. (b) t = 2 s. (c) 24 m. (d) −16 m. (e) 3.0, 0.0, −9.0, −24.0 m/s. (f) 0.0, −6.0, −12.0, −18.0 m/s². (g) −10 m/s.

SECTION 3-11

30. (a) With what speed must a ball be thrown vertically upward in order to rise to a height of 50 ft? (b) How long will it be in the air?

31. A tennis ball is dropped onto the floor from a height of 4.0 ft. It rebounds to a height of 3.0 ft. If the ball was in contact with the floor for 0.010 s, what was its average acceleration during contact?   Answer: 3000 ft/s²

32. While thinking of Isaac Newton, a person standing on a bridge overlooking a highway inadvertently drops an apple over the railing just as the front end of a truck passes directly below the railing. If the vehicle is moving at 55 km/h (34 mi/h) and is 12 m (39 ft) long, how far above the truck must the railing be if the apple just misses hitting the rear end of the truck?

33. A lead ball is dropped into a lake from a diving board 16 ft above the water. It hits the water with a certain velocity and then sinks to the bottom with this same constant velocity. It reaches the bottom 5.0 s after it is dropped. (a) How deep is the lake? (b) What is the average velocity of the ball? (c) Suppose all the water is drained from the lake. The ball is thrown from the diving board so that it again reaches the bottom in 5.0 s. What is the initial velocity of the ball?   Answer: (a) 128 ft. (b) 29 ft/s. (c) 51 ft/s upward.

34. A rocket is fired vertically and ascends with a constant vertical acceleration of 64 ft/s² for 1.0 min. Its fuel is then all used and it continues as a free particle. (a) What is the maximum altitude reached? (b) What is the total time elapsed from take-off until the rocket strikes the earth?

35. A balloon is ascending at the rate of 12 m/s at a height 80 m above the ground when a package is dropped. How long does it take the package to reach the ground?   Answer: 5.4 s.

36. A stone is dropped into the water from a bridge 144 ft (44 m) above the water. Another stone is thrown vertically down 1.0 s after the first is dropped. Both stones strike the water at the same time. (a) What was the initial speed of the second stone? (b) Plot speed versus time on a graph for each stone, taking zero time as the instant the first stone was released.

37. An open elevator is ascending with a constant speed v (32 ft/s). A ball is thrown straight up by a boy on the elevator when it is a height h (100 ft) above the ground. The initial speed of the ball with respect to the elevator is V_0 (64 ft/s). (a) What is the maximum height attained by the ball? (b) How long does it take for the ball to return to the elevator?
Answer: (a) 244 ft. (b) 4.0 s.

38. An arrow is shot straight up in the air with an initial speed of 250 ft/s. If on striking the ground it imbeds itself 6.0 in. into the ground, find (a) the acceleration (assumed constant) required to stop the arrow and (b) the time required for it to come to rest. Neglect air resistance during the arrow's flight.

39. A parachutist after bailing out falls 50 m without friction. When the parachute opens, he decelerates downward 2.0 m/s². He reaches the ground with a speed 3.0 m/s. (a) How long is the parachutist in the air? (b) At what height did he bail out?   *Answer:* (a) 17 s. (b) 290 m.

40. A shell is fired directly up from a gun; a rocket, propelled by burning fuel, takes off vertically from a launching area. Plot qualitatively (numbers not required) possible graphs of a_y versus t, of v_y versus t, and of y versus t for each. Take $t = 0$ at the instant the shell leaves the gun barrel or the rocket leaves the ground. Continue the plots until the rocket and the shell fall back to earth; neglect air resistance; assume that up is positive and down is negative.

41. If a body travels half its total path in the last second of its fall from rest, find (a) the time and (b) height of its fall. (c) Explain the physically unacceptable solution of the quadratic time equation.
Answer: (a) 3.4 s. (b) 57 m.

42. Two bodies begin a free fall from rest from the same height 1.0 s apart. How long after the first body begins to fall will the two bodies be 10 m apart?

43. A steel ball bearing is dropped from the roof of a building (the initial velocity of the ball is zero). An observer standing in front of a window 4.0 ft high notes that the ball takes $\frac{1}{8}$ s to fall from the top to the bottom of the window. The ball bearing continues to fall, makes a completely elastic collision with a horizontal sidewalk, and reappears at the bottom of the window 2.0 s after passing it on the way down. How tall is the building? (The ball will have the same speed at a point going up as it had going down after a completely elastic collision.)   *Answer:* 68 ft.

44. Water drips from the nozzle of a shower onto the floor 81 in. below. The drops fall at regular intervals of time, the first drop striking the floor at the instant the fourth drop begins to fall. Find the location of the individual drops when a drop strikes the floor.

45. An elevator ascends with an upward acceleration of 4.0 ft/s². At the instant its upward speed is 8.0 ft/s, a loose bolt drops from the ceiling of the elevator 9.0 ft from the floor. Calculate (a) the time of flight of the bolt from ceiling to floor and (b) the distance it has fallen relative to the elevator shaft.   *Answer:* (a) 0.71 s. (b) 2.3 ft.

46. A dog sees a flowerpot sail up and then back past a window 5.0 ft (1.5 m) high. If the total time the pot is in sight is 1.0 s, find the height above the window that the pot rises.

motion in a plane
4

In this chapter we return to a consideration of motion in two dimensions taken, for convenience, to be the x-y plane. Figure 4-1 shows a particle at time t moving along a curved path in this plane. Its *position*, or displacement from the origin, is measured by the vector **r**; its *velocity* is indicated by the vector **v** which, as we have seen in Section 3-4, must be tangent to the path of the particle. The *acceleration* is indicated by the vector **a**; the direction of **a**, as we shall see more explicitly later, does not bear any unique relationship to the path of the particle but depends rather on the rate at which the velocity **v** changes with time as the particle moves along its path.

4-1
DISPLACEMENT, VELOCITY, AND ACCELERATION

(a)

(b)

(c)

The vectors **r**, **v**, and **a** are interrelated (see Eqs. 3-4, 3-5, and 3-10) and can be expressed in terms of their components, using unit vector notation, as

$$\mathbf{r} = \mathbf{i}x + \mathbf{j}y, \tag{4-1}$$

$$\mathbf{v} = \frac{d\mathbf{r}}{dt} = \mathbf{i}v_x + \mathbf{j}v_y, \tag{4-2}$$

figure 4-1
A particle moves along a curved path in the x-y plane. (a) Its position **r**, (b) its velocity **v**, and (c) its acceleration **a** are shown at time t, along with the vector components of those vectors. Note that x, y, v_x, v_y, and a_x are positive but that a_y is negative. Compare to Fig. 3-3.

53

and
$$\mathbf{a} = \frac{d\mathbf{v}}{dt} = \mathbf{i}a_x + \mathbf{j}a_y. \qquad (4\text{-}3)$$

These equations can easily be extended to three dimensions by adding to them the terms $\mathbf{k}z$, $\mathbf{k}v_z$, and $\mathbf{k}a_z$, respectively in which $\mathbf{k}$ is a unit vector in the z-direction.

In Chapter 3 we considered the special case in which the particle moved in one dimension only, say along the x-axis, where the vectors $\mathbf{r}$, $\mathbf{v}$, and $\mathbf{a}$ were directed along this axis, either in the positive x-direction or the negative x-direction. The components y, v_y, and a_y were zero and we described the motion in terms of equations relating the scalar quantities x, v_x, and a_x. Or, when the particle moved along the y-axis, the components x, v_x, and a_x were zero and the motion was described in terms of equations relating the scalar quantities y, v_y, and a_y. In this chapter we consider motion in the x-y plane so that, in general, both sets of components have nonzero values.

Let us consider first the special case of motion in a plane with constant acceleration. Here, as the particle moves, the acceleration $\mathbf{a}$ does not vary either in magnitude or in direction. Hence the components of $\mathbf{a}$ also will not vary, that is, $a_x =$ constant and $a_y =$ constant. We then have a situation which can be described as the sum of two component motions occurring simultaneously with constant acceleration along each of two perpendicular directions. The particle will move, in general, along a curved path in the plane. This may be so even if one component of the acceleration, say a_x, is zero, for then the corresponding component of the velocity, say v_x, may have a constant, nonzero value. An example of this latter situation is the motion of a projectile which follows a curved path in a vertical plane and, neglecting the effects of air resistance, is subject to a constant acceleration $\mathbf{g}$ directed down along the y-axis only.

We can obtain the general equations for plane motion with constant $\mathbf{a}$ simply by setting

$$a_x = \text{constant} \quad \text{and} \quad a_y = \text{constant}.$$

The equations for constant acceleration, summarized in Table 3-1, then apply to both the x- and y-components of the position vector $\mathbf{r}$, the velocity vector $\mathbf{v}$, and the acceleration vector $\mathbf{a}$ (see Table 4-1).

4-2
MOTION IN A PLANE WITH CONSTANT ACCELERATION

Table 4-1
Motion with constant acceleration in the x-y plane

Equation No.	x-Motion Equations	Equation No.	y-Motion Equations
4-4a	$v_x = v_{x0} + a_x t$	4-4a'	$v_y = v_{y0} + a_y t$
4-4b	$x = x_0 + \frac{1}{2}(v_{x0} + v_x)t$	4-4b'	$y = y_0 + \frac{1}{2}(v_{y0} + v_y)t$
4-4c	$x = x_0 + v_{x0}t + \frac{1}{2}a_x t^2$	4-4c'	$y = y_0 + v_{y0}t + \frac{1}{2}a_y t^2$
4-4d	$v_x{}^2 = v_{x0}{}^2 + 2a_x(x - x_0)$	4-4d'	$v_y{}^2 = v_{y0}{}^2 + 2a_y(y - y_0)$

The two sets of equations in Table 4-1 are related in that the time parameter t is the same for each, since t represents the time at which the particle, moving in a curved path in the x-y plane, occupied a position described by the position components x and y.

The equations of motion in Table 4-1 may also be expressed in vector form. For example, substituting Eqs. 4-4a, 4a′ into Eq. 4-2 yields

$$\mathbf{v} = \mathbf{i}v_x + \mathbf{j}v_y$$
$$= \mathbf{i}(v_{x0} + a_x t) + \mathbf{j}(v_{y0} + a_y t)$$
$$= (\mathbf{i}v_{x0} + \mathbf{j}v_{y0}) + (\mathbf{i}a_x + \mathbf{j}a_y)t.$$

The first quantity in parentheses is the initial velocity vector $\mathbf{v}_0$ (see Eq. 4-2) and the second is the (constant) acceleration vector $\mathbf{a}$ (see Eq. 4-3). Thus the vector relation

$$\mathbf{v} = \mathbf{v}_0 + \mathbf{a}t \tag{4-5a}$$

is equivalent to the two scalar relations Eqs. 4-4a, a′ in Table 4-1. It shows simply and compactly that the velocity $\mathbf{v}$ at time t is the sum of the initial velocity $\mathbf{v}_0$ which the particle would have in the absence of acceleration plus the (vector) change in velocity, $\mathbf{a}t$, acquired during the time t under the constant acceleration $\mathbf{a}$. Similarly, the scalar equations 4-4c, c′ are equivalent to the single vector equation

$$\mathbf{r} = \mathbf{r}_0 + \mathbf{v}_0 t + \tfrac{1}{2}\mathbf{a}t^2, \tag{4-5b}$$

which is also easily interpreted. The proof of this and other relations is left to Problem 3.

4-3
PROJECTILE MOTION

An example of curved motion with constant acceleration is projectile motion. This is the two-dimensional motion of a particle thrown obliquely into the air. The ideal motion of a baseball or a golf ball is an example of projectile motion.* We assume that we can neglect the effect of the air on this motion.

The motion of a projectile is one of constant acceleration $\mathbf{g}$, directed downward, and thus should be described by the equations in Table 4-1. There is no horizontal component of acceleration. If we choose a coordinate system with the positive y-axis vertically upward, we may put $a_y = -g$ and $a_x = 0$ in these equations.

Let us further choose the origin of our coordinate system to be the point at which the projectile begins its flight (see Fig. 4-2). Hence the origin will be the point at which the ball leaves the thrower's hand or the fuel in the rocket burns out, for example. In Table 4-1 this choice of origin implies that $x_0 = y_0 = 0$. The velocity at $t = 0$, the instant the projectile begins its flight, is $\mathbf{v}_0$, which makes an angle θ_0 with the positive x-direction. The x- and y-components of $\mathbf{v}_0$ (see Fig. 4-2) are then

$$v_{x0} = v_0 \cos \theta_0 \quad \text{and} \quad v_{y0} = v_0 \sin \theta_0.$$

Because there is no horizontal component of acceleration, the horizontal component of the velocity will be constant. In Eq. 4-4a we set $a_x = 0$ and $v_{x0} = v_0 \cos \theta_0$, so that

$$v_x = v_0 \cos \theta_0. \tag{4-6a}$$

The horizontal velocity component retains its initial value throughout the flight.

The vertical component of the velocity will change with time in accordance with vertical motion with constant downward acceleration.

---

* See Galileo Galilei, *Dialogues Concerning Two New Sciences*, the "Fourth Day," for a fascinating discussion of Galileo's research on projectiles.

In Eq. 4-4a' we set

$$a_y = -g \quad \text{and} \quad v_{y0} = v_0 \sin \theta_0,$$

so that

$$v_y = v_0 \sin \theta_0 - gt. \tag{4-6a'}$$

The vertical velocity component is that of free fall. Indeed, if we view the motion of Fig. 4-2 from a reference frame that moves to the right with a speed v_{x0}, the motion will be that of an object thrown vertically upward with an initial speed $v_0 \sin \theta_0$.

The magnitude of the resultant velocity vector at any instant is

$$v = \sqrt{v_x^2 + v_y^2}. \tag{4-7}$$

The angle θ that the velocity vector makes with the horizontal at that instant is given by

$$\tan \theta = \frac{v_y}{v_x}.$$

The velocity vector is tangent to the path of the particle at every point, as shown in Fig. 4-2.

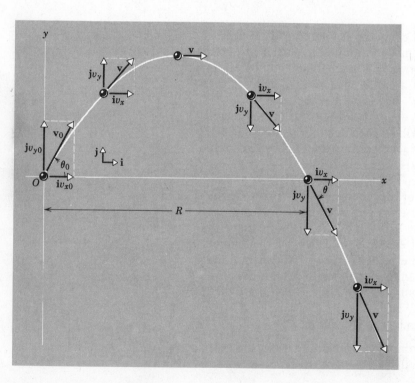

figure 4-2
The trajectory of a projectile, showing the initial velocity $\mathbf{v}_0$ and its vector components and also the velocity $\mathbf{v}$ and its vector components at five later times. Note that $v_x = v_{x0}$ throughout the flight. The distance R is called the range.

The x-coordinate of the particle's position at any time, obtained from Eq. 4-4c with $x_0 = 0$, $a_x = 0$, and $v_{x0} = v_0 \cos \theta_0$, is

$$x = (v_0 \cos \theta_0)t. \tag{4-6c}$$

The y-coordinate, obtained from Eq. 4-4c' with $y_0 = 0$, $a_y = -g$, and $v_{y0} = v_0 \sin \theta_0$, is

$$y = (v_0 \sin \theta_0)t - \tfrac{1}{2}gt^2. \tag{4-6c'}$$

Equations 4-6c, c' give us x and y as functions of the common parameter t, the time of flight. By combining and eliminating t from them, we obtain

$$y = (\tan \theta_0)x - \frac{g}{2(v_0 \cos \theta_0)^2}\, x^2, \qquad (4\text{-}8)$$

which relates y to x and is the equation of the trajectory of the projectile. Since v_0, θ_0, and g are constants, this equation has the form

$$y = bx - cx^2,$$

the equation of a parabola. Hence the trajectory of a projectile is parabolic.*

EXAMPLE 1

A plane is flying at a constant horizontal velocity of 500 km/h at an elevation of 5.0 km toward a point directly above its target. At what angle of sight ϕ should a survival package be released to strike the target (Fig. 4-3)?

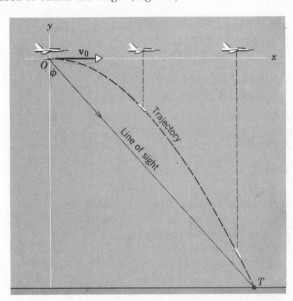

figure 4-3
Example 1. A survival package is released from an airplane with horizontal velocity v_0.

We choose a reference frame fixed with respect to the earth, its origin O being the release point. The motion of the package at the moment of release is the same as that of the plane. Hence the initial package velocity v_0 is horizontal and its magnitude is 500 km/h. The angle of projection θ_0 is zero.

We find the time of fall from Eq. 4-6c'. With $\theta_0 = 0$ and $y = 5.0$ km this gives

$$t = \sqrt{\frac{2y}{g}} = \sqrt{-\frac{(2)(-5.0 \times 10^3 \text{ m})}{(9.8 \text{ m/s}^2)}} = 31.9 \text{ s}$$

Note that the time of fall does not depend on the speed of the plane for a horizontal projection. (See, however, Problem 11.)

The horizontal distance traveled by the package in this time is given by Eq. 4-6c, $x = (v_0 \cos \theta_0)t$, or

$$x = (500 \text{ km/h}) \times 10^3 \text{m/km}) \times (1 \text{ h}/3600 \text{ s}) \times (31.9 \text{ s}) = 4430 \text{ m}.$$

so that the angle of sight (Fig. 4-3) should be

$$\phi = \tan^{-1} \frac{x}{|y|} = \tan^{-1} \frac{4430 \text{ m}}{5000 \text{ m}} = 42°.$$

Does the motion of the package appear to be parabolic when viewed from a reference frame fixed with respect to the plane?

---

* See "Galileo's Discovery of the Parabolic Trajectory" by Stillman Drake and James MacLachlan in *Scientific American*, March 1975.

A soccer player kicks a ball at an angle of 37° from the horizontal with an initial speed of 50 ft/s. (A right triangle, one of whose angles is 37°, has sides in the ratio 3:4:5, or 6:8:10.) Assuming that the ball moves in a vertical plane:

EXAMPLE 2

(a) Find the time t_1 at which the ball reaches the highest point of its trajectory.

At the highest point, the vertical component of velocity v_y is zero. Solving Eq. 4-6a' for t, we obtain

$$t = \frac{v_0 \sin \theta_0 - v_y}{g}.$$

With

$$v_y = 0, \qquad v_0 = 50 \text{ ft/s}, \qquad \theta_0 = 37°, \qquad g = 32 \text{ ft/s}^2,$$

we have

$$t_1 = \frac{[50(\tfrac{6}{10}) - 0] \text{ ft/sec}}{32 \text{ ft/sec}^2} = \frac{15}{16} \text{ s}.$$

(b) How high does the ball go?

The maximum height is reached at $t = 15/16$ s. By using Eq. 4-6c',

$$y = (v_0 \sin \theta_0)t - \tfrac{1}{2}gt^2,$$

we have

$$y_{max} = (50 \text{ ft/s})(\tfrac{6}{10})(\tfrac{15}{16} \text{ s}) - \tfrac{1}{2}(32 \text{ ft/s}^2)(\tfrac{15}{16})^2 \text{ s}^2 = 14 \text{ ft}.$$

(c) What is the range of the ball and how long is it in the air?

The horizontal distance from the starting point at which the ball returns to its original elevation (ground level) is the *range* R. We set $y = 0$ in Eq. 4-6c' and find the time t_2 required to transverse this range. We obtain

$$t_2 = \frac{2v_0 \sin \theta_0}{g} = \frac{2(50 \text{ ft/s})(\tfrac{6}{10})}{32 \text{ ft/s}^2} = \frac{15}{8} \text{ sec}.$$

Notice that $t_2 = 2t_1$. This corresponds to the fact that the same time is required for the ball to go up (reach its maximum height from ground) as is required for the ball to come down (reach the ground from its maximum height).

The range R can then be obtained by inserting this value t_2 for t in Eq. 4-6c. We obtain, from $x = (v_0 \cos \theta_0)t$,

$$R = (v_0 \cos \theta_0)t_2 = (50 \text{ ft/s})(\tfrac{8}{10})(\tfrac{15}{8} \text{ s}) = 75 \text{ ft}.$$

(d) What is the velocity of the ball as it strikes the ground? From Eq. 4-6a we obtain

$$v_x = v_0 \cos \theta_0 = (50 \text{ ft/s})(\tfrac{8}{10}) = 40 \text{ ft/s}.$$

From Eq. 4-6a' we obtain for $t = t_2 = \tfrac{15}{8}$ s,

$$v_y = v_0 \sin \theta_0 - gt = (50 \text{ ft/s})(\tfrac{6}{10}) - (32 \text{ ft/s}^2)(\tfrac{15}{8} \text{ s}) = -30 \text{ ft/s}.$$

Hence, from Eq. 4-7,

$$v = \sqrt{v_x^2 + v_y^2} = \sqrt{(40 \text{ ft/s})^2 + (-30 \text{ ft/s})^2} = 50 \text{ ft/s},$$

and

$$\tan \theta = v_y/v_x = -\tfrac{30}{40},$$

so that $\theta = -37°$, or 37° *clockwise* from the x-axis. Notice that $\theta = -\theta_0$, as we expect from symmetry (Fig. 4-2).

In a favorite lecture demonstration an air gun is sighted at an elevated target which is released in free fall by a trip mechanism as the "bullet" leaves the muzzle. No matter what the initial speed of the bullet, it always hits the falling target.

EXAMPLE 3

The simplest way to understand this is the following. If there were no acceleration due to gravity, the target would not fall and the bullet would move

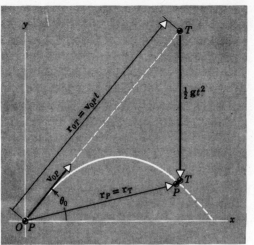

figure 4-4
Example 3. In the motion of a projectile, its displacement from the origin at any time t can be thought of as the sum of two vectors: $\mathbf{v}_{0P}t$, directed along $\mathbf{v}_{0P}$, and $\frac{1}{2}\mathbf{g}t^2$, directed down.

along the line of sight directly into the target (Fig. 4-4). The effect of gravity is to cause each body to accelerate down at the same rate from the position it would otherwise have had. Therefore, in the time t, the bullet will fall a distance $\frac{1}{2}gt^2$ from the position it would have had along the line of sight and the target will fall the same distance from its starting point. When the bullet reaches the line of fall of the target, it will be the same distance below the target's initial position as the target is and hence the collision. If the bullet moves faster than shown in the figure (v_0 larger), it will have a greater range and will cross the line of fall at a higher point; but since it gets there sooner, the target will fall a correspondingly smaller distance in the same time and collide with it. A similar argument holds for slower speeds.

For an equivalent analysis, let us use Eq. 4-5b.

$$\mathbf{r} = \mathbf{r}_0 + \mathbf{v}_0 t + \tfrac{1}{2}\mathbf{a}t^2$$

to describe the positions of the projectile and the target at any time t. For the projectile P, $\mathbf{r}_0 = 0$ and $\mathbf{a} = \mathbf{g}$, and we have

$$\mathbf{r}_P = \mathbf{v}_{0P}t + \tfrac{1}{2}\mathbf{g}t^2.$$

For the target T, $\mathbf{r}_0 = \mathbf{r}_{0T}$, $\mathbf{v}_0 = 0$, and $\mathbf{a} = \mathbf{g}$, leading to

$$\mathbf{r}_T = \mathbf{r}_{0T} + \tfrac{1}{2}\mathbf{g}t^2.$$

If there is a collision, we must have $\mathbf{r}_P = \mathbf{r}_T$. Inspection shows that this will always occur at a time t given by $\mathbf{r}_{0T} = \mathbf{v}_{0P}t$, that is, in the time $t (= r_{0T}/v_{0P})$ required for the projectile to travel to the target position along the line of sight, assuming that its initial velocity remains unchanged.

In Section 3-6 we saw that acceleration arises from a change in velocity. In the simple case of free fall the velocity changed in magnitude only, but not in direction. For a particle moving in a circle with constant speed, called *uniform circular motion*, the velocity vector changes continuously in direction but not in magnitude. We seek now to obtain the acceleration in uniform circular motion.

The situation is shown in Fig. 4-5a. Let P be the position of the particle at the time t and P' its position at the time $t + \Delta t$. The velocity at P is $\mathbf{v}$, a vector tangent to the curve at P. The velocity at P' is $\mathbf{v}'$, a vector tangent to the curve at P'. Vectors $\mathbf{v}$ and $\mathbf{v}'$ are equal in magnitude, the speed being constant, but their directions are different. The length of path traversed during Δt is the arc length PP', which is equal to $v \Delta t$, v being the constant speed.

Now redraw the vectors $\mathbf{v}$ and $\mathbf{v}'$, as in Fig. 4-5b, so that they originate

4-4
UNIFORM CIRCULAR MOTION

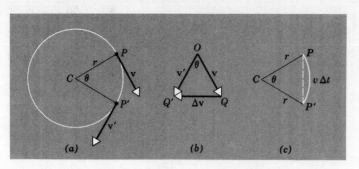

at a common point. We are free to do this as long as the magnitude and direction of each vector are the same as in Fig. 4-5a. This diagram (Fig. 4-5b) enables us to see clearly the *change in velocity* as the particle moved from P to P'. This change, $\mathbf{v}' - \mathbf{v} = \Delta\mathbf{v}$, is the vector which must be added to $\mathbf{v}$ to get $\mathbf{v}'$. Notice that it points inward, approximately toward the center of the circle.

Now the triangle OQQ' formed by $\mathbf{v}$, $\mathbf{v}'$, and $\Delta\mathbf{v}$ is similar to the triangle CPP' (Fig. 4-5c) formed by the chord PP' and the radii CP and CP'. This is so because both are isosceles triangles having the same vertex angle; the angle θ between $\mathbf{v}$ and $\mathbf{v}'$ is the same as the angle PCP' because $\mathbf{v}$ is perpendiculat to CP and $\mathbf{v}'$ is perpendicular to CP'. We can therefore write

$$\frac{\Delta v}{v} = \frac{v\,\Delta t}{r}, \qquad \text{approximately,}$$

the chord PP' being taken equal to the arc length PP'. This relation becomes more nearly exact as Δt is diminished, since the chord and the arc then approach each other. Notice also that $\Delta\mathbf{v}$ approaches closer and closer to a direction perpendicular to $\mathbf{v}$ and $\mathbf{v}'$ as Δt is diminished and therefore approaches closer and closer to a direction pointing to the exact center of the circle. It follows from this relation that

$$\frac{\Delta v}{\Delta t} = \frac{v^2}{r}, \qquad \text{approximately,}$$

and in the limit when $\Delta t \to 0$ this expression becomes exact. We therefore obtain

$$a = \lim_{\Delta t \to 0} \frac{\Delta v}{\Delta t} = \frac{v^2}{r} \qquad (4\text{-}9)$$

as the magnitude of the acceleration. The direction of $\mathbf{a}$ is instantaneously along a radius inward toward the center of the circle.

Figure 4-6 shows the instantaneous relation between $\mathbf{v}$ and $\mathbf{a}$ at various points of the motion. The magnitude of $\mathbf{v}$ is constant, but its direction changes continuously. This gives rise to an acceleration $\mathbf{a}$ which is also constant in magnitude (but not zero) but continuously changing in direction. The velocity $\mathbf{v}$ is always tangent to the circle in the direction of motion; the acceleration $\mathbf{a}$ is always directed radially inward. Because of this, $\mathbf{a}$ is called a radial, or *centripetal*, acceleration. Centripetal means "seeking a center."

Both in free fall and in projectile motion $\mathbf{a}$ is constant in direction and magnitude and we can use the equations developed for constant acceleration (see Table 4-1). We cannot use these equations for uniform circular motion because $\mathbf{a}$ varies in direction and is therefore not constant.

The units of centripetal acceleration are the same as those of an acceleration resulting from a change in the magnitude of a velocity.

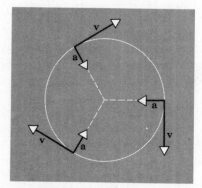

figure 4-6
In uniform circular motion the acceleration $\mathbf{a}$ is always directed toward the center of the circle and hence is perpendicular to $\mathbf{v}$.

Dimensionally, we have

$$\frac{v^2}{r} = \left(\frac{\text{length}}{\text{time}}\right)^2 \Big/ \text{length} = \frac{\text{length}}{\text{time}^2} \quad \text{or} \quad \frac{L}{T^2},$$

which are the dimensions of acceleration. The units therefore may be ft/s² and m/s², among others.

The acceleration resulting from a change in direction of a velocity is just as real and just as much an acceleration in every sense as that arising from a change in magnitude of a velocity. By definition, acceleration is the time rate of change of velocity, and velocity, being a vector, can change in direction as well as magnitude. If a physical quantity is a vector, its directional aspects cannot be ignored, for their effects will prove to be every bit as important and real as those produced by changes in magnitude.

It is worth emphasizing at this point that there need not be any motion in the direction of an acceleration and that there is no fixed relation in general between the directions of **a** and **v**. In Fig. 4-7 we give examples in which the angle between **v** and **a** varies from 0 to 180°. Only in one case, $\theta = 0°$, is the motion in the direction of **a**.

figure 4-7
Showing the relation between **v** and **a** for various motions.

The moon revolves about the earth, making a complete revolution in 27.3 days. Assume that the orbit is circular and has a radius of 239,000 miles. What is the magnitude of the acceleration of the moon toward the earth?

We have $r = 239{,}000$ mi $= 3.85 \times 10^8$ m. The time for one complete revolution, called the period, is $T = 27.3$ d $= 2.36 \times 10^6$ s. The speed of the moon (assumed constant) is therefore

$$v = 2\pi r/T = 1020 \text{ m/s}.$$

The centripetal acceleration is

$$a = \frac{v^2}{r} = \frac{(1020 \text{ m/s})^2}{3.85 \times 10^8 \text{ m}} = 0.00273 \text{ m/s}^2, \quad \text{or} \quad \text{only } 2.78 \times 10^{-4} \, g.$$

EXAMPLE 4

Calculate the speed of an earth satellite, assuming that it is traveling at an altitude h of 140 miles above the surface of the earth where $g = 30$ ft/s². The radius R of the earth is 3960 mi.

Like any free object near the earth's surface the satellite has an acceleration g toward the earth's center. It is this acceleration that causes it to follow the circular path. Hence the centripetal acceleration is g, and from Eq. 4-9, $a = v^2/r$, we have

$$g = v^2/(R + h),$$

or

$$v = \sqrt{(R + h)g} = \sqrt{(3960 \text{ mi} + 140 \text{ mi})(5280 \text{ ft/mi})(30 \text{ ft/s}^2)}$$
$$= 2.55 \times 10^4 \text{ ft/s} = 17{,}400 \text{ mi/h}.$$

EXAMPLE 5

Let us now derive Eq. 4-9 using vector methods. Figure 4-8a shows a particle in uniform circular motion about the origin O of a reference frame. For this motion the polar coordinates r, θ are more useful than the rectangular coordinates x, y because r remains constant throughout the motion and θ increases in a simple linear way with time; the behavior of x and y during such motion is more complex. The two sets of coordinates are related by

$$r = \sqrt{x^2 + y^2} \qquad \text{and} \qquad \theta = \tan^{-1} y/x \qquad (4\text{-}10a)$$

or by the reciprocal relations

$$x = r \cos \theta \qquad \text{and} \qquad y = r \sin \theta. \qquad (4\text{-}10b)$$

In rectangular coordinate systems we used the unit vectors $\mathbf{i}$ and $\mathbf{j}$ to describe motion in the x-y plane. Here we find it more convenient to introduce two new unit vectors $\mathbf{u}_r$ and $\mathbf{u}_\theta$. These, like $\mathbf{i}$ and $\mathbf{j}$, have unit length and are dimensionless; they designate direction only.

The unit vector $\mathbf{u}_r$ at any point is in the direction of increasing $\mathbf{r}$ at that point; it is directed radially outward from the origin. The unit vector $\mathbf{u}_\theta$ at any point is in the direction of increasing θ at that point; it is always tangent to a circle through the point in a counterclockwise direction. As Fig. 4-8a shows, $\mathbf{u}_r$ and $\mathbf{u}_\theta$ are at right angles to each other. The unit vectors $\mathbf{u}_r$ and $\mathbf{u}_\theta$ differ from the unit vectors $\mathbf{i}$ and $\mathbf{j}$ in that the directions of $\mathbf{u}_r$ and $\mathbf{u}_\theta$ vary from point to point in the plane; the unit vectors $\mathbf{u}_r$ and $\mathbf{u}_\theta$ are thus *not* constant vectors.

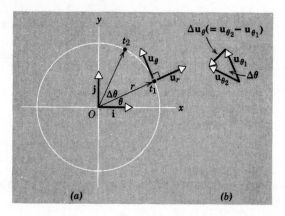

(a)   (b)

figure 4-8
(a) A particle moving counterclockwise in a circle of radius r. (b) The unit vectors $\mathbf{u}_{\theta 1}$ and $\mathbf{u}_{\theta 2}$ at times t_1 and t_2 respectively, and the change $\Delta\mathbf{u}_\theta\ (= \mathbf{u}_{\theta 2} - \mathbf{u}_{\theta 1})$.

In terms of $\mathbf{u}_r$ and $\mathbf{u}_\theta$ the motion of a particle moving counterclockwise at uniform speed v in a circle about the origin in Fig. 4-8a can be described by the vector equation

$$\mathbf{v} = \mathbf{u}_\theta v. \qquad (4\text{-}11)$$

This relation tells us, correctly, that the direction of $\mathbf{v}$ (which is the same as the direction of $\mathbf{u}_\theta$) is tangent to the circle and that the magnitude of $\mathbf{v}$ is the constant quantity v (because the magnitude of $\mathbf{u}_\theta$ is unity).

To find the acceleration we combine Eqs. 4-3 and 4-11, yielding

$$\mathbf{a} = \frac{d\mathbf{v}}{dt} = \frac{d\mathbf{u}_\theta}{dt}\, v. \qquad (4\text{-}12)$$

Note that v in Eq. 4-11 is a constant, but $\mathbf{u}_\theta$ is not since its direction changes as the particle moves. To evaluate $d\mathbf{u}_\theta/dt$, consider Fig. 4-8b which shows the unit vectors $\mathbf{u}_{\theta 1}$ and $\mathbf{u}_{\theta 2}$ corresponding to an elapsed time $\Delta t\ (= t_2 - t_1)$ for the moving particle. The vector $\Delta\mathbf{u}_\theta\ (= \mathbf{u}_{\theta 2} - \mathbf{u}_{\theta 1})$ points radially inward toward the origin in the limiting case as $\Delta t \to 0$. In other words, $d\mathbf{u}_\theta$ at any point has the direction of $-\mathbf{u}_r$. The angle between $\mathbf{u}_{\theta 2}$ and $\mathbf{u}_{\theta 1}$ in the figure is $\Delta\theta$, which is the angle swept out by a radial line from the origin to the particle in time Δt. The magnitude of $\Delta\mathbf{u}_\theta$ is simply $\Delta\theta$; bear in mind that the vectors $\mathbf{u}_{\theta 1}$ and $\mathbf{u}_{\theta 2}$ in Fig. 4-8b have the magnitude unity. Thus

$$\frac{d\mathbf{u}_\theta}{dt} = -\mathbf{u}_r \lim_{\Delta t \to 0} \frac{\Delta\theta}{dt} = -\mathbf{u}_r \frac{d\theta}{dt}$$

and, from Eq. 4-12,

$$\mathbf{a} = \frac{d\mathbf{u}_\theta}{dt}\, v = -\mathbf{u}_r \frac{d\theta}{dt}\, v. \qquad (4\text{-}13)$$

Now, $d\theta/dt$ is the uniform angular rotation rate of the particle and is given by

$$\frac{d\theta}{dt} = \frac{2\pi \text{ radians}}{\text{time for one revolution}} = \frac{2\pi}{2\pi r/v} = \frac{v}{r}.$$

Putting this into Eq. 4-13 leads us finally to

$$\mathbf{a} = -\mathbf{u}_r \frac{v^2}{r} \qquad (4\text{-}14)$$

which tells us that the acceleration in uniform circular motion has a magnitude v^2/r (see Eq. 4-9) and points radially inward (note the factor $-\mathbf{u}_r$). The vector relation Eq. 4-14 thus tells us *both* the magnitude *and* the direction of the centripetal acceleration **a.** Note that, as expected, **a** has a constant magnitude but changes continually in direction because $\mathbf{u}_r$ changes continually in direction.

We now consider the more general case of circular motion in which the speed v of the moving particle is *not* constant. We shall use vector methods in polar coordinates.

As before, the velocity is given by Eq. 4-11, or

$$\mathbf{v} = \mathbf{u}_\theta v$$

except that, in this case, not only $\mathbf{u}_\theta$ but also v varies with time. Recalling the formula for the derivative of a product, we obtain for the acceleration

$$\mathbf{a} = \frac{d\mathbf{v}}{dt} = \mathbf{u}_\theta \frac{dv}{dt} + v \frac{d\mathbf{u}_\theta}{dt} \qquad (4\text{-}15)$$

In Eq. 4-12 the first term in this equation was not present because, v being there assumed to be constant, dv/dt was zero. The last term in Eq. 4-15 reduces, as we saw in the last section, to $-\mathbf{u}_r(v^2/r)$. We can now write Eq. 4-15 as

$$\mathbf{a} = \mathbf{u}_\theta a_T - \mathbf{u}_r a_R, \qquad (4\text{-}16)$$

in which $a_T = dv/dt$ and $a_R = v^2/r$. The first term, $\mathbf{u}_\theta a_T$, is the vector component of **a** that is tangent to the path of the particle and arises from a change in the *magnitude* of the velocity in circular motion (see Fig. 4-9). This term and a_T are called the *tangential acceleration*. The second term $-\mathbf{u}_r a_R$ is the vector component of **a** directed radially in toward the center of the circle and arises from a change in the *direction* of the velocity in circular motion (see Fig. 4-9). This term and a_R are called the *centripetal acceleration*.

The magnitude of the instantaneous acceleration is

$$a = \sqrt{a_T{}^2 + a_R{}^2} \qquad (4\text{-}17)$$

If the speed is constant, then $a_T = dv/dt = 0$ and Eq. 4-16 reduces to Eq. 4-14.

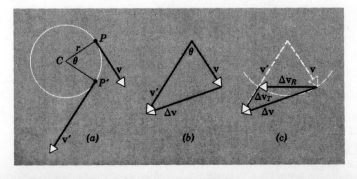

figure 4-9

(a) In nonuniform circular motion the speed is variable. *(b)* The change in velocity $\Delta\mathbf{v}$ in going from P to P' is made up of two parts: *(c)* $\Delta\mathbf{v}_R$ caused by the change in direction of **v**, and $\Delta\mathbf{v}_T$ caused by the change in magnitude of **v**. In the limit as $\Delta t \to 0$, $\Delta\mathbf{v}_R$ points toward the center C of the circle and $\Delta\mathbf{v}_T$ is tangent to the circular path.

When the speed v is not constant, a_T is not zero and a_R varies from point to point. If the speed changes at a rate that is not constant, then a_T will also vary from point to point.

If the motion is not circular, the formulas for a_T ($= dv/dt$) and for a_R ($= v^2/r$) can still be applied if instead of using for r the magnitude of the radius vector from the origin, we substitute the radius of curvature of the path at the instantaneous position of the particle. Then a_T gives the component of acceleration tangent to the curve at that position, and a_R gives the component of acceleration normal to the curve at that position. Figure 4-10 shows the track left in a liquid-hydrogen-filled bubble chamber by an energetic electron that spirals inward. The electron loses energy as it traverses the liquid in the chamber so that its speed v is being reduced steadily. Thus there is at every point a tangential acceleration a_T given by dv/dt. The centripetal acceleration a_R at any point is given by v^2/r, where r is the radius of curvature of the track at the point in question; both v and r become smaller as the particle loses energy. The force causing the electron to spiral is produced by a magnetic field present in the bubble chamber and at right angles to the plane of Fig. 4-10 (see Chapter 33).

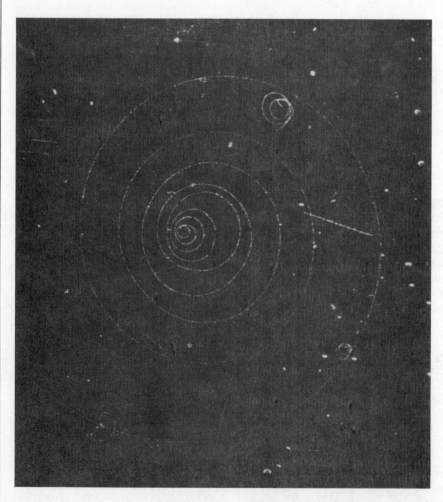

figure 4-10
A track left in a 10-in. liquid-hydrogen-filled bubble chamber by an energetic spiralling electron. (Courtesy Lawrence Radiation Laboratory.) This picture is one of a number in a collection prepared for easy stereoscopic viewing and published, with explanatory material, as *Introduction to the Detection of Nuclear Particles in a Bubble Chamber*, The Ealing Press, Cambridge, Massachusetts (1964). When viewed stereoscopically the electron is seen to be moving toward the reader as it moves in along the spiral. Its velocity vector at any point, thus, does not lie in the plane of the figure, but tilts up out of it; its motion is thus three-dimensional, rather than two-dimensional as we assumed for other examples in this chapter.

4-6
RELATIVE VELOCITY AND ACCELERATION

In earlier sections we considered the addition of velocities in a particular reference frame. Let us now consider the relation between the velocity of an object as determined by one observer S (= reference frame S) and the velocity of the same object as determined by another observer S' (= reference frame S') who is moving with respect to the first.

Consider observer S fixed to the earth, so that his reference frame is the earth. The other observer S' is moving on the earth—for example, a passenger sitting on a moving train—so that his reference frame is the train. They each follow the motion of the same object, say an automobile on a road or a man walking through the train. Each observer will record a displacement, a velocity, and an acceleration for this object measured *relative to his reference frame.* How will these measurements compare? In this section we consider only the case in which the second frame is in motion with respect to the first with a *constant* velocity **u.**

In Fig. 4-11 the reference frame S represented by the x- and y-axes can be thought of as fixed to the earth. The shaded region indicates another reference frame S', represented by x'- and y'-axes, which moves along the x-axis with a constant velocity **u,** as measured in the S-system; it can be thought of as drawn on the floor of a railroad flatcar.

Initially, a particle (say a ball on the flatcar) is at a position called A in the S-frame and called A' in the S'-frame. At a time t later the flatcar and its S' reference frame have moved a distance ut to the right and the particle has moved to B. The displacement of the particle from its initial position *in the S-frame* is the vector **r** from A to B. The displacement of the particle from its initial position *in the S'-frame* is the vector **r′** from A' to B. These are different vectors because the reference point A' of the moving frame has been displaced a distance ut along the x-axis during the motion. From the figure we see that **r** is the vector sum of **r′** and **u**t:

$$\mathbf{r} = \mathbf{r}' + \mathbf{u}t. \tag{4-18}$$

Differentiating Eq. 4-18 leads to

$$\frac{d\mathbf{r}}{dt} = \frac{d\mathbf{r}'}{dt} + \mathbf{u}.$$

But $d\mathbf{r}/dt = \mathbf{v}$, the instantaneous velocity of the particle measured in the S-frame, and $d\mathbf{r}'/dt = \mathbf{v}'$, the instantaneous velocity of the same particle measured in the S' frame, so that

$$\mathbf{v} = \mathbf{v}' + \mathbf{u}. \tag{4-19}$$

Hence the velocity of the particle relative to the S-frame, **v**, is the vector sum of the velocity of the particle relative to the S'-frame, **v′**, and the velocity **u** of the S'-frame relative to the S-frame.

figure 4-11
Two reference frames, S (= x, y) and S' (= x', y'); S' moves to the right, relative to S, with speed u.

EXAMPLE 6

(*a*) The compass of an airplane indicates that it is heading due east. Ground information indicates a wind blowing due north. Show on a diagram the velocity of the plane with respect to the ground.

The object is the airplane. The earth is one reference frame (S) and the air is the other reference frame (S') moving with respect to the first. Then

u is the velocity of the air with respect to the ground.
v′ is the velocity of the plane with respect to the air.
v is the velocity of the plane with respect to the ground.

In this case **u** points north and **v′** points east. Then the relation $\mathbf{v} = \mathbf{v}' + \mathbf{u}$ determines the velocity of the plane with respect to the ground, as shown in Fig. 4-12a.

The angle α is the angle N of E of the plane's course with respect to the ground and is given by

$$\tan \alpha = u/v'.$$

The airplane's speed with respect to the ground is given by

$$v = \sqrt{(v')^2 + u^2}.$$

For example, if the air-speed indicator shows that the plane is moving relative to the air at a speed of 200 mi/h, and if the speed of the wind with respect to the ground is 40.0 mi/h, then

$$v = \sqrt{(200)^2 + (40.0)^2} \text{ mi/h} = 204 \text{ mi/h}$$

is the ground speed of the plane and

$$\alpha = \tan^{-1} \frac{40.0}{200} = 11° 20'$$

gives the course of the plane N of E.

(b) Now draw the vector diagram showing the direction the pilot must steer the plane through the air for the plane to travel due east with respect to the ground.

He would naturally head partly into the wind. His speed relative to the earth will therefore be less than before. The vector diagram is shown in Fig. 4-12b. You should calculate θ and v, using the previous data for **u** and **v**'.

figure 4-12
Example 6

We have seen that different velocities are assigned to a particle by different observers when the observers are in relative motion. These velocities always differ by the relative velocity of the two observers, which here is a constant velocity. It follows that when the particle velocity changes, the change will be the same for both observers. Hence they each measure the same acceleration for the particle. *The acceleration of a particle is the same in all reference frames moving relative to one another with constant velocity;* that is, **a = a'**. This result follows in a formal way if we differentiate Eq. 4-19. Thus $d\mathbf{v}/dt = d\mathbf{v}'/dt + d\mathbf{u}/dt$; but $d\mathbf{u}/dt = 0$ when **u** is constant, so that **a = a'**.

questions

1. In projectile motion when air resistance is negligible, is it ever necessary to consider three-dimensional motion rather than two-dimensional?

2. In broad jumping does it matter how high you jump? What factors determine the span of the jump?

3. Why doesn't the electron in the beam from an electron gun fall as much because of gravity as a water molecule in the stream from a hose? Assume horizontal motion initially in each case.

4. At what point in the path of a projectile does it have its minimum speed? its maximum?

5. Suppose you could vary the angle of incline θ of a plane surface that is fixed at a hinge line to a horizontal table top. How should you choose θ so that the balls dropped vertically and rebounding elastically from the incline have the maximum range?

6. What advantage is there, if any, in measuring angles in radians rather than in degrees?

7. An aviator, pulling out of a dive, follows the arc of a circle. He was said to have "experienced 3g's" in pulling out of the dive. Explain what this statement means.

8. Describe qualitatively the acceleration acting on a bead which, sliding along a frictionless wire, moves inward with constant speed along a spiral.

9. Could the acceleration of a projectile be represented in terms of a radial and a tangential component at each point of the motion? If so, is there any advantage to this representation?

10. Over a short distance a circular arc is a good approximation to a parabola. What then is the radius r of the circular arc approximating the motion of a projectile, of initial speed v_0 and angle θ_0, near the top of its path?

11. A boy sitting in a railroad car moving at constant velocity throws a ball

straight up into the air. Will the ball fall behind him? In front of him? Into his hand? What happens if the car accelerates forward or goes around a curve while the ball is in the air?

12. A man on the observation platform of a train moving with constant velocity drops a coin while leaning over the rail. Describe the path of the coin as seen by (a) the man on the train, (b) a person standing on the ground near the track, and (c) a person in a second train moving in the opposite direction to the first train on a parallel track.

13. A bus with a vertical windshield moves along in a rainstorm at speed v_b. The raindrops fall vertically with a terminal speed v_r. At what angle do the raindrops strike the windshield?

14. Drops are falling vertically in a steady rain. In order to go through the rain from one place to another in such a way as to encounter the least number of raindrops, should you move with the greatest possible speed, the least possible speed, or some intermediate speed?

15. What is wrong with this picture (Fig. 4-13)? The sailor is running with the wind.

16. An elevator is descending at a constant speed. A passenger takes a coin from his pocket and drops it to the floor. What accelerations would (a) the passenger and (b) a person at rest with respect to the elevator shaft observe for the falling coin?

figure 4-13
Question 15

problems

SECTION 4-1

1. Prove that for a vector **a** defined by

$$\mathbf{a} = \mathbf{i}a_x + \mathbf{j}a_y + \mathbf{k}a_z$$

the scalar components are given by

$$a_x = \mathbf{i} \cdot \mathbf{a}, \ a_y = \mathbf{j} \cdot \mathbf{a}, \text{ and } a_z = \mathbf{k} \cdot \mathbf{a}.$$

SECTION 4-2

2. A particle moves so that its position as a function of time is

$$\mathbf{r}(t) = \mathbf{i} + 4t^2\mathbf{j} + t\mathbf{k}.$$

(a) Write expressions for its velocity and acceleration as functions of time. (b) What is the shape of the particle's trajectory?

3. Show (a) that Eqs. 4-4b, b′ can be expressed in vector form as

$$\mathbf{r} = \mathbf{r}_0 + \tfrac{1}{2}(\mathbf{v}_0 + \mathbf{v})t,$$

and (b) Eqs. 4-4 c, c′ as

$$\mathbf{r} = \mathbf{r}_0 + \mathbf{v}_0 t + \tfrac{1}{2}\mathbf{a}t^2.$$

Also, show (c) that Eqs. 4-4d, d′ can be combined to give

$$\mathbf{v} \cdot \mathbf{v} = \mathbf{v}_0 \cdot \mathbf{v}_0 + 2\mathbf{a} \cdot (\mathbf{r} - \mathbf{r}_0).$$

SECTION 4-3

4. Consider a projectile at the top of its trajectory. (a) What is its speed in terms of v_0 and θ_0? (b) What is its acceleration? (c) How is the direction of its acceleration related to that of its velocity? (See Question 10.)

5. A ball rolls off the edge of a horizontal table top 4.0 ft high. If it strikes the floor at a point 5.0 ft horizontally away from the edge of the table, what was its speed at the instant it left the table?   *Answer:* 10 ft/s.

6. A rifle with a muzzle velocity of 1500 ft/s shoots a bullet at a target 150 ft away. How high above the target must the rifle be aimed so that the bullet will hit the target?

7. (a) Show that the range of a projectile having an initial speed v_0 and angle

of projection θ_0 is $R = (v_0{}^2/g)\sin 2\theta_0$. Then show that a projection angle of 45° gives the maximum range (Fig. 4-14). (b) Show that the maximum height reached by the projectile is $y_{max} = (v_0 \sin \theta_0)^2/2g$. (c) Find the angle of projection at which the range and the maximum height of a projectile are equal.   *Answer:* (c) 76°.

8. A projectile is fired horizontally from a gun located 144 ft (44 m) above a horizontal plain with a muzzle speed of 800 ft/s (240 m/s). (a) How long does the projectile remain in the air? (b) At what horizontal distance does it strike the ground? (c) What is the magnitude of the vertical component of its velocity as it strikes the ground?

9. A ball is thrown from the ground into the air. At a height of 9.1 m the velocity is observed to be $\mathbf{v} = 7.6\mathbf{i} + 6.1\mathbf{j}$ in m/s (x-axis horizontal, y-axis vertical). (a) To what maximum height will the ball rise? (b) What will be the total horizontal distance traveled by the ball? (c) What is the velocity of the ball (magnitude and direction) the instant before it hits the ground? *Answer:* (a) 11 m. (b) 23 m. (c) 17 m/s, 63° below the horizontal.

10. Electrons, nuclei, atoms, and molecules, like all forms of matter, will fall under the influence of gravity. Consider separately a beam of electrons, of nuclei, of atoms, and of molecules traveling a horizontal distance of 1.0 m. Let the average speed be for an electron 3.0×10^7 m/s, for a thermal neutron 2.2×10^3 m/s, for a neon atom 5.8×10^2 m/s, and for an oxygen molecule 4.6×10^2 m/s. Let the beams move through vacuum with initial horizontal velocities and find by how much their paths deviate from a straight line (vertical displacement in 1.0 m) due to gravity. How do these results compare to that for a beam of golf balls (use reasonable data)? What is the controlling factor here?

11. A dive bomber, diving at an angle of 53° with the vertical, releases a bomb at an altitude of 730 m. The bomb hits the ground 5.0 s after being released. (a) What is the speed of the bomber? (b) How far did the bomb travel horizontally during its flight? (c) What were the horizontal and vertical components of its velocity just before striking the ground? *Answer:* (a) 200 m/s. (b) 810 m. (c) $v_h = 160$ m/s, $v_r = 170$ m/s.

12. A football is kicked off with an initial speed of 64 ft/s at a projection angle of 45°. A receiver on the goal line 60 yd away in the direction of the kick starts running to meet the ball at that instant. What must be his minimum speed if he is to catch the ball before it hits the ground? (See, in this connection, "Catching a Baseball" by Seville Chapman in *American Journal of Physics,* October 1968.)

13. In a cathode-ray tube a beam of electrons is projected horizontally with a speed of 1.0×10^9 cm/s into the region between a pair of horizontal plates 2.0 cm long. An electric field between the plates exerts a constant downward acceleration on the electrons of magnitude 1.0×10^{17} cm/s². Find (a) the vertical displacement of the beam in passing through the plates and (b) the velocity of the beam (direction and magnitude) as it emerges from the plates. *Answer:* (a) 2.0 mm. (b) $v_x = 1.0 \times 10^9$ cm/s, $v_y = 0.2 \times 10^9$ cm/s down.

14. A batter hits a pitched ball at a height of 4.0 ft above the ground so that its angle of projection is 45° and its initial speed is 110 ft/s. The ball is hit fair down the left field line where a 24-ft high fence is located 320 ft from home plate. Will the ball clear the fence?

15. Galileo, in his *Two New Sciences,* states that "for elevations (angles of projection) which exceed or fall short of 45° by equal amounts, the ranges are equal. . . ." Prove this statement. See Fig. 4-14.

16. A ball rolls off the top of a stairway with a horizontal velocity of magnitude 5.0 ft/s. The steps are 8.0 in. high and 8.0 in. wide. Which step will the ball hit first?

17. (a) Show that if the acceleration due to gravity changes by an amount dg, the range of a projectile (see Problem 7) of given initial speed v_0 and angle

figure 4-14
Problems 7, and 15

of projection θ_0 changes by dR where $dR/R = -dg/g$. (b) If the acceleration due to gravity changes by a small amount Δg (say by going from one place to another), the range for a given projectile system will change as well. Let the change in range by ΔR. If Δg, ΔR are small enough, we may write $\Delta R/R = -\Delta g/g$. In 1936, Jesse Owens established a world's running broad jump record of 8.09 m at the Olympic Games at Berlin ($g = 9.8128$ m/s²). By how much would his record have differed if he had competed instead in 1956 at Melbourne ($g = 9.7999$ m/s²)? (In this connection see "Bad Physics in Athletic Measurements," by P. Kirkpatrick, *American Journal of Physics*, February 1944.)
Answer: His record would have been longer by about 1 cm.

18. A juggler manages to keep five balls in motion, throwing each sequentially up a distance of 3.0 m. (a) Determine the time interval between successive throws. (b) Give the positions of the other balls at the instant when one reaches his hand. (Neglect the time taken to transfer balls from one hand to the other.)

19. A cannon is arranged to fire projectiles, with initial speed v_0, directly up the face of a hill of elevation angle α, as shown in Fig. 4-15. At what angle from the horizontal should the cannon be aimed to obtain the maximum possible range R up the face of the hill?   *Answer:* $\pi/4 + \alpha/2$.

figure 4-15
Problem 19

20. The kicker on a football team can give the ball an initial speed of 25 m/s. Within what angular range must he kick the ball if he is to just score a field goal from a point 50 m in front of the goalposts whose horizontal bar is 3.44 m above the ground?

21. A radar observer on the ground is "watching" an approaching projectile. At a certain instant he has the following information: the projectile has reached maximum altitude and is moving horizontally with a speed v; the straight-line distance to the projectile is l; the line of sight to the projectile is an angle θ above the horizontal. (a) Find the distance D between the observer and the point of impact of the projectile. D is to be expressed in terms of the observed quantities v, l, and θ and the known value of g. Assume a flat earth; assume also that the observer lies in the plane of the projectile's trajectory. (b) Does the projectile pass over his head or strike the ground before reaching him?
Answer: (a) $D = v\sqrt{(2l/g)\sin\theta} - l\cos\theta$. (b) The projectile will pass over the observer's head if D is positive and will fall short if D is negative.

22. Projectiles are hurled at a horizontal distance R from the edge of a cliff of height h in such a way as to land a horizontal distance x from the bottom of the cliff. If you want x to be as small as possible, how would you adjust θ_0 and v_0, assuming that v_0 can be varied from zero to some finite maximum value and that θ_0 can be varied continuously? Only one collision with the ground is allowed (see Fig. 4-16).

figure 4-16
Problem 22

SECTION 4-4

23. Certain neutron stars (extremely dense stars) are believed to be rotating at about 1 rev/s. If such a star has a radius of 20 km, what is the acceleration of an object on the equator of the star?   *Answer:* 8×10^5 m/s².

24. A magnetic field will deflect a charged particle perpendicular to its direction of motion. An electron experiences a radial acceleration of 3.0×10^{14} m/s² in one such field. What is its speed if the radius of its curved path is 0.15 m?

25. In Bohr's model of the hydrogen atom an electron revolves around a proton in a circular orbit of radius 5.28×10^{-11} m with a speed of 2.18×10^6 m/s. What is the acceleration of the electron in the hydrogen atom?
Answer: 9.00×10^{22} m/s².

26. A particle rests on the top of a hemisphere of radius R. Find the smallest horizontal velocity that must be imparted to the particle if it is to leave the hemisphere without sliding down it.

27. What is the acceleration of an object (a) on the equator and (b) at latitude

60°, due to rotation of the earth? (c) By what factor would the speed of the earth's rotation have to increase for a body on the equator to require an acceleration of g to keep it on the earth?
Answer: (a) 3.4×10^{-2} m/s². (b) 1.7×10^{-2} m/s². (c) 17.

28. A boy whirls a stone in a horizontal circle 6.0 ft (1.8 m) above the ground by means of a string 4.0 ft (1.2 m) long. The string breaks, and the stone flies off horizontally, striking the ground 30 ft (9.1 m) away. What was the centripetal acceleration during circular motion?

29. A particle P travels with constant speed counterclockwise on a circle of radius 3.0 m and completes 1.0 rev in 20 s (Fig. 4-17). The particle passes through O at $t = 0$. Starting from the origin O, find (a) the magnitude and direction of the position vectors 5.0 s, 7.5 s, and 10 s later; (b) the magnitude and direction of the displacement in the 5.0-s interval from the fifth to the tenth second; (c) the average velocity vector in this interval; (d) the instantaneous velocity vector at the beginning and at the end of this interval; (e) the average acceleration vector in this interval; and (f) the instantaneous acceleration vector at the beginning and at the end of this interval. (Measure directions counterclockwise from the x-axis in Fig. 4-17.)
Answer: (a) 4.2 m, 45°; 5.5 m, 68°; 6.0 m, 90°. (b) 4.2 m, 135°. (c) 0.85 m/s, 135°. (d) 0.94 m/s, 90°; 0.94 m/s, 180°. (e) 0.27 m/s², 225°. (f) 0.30 m/s², 180°; 0.30 m/s², 270°.

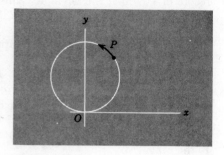

figure 4-17
Problem 29

30. (a) Write an expression for the position vector **r** for a particle describing uniform circular motion, using rectangular coordinates and the unit vectors **i** and **j**. (b) From (a) derive vector expressions for the velocity **v** and the acceleration **a**. (c) Prove that the acceleration is directed toward the center of the circular motion.

31. (a) Express the unit vectors $\mathbf{u}_r$ and $\mathbf{u}_\theta$ in terms of **i**, **j**, and the angle θ in Fig. 4-8. (b) Write an expression, using the unit vectors $\mathbf{u}_\theta$ and $\mathbf{u}_r$, for the position vector **r** for a particle describing uniform circular motion and from it derive Eq. 4-11, $\mathbf{v} = \mathbf{u}_\theta v$.

32. A particle in uniform circular motion about the origin O has a speed v. (a) Show that the time Δt required for it to pass through an angular displacement $\Delta \theta$ is given by

$$\Delta t = \frac{2\pi r}{v} \, \Delta\theta / 360°,$$

where $\Delta\theta$ is in degrees and r is the radius of the circle. (b) Refer to Fig. 4-18, and by taking x and y components of the velocities at points 1 and 2 show that $\overline{a_x} = 0$ and $\overline{a_y} = -0.9 \, v^2/r$, for a pair of points symmetric about the y-axis with $\Delta\theta = 90°$. (c) Show that if $\Delta\theta = 30°$, $\overline{a_x} = 0$ and $\overline{a_y} = -0.99 \, v^2/r$. (d) Argue that $\overline{a_y} \to -v^2/r$ as $\Delta\theta \to 0$ and that circular symmetry requires this answer for each point on the circle.

figure 4-18
Problem 32

SECTION 4-5

33. A particle moves in a plane according to

$$x = R \sin \omega t + \omega R t,$$
$$y = R \cos \omega t + R,$$

where ω and R are constants. This curve, called a *cycloid*, is the path traced out by a point on the rim of a wheel which rolls without slipping along the x-axis. (a) Sketch the path. (b) Calculate the instantaneous velocity and acceleration when the particle is at its maximum and minimum value of y.
Answer: (b) At minimum y: $v_x = v_y = a_x = 0$; $a_y = +\omega^2 R$. At maximum y: $v_x = 2\omega R$; $v_y = a_x = 0$; $a_y = -\omega^2 R$.

SECTION 4-6

34. Snow is falling vertically at a constant speed of 8.0 m/s. (a) At what angle from the vertical and (b) with what speed do the snowflakes appear to be falling as viewed by the driver of a car traveling on a straight road with a speed of 50 km/h?

35. A train travels due south at 88.2 ft/s (relative to ground) in a rain that is blown toward the south by the wind. The path of each raindrop makes the angle 21.6° with the vertical, as measured by an observer stationary on the earth. An observer seated in the train, however, sees perfectly vertical tracks of rain on the windowpane. Determine the speed of each raindrop relative to the earth. *Answer:* 240 ft/s.

36. A helicopter is flying in a straight line over a level field at a constant speed of 4.9 m/s and at a constant altitude of 4.9 m. A package is ejected horizontally from the helicopter with an initial velocity of 12 m/s relative to the helicopter, and in a direction opposite to the helicopter's motion. (*a*) Find the initial velocity of the package relative to the ground. (*b*) What is the horizontal distance between the helicopter and the package at the instant the package strikes the ground? (*c*) What angle does the velocity vector of the package make with the ground at the instant before impact?

37. Find the speeds of two objects if, when they move uniformly toward each other, they get 4.0 m closer each second, and, when they move uniformly in the same direction with the original speeds, they get 4.0 m closer each 10 seconds. *Answer:* 2.2 m/s, 1.8 m/s.

38. A man can row a boat 4.0 mi/h in still water. (*a*) If he is crossing a river where the current is 2.0 mi/h, in what direction will his boat be headed if he wants to reach a point directly opposite from his starting point? (*b*) If the river is 4.0 mi wide, how long will it take him to cross the river? (*c*) How long will it take him to row 2.0 mi *down* the river and then back to his starting point? (*d*) How long will it take him to row 2.0 mi *up* the river and then back to his starting point? (*e*) In what direction should he head the boat if he wants to cross in the smallest possible time?

39. An airplane has a speed of 135 mi/h in still air. It is flying straight north so that it is at all times directly above a north-south highway. A ground observer tells the pilot by radio that a 70 mi/h wind is blowing, but neglects to tell him the wind direction. The pilot observes that in spite of the wind he can travel 135 miles along the highway in one hour. In other words, his ground speed is the same as if there were no wind. (*a*) What is the direction of the wind? (*b*) What is the heading of the plane, that is, the angle between its axis and the highway?
Answer: (*a*) From 75° E of S. (*b*) 30° E of N. Substituting W for E gives a second solution.

40. A pilot is supposed to fly due east from A to B and then back again to A due west. The velocity of the plane in air is **v′** and the velocity of the air with respect to the ground is **u**. The distance between A and B is *l* and the plane's air speed *v′* is constant. (*a*) If $u = 0$ (still air), show that the time for the round trip is $t_0 = 2l/v'$. (*b*) Suppose that the air velocity is due east (or west). Show that the time for a round trip is then

$$t_E = \frac{t_0}{1 - u^2/(v')^2}.$$

(*c*) Suppose that the air velocity is due north (or south). Show that the time for a round trip is then

$$t_N = \frac{t_0}{\sqrt{1 - u^2/(v')^2}}.$$

(*d*) In parts (*b*) and (*c*) one must assume that $u < v'$. Why?

41. A person walks up a stalled escalator in 90 s. When standing on the same escalator, now moving, he is carried up in 60 s. How much time would it take him to walk up the moving escalator? *Answer:* 36 s.

42. A man wants to cross a river 500 m wide. His rowing speed (relative to the water) is 3000 m/h. The river flows with a speed of 2000 m/h. If the man's walking speed on shore is 5000 m/h, (*a*) find the path (combined rowing and walking) he should take to get to the point directly opposite his starting point in the shortest time. (*b*) How long does it take?

5
particle dynamics—I

In Chapters 3 and 4, we studied the motion of a particle, with emphasis on motion along a straight line or in a plane. We did not ask what "caused" the motion; we simply described it in terms of the vectors **r**, **v**, and **a.** Our discussion was thus largely geometrical. In this chapter and the next we discuss the causes of motion, an aspect of mechanics called *dynamics.* As before, bodies will be treated as though they were single particles. Later in the book we shall treat groups of particles and rigid bodies as well.

The motion of a given particle is determined by the nature and the arrangement of the other bodies that form its *environment.* Table 5-1 shows some "particles" and possible environments for them.

In what follows, we limit ourselves to the very important special case of gross objects moving at speeds that are small compared to c, the speed of light; this is the realm of *classical mechanics.* Specifically, we shall not inquire here into such questions as the motion of an electron in a uranium atom or the collision of two protons whose speeds are, say, $0.90c$. The first inquiry would involve us with the quantum theory and the second with the theory of relativity. We leave consideration of these theories, of which classical mechanics is a special case (see Section 6-4), to later.

The central problem of classical mechanics is this: (1) We are given a particle whose characteristics (mass, charge, magnetic dipole moment, etc.) we know. (2) We place this particle, with a known initial velocity, in an environment of which we have a complete description. (3) Problem: what is the subsequent motion of the particle?

This problem was solved, at least for a large variety of environments, by Isaac Newton (1642–1727) when he put forward his laws of motion and formulated his law of universal gravitation. The program for solving this problem, in terms of our present understanding of classical mechanics,* is: (1) We introduce the concept of *force* **F** and define it in terms of the acceleration **a** experienced by a particular standard body. (2) We develop a procedure for assigning a *mass m* to a body so that we may understand the fact that different particles of the same kind experience different accelerations in the same environment. (3) Finally, we try to find ways of calculating the forces that act on particles from the properties of the particle and of its environment; that is, we look for *force laws.* Force, which is at root a technique for relating the environment to the motion of the particle, appears both in the laws of motion (which tell us what acceleration a given body will experience under the action of a given force) and in the force laws (which tell us how to calculate the force that will act on a given body in a given environment). The laws of motion and the force laws, taken together, constitute the laws of mechanics, as the sketch suggests.

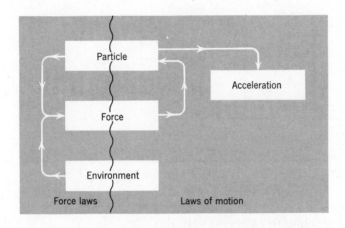

The program of mechanics cannot be tested piecemeal. We must view it as a unit and we shall judge it to be successful if we can say "yes" to these two questions. (1) Does the program yield results that agree with experiment? (2) Are the force laws simple in form? It is the crowning glory of Newtonian mechanics that we can indeed answer each of these questions in the affirmative.

In this section we have used the terms *force* and *mass* rather unprecisely, having identified force with the influence of the environment, and mass with the resistance of a body to be accelerated when a force acts on it, a property often called inertia. In later sections we shall refine these primitive ideas about force and mass.

5-2
NEWTON'S FIRST LAW

For centuries the problem of motion and its causes was a central theme of natural philosophy, an early name for what we now call physics. It was not until the time of Galileo and Newton, however, that dramatic progress was made. Isaac Newton, born in England in the year of Gali-

---

* See "Presentation of Newtonian Mechanics" by Norman Austern, *American Journal of Physics,* September 1961, "On the Classical Laws of Motion" by Leonard Eisenbud, *American Journal of Physics,* March 1958, and "The Laws of Classical Motion: What's *F?* What's *m?* What's *a?*" by Robert Weinstock, *American Journal of Physics,* October 1961, for expositions of the laws of classical mechanics as we now view them.

Table 5-1

System	The Particle	The Environment
1.	A block	The spring; the rough surface
2.	A golf ball	The earth
3.	A satellite	The earth
4.	An electron	A large uniformly charged sphere
5.	A bar magnet	A second bar magnet

leo's death, is the principal architect of classical mechanics.* He carried to full fruition the ideas of Galileo and others who preceded him. His three laws of motion were first presented (in 1686) in his *Philosophiae Naturalis Principia Mathematica*, usually called the *Principia*.

Before Galileo's time most philosophers thought that some influence or "force" was needed to keep a body moving. They thought that a body was in its "natural state" when it was at rest. For a body to move in a straight line at constant speed, for example, they believed that some external agent had to continually propel it; otherwise it would "naturally" stop moving.

If we wanted to test these ideas experimentally, we would first have to find a way to free a body from all influences of its environment or from all forces. This is hard to do, but in certain cases we can make the forces very small. If we study the motions as we make the forces smaller and smaller, we shall have some idea of what the motion would be like if the external forces were truly zero.

Let us place our test body, say a block, on a rigid horizontal plane. If we let the block slide along this plane, we notice that it gradually slows down and stops. This observation was used, in fact, to support the idea that motion stopped when the external force, in this case the hand initially pushing the block, was removed. We can argue against this idea, however, reasoning as follows: Let us repeat our experiment, now using a smoother block and a smoother plane and providing a lubricant. We notice that the velocity decreases more slowly than before. Let us use still smoother blocks and surfaces and better lubricants. We find that the block decreases in velocity at a slower and slower rate and travels

---

* Newton also invented the (fluxional) calculus, conceived the idea of universal gravitation and formulated its law, and discovered the composition of white light. He was a skillful experimenter and a mathematician of first rank as well as what today would be called theoretical physicist.

farther each time before coming to rest.* We can now extrapolate and say that if all friction could be eliminated, the body would continue indefinitely in a straight line with constant speed. Some external force is necessary to *change* the velocity of a body but no external force is necessary to *maintain* the velocity of a body. Our hand, for example, exerts a force on the block when it sets it in motion. The rough plane exerts a force on it when it slows it down. Both of these forces produce a change in the velocity, that is, they produce an acceleration.

This principle was adopted by Newton as the first of his three laws of motion. Newton stated his first law in these words: *"Every body persists in its state of rest or of uniform motion in a straight line unless it is compelled to change that state by forces impressed on it."*

Newton's first law is really a statement about reference frames. For, in general, the acceleration of a body depends on the reference frame relative to which it is measured. The first law tells us that, if there are no nearby objects (and by this we mean that there are no forces because every force must be associated with an object in the environment), then it is possible to find a family of reference frames in which a particle has no acceleration. The fact that bodies stay at rest or retain their uniform linear motion in the absence of applied forces is often described by assigning a property to matter called inertia. Newton's first law is often called the law of inertia and the reference frames to which it applies are called inertial frames. Such frames are assumed to be fixed with respect to the distant stars.

In nearly all cases in this book we will apply the laws of classical mechanics from the point of view of an observer in an inertial frame. It is possible to solve problems in mechanics using a noninertial frame, such as a frame rotating with respect to the fixed stars, but to do so we have to introduce forces that cannot be associated with objects in the environment. We will discuss this in Chapters 6, 11, and 16. A reference frame attached to the earth can be considered to be an inertial frame for most practical purposes. We shall see in Chapter 16 how good an approximation this is.

Notice that there is no distinction in the first law between a body at rest and one moving with a constant velocity. Both motions are "natural" in the absence of forces. That this is so becomes clear when a body at rest in one inertial frame is viewed from a second inertial frame, that is, a frame moving with constant velocity with respect to the first. An observer in the first frame finds the body to be at rest; an observer in the second frame finds the same body to be moving with uniform velocity. Both observers find the body to have no acceleration, that is, no change in velocity, and both may conclude from the first law that no force acts on the body.

Notice, too, that by implication there is no distinction in the first law between the absence of all forces and the presence of forces whose resultant is zero. For example, if the push of our hand on the book exactly counteracts the force of friction on it, the book will move with uniform velocity. Hence another way of stating the first law is: *If no net force acts on a body, its acceleration* **a** *is zero.*

If there *is* an interaction between the body and objects present in the

---

* You may have experimented in the laboratory with a dry ice puck. This is a device which can be made to move over a smooth horizontal surface, floating on a layer of CO_2 gas. The friction between the puck and the surface is very low indeed and it is hard to measure any reduction in speed for path lengths of practical dimensions.

environment, the effect may be to change the "natural" state of the body's motion. To investigate this we must now examine carefully the concept of force.

5-3
FORCE

Let us refine our concept of force by defining it operationally. In our everyday language force is associated with a push or a pull, perhaps exerted by our muscles. In physics, however, we need a more precise definition. We define force here in terms of the acceleration that a given standard body experiences when placed in a suitable environment.

As a standard body we find it convenient to use (or rather to imagine that we use!) the standard kilogram (see Fig. 1-2). This body has been selected as our standard of mass and has been assigned, by definition, a mass m_0 of exactly 1 kg. Later we will describe how masses are assigned to other bodies.

As for an environment we place the standard body on a horizontal table having negligible friction and we attach a spring to it. We hold the other end of the spring in our hand, as in Fig. 5-1a. Now we pull the spring horizontally to the right so that by trial and error the standard body experiences a measured uniform acceleration of 1.0 m/s². We then declare, as a matter of definition, that the spring (which is the significant body in the environment) is exerting a constant force whose magnitude we will call "1.00 newton," or in SI notation: 1.00 N, on the standard body. We note that, in imparting this force, the spring is kept stretched an amount Δl beyond its normal unextended length, as Fig. 5-1b shows.

(a)

(b)

figure 5-1
(a) A "particle" P (the standard kilogram) at rest on a horizontal frictionless surface. (b) The body is accelerated by pulling the spring to the right.

We can repeat the experiment, either stretching the spring more or using a stiffer spring, so that we measure an acceleration of 2.00 m/s² for the standard body. We now declare that the spring is exerting a force of 2.00 N on the standard body. In general, if we observe this particular standard body to have an acceleration a in a particular environment, we then say that the environment is exerting a force F on the standard body, where F (in newtons) is numerically equal to a (in m/s²).

Now let us see whether force, as we have defined it, is a *vector* quantity. In Fig. 5-2b we assigned a magnitude to the force F, and it is a simple matter to assign a direction to it as well, namely, the direction of the acceleration that the force produces. However, to be a vector it is not enough for a quantity to have magnitude and direction; it must also obey the laws of vector addition described in Chapter 2. We can learn only from experiment whether forces, as we defined them, do indeed obey these laws.

Let us arrange to exert a 4.00-N force along the x-axis and a 3.00-N force along the y-axis and let us apply these forces simultaneously to the standard body placed, as before, on a horizontal, frictionless surface. What will be the acceleration of the standard body? We would find by experiment that it was 5.00 m/s², directed along a line that makes an angle of 37° with the x-axis. In other words, we would say that the standard body was experiencing a force of 5.00 N in this same direction. This same result can be obtained by adding the 4.00-N and 3.00-N forces vectorially according to the parallelogram method. Experiments of this kind show conclusively that forces are vectors; they have magnitude; they have direction; they add according to the parallelogram law.

The result of experiments of this general type is often stated as follows: *When several forces act on a body, each produces its own acceleration independently. The resulting acceleration is the vector sum of the several independent accelerations.*

5-4
MASS; NEWTON'S SECOND LAW

In Section 5-3 we considered only the accelerations given to one particular object, the standard kilogram. We were able thereby to define forces quantitatively. What effect would these forces have on other objects? Because our standard body was chosen arbitrarily in the first place, we know that for any given object the acceleration will be directly proportional to the force applied. The significant question remaining then is: What effect will the *same force* have on *different objects?* Everyday experience gives us a qualitative answer. The same force will produce different accelerations on different bodies. A baseball will be accelerated more by a given force than will an automobile. In order to obtain a quantitative answer to this question we need a method to measure mass, the property of a body which determines its resistance to a change in its motion.

Let us attach a spring to our standard body (the standard kilogram, to which we have arbitrarily assigned a mass m_0 = one kg, exactly) and arrange to give it an acceleration a_0 of, say 2.00 m/s², using the method of Fig. 5-1b. Let us measure carefully the extension Δl of the spring associated with the force that the spring is exerting on the block.

Now we remove the standard kilogram and substitute an arbitrary body, whose mass we label m_1. We apply the same force (the one that accelerated the standard kilogram 2.00 m/s²) to the arbitrary body (by stretching the spring by the same amount) and we measure an acceleration a_1 of, say, 0.50 m/s².

We *define* the ratio of the masses of the two bodies to be the inverse ratio of the accelerations given to these bodies by the same force, or

$$m_1/m_0 = a_0/a_1 \qquad \text{(same force } \mathbf{F} \text{ acting).}$$

In this example we have, numerically,

$$m_1 = m_0(a_0/a_1) = 1.00 \text{ kg } [(2.00 \text{ m/s}^2)/(0.50 \text{ m/s}^2)]$$
$$= 4.00 \text{ kg.}$$

The second body, which has only one-fourth the acceleration of the first body when the same force acts on it, has, by definition, four times the mass of the first body. Hence mass may be regarded as a quantitative measure of inertia.

If we repeat the preceding experiment with a different common force acting, we find the ratio of the accelerations, a_0'/a_1', to be the same as in the previous experiment, or

$$m_1/m_0 = a_0/a_1 = a_0'/a_1'.$$

The ratio of the masses of two bodies is thus independent of the common force used.

Furthermore, experiment shows that we can consistently assign masses to any body by this procedure. For example, let us compare a second arbitrary body with the standard body, and thus determine its mass, say m_2. We can now compare the two arbitrary bodies, m_2 and m_1, directly, obtaining accelerations a_2'' and a_1'' when the same force is applied. The mass ratio, defined as usual from

$$m_2/m_1 = a_1''/a_2'', \quad \text{(same force acting)}$$

turns out to have the same value that we obtain by using the masses m_2 and m_1 previously determined by direct comparison with the standard.

We can show, in still another experiment of this type, that if objects of mass m_1 and m_2 are fastened together, they behave mechanically as a single object of mass $(m_1 + m_2)$. In other words, masses add like (and are) scalar quantities.

We can now summarize all the experiments and definitions described above in one equation, the fundamental equation of classical mechanics,

$$\mathbf{F} = m\mathbf{a}. \tag{5-1}$$

In this equation $\mathbf{F}$ is the (vector) *sum* of *all* the forces acting *on* the body, m is the mass of the body, and $\mathbf{a}$ is its (vector) acceleration. Equation 5-1 may be taken as a statement of Newton's second law. If we write it in the form $\mathbf{a} = \mathbf{F}/m$, we can easily see that the acceleration of the body is directly proportional to the resultant force acting on it and parallel in direction to this force and that the acceleration, for a given force, is inversely proportional to the mass of the body.

Notice that the first law of motion is contained in the second law as a special case, for if $\mathbf{F} = 0$, then $\mathbf{a} = 0$. In other words, if the resultant force on a body is zero, the acceleration of the body is zero. Therefore in the absence of applied forces a body will move with constant velocity or be at rest (zero velocity), which is what the first law of motion says. Therefore of Newton's three laws of motion only two are independent, the second and the third (Section 5-5). The division of translational particle dynamics that includes only systems for which the resultant force $\mathbf{F}$ is zero is called *statics*.

Equation 5-1 is a vector equation. We can write this single vector equation as three scalar equations,

$$F_x = ma_x, \quad F_y = ma_y, \quad \text{and} \quad F_z = ma_z, \tag{5-2}$$

relating the x, y, and z components of the resultant force (F_x, F_y, and F_z) to the x, y, and z components of acceleration (a_x, a_y, and a_z) for the mass m. It should be emphasized that F_x is the sum of the x-components of *all* the forces, F_y is the sum of the y-components of *all* the forces, and F_z is the *sum* of the z-components of *all* the forces acting on m.

Forces acting on a body originate in other bodies that make up its environment. Any single force is only one aspect of a mutual interaction between *two* bodies. We find by experiment that when one body exerts a force on a second body, the second body always exerts a force on the first. Furthermore, we find that these forces are equal in magnitude but

5-5
NEWTON'S THIRD LAW OF MOTION

opposite in direction. A single isolated force is therefore an impossibility.

If one of the two forces involved in the interaction between two bodies is called an "action" force, the other is called the "reaction" force. *Either* force may be considered the "action" and the other the "reaction." Cause and effect is *not* implied here, but a mutual simultaneous interaction *is* implied.

This property of forces was first stated by Newton in his third law of motion: *"To every action there is always opposed an equal reaction; or, the mutual actions of two bodies upon each other are always equal, and directed to contrary parts."*

In other words, if body A exerts a force on body B, body B exerts an equal but oppositely directed force on body A; and furthermore the forces lie along the line joining the bodies. Notice that the action and reaction forces, which always occur in pairs, act on *different* bodies. If they were to act on the same body, we could never have accelerated motion because the resultant force on every body would always be zero.

Imagine a boy kicking open a door. The force exerted by the boy B on the door D accelerates the door (it flies open); at the same time, the door D exerts an equal but opposite force on the boy B, which decelerates the boy (his foot loses forward velocity). The boy will be painfully aware of the "reaction" force to his "action," particularly if his foot is bare.

The following examples illustrate the application of the third law and clarify its meaning.

EXAMPLE 1

Consider a man pulling horizontally on a rope attached to a block on a horizontal table as in Fig. 5-2. The man pulls on the rope with a force $\mathbf{F}_{MR}$. The rope exerts a reaction force $\mathbf{F}_{RM}$ on the man. According to Newton's third law, $\mathbf{F}_{MR} = -\mathbf{F}_{RM}$. Also, the rope exerts a force $\mathbf{F}_{RB}$ on the block, and the block exerts a reaction force $\mathbf{F}_{BR}$ on the rope. Again according to the third law, $\mathbf{F}_{RB} = -\mathbf{F}_{BR}$.

Suppose that the rope has a mass m_R. Then, in order to start the block and rope moving from rest, we must have an acceleration, say $\mathbf{a}$. The only forces acting *on the rope* are $\mathbf{F}_{MR}$ and $\mathbf{F}_{BR}$, so that the resultant force on it is $\mathbf{F}_{MR} + \mathbf{F}_{BR}$, and this must be different from zero if the rope is to accelerate. In fact, from the

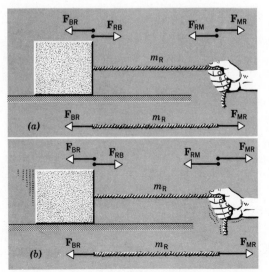

figure 5-2
Example 1. A man pulls on a rope attached to a block. *(a)* The forces exerted on the rope by the block and by the man are equal and opposite. Thus the resultant horizontal force on the rope is zero, as is shown in the free-body diagram. The rope does not accelerate. *(b)* The force exerted on the rope by the man exceeds that exerted by the block. The net horizontal force has magnitude $F_{MR} - F_{BR}$ and points to the right. Thus the rope is accelerated to the right. The block is also acted upon by a frictional force not shown here.

second law we have

$$\mathbf{F}_{MR} + \mathbf{F}_{BR} = m_R \mathbf{a}$$

Since the forces and the acceleration are along the same line, we can drop the vector notation and write the relation between the *magnitudes* of the vectors, namely

$$F_{MR} - F_{BR} = m_R a.$$

We see therefore that in general $\mathbf{F}_{MR}$ does not have the same magnitude as $\mathbf{F}_{BR}$ (Fig. 5-2b). These two forces act on the *same* body and are *not* action and reaction pairs.

According to Newton's third law the magnitude of $\mathbf{F}_{MR}$ always equals the magnitude of $\mathbf{F}_{RM}$, and the magnitude of $\mathbf{F}_{RB}$ always equals the magnitude of $\mathbf{F}_{BR}$. However, only if the acceleration $\mathbf{a}$ of the system is zero will we have the pair of forces $\mathbf{F}_{MR}$ and $\mathbf{F}_{RM}$ equal in magnitude to the pair of forces $\mathbf{F}_{RB}$ and $\mathbf{F}_{BR}$ (Fig. 5-2a). In this special case only, we could imagine that the rope merely transmits the force exerted by the man to the block without change. This same result holds in principle if $m_R = 0$. In practice, we never find a massless rope. However, we can often neglect the mass of a rope, in which case the rope is assumed to transmit a force unchanged. The force exerted at any point in the rope is called the *tension* at that point. We may measure the tension at any point in the rope by cutting a suitable length from it and inserting a spring scale; the tension is the reading of the scale. The tension is the same at all points in the rope only if the rope is unaccelerated or assumed to be massless.

Consider a spring attached to the ceiling and at the other end holding a block at rest (Fig. 5-3a). Since no body is accelerating, all the forces on any body will add vectorially to zero. For example, the forces on the suspended block are $\mathbf{T}$, the tension in the stretched spring, pulling vertically up on the mass, and $\mathbf{W}$, the pull of the earth acting vertically down on the body, called its weight. These are drawn in Fig. 5-3b, where we show only the block for clarity. There are no other forces on the block.

In Newton's second law, $\mathbf{F}$ represents the *sum* of *all* the forces acting *on* a body, so that for the block

$$\mathbf{F} = \mathbf{T} + \mathbf{W}.$$

The block is at rest so that its acceleration is zero, or $\mathbf{a} = 0$. Hence, from the relation $\mathbf{F} = m\mathbf{a}$, we obtain $\mathbf{T} + \mathbf{W} = 0$, or

$$\mathbf{T} = -\mathbf{W}.$$

The forces act along the same line, so that their magnitudes are equal, or

$$T = W.$$

Therefore the tension in the spring is an exact measure of the weight of the block. We shall use this result later in presenting a static procedure for measuring forces.

It is instructive to examine the forces exerted on the spring; they are shown in Fig. 5-3c. $\mathbf{T}'$ is the pull of the block on the spring and is the reaction force of the action force $\mathbf{T}$. $\mathbf{T}'$ therefore has the same magnitude as $\mathbf{T}$, which is W. $\mathbf{P}$ is the upward pull of the ceiling on the spring, and $\mathbf{w}$ is the weight of the spring, that is, the pull of the earth on it. Since the spring is at rest and all forces act along the same line, we have

$$\mathbf{P} + \mathbf{T}' + \mathbf{w} = 0,$$

or

$$P = W + w.$$

The ceiling therefore pulls up on the spring with a force whose magnitude is the sum of the weights of the block and spring.

EXAMPLE 2

figure 5-3

Example 2. *(a)* A block is suspended by a spring. *(b)* A free-body diagram showing all the vertical forces exerted on the block. *(c)* A similar diagram for the vertical forces on the spring.

From the third law of motion, the force exerted by the spring on the ceiling, P', must be equal in magnitude to P, which is the reaction force to the action force P'. P' therefore has a magnitude $W + w$.

In general, the spring exerts different forces on the bodies attached at its different ends, for $P' \neq T$. In the special case in which the weight of the spring is negligible, $w = 0$ and $P' = W = T$. Therefore a weightless spring (or cord) may be considered to transmit a force from one end to the other without change.

It is instructive to classify all the forces in this problem according to action and reaction pairs. The reaction to W, a force exerted by the earth on the block, must be a force exerted by the block on the earth. Similarly, the reaction to w is a force exerted by the spring on the earth. Because the earth is so massive, we do not expect these forces to impart a noticeable acceleration to the earth. Since the earth is not shown in our diagrams, these forces have not been shown. The forces T and T' are action-reaction pairs, as are P and P'. Notice that although $T = -W$ in our problem, these forces are *not* an action-reaction pair because they act on the *same* body.

5-6
SYSTEMS OF MECHANICAL UNITS

Unit force is defined as a force that causes a unit of acceleration when applied to a unit mass. In SI terms unit force is the force that will accelerate a one-kg mass at one m/s^2; we have seen that this unit is called the newton (abbreviation, N). In the cgs (centimeter, gram, second) system unit force is the force that will accelerate a one-g mass at one cm/s^2; this unit is called the *dyne*. Since $1 \text{ kg} = 10^3 \text{ g}$ and $1 \text{ m/s}^2 = 10^2 \text{ cm/s}^2$, it follows that $1 \text{ N} = 10^5 \text{ dynes}$.

In each of our systems of units we have chosen mass, length, and time as our fundamental quantities. Standards were adopted for these fundamental quantities and units defined in terms of these standards. Force appears as a derived quantity, determined from the relation $F = ma$.

In the BE (British engineering) system of units, however, *force*, length, and time are chosen as the fundamental quantities and mass is a derived quantity. In this system, mass is determined from the relation $m = F/a$. The standard and unit of force in this system is the *pound*. Actually, the pound of force was originally defined to be the pull of the earth on a certain standard body at a certain place on the earth. We can get this force in an operational way by hanging the standard body from a spring at the particular point where the earth's pull on it is defined to be one lb of force. If the body is at rest, the earth's pull on the body, its weight W, is balanced by the tension in the spring. Therefore $T = W =$ one lb, in this instance. We can now use this spring (or any other one thus calibrated) to exert a force of one lb on any other body; to do this we simply attach the spring to another body and stretch it the same amount as the pound force had stretched it. The standard body can be compared to the kilogram and it is found to have the mass 0.45359237 kg. The acceleration due to gravity at the certain place on the earth is found to be 32.1740 ft/s^2. The pound of force can therefore be defined from $F = ma$ as the force that accelerates a mass of 0.45359237 kg at the rate of 32.1740 ft/s^2.

This procedure enables us to compare the pound-force with the newton. Using the fact that 32.1740 ft/s^2 equals 9.8066 m/s^2, we find that

$$1 \text{ lb} = (0.45359237 \text{ kg})(32.1740 \text{ ft/s}^2)$$
$$= (0.45359237 \text{ kg})(9.8066 \text{ m/s}^2)$$
$$\cong 4.45 \text{ N}.$$

The unit of mass in the British engineering system can now be derived. It is defined as the mass of a body whose acceleration is 1 ft/s^2

when the force on it is 1 lb; this mass is called the slug. Thus, in this system

$$F[\text{lb}] = m[\text{slugs}] \times a[\text{ft/s}^2].$$

Legally the pound is a unit of mass but in engineering practice the pound is treated as a unit of force or weight. This has given rise to the terms pound-mass and pound-force. The pound-mass is a body of mass 0.45359237 kg; no standard block of metal is preserved as the pound-mass, but, like the yard, it is defined in terms of the SI standard. The pound-force is the force that gives a standard pound an acceleration equal to the standard acceleration of gravity, 32.1740 ft/s². As we shall see later, the acceleration of gravity varies with distance from the center of the earth, and this "standard acceleration" is, therefore, the value at a particular distance from the center of the earth. (Any point at sea level and 45°N latitude is a good approximation.)

In this book only forces will be measured in pounds. Thus the corresponding unit of mass is the slug. The units of force, mass, and acceleration in the three systems are summarized in Table 5-2.

Table 5-2
Units in $F = ma$

Systems of Units	Force	Mass	Acceleration
SI	newton (N)	kilogram (kg)	m/s²
Cgs	dyne	gram (g)	cm/s²
BE	pound (lb)	slug	ft/s²

The *dimensions* of force are the same as those of mass times acceleration. In a system in which mass, length, and time are the fundamental qualities, the dimensions of force are, therefore, mass × length/time², or MLT^{-2}. We shall arbitrarily adopt mass, length, and time as our fundamental mechanical quantities.

The three laws of motion that we have described are only part of the program of mechanics that we outlined in Section 5-1. It remains to investigate the *force laws*, which are the procedures by which we calculate the force acting on a given body in terms of the properties of the body and its environment. Newton's second law

$$\mathbf{F} = m\mathbf{a} \qquad (5\text{-}3)$$

is essentially not a law of nature but a definition of force. We need to identify various functions of the type:

$$\mathbf{F} = \text{a function of the properties of the particle and of the environment} \qquad (5\text{-}4)$$

so that we can, in effect, eliminate **F** between Eqs. 5-3 and 5-4, thus obtaining an equation that will let us calculate the acceleration of a particle in terms of the properties of the particle and its environment. We see here clearly that force is a concept that connects the acceleration of the particle on the one hand with the properties of the particle and its environment on the other. We indicated earlier that one criterion for declaring the program of mechanics to be successful would be the

5-7
THE FORCE LAWS

discovery that *simple* laws of the type of Eq. 5-4 do indeed exist. This turns out to be the case, and this fact constitutes the essential reason that we "believe" the laws of classical mechanics. If the force laws had turned out to be very complicated, we would not be left with the feeling that we had gained much insight into the workings of nature.

The number of possible environments for an accelerated particle is so great that a detailed discussion of all the force laws is not feasible in this chapter. We shall, however, indicate in Table 5-3 the force laws that apply to the five particle-plus-environment situations of Table 5-1. At appropriate places throughout the text we will discuss these and other force laws in detail; several of the laws in Table 5-3 are approximations or special cases.

Table 5-3
The force laws for the systems of table 5-1

System	Force Law
1. A block propelled by a stretched spring over a rough horizontal surface	(a) Spring force: $F = -kx$, where x is the extension of the spring and k is a constant that describes the spring; **F** points to the right; see Chapter 15
	(b) Friction force: $F = \mu mg$, where μ is the coefficient of friction and mg is the weight of the block; **F** points to the left; see Chapter 6
2. A golf ball in flight	$F = mg$; **F** points down (see Section 5-8)
3. An artificial satellite	$F = GmM/r^2$, where G is the *gravitational constant*, M the mass of the earth, and r the orbit radius; **F** points toward the center of the earth; see Chapter 16. This is *Newton's law of universal gravitation*
4. An electron near a positively charged sphere	$F = (1/4\pi\epsilon_0)eQ/r^2$, where ϵ_0 is a constant, e is the electron charge, Q is the charge on the sphere, and r is the distance from the electron to the center of the sphere; **F** points to the right; see Chapter 26. This is *Coulomb's law of electrostatics*
5. Two bar magnets	$F = (3\mu_0/2\pi)\mu^2/r^4$, where μ_0 is a constant, μ is the magnetic dipole moment of each magnet, and r is the center-to-center separation of the magnets; we assume that $r \gg l$, where l is the length of each magnet; **F** points to the right

5-8
WEIGHT AND MASS

The *weight* of a body is the gravitational force exerted on it by the earth. Weight, being a force, is a vector quantity. The direction of this vector is the direction of the gravitational force, that is, toward the center of the earth. The magnitude of the weight is expressed in force units, such as pounds or newtons.

When a body of mass m is allowed to fall freely, its acceleration is that of gravity **g** and the force acting on it is its weight **W.** Newton's second law, **F** = m**a,** when applied to a freely falling body, gives us **W** = m**g.** Both **W** and **g** are vectors directed toward the center of the earth. We can therefore write

$$W = mg, \qquad\qquad (5\text{-}5)$$

where W and g are the magnitudes of the weight and acceleration vectors. To keep an object from falling we have to exert on it an upward force equal in magnitude to W, so as to make the net force zero. In Fig. 5-3a the tension in the spring supplies this force.

We stated previously that g is found experimentally to have the same value for all objects *at the same place*. From this it follows that the ratio of the weights of two objects must be equal to the ratio of their masses. Therefore a chemical balance, which actually is an instrument for comparing two downward forces, can be used in practice to compare masses. If a sample of salt in one pan of a balance is pulling down on that pan with the same force as is a standard one gram-mass on the other pan, we know* that the mass of salt is equal to one gram. We are likely to say that the salt "weighs" one gram, although a gram is a unit of mass, not weight. However, it is always important to distinguish carefully between weight and mass.

We have seen that the weight of a body, the downward pull of the earth on that body, is a vector quantity. The mass of a body is a scalar quantity. The quantitative relation between weight and mass is given by $\mathbf{W} = m\mathbf{g}$. Because g varies from point to point on the earth, $\mathbf{W}$, the weight of a body of mass m, is different in different localities. Thus, the weight of a one kg-mass in a locality where g is 9.80 m/s² is 9.80 N; in a locality where g is 9.78 m/s², the same one kg-mass weighs 9.78 N. If these weights were determined by measuring the amount of stretch required in a spring to balance them, the difference in weight of the same one kg-mass at the two different localities would be evident in the slightly different stretch of the spring at these two localities. Hence, unlike the mass of a body, which is an intrinsic property of the body, the weight of a body depends on its location relative to the center of the earth. Spring scales read differently, balances the same, at different parts of the earth.

We shall generalize the concept of weight in Chapter 16 in which we discuss universal gravitation. There we shall see that the weight of a body is zero in regions of space where the gravitational effects are nil, although the inertial effects, and hence the mass of the body, remain unchanged from those on earth. In a space ship free from the influence of gravity it is a simple matter to lift a large block of lead ($\mathbf{W} = 0$), but the astronaut would still stub his toe if he were to kick the block ($m \neq 0$).

It takes the same force to accelerate a body in gravity-free space as it does to accelerate it along a horizontal frictionless surface on earth, for its mass is the same in each place. But it takes a greater force to hold the body up against the pull of the earth on the earth's surface than it does high up in space, for its weight is different in each place.

Often, instead of being given the mass, we are given the weight of a body on which forces are exerted. The acceleration $\mathbf{a}$ produced by the force $\mathbf{F}$ acting on a body whose weight has a magnitude W can be obtained by combining Eq. 5-3 and Eq. 5-5. Thus from $\mathbf{F} = m\mathbf{a}$ and $W = mg$ we obtain

$$m = W/g, \qquad \text{so that} \qquad \mathbf{F} = (W/g)\mathbf{a}. \qquad (5\text{-}6)$$

The quantity W/g plays the role of m in the equation $F = ma$ and is, in fact, the mass of a body whose weight has the magnitude W. For example, a man whose weight is 160 lb at a point where $g = 32.0$ ft/s² has a mass $m = W/g = (160 \text{ lb})/(32.0 \text{ ft/s}^2) = 5.00$ slugs. Notice that his weight at another point where $g = 32.2$ ft/s² is $W = mg = (5.00 \text{ slugs})(32.2 \text{ ft/s}^2) = 161$ lb.

---

* Corrections for buoyancy, owing to the different volumes of air displaced by the salt and the standard, must be made. We discuss these in Chapter 17.

In Section 5-3 we defined force by measuring the acceleration imparted to a standard body by pulling on it with a stretched spring. That may be called a dynamic method for measuring force. Although convenient for the purposes of definition, it is not a particularly practical procedure for the measurement of forces. Another method for measuring forces is based on measuring the change in shape or size of a body (a spring, say) on which the force is applied when the body is unaccelerated. This may be called the static method of measuring forces.

The idea of the static method is to use the fact that when a body, under the action of several forces, has zero acceleration, the vector sum of all the forces acting on the body must be zero. This is, of course, just the content of the first law of motion. A single force acting on a body would produce an acceleration; this acceleration can be made zero if we apply another force to the body equal in magnitude but oppositely directed. In practice we seek to keep the body at rest. If now we choose some force as our unit force, we are in a position to measure forces. The pull of the earth on a standard body at a particular point can be taken as the unit force, for example.

The instrument most commonly used to measure forces in this way is the spring balance. It consists of a coiled spring having a pointer at one end that moves over a scale. A force exerted on the balance changes the length of the spring. If a body weighing 1.00 N is hung from the spring, the spring stretches until the pull of the spring on the body is equal in magnitude but opposite in direction to its weight. A mark can be made on the scale next to the pointer and labeled "1.00-N force." Similarly, 2.00-N, 3.00-N, etc., weights may be hung from the spring and corresponding marks can be made on the scale next to the pointer in each case. In this way the spring is calibrated. We assume that the force exerted on the spring is always the same when the pointer stands at the same position. The calibrated balance can now be used to measure any suitable unknown force, not merely the pull of the earth on some body.

The third law is tacitly used in our static procedure because we assume that the force exerted by the spring on the body is the same in magnitude as the force exerted by the body on the spring. This latter force is the force we wish to measure. The first law is used too, because we assume $\mathbf{F}$ is zero when $\mathbf{a}$ is zero. It is worth noting again here that if the acceleration were not zero, the body of weight W would not stretch the spring to the same length as it did with $\mathbf{a} = 0$. In fact, if the spring and attached body of weight W were to fall freely under gravity so that $\mathbf{a} = \mathbf{g}$, the spring would not stretch at all and its tension would be zero.

5-9
A STATIC PROCEDURE FOR MEASURING FORCES

5-10
SOME APPLICATIONS OF NEWTON'S LAWS OF MOTION

It will be helpful to write down some procedures for solving problems in classical mechanics and to illustrate them by several examples. Newton's second law states that the vector sum of all the forces acting on a body is equal to its mass times its acceleration. The first step in problem solving is therefore: (1) Identify the body to whose motion the problem refers. Lack of clarity on the point as to what has been or should be picked as "the body" is a major source of mistakes. (2) Having selected "the body," we next turn our attention to the objects in "the environment" because these objects (inclined planes, springs, cords, the earth, etc.) exert forces on the body. We must be clear as to the nature of these forces. (3) The next step is to select a suitable (inertial) reference frame. We should position the origin and orient the coordinate axes so as to simplify the task of our next step as much as possible. (4) We now make a separate diagram of the body alone, showing the reference frame and *all* of the forces acting *on* the body. This is called a *free-body* diagram. (5) Finally we apply Newton's second law, in the form of Eq. 5-2, to each component of force and acceleration.

The following examples illustrate the method of analysis used in applying Newton's laws of motion. Each body is treated as if it were a particle of definite mass, so that the forces acting on it may be assumed to act at a point. Strings and pulleys are considered to have negligible mass. Although some of the situations picked for analysis may seem simple and artificial, they are the prototypes for many interesting real situations; but, more important, the method of analysis—which is the chief thing to understand—is applicable to all the modern and sophisticated situations of classical mechanics, even sending a spaceship to Mars.

Fig. 5-4a shows a weight W hung by strings. Consider the knot at the junction of the three strings to be "the body." The body remains at rest under the action of the three forces shown in Fig. 5-4b. Suppose we are given the magnitude of one of these forces. How can we find the magnitude of the other forces?

EXAMPLE 3

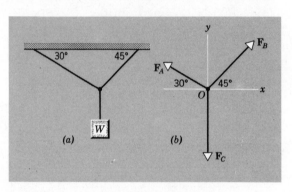

(a)   (b)

figure 5-4
Example 3. *(a)* A weight is suspended by strings. *(b)* A free-body diagram showing all the forces acting on the knot. The strings are assumed to be weightless.

$\mathbf{F}_A$, $\mathbf{F}_B$, and $\mathbf{F}_C$ are *all* the forces acting *on* the body. Since the body is unaccelerated (actually at rest), $\mathbf{F}_A + \mathbf{F}_B + \mathbf{F}_C = 0$. Choosing the x- and y-axes as shown, we can write this vector equation as three scalar equations:

$$F_{Ax} + F_{Bx} = 0,$$

$$F_{Ay} + F_{By} + F_{Cy} = 0,$$

using Eq. 5-2. The third scalar equation for the z-axis is simply

$$F_{Az} = F_{Bz} = F_{Cz} = 0.$$

That is, the vectors all lie in the x-y plane so that they have no z-components.
From the figure we see that

$$F_{Ax} = -F_A \cos 30° = -0.866 F_A,$$

$$F_{Ay} = F_A \sin 30° = 0.500 F_A,$$

and

$$F_{Bx} = F_B \cos 45° = 0.707 F_B,$$

$$F_{By} = F_B \sin 45° = 0.707 F_B.$$

Also,

$$F_{Cy} = -F_C = -W,$$

because the string C merely serves to transmit the force on one end to the junction at its other end. Substituting these results into our original equations, we obtain

$$-0.866F_A + 0.707F_B = 0,$$
$$0.500F_A + 0.707F_B - W = 0.$$

If we are given the magnitude of any one of these three forces, we can solve these equations for the other two. For example, if $W = 100$ N, we obtain $F_A = 73.3$ N and $F_B = 89.6$ N.

EXAMPLE 4

We wish to analyze the motion of a block on a smooth incline.

(a) *Static case.* Figure 5-5a shows a block of mass m kept at rest on a smooth plane, inclined at an angle θ with the horizontal, by means of a string attached to the vertical wall. The forces acting *on* the block are shown in Fig. 5-5b. F_1 is the force exerted *on* the block by the string; $m\mathbf{g}$ is the force exerted *on* the block by the earth, that is, its weight; and F_2 is the force exerted *on* the block by the inclined surface. F_2, called the normal force, is normal to the surface of contact because there is no frictional force between the surfaces.* If there were a frictional force, F_2 would have a component parallel to the incline. Because we wish to analyze the motion of *the block*, we choose *ALL the forces acting ON the block.* You will note that the block will exert forces on other bodies in its environment (the string, the earth, the surface of the incline) in accordance with the action-reaction principle; these forces, however, are not needed to determine the motion of the block because they do not act on the block.

Suppose θ and m are given. How do we find F_1 and F_2? Since the block is unaccelerated, we obtain

$$\mathbf{F}_1 + \mathbf{F}_2 + m\mathbf{g} = 0.$$

It is convenient to choose the x-axis of our reference frame to be along the incline and the y-axis to be normal to the incline (Fig. 5-5b). With this choice of coordinates, only one force, $m\mathbf{g}$, must be resolved into components in solving the problem. The two scalar equations obtained by resolving $m\mathbf{g}$ along the x- and y-axes are

$$F_1 - mg \sin \theta = 0, \quad \text{and} \quad F_2 - mg \cos \theta = 0,$$

from which F_1 and F_2 can be obtained if θ and m are given.

(b) *Dynamic case.* Now suppose that we cut the string. Then the force $\mathbf{F}_1$, the pull of the string on the block, will be removed. The resultant force on the block will no longer be zero, and the block will accelerate. What is its acceleration?

From Eq. 5-2 we have $F_x = ma_x$ and $F_y = ma_y$. Using these relations we obtain

$$F_2 - mg \cos \theta = ma_y = 0,$$

and

$$-mg \sin \theta = ma_x,$$

which yield

$$a_y = 0, \quad a_x = -g \sin \theta.$$

The acceleration is directed down the incline with a magnitude of $g \sin \theta$.

figure 5-5
Example 4. (a) A block is held on a smooth inclined plane by a string. (b) A free-body diagram showing all the forces acting on the block.

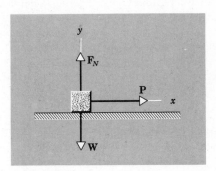

figure 5-6
Example 5. A block is being pulled along a smooth table. The forces acting on the block are shown.

EXAMPLE 5

Consider a block of mass m pulled along a smooth horizontal surface by a horizontal force $\mathbf{P}$, as shown in Fig. 5-6. F_N is the normal force exerted on the block by the frictionless surface and $\mathbf{W}$ is the weight of the block.

(a) If the block has a mass of 2.0 kg, what is the normal force?

From the second law of motion with $a_y = 0$, we obtain

$$F_y = ma_y \quad \text{or} \quad F_N - W = 0.$$

Hence, $F_N = W = mg = (2.0 \text{ kg})(9.8 \text{ m/s}^2) = 20$ N.

---

* The normal force is an example of a constraining force, one which limits the freedom of movement a body might otherwise have. It is an elastic force arising from small deformations of the bodies in contact, which are never perfectly rigid as we often tacitly assume.

(*b*) What force P is required to give the block a horizontal velocity of 4.0 m/s in 2.0 s starting from rest?

The acceleration a_x follows from

$$a_x = \frac{v_x - v_{x0}}{t} = \frac{4.0 \text{ m/s} - 0}{2.0 \text{ s}} = 2.0 \text{ m/s}^2.$$

From the second law, $F_x = ma_x$ or $P = ma_x$. The force P is then

$$P = ma_x = (2.0 \text{ kg})(2.0 \text{ m/s}^2) = 4.0 \text{ N}.$$

Figure 5-7*a* shows a block of mass m_1 on a smooth horizontal surface pulled by a massless string which is attached to a block of mass m_2 hanging over a pulley. We assume that the pulley has no mass and is frictionless and that it merely serves to change the direction of the tension in the string at that point. The magnitude of the tension is the same throughout a massless string (see Example 2). Find the acceleration of the system and the tension in the string.

Suppose we choose the block of mass m_1 as the body whose motion we investigate. The forces on this block, taken to be a particle, are shown in Fig. 5-7*b*. **T**, the tension in the string, pulls on the block to the right; m_1g is the downward pull of the earth on the block and $\mathbf{F}_N$ is the vertical force exerted on the block by the smooth table. The block will accelerate in the *x*-direction only, so that $a_{1y} = 0$. We, therefore, can write

$$F_N - m_1g = 0 = m_1a_{1y},$$

$$\text{(5-7)}$$

and

$$T = m_1a_{1x}.$$

From these equations we conclude that $F_N = m_1g$. We do not know T, so we cannot solve for a_{1x}.

To determine T we must consider the motion of the block m_2. The forces acting on m_2 are shown in Fig. 5-7*c*. Because the string and block are accelerating, we cannot conclude that T equals m_2g. In fact, if T were to equal m_2g, the resultant force on m_2 would be zero, a condition holding only if the system is not accelerated.

The equation of motion for the suspended block is

$$m_2g - T = m_2a_{2y}. \qquad \text{(5-8)}$$

The direction of the tension in the string changes at the pulley and, because the string has a fixed length, it is clear that

$$a_{2y} = a_{1x},$$

so that we can represent the acceleration of the system as simply *a*. We then obtain from Eqs. 5-7 and 5-8

$$m_2g - T = m_2a, \qquad \text{(5-9)}$$

and

$$T = m_1a.$$

These yield

$$m_2g = (m_1 + m_2)a, \qquad \text{(5-10)}$$

or

$$a = \frac{m_2}{m_1 + m_2}\, g,$$

and

$$T = \frac{m_1m_2}{m_1 + m_2}\, g, \qquad \text{(5-11)}$$

which gives us the acceleration of the system *a* and the tension in the string *T*.

Notice that the tension in the string is always less than m_2g. This is clear from Eq. 5-11, which can be written

$$T = m_2g\,\frac{m_1}{m_1 + m_2}.$$

EXAMPLE 6

figure 5-7
Example 6. (*a*) Two masses are connected by a string; m_1 lies on a smooth table, m_2 hangs freely. (*b*) A free-body diagram showing all the forces acting on m_1. (*c*) A similar diagram for m_2.

Notice also that a is always less than g, the acceleration due to gravity. Only when m_1 equals zero, which means that there is no block at all on the table, do we obtain $a = g$ (from Eq. 5-10). In this case $T = 0$ (from Eq. 5-9).

We can interpret Eq. 5-10 in a simple way. The net unbalanced force on the system of mass $m_1 + m_2$ is represented by $m_2 g$. Hence, from $F = ma$, we obtain Eq. 5-10 directly.

To make the example specific, suppose $m_1 = 2.0$ kg and $m_2 = 1.0$ kg. Then

$$a = \frac{m_2}{m_1 + m_2} g = \tfrac{1}{3} g = 3.3 \text{ m/s}^2,$$

and

$$T = \frac{m_1 m_2}{m_1 + m_2} g = (\tfrac{2}{3})(9.8) \text{ kg m/s}^2 = 6.5 \text{ N}.$$

EXAMPLE 7

Consider two unequal masses connected by a string which passes over a frictionless and massless pulley, as shown in Fig. 5-8a. Let m_2 be greater than m_1. Find the tension in the string and the acceleration of the masses.

We consider an *upward* acceleration *positive*. If the acceleration of m_1 is a, the acceleration of m_2 must be $-a$. The forces acting on m_1 and on m_2 are shown in Fig. 5-8b in which T represents the tension in the string.

The equation of motion for m_1 is

$$T - m_1 g = m_1 a$$

and for m_2 is

$$T - m_2 g = -m_2 a.$$

Combining these equations, we obtain

$$a = \frac{m_2 - m_1}{m_2 + m_1} g, \tag{5-12}$$

and

$$T = \frac{2 m_1 m_2}{m_1 + m_2} g.$$

For example, if $m_2 = 2.0$ slugs and $m_1 = 1.0$ slug,

$$a = (32/3.0) \text{ ft/s}^2 = g/3,$$

$$T = (\tfrac{4}{3})(32) \text{ slug ft/s}^2 = 43 \text{ lb}.$$

Notice that the magnitude of T is always intermediate between the weight of the mass m_1 (32 lb in our example) and the weight of the mass m_2 (64 lb in our example). This is to be expected, since T must exceed $m_1 g$ to give m_1 an upward acceleration, and $m_2 g$ must exceed T to give m_2 a downward acceleration. In the special case when $m_1 = m_2$, we obtain $a = 0$ and $T = m_1 g = m_2 g$, which is the static result to be expected.

Figure 5-8c shows the forces acting on the massless pulley. If we treat the pulley as a particle, all the forces can be taken to act through its center. P is the

figure 5-8
Example 7. *(a)* Two unequal masses are suspended by a string from a pulley (Atwood's machine). *(b)* Free-body diagrams for m_1 and m_2. *(c)* Free-body diagram for the pulley, assumed massless.

upward pull of the support on the pulley and T is the downward pull of each segment of the string on the pulley. Since the pulley has no translational motion,

$$P = T + T = 2T.$$

If we were to drop our assumption of a massless pulley and assign a mass m to it, we would then be required to include a downward force mg on the support. Also, as we shall see later, the rotational motion of the pulley results in a different tension in each segment of the string. Friction in the bearings also affects the rotational motion of the pulley and the tension in the strings.

EXAMPLE 8

Consider an elevator moving vertically with an acceleration **a.** We wish to find the force exerted by a passenger on the floor of the elevator.

Acceleration will be taken *positive upward* and *negative downward.* Thus positive acceleration in this case means that the elevator is either moving upward with increasing speed or is moving downward with decreasing speed. Negative acceleration means that the elevator is moving upward with decreasing speed or downward with increasing speed.

From Newton's third law the force exerted by the passenger on the floor will always be equal in magnitude but opposite in direction to the force exerted by the floor on the passenger. We can therefore calculate either the action force or the reaction force. When the forces acting on the passenger are used, we solve for the latter force. When the forces acting on the floor are used, we solve for the former force.

The situation is shown in Fig. 5-9: The passenger's true weight is **W** and the force exerted on him by the floor, called **P,** is his *apparent* weight in the accelerating elevator. The resultant force acting on him is **P + W.** Forces will be taken as positive when directed upward. From the second law of motion we have

$$F = ma,$$

or

$$P - W = ma, \tag{5-13}$$

where m is the mass of the passenger and a is his (and the elevator's) acceleration.

Suppose, for example, that the passenger weighs 160 lb and the acceleration is 2.0 ft/s² upward. We have

$$m = \frac{W}{g} = \frac{160 \text{ lb}}{32 \text{ ft/s}^2} = 5.0 \text{ slugs},$$

and from Eq. 5-13,

$$P - 160 \text{ lb} = (5.0 \text{ slugs})(2.0 \text{ ft/s}^2)$$

or

$$P = \text{apparent weight} = 170 \text{ lb}.$$

If we were to measure this force directly by having the passenger stand on a spring scale fixed to the elevator floor (or suspended from the ceiling), we would find the scale reading to be 170 lb for a man whose weight is 160 lb. The passenger feels himself pressing down on the floor with greater force (the floor is pressing upward on him with greater force) than when he and the elevator are at rest. Everyone experiences this feeling when an elevator starts upward from rest.

If the acceleration were taken as 2.0 ft/s² downward, then $a = -2.0$ ft/s² and $P = 150$ lb for the passenger. The passenger who weighs 160 lb feels himself pressing down on the floor with less force than when he and the elevator are at rest.

If the elevator cable were to break and the elevator were to fall freely with an acceleration $a = -g$, then P would equal $W + (W/g)(-g) = 0$. Then the passenger and floor would exert no forces on each other. The passenger's apparent weight, as indicated by the spring scale on the floor, would be zero. Such a situation is

figure 5-9

Example 8. *(a)* A passenger stands on the floor of an elevator. *(b)* A free-body diagram for the passenger.

popularly referred to as "weightlessness." The passenger's weight (the pull of gravity on him) has not changed, of course, but the force he exerts on the floor and the reaction force of the floor on him are zero.

questions

1. What is your mass in slugs? Your weight in newtons?

2. Why do you fall forward when a moving train decelerates to a stop and fall backward when a train accelerates from rest? What would happen if the train rounded a curve at constant speed?

3. A block of mass m is supported by a cord C from the ceiling, and another cord D is attached to the bottom of the block (Fig. 5-10). Explain this: If you give a sudden jerk to D, it will break, but if you pull on D steadily, C will break.

4. A horse is urged to pull a wagon. The horse refuses to try, citing Newton's third law as his defense: " 'The pull of the horse on the wagon is equal but opposite to the pull of the wagon on the horse.' If I can never exert a greater force on the wagon than it exerts on me, how can I ever start the wagon moving?" asks the horse. How would you reply?

5. Comment on whether the following pairs of forces are examples of action-reaction: (a) the earth attracts a brick; the brick attracts the earth; (b) a propellered airplane pulls air in toward the plane; the air pushes the plane forward; (c) a horse pulls forward on a cart, accelerating it; the cart pulls backward on the horse; (d) a horse pulls forward on a cart without moving it; the cart pulls back on the horse; (e) a horse pulls forward on a cart without moving it; the earth exerts an equal and opposite force on the cart.

6. Criticize the statement, often made, that the mass of a body is a measure of the "quantity of matter" in it.

7. Using force, length, and time as fundamental quantities, what are the dimensions of mass?

8. Is the definition of mass that we have given limited to objects initially at rest?

9. Comment on the following statements about mass and weight taken from examination papers. (a) Mass and weight are the same physical quantities expressed in different units; (b) mass is a property of one object alone whereas weight results from the interaction of two objects; (c) the weight of an object is proportional to its mass; (d) the mass of a body varies with changes in its local weight.

10. A horizontal force acts on a mass which is free to move. Can it produce an acceleration if the force is less than the weight of that mass?

11. Does the acceleration of a freely falling body depend upon the weight of the body?

12. A bird alights on a stretched telegraph wire. Does this change the tension in the wire? If so, by an amount less than, equal to, or greater than the weight of the bird?

13. In Fig. 5-11, we show four forces which are about equal in magnitude. What combination of three forces, acting together on the same body, might keep that body in translational equilibrium?

14. Why do raindrops fall with constant speed during the later stages of their descent?

15. In a tug of war, three men pull on a rope to the left at A and three men pull to the right at B with forces of equal magnitude. Now a weight of 5.0 lb is hung vertically from the center of the rope. (a) Can the men get the rope AB to be horizontal? (b) If not, explain. If so, determine the magnitude of the forces required at A and B to do this.

16. Both the following statements are true; explain them. Two teams having a tug of war must always pull equally hard on one another. The team that pushes harder against the ground wins.

figure 5-10
Question 3

figure 5-11
Question 13

17. A massless rope is strung over a frictionless pulley. A monkey holds onto one end of the rope and a mirror, having the same weight as the monkey, is attached to the other end of the rope at the monkey's level. Can the monkey get away from his image seen in the mirror (a) by climbing up the rope, (b) by climbing down the rope, (c) by releasing the rope?

18. Two objects of equal mass rest on opposite pans of a trip scale. Does the scale remain balanced when it is accelerated up or down in an elevator?

19. You stand on the large platform of a spring scale and note your weight. You then take a step on this platform and notice that the scale reads less than your weight at the beginning of the step and more than your weight at the end of the step. Explain.

20. A weight is hung by a cord from the ceiling of an elevator. From the following conditions, choose the one in which the tension in the cord will be greatest . . . least? (a) elevator at rest; (b) elevator rising with uniform speed; (c) elevator descending with decreasing speed; (d) elevator descending with increasing speed.

21. A woman stands on a spring scale in an elevator. In which case below will the scale record the minimum reading . . . the maximum reading? (a) elevator stationary; (b) elevator cable breaks, free fall; (c) elevator accelerating upward; (d) elevator accelerating downward; (e) elevator moving at constant velocity.

22. Under what circumstances would your weight be zero? Does your answer depend on the choice of a reference system?

problems

SECTION 5-4

1. Two blocks, mass m_1 and m_2, are connected by a light spring on a horizontal frictionless table. Find the ratio of their accelerations a_1 and a_2 after they pulled apart and then released. *Answer: $a_1/a_2 = m_2/m_1$.*

SECTION 5-5

2. (a) Two 10-lb weights are attached to a spring scale as shown in Fig. 5-12(a). What is the reading of the scale? (b) A single 10-lb weight is attached to a spring scale which itself is attached to a wall, as shown in Fig. 5-12(b). What is the reading of the scale?

3. Two blocks are in contact on a frictionless table. A horizontal force is applied to one block, as shown in Fig. 5-13. (a) If $m_1 = 2.0$ kg, $m_2 = 1.0$ kg, and $F = 3.0$ N, find the force of contact between the two blocks. (b) Show that if the same force F is applied to m_2 rather than to m_1, the force of contact between the blocks is 2.0 N, which is not the same value derived in (a). Explain. *Answer: (a) 1.0 N.*

SECTION 5-8

4. A space traveler whose mass is 75 kg leaves the earth. Compute his weight (a) on the earth, (b) on Mars, where $g = 3.8$ m/s², and (c) in interplanetary space. (d) What is his mass at each of these locations?

SECTION 5-10

5. A car moving initially at a speed of 50 mi/h (80 km/h) and weighing 3000 lb (13,000 N) is brought to a stop in a distance of 200 ft (61 m). Find (a) the braking force, and (b) the time required to stop. Assuming the same braking force, find (c) the distance, and (d) the time required to stop if the car was going 25 mi/h (40 km/h) initially.
Answer: (a) 1300 lb (5400 N). (b) 5.5 s (5.5 s). (c) 50 ft (15 m). (d) 2.7 s (2.7 s).

6. A body of mass m is acted on by two forces $\mathbf{F}_1$ and $\mathbf{F}_2$, as shown in Fig. 5-14. If $m = 5.0$ kg, $F_1 = 3.0$ N, and $F_2 = 4.0$ N, find the vector acceleration of the body.

7. An electron is projected horizontally at a speed of 1.2×10^7 m/s into an

figure 5-12(a)
Problem 2(a)

figure 5-12(b)
Problem 2(b)

electric field which exerts a constant vertical force of 4.5×10^{-16} N on it. The mass of the electron is 9.1×10^{-31} kg. Determine the vertical distance the electron is deflected during the time it has moved forward 3.0 cm horizontally. *Answer:* 1.5 mm.

8. A body of mass 2.0 slugs is acted on by the downward force of gravity and a horizontal force of 130 lb. Find (a) its acceleration and (b) its velocity as functions of time, assuming it starts from rest.

figure 5-13
Problem 3

9. An electron travels in a straight line from the cathode of a vacuum tube to its anode, which is exactly 1.0 cm away. It starts with zero speed and reaches the anode with a speed of 6.0×10^6 m/s. (a) Assume constant acceleration and compute the force on the electron. Take the electron's mass to be 9.1×10^{-31} kg. This force is electrical in origin. (b) Compare it with the gravitational force on the electron, which we neglected when we assumed straight line motion. *Answer:* (a) 1.6×10^{-15} N. (b) 8.9×10^{-30} N.

figure 5-14
Problem 6

10. A man of mass 80 kg (weight $mg = 176$ lb) jumps down to a concrete patio from a window ledge only 0.50 m (1.6 ft) above the ground. He neglects to bend his knees on landing, so that his motion is arrested in a distance of about 2.0 cm (0.79 in). (a) What is the average acceleration of the man from the time his feet first touch the patio to the time he is brought fully to rest? (b) With what average force does this jump jar his bone structure?

11. Let the only forces acting on two bodies be their mutual interactions. If both bodies start from rest, show that the distances traveled by each are inversely proportional to the respective masses of the bodies.

12. Determine the frictional force of the air on a body of mass 0.25 kg falling with an acceleration of 9.2 m/s².

13. A charged sphere of mass 3.0×10^{-4} kg is suspended from a string. An electric force acts horizontally on the sphere so that the string makes an angle of 37° with the vertical when at rest. Find (a) the magnitude of the electric force and (b) the tension in the string.
Answer: (a) 2.2×10^{-3} N. (b) 3.7×10^{-3} N.

14. A block of mass M is pulled along a horizontal frictionless surface by a rope of mass m, as shown in Fig. 5-15. A horizontal force **P** is applied to one end of the rope. (a) Show that the rope *must* sag, even if only by an imperceptible amount. Then, assuming that the sag is negligible, find (b) the acceleration of rope and block, (c) the force that the rope exerts on the block M, and (d) the tension in the rope at its midpoint.

figure 5-15
Problem 14

figure 5-16
Problem 15

15. Three blocks are connected, as shown in Fig. 5-16, on a horizontal frictionless table and pulled to the right with a force $T_3 = 60$ N. If $m_1 = 10$ kg, $m_2 = 20$ kg, and $m_3 = 30$ kg, find the tensions T_1 and T_2. Draw an analogy to bodies being pulled in tandem, such as an engine pulling a train of coupled cars. *Answer:* $T_1 = 10$ N, $T_2 = 30$ N.

16. A rocket and its payload have a total mass of 50,000 kg (weight $mg = 110,250$ lb). How large is the thrust of the rocket engine when (a) the rocket is "hovering" over the launch pad, just after ignition, and (b) when the rocket is accelerating upward at 20 m/s² (66 ft/s²)?

17. How could a 100-lb object be lowered from a roof using a cord with a breaking strength of 87 lb without breaking the cord?
Answer: Lower object with an acceleration ≥ 4.2 ft/s².

18. A block is released from rest at the top of a frictionless inclined plane 16 m long. It reaches the bottom 4.0 s later. A second block is projected up the plane from the bottom at the instant the first block is released in such a way that it returns to the bottom simultaneously with the first block. (a) Find the acceleration of each block on the incline. (b) What is the initial velocity of the second block? (c) How far up the incline does it travel? (d) What angle does the plane make with the horizontal?

19. A block of mass $m_1 = 3.0$ slugs on a smooth inclined plane of angle 30° is connected by a cord over a small frictionless pulley to a second block of mass $m_2 = 2.0$ slugs hanging vertically (Fig. 5-17). (a) What is the acceleration of each body? (b) What is the tension in the cord? *Answer: (a)* 3.2 ft/s². *(b)* 58 lb.

20. A block is projected up a frictionless inclined plane with a speed v_0. The angle of incline is θ. (a) How far up the plane does it go? (b) How long does it take to get there? (c) What is its speed when it gets back to the bottom? Find numerical answers for $\theta = 30°$ and $v_0 = 8.0$ ft/s.

21. An elevator weighing 6000 lb is pulled upward by a cable with an acceleration of 4.0 ft/s². (a) What is the tension in the cable? (b) What is the tension when the elevator is accelerating downward at 4.0 ft/s², but is still moving upward? *Answer: (a)* 6800 lb. *(b)* 5300 lb.

22. A lamp hangs vertically from a cord in a descending elevator. The elevator has a deceleration of 8.0 ft/s² (2.4 m/s²) before coming to a stop. (a) If the tension in the cord is 20 lb (89 N), what is the mass of the lamp? (b) What is the tension in the cord when the elevator ascends with an acceleration of 8.0 ft/s² (2.4 m/s²)?

23. An 80-kg man is parachuting and experiencing a downward acceleration of 2.5 m/s². The mass of the parachute is 5.0 kg. (a) What is the value of the upward force exerted on the parachute by the air? (b) What is the value of the downward force exerted by the man on the parachute? *Answer: (a)* 620 N. *(b)* 580 N.

24. A research balloon of total mass M is descending vertically with downward acceleration a. How much ballast must be thrown from the car to give the balloon an *upward* acceleration a?

25. An elevator consists of the elevator cage (A), the counterweight (B), the driving mechanism (C), and the cable and pulleys as shown in Fig. 5-18. The mass of the cage is 1100 kg and the mass of the counterweight is 1000 kg. Neglect friction and the mass of the cable and pulleys. The elevator accelerates upward at 2.0 m/s² and the counterweight accelerates downward at the same rate. (a) What is the value of the tension T_1? (b) T_2? (c) What force is exerted on the cable by the driving mechanism? *Answer: (a)* 1.3×10^4 N. *(b)* 0.78×10^4 N. *(c)* 5.2×10^3 N, toward the counterweight.

26. A 100-kg man lowers himself to the ground from a height of 10 m by means of a rope passed over a frictionless pulley and attached to a 70-kg sandbag. (a) With what speed does the man hit the ground? (b) Is there anything he could do to reduce the speed with which he hits the ground?

27. Someone exerts a force **F** directly up on the axle of the pulley shown in Fig. 5-19. Consider the pulley and string to be massless and the bearing frictionless. Two bodies, m_1 of mass 1.0 kg and m_2 of mass 2.0 kg, are attached, as shown, to the opposite ends of the string which passes over the pulley. The body m_2 is in contact with the horizontal floor. (a) Draw a free body diagram for the pulley and for each of the masses. (b) What is the largest value the force F may have so that m_2 will remain at rest on the floor? (c) What is the tension in the string if the upward force **F** is 100 N? (d) With the tension determined in part (c), what is the acceleration of m_1? *Answer: (b)* 39 N. *(c)* 50 N. *(d)* 40 m/s², upward.

28. A 10-kg monkey is climbing a massless rope attached to a 15-kg mass over a (frictionless!) tree limb. (a) Explain quantitatively how the monkey can

figure 5-17
Problem 19

figure 5-18
Problem 25

figure 5-19
Problem 27

climb up the rope so that he can raise the 15-kg mass off the ground. If, after the mass has been raised off the ground, the monkey stops climbing and holds on to the rope, what will now be (b) his acceleration and (c) the tension in the rope?

29. A plumb bob hanging from the ceiling of a railroad car acts as an accelerometer. (a) Derive the general expression relating the horizontal acceleration a of the car to the angle θ made by the bob with the vertical. (b) Find a when θ = 20°. (c) Find θ when a = 5.0 ft/s².
Answer: (a) $a = g \tan \theta$. (b) 12 ft/s². (c) 8.9°.

30. A uniform flexible chain of length *l*, with weight per unit length λ, passes over a small, frictionless, massless pulley. It is released from a rest position with a length of chain *x* hanging from one side and a length *l* − *x* from the other side. Find the acceleration *a* as a function of *x*.

31. Two particles, each of mass *m*, are connected by a light string of length 2*l*, as shown in Fig. 5-20. A continuous force **F** is applied at the midpoint of the string (*x* = 0) at right angles to the initial position of the string. Show that the acceleration of *m* in the direction at right angles to **F** is given by

$$a_r = \frac{F}{2m} \frac{x}{\sqrt{l^2 - x^2}}$$

figure 5-20
Problem 31

in which *x* is the perpendicular distance of one of the particles from the line of action of **F**. Discuss the situation when *x* = *l*.

32. A chain consisting of five links, each of mass 0.10 kg, is lifted vertically with a constant acceleration of 2.5 m/s², as shown in Fig. 5-21. Find (a) the forces acting between adjacent links, (b) the force **F** exerted on the top link by the agent lifting the chain, and (c) the *net* force acting on each link.

33. *Terminal velocity.* The resistance of the air to the motion of bodies in free fall depends on many factors, such as the size of the body and its shape, the density and temperature of the air, and the velocity of the body through the air. A useful assumption, only approximately true, is that the resisting force f_R is proportional to the velocity and oppositely directed; that is, $f_R = -k\mathbf{v}$, where *k* is a constant whose value in any particular case is determined by factors other than velocity.

figure 5-21
Problem 32

Consider free fall of an object from rest through the air.
(a) Show that Newton's second law gives

$$mg - kv = ma \quad \text{or} \quad mg - k\frac{dy}{dt} = m\frac{d^2y}{dt^2}.$$

(b) Show that the body ceases to accelerate when it reaches a velocity $v_T = mg/k$, called the *terminal velocity.*
(c) Prove, by substituting it in the equation of motion of part (a), that the velocity varies with time as

$$v = v_T(1 - e^{-kt/m})$$

and plot *v* versus *t*.
(d) Sketch qualitatively curves of *y* versus *t* and *a* versus *t* for this motion, noting that the initial acceleration is *g* and the final acceleration is zero.

34. A right triangular wedge of mass *M* and angle θ, supporting a cubical block of mass *m* on its side, rests on a horizontal table, as shown in Fig. 5-22. (a) What horizontal acceleration *a* must *M* have relative to the table to keep *m* stationary relative to the wedge, assuming frictionless contacts? (b) What horizontal force **F** must be applied to the system to achieve this result, assuming a frictionless table top? (c) Suppose no force is supplied to *M* and both surfaces are frictionless. Describe the resulting motion.

figure 5-22
Problem 34

35. A block, mass *m*, slides down a frictionless incline making an angle θ with an elevator floor. Find its acceleration relative to the incline in the following cases. (a) Elevator descends at constant speed *v*. (b) Elevator ascends at constant speed *v*. (c) Elevator descends with acceleration *a*. (d) Elevator

descends with deceleration *a*. (*e*) Elevator cable breaks. (*f*) In part (*c*) above, what is the force exerted on the block by the incline?

Answer: (*a*) $g \sin \theta$ down the incline. (*b*) $g \sin \theta$ down the incline. (*c*) $(g - a)$ $\sin \theta$ down the incline. (*d*) $(g + a) \sin \theta$ down the incline. (*e*) Zero. (*f*) $m(g - a) \cos \theta$.

6
particle
dynamics—II

In Chapter 5 we considered particle dynamics for bodies subject to a force that was constant in both magnitude and direction. The forces that we dealt with were exerted by the earth or by taut cords, that is, they were either gravitational or elastic. In this chapter we consider another kind of force, that resulting from friction.

We shall also discuss the dynamics of uniform circular motion, in which the force, although constant in magnitude, changes in direction with time. In Chapter 10 we shall consider problems in which the force, although constant in direction, changes in magnitude with time, as when one body exerts a transient force on another during a collision. Finally, in Chapter 15, we shall consider problems in which the force changes in *both* magnitude *and* direction with time, such as the force exerted by a spring on an oscillating mass suspended from it.

If we project a block of mass m with initial velocity $\mathbf{v}_0$ along a long horizontal table, it eventually comes to rest. This means that, while it is moving, it experiences an average acceleration $\bar{\mathbf{a}}$ that points in the direction opposite to its motion. If (in an inertial frame) we see that a body is being accelerated, we always associate a force, defined from Newton's second law, with the motion. In this case we declare that the table exerts a force of friction, whose average value is $m\bar{\mathbf{a}}$, on the sliding block.

Actually, whenever the surface of one body slides over that of an-

* See "The Friction of Solids" by E. H. Freitag, in *Contemporary Physics*, Vol. 2, 1961, p. 198, for a good general reference; see also the article "Friction" in Britannica 3.

other, each body exerts a frictional force on the other. The frictional force on each body is in a direction opposite to its motion relative to the other body. Frictional forces automatically oppose the motion and never aid it. Even when there is no relative motion, frictional forces may exist between surfaces.

Although we have ignored its effects up to now, friction is very important in our daily lives. Left to act alone it brings every rotating shaft to a halt. In an automobile, about 20% of the engine power is used to counteract frictional forces. Friction causes wear and seizing of moving parts and many engineering man-hours are devoted to reducing it. On the other hand, without friction we could not walk; we could not hold a pencil and if we could it would not write; wheeled transport as we know it would not be possible.

We want to know how to express frictional forces in terms of the properties of the body and its environment; that is, we want to know the force law for frictional forces. In what follows we consider the sliding (not rolling) of one dry (unlubricated) surface over another. As we shall see later, friction, viewed at the microscopic level, is a very complicated phenomenon* and the force laws for dry, sliding friction are empirical in character and approximate in their predictions. They do not have the elegant simplicity and accuracy that we find for the gravitational force law (Chapter 16) or for the electrostatic force law (Chapter 26). It is remarkable, however, considering the enormous diversity of surfaces one encounters, that many aspects of frictional behavior can be understood qualitatively on the basis of a few simple mechanisms.

Consider a block at rest on a horizontal table as in Fig. 6-1. Attach a spring to it to measure the force required to set the block in motion. We find that the block will not move even though we apply a small force. We say that our applied force is balanced by an opposite frictional force exerted on the block by the table, acting along the surface of contact. As we increase the applied force we find some definite force at which the block just begins to move. Once motion has started, this same force produces accelerated motion. By reducing the force once motion has started, we find that it is possible to keep the block in uniform motion without acceleration; this force may be small, but it is never zero.

The frictional forces acting between surfaces at rest with respect to each other are called forces of *static friction*. The maximum force of static friction will be the same as the smallest force necessary to start motion. Once motion is started, the frictional forces acting between the surfaces usually decrease so that a smaller force is necessary to maintain uniform motion. The forces acting between surfaces in relative motion are called forces of *kinetic friction*.

The maximum force of static friction between any pair of dry unlubricated surfaces follows these two empirical laws. (1) It is approximately independent of the area of contact, over wide limits and (2) it is proportional to the normal force. The normal force, sometimes called the loading force, is the one which either body exerts on the other at right angles to their mutual interface. It arises from the elastic deformation of the bodies in contact, such bodies never really being entirely rigid. For a block resting on a horizontal table or sliding along it, the normal force is equal in magnitude to the weight of the block. Because

* See, for example, "Stick and Slip" by Ernest Rabinowicz, in *Scientific American*, May 1956.

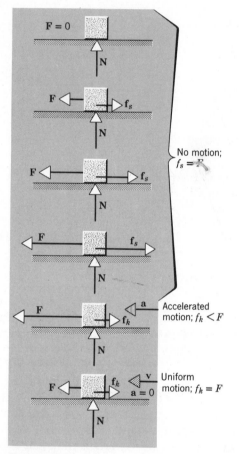

figure 6-1

A block being put into motion as applied force **F** overcomes frictional forces. In the first four drawings the applied force is gradually increased from zero to magnitude $\mu_s N$. No motion occurs until this point because the frictional force always just balances the applied force. The instant F becomes greater than $\mu_s N$, the block goes into motion, as is shown in the fifth drawing. In general, $\mu_k N < \mu_s N$; this leaves an unbalanced force to the left and the block accelerates. In the last drawing F has been reduced to equal $\mu_k N$. The net force is zero, and the block continues with constant velocity.

the block has no vertical acceleration, the table must be exerting a force on the block that is directed upward and is equal in magnitude to the downward pull of the earth on the block, that is, equal to the block's weight.

The ratio of the magnitude of the maximum force of static friction to the magnitude of the normal force is called the *coefficient of static friction* for the surfaces involved. If f_s represents the magnitude of the force of static friction, we can write

$$f_s \leq \mu_s N, \tag{6-1}$$

where μ_s is the coefficient of static friction and N is the magnitude of the normal force. The equality sign holds only when f_s has its maximum value.

The force of kinetic friction f_k between dry, unlubricated surfaces follows the same two laws as those of static friction. (1) It is approximately independent of the area of contact over wide limits and (2) it is proportional to the normal force. The force of kinetic friction is also reasonably independent of the relative speed with which the surfaces move over each other.

The two laws of friction above were first discovered experimentally by Leonardo da Vinci (1452–1519). Leonardo's statement of the two laws was remarkable, coming as it did about two centuries before the concept of force was developed by Newton. Leonardo's formulation was: (1) "Friction made by the same weight will be of equal resistance at the beginning of the movement though the contact may be of different breadths or lengths" and (2) "Friction produces double the amount of effort if the weight be doubled." The French scientist, Charles A. Coulomb, (1736–1806) did many experiments on friction and pointed out the difference between static and kinetic friction.

The ratio of the magnitude of the force of kinetic friction to the magnitude of this normal force is called the *coefficient of kinetic friction*. If f_k represents the magnitude of the force of kinetic friction,

$$f_k = \mu_k N, \tag{6-2}$$

where μ_k is the coefficient of kinetic friction.

Both μ_s and μ_k are dimensionless constants, each being the ratio of (the magnitudes of) two forces. Usually, for a given pair of surfaces $\mu_s > \mu_k$. The actual values of μ_s and μ_k depend on the nature of both the surfaces in contact. Both μ_s and μ_k can exceed unity, although commonly they are less than one. Notice that Eqs. 6-1 and 6-2 are relations between the *magnitudes only* of the normal and frictional forces. These forces are always directed perpendicularly to one another.

On the atomic scale even the most finely polished surface is far from plane. Figure 6-2, for example, shows an actual profile, highly magnified, of a steel surface that would be considered to be highly polished. One can readily believe that when two bodies are placed in contact, the actual microscopic area of contact is much less than apparent macroscopic area of contact; in a particular case these areas can be easily in the ratio of 1 to 10^4.

The actual (microscopic) area of contact is proportional to the normal force, because the contact points deform plastically under the great stresses that develop at these points. Many contact points actually become "cold-welded" together. This phenomenon, *surface adhesion*, occurs because at the contact points the molecules on opposite sides of the surface are so close together that they exert strong intermolecular forces on each other.

When one body (a metal, say) is pulled across another, the frictional resistance is associated with the rupturing of these thousands of tiny welds,

figure 6-2
A highly magnified view of a section of a finely polished steel surface. The section was cut at an angle so that vertical distances are exaggerated by a factor of ten with respect to horizontal distances. The surface irregularities are several thousand atomic diameters high. From *Friction and Lubrication of Solids*, by F. P. Bowden and D. Tabor, Clarendon Press, 1950.

which continually reform as new chance contacts are made (see Fig. 6-3). Radio-active tracer experiments have shown that, in the rupturing process, small fragments of one metallic surface may be sheared off and adhere to the other surface. If the relative speed of the two surfaces is great enough, there may be local melting at certain contact areas even though the surface as a whole may feel only moderately warm.

(b)   (a)

figure 6-3
Sliding friction. *(a)* The upper body is sliding to the right over the lower body in this enlarged diagram. *(b)* A further enlarged view showing two spots where surface adhesion has occurred. Force is required to break these welds apart and maintain the motion.

The coefficient of friction depends on many variables, such as the nature of the materials, surface finish, surface films, temperature, and extent of contamination. For example, if two carefully cleaned metal surfaces are placed in a highly evacuated chamber so that surface oxide films do not form, the coefficient of friction rises to enormous values and the surfaces actually become firmly "welded" together. The admission of a small amount of air to the chamber so that oxide films may form on the opposing surfaces reduces the coefficient of friction to its "normal" value.

With these complications it is not surprising that there is no exact theory of dry friction and that the laws of friction are empirical. The surface adhesion theory of friction for metals leads to a ready understanding of the two laws of friction mentioned above however. (1) The microscopic contact area, which determines the frictional force f_k, is proportional to the normal force N and thus f_k is proportional to N, as Eq. 6-2 shows. (2) The fact that the frictional force is independent of the apparent area of contact means, for example, that the force required to drag a metal "brick" along a metal table is the same no matter which face of the brick is in contact with the table. We can understand this only if the microscopic area of contact is the same for all positions of the brick, and this is indeed the case. With the largest face down, there are a relatively large number of relatively small area contacts supporting the load; with the smallest face down there are fewer contacts (because the apparent contact area is smaller), but the area of individual contact is larger by just the same factor because of the higher pressure exerted by the up-ended brick on this smaller number of contacts supporting the same load.

The frictional force that opposes one body *rolling* over another is much less than that for a sliding motion and this, indeed, is the advantage of the wheel over the sledge. This reduced friction is due in large part to the fact that, in rolling, the microscopic contact welds are "peeled" apart rather than "sheared" apart as in sliding friction. This will reduce the frictional force by a large factor.

Frictional resistance in dry, sliding, friction can be considerably reduced by lubrication. A mural in a grotto in Egypt dating back to 1900 B.C. shows a large stone statue being pulled on a sledge while a man in front of the sledge pours lubricating oil in its path. A still more effective technique is to introduce a layer of gas between the sliding surfaces; the dry ice puck and the gas-supported bearing are two examples. Friction can be reduced still further by suspending a rotating object in an evacuated space by means of magnetic forces. J. W. Beams, for example, has spun a 30-lb rotor of this type at 1000 rev/s; when the drive was cut off, the rotor lost speed at the rate of only 1 rev/s in a day.*

* See "Ultrahigh-Speed Rotation," Jesse W. Beams in *Scientific American*, April 1961.

Examples of the application of the empirical force law for friction follow. The coefficients of friction given are assumed to be constant. Actually μ_k can be regarded as a good average value that is not greatly different from the value at any particular speed in the range.

EXAMPLE 1

A block is at rest on an inclined plane making an angle θ with the horizontal, as in Fig. 6-4a. As the angle of incline is raised, it is found that slipping just begins at an angle of inclination θ_s. What is the coefficient of static friction between block and incline?

figure 6-4
Example 1. (a) A block at rest on a rough inclined plane. (b) A free-body force diagram for the block.

The forces acting on the block, considered to be a particle, are shown in Fig. 6-4b. $\mathbf{W}$ is the weight of the block, $\mathbf{N}$ the normal force exerted on the block by the inclined surface, and $\mathbf{f}_s$ the tangential force of friction exerted by the inclined surface on the block. Notice that the resultant force exerted by the inclined surface on the block, $\mathbf{N} + \mathbf{f}_s$, is no longer perpendicular to the surface of contact, as was true for smooth surfaces ($\mathbf{f}_s = 0$). The block is at rest, so that

$$\mathbf{N} + \mathbf{f}_s + \mathbf{W} = 0.$$

Resolving our forces into x- and y-components, along the plane and the normal to the plane, respectively, we obtain

$$N - W \cos \theta = 0,$$
$$f_s - W \sin \theta = 0. \tag{6-3}$$

However, $f_s \leq \mu_s N$. If we increase the angle of incline slowly until slipping just begins, then for that angle, $\theta = \theta_s$ and we can use $f_s = \mu_s N$. Substituting this into Eqs. 6-3, we obtain

$$N = W \cos \theta_s$$

and

$$\mu_s N = W \sin \theta_s,$$

so that

$$\mu_s = \tan \theta_s.$$

Hence measurement of the angle of inclination at which slipping just starts provides a simple experimental method for determining the coefficient of static friction between two surfaces.

You can use similar arguments to show that the angle of inclination θ_k required to maintain a *constant speed* for the block as it slides down the plane, once it has been started by tapping, is given by

$$\mu_k = \tan \theta_k,$$

where $\theta_k < \theta_s$. With the aid of a ruler you can now determine μ_s and μ_k for a coin sliding down your textbook.

EXAMPLE 2

Consider an automobile moving along a straight horizontal road with a speed v_0. If the coefficient of static friction between the tires and the road is μ_s, what is the shortest distance in which the automobile can be stopped?

The forces acting on the automobile, considered to be a particle, are shown in Fig. 6-5. The car is assumed to be moving in the positive x-direction. If we assume that f_s is a constant force, we have uniformly decelerated motion.

From the relation (see Eq. 3-16)

$$v^2 = v_0^2 + 2ax,$$

with the final speed $v = 0$, we obtain

$$x = -v_0^2/2a,$$

where the minus sign means that **a** points in the negative x-direction.

To determine a, apply the second law of motion to the x-component of the motion:

$$-f_s = ma = (W/g)a \qquad \text{or} \qquad a = -g(f_s/W).$$

From the y components we obtain

$$N - W = 0 \qquad \text{or} \qquad N = W,$$

so that

$$\mu_s = f_s/N = f_s/W$$

and

$$a = -\mu_s g.$$

Then the distance of stopping is

$$x = -v_0^2/2a = v_0^2/2g\mu_s. \qquad (6\text{-}4)$$

The greater the initial speed, the longer the distance required to come to a stop; in fact, this distance varies as the square of the initial velocity. Also, the greater the coefficient of static friction between the surfaces, the less the distance required to come to a stop.

We have used the coefficient of static friction in this problem, rather than the coefficient of sliding friction, because we assume there is no sliding between the tires and the road. We have neglected rolling friction. Furthermore, we have assumed that the maximum force of static friction ($f_s = \mu_s N$) operates because the problem seeks the shortest distance for stopping. With a smaller static frictional force the distance for stopping would obviously be greater. The correct braking technique required here is to keep the car just on the verge of skidding. If the surface is smooth and the brakes are fully applied, sliding may occur. In this case μ_k replaces μ_s, and the distance required to stop is seen to increase from Eq. 6-4.

The assumption that the car is a particle is valid if the wheels are locked (skidding). When the wheels rotate, internal forces (and torques) in the brake drums must be considered to understand work and energy ideas (see Questions 3, 4, and 5 of Chapter 8), though the result (Eq. 6-4) is correct. The rotation of the wheels is explicitly considered in Chapter 13.

As a specific example, if $v_0 = 60$ mi/h $= 88$ ft/s $= 97$ km/h, and $\mu_s = 0.60$ (a typical value), we obtain

$$x = \frac{v_0^2}{2\mu_s g} = \frac{(88 \text{ ft/s})^2}{2(0.60)(32 \text{ ft/s}^2)} = 200 \text{ ft} = 61 \text{ m}.$$

Notice that the mass of the car does not appear in Eq. 6-4. How can you explain the practice of "weighing down" a car in order to increase safety in driving on icy roads? (Hint: See Prob. 6-2.)

How do the forces of friction modify the results of the examples of Section 5-10?

figure 6-5
Example 2. The forces on a decelerating automobile.

6-3
THE DYNAMICS OF UNIFORM CIRCULAR MOTION

In Section 4-4 we pointed out that if a body is moving at uniform speed v in a circle of radius r, it experiences a centripetal acceleration **a** whose magnitude is v^2/r. The direction of **a** is always radially inward toward the center of rotation. Thus **a** is a variable vector because, even though its magnitude remains constant, its direction changes continuously as the motion progresses.

Recall that there need not be any motion in the direction of an acceleration. In general, there is no fixed relation between the directions of the acceleration **a** and the velocity **v** of a particle, as Fig. 4-7 shows. As it happens, for a particle in uniform circular motion the acceleration **a** and velocity **v** are always at right angles to each other.

Every accelerated body must have a force **F** acting on it, defined by Newton's second law (**F** = m**a**). Thus (assuming that we are in an inertial frame), if we see a body undergoing uniform circular motion, we can be certain that a net force **F**, given in magnitude by

$$F = ma = mv^2/r$$

must be acting on the body; the body is *not* in equilibrium. The direction of **F** at any instant must be the direction of **a** at that instant, namely, radially inward. We must always be able to account for this force by pointing to a particular object in the environment that is exerting the force on the circulating, accelerating body.

If the body in uniform circular motion is a disc on the end of a string moving in a circle on a frictionless horizontal table as in Fig. 6-6, the force **F** on the disc is provided by the tension **T** in the string. This force **T** is the net force acting on the disc. It accelerates the disc by constantly changing the direction of its velocity so that the disc moves in a circle. **T** is always directed toward the pin at the center and its magnitude is mv^2/R. If the string were to be cut where it joins the disc, there would be no net force exerted on the disc. The disc would then move with constant speed in a straight line along the direction of the tangent to the circle at the point at which the string was cut. Hence, to keep the disc moving in a circle, a force must be supplied to it pulling it *inward* toward the center.

Forces responsible for uniform circular motion are called *centripetal* forces because they are directed "toward the center" of the circular motion. To label a force as "centripetal," however, simply means that it always points radially inward; the name tells us nothing about the nature of the force or about the body that is exerting it. Thus, for the revolving disc of Fig. 6-6, the centripetal force is an elastic force provided by the string; for the moon revolving around the earth the centripetal force is the gravitational pull of the earth on the moon; for an electron circulating about an atomic nucleus the centripetal force is electrostatic. A centripetal force is not a new kind of force but simply a way of describing the behavior with time of forces that are attributable to specific bodies in the environment. Thus a force can be centripetal *and* elastic, centripetal *and* gravitational, or centripetal *and* electrostatic, among other possibilities.

Let us consider some examples of forces that act centripetally.

figure 6-6
A disk m moves with constant speed in a circular path on a horizontal frictionless surface. The only horizontal force acting on m is the centripetal force **T** with which the string pulls on the body.

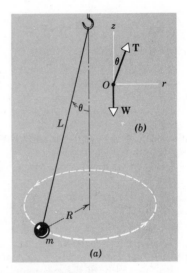

figure 6-7
Example 3. *(a)* A mass m suspended from a string of length L swings so as to describe a circle. The string describes a right circular cone of semiangle θ. *(b)* A free-body force diagram for m.

EXAMPLE 3

The Conical Pendulum. Figure 6-7a shows a small body of mass m revolving in a horizontal circle with constant speed v at the end of a string of length L. As the body swings around, the string sweeps over the surface of a cone. This device is called a *conical pendulum*. Find the time required for one complete revolution of the body.

If the string makes an angle θ with the vertical, the radius of the circular path is $R = L \sin \theta$. The forces acting on the body of mass m are **W**, its weight, and **T**, the pull of the string, as shown in Fig. 6-7b. It is clear that **T** + **W** $\neq$ 0. Hence, the resultant force acting on the body is nonzero, which is as it should be because a force is required to keep the body moving in a circle with constant speed.

We can resolve **T** at any instant into a radial and a vertical component

$$T_r = T \sin \theta \qquad \text{and} \qquad T_z = T \cos \theta.$$

Since the body has no vertical acceleration,

$$T_z - W = 0.$$

But

$$T_z = T \cos \theta \qquad \text{and} \qquad W = mg,$$

so that

$$T \cos \theta = mg.$$

The radial acceleration is v^2/R. This acceleration is supplied by T_r, the radial component of **T**, which is the centripetal force acting on m. Hence

$$T_r = T \sin \theta = mv^2/R.$$

Dividing this equation by the preceding one, we obtain

$$\tan \theta = v^2/Rg, \qquad \text{or} \qquad v^2 = Rg \tan \theta,$$

which gives the constant speed of the bob. If we let τ represent the time for one complete revolution of the body, then

$$v = \frac{2\pi R}{\tau} = \sqrt{Rg \tan \theta}$$

or

$$\tau = \frac{2\pi R}{v} = \frac{2\pi R}{\sqrt{Rg \tan \theta}} = 2\pi \sqrt{R/(g \tan \theta)}.$$

But $R = L \sin \theta$, so that

$$\tau = 2\pi \sqrt{(L \cos \theta)/g}.$$

This equation gives the relation between τ, L, and θ. Notice that τ, called the *period* of motion, does not depend on m.

If $L = 1.0$ m and $\theta = 30°$, what is the period of the motion? We have

$$\tau = 2\pi \sqrt{\frac{(1.0 \text{ m})(0.866)}{9.8 \text{ m/s}^2}} = 1.9 \text{ s}.$$

EXAMPLE 4

The Rotor. In many amusement parks* we find a device called the rotor. The rotor is a hollow cylindrical room which can be set rotating about the central vertical axis of the cylinder. A person enters the rotor, closes the door, and stands up against the wall. The rotor gradually increases its rotational speed from rest until, at a predetermined speed, the floor below the person is opened downward, revealing a deep pit. The person does not fall but remains "pinned up" against the wall of the rotor. Find the coefficient of friction necessary to prevent falling.

The forces acting on the person are shown in Fig. 6-8. **W** is the person's weight, $\mathbf{f}_s$ is the force of static friction between person and rotor wall, and **P** is the centripetal force exerted by the wall on the person necessary to keep him moving in a circle. Let the radius of the rotor be R and the final speed of the passenger be v. Since the person does not move vertically, but experiences a radial acceleration v^2/R at any instant, we have

$$f_s - W = 0$$

and

$$P(= ma) = (W/g)(v^2/R).$$

If μ_s is the coefficient of static friction between person and wall necessary to

* See "Physics and the Amusement Park" by John L. Roeder in *The Physics Teacher*, September 1975.

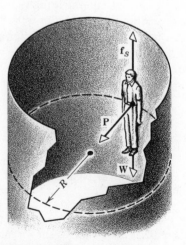

figure 6-8
Example 4. The forces on a person in a "rotor" of radius R.

prevent slipping, then $f_s = \mu_s P$ and

$$f_s = W = \mu_s P$$

or

$$\mu_s = \frac{W}{P} = \frac{gR}{v^2}.$$

This equation gives the minimum coefficient of friction necessary to prevent slipping for a rotor of radius R when a particle on its wall has a speed v. Notice that the result does not depend on the person's weight.

As a practical matter the coefficient of friction between the textile material of clothing and a typical rotor wall (canvas) is about 0.40. For a typical rotor the radius is 2.0 m, so that v must be about 7.0 m/s or 25 km/h or more.

Let the block in Fig. 6-9a represent an automobile or railway car moving at constant speed v on a *level* road-bed around a curve having a radius of curvature R. In addition to two vertical forces, namely, the force of gravity **W** and a normal force **N**, a horizontal centripetal force **P** acts on the car. In the case of the automobile this centripetal force is supplied by a sidewise frictional force exerted by the road on the tires; in the case of the railway car the centripetal force is supplied by the rails exerting a sidewise force on the inner rims of the car's wheels. Neither of these sidewise forces can be safely relied upon to be large enough at all times and both cause unnecessary wear. Hence, the roadbed is *banked* on curves, as shown in Fig. 6-9b. In this case, the normal force **N** has not only a vertical component, as before, but also a horizontal component which supplies the centripetal force necessary for uniform circular motion; no additional sidewise forces are needed, therefore, with a properly banked roadbed.

The correct angle θ of banking can be obtained as follows. There is no vertical acceleration, so that

$$N \cos \theta = W.$$

The centripetal force is $N \sin \theta$, so that $N \sin \theta = mv^2/R$. Dividing the latter equation by the former and setting $W = mg$, we obtain

$$\tan \theta = v^2/Rg$$

Notice that the proper angle of banking depends upon the speed of the car and

EXAMPLE 5

figure 6-9
Examples.
(a) a level roadbed.
(b) a banked roadbed.

(a)

(b)

the curvature of the road. For a given curvature, the road is banked at an angle corresponding to an expected average speed. Often curves are marked by signs giving the proper speed for which the road was banked.

Check the banking formula for the limiting cases $v = 0$; $R \to \infty$; v large; and R small. Also note the similarity between Fig. 6-7 of Example 3 and Fig. 6-9b of this example.

6-4
CLASSIFICATION OF FORCES; INERTIAL FORCES

All forces in nature can be classified under four headings, each with a different relative strength: (1) gravitational forces, which are relatively very weak, (2) electromagnetic forces, which are of intermediate strength, (3) nuclear forces which bind neutrons and protons in the nucleus and are the strongest of all, and (4) the weak interaction force, which is involved in the β-decay of nuclei and in the interactions of many elementary particles (see Appendix F).

These forces are "real" in the sense that we can associate them with specific objects in the environment. Such forces as the tension in a rope, the force of friction, the force that we exert on a wall by pushing on it, or the force exerted by a compressed spring are electromagnetic forces; all are macroscopic manifestations of the (electromagnetic) attractions and repulsions between atoms.

In our treatment of classical mechanics so far we have assumed that our measurements and observations were made from an inertial frame. This, we recall, is a reference frame that is either at rest or is moving at constant velocity with respect to the average positions of the fixed stars; it is the set of reference frames defined by Newton's first law, namely, that set of frames in which a body will not be accelerated ($\mathbf{a} = 0$) if there are no identifiable force-producing bodies in its environment ($\mathbf{F} = 0$). The choice of a reference frame is always ours to make, so that if we choose to select only inertial frames, we do not restrict in any way our ability to apply classical mechanics to natural phenomena.

Nevertheless we can, if we find it convenient, apply classical mechanics from the point of view of an observer in a *noninertial frame*. Such a frame might be one that is attached to a falling body or one that is rotating (and therefore accelerating) with respect to the fixed stars. We sometimes choose a noninertial reference frame when we consider, for example, the separation of liquids of different density in a spinning centrifuge, the global circulation of the winds on the rotating earth, or the experiences of an astronaut in an orbiting satellite.

We can apply classical mechanics in noninertial frames if we introduce non-Newtonian forces called *inertial forces*. Unlike the forces that we have examined so far, we cannot associate inertial forces with any particular body in the environment of the particle on which they act and we cannot classify them into any of the categories listed in the first paragraph of this section. Moreover, if we view the particle from an inertial frame, the inertial forces disappear. These forces are, then, simply a technique that permits us to apply classical mechanics in the normal way to events if we insist on viewing the events from a noninertial reference frame.

Consider a rotating merry-go-round on which a marble is lodged against a raised rim at the outer edge. An observer on the merry-go-round is in a noninertial frame. As he kneels down and examines the marble he sees that, with respect to him, it is not moving; if he pulls it away a bit from the rim toward the center of rotation, he observes that it moves back again, as if under the influence of a force directed radially outward. He would declare the marble to be in equilibrium under the action of this outward force (an inertial force called, in this case, a *centrifugal force*) and the radially inward force exerted by the rim.

An observer on the ground (an inertial frame) watching the marble would describe it differently. He would declare the marble to be in uniform circular motion, accelerated radially inward with $a = v^2/R$. The inward force $\mathbf{F}$ exerted by the rim on the marble accounts for this acceleration from Newton's second law, or $F = ma = mv^2/R$. The marble is definitely *not* in equilibrium from the point of view of this observer or of an observer in any inertial frame. Only if the rim were *not* exerting this inward force would the marble move with uniform speed in a straight line and be in equilibrium. This observer would find no trace of a force directed radially outward on the marble (the inertial force) and, indeed, there is no room for such a force in his analysis of the motion.

It is clear from this simple example that the radially outward inertial force (or centrifugal force) noted by the observer on the rotating merry-go-round must have a magnitude mv^2/R. Thus the magnitude of the inertial force depends on the speed of the particle *as seen from another reference frame*, namely, the ground; the speed of the particle in its own (rotating) reference frame is zero.

This example illustrates why inertial forces are non-Newtonian, namely, Newton's third law of motion does not apply to them. That is, there is no reaction force to the inertial (action) force. In the rotating frame, if the rim were *not* present we would have an inertial (centrifugal) force acting on the marble without any reaction force of the marble on another body. When the rim *is* present we have two forces acting on the *same* body, the centripetal force due to the rim and the inertial (centrifugal) force each acting on the marble. The marble is viewed as being in equilibrium under the influence of two forces acting on it but, as we have seen, we can have one force without the other. In an inertial frame, on the other hand, the (action) force of the rim is the only force on the marble and the marble exerts a (reaction) force on the rim, equal in magnitude but oppositely directed. If one wished to use the terms *centripetal* and *centrifugal* here, he would have an action-reaction pair acting on *different* bodies, consistent with Newton's third law. But in the accelerated frame, the forces called by these names act on the same body and are not an action-reaction pair.

In more general terms we might say that the expression $\mathbf{F} = m\mathbf{a}$ used in an inertial frame is changed to $\mathbf{F} - m\mathbf{a} = 0$ in a noninertial frame and that the interpretation given to the term $-m\mathbf{a}$ in the latter case is that it is an (inertial) force existing only in the accelerated frame which permits one to regard the object acted upon as always being in equilibrium. In this sense it is sometimes simpler to use a noninertial frame to describe motion, such as circular motion, the object being regarded as at rest in such a frame.

In mechanical problems, then, we have two choices: (1) select an *inertial frame* as a reference frame and consider only "real" forces, that is, forces that we can associate with definite bodies in the environment or (2) select a *noninertial frame* as a reference frame and consider not only the "real" forces but suitably defined inertial forces. Although we usually choose the first alternative, we sometimes choose the second; both are completely equivalent and the choice is a matter of convenience. We shall discuss noninertial frames and inertial forces further in Chapters 11 and 16.

6-5
CLASSICAL MECHANICS, RELATIVISTIC MECHANICS, AND QUANTUM MECHANICS

In these first chapters we have laid the groundwork of classical mechanics. We have presented the laws of motion and have given several examples of the force laws. In later chapters we shall discuss other kinds of forces and shall continue to develop the structure of the theory. Here we want to point out where classical mechanics stands in the framework of modern physics.

Physics is not a static body of doctrine but a developing science. Historically there have been long periods of deep concern with a certain class of problem, culminating, often rather suddenly and in unexpected ways, in a "breakthrough" in the form of a new, more comprehensive theory.* This occurred about 1690 (Newtonian mechanics), about 1870 (Maxwell's theory of electromagnetism), 1905 (Einstein's theory of relativity), and about 1925 (quantum

*See "The Structure of Scientific Revolutions" by Thomas Kuhn, The University of Chicago Press, 1970.

mechanics). Some physicists believe that our present concern for problems in the area of elementary particles (see Appendix F) will lead us eventually to another major "breakthrough."

As physics has evolved, many things have changed, such as the problems to be solved and the tools we use to investigate them. But through it all the general method of inquiry or process of solution remains basically the same. Thus earlier theories of physics are found to have limited ranges of validity and to be special cases of more comprehensive theories, which in turn are found to have limitations, and so on. However, independent of any particular area or problem in physics, we always demand that theory meet the test of experiment, we search for quantities that are invariant, we are guided by a belief in the simplicity and symmetry of nature, and we seek and use analogies and models. Major unifying concepts arise which are valid in all domains of physics, such as the conservation laws. All this is important to understand for its own sake, independent of mastery of any particular special topic, and is exemplified throughout the book. If, in addition to mastering classical mechanics, the student comes to understand this process, he will find it much easier to understand and master such theories as relativity theory and quantum theory, wherein the same method of inquiry applies but whose areas of application, unlike those of classical mechanics, are not a familiar part of his daily life experience.

Classical mechanics, like all theories in physics, is based on observations of things that happen in nature. It will help to point out how limited are our normal experiences of natural phenomena. This is particularly true during our formative years which is the period when we develop our intuitive notions (often false!) of what is "common sense" in natural events and what is not.

For example, the highest speed that can be used to transmit signals from one point to another is the speed of light ($c = 186{,}000$ mi/s $= 3.00 \times 10^8$ m/s) and this seems to set an upper limit to the speeds of material objects. However, gross objects, even the fastest of them, such as jet planes or earth satellites, have speeds v that are very much less than c. For an earth satellite moving at 17,000 mi/h, v/c is only 0.00025. Classical mechanics was built up over several centuries on a body of observations of relatively slow-moving objects such as planets, balls rolling down inclined planes, and falling bodies. Our experience with moving objects has indeed been limited, until the last few decades, to a tiny fraction of the range of possible speeds.

During these last decades it has become possible to make measurements on small particles, of potentially high speed, such as electrons, protons, and other fundamental particles. A proton accelerated in the 30-billion electron volt accelerator at the Brookhaven National Laboratories has, for example, $v/c = 0.98$. Are we to expect that the laws of classical mechanics, which work so beautifully when $v/c \ll 1$, will also describe correctly the collisions, decays, and interactions of these elementary particles moving at such high speeds? This is the grossest kind of extrapolation and indeed we find by experiment that it simply does not work; classical mechanics gives answers that do not agree with experiment if the speeds of the objects involved are appreciable compared to the speed of light. This does not make us think less of classical mechanics, which serves so well in the region of low speed, precisely the very important region of our daily experiences. We are led, however, to view classical mechanics as a special case of a more general theory which would hold for all speeds up to the speed of light.

Einstein, in 1905, first proposed this more general theory, the *special theory of relativity*.* We shall discuss it again later but will state here its fundamental postulate. This is that the speed of light c is the *same* for all observers in inertial frames, no matter what the motion of the light source may be. In other words, if a light source is moving directly toward you at a speed v, you would measure the same value for c, if you observed a light pulse passing you, no matter what the value of v; you would also obtain speed c for the light pulse if the source were rushing away from you at speed v. If this basic assumption seems to violate "common sense," we must realize that our intuitive feelings are based on

* For a summary of special relativity, see Supplementary Topic V.

"common sense at low speeds." We have no direct experience in our daily activities about what really happens in nature at high speeds. Furthermore, all of Einstein's predictions (1) agree with experiment and (2) reduce to the predictions of classical mechanics at low speeds.

We list here just one of the predictions of the theory of relativity that is at variance with classical mechanics. If two observers watch an object moving parallel to the common $x - x'$-axis in Fig. 4-11, they will find, from Eq. 4-19,

$$v = v' + u, \qquad (6\text{-}5)$$

where v' is the speed as measured by observer S', v is that measured by observer S, and u is the relative speed of separation of the two reference frames. Note that there is nothing in Eq. 6-5 to prevent v from exceeding c if v' and u are large enough.

The theory of relativity predicts that Eq. 6-5 is a special case of a more general formula, namely,

$$v = \frac{v' + u}{1 + v'u/c^2}. \qquad (6\text{-}6)$$

Note that for $v' \ll c$ and $u \ll c$ Eq. 6-6 does indeed reduce to Eq. 6-5. Also, if $v' < c$ and $u < c$, then v cannot exceed c. If $v' = u = 0.8\ c$, for example, Eq. 6-6 yields $v = 0.975\ c$; Eq. 6-5, on the other hand, yields $v = 1.6\ c$, which is contrary to experience.

For gross objects, Eqs. 6-5 and 6-6 give the same results within experimental error, so that we naturally use the simpler, Eq. 6-5. If two satellites moving in opposite directions have speeds $v' = u = 17{,}000$ mi/h, the denominator in Eq. 6-6 has the value 1.0000000007, so that the speed v of one satellite as seen from the other differs very slightly from the value $v' + u$ predicted by Eq. 6-5. It would take speeds almost 3000 *times* as great as above, nearly 50 million mi/h, generally achievable only in the subatomic domain, to obtain a difference as great as one-half of one percent in the two formulas.

We point out a second way in which our daily experiences are limited, namely, that all the objects that we normally deal with have masses that greatly exceed, for example, the electron mass ($m = 9.11 \times 10^{-31}$ kg). This turns out to have an interesting consequence, closely related to the very concept of "particle" on which classical mechanics is based. We have not hesitated to assign a mass m, a position x, and a velocity v_x to a particle, assumed to be moving along the x-axis.* If we are asked within what accuracy Δx and Δv_x we could measure the position x and the velocity v_x respectively, we would be inclined to say that, although there might be limits in practice there are none in principle and, with sufficient attention to methods of measurement, we can specify x and v_x as closely as we wish. Experiment seems to confirm this view for large objects like golf balls.

When we deal with objects of very small mass, however, such as electrons, we learn that the very procedures of measurement introduce fundamental uncertainties and that, in fact, the more precise our knowledge of x becomes the less precise is our knowledge of v_x and conversely. We can express this in terms of the famous Heisenberg uncertainty relation, which we write as

$$\Delta x \cong \frac{h}{m\ \Delta v_x} \qquad (6\text{-}7)$$

in which h (Planck's constant) is a fundamental constant of nature and has the value $h = 6.63 \times 10^{-34}$ kg m²/s. Equation 6-7 shows clearly that if Δv_x is very small (which means that we know v_x very precisely), then Δx must be relatively large (which means that we do not know x very precisely). Thus it does not seem possible to measure *both* the position *and* the velocity of a particle to any given precision at the same time. If we cannot do this, then our whole concept of a particle as a mass point following a trajectory, which is a basic concept of classical mechanics, is open to question.

---

* We assume $v_x \ll c$ so that considerations of relativity do not enter this new discussion.

Just as for relativity theory, these considerations of quantum mechanics simply do not make any difference for the gross objects of our daily experience. Consider a ball bearing with a speed of 10^3 m/s and a mass of 1.0 g ($= 10^{-3}$ kg). Let us assume that we know the speed to be accurate to 0.1%, which means that $\Delta v_x = 0.001 \times 10^3 = 1$ m/s. The uncertainty in the position of the ball bearing is now given by Eq. 6-7 as

$$\Delta x \cong \frac{6.63 \times 10^{-34} \text{ kg m}^2/\text{s}}{(10^{-3} \text{ kg})(1 \text{ m/s})} \cong 7 \times 10^{-31} \text{ m}$$

This is such a small distance (being 10^{-15} times smaller than an atomic nucleus!) that we could not possibly detect any limitation on the measurement of x set by Eq. 6-7.

Consider, however, not a ball bearing but an electron ($m = 9.11 \times 10^{-31}$ kg) whose speed is measured to be 2×10^6 m/s, which is about the speed of an electron in a hydrogen atom. If we assume that we know this speed to be accurate to, say, 1%, then $\Delta v_x = 0.01 \times 2 \times 10^6$ m/s $= 2 \times 10^4$ m/s. The uncertainty in position predicted by Eq. 6-7 is then

$$\Delta x \cong \frac{6.63 \times 10^{-34} \text{ kg m}^2/\text{s}}{(9.11 \times 10^{-31} \text{ kg})(2 \times 10^{-4} \text{ m/s})} = 3 \times 10^{-8} \text{ m}.$$

Since the radius of a hydrogen atom is about 5×10^{-11} m we see that the uncertainty with which we can locate the electron in the hydrogen atom, assuming that we have measured its speed as accurately as we claim, is 600 times the radius of the atom! The concept of "particle" does not mean much under these circumstances. This simply means that we cannot use classical mechanics to describe the motions of electrons in atoms; we need quantum mechanics.

The situation is very much like that of relativity theory. Ideas that we find acceptable in a certain region of experience (ball bearings) fall down when we apply them to a region outside our direct normal experience (electrons in atoms). Once again the solution is the same: Classical mechanics turns out to be an important special case of a more general theory. In this case the general theory is that of quantum mechanics developed about 1925 to 1926 by Heisenberg, Schrödinger, Born, and others. Once again, quantum mechanics does not detract from the merit of classical mechanics, which continues to give results that agree admirably with experiment for particles of relatively large mass.

The situation most remote from our daily experience deals with particles that have *both* small mass *and* high speed. Here we must use a still more general theory, *relativistic quantum mechanics*, which combines both relativity theory and quantum mechanics; such a theory was first developed by Dirac in 1927.

In the rest of our treatment of mechanics we return to the familiar special case of our daily experience, that of relatively massive and relatively slow-moving objects (classical mechanics). From time to time we will point out parenthetically how the predictions of classical mechanics must be modified when we depart from this region of experience.

questions

1. There is a limit beyond which further polishing of a surface *increases* rather than decreases frictional resistance. Can you explain this?

2. Is it unreasonable to expect a coefficient of friction to exceed unity?

3. How could a person who is at rest on completely frictionless ice covering a pond reach shore? Could he do this by walking, rolling, swinging his arms, or kicking his feet? How could a person be placed in such a position in the first place?

4. Explain how the range of your car's headlights limits the safe driving speed at night.

5. Your car skids across the center line on an icy highway. Should you turn the front wheels in the direction of skid or in the opposite direction (a) when

you want to avoid a collision with an oncoming car, (b) when no other car is near but you want to regain control of the steering?

6. If you want to stop the car in the shortest distance on an icy road, should you (a) push hard on the brakes to lock the wheels, (b) push just hard enough to prevent slipping, or (c) "pump" the brakes?

7. Discuss how the choice of angle for maximum range of a projectile would be affected by the resistance of the air to motion of the projectile through it.

8. Why are the train roadbeds and highways banked on curves?

9. How does the earth's rotation affect the apparent weight of a body at the equator?

10. Explain why a plumb bob will not hang exactly in the direction of the earth's gravitational attraction at most latitudes.

11. Suppose you need to measure whether a table top in a train is truly horizontal. If you use a spirit level, can you determine this when the train is moving down or up a grade? When the train is moving along a curve? (Hint: there are two horizontal components.)

12. In the conical pendulum of Example 3, what happens to the period τ and the speed v when $\theta = 90°$? Why is this angle not achievable physically? Discuss the case for $\theta = 0°$.

13. A coin is put on a photograph turntable. The motor is started, but before the final speed of rotation is reached, the coin flies off. Explain.

14. Suppose that a body that is acted upon by exactly two forces is accelerated. Does it then follow that (a) the body cannot move with constant speed? (b) the velocity can never be zero? (c) the sum of the two forces cannot be zero? (d) the two forces must act in the same line?

15. A car is riding on a country road that resembles a roller coaster track. If the car travels with uniform speed, compare the force it exerts on a horizontal section of the road to the force it exerts on the road at the top of a hill and at the bottom of a hill. Explain.

16. A passenger in the front seat of a car finds himself sliding toward the door as the driver makes a sudden left turn. Describe the forces on the passenger and on the car at this instant if (a) the motion is viewed from a reference frame attached to the earth and (b) if attached to the car.

17. Astronauts in the orbiting Skylab spacecraft want to keep a daily record of their weight. Can you think how they might do it, considering that they are 'weightless'?

18. What conclusion might a physicist draw if, while standing in an elevator, he observes that unequal masses hung over a pulley remain balanced, that is, there is no tendency for the pulley to turn?

19. Explain how the question "What is the linear velocity of a point on the equator?" requires an assumption about the reference frame used. Show how the answer changes as you change reference frames.

20. What is the distinction between inertial reference frames and those differing only by a translation or rotation of the axes?

SECTION 6-2

1. A hockey puck weighing 0.25 lb (1.1 N) slides on the ice for 50 ft (15 m) before it stops. (a) If its initial speed was 20 ft/s (6.1 m/s), what is the force of friction between puck and ice? (b) What is the coefficient of kinetic friction?   *Answer:* (a) 0.031 lb (0.14 N). (b) 0.12 (0.13).

2. Suppose that only the rear wheels of an automobile can accelerate it, and that half the total weight of the automobile is supported by those wheels. (a) What is the maximum acceleration attainable if the coefficient of static friction between tires and road is μ_s? (b) Take $\mu_s = 0.35$ and get a numerical value for this acceleration.

problems

3. Frictional heat generated by the moving ski is the chief factor promoting sliding in skiing. The ski sticks at the start, but once in motion will melt the snow beneath it. Waxing the ski makes it water repellent and reduces friction with the film of water. A magazine reports that a new type of plastic ski is even more water repellent and that on a gentle 700-ft slope in the Alps, a skier reduced his time from 61 to 42 s with new skis. (a) Determine the average accelerations for each pair of skis. (b) Assuming a 3°-slope compute the coefficient of kinetic friction for each case.
   Answer: (a) 0.38 ft/s²; 0.79 ft/s². (b) 0.041; 0.028.

4. A fireman weighing 160 lb (710 N) slides down a vertical pole with an average acceleration of 10 ft/s² (3 m/s²). What is the average vertical force he exerts on the pole?

5. A man drags a 150-lb crate across a floor by pulling on a rope inclined 15° above the horizontal. (a) If the coefficient of static friction is 0.50, what tension in the rope is required to start the crate moving? (b) If $\mu_k = 0.35$, what is the initial acceleration of the crate?
   Answer: (a) 68 lb. (b) 4.2 ft/s².

6. A cube of weight W rests on a rough inclined plane which makes an angle θ with the horizontal. (a) What is the minimum force necessary to start the cube moving *down* the plane? (b) What is the minimum force necessary to start the cube moving *up* the plane? (c) What is the minimum *horizontal* (transverse to the slope) force necessary to start the cube moving down the plane?

7. The handle of a floor mop of mass m makes an angle θ with the vertical direction. Let μ_k be the coefficient of kinetic friction between mop and floor, and μ_s be the coefficient of static friction between mop and floor. Neglect the mass of the handle. (a) Find the magnitude of the force F directed along the handle required to slide the mop with uniform velocity across the floor. (b) Show that if θ is smaller than a certain angle θ_0, the mop cannot be made to slide across the floor no matter how great a force is directed along the handle. (c) What is the angle θ_0?
   Answer: (a) $\mu_k mg/(\sin\theta - \mu_k \cos\theta)$. (c) $\theta_0 = \tan^{-1}\mu_s$.

8. A piece of ice slides down a 45°-incline in twice the time it takes to slide down a frictionless 45°-incline. What is the coefficient of kinetic friction between the ice and the incline?

9. A block slides down an inclined plane of slope angle φ with constant velocity. It is then projected up the same plane with an initial speed v_0. (a) How far up the incline will it move before coming to rest? (b) Will it slide down again?   *Answer:* (a) $v_0^2/4g \sin\varphi$. (b) No.

10. A student wants to determine the coefficients of static friction and kinetic friction between a box and a plank. He places the box on the plank and gradually raises the plank. When the angle of inclination with the horizontal reaches 30°, the box starts to slip and slides 4.0 m down the plank in 4.0 s. What are the coefficients of friction?

figure 6-10
Problem 11

11. A horizontal force F of 12 lb pushes a block weighing 5.0 lb against a vertical wall (Fig. 6-10). The coefficient of static friction between the wall and the block is 0.60 and the coefficient of kinetic friction is 0.40. Assume the block is not moving initially. (a) Will the block start moving? (b) What is the force exerted on the block by the wall?
   Answer: (a) No. (b) A 12-lb force to the left and a 5.0-lb force up.

12. A 10-lb block of steel is at rest on a horizontal table. The coefficient of static friction between block and table is 0.50. (a) What is the magnitude of the horizontal force that will just start the block moving? (b) What is the magnitude of a force acting upward 60° from the horizontal that will just start the block moving? (c) If the force acts down at 60° from the horizontal, how large can it be without causing the block to move?

13. Block B in Fig. 6-11 weighs 160 lb (710 N). The coefficient of static friction

figure 6-11
Problem 13

between block and table is 0.25. Find the maximum weight of block A for which the system will be in equilibrium.   *Answer:* 40 lb (180 N).

14. Two masses, $m_1 = 1.65$ kg and $m_2 = 3.30$ kg, attached by a massless rod parallel to the incline on which both slide, as shown in Fig. 6-12, travel down along the plane with m_1 trailing m_2. The angle of incline is $\theta = 30°$. The coefficient of kinetic friction between m_1 and the incline is $\mu_1 = 0.226$; between m_2 and the incline the corresponding coefficient is $\mu_2 = 0.113$. Compute (a) the tension in the rod linking m_1 and m_2 and (b) the common acceleration of the two masses. (c) Would the answers to (a) and (b) be changed if m_2 trails m_1?

figure 6-12
Problem 14

15. A 4.0-kg block is put on top of a 5.0-kg block. In order to cause the top block to slip on the bottom one, held fixed, a horizontal force of 12 N must be applied to the top block. The assembly of blocks is now placed on a horizontal, frictionless table (Fig. 6-13). Find (a) the maximum horizontal force F which can be applied to the lower block so that the blocks will move together, and (b) the resulting acceleration of the blocks.
Answer: (a) 27 N. (b) 3.0 m/s².

figure 6-13
Problem 15

16. A railroad flatcar is loaded with crates having a coefficient of static friction 0.25 with the floor. If the train is moving at 30 mi/h (48 km/h), in how short a distance can the train be stopped without letting the crates slide?

17. A 40-kg slab rests on a frictionless floor. A 10-kg block rests on top of the slab (Fig. 6-14). The static coefficient of friction between the block and the slab is 0.60 while the kinetic coefficient is 0.40. The 10-kg block is acted upon by a horizontal force of 100 N. What are the resulting accelerations of (a) the block, and (b) the slab?   *Answer:* (a) 6.1 m/s². (b) 0.98 m/s².

figure 6-14
Problem 17

18. In Fig. 6-15, A is a 10-lb (44-N) block and B is a 5.0-lb (22-N) block. (a) Determine the minimum weight (block C) which must be placed on A to keep it from sliding, if μ_s between A and the table is 0.20. (b) The block C is suddenly lifted off A. What is the acceleration of block A, if μ_k between A and the table is 0.20?

19. An 8.0-lb block and a 16-lb block connected together by a string slide down a 30° inclined plane. The coefficient of kinetic friction between the 8.0-lb block and the plane is 0.10; between the 16-lb block and the plane it is 0.20. Find (a) the acceleration of the blocks and (b) the tension in the string, assuming that the 8.0-lb block leads. (c) Describe the motion if the blocks are reversed.
Answer: (a) 11 ft/s². (b) 0.46 lb. (c) Blocks move independently, unless they subsequently collide.

figure 6-15
Problem 18

20. Body B weighs 100 lb and body A weighs 32 lb (Fig. 6-16). Given $\mu_s = 0.56$ and $\mu_k = 0.25$, (a) find the acceleration of the system if B is initially at rest and (b) find the acceleration if B is moving initially.

21. A block of mass m slides in an inclined right-angled trough as in Fig. 6-17. If the coefficient of kinetic friction between the block and the material composing the trough is μ_k, find the acceleration of the block.
Answer: $g(\sin \theta - \sqrt{2}\,\mu_k \cos \theta)$.

SECTION 6-3

22. In the Bohr model of the hydrogen atom, the electron revolves in a circular orbit around the nucleus. If the radius is 5.3×10^{-11} meters and the electron makes 6.6×10^{15} rev/s, find (a) the acceleration (magnitude and direction) of the electron and (b) the centripetal force acting on the electron. (This force is due to the attraction between the positively charged nucleus and the negatively charged electron.) The mass of the electron is 9.1×10^{-31} kg.

figure 6-16
Problem 20

figure 6-17
Problem 21

23. A mass m on a frictionless table is attached to a hanging mass M by a cord through a hole in the table (Fig. 6-18). Find the condition (v and r) with which m must spin for M to stay at rest.   *Answer:* $v^2/r = Mg/m$.

24. Show that the periods of two conical pendula of different lengths which are hung from a ceiling and rotate with their bobs an equal distance below the ceiling are equal.

25. A small coin is placed on a flat, horizontal turntable. The turntable is observed to make three revolutions in 3.14 s. (*a*) What is the speed of the coin when it rides without slipping at a distance 5.0 cm from the center of the turntable? (*b*) What is the acceleration (magnitude and direction) of the coin in part (*a*)? (*c*) What is the frictional force acting on the coin in part (*a*) if the coin has a mass of 2.0 g? (*d*) What is the coefficient of static friction between the coin and the turntable if the coin is observed to slide off the turntable when it is more than 10 cm from the center of the turntable?
Answer: (*a*) 30 cm/s. (*b*) 180 cm/s², radially inward. (*c*) 3.6×10^{-3} N. (*d*) 0.37.

figure 6-18
Problem 23

26. A block of mass m at the end of a string is whirled around in a vertical circle of radius R. Find the critical speed below which the string would become slack at the highest point?

27. A circular curve of highway is designed for traffic moving at 40 mi/h. (*a*) If the radius of the curve is 400 ft, what is the correct angle of banking of the road? (*b*) If the curve is not banked, what is the minimum coefficient of friction between tires and road that would keep traffic from skidding at this speed?   *Answer:* (*a*) 15°. (*b*) 0.27.

28. A driver's manual states that a driver traveling at 30 mi/h (48 km/h) and desiring to stop as quickly as possible travels 33 ft (10 m) before his foot reaches the brake. He travels an additional 68 ft (21 m) before coming to rest. (*a*) What coefficient of friction is assumed in these calculations? (*b*) What is the minimum radius for turning a corner at 30 mi/h (48 km/h) without skidding?

29. A 5000-lb airplane loops at a speed of 200 mi/h. Find (*a*) the radius of the largest circular loop possible, (*b*) the net force on the plane at the bottom of this loop, and (*c*) the lift on the plane at the bottom of this loop.
Answer: (*a*) 2700 ft. (*b*) 5000 lb. (*c*) 10,000 lb.

30. A 150-lb student on a steadily rotating Ferris wheel has an apparent weight of 125 lb at his highest point. (*a*) What is his apparent weight at the lowest point? (*b*) What would be his apparent weight at the highest point if the speed of the Ferris wheel were doubled?

31. Assume that the standard kilogram would weigh exactly 9.80 N at sea level on the earth's equator if the earth did not rotate. Then take into account the fact that the earth does rotate so that this mass moves in a circle of radius 6.40×10^6 m (earth's radius) at a constant speed of 465 m/s. (*a*) Determine the centripetal force needed to keep the standard moving in its circular path. (*b*) Determine the force exerted by the standard kilogram on a spring balance from which it is suspended at the equator (its weight).
Answer: (*a*) 0.0338 N. (*b*) 9.77 N.

32. An old streetcar rounds a corner on unbanked tracks. (*a*) If the radius of the tracks is 30 ft and the car's speed is 10 mi/h, what angle with the vertical will be made by the loosely hanging hand straps? (*b*) Is there a force acting on these straps? If so, is it a centripetal or centrifugal force? Do your answers depend on what reference frame you choose?

33. A particle of mass $M = 0.305$ kg moves counterclockwise in a horizontal circle of radius $r = 2.63$ m with uniform speed $v = 0.754$ m/s as in Fig. 6-19. Determine at the instant $\theta = 322°$ (measured counterclockwise from the positive x-direction) the following quantities: (*a*) the x-component of the velocity; (*b*) the y-component of the acceleration; (*c*) the total force on the particle; (*d*) the component of the total force on the particle in the direction of its velocity.
Answer: (*a*) 0.464 m/s. (*b*) 0.133 m/s². (*c*) 6.59×10^{-2} N. (*d*) Zero.

figure 6-19
Problem 33

34. A 1.0-kg ball is attached to a rigid vertical rod by means of two massless strings each 1.0 m long. The strings are attached to the rod at points 1.0 m apart. The system is rotating about the axis of the rod, both strings being taut and forming an equilateral triangle with the rod, as shown in Fig. 6-20. The tension in the upper string is 25 N. (a) Draw the free-body diagram for the ball. (b) What is the tension in the lower string? (c) What is the net force on the ball at the instant shown in the figure? (d) What is the speed of the ball?

35. An airplane is flying in a horizontal circle at a speed of 300 mi/h (480 km/h). If the wings of the plane are tilted 45° to the vertical, what is the radius of the circle the plane is flying?   *Answer:* 1.1 mi (1.8 km).

36. Because of the rotation of the earth, a plumb bob may not hang exactly along the direction of the earth's gravitational pull (its weight) but deviate slightly from that direction. Calculate the deviation (a) at 40° latitude, (b) at the poles, and (c) at the equator.

37. Imagine that the disc of Fig. 6-6 is attached to a spring rather than a string. The unstretched length of the spring is l_o and the tension in the spring increases in direct proportion to its elongation, the tension per unit elongation being k. If the disc revolves with a frequency ν (revolutions per unit time), show that (a) the radius R of the uniform circular motion is $kl_o/(k - 4\pi^2m\nu^2)$ and (b) the tension T in the spring is $4\pi^2mkl_o\nu^2/(k - 4\pi^2m\nu^2)$.

38. A very small cube of mass m is placed on the inside of a funnel (Fig. 6-21) rotating about a vertical axis at a constant rate of ν rev/s. The wall of the funnel makes an angle θ with the horizontal. If the coefficient of static friction between the cube and the funnel is μ and the center of the cube is a distance r from the axis of rotation, what are (a) the largest and (b) the smallest values of ν for which the cube will not move with respect to the funnel?

figure 6-20
Problem 34

figure 6-21
Problem 38

$$7$$

work and energy

A fundamental problem of particle dynamics is to find how a particle will move when we know the forces that act on it. By "how a particle will move" we mean how its position varies with time. If the motion is one-dimensional, the problem is to find x as a function of time, $x(t)$. In the previous two chapters we solved this problem for the special case of a constant force. The method used is this. We find the resultant force $\mathbf{F}$ acting on the particle from the appropriate force law. We then substitute $\mathbf{F}$ and the particle mass m into Newton's second law of motion. This gives us the acceleration $\mathbf{a}$ of the particle; or

$$\mathbf{a} = \mathbf{F}/m.$$

If the force $\mathbf{F}$ and the mass m are constant, the acceleration $\mathbf{a}$ must be constant. Let us choose the x-axis to be along the direction of this constant acceleration. We can then find the speed of the particle from Eq. 3-12,

$$v = v_0 + at,$$

and the position of the particle from Eq. 3-15 (with $x_0 = 0$), or

$$x = v_0 t + \tfrac{1}{2}at^2;$$

note that, for simplicity and convenience, we have dropped the subscript x in these equations. The last equation gives us directly what we usually want to know, namely $x(t)$, the position of the particle as a function of time.

The problem is more difficult, however, when the force acting on a particle is *not constant*. In such a case we still obtain the acceleration of the particle, as before, from Newton's second law of motion. However, in order to get the speed or position of the particle, we can no longer use the formulas previously developed for constant acceleration because the acceleration now is *not* constant. To solve such problems, we use the mathematical process of integration, which we consider in this chapter.

We confine our attention to forces that vary with the position of the particle in its environment. This type of force is common in physics. Some examples are the gravitational forces between bodies, such as the sun and earth or earth and moon, and the force exerted by a stretched spring on a body to which it is attached. The procedure used to determine the motion of a particle subject to such a force leads us to the concepts of work and kinetic energy and to the development of the work-energy theorem, which is the central feature of this chapter. In Chapter 8 we consider a broader view of energy, embodied in the law of conservation of energy, a concept which has played a major role in the development of physics.

7-2
WORK DONE BY A CONSTANT FORCE

Consider a particle acted on by a force. In the simplest case the force **F** is constant and the motion takes place in a straight line in the direction of the force. In such a situation we define the work done by the force on the particle as the product of the magnitude of the force F and the distance d through which the particle moves. We write this as

$$W = Fd.$$

However, the constant force acting on a particle may not act in the direction in which the particle moves. In this case we define the work done by the force on the particle as the product of the component of the force along the line of motion by the distance d the body moves along that line. In Fig. 7-1 a constant force **F** makes an angle ϕ with the x-axis and acts on a particle whose displacement along the x-axis is **d.** If W represents the work done by **F** during this displacement, then according to our definition

$$W = (F \cos \phi)d. \tag{7-1}$$

figure 7-1
A force **F** makes the block undergo a displacement **d**. The component of **F** that does the work has magnitude $F \cos \phi$; the work done is $Fd \cos \phi \ (= \mathbf{F} \cdot \mathbf{d})$.

Of course, other forces must act on a particle that moves in this way (its weight and the frictional force exerted by the plane, to name two). A particle acted on by only a single force may have a displacement in a direction other than that of this single force, as in projectile motion. But it cannot move in a straight line unless the line has the same direction as that of the single force applied to it. *Equation 7-1 refers only to the work done on the particle by the particular force* **F.** *The work done on the particle by the other forces must be calculated separately.* The total work done on the particle is the sum of the works done by the separate forces.

When ϕ is zero, the work done by **F** is simply Fd, in agreement with our previous equation. Thus, when a horizontal force draws a body horizontally, or when a vertical force lifts a body vertically, the work done by the force is the product of the magnitude of the force by the distance moved. When ϕ is 90°, the force has no component in the direction of motion. That force then does no work on the body. For instance, the vertical force holding a body a fixed distance off the ground does no work on the body, even if the body is moved horizontally over the ground. Also, the centripetal force acting on a body in motion does no work on that body because the force is always at right angles to the direction in which the body is moving. Of course, a force does no work on a body that does not move, for its displacement is then zero. In Fig. 7-2 we illustrate common examples in which a force applied to a body does no work on that body.

figure 7-2
Work is not always done by a force that is applied to a body. *(a)* The block is moving to the right at constant speed v over a frictionless surface. Work is not done by either the weight **W** or the normal force **N**. *(b)* The ball moves in a circle under the influence of a centripetal force **T**. There is a centripetal acceleration **a** but no work is done by **T**. In both *(a)* and *(b)* the forces being considered (**W**, **N**, and **T**) are at right angles to the displacement so that $W = \mathbf{F} \cdot \mathbf{d} = Fd \cos \phi = Fd \cos 90° = 0$. *(c)* A cylinder hangs from a cord. No work is done either by **T**, the tension in the cord, or by **W** the weight of the cylinder. *(d)* A cylinder rests in a groove; no work is done by **W**, N_1 or N_2. In both *(c)* and *(d)* the work done by the individual forces is zero because the displacement is zero.

Notice that we can write Eq. 7-1 either as $(F \cos \phi)d$ or $F(d \cos \phi)$. This suggests that the work can be calculated in two different ways: Either we multiply the magnitude of the displacement by the component of the force in the direction of the displacement or we multiply the magnitude of the force by the component of the displacement in the direction of the force. These two methods always give the same result.

Work is a *scalar*, although the two quantities involved in its definition, force and displacement, are vectors. In Section 2-4 we defined the *scalar product* of two vectors as the scalar quantity that we find when we multiply the magnitude of one vector by the component of a second vector along the direction of the first. We promised in that section that we would soon run across physical quantities that behave like scalar products. Equation 7-1 shows that work is such a quantity. In the terminology of vector algebra we can write this equation as

$$W = \mathbf{F} \cdot \mathbf{d}, \tag{7-2}$$

where the dot indicates a scalar (or dot) product. Equation 7-2 for **F** and **d** corresponds to Eq. 2-11 for **a** and **b**.

Work can be either positive or negative. If the particle on which a force acts has a component of motion opposite to the direction of the force, the work done by that force is negative. This corresponds to an obtuse angle between the force and displacement vectors. For example, when a person lowers an object to the floor, the work done on the object by the upward force of his hand holding the object is negative. In this case ϕ is 180°, for **F** points up and **d** points down.

Work as we have defined it (Eq. 7-2) proves to be a very useful concept in physics. Our special definition of the word "work" does not correspond to the colloquial usage of the term. This may be confusing. A person holding a heavy weight at rest in the air may say that he is doing hard work — and he may work hard in the physiological sense — but from the point of view of physics we say that he is not doing any work. We say this because the applied force causes no displacement. The word *work* is used only in the strict sense of Eq. 7-2. In many scientific fields words are borrowed from our everyday language and are used to name a very specific concept. The words "basic" and "cell," for example, mean quite different things in chemistry and biology than in everyday language.

The *unit* of work is the work done by a unit force in moving a body a unit distance in the direction of the force. In SI units the unit of work is 1 *newton-meter*, called 1 *joule* (abbreviation J). In the British engineering system the unit of work is the *foot-pound*. In cgs systems the unit of work is 1 *dyne-centimeter*, called 1 *erg*. Using the relations between the newton, the dyne, and the pound, and the meter, the centimeter, and foot, we obtain 1 joule = 10^7 ergs = 0.7376 ft · lb.

EXAMPLE 1

A block of mass 10.0 kg is to be raised from the bottom to the top of an incline 5.00 m long and 3.00 m off the ground at the top. Assuming frictionless surfaces, how much work must be done by a force parallel to the incline pushing the block up at *constant speed* at a place where $g = 9.80$ m/s².

The situation is shown in Fig. 7-3a. The forces acting on the block are shown in Fig. 7-3b. We must first find P, the magnitude of the force pushing the block up the incline. Because the motion is not accelerated, the resultant force parallel to the plane must be zero. Thus

$$P - mg \sin \theta = 0,$$

or

$$P = mg \sin \theta = (10.0 \text{ kg})(9.80 \text{ m/s}^2)(\tfrac{3}{5}) = 58.8 \text{ N}.$$

Then the work done by **P**, from Eq. 7-1 with $\phi = 0°$, is

$$W = \mathbf{P} \cdot \mathbf{d} = Pd \cos 0° = Pd = (58.8 \text{ N})(5.00 \text{ m}) = 294 \text{ J}.$$

If a man were to raise the block vertically without using the incline, the work he would do would be the vertical force mg times the vertical distance or

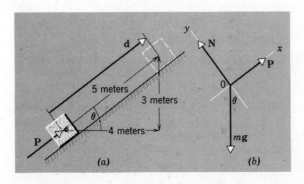

figure 7-3
Example 1. *(a)* A force **P** displaces a block a distance **d** up an inclined plane which makes an angle θ with the horizontal. *(b)* A free-body force diagram for the block.

$$(98.0 \text{ N})(3.00 \text{ m}) = 294 \text{ J},$$

the same as before. The only difference is that with the incline he could apply a smaller force ($P = 58.8$ N) to raise the block than is required without the incline ($mg = 98.0$ N); on the other hand, he had to push the block a greater distance (5.00 m) up the incline than he had to raise the block directly (3.00 m).

A boy pulls a 10-lb sled 30 ft along a horizontal surface at a *constant speed*. What work does he do on the sled if the coefficient of kinetic friction is 0.20 and his pull makes an angle of 45° with the horizontal?

EXAMPLE 2

figure 7-4
Example 2. *(a)* A boy displaces a sled an amount **d** by pulling with a force **P** on a rope that makes an angle ϕ with the horizontal. *(b)* A free-body diagram for the sled.

The situation is shown in Fig. 7-4*a* and the forces acting on the sled are shown in Fig. 7-4*b*. **P** is the boy's pull, **w** the sled's weight, **f** the frictional force, and **N** the normal force exerted by the surface on the sled. The work done by the boy on the sled is

$$W = \mathbf{P} \cdot \mathbf{d} = Pd \cos \phi.$$

To evaluate this we first must determine P, whose value has not been given. To obtain P we refer to the force diagram.

The sled is unaccelerated, so that from the second law of motion we obtain

$$P \cos \phi - f = 0,$$

and

$$P \sin \phi + N - w = 0.$$

We know that f and N are related by

$$f = \mu_k N.$$

These three equations contain three unknown quantities, P, f, and N. To find P we eliminate f and N from these equations and solve the remaining equation for P. You should verify that

$$P = \mu_k w/(\cos \phi + \mu_k \sin \phi).$$

With $\mu_k = 0.20$, $w = 10$ lb, and $\phi = 45°$ we obtain

$$P = (0.20)(10 \text{ lb})/(0.707 + 0.141) = 2.4 \text{ lb}.$$

Then with $d = 30$ ft, the work done by the boy on the sled is

$$W = Pd \cos \phi = (2.4 \text{ lb})(30 \text{ ft})(0.707) = 51 \text{ ft} \cdot \text{lb}.$$

The vertical component of the boy's pull **P** does no work on the sled. Notice, however, that it reduces the normal force between the sled and the surface ($N = w - P \sin \phi$) and thereby reduces the magnitude of the force of friction ($f = \mu_k N$).

Would the boy do more work, less work, or the same amount of work on the sled if he pulled horizontally instead of at 45° from the horizontal? Do any of the other forces acting on the sled do work on it?

Let us now consider the work done by a force that is not constant. We consider first a force that varies in magnitude only. Let the force be given as a function of position $F(x)$ and assume that the force acts in the x-direction. Suppose a body is moved along the x-direction by this force. What is the work done by this variable force in moving the body from x_1 to x_2?

In Fig. 7-5 we plot F versus x. Let us divide the total displacement into a large number of small equal intervals Δx (Fig. 7-5a). Consider the small displacement Δx from x_1 to $x_1 + \Delta x$. During this small displacement the force F has a nearly constant value and the work it does, ΔW, is approximately

$$\Delta W = F \Delta x, \qquad (7\text{-}3)$$

where F is the value of the force at x_1. Likewise, during the small displacement from $x_1 + \Delta x$ to $x_1 + 2\Delta x$, the force F has a nearly constant value and the work it does is approximately $\Delta W = F \Delta x$, where F is the value of the force at $x_1 + \Delta x$. The total work done by F in displacing the body from x_1 to x_2, W_{12}, is approximately the sum of a large number of terms like that of Eq. 7-3, in which F has a different value for each term. Hence

$$W_{12} = \sum_{x_1}^{x_2} F \Delta x, \qquad (7\text{-}4)$$

where the Greek letter sigma (Σ) stands for sum over all intervals from x_1 to x_2.

To make a better approximation we can divide the total displacement from x_1 to x_2 into a larger number of equal intervals, as in Fig. 7-5b, so that Δx is smaller and the value of F at the beginning of each interval is more typical of its values within the interval. It is clear that we can obtain better and better approximations by taking Δx smaller and smaller so as to have a larger and larger number of intervals. We can obtain an exact result for the work done by F if we let Δx go to zero and the number of intervals go to infinity. Hence the exact result is

$$W_{12} = \lim_{\Delta x \to 0} \sum_{x_1}^{x_2} F \Delta x. \qquad (7\text{-}5)$$

The relation

$$\lim_{\Delta x \to 0} \sum_{x_1}^{x_2} F \Delta x = \int_{x_1}^{x_2} F \, dx,$$

as you may have learned in your calculus course, *defines the integral of F with respect to x from x_1 to x_2*. Numerically, this quantity is exactly equal to the area between the force curve and the x-axis between the limits x_1 and x_2 (Fig. 7-5c). Hence, graphically an integral can be interpreted as an area. The symbol $\int$ is a distorted S (for *sum*) and symbolizes the integration process. We can write the total work done by F in displacing a body from x_1 to x_2 as

$$W_{12} = \int_{x_1}^{x_2} F(x) \, dx. \qquad (7\text{-}6)$$

As an example, consider a spring attached to a wall. Let the (horizontal) axis of the spring be chosen as an x-axis, and let the origin, $x = 0$, coincide with the endpoint of the spring in its normal, unstretched state. We assume that the positive x-direction points away from the

7-3
WORK DONE BY A VARIABLE FORCE— ONE-DIMENSIONAL CASE

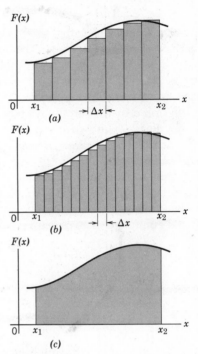

figure 7-5

Computing $\int_{x_1}^{x_2} F(x) \, dx$ amounts to finding the area under the curve $F(x)$ between the limits x_1 and x_2. This can be done approximately as in the top drawing *(a)* by dividing the area into a few strips, each of width Δx. The areas of the rectangles are then summed to give a rough value of the area. In the middle drawing *(b)* the strips are narrower and the value for the area becomes more exact as the errors at the tops of the rectangles become smaller. In the bottom drawing *(c)* the strips are only infinitesimal in width. The measurement of area is exact, since the errors at the tops of the rectangles go to zero as the strip width dx goes to zero.

wall. In what follows we imagine that we stretch the spring so slowly that it is essentially in equilibrium at all times ($\mathbf{a} = 0$).

If we stretch the spring so that its endpoint moves to a position x, the spring will exert a force on the agent doing the stretching given to a good approximation by

$$F = -kx, \tag{7-7}$$

where k is a constant called the *force constant* of the spring. Equation 7-7 is the *force law* for springs. The direction of the force is always opposite to the displacement of the endpoint from the origin. When the spring is stretched, $x > 0$ and F is negative; when the spring is compressed, $x < 0$ and F is positive. The force exerted by the spring is a *restoring force* in that it always points toward the origin. Real springs will obey Eq. 7-7, known as *Hooke's law*, if we do not stretch them beyond a limited range. We can think of k as the magnitude of the force per unit elongation. Thus very stiff springs have large values of k.

To stretch a spring we must exert a force F' on it equal but opposite to the force F exerted by the spring on us. The applied force* is therefore $F' = kx$ and the work done by the applied force in stretching the spring so that its endpoint moves from x_1 to x_2 is†

$$W_{12} = \int_{x_1}^{x_2} F'(x)\ dx = \int_{x_1}^{x_2} (kx)\ dx = \tfrac{1}{2}kx_2{}^2 - \tfrac{1}{2}kx_1{}^2.$$

If we let $x_1 = 0$ and $x_2 = x$, we obtain   VARIABLE FORCE 1-D.

$$W = \int_0^x (kx)\ dx = \tfrac{1}{2}kx^2. \qquad \text{FOR A SPRING} \tag{7-8}$$

This is the work done in stretching a spring so that its endpoint moves from its unstretched position to x. Note that the work to *compress* a spring by x is the same as that to stretch it by x because the displacement x is squared in Eq. 7-8; either sign for x gives a positive value for W.

We can also evaluate this integral by computing the area under the force-displacement curve and the x-axis from $x = 0$ to $x = x$. This is drawn as the white area in Fig. 7-6. The area is a triangle of base x and altitude kx. The white area is therefore

$$\tfrac{1}{2}(x)(kx) = \tfrac{1}{2}kx^2.$$

in agreement with Eq. 7-8.

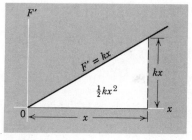

figure 7-6
The force exerted in stretching a spring is $F' = kx$. The area under the force curve is the work done in stretching the spring a distance x and can be found by integrating or by using the formula for the area of a triangle.

The force $\mathbf{F}$ acting on a particle may vary in direction as well as in magnitude, and the particle may move along a curved path. To compute the work in this general case we divide the path into a large number of small displacements $\Delta \mathbf{r}$, each pointing along the path in the direction of motion. Figure 7-7 shows two selected displacements for a particular situation; it also shows the value of $\mathbf{F}$ and the angle ϕ between $\mathbf{F}$ and $\Delta \mathbf{r}$ at each location. We can find the amount of

7-4

WORK DONE BY A VARIABLE FORCE— TWO-DIMENSIONAL CASE

---

* If the applied force were different from $F' = kx$, we would have a net unbalanced force acting on the spring and its motion would be accelerated. To compute the work done we would have to specify exactly what the applied force is at each point. No matter what the force turned out to be, the work done would always be the same for the same displacement x_1 to x_2, providing the spring has the same speed initially and finally. However, it is much easier to use the simple force $F' = kx$ in calculating the work done. Such an applied force leads to unaccelerated motion. It is in order to be able to use this simple force that we specified unaccelerated motion in the first place.

† The student just becoming familiar with calculus should consult the list of integrals in Appendix I.

work done on the particle during a displacement $\Delta \mathbf{r}$ from

$$dW = \mathbf{F} \cdot \Delta \mathbf{r} = F \cos \phi \, \Delta r \qquad (7\text{-}9)$$

The work done by the variable force $\mathbf{F}$ on the particle as the particle moves, say, from a to b in Fig. 7-7 is found very closely by adding up (summing) the elements of work done over each of the line segments that make it up. As the line segments $\Delta \mathbf{r}$ become smaller they may be replaced by differentials $d\mathbf{r}$ and the sum over the line segments may be replaced by an intergral, as in Eq. 7-6. The work is then found from

$$W_{ab} = \int_a^b \mathbf{F} \cdot d\mathbf{r} = \int_a^b F \cos \phi \, dr. \qquad (7\text{-}10a)$$

We cannot evaluate this integral until we are able to say how F and ϕ in Eq. 7-10a vary from point to point along the path; both are functions of the x- and y-coordinates of the particle in Fig. 7-7.

We can obtain another equivalent expression for Eq. 7-10a by expressing $\mathbf{F}$ and $d\mathbf{r}$ in terms of their components. Thus $\mathbf{F} = \mathbf{i} F_x + \mathbf{j} F_y$ and $d\mathbf{r} = \mathbf{i} \, dx + \mathbf{j} \, dy$, so that $\mathbf{F} \cdot d\mathbf{r} = F_x \, dx + F_y \, dy$. In this evaluation recall (see Problem 21, Chapter 2) that $\mathbf{i} \cdot \mathbf{i} = \mathbf{j} \cdot \mathbf{j} = 1$ and $\mathbf{i} \cdot \mathbf{j} = \mathbf{j} \cdot \mathbf{i} = 0$. Substituting this result into Eq. 7-10a, we obtain

Z.D. VARIABLE FORCE

$$W_{ab} = \int_a^b (F_x \, dx + F_y \, dy) \qquad (7\text{-}10b)$$

Integrals such as those in Eqs. 7-10a and 7-10b are called *line integrals.*

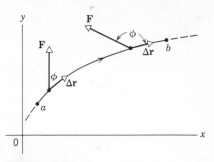

figure 7-7
How F and ϕ might change along a path. As $\Delta \mathbf{r} \to 0$ we may replace it by the differential $d\mathbf{r}$, which always points in the direction of the velocity of the moving object, since $\mathbf{v} = d\mathbf{r}/dt$, and hence is tangent to the path at all points.

EXAMPLE 3

As an example of a variable force consider a particle of mass m suspended from a weightless cord of length l. This is called a simple pendulum. Let us displace the particle along a circular path of radius l from $\phi = 0$ to $\phi = \phi_0$ by applying a force that is always *horizontal*. We can apply such a force by pulling horizontally on the particle with an attached string, for example. The particle will then have been displaced a vertical distance h. Figure 7-8a shows the situation and Fig. 7-8b shows the forces acting on the particle in the arbitrary position ϕ. The applied force is $\mathbf{F}$, $\mathbf{T}$ is the tension in the cord, and mg the weight of the particle.

Again we assume that there is no acceleration (the reason is the same as before), so that in practice the motion must be very slow. The force $\mathbf{F}$ is always horizontal, but the displacement $d\mathbf{r}$ is along the arc. The direction of $d\mathbf{r}$ depends on the value of ϕ and is tangent to the circle at each point. $\mathbf{F}$ will vary in magnitude in such a way as to balance the horizontal component of the tension. Notice that the angle between $\mathbf{F}$ and $d\mathbf{r}$ is equal to the angular displacement ϕ in this case.

The work done as the mass m moves from $\phi = 0$ to $\phi = \phi_0$ under the action of the force $\mathbf{F}$ is

$$W = \int_{\phi=0}^{\phi=\phi_0} \mathbf{F} \cdot d\mathbf{r} = \int_{\phi=0}^{\phi=\phi_0} F \cos \phi \, dr \qquad (7\text{-}10a)$$

or

$$W = \int_{x=0, y=0}^{x=(l-h)\tan \phi_0, y=h} (F_x \, dx + F_y \, dy). \qquad (7\text{-}10b)$$

Let us evaluate Eq. 7-10b.

Note that, from Newton's first law (see Fig. 7-8b)

$$F_x = T \sin \phi \qquad \text{and} \qquad mg = T \cos \phi.$$

Eliminating T between these relations gives us

$$F_x = mg \tan \phi.$$

We also note in Fig. 7-8b that $F_y = 0$. Substituting these values for F_x and F_y into Eq. 7-10b yields

figure 7-8
Example 3. *(a)* A simple pendulum. A mass point m is suspended on a string of length l. Its maximum displacement is ϕ_0. *(b)* A free-body force diagram for the mass subjected to an applied horizontal force.

$$W = \int_{x=0, y=0}^{x=(l-h)\tan\phi_0, y=h} mg \tan\phi \, dx.$$

Now from Fig. 7-8a we see that

$$\tan\phi = dy/dx \quad \text{or} \quad \tan\phi \, dx = dy.$$

Making this substitution and noting that the integral depends only on the variable y, we obtain finally

$$W = \int_{y=0}^{y=h} (mg) \, dy = mg \int_0^h dy = mgh.$$

You should now try to compute the work done in displacing the particle along the arc with constant speed by applying a force that is always directed *along the arc*. Here it will be simpler to work with Eq. 7-10a, using the tantential force and taking $dr = l \, d\phi$. The result will be the same as before, $W = mgh$. Notice that both these results are the same as the work that would be done in raising a mass m vertically through a height h.

What work has been done on the particle by the tension **T** in the string?

In our previous examples of work done by forces, we dealt with *unaccelerated* objects. In such cases the *resultant force* acting on the object is *zero*. Let us suppose now that the *resultant force* acting on an object is *not zero*, so that the object is *accelerated*. The conditions are the same in all respects to those that exist when a single unbalanced force acts on the object.

The simplest situation to consider is that of a *constant resultant force* **F**. Such a force, acting on a particle of mass m, will produce a constant acceleration **a.** Let us choose the x-axis to be in the common direction of **F** and **a.** What is the work done by this force on the particle in causing a displacement x? We have (for constant acceleration) the relations

$$a = \frac{v - v_0}{t}$$

and

$$x = \frac{v + v_0}{2} \cdot t,$$

which are Eqs. 3-12 and 3-14 respectively (in which we have dropped the subscript x, for convenience, and chosen $x_0 = 0$ in the last equation). Here v_0 is the particle's speed at $t = 0$ and v its speed at the time t. Then the work done is

$$W = Fx = max$$

$$= m\left(\frac{v - v_0}{t}\right)\left(\frac{v + v_0}{2}\right)t = \tfrac{1}{2}mv^2 - \tfrac{1}{2}mv_0^2. \qquad (7\text{-}11)$$

We call one-half the product of the mass of a body and the square of its speed the kinetic energy of the body. If we represent kinetic energy by the symbol K, then

$$K = \tfrac{1}{2}mv^2. \qquad (7\text{-}12)$$

We may then state Eq. 7-11 in this way: *The work done by the resultant force acting on a particle is equal to the change in the kinetic energy of the particle.*

Although we have proved this result for a constant force only, it holds whether the resultant force is constant or variable. Let the resultant force vary in magnitude (but not in direction), for example. Take the displacement to be in the direction of the force. Let this direction be the x-axis. The work done by the resultant force in displacing the particle from x_0 to x is

$$W = \int \mathbf{F} \cdot d\mathbf{r} = \int_{x_0}^{x} F \, dx.$$

But from Newton's second law we have $F = ma$, and we can write the acceleration a as

$$a = \frac{dv}{dt} = \frac{dv}{dx} \cdot \frac{dx}{dt} = \frac{dv}{dx} v = v \frac{dv}{dx}.$$

Hence

$$W = \int_{x_0}^{x} F \, dx = \int_{x_0}^{x} mv \frac{dv}{dx} \, dx = \int_{v_0}^{v} mv \, dv = \tfrac{1}{2}mv^2 - \tfrac{1}{2}mv_0^2. \quad (7\text{-}13)$$

A more general case is that in which the force varies both in direction and magnitude and the motion is along a curved path, as in Fig. 7-7. (See Problem 8.) Once again we find that the work done on a particle by the resultant force is equal to the change in the kinetic energy of the particle.

The work done *on* a particle by the *resultant* force is *always* equal to the change in the kinetic energy of the particle:

$$W \text{ (of the } \textit{resultant} \text{ force)} = K - K_0 = \Delta K. \quad (7\text{-}14)$$

Equation 7-14 is known as the *work-energy theorem* for a particle.

Notice that when the speed of the particle is constant, there is no change in kinetic energy and the work done by the resultant force is zero. With uniform circular motion, for example, the speed of the particle is constant and the centripetal force does no work on the particle. A force at right angles to the direction of motion merely changes the *direction* of the velocity and not its magnitude. Only when the *resultant* force has a component along the direction of motion does it change the speed of the particle or its kinetic energy. Work is done on a particle by that component of the resultant force along the line of motion. This agrees with our definition of work in terms of a scalar product, for in $\mathbf{F} \cdot d\mathbf{r}$ only the component of $\mathbf{F}$ along $d\mathbf{r}$ contributes to the product.

If the kinetic energy of a particle decreases, the work done on it by the resultant force is negative. The displacement and the component of the resultant force along the line of motion are oppositely directed. The work done *on* the particle by the force is the negative of the work done *by* the particle on whatever produced the force. This is a consequence of Newton's third law of motion. Hence Eq. 7-14 can be interpreted to say that the kinetic energy of a particle *decreases* by an amount just equal to the amount of work which the particle *does*. A body is said to have energy stored in it because of its motion; as it does work it slows down and loses some of this energy. Therefore, *the kinetic energy of a body in motion is equal to the work it can do in being brought to rest.* This result holds whether the applied forces are constant or variable.

The units of kinetic energy and of work are the same. Kinetic energy, like work, is a scalar quantity. The kinetic energy of a group of particles is simply the (scalar) sum of the kinetic energies of the individual particles in the group.

EXAMPLE 4

A neutron, one of the constituents of a nucleus, is found to pass two points 6.0 m apart in a time interval of 1.8×10^{-4} s. Assuming its speed was constant, find its kinetic energy. The mass of a neutron is 1.7×10^{-27} kg.

We find the speed from

$$v = \frac{d}{t} = \frac{6.0 \text{ m}}{1.8 \times 10^{-4} \text{ s}} = 3.3 \times 10^{4} \text{ m/s}.$$

The kinetic energy is

$$K = \tfrac{1}{2}mv^2 = (\tfrac{1}{2})(1.7 \times 10^{-27} \text{ kg})(3.3 \times 10^{4} \text{ m/s})^2 = 9.3 \times 10^{-19} \text{ J}.$$

For purposes of nuclear physics the joule is a very large energy unit. A unit more commonly used is the *electron volt* (eV), which is equal to 1.60×10^{-19} J. The kinetic energy of the neutron in our example can then be expressed as

$$K = (9.3 \times 10^{-19} \text{ J})\left(\frac{1 \text{ eV}}{1.60 \times 10^{-19} \text{ J}}\right) = 5.8 \text{ eV}.$$

EXAMPLE 5

Assume the force of gravity to be constant for small distances above the surface of the earth. A body is dropped from rest at a height h above the earth's surface. What will its kinetic energy be just before it strikes the ground?

The gain in kinetic energy is equal to the work done by the resultant force, which here is the force of gravity. This force is constant and directed along the line of motion. so that the work done by gravity is

$$W = \mathbf{F} \cdot \mathbf{d} = mgh.$$

Initially the body has a speed $v_0 = 0$ and finally a speed v. The gain in kinetic energy of the body is

$$\tfrac{1}{2}mv^2 - \tfrac{1}{2}mv_0^2 = \tfrac{1}{2}mv^2 - 0.$$

Equating these two equivalent terms we obtain

$$K = \tfrac{1}{2}mv^2 = mgh$$

as the kinetic energy of the body just before it strikes the ground.

The speed of the body is then

$$v = \sqrt{2gh}.$$

You should show that in falling from a height h_1 to a height h_2 a body will increase its kinetic energy from $\tfrac{1}{2}mv_1^2$ to $\tfrac{1}{2}mv_2^2$, where

$$\tfrac{1}{2}mv_2^2 - \tfrac{1}{2}mv_1^2 = mg(h_1 - h_2).$$

In this example we are dealing with a constant force and a constant acceleration. The methods developed in previous chapters should be useful here too. Can you show how the results obtained by energy considerations could be obtained directly from the laws of motion for uniformly accelerated bodies?

EXAMPLE 6

A block weighing 8.0 lb (= 35.6 N) slides on a horizontal frictionless table with a speed of 4.0 ft/s (= 1.22 m/s). It is brought to rest in compressing a spring in its path. By how much is the spring compressed if its force constant is 0.25 lb/ft (= 1.35 N/m)?

The kinetic energy of the block is

$$K = \tfrac{1}{2}mv^2 = \tfrac{1}{2}(w/g)v^2.$$

This kinetic energy is equal to the work W that the block can do before it is brought to rest. The work done in compressing the spring a distance x beyond its unstretched length is

$$W = \tfrac{1}{2}kx^2,$$

so that

$$\tfrac{1}{2}kx^2 = \tfrac{1}{2}(w/g)v^2$$

or

$$x = \sqrt{\frac{w}{gk}}\,v = \sqrt{\frac{8.0}{(32)(0.25)}}\,4.0 \text{ ft} = 4.0 \text{ ft } (= 1.22 \text{ m}).$$

7-6
SIGNIFICANCE OF THE WORK-ENERGY THEOREM

The work-energy theorem does *not* represent a new, independent law of classical mechanics. We have simply *defined* work and kinetic energy and *derived* the relation between them directly from Newton's second law. The work-energy theorem is useful, however, for solving problems in which the work done by the resultant force is easily computed and in which we are interested in finding the particle's speed at certain positions. Of greater significance, perhaps, is the fact that the work-energy theorem is the starting point for a sweeping generalization in physics. It has been emphasized that the work-energy theorem is valid when *W* is interpreted as the work done by the *resultant* force acting on the particle. However, it is helpful in many problems to compute separately the work done by certain types of force and give special names to the work done by each type. This leads to the concepts of different types of energy and the principle of the conservation of energy, which is the subject of the next chapter.

7-7
POWER

Let us now consider the time involved in doing work. The same amount of work is done in raising a given body through a given height whether it takes one second or one year to do so. However, the *rate at which work is done* is often more interesting to us than the total work performed.

We define *power* as the time rate at which work is done. The average power delivered by an agent is the total work done by the agent divided by the total time interval, or

$$\bar{P} = W/t.$$

The instantaneous power delivered by an agent is

$$P = dW/dt. \qquad (7\text{-}15)$$

If the power is constant in time, then $P = \bar{P}$ and

$$W = Pt.$$

In the International System of units, the unit of power is 1 joule/sec, which is called 1 *watt* (abbreviation W). This unit of power is named in honor of James Watt who made major improvements to the steam engines of his day that pointed the way toward today's more efficient engines. In the British engineering system, the unit of power is 1 ft · lb/sec. Because this unit is quite small for practical purposes, a larger unit, called the *horsepower* (abbreviation hp), has been adopted. Actually Watt himself suggested as a unit of power the power delivered by a horse as an engine. One horsepower was chosen to equal 550 ft · lb/sec. One horsepower is equal to about 746 watts or about three-fourths of a kilowatt. A horse would not last very long at that rate.

Work can also be expressed in units of power × time. This is the origin of the term *kilowatt-hour*, for example. One kilowatt-hour is the work done in 1 hour by an agent working at a constant rate of 1 kW.

An automobile uses 100 hp and moves at a uniform speed of 60 mi/h (= 88 ft/s). What is the forward thrust exerted by the engine on the car?

EXAMPLE 7

$$P = \frac{W}{t} = \frac{\mathbf{F} \cdot \mathbf{d}}{t} = \mathbf{F} \cdot \mathbf{v}.$$

The forward thrust $\mathbf{F}$ is in the direction of motion given by $\mathbf{v}$, so that

$$P = Fv,$$

and

$$F = \frac{P}{v} = \left(\frac{100 \text{ hp}}{88 \text{ ft/s}}\right)\left(\frac{550 \text{ ft} \cdot \text{lb/sec}}{1 \text{ hp}}\right) = 630 \text{ lb}.$$

Why doesn't the car accelerate?

questions

1. Can you think of other words like "work" whose colloquial meanings are often different from their scientific meanings?

2. Suppose that three constant forces act on a particle as it moves from one position to another. Is the work done on the particle by the resultant of these three forces equal to the sum of the work done by each of the three forces acting separately?

3. The inclined plane (see Example 1) is a simple machine which enables us to do work with the application of a smaller force than is otherwise necessary. The same statement applies to a wedge, a lever, a screw, a gear wheel, and a pulley. Do such machines save us work?

4. In a tug of war one team is slowly giving way to the other. What work is being done and by whom?

5. The work done by frictional forces is always negative. Can you explain this?

6. A man exerts a constant force on a fixed wall and does no mechanical work on it. Why does he feel tired doing this?

7. You lift a bowling ball from the floor and put it on a table. Two forces act on the ball: its weight $-m\mathbf{g}$, and your upward force $+m\mathbf{g}$. These two forces cancel each other so that it would seem that no work is done. On the other hand you know that you have done some work. What is wrong?

8. You cut a spring in half. What is the relationship of the spring constant k for the original spring to that for either of the half-springs?

9. Springs A and B are identical except that A is stiffer than B, that is, $k_A > k_B$. On which spring is more work expended if (a) they are stretched by the same amount, (b) they are stretched by the same force?

10. Does kinetic energy depend on the direction of the motion involved? Can it be negative?

11. The work done by the resultant force is always equal to the change in kinetic energy. Can it happen that the work done by one of the component forces alone will be greater than the change in kinetic energy? If so, give examples.

12. You throw a ball vertically in the air and catch it. What does the work-energy theorem say qualitatively about the free flight during this round trip? Answer the question first neglecting air resistance and second, taking it into account.

13. When two children play catch on a train, does the kinetic energy of the ball depend on the speed of the train? Does the reference frame chosen affect your answer? If so, would you call kinetic energy a scalar quantity? (See Problem 21.)

14. Does the work done by the resultant force acting on a particle depend on the (inertial) reference frame of the observer? Does the change in kinetic energy so depend?

15. A man rowing a boat upstream is at rest with respect to the shore. (a) Is he

doing any work? (b) If he stops rowing and moves down with the stream, is any work being done on him?

16. Does the power needed to raise a box onto a platform depend on how fast it is raised?

17. You lift some books from a library shelf to a higher shelf in time t. Does the work that you do depend on (a) the mass of the books, (b) the weight of the books, (c) the height of the upper shelf above the floor, (d) the time t, and (e) whether you lift the books sideways or directly upward?

18. We hear a lot about the "energy crisis." Would it be more accurate to speak of a "power crisis"?

problems

SECTION 7-2

1. A man pushes a 60-lb (270-N) block 30 ft (9.1 m) along a level floor at constant speed with a force directed 45° below the horizontal. If the coefficient of kinetic friction is 0.20, how much work does the man do on the block? *Answer:* 450 ft · lb (610 J).

2. A block of mass $m = 3.57$ kg is drawn at constant speed a distance $d = 4.06$ m along a horizontal floor by a rope exerting a constant force of magnitude $F = 7.68$ N making an angle $\theta = 15.0°$ above the horizontal. Compute (a) the total work done on the block, (b) the work done by the rope on the block, (c) the work done by friction on the block, and (d) the coefficient of kinetic friction between block and floor.

3. A 100-lb block of ice slides down an incline 5.0 ft long and 3.0 ft high. A man pushes up on the ice parallel to the incline so that it slides down at constant speed. The coefficient of friction between the ice and the incline is 0.10. Find (a) the force exerted by the man, (b) the work done by the man on the block, (c) the work done by gravity on the block, (d) the work done by the surface of the incline on the block, (e) the work done by the resultant force on the block, and (f) the change in kinetic energy of the block. Answer: (a) 52 lb. (b) −260 ft · lb. (c) 300 ft · lb. (d) −40 ft · lb. (e) Zero. (f) Zero.

4. A crate weighing 500 lb (2200 N) is suspended from a rope 40 ft (12 m) long. The crate is then pushed aside 4.0 ft (1.2 m) from the vertical and held there. (a) What force directed along the arc is needed to keep the crate in this position? (b) Is work being done in holding it there? (c) Was work done in moving it aside? If so, how much? (d) Does the tension in the rope perform any work on the crate?

5. A cord is used to lower vertically a block of mass M a distance d at a constant downward acceleration of $g/4$. Find the work done by the cord on the block. *Answer:* $-3Mgd/4$.

SECTION 7-3

6. (a) Estimate the work done by the force shown on the graph (Fig. 7-9) in displacing a particle from $x = 1$ m to $x = 3$ m. Refine your method to see how

figure 7-9
Problem 6

close you can come to the exact answer of 6 J. (*b*) The curve is given analytically by $F = a/x^2$ where $a = 9$ N · m². Show how to get the work done by the
rules of integration.

7. A single force acts on a body in rectilinear motion. A plot of velocity versus
time for the body is shown in Fig. 7-10. Find the sign (positive or negative)
of the work done *by* the force *on* the body in each of the intervals AB, BC,
CD, and DE.

   Answer: AB, BC, CD, DE.

   $$+ \quad 0 \quad -. \quad +$$

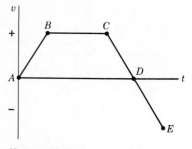

figure 7-10
Problem 7

SECTION 7-4

8. When the force **F** varies both in direction and magnitude and the motion is
along a curved path, the work done by **F** is obtained from $dW = \mathbf{F} \cdot d\mathbf{r}$, the
subsequent integration being taken along the curved path. Notice that both
F and ϕ, the angle between **F** and $d\mathbf{r}$, may vary from point to point (see Fig.
7-7). Show that for two- or three-dimensional motion

$$W = \tfrac{1}{2}mv^2 - \tfrac{1}{2}mv_0^2,$$

where v is the final speed and v_0 the initial speed.

SECTION 7-5

9. From what height would an automobile have to fall to gain the kinetic
energy equivalent to what it would have when going 60 mi/h (97 km/h)?
Answer: 120 ft (37 m).

10. A running man has half the kinetic energy of a boy half his mass. The man
speeds up by 1.0 m/s and then has the same kinetic energy as the boy. What
were the original speeds of (*a*) man and (*b*) boy?

11. If a 2.9×10^5 kg (weight $mg = 6.4 \times 10^5$ lb) Saturn V rocket with an Apollo
spacecraft attached must achieve an escape velocity of 11.2 km/s (25,000
mi/h) near the surface of the earth, how much energy must the fuel contain? Would the system actually need as much, or would it need more?
Why?   *Answer:* 1.8×10^{13} J (1.3×10^{13} ft · lb).

12. A proton starting from rest is accelerated in a cyclotron to a final speed of
3.0×10^7 m/s (about one-tenth the speed of light). How much work, in
electron volts, is done on the proton by the electrical force of the cyclotron
that accelerates it? 1 eV $= 1.6 \times 10^{-19}$ J.

13. A 30-g bullet initially traveling 500 m/s penetrates 12 cm into a wooden
block. What average force does it exert on the block?
Answer: 3.1×10^4 N.

14. An outfielder throws a baseball with an initial speed of 60 ft/s. An infielder
catches the ball at the same level when its speed is reduced to 40 ft/s. What
work was done in overcoming the resistance of the air? The weight of a
baseball is 9.0 oz.

15. A proton (nucleus of the hydrogen atom) is being accelerated in a linear
accelerator. In each stage of such an accelerator the proton is accelerated
along a straight line at 3.6×10^{15} m/s². If a proton enters such a stage moving
initially with a speed of 2.4×10^7 m/s, and the stage is 3.5 cm long, compute
(*a*) its speed at the end of the stage and (*b*) the gain in kinetic energy resulting from the acceleration. Take the mass of the proton to be 1.67×10^{-27} kg
and express the energy in electron volts; 1 eV $= 1.6 \times 10^{-19}$ J.
Answer: (*a*) 2.9×10^7 m/s. (*b*) 1.3×10^6 eV.

16. Show *from considerations of work and kinetic energy* that, assuming the
driver jams on the brakes, the stopping distance for a car of mass m moving
with speed v along a level road is $v^2/2\mu_k g$, where μ_k is the coefficient of
kinetic friction between tires and road. (See Example 2, Chapter 6, and
Questions 3, 4, 5 of Chapter 8.)

17. A 5.0-kg block moves in a straight line on a horizontal frictionless surface
under the influence of a force that varies with position as shown in Fig.
7-11. (*a*) How much work is done by the force as the block moves from the

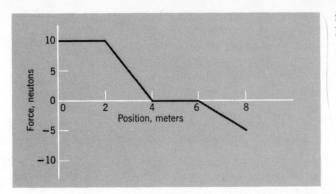

figure 7-11
Problem 17

origin to $x = 8.0$ m? (b) If the particle's speed passing through the origin was 4.0 m/s, with what speed does it pass the point $x = 8.0$ m?
Answer: (a) 25 J. (b) 5.1 m/s.

18. A helicopter is used to lift a 160-lb (710-N) astronaut 50 ft (15 m) vertically from the ocean by means of a cable. The acceleration of the astronaut is $g/10$. (a) How much work is done by the helicopter on the astronaut? (b) How much work is done by the gravitational force on the astronaut? (c) With what speed does the astronaut reach the helicopter?

figure 7-12
Problem 19

19. The block of mass M shown in Fig. 7-12 initially has a velocity v_0 to the right and its position is such that the spring exerts no force on it, that is, the spring is not stretched or compressed. The block moves to the right a distance l before stopping in the dotted position shown. The spring constant is k and the coefficient of kinetic friction between block and table is μ_k. As the block moves the distance l, (a) what is the work done on it by the friction force? (b) What is the work done on it by the spring force? (c) Are there other forces acting on the block, and, if so, what work do they do? (d) What is the total work done on the block? (e) Use the work-energy theorem to find the value of l in terms of M, v_0, μ_k, g. and k.
Answer: (a) $-\mu_k Mgl$. (b) $-kl^2/2$. (c) Gravity and the vertical thrust of the table, which do no work. (d) $-(\mu_k Mgl + kl^2/2)$.
(e) $(\sqrt{\mu_k^2 M^2 g^2 + v_0^2 kM} - \mu_k Mg)/k$.

20. (a) A mass of 0.675 kg on a frictionless table is attached to a string which passes through a hole in the table at the center of the horizontal circle in which the mass moves with constant speed. If the radius of the circle is 0.500 m and the speed is 10.0 m/s, compute the tension in the string. (b) It is found that drawing an additional 0.200 m of the string down through the hole, thereby reducing the radius of the circle to 0.300 m, has the effect of multiplying the original tension in the string by 4.63. Compute the total work done by the string on the revolving mass during the reduction of the radius.

21. *Work and Kinetic Energy in Moving Reference Frames.* Consider two observers, one whose frame is attached to the ground and another whose frame is attached, say, to a train moving with uniform velocity **u** with respect to the ground. Each observes that a particle, initially at rest with respect to the train, is accelerated by a constant force applied to it for time t in the forward direction.
(a). Show that for each observer the work done by the force is equal to the

gain in kinetic energy of the particle, but that one observer measures these quantities to be $\frac{1}{2}ma^2t^2$, whereas the other observer measures them to be $\frac{1}{2}ma^2t^2 + maut$. Here a is the common acceleration of the particle of mass m.

(b). Explain the differences in work done by the same force in terms of the different distances through which the observers measure the force to act during the time t. Explain the different final kinetic energies measured by each observer in terms of the work the particle could do in being brought to rest relative to each observer's frame.

SECTION 7-7

22. If a 128-lb (570 N) woman runs up a flight of stairs having a rise of 14 ft (4.3 m) in 3.5 s, what average power must she supply?

23. 1200 m³ of water passes each second over a waterfall 100 m high. Assuming that three-fourths of the kinetic energy gained by the water in falling is converted to electrical energy by a hydroelectric generator, what is the power output of the generator?   *Answer:* 8.8×10^5 kW.

24. The loaded cab of an elevator has a mass m of 3.0×10^3 kg and moves 200 m up the shaft in 20 s. At what average rate does the cable do work on the cab?

25. A horse pulls a cart horizontally with a force of 40 lb at an angle of 30° above the horizontal and moves along at a speed of 6.0 mi/h. (a) How much work does the horse do in 10 minutes? (b) What is the power output of the horse? *Answer:* (a) 1.8×10^5 ft · lb. (b) 0.55 hp.

26. What power is developed by a grinding machine whose wheel has a radius of 8.0 in. and runs at 2.5 rev/s when the tool to be sharpened is held against the wheel with a force of 40 lb? The coefficient of friction between the tool and the wheel is 0.32.

27. A satellite rocket weighing 100,000 lb acquires a speed of 4000 mi/h in 1.0 min after launching. (a) What is its kinetic energy at the end of the first minute? (b) What is the average power expended during this time, neglecting frictional and gravitational forces?
Answer: (a) 5.4×10^{10} ft · lb. (b) 1.6×10^6 hp.

28. A net force of 5.0 N acts on a 15 kg-body initially at rest. Compute (a) the work done by the force in the first, second, and third second and (b) the instantaneous power exerted by the force at the end of the third second.

29. A force acts on a 3.0-kg particle in such a way that the position of the particle as a function of time is given by $x = 3t - 4t^2 + t^3$, where x is in meters and t is in seconds. (a) Find the work done by the force during the first 4.0 s. (b) At what instantaneous rate is the force doing work on the particle at the instant $t = 1.0$ s?   *Answer:* (a) 530 J. (b) 12 W.

30. The force required to tow a boat at constant velocity is proportional to the speed. If it takes 10 hp (7500 W) to tow a certain boat at a speed of 2.5 mi/h (4.0 km/h), how much power does it take to tow it at a speed of 7.5 mi/h (12 km/h)?

31. A body of mass m accelerates uniformly from rest to a speed v_f in time t_f. (a) Show that the work done on the body as a function of time t, in terms of v_f and t_f, is

$$\tfrac{1}{2}m\,\frac{v_f^2}{t_f^2}\,t^2.$$

(b) As a function of time t, what is the instantaneous power delivered to the body? (c) What is the instantaneous power at the end of 5.0 s delivered to a 3200-lb body which accelerates to 60 mi/h in 10 s?
Answer: (b) mv_f^2t/t_f^2. (c) 70 hp.

32. A truck can move up a road having a grade of 1.0-ft rise every 50 ft with a speed of 15 mi/h. The resisting force is equal to 1/25 the weight of the truck. How fast will the same truck move down the hill with the same horsepower?

33. A 1.5×10^6 W railroad locomotive accelerates a train from a speed of 10 m/s to 25 m/s at full power in 6.0 minutes. (a) Neglecting friction, calculate the mass of the train. (b) Find the speed of the train as a function of time during the interval. (c) Find the force accelerating the train as a function of time during the interval. (d) Find the distance moved by the train during the interval.

Answer: (a) 2.1×10^6 kg. (b) $\sqrt{100 + 1.4\,t}$ m/s.
(c) $(1.5 \times 10^6)/\sqrt{100 + 1.4\,t}$ N. (d) 6.9 km.

8
the conservation of energy

In Chapter 7 we derived the work-energy theorem from Newton's second law of motion. This theorem says that the work W done by the resultant force $\mathbf{F}$ acting on a particle as it moves from one point to another is equal to the change ΔK in the kinetic energy of the particle, or

$$W = \Delta K. \qquad (8\text{-}1)$$

Often several forces act on a particle, the resultant force $\mathbf{F}$ being their vector sum, that is, $\mathbf{F} = \mathbf{F}_1 + \mathbf{F}_2 + \cdots \mathbf{F}_n$, in which we assume that n forces act. The work done by the resultant force $\mathbf{F}$ is the algebraic sum of the work done by these individual forces, or $W = W_1 + W_2 + \cdots W_n$. Thus we can write the work-energy theorem (Eq. 8-1) as

$$W_1 + W_2 + \cdots + W_n = \Delta K. \qquad (8\text{-}2)$$

In this chapter we shall consider systems in which a particle is acted upon by various kinds of forces and we shall compute W_1, W_2, and so on, for these forces; this will lead us to define different kinds of energy such as potential energy and heat energy. The process culminates in the formulation of one of the great principles of science, the *conservation of energy principle*.

Let us first distinguish between two types of forces, *conservative* and *nonconservative*. We shall consider an example of each type and we discuss each example from several different, but related, points of view.

Imagine a spring fastened at one end to a rigid wall as in Fig. 8-1. Let us slide a block of mass m with velocity $\mathbf{v}$ directly toward the spring; we assume that the horizontal plane is frictionless and that the spring is ideal, that is, that it obeys Hooke's law (Eq. 7-7)

$$F = -kx, \qquad (8\text{-}3)$$

where F is the force exerted by the spring when its free end is displaced through a distance x; we assume further that the mass of the spring is so small compared with that of the block that we can neglect the kinetic energy of the spring. Thus, in the system (mass + spring), all the kinetic energy is concentrated in the mass.

After the block touches the spring, the speed and hence the kinetic energy of the block decrease until finally the block is brought to rest by the action of the spring force, as in Fig. 8-1b. The block now reverses its motion as the compressed spring expands. It gains speed and kinetic energy and, when it comes once again to its position of initial contact with the spring, we find that it has the same speed and kinetic energy as it had originally; only the direction of motion has changed. The block loses kinetic energy during one part of its motion but gains it all back during the other part of its motion as it returns to its starting point (Fig. 8-1c).

We have interpreted the kinetic energy of a body as its ability to do work by virtue of its motion. It is clear that at the completion of a round trip the ability of the block in Fig. 8-1 to do work remains the same; it has been *conserved*. The elastic force exerted by an ideal spring, and other forces that act in this same way, are called *conservative*. The force of gravity is also conservative; if we throw a ball vertically upward, it will (if we assume air resistance to be negligible) return to our hand with the same kinetic energy that it had when it left our hand.

If, however, a particle on which one or more forces act returns to its initial position with either more or less kinetic energy than it had initially, then in a round trip its ability to do work has been changed. In this case the ability to do work has *not* been conserved and at least one of the forces acting is labeled *nonconservative*.

To illustrate a nonconservative force let us assume that the surfaces of the block and the plane in Fig. 8-1 are not frictionless but rather that a force of friction $\mathbf{f}$ is exerted by the plane on the block. The frictional force opposes the motion of the block no matter which way the block is moving and we find that the block returns to its starting point with *less* kinetic energy than it had initially. Since we showed in our first experiment that the spring force was conservative, we must attribute this new result to the action of the friction force.* We say that this force, and other forces that act in this same way, are *nonconservative*. The induction force in a betatron (Section 35-6) is also a nonconservative force. Instead of dissipating kinetic energy, however, it generates it, so that an electron moving in the circular betatron orbit will return to its initial position with *more* kinetic energy than it had there originally. In a round trip the electron gains kinetic energy, as it must do if the betatron is to be effective.

We can define conservative force from another point of view, that of the work done by the force on the particle. In our first example above, the work done by the elastic spring force on the block while the spring was being compressed was negative, because the force exerted on the block by the spring (to the left in Fig. 8-1a) was directed opposite to the

figure 8-1
(a) A block of mass m is projected with speed v against a spring. (b) The block is brought to rest by the action of the spring force. (c) The block has regained its initial speed v as it returns to its starting point.

---

* Actually two other forces act on the block in Fig. 8-1, its weight $\mathbf{W}$ and the normal force $\mathbf{N}$ exerted by the plane. Because these act at right angles to the motion, they cannot change the kinetic energy of the block and hence do not enter into this discussion.

displacement of the block (to the right in Fig. 8-1a). While the spring was being extended the work that the spring force did on the block was positive (force and displacement in the same direction). In our first example the net work done on the block by the spring force during a complete cycle, or round trip, is zero.

In our second example we considered the effect of the frictional force. The work done on the block by this force was negative for each portion of the cycle because the frictional force always opposed the motion. Hence the work done by friction in a round trip cannot be zero. In general, then: *A force is conservative if the work done by the force on a particle that moves through any round trip is zero. A force is non-conservative if the work done by the force on a particle that moves through any round trip is not zero.*

The work-energy theorem shows that this second way of defining conservative and nonconservative forces is fully equivalent to our first definition. If there is no change in the kinetic energy of a particle moving through any round trip then $\Delta K = 0$ and, from Eq. 8-1, $W = 0$ and the resultant force acting must be conservative. Similarly, if $\Delta K \neq 0$, then from Eq. 8-1, $W \neq 0$ and at least one of the forces acting must be non-conservative.

We can look into this matter in a little more detail. When friction is present in the system of Fig. 8-1, four forces act on the block, the resultant force being

$$\mathbf{F} = \mathbf{F}_s + \mathbf{W} + \mathbf{N} + \mathbf{f}$$

in which the forces are the spring force $\mathbf{F}_s$, the weight of the block $\mathbf{W}$, the normal force exerted on the block by the plane $\mathbf{N}$, and the frictional force $\mathbf{f}$. We can write Eq. 8-2, the work-energy theorem, as

$$W_s + W_W + W_N + W_f = \Delta K,$$

where the terms on the left are the work done on the block by the four forces above. We have seen that for a round trip $W_s = 0$. Similarly, $W_W = W_N = 0$ because the corresponding forces are at right angles to the displacement of the block. Thus the change in kinetic energy is due entirely to W_f, the work done by the frictional force.

We can consider the difference between conservative and nonconservative forces in still a third way. Suppose a particle goes from a to b along path 1 and back from b to a along path 2 as in Fig. 8-2a. Several forces may act on the particle during this round trip; we consider each force separately. If the force being considered is conservative, the work done on the particle by that particular force for the round trip is zero, or

$$W_{ab,1} + W_{ba,2} = 0,$$

which we can write as

$$W_{ab,1} = -W_{ba,2}.$$

That is, the work in going from a to b along path 1 is the negative of the work in going from b to a along path 2. However, if we cause the particle to go from a to b along path 2, as shown in Fig. 8-2b, we merely reverse the direction of the previous motion along 2, so that

$$W_{ab,2} = -W_{ba,2}.$$

Hence

$$W_{ab,1} = W_{ab,2},$$

which tells us that the work done on the particle by a conservative force in going from a to b is the same for either path.

Paths 1 and 2 can be any paths at all as long as they go from a to b;

(a)   (b)

figure 8-2

and *a* and *b* can be chosen to be any two points at all. We always find the same result if the force is conservative. Hence, we have another equivalent definition of conservative and nonconservative forces: *A force is conservative if the work done by it on a particle that moves between two points depends only on these points and not on the path followed. A force is nonconservative if the work done by that force on a particle that moves between two points depends on the path taken between those points.*

To illustrate this third (equivalent) definition of conservative forces, let us consider a second kind of conservative force, that due to gravity. Suppose that we take a stone of mass *m* in our hand and raise it to a height *h* above the ground, going from *a* to *b* by several different paths as in Fig. 8-3. We already know that in a round trip the total work done by a conservative force is zero and that the gravitational force is conservative. The work done *on* the stone by gravity along the return path *bca* is simply *mgh*. Hence, because gravity is a conservative force, the work done by gravity *on* the stone along any of the paths from *a* to *b* must be −*mgh*, for only if this is true can the total work done by gravity in a round trip be zero. This means that gravity does negative work on the stone as it moves from *a* to *b*, or, to put it another way, work must be done *against* gravity along any of the paths *ab*. You can compute directly the result that the work done by gravity along any path from *a* to *b* equals −*mgh*. For any of these paths can be decomposed into infinitesimal displacements which are alternately horizontal and vertical; no work is done by gravity in horizontal displacements, and the net vertical displacement is the same in all cases. Hence the work done by gravity on the stone moving from *a* to *b* depends only on the positions of *a* and *b* and not at all on the path taken.

For a nonconservative force, such as friction, the work done is *not* independent of the path taken between two fixed points. We need only point out that as we push a block over a (rough) table between any two points *a* and *b* by various paths, the distance traversed varies and so does the work done by the frictional force. It depends on the path.

The definitions of conservative force which we have given are equivalent to one another. Which one we use depends only on convenience. The round-trip approach shows clearly that kinetic energy is conserved when conservative forces act. If we wish to develop the idea of potential energy, however, the path independence statement is preferable.

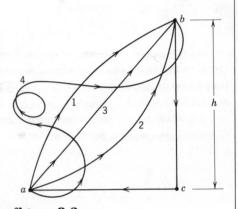

figure 8-3
A stone is raised from *a* to *b* via various paths 1, 2, 3, and 4.

8-3
POTENTIAL ENERGY

In this section we shall focus attention not on the moving block of Fig. 8-1 but rather on the isolated system (block + spring). Instead of saying that the block is moving we prefer, from this point of view, to say that the configuration of the system is changing. We measure both the position of the block and the configuration of the system at any instant by the same parameter *x*, namely, the displacement of the free end of the spring from its normal position, corresponding to an unstretched spring. The kinetic energy of the system is the same as that of the block because we have assumed the spring to be massless.

We have seen that the kinetic energy of the system of Fig. 8-1 decreases during the first half of the motion, becomes zero, and then increases during the second half of the motion. If there is no friction, the kinetic energy of the system when it has regained its initial configuration returns to its initial value.

Under these circumstances (conservative forces acting) it makes sense to introduce the concept of *energy of configuration*, or *potential energy U*, and to say that if K for the system changes by ΔK as the configuration of the system changes (that is, as the block moves in the system of Fig. 8-1), then U for the system must change by an equal but opposite amount so that the sum of the two changes is zero, or

$$\Delta K + \Delta U = 0. \qquad (8\text{-}4a)$$

Alternatively, we can say that any change in kinetic energy K of the system is compensated for by an equal but opposite change in the potential energy U of the system so that their sum remains constant throughout the motion, or

$$K + U = \text{a constant.} \qquad (8\text{-}4b)$$

The potential energy of a system represents a form of stored energy which can be fully recovered and converted into kinetic energy. We cannot associate a potential energy with a nonconservative force such as the force of friction because the kinetic energy of a system in which such forces act does *not* return to its initial value when the system returns to its initial configuration.

Equations 8-4 apply to a closed system of interacting objects, such as the (mass + spring) system of Fig. 8-1. In this example, because we have taken the spring to be effectively massless, the kinetic energy may be associated with the moving mass alone. The block slows down (or speeds up) because a force is exerted on it *by the spring*; it is appropriate, then, to associate the potential energy of the system with this force, that is to say, with the spring. Thus in this simple case we say that kinetic energy, localized in the mass, decreases during the first part of the motion whereas potential energy, localized in the spring, increases during this same time.*

Equations 8-4 are essentially bookkeeping statements about energy. They, and the concept of potential energy, have no real meaning, however, until we have shown how to calculate U as a function of the configuration of the system within which the conservative forces act; in the example of Fig. 8-1 this means that we must be able to calculate $U(x)$, where x is the spring displacement.

To refine our concept of potential energy U let us consider the work-energy theorem, $W = \Delta K$, in which W is the work done by the resultant force on a particle as it moves from a to b. For simplicity let us assume that only a single force $\mathbf{F}$ acts on the particle; this is effectively true in the system of Fig. 8-1. If $\mathbf{F}$ is conservative, we can combine the work-energy theorem (Eq. 8-1) with Eq. 8-4a, obtaining

$$W = \Delta K = -\Delta U. \qquad (8\text{-}5a)$$

The work W done by a conservative force depends only on the starting and the end points of the motion and not on the path followed between them. Such a force can depend only on the position of a particle; it does not depend on the velocity of the particle or on the time, for example.

For motion in one dimension, Eq. 8-5a becomes

* Just as we assumed the spring to be effectively massless we also assume the block to be rigid, that is, effectively "springless." In a more general system, kinetic and potential energy could each be present in various parts of the system, in varying proportions as the system configuration changes.

$$\Delta U = -W = -\int_{x_0}^{x} F(x)\ dx, \qquad (8\text{-}5b)$$

the particle moving from x_0 to x. Equation 8-5b shows how to calculate the change in potential energy ΔU when a particle, acted on by a conservative force $F(x)$, moves from point a, described by x_0, to point b, described by x. The equation shows that we can only calculate ΔU if the force **F** depends only on the position of the particle (that is, on the system configuration), which is equivalent to saying that potential energy has meaning only for conservative forces.

Now that we know that the potential energy U depends on the position of the particle only, we can write Eq. 8-4b as

$$\tfrac{1}{2}mv^2 + U(x) = E \qquad \text{(one-dimension)} \qquad (8\text{-}6a)$$

in which E, which remains constant as the particle moves, is called the *total mechanical energy*. Suppose that the particle moves from point a (where its position is x_0 and its speed is v_0) to point b (where its position is x and its speed is v); the total mechanical energy E must be the same for each system configuration when the force is conservative, or, from Eq. 8-6a,

$$\tfrac{1}{2}mv^2 + U(x) = \tfrac{1}{2}mv_0^2 + U(x_0). \qquad (8\text{-}6b)$$

The quantity on the right depends only on the initial position x_0 and the initial speed v_0, which have definite values; it is, therefore, *constant during the motion*. This is the constant total mechanical energy E. Notice that force and acceleration do not appear in this equation, only position and speed. Equations 8-6 are often called the *law of conservation of mechanical energy* for conservative forces.

In many problems we find that although some of the individual forces are not conservative, they are so small that we can neglect them. In such cases we can use Eqs. 8-6 as a good approximation. For example, air resistance may be present but may have so little effect on the motion that we can ignore it.

Notice that, instead of starting with Newton's laws, we can simplify problem solving when conservative forces alone are involved by starting with Eqs. 8-6. This relation is derived from Newton's laws, of course, but it is one step closer to the solution (the so-called first integral of the motion). We often solve problems without analyzing the forces or writing down Newton's laws by looking instead for something in the motion that is constant; here the mechanical energy is constant and we can write down Eqs. 8-6 as the first step.

For one-dimensional motion we can also write the relation between force and potential energy (Eq. 8-5b) as

$$F(x) = -\frac{dU(x)}{dx}. \qquad (8\text{-}7)$$

To show this, substitute this expression for $F(x)$ into Eq. 8-5b and observe that you get an identity. Equation 8-7 gives us another way of looking at potential energy. *The potential energy is a function of position whose negative derivative gives the force.*

You may have noticed that we wrote down the quantity $U(x)$ in Eqs. 8-6 although we are only able to calculate *changes* in U (from Eq. 8-5b) and not U itself. Let us imagine that a particle moves from a to b along the x-axis and that a single conservative force $F(x)$ acts on it. To assign a value to U_b, the potential energy at point b, let us write

$$\Delta U = U_b - U_a,$$

or (see Eq. 8-5b),

$$U_b = \Delta U + U_a = -\int_{x_a}^{x_b} F(x)\ dx + U_a. \qquad (8\text{-}8)$$

We cannot assign a value to U_b until we have assigned one to U_a. If point b is any arbitrary position x, so that $U_b = U(x)$, we give meaning to $U(x)$ by choosing point a to be some convenient reference position, described by $x_a = x_0$, and by arbitrarily assigning a value to the potential energy $U_a = U(x_0)$ when the body is at that point. Thus Eq. 8-8 becomes

$$U(x) = -\int_{x_0}^{x} F(x)\ dx + U(x_0). \qquad (8\text{-}9)$$

The potential energy when the body is at the reference position, that is, $U(x_0)$, is usually given the arbitrary value zero.

It is often convenient to choose the reference position x_0 to be that at which the force acting on the particle is zero. Thus the force exerted by a spring is zero when the spring has its normal unstretched length; we usually say that the potential energy is also zero for this condition. Also, the attraction of the earth on a body decreases as the body moves away from the earth, becoming zero at an infinite distance. We usually take infinity as our reference position and assign the value zero to the potential energy associated with the gravitational force at that position (see Chapter 16). So far, however, we have been more concerned with the gravitational pull on bodies such as baseballs, etc., which, in comparison to the earth's radius, never move very far from the earth's surface. Here the gravitational force $(= m\mathbf{g})$ is essentially constant and we find it convenient to take the zero of potential energy to be, not at infinity, but at the surface of the earth.

The effect of changing the coordinate of the standard reference position x_0, or of the arbitrary value assigned to $U(x_0)$, is simply to change the value of $U(x)$ by an added constant. The presence of an arbitrary added constant in the potential energy expression (Eq. 8-9) makes no difference to the equations that we have written so far. This simply adds the same constant term to each side of Eq. 8-6b, for example, leaving that equation unchanged. Furthermore, changing $U(x)$ by an added constant does not change the force calculated from Eq. 8-7 because the derivative of a constant is zero. All this simply means that the choice of a reference point for potential energy is immaterial because we are always concerned with *differences* in potential energy, rather than with any absolute value of potential energy at a given point.

There is a certain arbitrariness in specifying kinetic energy also. In order to determine speed, and hence kinetic energy, we must specify a reference frame. The speed of a man sitting on a train is zero if we take the train as a reference frame, but it is not zero to an observer on the ground who sees the man move by with uniform velocity. The value of the kinetic energy depends on the reference frame used by the observer. Hence the important thing about mechanical energy E, which is the sum of the kinetic and the potential energies, is *not* its actual value during a given motion (this depends on the observer) but the fact that this value *does not change* during the motion for any particular observer when the forces are conservative.

Let us now calculate the potential energy in one-dimensional motion for two examples of conservative forces, the force of gravity for motions near the earth's surface and the elastic restoring force of an (ideal) stretched spring.

For the force of gravity we take the one-dimensional motion to be vertical, along the y-axis. We take the positive direction of the y-axis to be upward; the force of gravity is then in the negative y-direction, or downward. We have $F(y) = -mg$, a constant. The potential energy at position y is found from Eq. 8-9, or

$$U(y) = -\int_0^y F(y)\ dy + U(0) = -\int_0^y (-mg)\ dy + U(0) = mgy + U(0).$$

The potential energy can be taken as zero where $y = 0$, so that $U(0) = 0$, and

$$U(y) = mgy. \tag{8-10}$$

The gravitational potential energy is then mgy. The relation $F(y) = -dU/dy$ (Eq. 8-7) is satisfied, for $-d(mgy)/dy = -mg$. We choose $y = 0$ to be at the surface of the earth for convenience, so that the gravitational potential energy is zero at the earth's surface and increases linearly with altitude y.

If we compare points y and $y = 0$, the conservation of kinetic plus potential energy, Eq. 8-6b, gives us the relation

$$\tfrac{1}{2}mv^2 + mgy = \tfrac{1}{2}mv_0{}^2.$$

This is mathematically equivalent to the well-known result (see Eq. 3-17),

$$v^2 = v_0{}^2 - 2gy.$$

If our particle moves from a height h_1 to a height h_2, we can use Eq. 8-6b to obtain

$$\tfrac{1}{2}mv_1{}^2 + mgh_1 = \tfrac{1}{2}mv_2{}^2 + mgh_2.$$

This result is equivalent to that of Example 5, Chapter 7. The total mechanical energy E is constant and is conserved during the motion, even though the kinetic energy and the potential energy vary as the configuration of the system (particle + earth) changes.

A second example of a conservative force is that exerted by an elastic spring on a body of mass m attached to it moving on a horizontal frictionless surface. If we take $x_0 = 0$ as the position of the end of the spring when unextended, the force exerted *on* the mass when the spring is stretched a distance x from its unextended length is $F = -kx$. The potential energy is obtained from Eq. 8-9,

$$U(x) = -\int_0^x F(x)\ dx + U(0) = -\int_0^x (-kx)\ dx + U(0).$$

If we choose $U(0) = 0$, the potential energy, as well as the force, is zero when the spring is unextended, and

$$U(x) = -\int_0^x (-kx)\ dx = \tfrac{1}{2}kx^2.$$

The result is the same whether we stretch or compress the spring, that is, whether x is plus or minus.

The relation $F(x) = -dU/dx$ (Eq. 8-7) is satisfied, for $-d(\tfrac{1}{2}kx^2)/dx = -kx$. The elastic potential energy of the spring is then

$$U(x) = \tfrac{1}{2}kx^2. \tag{8-11}$$

The body of mass m will undergo a motion in which the total energy E is conserved (Fig. 8-4). From Eq. 8-6b we have

$$\tfrac{1}{2}mv^2 + \tfrac{1}{2}kx^2 = \tfrac{1}{2}mv_0^2.$$

figure 8-4
A mass attached to a spring slides back and forth on a frictionless surface. The system is called a harmonic oscillator. The motion of the mass through one cycle is illustrated. Starting at the left (9 o'clock) the mass is in its extreme left position and momentarily at rest: $K = 0$. The spring is extended to its maximum length: $U = U_{\text{max}}$. (K and U are illustrated in the bar graphs below each sketch.) An eighth-cycle later (next drawing) the mass has gained kinetic energy, but the spring is no longer so elongated; K and U have here the same value, $K = U = U_{\text{max}}/2$. At the top this spring is neither elongated nor compressed and the speed is a maximum: $U = 0$, $K = K_{\text{max}} = U_{\text{max}}$. The cycle continues, with the total energy $E = K + U$ always the same: $E = K_{\text{max}} = U_{\text{max}}$. The harmonic oscillator will be analyzed more closely in Chapter 15.

Here v_0 is the speed of the particle for $x = 0$. Physically we achieve such a result by stretching the spring with an applied force to some position, x_m, and then releasing the spring. Notice that at $x = 0$ the energy of the system (particle + spring) is all kinetic. At $x = x_m$ (the maximum value of x), v must be zero, so that here the system energy is all potential. At $x = x_m$, we have

$$\tfrac{1}{2}kx_m^2 = \tfrac{1}{2}mv_0^2$$

or

$$x_m = \sqrt{m/k}\ v_0.$$

For positions between x_1 and x_2, Eq. 8-6b gives

$$\tfrac{1}{2}kx_1^2 + \tfrac{1}{2}mv_1^2 = \tfrac{1}{2}kx_2^2 + \tfrac{1}{2}mv_2^2.$$

We have seen that *the kinetic energy of a body is the work that a body can do by virtue of its motion*. We express the kinetic energy by the formula $K = \tfrac{1}{2}mv^2$. We cannot give a similar universal formula by which potential energy can be expressed. *The potential energy of a system of bodies is the work that the system of bodies can do by virtue of the relative position of its parts, that is, by virtue of its configuration.* In each case we must determine how much work the system can do in passing from one configuration to another and then take this as the difference in potential energy of the system between these two configurations.

The potential energy of the spring depends on the relative position of the parts of the spring. Work can be obtained by allowing the spring to return from its extended to its unextended length, during which time it exerts a force through a distance. If a mass is attached to the spring, as in our example, the mass will be accelerated by this force and the potential energy will be converted to kinetic energy. In the gravitational case an object occupies a position with respect to the earth. The potential energy is a property of the object and the earth, considered as a system of bodies. It is the relative position of the parts of this system that determines its potential energy. The potential energy is greater when the parts are far apart than when they are close together. The loss of potential energy is equal to the work done in this process. This work is converted into kinetic energy of the bodies. In our example we ignored the kinetic energy acquired by the earth itself as an object fell toward it. In principle, this object exerts a force on the earth and causes it to acquire an acceleration, relative to some inertial frame. The resulting change in speed, however, is extremely small, and in spite of the enormous mass of the earth, its additional kinetic energy is negligible compared to that acquired by the falling object. This will be proved in a later chapter. In other cases, such as in planetary motion where the masses of the objects in our system may be comparable, we cannot ignore any part of the system. In general, *potential energy* is not assigned to either body separately but *is considered a joint property of the system.*

EXAMPLE 1

What is the change in gravitational potential energy when a 1600-lb (= 7117 N) elevator moves from street level to the top of the Empire State Building, 1250 ft (= 381 m) above street level?

The gravitational potential energy of the system (elevator + earth) is $U = mgy.$ Then

$$\Delta U = U_2 - U_1 = mg(y_2 - y_1).$$

But

$$mg = W = 1600 \text{ lb} \quad \text{and} \quad y_2 - y_1 = 1250 \text{ ft},$$

so that

$$\Delta U = 1600 \times 1250 \text{ ft} \cdot \text{lb} = 2.00 \times 10^6 \text{ ft} \cdot \text{lb} = 2.71 \times 10^6 \text{ J}.$$

EXAMPLE 2

As an example of the simplicity and usefulness of the energy method of solving dynamical problems, consider the problem illustrated in Fig. 8-5. A block of mass m slides down a curved frictionless surface. The force exerted by the surface on the block is always normal to the surface and to the direction of the

figure 8.5
Example 2. A block sliding down a frictionless curved surface.

motion of the block, so that this force does no work. Only the gravitational force does work on the block and that force is conservative. The mechanical energy E is, therefore, conserved and we can write at once

$$\tfrac{1}{2}mv_1^2 + mgy_1 = \tfrac{1}{2}mv_2^2 + mgy_2.$$

This gives

$$v_2^2 = v_1^2 + 2g(y_1 - y_2).$$

The speed at the bottom of the curved surface depends only on the initial speed and the change in vertical height but does not depend at all on the shape of the surface. In fact, if the block is initially at rest at $y_1 = h$, and if we set $y_2 = 0$, we obtain

$$v_2 = \sqrt{2gh}.$$

At this point you should recall the independence of path feature of work done by conservative forces and should be able to justify applying the ideas developed for one-dimensional motion to this two-dimensional example.

In this problem the value of the force depends on the slope of the surface at each point. Hence, the acceleration is not constant but is a function of position. To obtain the speed by starting with Newton's laws we would first have to determine the acceleration at each point and then integrate the acceleration over the path. We avoid all this labor by starting at once from the fact that the mechanical energy is constant throughout the motion.

The spring in a spring gun has a force constant of 4.0 lb/in. ($= 7.0$ N/cm). It is compressed 2.0 in. ($= 5.1$ cm) from its natural length, and a ball weighing 0.030 lb ($= 0.133$ N) is put into the barrel against it. Assuming no friction and a horizontal gun barrel, with what speed will the ball leave the gun when released?

The force is conservative so that mechanical energy is conserved in the process. The initial mechanical energy is the elastic potential energy of the spring, $\tfrac{1}{2}kx^2$, and the final mechanical energy is the kinetic energy of the ball, $\tfrac{1}{2}mv^2$. Hence,

$$\tfrac{1}{2}kx^2 = \tfrac{1}{2}mv^2$$

or

$$v = \sqrt{\frac{k}{m}}\,x = \sqrt{\frac{48\ \text{lb/ft}}{(0.030\ \text{lb})/(32\ \text{ft/s}^2)}}\ (\tfrac{1}{6}\ \text{ft}) = 38\ \text{ft/s}\ (= 11.6\ \text{m/s}).$$

EXAMPLE 3

Equation 8-6a gives the relation between coordinate and speed for one-dimensional motion when the force depends on position only. The force and the acceleration have been eliminated in arriving at this equation. To complete the solution of the dynamical problem we must eliminate the speed and determine position as a function of time.

We can do this in a formal way, as follows. From Eq. 8-6a we have

$$\tfrac{1}{2}mv^2 + U(x) = E.$$

Solving for v, we obtain

$$v = \frac{dx}{dt} = \sqrt{\frac{2}{m}\,[E - U(x)]}, \qquad (8\text{-}12)$$

or

$$\frac{dx}{\sqrt{\dfrac{2}{m}\,[E - U(x)]}} = dt.$$

Then the function $x(t)$ may be found by solving for x the equation

8-5
THE COMPLETE SOLUTION OF THE PROBLEM FOR ONE-DIMENSIONAL FORCES DEPENDING ON POSITION ONLY

$$\int_{x_0}^{x} \frac{dx}{\sqrt{\dfrac{2}{m}\left[E - U(x)\right]}} = \int_{t_0}^{t} dt = t - t_0. \qquad (8\text{-}13)$$

Here the particle is taken to be at x_0 at the time t_0 and E is the constant total energy. In applying this equation, the sign of the square root taken corresponds to whether **v** points in the positive or in the negative x-direction. When **v** changes direction during the motion it may be necessary to carry out the integration separately for each part of the motion.

Even when this integral cannot be evaluated or when the resulting equation cannot be solved to give an explicit solution for $x(t)$, the equation of energy conservation gives us useful information about the solution. For example, for a given total energy E, Eq. 8-12 tells us that the particle is restricted to those regions on the x-axis where $E > U(x)$. We cannot have an imaginary speed or a negative kinetic energy physically, so that $E - U(x)$ must be zero or greater. Furthermore, we can obtain a good qualitative description of the types of motion possible by plotting $U(x)$ versus x. This description depends on the fact that the speed is proportional to the square root of the difference between E and U.

For example, consider the potential energy function shown in Fig. 8-6. This could be thought of as an actual profile of a frictionless roller coaster, but in general it can represent the potential energy of a nongravitational system. Since we must have $E \geq U(x)$ for real motion, the lowest total energy possible is E_0. At this value of the total energy, $E_0 = U$ and the kinetic energy must be zero. The particle must be at rest at the point x_0. At a slightly higher energy E_1 the particle can move between x_1 and x_2 only. As it moves from x_0 its speed decreases on approaching either x_1 or x_2. At x_1 or x_2 the particle stops and reverses its direction. These points x_1 and x_2 are, therefore, called *turning points* of the motion. At a total energy E_2 there are four turning points, and the particle can oscillate in either one of the two potential valleys. At the total energy E_3 there is only one turning point of the motion, at x_3. If the particle is initially moving in the negative x-direction, it will stop at x_3 and then move in the positive x-direction. It will speed up as U decreases and slow down as U increases. At energies above E_4 there are no turning points, and the particle will not reverse direction. Its speed will change according to the value of the potential at each point.

At a point where $U(x)$ has a minimum value, such as at $x = x_0$, the slope of the curve is zero so that the force is zero, that is, $F(x_0) = -(dU/dx)_{x=x_0} = 0$. A particle at rest at this point will remain at rest. Furthermore, if the particle is displaced slightly in either direction, the force, $F(x) = -dU/dx$, will tend to return it, and it will oscillate about the equilibrium point. This equilibrium point is, therefore, called a point of *stable equilibrium*.

At a point where $U(x)$ has a maximum value, such as at $x = x_4$, the slope of the curve is zero so that the force is again zero, that is, $F(x_4) = -(dU/dx)_{x=x_4} = 0$. A particle at rest at this point will remain at rest. However, if the particle is dis-

figure 8-6
A potential energy curve.

placed even the slightest distance from this point, the force, $F(x) = -dU/dx$, will tend to push it farther away from the equilibrium position. Such an equilibrium point is, therefore, called a point of *unstable equilibrium*.

In an interval in which $U(x)$ is constant, such as near $x = x_5$, the slope of the curve is zero so that the force is zero, that is, $F(x_5) = -(dU/dx)_{x=x_5} = 0$. Such an interval is called one of *neutral equilibrium*, since a particle can be displaced slightly without experiencing either a repelling or a restoring force.

From this it is clear that if we know the potential energy function for the region of x in which the body moves, we know a great deal about the motion of the body.

EXAMPLE 4

The potential energy function for the force between two atoms in a diatomic molecule can be expressed approximately as follows:

$$U(x) = \frac{a}{x^{12}} - \frac{b}{x^6}$$

where a and b are positive constants and x is the distance between atoms.

(a) At what values of x is $U(x)$ equal to zero? At what value of x is $U(x)$ a minimum?

In Fig. 8-7a we show $U(x)$ versus x. The values of x at which $U(x)$ equals zero are found from

$$\frac{a}{x^{12}} - \frac{b}{x^6} = 0.$$

Hence

$$x^6 = \frac{a}{b} \qquad x = \sqrt[6]{\frac{a}{b}}.$$

$U(x)$ also becomes zero as $x \to \infty$ [see figure or put $x = \infty$ into equation for $U(x)$], so that $x = \infty$ is also a solution.

The value of x at which $U(x)$ is a minimum is found from

$$\frac{d}{dx} U(x) = 0.$$

That is,

$$\frac{-12a}{x^{13}} + \frac{6b}{x^7} = 0$$

or

$$x^6 = \frac{2a}{b} \qquad x = \sqrt[6]{\frac{2a}{b}}.$$

(b) Determine the force between the atoms.

From Eq. 8-7

$$F(x) = -\frac{d}{dx} U(x),$$

$$F = \frac{-d}{dx}\left(\frac{a}{x^{12}} - \frac{b}{x^6}\right) = \frac{12a}{x^{13}} - \frac{6b}{x^7}.$$

We plot the force as a function of the separation between the atoms in Fig. 8-7b. When the force is positive (from $x = 0$ to $x = \sqrt[6]{2a/b}$), the atoms are repelled from one another (force directed toward increasing x). When the force is negative (from $x = \sqrt[6]{2a/b}$ to $x = \infty$), the atoms are attracted to one another (force directed toward decreasing x). At $x = \sqrt[6]{2a/b}$ the force is zero; this is the equilibrium point and is a point of stable equilibrium.

(c) Assume that one of the atoms remains at rest and that the other moves along x. Describe the possible motions.

From the analysis of this section it is clear that the atom oscillates about the equilibrium separation at $x = \sqrt[6]{2a/b}$, much as a particle sliding up and down the frictionless hills of the potential valley.

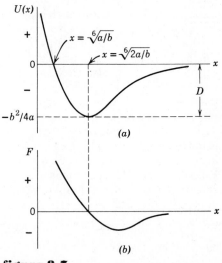

figure 8-7

Example 4. *(a)* The potential energy and *(b)* the force between two atoms in a diatomic molecule as a function of the distance x between atoms.

(d) The energy needed to break up the molecule into separate atoms is called the dissociation energy. What is the dissociation energy of the molecule?

If one atom has enough kinetic energy to get over the potential hill, it will no longer be bound to the other atom. Hence, the dissociation energy D equals the change in potential energy from the minimum value at $x = \sqrt[6]{2a/b}$ to the value at $x = \infty$. This is simply

$$U(x = \infty) - U\left(x = \sqrt[6]{\frac{2a}{b}}\right) = 0 - \left(\frac{a}{4a^2/b^2} - \frac{b}{2a/b}\right) = \frac{b^2}{4a}.$$

If the kinetic energy at the equilibrium position is equal to or greater than this value, the molecule will dissociate.

8-6
TWO- AND THREE-DIMENSIONAL CONSERVATIVE SYSTEMS

So far we have discussed potential energy and energy conservation for one-dimensional systems in which the force was directed along the line of motion. We can easily generalize the discussion to three-dimensional motion.

If the work done by the force **F** depends only on the end points of the motion and is independent of the path taken between these points, the force is conservative. We define the potential energy U by analogy with the one-dimensional system and find that it is a function of three space coordinates, that is, $U = U(x,y,z)$. Again we obtain an expression for the conservation of mechanical energy.

The generalization of Eq. 8-5b to motion in three dimensions is

$$\Delta U = -\int_{x_0}^{x} F_x \, dx - \int_{y_0}^{y} F_y \, dy - \int_{z_0}^{z} F_z \, dz \qquad (8\text{-}5c)$$

or, more compactly in vector notation,

$$\Delta U = -\int_{\mathbf{r}_0}^{\mathbf{r}} \mathbf{F}(\mathbf{r}) \cdot d\mathbf{r} \qquad (8\text{-}5d)$$

in which ΔU is the change in potential energy for the system as the particle moves from the point (x_0, y_0, z_0), described by the position vector $\mathbf{r}_0$, to the point (x,y,z), described by the position vector $\mathbf{r}$. F_x, F_y, and F_z are the components of the conservative force $\mathbf{F}(\mathbf{r}) = \mathbf{F}(x,y,z)$.

The generalization of Eq. 8-6b to three-dimensional motion is

$$\tfrac{1}{2}mv^2 + U(x,y,z) = \tfrac{1}{2}mv_0^2 + U(x_0, y_0, z_0) \qquad (8\text{-}6c)$$

which can be written in vector notation as

$$\tfrac{1}{2}m\mathbf{v} \cdot \mathbf{v} + U(\mathbf{r}) = \tfrac{1}{2}m\mathbf{v}_0 \cdot \mathbf{v}_0 + U(\mathbf{r}_0) \qquad (8\text{-}6d)$$

in which $\mathbf{v} \cdot \mathbf{v} = v_x^2 + v_y^2 + v_z^2 = v^2$ and $\mathbf{v}_0 \cdot \mathbf{v}_0 = v_{0x}^2 + v_{0y}^2 + v_{0z}^2 = v_0^2$. Likewise Eq. 8-6a becomes

$$\tfrac{1}{2}mv^2 + U(x,y,z) = E$$

in three dimensions, E being the constant total mechanical energy.

Finally, the generalization of Eq. 8-7 to three dimensions is

$$\mathbf{F}(\mathbf{r}) = -\mathbf{i}\frac{\partial U}{\partial x} - \mathbf{j}\frac{\partial U}{\partial y} - \mathbf{k}\frac{\partial U}{\partial z}.$$

If we substitute this expression for **F** into Eq. 8-5d, we again obtain an identity. In vector language the conservative force **F** is said to be the negative of the *gradient* of the potential energy $U(x,y,z)$.

You can show that all these expressions reduce to the correct one-dimensional equations for motion along the x-axis.

EXAMPLE 5

Consider the single pendulum, Section 7-4, Fig. 7-8a. The motion of the system is in the x-y plane, that is, it is a two-dimensional motion. The tension in the cord is always at right angles to the motion of the suspended particle, so that this force does no work on the particle. If the pendulum is displaced through some angle and is then released, only the gravitational force of attraction exerted on the particle by the earth does work on it. Since this force is conservative, we can use the equation of energy conservation in two dimensions,

$$\tfrac{1}{2}mv^2 + U(x,y) = E.$$

But $U(x,y)$ equals mgy, where y is taken as zero at the lowest point of the arc $(\phi = 0°)$. Then,

$$\tfrac{1}{2}mv^2 + mgy = E.$$

The particle is pulled through an angle ϕ_0 before being released. The potential energy there is mgh, corresponding to a height $y = h$ above the reference point. At the release point $(\phi = \phi_0)$ the speed and the kinetic energy are zero so that the potential energy equals the total mechanical energy at that point.

Hence,

$$E = mgh$$

and

$$\tfrac{1}{2}mv^2 + mgy = mgh,$$

or

$$\tfrac{1}{2}mv^2 = mg(h - y).$$

The maximum speed occurs at $y = 0$, where $v = \sqrt{2gh}$.

The minimum speed occurs at $y = h$, where $v = 0$.

At $y = 0$ the energy is entirely kinetic, the potential energy being zero.

At $y = h$ the energy is entirely potential, the kinetic energy being zero.

At intermediate positions the energy is partly kinetic and partly potential.

Notice that $U \leq E$ at all points of the motion; the pendulum cannot rise higher than $y = h$, its initial release point.

8-7 NONCONSERVATIVE FORCES

So far we have considered only the action of a single conservative force on a particle. Starting from the work-energy theorem, or

$$W_1 + W_2 + \cdots + W_n = \Delta K \qquad (8\text{-}2)$$

we saw that, if only one force, say $\mathbf{F}_1$, was acting and if it was conservative, then we could represent the work W_1 that it did on the particle as a decrease in potential energy ΔU_1 of the system (see Eq. 8-5a), or

$$W_1 = -\Delta U_1.$$

Combining this with Eq. 8-2 yielded

$$\Delta K + \Delta U_1 = 0.$$

If several conservative forces such as gravity, an elastic spring force, an electrostatic force, etc., are acting, we can easily extend these two equations to

$$\Sigma\, W_c = -\Sigma\, \Delta U \qquad (8\text{-}14a)$$

and

$$\Delta K + \Sigma\, \Delta U = 0 \qquad (8\text{-}14b)$$

in which $\Sigma\, W_c$ is the sum of the work done by the various (conservative) forces and the ΔU's are the changes in the potential energy of the system associated with these forces. The quantity on the left of Eq. 8-14b is simply ΔE, the change in the total mechanical energy, for the case in which several conservative forces are acting on a particle. We can write this equation then as

$$\Delta E = 0 \qquad \text{(conservative forces)}, \qquad (8\text{-}15)$$

which tells us that, as the system configuration changes the total mechanical energy E for the system remains constant.

Let us now suppose that, in addition to the several conservative forces, a single nonconservative force due to friction acts on the particle. We can then write Eq. 8-2 as

$$W_f + \Sigma \ W_c = \Delta K,$$

where $\Sigma \ W_c$ is again the sum of the work done by the conservative forces and W_f is the work done by friction. We can recast this (see Eq. 8-14a) as

$$\Delta K + \Sigma \ \Delta U = W_f. \qquad (8\text{-}16)$$

Equation 8-16 shows that, if a frictional force acts, the total mechanical energy is *not* constant, but changes by the amount of work done by the frictional force. We can write Eq. 8-16 as

$$\Delta E = E - E_0 = W_f. \qquad (8\text{-}17)$$

Since W_f, the work done by friction *on* the particle, is always negative, it follows from Eq. 8-17 that the final mechanical energy $E \ (= K + \Sigma \ U)$ is less than the initial mechanical energy $E_0 \ (= K_0 + \Sigma \ U_0)$.

Friction is an example of a dissipative force, one which does negative work on a body and tends to diminish the total mechanical energy of the system. If we had used another nonconservative force, then W_f in Eqs. 8-16 and 8-17 would be replaced by a term W_{nc}, showing again that the total mechanical energy E of the system is *not* constant, but changes by the amount of work done by the nonconservative force. Hence, *only when there are no nonconservative forces, or when we neglect the work they do, can we assume conservation of mechanical energy.*

What happened to the "lost" mechanical energy in the case of friction? It is transformed into internal energy U_{int}, resulting in a temperature rise. The internal energy developed is exactly equal to the mechanical energy dissipated. We shall have much more to say about internal energy in later chapters.

Just as the work done by a conservative force *on* an object is the negative of the potential energy gain, so the work done by a frictional force *on* an object is the negative of the internal energy gained. In other words, the internal energy produced is equal to the work done *by* the object. Then we can replace W_f in Eq. 8-17 by $-U_{int}$, in which U_{int} is the internal energy produced, or

$$\Delta E + U_{int} = 0. \qquad (8\text{-}18)$$

This asserts that there is no change in the sum of the mechanical and internal energy of the system when only conservative and frictional forces act on the system. Writing this equation as $U_{int} = -\Delta E$ we see that the loss of mechanical energy equals the gain in internal energy.

EXAMPLE 6

An object with an initial velocity v_0 of 14 m/s falls from a height of 240 m and buries itself in 0.20 m of sand. The mass of the body is 1.0 kg. Find the average resistive force exerted by the sand on the body. Neglect air resistance and solve the problem by considerations of work and energy.

The kinetic energy of the body just as it enters the sand is

$$K = \tfrac{1}{2}mv_0^2 + mgh,$$

where m is the mass of the body and h is the height of fall.

Also, from the work-energy principle, we have (approximately)

$$K = \overline{F} s,$$

where $\overline{F}$ is the average resistive force and s is the distance of penetration into the ground.

Equating and solving for $\overline{F}$ gives

$$\overline{F} = \frac{mv_0^2}{2\,s} + \frac{mgh}{s}$$

$$= \frac{(1.0 \text{ kg})(14 \text{ m/s})^2}{2(0.20 \text{ m})} + \frac{(1.0 \text{ kg})(9.8 \text{ m/s}^2)(240 \text{ m})}{(0.20 \text{ m})}$$

$$= 12,250 \text{ N.}$$

For what equations in this chapter are the first two equations of this example special cases?

What error do we make by neglecting (in comparison to h) the extra distance of fall s before the object is brought to rest? Show that this is equivalent to neglecting mg in comparison to $\overline{F}$ in arriving at the resultant force to be used in the work-energy theorem. Such terms are not always negligible in practice (see Problem 19, for example).

EXAMPLE 7

A 44-N block is thrust up a 30° inclined plane with an initial speed of 5.0 m/s. It is found to travel 1.5 m along the plane, stop, and slide back to the bottom. Compute the force of friction **f** (assumed to have a constant magnitude) acting on the block and find the speed v of the block when it returns to the bottom of the inclined plane.

Consider first the upward motion. At the top, where this motion ends,

$$E = K + U = 0 + (44 \text{ N})(1.5 \text{ m})(\sin 30°) = 33 \text{ J.}$$

At the bottom, where this motion begins,

$$E_0 = K_0 + U_0 = \tfrac{1}{2}\left(\frac{44 \text{ N}}{9.8 \text{ m/s}^2}\right)(5.0 \text{ m/s})^2 + 0 = 57 \text{ J.}$$

But

$$W_f = -fs = -f(1.5 \text{ m})$$

and

$$E - E_0 = W_f,$$

so that

$$33 \text{ J} - 57 \text{ J} = -f(1.5 \text{ m})$$

and

$$f = 16 \text{ N.}$$

Now consider the downward motion. The block returns to the bottom of the inclined plane with a speed v. Then, at the bottom, where this motion ends,

$$E = K + U = \tfrac{1}{2}\left(\frac{44 \text{ N}}{9.8 \text{ m/s}^2}\right)v^2 + 0 = \left(\frac{22}{9.8} \text{ N s}^2/\text{m}\right)v^2.$$

At the top, where this motion begins,

$$E_0 = K_0 + U_0 = 0 + (44 \text{ N})(1.5 \text{ m})(\sin 30°) = 33 \text{ J.}$$

But

$$W_f = -(16 \text{ N})(1.5 \text{ m}) = -24 \text{ J}$$

and

$$E - E_0 = W_f,$$

so that

$$\left(\frac{22}{9.8}\ \text{N s}^2/\text{m}\right)v^2 - 33\ \text{J} = -24\ \text{J}$$

and

$$v = 2.0\ \text{m/s}.$$

We can extend the discussion of the previous section by considering not only conservative forces and the force of friction but also other, nonfrictional, nonconservative forces. We can regroup the work-energy theorem (Eq. 8-2)

$$W_1 + W_2 + \cdots + W_n = \Delta K$$

as

$$\Sigma\ W_c + W_f + \Sigma\ W_{nc} = \Delta K \qquad\qquad (8\text{-}19)$$

in which $\Sigma\ W_c$ is the total work done on the particle by conservative forces, W_f is the work done by friction, and $\Sigma\ W_{nc}$ is the total work done by nonconservative forces other than friction. We have seen that each conservative force can be associated with a potential energy and that friction is associated with internal energy, or

$$\Sigma\ W_c = -\Sigma\ \Delta U$$

and

$$W_f = -U_{int},$$

so that Eq. 8-19 becomes

$$\Sigma\ W_{nc} = \Delta K + \Sigma\ \Delta U + U_{int}.$$

Now whatever the W_{nc} are, it has always been possible to find new forms of energy which corresponds to this work. We can then represent $\Sigma\ W_{nc}$ by another change of energy term on the right-hand side of the equation, with the result that we can always write the work-energy theorem as

$$0 = \Delta K + \Sigma\ \Delta U + U_{int} + \text{(change in other forms of energy)}.$$

In other words, the total energy—kinetic plus potential plus internal plus all other forms—does not change. *Energy may be transformed from one kind to another, but it cannot be created or destroyed; the total energy is constant.*

This statement is a generalization from our experience, so far not contradicted by observation of nature. It is called the *principle of the conservation of energy.* Often in the history of physics this principle seemed to fail. But its apparent failure stimulated the search for the reasons. Experimentalists searched for physical phenomena besides motion that accompany the forces of interaction between bodies. Such phenomena have always been found. With work done against friction, internal energy is produced; in other interactions energy in the form of sound, light, electricity, etc., may be produced. Hence the concept of energy was generalized to include forms other than kinetic and potential energy of directly observable bodies. This procedure, which relates the mechanics of bodies observed to be in motion to phenomena which are not mechanical or in which motion is not directly detected, has linked mechanics to all other areas of physics. The energy concept now permeates all of physical science and has become one of the unifying ideas of physics.*

* See for example, "Concept of Energy in Mechanics," by R. B. Lindsay in *The Scientific Monthly*, October 1957.

In subsequent chapters we shall study various transformations of energy — from mechanical to internal, mechanical to electrical, nuclear to internal, etc. It is during such transformations that we measure the energy changes in terms of work, for it is during these transformations that forces arise and do work.

Although the principle of the conservation of kinetic plus potential energy is often useful, we see that it is a restricted case of the more general principle of the conservation of energy. Kinetic and potential energy are conserved only when conservative forces act. Total energy is *always* conserved.

One of the great conservation laws of science had been the law of conservation of matter. From a philosophical point of view an early statement of this general principle was given by the Roman poet Lucretius, a contemporary of Julius Caesar, in his celebrated work *De Rerum Natura.* Lucretius wrote "Things cannot be born from nothing, cannot when begotten be brought back to nothing." It was a long time before this concept was established as a firm scientific principle. The principal experimental contribution was made by Antoine Lavoisier (1743–1794), regarded by many as the father of modern chemistry. He wrote in 1789 "We must lay it down as an incontestable axiom, that in all the operations of art and nature, nothing is created; an equal quantity of matter exists both before and after the experiment . . . and nothing takes place beyond changes and modifications in the combinations of these elements."

This principle, subsequently called the conservation of mass, proved extremely fruitful in chemistry and physics. Serious doubts as to the general validity of this principle were raised by Albert Einstein in his papers introducing the theory of relativity. Subsequent experiments on fast-moving electrons and on nuclear matter confirmed his conclusions.

Einstein's findings suggested that, if certain physical laws were to be retained, the mass of a particle had to be redefined as

$$m = \frac{m_0}{\sqrt{1 - v^2/c^2}}. \tag{8-20}$$

Here m_0 is the mass of the particle when at rest with respect to the observer, called the *rest mass*; m is the mass of the particle measured as it moves at a speed v relative to the observer; and c is the speed of light, having a constant value of approximately 3×10^8 m/s. Experimental checks of this equation can be made, for example, by deflecting high-speed electrons in magnetic fields and measuring the radii of curvature of their path. The paths are circular and the magnetic force a centripetal one ($F = mv^2/r$, F and v being known). At ordinary speeds the difference between m and m_0 is too small to be detectable. Electrons, however, can be emitted from radioactive nuclei with speeds greater than nine-tenths that of light. In such cases the results (Fig. 8-8 shows early data) confirm Eq. 8-20.

It is convenient to let the ratio v/c be represented by β. Then Eq. 8-20 becomes

$$m = m_0(1 - \beta^2)^{-1/2}.$$

To find the kinetic energy of a body, we compute the work done by the resultant force in setting the body in motion. In Section 7-5 we obtained

8-9
MASS AND ENERGY

$$K = \int_0^v \mathbf{F} \cdot d\mathbf{r} = \tfrac{1}{2}m_0 v^2$$

for kinetic energy, when we assumed a constant mass m_0. Suppose now instead we take into account the variation of mass with speed and use $m = m_0(1 - \beta^2)^{-1/2}$ in our previous equation. We find (Problem 29, Chapter 9) that the kinetic energy is no longer given by $\tfrac{1}{2}m_0 v^2$ but instead is

$$K = mc^2 - m_0 c^2 = (m - m_0)c^2 = \Delta mc^2. \qquad (8\text{-}21)$$

The kinetic energy of a particle is, therefore, the product of c^2 and the *increase in mass* Δm resulting from the motion.

Now, at small speeds we expect the relativistic result to agree with the classical result. By the binomial theorem we can expand $(1 - \beta^2)^{-1/2}$ as

$$(1 - \beta^2)^{-1/2} = 1 + \tfrac{1}{2}\beta^2 + \tfrac{3}{8}\beta^4 + \tfrac{5}{16}\beta^6 + \cdots.$$

At small speeds $\beta = v/c \ll 1$ so that all terms beyond β^2 are negligible. Then

$$K = (m - m_0)c^2 = m_0 c^2[(1 - \beta^2)^{-1/2} - 1]$$
$$= m_0 c^2(1 + \tfrac{1}{2}\beta^2 + \cdots - 1) \cong \tfrac{1}{2}m_0 c^2 \beta^2 = \tfrac{1}{2}m_0 v^2,$$

which is the classical result. Notice also that when K equals zero, $m = m_0$ as expected.

The basic idea that energy is equivalent to mass can be extended to include energies other than kinetic. For example, when we compress a spring and give it elastic potential energy U, its mass increases from m_0 to $m_0 + U/c^2$. When we add heat energy in amount Q to an object, its mass increases by an amount Δm, where Δm is Q/c^2. We arrive at a principle of *equivalence of mass and energy:* For every unit of energy E of *any* kind supplied to a material object, the mass of the object increases by an amount

$$\Delta m = E/c^2.$$

This is the famous Einstein formula

$$E = \Delta mc^2. \qquad (8\text{-}22)$$

In fact, since rest mass itself is just one form of energy, we can now assert that a body at rest has an energy $m_0 c^2$ by virtue of its rest mass.

figure 8-8
The way an electron's mass increases as its speed relative to the observer increases. The solid line is a plot of $m = m_0(1 - v^2/c^2)^{-1/2}$, and the circles are adapted from experimental values obtained by Bucherer and Neumann in 1914. The curve tends toward infinity as $v \to c$.

This is called its *rest energy*. If we now consider a closed system, the principle of the conservation of energy, as generalized by Einstein, becomes

$$\Sigma \,(m_0 c^2 + \mathscr{E}) = \text{constant}$$

or

$$\Delta(\Sigma \, m_0 c^2 + \Sigma \, \mathscr{E}) = 0,$$

where $\Sigma \, m_0 c^2$ is the total rest energy and $\Sigma \, \mathscr{E}$ is the total energy of *all other* kinds. As Einstein wrote, "Pre-relativity physics contains two conservation laws of fundamental importance, namely the law of conservation of energy and the law of conservation of mass; these two appear there as completely independent of each other. Through relativity theory they melt together into *one* principle."

Because the factor c^2 is so large, we would not expect to be able to detect changes in mass in ordinary mechanical experiments. A change in mass of 1 g would require an energy of 9×10^{13} joules. But when the mass of a particle is quite small to begin with and high energies can be imparted to it, the relative change in mass may be readily noticeable. This is true in nuclear phenomena, and it is in this realm that classical mechanics breaks down and relativistic mechanics receives its most striking verification.

A beautiful example of exchange of energy between rest mass and other forms is given by the phenomenon of pair annihilation or pair production. In this phenomenon an electron and a positron, elementary material particles differing only in the sign of their electric charge, can combine and literally disappear. In their place we find high-energy radiation, called γ-radiation, whose radiant energy is exactly equal to the rest mass plus kinetic energies of the disappearing particles. The process is reversible, so that a materialization of rest mass from radiant energy can occur when a high enough energy γ-ray, under proper conditions, disappears; in its place appears a positron-electron pair whose total energy (rest mass + kinetic) is equal to the radiant energy lost.

EXAMPLE 8

Consider a quantitative example. On the atomic mass scale the unit of mass is 1.66×10^{-27} kg approximately. On this scale the rest mass of the proton (the nucleus of a hydrogen atom) is 1.00731 and the rest mass of the neutron (a neutral particle, one of the constituents of all nuclei except hydrogen) is 1.00867. A deuteron (the nucleus of heavy hydrogen) is known to consist of a neutron and a proton; the rest mass of the deuteron is found to be 2.01360. The rest mass of the deuteron is *less than* the combined rest masses of neutron and proton by 0.00238 atomic mass units. The discrepancy is equivalent in energy to

$$E = \Delta m c^2 = (0.00238 \times 1.66 \times 10^{-27} \text{ kg})(3.00 \times 10^8 \text{ m/s})^2$$

$$= 3.57 \times 10^{-13} \text{ joules} = 2.22 \times 10^6 \text{ eV}.$$

When a neutron and a proton combine to form a deuteron, this exact amount of energy is given off in the form of γ-radiation. Similarly, it is found that the same amount of energy must be *added* to the deuteron to break it up into a proton and a neutron. This energy is therefore called the *binding energy* of the deuteron.

---

questions

1. Mountain roads rarely go straight up the slope but wind up gradually. Explain why.

2. Is any work being done on a car moving with constant speed along a straight level road?

3. An automobile of mass m and speed v is moving along a highway. The

driver jams on the brakes and the car skids to a halt. In what form does the lost kinetic energy of the car appear?

4. In the above question, assume that the driver "pumps" the brakes in such a way that there is no skidding or sliding. In this case, in what form does the lost kinetic energy of the car appear?

5. An automobile accelerates from rest to a speed v, under conditions such that no slipping of the driving wheels occurs. Where does the kinetic energy of the car come from? In particular, is it true that it is provided by the work done on the car by the (static) frictional force exerted by the road on the car?

6. If it takes no work to hold up a heavy object, why is it tiring?

7. What happens to the potential energy an elevator loses in coming down from the top of a building to a stop at the ground floor?

8. In Example 2 (see Fig. 8-5) we asserted that the speed at the bottom does not depend at all on the *shape* of the surface. Would this still be true if friction were present?

9. Give physical examples of unstable equilibrium. Of neutral equilibrium. Of stable equilibrium.

10. Explain, using work and energy ideas, how a child pumps a swing up to large amplitudes from a rest position. (See "How to Make a Swing Go" by R. V. Hesheth, *Physics Education*, July 1975.)

11. A swinging pendulum eventually comes to rest. Is this a violation of the law of conservation of energy?

12. A scientific article ("The Energetic Cost of Moving About" by V. A. Tucker, *American Scientist* July–August 1975) asserts that walking and running are extremely inefficient forms of locomotion and that much greater efficiency is achieved by birds, fish, and bicyclists. Can you suggest an explanation?

13. Two disks are connected by a stiff spring. Can one press the upper disk down enough so that when it is released it will spring back and raise the lower disk off the table (see Fig. 8-9)? Can mechanical energy be conserved in such a case?

14. In the case of work done against friction, the amount of heat energy generated is independent of the velocity (or inertial reference frame) of the observer. That is, different observers would assign the same quantity of mechanical energy transformed into heat energy due to friction. How can this be explained, considering that such observers measure different quantities of total work done and different changes in kinetic energy in general (see Problem 21, Chapter 7)?

15. Must *all* nonconservative forces be dissipative, as friction is? Could ΣW_{nc} be greater than zero, in principle?

16. An object is dropped and observed to bounce to one and one-half times its original height. What conclusion can you draw from this observation?

17. The driver of an automobile traveling at speed v suddenly sees a brick wall at a distance d directly in front of him. To avoid crashing, is it better for him to slam on the brakes or to turn the car sharply away from the wall? (Hint: Consider the force required in each case.)

18. A spring is kept compressed by tying its ends together tightly. It is then placed in acid and dissolves. What happened to its stored potential energy?

figure 8-9
Question 13

SECTION 8-3

1. A body moving along the x-axis is subject to a force repelling it from the origin, given by $F = kx$. (a) Find the potential energy function $U(x)$ for the motion and write down the conservation of energy condition. (b) Describe the motion of the system and show that this is the kind of motion we would expect near a point of unstable equilibrium. *Answer:* (a) $-kx^2/2$.

2. If the magnitude of the force of attraction between a particle of mass m_1 and one of mass m_2 is given by

problems

$$F = k \frac{m_1 m_2}{x^2}$$

where k is a constant and x is the distance between the particles, find (a) the potential energy function and (b) the work required to increase the separation of the masses from $x = x_1$ to $x = x_1 + d$.

3. A chain is held on a frictionless table with one-fifth of its length hanging over the edge. If the chain has a length l and a mass m, how much work is required to pull the hanging part back on the table?   *Answer: mgl/50.*

SECTION 8-4

4. A 2.0-g (weight $mg = 0.071$-oz) penny is pushed down on a vertical spring, compressing the spring by 1.0 cm (0.39 in.). The force constant of the spring is 40 N/m (2.7 lb/ft). How far above this original position will the penny fly if it is released?

5. A 200-lb man jumps out a window into a fire net 30 ft below. The net stretches 6.0 ft before bringing him to rest and tossing him back into the air. What is the potential energy of the stretched net if no energy is dissipated by nonconservative forces?   *Answer: 7200 ft · lb.*

6. A 2.0-kg (0.14-slug) block is dropped from a height of 0.40 m (1.3 ft) onto a spring of force constant $k = 1960$ N/m (134 lb/ft). Find the maximum distance the spring will be compressed (neglect friction).

7. Show that for the same initial speed v_0, the speed v of a projectile will be the same at all points at the same elevation, regardless of the angle of projection.

8. A certain peculiar spring is found *not* to conform to Hooke's law. The force (in newtons) it exerts when stretched a distance x (in meters) is found to have magnitude $52.8x + 38.4x^2$ in the direction opposing the stretch. (a) Compute the total work required to stretch the spring from $x = 0.50$ to $x = 1.00$ m. (b) With one end of the spring fixed, a particle of mass 2.17 kg is attached to the other end of the spring when it is extended by an amount $x = 1.00$ m. If the particle is then released from rest, compute its *speed* at the instant the spring has returned to the configuration in which the extension is $x = 0.50$ m. (c) Is the force exerted by the spring conservative or nonconservative? Explain.

9. It is claimed that large trees can evaporate as much as 1 ton (910 kg mass) of water per day. (a) Assuming the average rise of water to be 30 ft (9.1 m) from the ground, how much energy (in kW · h) must be supplied to do this? (b) What is the average power if the evaporation is assumed to occur during 12 hours of the day?   *Answer: (a) 2.3×10^{-2} kW · h. (b) 1.9 W.*

10. An object is attached to a vertical spring and slowly lowered to its equilibrium position. This stretches the spring by an amount d. If the same object is attached to the same vertical spring but permitted to fall instead, through what maximum distance does it stretch the spring?

11. A body falls from rest from a height h. Determine the kinetic energy and the potential energy of the body as a function (a) of time and (b) of height. Graph the expressions and show that their sum, the total energy, is constant in each case.

SECTION 8-5

12. A particle moves along a line in a region in which its potential energy varies as in Fig. 8-10. (a) Sketch, with the same scale on the abscissa, the force $F(x)$ acting on the particle. Indicate on the graph the approximate numerical scale for $F(x)$. (b) If the particle has a constant total energy of 4.0 joules, sketch the graph of its kinetic energy. Indicate the numerical scale on the $K(x)$ axis.

13. An α-particle (helium atom nucleus) in a large nucleus is bound by a potential like that shown in Fig. 8-11. (a) Construct a function of x, which has

figure 8-10
Problem 12

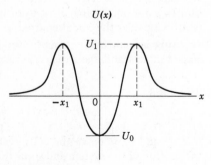

figure 8-11
Problem 13

this general shape, with a minimum value U_0 at $x=0$ and a maximum value U_1 at $x=x_1$ and $x=-x_1$. (b) Determine the force between the α-particle and the nucleus as a function of x. (c) Describe the possible motions.

SECTION 8-6

14. The string in Fig. 8-12 has a length $l = 4.0$ ft. When the ball is released, it will swing down the dotted arc. How fast will it be going when it reaches the lowest point in its swing?

15. A frictionless roller coaster of mass m starts at point A with speed v_0, as in Fig. 8-13. Assume that the roller coaster can be considered as a particle and that it always remains on the track. (a) What will be the speed of the roller coaster at points B and C? (b) What constant deceleration is required to stop it at E if the brakes are applied at point D?

    Answer: (a) $v_B = v_0$; $v_C = \sqrt{v_0{}^2 + gh}$. (b) $(v_0{}^2 + 2gh)/2L$.

figure 8-12
Problems 14, 27, 30

figure 8-13
Problem 15

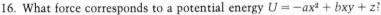

16. What force corresponds to a potential energy $U = -ax^2 + bxy + z$?

17. The potential energy corresponding to a certain two-dimensional force field is given by $U(x,y) = \frac{1}{2}k(x^2 + y^2)$. (a) Derive F_x and F_y and describe the vector force at each point in terms of its coordinates x and y. (b) Derive F_r and F_θ and describe the vector force at each point in terms of the polar coordinates r and θ of the point. (c) Can you think of a physical model of such a force?

    Answer: (a) $F_x = -kx$; $F_y = -ky$; $\mathbf{F}$ points toward the origin. (b) $F_r = -kr$; $F_\theta = 0$.

18. The so-called Yukawa potential

$$U(r) = -\frac{r_0}{r} U_0 e^{-r/r_0}$$

gives a fairly accurate description of the interaction between nucleons (i.e., neutrons and protons, the constituents of the nucleus). The constant r_0 is about 1.5×10^{-15} meter and the constant U_0 is about 50 MeV. (a) Find the corresponding expression for the force of attraction. (b) To show the short range of this force, compute the ratio of the force at $r = 2r_0$, $4r_0$, and $10r_0$ to the force at $r = r_0$.

19. An ideal massless spring S can be compressed 1.0 m by a force of 100 N. This same spring is placed at the bottom of a frictionless inclined plane which makes an angle of $\theta = 30°$ with the horizontal (see Fig. 8-14). A 10-kg mass M is released from the top of the incline and is brought to rest momentarily after compressing the spring 2.0 m. (a) Through what distance does the mass slide before coming to rest? (b) What is the speed of the mass just before it reaches the spring?   *Answer:* (a) 4.1 m. (b) 4.5 m/s.

20. The magnitude of the force of attraction between the positively charged nucleus and the negatively charged electron in the hydrogen atom is given by

$$F = k\frac{e^2}{r^2}$$

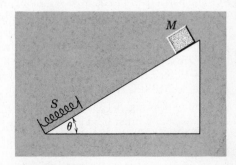

figure 8-14
Problem 19

where e is the charge of the electron, k is a constant, and r is the separation between electron and nucleus. Assume that the nucleus is fixed. The electron, initially moving in a circle of radius R_1 about the nucleus, jumps suddenly into a circular orbit of smaller radius R_2. (a) Calculate the change in kinetic energy of the electron, using Newton's second law. (b) Using the relation between force and potential energy, calculate the change in potential energy of the atom. (c) Show by how much the mechanical energy of the atom has changed in this process. (This energy is given off in the form of radiation.)

21. A light rigid rod of length l has a mass m attached to its end, forming a simple pendulum. It is inverted and then released. What are (a) the speed v at its lowest point and (b) the tension T in the suspension at that instant? (c) The same pendulum is next put in a horizontal position and released from rest. At what angle from the vertical will the tension in the suspension equal the weight in magnitude?   Answer: (a) $2\sqrt{gl}$. (b) 5 mg. (c) 71°.

22. A simple pendulum of length l, the mass of whose bob is m, is observed to have a speed v_0 when the cord makes an angle θ_0 with the vertical $(0 < \theta_0 < \pi/2)$, as in Fig. 8-15. In terms of g and the foregoing given quantities, determine (a) the speed v_1 of the bob when it is at its lowest position; (b) the least value v_2 that v_0 could have if the cord is to achieve a horizontal position during the motion; (c) the speed v_3 such that if $v_0 > v_3$, the pendulum will not oscillate but rather will continue to move around in a vertical circle.

figure 8-15
Problem 22

23. A simple pendulum is made by tying a 2.0-kg stone to a string 4.0 m long. The stone is projected perpendicular to the string, away from the ground, with the string at an angle of 60° with the vertical. It is observed to have a speed of 8.0 m/s when it passes its lowest point. (a) What was the speed of the stone at the moment of release? (b) What is the largest angle with the vertical that the string will reach during the stone's motion? (c) Using the lowest point of the swing as the zero of gravitational potential energy, what is the total mechanical energy of the system?
Answer: (a) 5.0 m/s. (b) 80°. (c) 64 J.

figure 8-16
Problem 24

24. A small block of mass m slides along the frictionless loop-the-loop track shown in Fig. 8-16. (a) If it starts from rest at P, what is the resultant force acting on it at Q? (b) At what height above the bottom of the loop should the block be released so that the force exerted on it by the track at the top of the loop is equal to its weight?

25. A point mass m starts from rest and slides down the surface of a frictionless solid sphere of radius r as in Fig. 8-17. Measure angles from the vertical and potential energy from the top. Find (a) the change in potential energy of the mass with angle; (b) the kinetic energy as a function of angle; (c) the radial and tangential accelerations as a function of angle; (d) the angle at which the mass flies off the sphere.
Answer: (a) $-mgr(1 - \cos\theta)$. (b) $mgr(1 - \cos\theta)$. (c) $2g(1 - \cos\theta)$; $g\sin\theta$. (d) $\cos^{-1} 2/3$.

figure 8-17
Problem 25

26. The particle m in Fig. 8-18 is moving in a vertical circle of radius R inside a track. There is no friction. When m is at its lowest position, its speed is v_0. (a) What is the minimum value v_m of v_0 for which m will go completely around the circle without losing contact with the track? (b) Suppose v_0 is $0.775v_m$. The particle will move up the track to some point at P at which it will lose contact with the track and travel along a path shown roughly by the dashed line. Find the angular position θ of point P.

27. The nail in Fig. 8-12 is located a distance d below the point of suspension. Show that d must be at least $0.6l$ if the ball is to swing completely around in a circle centered on the nail.

28. Two children are playing a game in which they attempt to hit a small box on the floor using a spring-loaded marble gun placed horizontally on a frictionless table (Fig. 8-19). The first child compresses the spring 1.0 cm and the

figure 8-18
Problem 26

figure 8-19
Problem 28

marble falls 20 cm short of the target, which is 2.0 m horizontally from the edge of the table. How far should the second child compress the spring so that the same marble falls into the box?

29. An escalator joins one floor with another one 25 ft (7.6 m) above. The escalator is 40 ft (12 m) long and moves along its length at 2.0 ft/s (0.61 m/s). (a) What power must its motor deliver if it is required to carry a maximum of 100 persons per minute, of average mass 5.0 slugs (73 kg)? (b) A 160-lb (710-N) man walks up the escalator in 10 s. How much work does the motor do on him? (c) If this man turned around at the middle and walked down the escalator so as to stay at the same level in space, would the motor do work on him? If so, what power does it deliver for this purpose? (d) Is there any (other?) way the man could walk along the escalator without consuming power from the motor?
Answer: (a) 6700 ft · lb/s (9100 W). (b) 2000 ft · lb (2700 J).

30. Suppose that the string of Fig. 8-12 is very elastic, made of rubber, say, and that the string is unextended at length l when the ball is released. (a) Explain why you would expect the ball to reach a low point greater than a distance l below the point of suspension. (b) Show, using dynamic and energy considerations, that if Δl is small compared to l, the string will stretch by an amount $\Delta l = 3 \, mg/k$, where k is the assumed force constant of the string. Notice that the larger k is, the smaller Δl is, and the better the approximation $\Delta l \ll l$. (c) Show, under these circumstances, that the speed of the ball at the bottom is $v = \sqrt{2g(l - 3mg/2k)}$, *less* than it would be for an inelastic string $(k = \infty)$. Give a physical explanation for this result using energy considerations.

SECTION 8-7

31. Two snow-covered peaks at elevations of 3500 m and 3400 m are separated by a valley. A ski-run extends from the top of the higher peak to the top of the lower one, with a total length of 3000 m. (a) A skier starts from rest on the higher peak. With what speed will he arrive at the top of the lower peak if he goes as fast as possible, never trying to slow down? Neglect friction. (b) Make a rough estimate of how large a coefficient of friction with the snow could be tolerated without preventing him from reaching the lower peak. *Answer:* (a) 44 m/s. (b) Approximately 1/10.

32. A projectile of mass 9.4 kg is fired straight up with an initial speed of 470 m/s. How much higher would it have gone if the air resistance did not dissipate the energy of 6.8×10^5 J that it does?

33. Show that when friction is present in an otherwise conservative mechanical system, the rate at which mechanical energy is dissipated equals the frictional force times the speed at that instant, or

$$\frac{d}{dt} (K + U) = -fv$$

34. A boy is seated on the top of a hemispherical mound of ice (Fig. 8-20). He is given a very small push and starts sliding down the ice. (a) Show that he leaves the ice at a point whose height is $2R/3$ if the ice is frictionless. (b) If there is friction between the ice and the boy, would he fly off at a greater or lesser height than in (a)?

figure 8-20
Problem 34

35. A 1.0-kg (weight $mg = 2.2$ lb) block collides with a horizontal weightless spring of force constant 2.0 N/m (0.14 lb/ft) (Fig. 8-21). The block compresses the spring 4.0 m (13 ft) from the rest position. Assuming that the coefficient of kinetic friction between block and horizontal surface is 0.25, what was the speed of the block at the instant of collision?
Answer: 7.2 m/s (23 ft/s).

36. A body of mass m starts from rest down a plane of length l inclined at an angle θ with the horizontal. (a) Take the coefficient of friction to be μ and find the body's speed at the bottom. (b) How far, d, will it slide horizontally on a similar surface after reaching the bottom of the incline? Solve by using energy methods and solve again using Newton's laws directly.

37. A 4.0-kg block starts up a 30° incline with 128 J of kinetic energy. How far will it slide up the plane if the coefficient of friction is 0.30?
Answer: 4.3 m.

38. A 40-lb body is pushed up a frictionless 30° inclined plane 10 ft long by a horizontal force **F**. (a) If the speed at the bottom is 2.0 ft/s and at the top is 10 ft/s, how much work is done by **F**? (b) What is the magnitude of the force **F**? (c) Suppose the plane is not frictionless, and that $\mu_k = 0.15$. How far up the plane goes the body go?

39. A particle slides along a track with elevated ends and a flat central part, as shown in Fig. 8-22. The flat part has a length $l = 2.0$ m. The curved portions of the track are frictionless. For the flat part the coefficient of kinetic friction is $\mu_k = 0.20$. The particle is released at point A which is at a height $h = 1.0$ m above the flat part of the track. Where does the particle finally come to rest?   *Answer:* In the center of the flat part.

40. A very light rigid rod whose length is l has a ball of mass m attached to one end (Fig. 8-23). The other end is pivoted frictionlessly in such a way that the ball moves in a vertical circle. The system is launched from the horizontal position A with downward initial velocity $\mathbf{v}_0$. The ball just reaches point D and then stops. (a) Derive an expression for v_0 in terms of l, m, and g. (b) What is the tension in the rod when the ball is at B? (c) A little sand is placed on the pivot, after which the ball just reaches C when launched from A with the same speed as before. How much work is done by friction during this motion? (d) How much total work is done by friction before the ball finally comes to rest at B after oscillating back and forth several times?

41. The cable of a 4000-lb elevator in Fig. 8-24 snaps when the elevator is at rest at the first floor so that the bottom is a distance $d = 12$ ft above a cushioning spring whose spring constant is $k = 10,000$ lb/ft. A safety device clamps the guide rails so that a constant friction force of 1000 lb opposes the motion of the elevator. (a) Find the speed of the elevator just before it hits the spring. (b) Find the distance s that the spring is compressed. (c) Find the distance that the elevator will "bounce" back up the shaft. (d) Using the conservation of energy principle, find approximately the total distance that the elevator will move before coming to rest. Why is the answer not exact?   *Answer:* (a) 24 ft/s. (b) 3.0 ft. (c) 9.0 ft. (d) 49 ft.

SECTION 8-9

42. A vacuum diode consists of a cylindrical anode enclosing a cylindrical cathode. An electron with a potential energy relative to the anode of 4.8×10^{-16} J leaves the surface of the cathode with zero initial speed. Assume that the electron does not collide with any air molecules and that the gravitational force is negligible. (a) What kinetic energy would the electron have when it strikes the anode? (b) Take 9.1×10^{-31} kg as the mass of the electron and find its final speed. (c) Were you justified in using classical relations for kinetic energy and mass rather than the relativistic ones?

43. What is the speed of an electron with a kinetic energy of (a) 1.0×10^5 eV $(1.2 \times 10^{-14}$ ft · lb), (b) 1.0×10^6 eV $(1.2 \times 10^{-13}$ ft · lb)?
Answer: (a) 1.6×10^8 m/s (103,000 mi/s). (b) 2.8×10^8 m/s (175,000 mi/s).

figure 8-21
Problem 35

figure 8-22
Problem 39

figure 8-23
Problem 40

figure 8-24
Problem 41

44. The United States consumed about 1.6×10^{12} kW $\cdot$ h of electrical energy in 1970. How many kilograms of matter would have to be completely destroyed to yield this energy?

45. A nuclear reactor generating plant supplies 60 MW (8.0×10^4 hp) of useful power steadily for a year. (a) How much energy, in joules (ft $\cdot$ lb), did it supply? (b) Assuming that, in addition, 90 MW (12×10^4 hp) of power is wasted in heat production, determine the mass (weight) converted to energy in a year at this plant.
 Answer: (a) 1.9×10^{15} J (1.4×10^{15} ft $\cdot$ lb). (b) 52 g (1.9 oz).

46. How much matter would have to be converted into energy in order to accelerate a 1.0-kiloton spaceship from rest to a speed of $(1/10)c$?

47. An electron (rest mass 9.1×10^{-31} kg) is moving with a speed 0.99 c. (a) What is its total energy? (b) Find the ratio of the Newtonian kinetic energy to the relativistic kinetic energy in this case?   *Answer:* (a) 5.8×10^{-13} J. (b) 0.08.

48. (a) The rest mass of a body is 0.010 kg. What is its mass when it moves at a speed of 3.0×10^7 m/s relative to the observer? At 2.7×10^8 m/s? (b) Compare the classical and relativistic kinetic energies for these cases. (c) What if the observer, or measuring apparatus, is riding on the body?

49. Equation (8-21), $K = (m - m_0)c^2$, is the usual relativistic equation for kinetic energy. (a) Show that, by using Eq. (8-20), $m = m_0(1 - \beta^2)^{-1/2}$, we can also express the relativistic kinetic energy as $K = \dfrac{m}{m + m_0} mv^2$. (b) Contrast the way these two expressions reduce to the classical result as $m \to m_0$ or $v/c \to 0$. (See "Parallels between Relativistic and Classical Dynamics for Introductory Courses" by Donald E. Fahnline, *American Journal of Physics*, June 1975.)

50. It is believed that the sun obtains its energy by a fusion process in which four hydrogen atoms are transformed into a helium atom with the emission of energy in various forms of radiation. If a hydrogen atom has a rest mass of 1.0081 atomic mass units (see Example 7) and a helium atom has a rest mass of 4.0039 atomic mass units, calculate the energy released in each fusion process.

9
conservation of linear momentum

So far we have treated objects as though they were particles, having mass but no size. In translational motion each point on a body experiences the same displacement as any other point as time goes on, so that the motion of one particle represents the motion of the whole body. But even when a body rotates or vibrates as it moves, there is one point on the body, called the *center of mass*, that moves in the same way that a single particle subject to the same external forces would move. Figure 9-1 shows the simple parabolic motion of the center of mass of an Indian club thrown from one performer to another; no other point in the club moves in such a simple way. Note that if the club were moving in pure translation (see Fig. 3-1), then *every* point in it would experience the same displacements as does the center of mass in Fig. 9-1. For this reason the motion of the center of mass of a body is called the translational motion of the body.

When the system with which we deal is not a rigid body, a center of mass (whose motion can also be described in a relatively simple way) can be assigned, even though the particles that make up the system may be changing their positions with respect to each other in a relatively complicated way as the motion proceeds. In this section we define the center of mass and show how to calculate its position. In the next section we discuss the properties that make it useful in describing the motion of extended objects or systems of particles.

Consider first the simple case of a system of two particles m_1 and m_2 at distances x_1 and x_2 respectively, from some origin O. We define a point C, the center of mass of the system, as a distance x_{cm} from the origin O, where x_{cm} is defined by

9-1
CENTER OF MASS

$$x_{cm} = \frac{m_1x_1 + m_2x_2}{m_1 + m_2}.$$ (9-1)

This point (Fig. 9-2) has the property that the product of the total mass of the system $M(= m_1 + m_2)$ times the distance of this point from the origin is equal to the sum of the products of the mass of each particle by its distance from the origin; that is,

$$(m_1 + m_2)x_{cm} = Mx_{cm} = m_1x_1 + m_2x_2.$$

In Eq. 9-1, x_{cm} can be regarded as the *mass-weighted mean* of x_1 and x_2.

An analogy might help to fix this idea. Suppose, for example, that we are given two boxes of nails. In one box we have n_1 nails all having the same length l_1; in the other box we have n_2 nails all having the same length l_2. We are asked to get the mean length of the nails. If $n_1 = n_2$, the mean length is simply $(l_1 + l_2)/2$. But if $n_1 \neq n_2$, we must allow for the fact that there are more nails of one length than another by a "weighting" factor for each length. For l_1 this factor is $n_1/(n_1 + n_2)$ and for l_2 this factor is $n_2/(n_1 + n_2)$, the fraction of the total number of nails in each box. Then the weighted-mean length is

$$\bar{l} = \left(\frac{n_1}{n_1 + n_2}\right)l_1 + \left(\frac{n_2}{n_1 + n_2}\right)l_2$$

or

$$\bar{l} = \frac{n_1l_1 + n_2l_2}{n_1 + n_2}.$$

figure 9-2
The center of mass of the two masses m_1 and m_2 lies on the line joining m_1 and m_2 at C, a distance x_{cm} from the origin.

The center of mass, defined in Eq. 9-1, is then a weighted-mean displacement where the "weighting" factor for each particle is the fraction of the total mass that each particle has.

If we have n particles, $m_1, m_2, \cdots, m_n$ along a straight line, by definition the center of mass of these particles relative to some origin is

$$X_{cm} = \frac{m_1 x_1 + m_2 x_2 + \cdots + m_n x_n}{m_1 + m_2 + \cdots + m_n} = \frac{\Sigma \, m_i x_i}{\Sigma \, m_i}, \qquad (9\text{-}2)$$

where $x_1, x_2, \cdots, x_n$ are the distances of the masses from the origin from which x_{cm} is measured. The symbol Σ represents a summation operation, in this case over all n particles. The sum

$$\Sigma \, m_i = M$$

is the total mass of the system. We can then rewrite Eq. 9-2 in the form

$$M x_{cm} = \Sigma \, m_i x_i. \qquad (9\text{-}2a)$$

Suppose now that we have three particles *not* in a straight line; they will lie in a *plane*, as in Fig. 9-3. The center of mass C is defined and located by the coordinates x_{cm} and y_{cm}, where

$$X_{cm} = \frac{m_1 x_1 + m_2 x_2 + m_3 x_3}{m_1 + m_2 + m_3}, \qquad (9\text{-}3)$$

$$y_{cm} = \frac{m_1 y_1 + m_2 y_2 + m_3 y_3}{m_1 + m_2 + m_3},$$

in which x_1, y_1 are the coordinates of the particle of mass m_1; x_2, y_2 are those of m_2; and x_3, y_3 are those of m_3. The coordinates x_{cm}, y_{cm} of the center of mass are measured from the same arbitrary origin.

For a large number of particles lying in a plane, the center of mass is at x_{cm}, y_{cm}, where

$$X_{cm} = \frac{\Sigma \, m_i x_i}{\Sigma \, m_i} = \frac{1}{M} \Sigma \, m_i x_i \quad \text{and} \quad y_{cm} = \frac{\Sigma \, m_i y_i}{\Sigma \, m_i} = \frac{1}{M} \Sigma \, m_i y_i \qquad (9\text{-}4)$$

in which $M \, (= \Sigma \, m_i)$ is the total mass of the system.

For a large number of particles not necessarily confined to a plane but *distributed in space*, the center of mass is at x_{cm}, y_{cm}, z_{cm}, where

$$X_{cm} = \frac{1}{M} \Sigma \, m_i x_i, \quad y_{cm} = \frac{1}{M} \Sigma \, m_i y_i, \quad z_{cm} = \frac{1}{M} \Sigma \, m_i z_i. \qquad (9\text{-}5a)$$

In vector notation each particle in the system can be described by a position vector $\mathbf{r}_i$ in a particular reference frame and the center of mass can be located by a position vector $\mathbf{r}_{cm}$. These vectors are related to $x_i, y_i, z_i,$ and x_{cm}, y_{cm}, z_{cm} in Eq. 9-5a by

$$\mathbf{r}_i = \mathbf{i} x_i + \mathbf{j} y_i + \mathbf{k} z_i$$

and

$$\mathbf{r}_{cm} = \mathbf{i} x_{cm} + \mathbf{j} y_{cm} + \mathbf{k} z_{cm}.$$

Thus the three scalar equations of Eq. 9-5a can be replaced by a single vector equation

$$\mathbf{r}_{cm} = \frac{1}{M} \Sigma \, m_i \mathbf{r}_i \qquad (9\text{-}5b)$$

in which the sum is a vector sum. You can prove that Eq. 9-5b is true by substituting the expressions given for $\mathbf{r}_i$ and $\mathbf{r}_{cm}$ just above into Eq. 9-5b. Note the economy of expression permitted by the use of vectors. Equation 9-5b shows that, if the origin of our reference frame is at the center of mass (which means that $\mathbf{r}_{cm} = 0$), then $\Sigma \, m_i \mathbf{r}_i = 0$ for the system.

Equations 9-5 are the most general case for a collection of particles. Equations 9-1 through 9-4 are special instances of this one. The location

figure 9-3
The center of mass of the three masses m_1, m_2, and m_3 lies at point C, with coordinates x_{cm}, y_{cm}. C lies in the same plane as that of the triangle formed by the three masses.

of the center of mass is independent of the reference frame used to locate it (see Problem 1). *The center of mass of a system of particles depends only on the masses of the particles and the positions of the particles relative to one another.*

A rigid body, such as a meter stick, can be thought of as a system of closely packed particles. Hence it also has a center of mass. The number of particles (atoms, for example) in the body is so large and their spacing so small, however, that we can treat such a body as though it has a continuous distribution of mass. To obtain the expression for the center of mass of a continuous body, let us begin by subdividing the body into n small elements of mass Δm_i located approximately at the points x_i, y_i, z_i. The coordinates of the center of mass are then given approximately by

$$x_{cm} = \frac{\Sigma \, \Delta m_i x_i}{\Sigma \, \Delta m_i}, \quad y_{cm} = \frac{\Sigma \, \Delta m_i y_i}{\Sigma \, \Delta m_i}, \quad z_{cm} = \frac{\Sigma \, \Delta m_i z_i}{\Sigma \, \Delta m_i}.$$

Now let the elements of mass be further subdivided so that the number of elements n tends to infinity. The points x_i, y_i, z_i will locate the mass elements more precisely as n is increased and will locate them exactly as n becomes infinite. The continuous body is then subdivided into an infinite number of infinitesimal mass elements. We can now give the coordinates of the center of mass precisely as

$$x_{cm} = \lim_{\Delta m_i \to 0} \frac{\Sigma \, \Delta m_i x_i}{\Sigma \, \Delta m_i} = \frac{\int x \, dm}{\int dm} = \frac{1}{M} \int x \, dm,$$

$$y_{cm} = \lim_{\Delta m_i \to 0} \frac{\Sigma \, \Delta m_i y_i}{\Sigma \, \Delta m_i} = \frac{\int y \, dm}{\int dm} = \frac{1}{M} \int y \, dm, \qquad (9\text{-}6a)$$

$$z_{cm} = \lim_{\Delta m_i \to 0} \frac{\Sigma \, \Delta m_i z_i}{\Sigma \, \Delta m_i} = \frac{\int z \, dm}{\int dm} = \frac{1}{M} \int z \, dm.$$

In these expressions dm is the differential element of mass at the point x, y, z, and $\int dm$ equals M, where M is the total mass of the body. For a continuous body the summation of Eq. 9-5a is replaced by the integration of Eq. 9-6a.

The vector expression that is equivalent to the three scalar expressions of Eq. 9-6a is

$$\mathbf{r}_{cm} = \frac{1}{M} \int \mathbf{r} \, dm \qquad (9\text{-}6b)$$

As before, the summation of Eq. 9-5b has been replaced by an integration. Once again we see that if the origin of our reference frame is at the center of mass (that is, if $\mathbf{r}_{cm} = 0$), then $\int \mathbf{r} \, dm = 0$ for the body. This integral, and the corresponding sum $\Sigma \, m_i \mathbf{r}_i$ of Eq. 9-5b, is called the *first moment of mass for the system.*

Often we deal with homogeneous objects having a point, a line, or a plane of symmetry. Then the center of mass will lie at the point, on the line, or in the plane of symmetry. For example, the center of mass of a homogeneous sphere (which has a point of symmetry) will be at the center of the sphere, the center of mass of a cone (which has a line of symmetry) will be on the axis of the cone, etc. We can understand that this is so because, from symmetry, the first moment of mass ($\int \mathbf{r} \, dm$) is zero at the center of a sphere, somewhere along the axis of a cone, etc. If follows from Eq. 9-6b that $\mathbf{r}_{cm} = 0$ for such points which means that the center of mass is located at these points.

Locate the center of mass of three particles of mass $m_1 = 1.0$ kg, $m_2 = 2.0$ kg, and $m_3 = 3.0$ kg at the corners of an equilateral triangle 1.0 m on a side.

Choose the x-axis along one side of the triangle as shown in Fig. 9-4. Note that m_3 is then $\sqrt{3}/2$ m along the y-axis. Then,

$$x_{\text{cm}} = \frac{\Sigma\, m_i x_i}{\Sigma\, m_i} = \frac{(1.0\ \text{kg})(0) + (2.0\ \text{kg})(1.0\ \text{m}) + (3.0\ \text{kg})(\frac{1}{2}\ \text{m})}{(1.0 + 2.0 + 3.0)\ \text{kg}} = \tfrac{7}{12}\ \text{m},$$

$$y_{\text{cm}} = \frac{\Sigma\, m_i y_i}{\Sigma\, m_i} = \frac{(1.0\ \text{kg})(0) + (2.0\ \text{kg})(0) + (3.0\ \text{kg})(\sqrt{3}/2\ \text{m})}{(1.0 + 2.0 + 3.0)\ \text{kg}} = \frac{\sqrt{3}}{4}\ \text{m}.$$

The center of mass C is shown in the figure. Why is it not at the geometric center of the triangle?

EXAMPLE 1

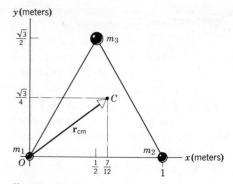

figure 9-4
Example 1. Finding the center of mass C of three unequal masses forming an equilateral triangle.

Find the center of mass of the triangular plate of Fig. 9-5.

If we can divide a body into parts such that the center of mass of each part is known, we can usually find the center of mass of the body simply. The triangular plate may be divided into narrow strips parallel to one side. The center of mass of each strip lies on the line which joins the middle of that side to the opposite vertex. But we can divide up the triangle in three different ways, using this process for each of three sides. Hence the center of mass lies at the intersection of the three lines which join the middle of each side with the opposite vertices. This is the only point that is common to the three lines.

EXAMPLE 2

figure 9-5
Example 2. Finding the center of mass C of a triangular plate.

Now we can discuss the physical importance of the center-of-mass concept. Consider the motion of a group of particles whose masses are $m_1, m_2, \ldots, m_n$ and whose total mass is M. For the time being we will assume that mass neither enters nor leaves the system so that the total mass M of the system remains constant with time. In Section 9-7 we shall consider systems in which M is not constant; a familiar example is a rocket, which expels hot gases as its fuel burns, thus reducing its mass.

From Eq. 9-5b we have, for our fixed system of particles,

$$M\mathbf{r}_{\text{cm}} = m_1\mathbf{r}_1 + m_2\mathbf{r}_2 + \cdots + m_n\mathbf{r}_n,$$

where $\mathbf{r}_{\text{cm}}$ is the position vector identifying the center of mass in a particular reference frame. Differentiating this equation with respect to time, we obtain

$$M\frac{d\mathbf{r}_{\text{cm}}}{dt} = m_1\frac{d\mathbf{r}_1}{dt} + m_2\frac{d\mathbf{r}_2}{dt} + \cdots + m_n\frac{d\mathbf{r}_n}{dt} \tag{9-7}$$

9-2
MOTION OF THE CENTER OF MASS

or

$$M\mathbf{v}_{cm} = m_1\mathbf{v}_1 + m_2\mathbf{v}_2 + \cdots + m_n\mathbf{v}_n,$$

where $\mathbf{v}_1$ is the velocity of the first particle, etc., and $d\mathbf{r}_{cm}/dt \ (= \mathbf{v}_{cm})$ is the velocity of the center of mass.

Differentiating Eq. 9-7 with respect to time, we obtain

$$M\frac{d\mathbf{v}_{cm}}{dt} = m_1\frac{d\mathbf{v}_1}{dt} + m_2\frac{d\mathbf{v}_2}{dt} + \cdots + m_n\frac{d\mathbf{v}_n}{dt} \qquad (9\text{-}8)$$

$$= m_1\mathbf{a}_1 + m_2\mathbf{a}_2 + \cdots + m_n\mathbf{a}_n,$$

where $\mathbf{a}_1$ is the acceleration of the first particle, etc., and $d\mathbf{v}_{cm}/dt \ (= \mathbf{a}_{cm})$ is the acceleration of the center of mass of the system. Now, from Newton's second law, the force $\mathbf{F}_1$ acting on the first particle is given by $\mathbf{F}_1 = m_1\mathbf{a}_1$. Likewise, $\mathbf{F}_2 = m_2\mathbf{a}_2$, etc. We can then write Eq. 9-8 as

$$M\mathbf{a}_{cm} = \mathbf{F}_1 + \mathbf{F}_2 + \cdots + \mathbf{F}_n. \qquad (9\text{-}9)$$

Hence *the total mass of the group of particles times the acceleration of its center of mass is equal to the vector sum of all the forces acting on the group of particles.*

Among all these forces will be *internal* forces exerted by the particles on each other. However, from Newton's third law, these internal forces will occur in equal and opposite pairs, so that they contribute nothing to the sum. Hence the internal forces can be removed from the problem. The right-hand sum in Eq. 9-9 represents the sum of only the *external* forces acting on all the particles. We can then rewrite Eq. 9-9 as simply

$$M\mathbf{a}_{cm} = \mathbf{F}_{ext}. \qquad (9\text{-}10)$$

This states that *the center of mass of a system of particles moves as though all the mass of the system were concentrated at the center of mass and all the external forces were applied at that point.*

Notice that we obtain this simple result without specifying the nature of the system of particles. The system can be a rigid body in which the particles are in fixed positions with respect to one another, or it can be a collection of particles in which there may be all kinds of internal motion. Whatever the system is, and however its individual parts may be moving, its center of mass moves according to Eq. 9-10.

Hence, instead of treating bodies as single particles as we have done in previous chapters, we can treat them as collections of particles. Then we can obtain the translational motion of the body, that is, the motion of its center of mass, by assuming that all the mass of the body is concentrated at its center of mass and all the external forces are applied at that point.* This, in fact, is the procedure that we followed implicitly in all our force diagrams and problem solving.

Aside from justifying and making more concrete our previous procedure, we have now found how to describe the translational motion of a system of particles and how to describe the translational motion of a body which may be rotating as well. In this chapter and the next we apply this result to the linear motion of a system of particles. In later chapters we shall see how it simplifies the analysis of rotational motion.

---

* When the external force is gravity, it acts through the *center of gravity* of the body. In every case we have considered, the center of gravity coincides with the center of mass, which is a more general concept. We will discuss the conditions under which these points are different for a body in Chapter 14.

Consider three particles of different masses acted on by external forces, as shown in Fig. 9-6. Find the acceleration of the center of mass of the system.

First we find the coordinates of the center of mass. From Eq. 9-3,

$$x_{cm} = \frac{(8.0 \times 4) + (4.0 \times -2) + (4.0 \times 1)}{16}\ m = 1.8\ m,$$

$$y_{cm} = \frac{(8.0 \times 1) + (4.0 \times 2) + (4.0 \times -3)}{16}\ m = 0.25\ m.$$

These are shown as C in Fig. 9-6.

To obtain the acceleration of the center of mass, we first determine the resultant external force acting on the system consisting of the three particles. The x-component of this force is

$$F_x = 14\ N - 6.0\ N = 8.0\ N,$$

and the y-component is

$$F_y = 16\ N.$$

Hence the resultant external force has a magnitude

$$F = \sqrt{(8.0)^2 + (16)^2}\ N = 18\ N,$$

and makes an angle θ with the x-axis given by

$$\tan \theta = \frac{16\ N}{8.0\ N} = 2.0 \quad \text{or} \quad \theta = 63°.$$

Then, from Eq. 9-10, the acceleration of the center of mass is

$$a_{cm} = \frac{F}{M} = \frac{18\ N}{16\ kg} = 1.1\ m/s^2,$$

making an angle of 63° with the x-axis.

Although the three particles will change their relative positions as time goes on, the center of mass will move, as shown, with this constant acceleration.

EXAMPLE 3

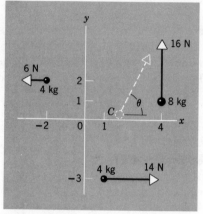

figure 9-6
Example 3. Finding the motion of the center of mass of three masses, each subjected to a different force. The forces all lie in the plane defined by the particles. The distances indicated along the axes are in meters.

The *momentum* of a single particle is a vector **p** defined as the product of its mass m and its velocity **v**. That is,

$$\mathbf{p} = m\mathbf{v}. \tag{9-11}$$

Momentum, being the product of a scalar by a vector, is itself a vector. Because it is proportional to **v**, the momentum **p** of a particular particle depends on the reference frame of the observer; we must always specify this frame.

Newton, in his famous *Principia*, expressed the second law of motion in terms of momentum (which he called "quantity of motion"). Expressed in modern terminology Newton's second law reads: *The rate of change of momentum of a body is proportional to the resultant force acting on the body and is in the direction of that force.* In symbolic form this becomes

$$\mathbf{F} = \frac{d\mathbf{p}}{dt}. \tag{9-12}$$

If our system is a single particle of (constant) mass m, this formulation of the second law is equivalent to the form $\mathbf{F} = m\mathbf{a}$, which we have used up to now. That is, if m is a constant, then

$$\mathbf{F} = \frac{d\mathbf{p}}{dt} = \frac{d}{dt}(m\mathbf{v}) = m\frac{d\mathbf{v}}{dt} = m\mathbf{a}.$$

9-3
LINEAR MOMENTUM OF A PARTICLE

The relations $\mathbf{F} = m\mathbf{a}$ and $\mathbf{F} = d\mathbf{p}/dt$ for single particles are completely equivalent in classical mechanics.

In relativity theory (See Supplementary Topic V) the second law for a single particle in the form $\mathbf{F} = m\mathbf{a}$ is not valid. However, it turns out that Newton's second law in the form $\mathbf{F} = d\mathbf{p}/dt$ *is* still a valid law *if* the momentum $\mathbf{p}$ for a single particle is defined not as $m_0\mathbf{v}$ but as

$$\mathbf{p} = \frac{m_0\mathbf{v}}{\sqrt{1 - v^2/c^2}}. \tag{9-13}$$

This result suggested a new definition of mass (compare Eqs. 9-11 and 9-13)

$$m = \frac{m_0}{\sqrt{1 - v^2/c^2}},$$

so that the momentum could still be written as $\mathbf{p} = m\mathbf{v}$; see Section 8-9. In this equation v is the speed of the particle, c is the speed of light, and m_0 is the "rest mass" of the body (its mass when $v = 0$). From this definition we must expect the mass of a particle to increase with its speed. Elementary particles such as electrons, protons, etc., may acquire enormous speeds, comparable to the speed of light. This concept can be put to a direct test in such cases because the increase in mass over the rest mass for such particles is large enough to measure accurately. Results of all such experiments indicate that this effect is real and given exactly by the equation above. (See for example, Fig. 8-8.)

Suppose that instead of a single particle we have a system of n particles, with masses m_1, m_2, etc. We shall continue to assume, as we did in Section 9-2, that no mass enters or leaves the system, so that the mass $M \,(= \Sigma\, m_i)$ of the system remains constant with time. The particles may interact with each other and external forces may act on them as well. Each particle will have a velocity and a momentum. Particle 1 of mass m_1 and velocity $\mathbf{v}_1$ will have a momentum $\mathbf{p}_1 = m_1\mathbf{v}_1$, for example. The system as a whole will have a *total momentum* $\mathbf{P}$ in a particular reference frame, which is defined to be simply the vector sum of the momenta of the individual particles in that same frame, or

$$\mathbf{P} = \mathbf{p}_1 + \mathbf{p}_2 + \cdots + \mathbf{p}_n \tag{9-14}$$
$$= m_1\mathbf{v}_1 + m_2\mathbf{v}_2 + \cdots + m_n\mathbf{v}_n.$$

If we compare this relation with Eq. 9-7, we see at once that

$$\mathbf{P} = M\mathbf{v}_{cm}, \tag{9-15}$$

which is an equivalent definition for the momentum of a system of particles. In words, Eq. 9-15 states: *The total momentum of a system of particles is equal to the product of the total mass of the system and the velocity of its center of mass.*

We have seen (Eq. 9-10) that Newton's second law for a system of particles can be written as

$$\mathbf{F}_{ext} = M\mathbf{a}_{cm} \tag{9-10}$$

in which $\mathbf{F}_{ext}$ is the vector sum of all the *external* forces acting on the system; we recall that the internal forces acting between particles cancel in pairs because of Newton's third law (see Fig. 9-7). If we differentiate Eq. 9-15 with respect to time we obtain, for an assumed constant mass M,

$$\frac{d\mathbf{P}}{dt} = M\frac{d\mathbf{v}_{cm}}{dt} = M\mathbf{a}_{cm}. \tag{9-16}$$

9-4
LINEAR MOMENTUM OF A SYSTEM OF PARTICLES

figure 9-7
Relationship between the forces acting on a system of three masses m_1, m_2, and m_3. *(a)* All the forces acting on each mass are shown here, as well as the location of the center of mass. On m_1 act forces $\mathbf{f}_{21}$ and $\mathbf{f}_{31}$ exerted by m_2 and m_3 respectively, as well as $\mathbf{F}_1$, a force from some external agent. Similar sets of forces act on m_2 and m_3. However, according to Newton's third law, internal forces $\mathbf{f}_{31}$ and $\mathbf{f}_{13}$ must be equal and opposite and must both lie along the line of centers of m_1 and m_3. Similar statements hold for the other two pairs of action-reaction forces. *(b)* If we are interested only in the motion of the system as a whole, we may consider all the forces to act on a mass $M = m_1 + m_2 + m_3$, located at the center of mass. Owing to the equality of the action-reaction pairs of internal forces as just stated, they cancel each other identically, leaving only the three external forces $\mathbf{F}_1$, $\mathbf{F}_2$, and $\mathbf{F}_3$. We add these graphically in *(c)* to yield a net force $\mathbf{F}_{\text{ext}}$ acting on the center of mass of the system.

Comparison of Eqs. 9-10 and 9-16 allows us to write Newton's second law for a system of particles in the form

$$\mathbf{F}_{\text{ext}} = \frac{d\mathbf{P}}{dt}. \tag{9-17}$$

This equation is the generalization of the single-particle equation $\mathbf{F} = d\mathbf{p}/dt$ (Eq. 9-12) to a system of many particles, no mass entering or leaving the system. Equation 9-17 reduces to Eq. 9-12 for the special case of a single particle, there being only external forces on a one-particle system.

Suppose that the sum of the external forces acting on a system is zero. Then, from Eq. 9-17,

$$\frac{d\mathbf{P}}{dt} = 0 \qquad \text{or} \quad \mathbf{P} = \text{a constant.}$$

When the resultant external force acting on a system is zero, the total vector momentum of the system remains constant. This simple but quite general result is called *the principle of the conservation of linear momentum.* We shall see that it is applicable to many important physical situations.

The conservation of linear momentum principle is the second of the great conservation principles that we have met so far, the first being the conservation of energy principle. Later we shall meet several others, among them the conservation of electric charge and of angular momentum. Conservation principles are of theoretical and practical importance in physics because they are simple and universal. They are all cast in the form: While the system is changing there is one aspect of the system that remains unchanged. Different observers, each in his own reference frame, would all agree, if they watched the same changing system, that the conservation laws applied to the system. For the con-

9-5
CONSERVATION OF LINEAR MOMENTUM

servation of linear momentum, for example, observers in different reference frames would assign different values of **P** to the linear momentum of the system, but each would agree (assuming $\mathbf{F}_{ext} = 0$) that his own value of **P** remained unchanged as the particles that make up the system move about.

The total momentum of a system can only be changed by external forces acting on the system. The internal forces, being equal and opposite, produce equal and opposite changes in momentum which cancel each other. For a system of particles

$$\mathbf{p}_1 + \mathbf{p}_2 + \cdots + \mathbf{p}_n = \mathbf{P},$$

so that when the total momentum **P** is constant we have

$$\mathbf{p}_1 + \mathbf{p}_2 + \cdots + \mathbf{p}_n = \text{a constant} = \mathbf{P}_0. \qquad (9\text{-}18)$$

The momenta of the individual particles may change, but their sum remains constant if there is no external force.

Momentum is a vector quantity. Equation 9-18 is therefore equivalent to three scalar equations, one for each coordinate direction. Hence the conservation of linear momentum supplies us with three conditions on the motion of a system to which it applies. The conservation of energy on the other hand supplies us with only one condition on the motion of a system to which it applies, because energy is a scalar.

The law of the conservation of linear momentum holds true even in atomic and nuclear physics, although Newtonian mechanics does not. Hence this conservation law must be more fundamental than the Newtonian principles. In our derivation of this principle we must have made more rigid assumptions than we needed to. This is true even in the framework of classical mechanics. Recall the key role played by Newton's third law in this deduction of momentum conservation. This law was used to justify the assumption that the sum of the internal forces acting on all the particles is zero. However, it is somewhat artificial to regard the internal forces in a piece of matter as resulting from pairs of equal and opposite forces between the various pairs of atoms. These internal forces are actually many-body forces, depending on not only the relative separation and orientation of two atoms but also on the positions and orientations of neighboring atoms. If it were possible to prove our assumption without using Newton's third law, the law of conservation of linear momentum would not depend on the validity of the third law of motion. Actually we can prove this assumption on the basis of a much less stringent requirement than that the third law should hold. The proof lies outside the scope of this text.*

9-6
SOME APPLICATIONS OF THE MOMENTUM PRINCIPLE

EXAMPLE 4

Consider first a problem in which an external force acts on a system of particles. Recall our previous discussion of projectile motion (Chapter 4). Now let us imagine that our projectile is a fireworks shell that explodes while in flight. The path of the shell is shown in Fig. 9-8. We assume that the air resistance is negligible. The system is the shell, the earth is our reference frame, and the external force is that of gravity. At the point x_1 the shell explodes and shell fragments are blown in all directions. What can we say about the motion of this system thereafter?

The forces of the explosion are all *internal forces*; they are forces exerted by part of the system on other parts of the system. These forces may change the momenta of all the individual fragments from the values they had when they made up the shell, but they cannot change the *total* vector momentum of the system. Only an external force can change the total momentum. The external

* See "On Newton's Third Law and the Conservation of Momentum" by E. Gerjuoy, *American Journal of Physics*, November 1949.

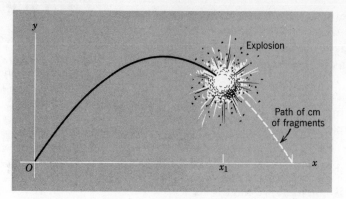

figure 9-8
Example 4. A projectile, following the usual parabolic trajectory, bursts at x_1. The center of mass of the fragments continues along the same parabolic path.

force, however, is simply that due to gravity. Because a system of particles as a whole moves as though all its mass were concentrated at the center of mass with the external force applied there, the center of mass of the fragments will continue to move in the parabolic trajectory that the unexploded shell would have followed. The change in the total momentum of the system attributable to gravity is the same whether the shell explodes or not.

What can you say about the *mechanical energy* of the system before and after the explosion?

Consider now two blocks A and B, of masses m_A and m_B, coupled by a spring and resting on a horizontal frictionless table. Let us pull the blocks apart and stretch the spring, as in Fig. 9-9, and then release the blocks. Describe the subsequent motion.

EXAMPLE 5

figure 9-9
Example 5. Two blocks A and B, resting on a frictionless surface, are connected by a spring. If they are held apart and then released, the sum of their momenta remains zero.

If the system consists of the two blocks and spring, then after we have released the blocks there is no net *external* force acting on the system. We can therefore apply the conservation of linear momentum to the motion. The momentum of the system before the blocks were released was zero in the reference frame shown attached to the table, so the momentum must remain zero thereafter. The total momentum can be zero even though the blocks move because momentum is a *vector* quantity. One block will have positive momentum (A moves in the $+x$ direction) and the other block will have negative momentum (B moves in the $-x$ direction). From the conservation of momentum we have

initial momentum = final momentum.

$$0 = m_B\mathbf{v}_B + m_A\mathbf{v}_A.$$

Therefore

$$m_B\mathbf{v}_B = -m_A\mathbf{v}_A$$

or

$$\mathbf{v}_A = -\frac{m_B}{m_A}\mathbf{v}_B.$$

For example, if m_A is 2 kg and m_B is 1 kg, then $\mathbf{v}_A$ will always be one-half $\mathbf{v}_B$ in magnitude and oppositely directed as the blocks move.

The kinetic energy of block A is $\frac{1}{2}m_A v_A^2$ and can be written as $(m_A v_A)^2/2m_A$ and that of block B is $\frac{1}{2}m_B v_B^2$ and can be written as $(m_B v_B)^2/2m_B$. But

$$\frac{K_A}{K_B} = \frac{2m_B(m_A v_A)^2}{2m_A(m_B v_B)^2} = \frac{m_B}{m_A},$$

in which $m_A v_A$ equals $m_B v_B$ because of momentum conservation. The kinetic energies of the blocks at any instant are inversely proportional to their respective masses. Because mechanical energy is conserved also, the blocks will continue to oscillate back and forth, the energy being partly kinetic and partly potential. What is the motion of the center of mass of this system?

If mechanical energy is not conserved, as would be true if friction were present, the motion will die out as the energy is dissipated. Can we apply the conservation of linear momentum in this case? Explain.

EXAMPLE 6

As an example of recoil, consider radioactive decay. An α-particle (the nucleus of a helium atom) is emitted from a uranium-238 nucleus, originally at rest, with a speed of 1.4×10^7 m/s and a kinetic energy of 4.1 MeV (million electron volts). Find the recoil speed of the residual nucleus (thorium-234).

We think of the system (thorium + α-particle) as initially bound and forming the uranium nucleus. The system then fragments into two separate parts. The momentum of the system before fragmentation is zero. In the absence of external forces, the momentum after fragmentation is also zero. Hence,

initial momentum = final momentum,

$$0 = M_\alpha \mathbf{v}_\alpha + M_{\text{Th}} \mathbf{v}_{\text{Th}},$$

$$\mathbf{v}_{\text{Th}} = -\frac{M_\alpha}{M_{\text{Th}}} \mathbf{v}_\alpha.$$

The ratio of the α-particle mass to the thorium nucleus mass, M_α/M_{Th}, is 4/234 and $v_\alpha = 1.4 \times 10^7$ m/s. Hence,

$$v_{\text{Th}} = -(4/234)(1.4 \times 10^7 \text{ m/s}) = -2.4 \times 10^5 \text{ m/s}.$$

The minus sign indicates that the residual thorium nucleus recoils in a direction exactly opposite to the motion of the α-particle, so as to give a resultant vector momentum of zero.

How can we compute the kinetic energy of the recoiling nucleus (see previous example)? Where does the energy of the fragments come from?

EXAMPLE 7

Consider now the apparently simple example of a ball thrown up from the earth by a person and then caught by him on its return. To simplify matters we can consider the person to be part of the earth since he does not lose contact with it. We also assume that air resistance is negligible.

The system being considered consists of the earth and the ball. The gravitational forces between the parts of the system are now internal forces. Let us choose a reference frame in which the system (earth + ball) is at rest. When the ball is thrown up, the earth must recoil as seen by an observer in this reference frame. The momentum of the system (earth + ball) is zero initially and no external forces act. Therefore, momentum is conserved and the total momentum remains zero throughout the motion. The upward momentum acquired by the ball is balanced by an equal and opposite downward momentum of the earth. We have

initial momentum = final momentum,

$$0 = m_B \mathbf{v}_B + m_E \mathbf{v}_E,$$

$$m_B \mathbf{v}_B = -m_E \mathbf{v}_E.$$

Here m_B and m_E are the masses of ball and earth respectively and $\mathbf{v}_B$ and $\mathbf{v}_E$ are the velocities of the ball and the earth in our selected reference frame. Owing to the enormous mass of the earth in comparison with the ball, the recoil speed of the earth is negligibly small.

As the ball and earth separate, the internal force of gravitational attraction pulls them together until they cease separating and begin to approach one another. As the ball falls toward the earth, the earth falls toward the ball with an equal but oppositely directed momentum. As the ball is caught, its momentum is neutralized by (and it neutralizes) the momentum of the earth. Both objects lose their relative motion, the total momentum is still zero, and the original situation before throwing is restored.

You will recall that when we discussed the conservation of energy in the presence of gravitational potential, we neglected to consider the motion of the earth itself. We took the surface of the earth as our zero level of gravitational potential energy. The reference position did not matter, because we were concerned only with *changes* in potential energy. However, in computing changes in kinetic energy, we assumed that the earth remained stationary, as in the case of the ball thrown up from the earth.

In principle, we cannot ignore the change in the kinetic energy of the earth itself. For example, when the ball falls toward the earth, the earth is slightly accelerated toward the ball. We neglected this fact before because we assumed that the change in kinetic energy of the earth is negligible. This result is not obvious, because although the earth's speed will certainly be small, its mass is enormous and the kinetic energy acquired might be significant. To settle the point we compute the ratio of the kinetic energy of the earth to that of the ball. Using $m_E v_E = m_B v_B$ from momentum conservation, we have

$$\frac{K_E}{K_B} = \frac{\frac{1}{2} m_E v_E{}^2}{\frac{1}{2} m_B v_B{}^2} = \frac{\frac{1}{2}(m_E v_E)^2}{\frac{1}{2}(m_B v_B)^2} \cdot \frac{m_B}{m_E} = \frac{m_B}{m_E}.$$

Since the mass of the ball m_B is negligibly small compared to the mass of the earth m_E, the kinetic energy acquired by the earth, K_E, is negligibly small compared to that of the ball, K_B. For example, if $m_B = 6$ kg (a rather massive ball), then, since $m_E = 6 \times 10^{24}$ kg, $K_E/K_B = 10^{-24}$!

Notice that this problem is identical in principle to Example 5. The differences are only those of detail; in one the potential energy is elastic and in the other the potential energy is gravitational; in one the masses are pictured as of the same order of magnitude, and in the other they are of very different orders of magnitude.

9-7 SYSTEMS OF VARIABLE MASS

So far we have dealt only with systems in which the total system mass M remained constant with time. Now we consider systems in which mass enters or leaves the system while we are observing it, dM/dt being positive in the former case and negative in the latter.

Figure 9-10a shows a system of mass M whose center of mass is moving with velocity $\mathbf{v}$ as seen from a particular reference frame. An external force $\mathbf{F}_{\text{ext}}$ acts on the system. At a time Δt later the configuration has changed to that shown in Fig. 9-10b. A mass ΔM has been ejected from the system, its center of mass moving with velocity $\mathbf{u}$ as seen by our observer. The system mass is reduced to $M - \Delta M$ and the velocity $\mathbf{v}$ of the center of mass of the system is changed to $\mathbf{v} + \Delta\mathbf{v}$.

The student may imagine the system of Fig. 9-10 to represent a rocket. It ejects hot gas from its orifice at a fairly high speed, decreasing its own mass and increasing its own speed. In a rocket the loss of mass is continuous during the burning process. The external force $\mathbf{F}_{\text{ext}}$ is *not* the thrust of the rocket but is the force of gravity on the rocket and the resisting force of the atmosphere.

To analyze the situation let us, for the time being, define the system to be one of constant mass. This means that in Fig. 9-10b, we shall include in our system not only the mass $M - \Delta M$ of the body but also the ejected mass ΔM, the total mass of the system being the M of Fig. 9-10a. Doing so permits us to apply

(a)   (b)

figure 9-10
A mass M moving with velocity $\mathbf{v}$ ejects a mass ΔM during a time interval Δt. An external force $\mathbf{F}_{ext}$ (not shown) acts on the system.

the results that we have derived so far for constant mass systems. We shall see that this approach leads us to the form of Newton's second law for systems in which the mass is not constant.

From Eq. 9-17

$$\mathbf{F}_{ext} = \frac{d\mathbf{P}}{dt} \qquad (9\text{-}17)$$

we can write, as an approximate result for the finite time interval Δt,

$$\mathbf{F}_{ext} \cong \frac{\Delta\mathbf{P}}{\Delta t} = \frac{\mathbf{P}_f - \mathbf{P}_i}{\Delta t}$$

in which $\mathbf{P}_f$ is the (final) system momentum in Fig. 9-10b and $\mathbf{P}_i$ is the (initial) system momentum for Fig. 9-10a. But $\mathbf{P}_f = (M - \Delta M)(\mathbf{v} + \Delta\mathbf{v}) + \Delta M\mathbf{u}$ and $\mathbf{P}_i = M\mathbf{v}$. This leads to

$$\mathbf{F}_{ext} \cong \frac{[(M - \Delta M)(\mathbf{v} + \Delta\mathbf{v}) + \Delta M\mathbf{u}] - [M\mathbf{v}]}{\Delta t}$$

$$= M\frac{\Delta\mathbf{v}}{\Delta t} + [\mathbf{u} - (\mathbf{v} + \Delta\mathbf{v})]\frac{\Delta M}{\Delta t}. \qquad (9\text{-}19)$$

Now, if we let Δt approach zero, the configuration of Fig. 9-10b approaches that of Fig. 9-10a; that is, $\Delta\mathbf{v}/\Delta t$ approaches $d\mathbf{v}/dt$, the acceleration of the body in Fig. 9-10a. The quantity ΔM is the mass ejected in Δt: this leads to a decrease in the mass M of the original body. Since dM/dt, the change in mass of the body with time, is intrinsically negative in this case, the positive quantity $\Delta M/\Delta t$ is replaced by $-dM/dt$ as Δt approaches zero. Finally, $\Delta\mathbf{v}$ goes to zero as Δt approaches zero. Making these changes in Eq. 9-19 leads to

$$\mathbf{F}_{ext} = M\frac{d\mathbf{v}}{dt} + \mathbf{v}\frac{dM}{dt} - \mathbf{u}\frac{dM}{dt} \qquad (9\text{-}20a)$$

or

$$\mathbf{F}_{ext} = \frac{d}{dt}(M\mathbf{v}) - \mathbf{u}\frac{dM}{dt}, \qquad (9\text{-}20b)$$

which is Newton's second law, defining the external forces on a body (like that of Fig. 9-10a) whose mass is changing.

Note that these equations reduce to the familiar forms $\mathbf{F}_{ext} = M\mathbf{a}$ and $\mathbf{F}_{ext} = (d/dt)(M\mathbf{v})$ respectively for the special case of a body of constant mass ($dM/dt = 0$). It is important to note that we *cannot* derive a general expression for Newton's second law for variable mass systems by treating the mass in $\mathbf{F}_{ext} = d\mathbf{P}/dt = d(M\mathbf{v})/dt$ as a variable. For this leads to

$$\mathbf{F}_{ext} = d(M\mathbf{v})/dt = M\, d\mathbf{v}/dt + \mathbf{v}\, dM/dt,$$

which is only a special case of the more general Eq. 9-20, namely, the case in which either (a) $dM/dt = 0$, a system of constant mass, or (b) $\mathbf{u} = 0$, a special choice of reference frame. We *can* use $\mathbf{F}_{ext} = d\mathbf{P}/dt$ to analyze variable mass systems *only* if we apply it to an *entire system of constant mass* having parts among which there is an interchange of mass. This indeed is what we have done in deriving Eqs. 9-20. The importance of the momentum formulation $\mathbf{F}_{ext} = d\mathbf{P}/dt$ in classical physics lies in the fact that it highlights momentum conservation and gives us a simple, physical way to treat complicated systems. Since the choice of what we will take as the system is ours to make, we can always choose a system of constant mass by defining our system broadly enough.

However, it is often convenient, as in rocket problems, to choose a system whose mass varies with time. In such cases we apply Newton's second law of Eqs. 9-20 in a form that is sometimes more convenient and interpretable more physically. The quantity $\mathbf{u} - (\mathbf{v} + \Delta\mathbf{v})$ in Eq. 9-19 is just $\mathbf{v}_{rel}$, the relative velocity of the ejected mass with respect to the main body. Therefore Eq. 9-20a may be written as

$$M\frac{d\mathbf{v}}{dt} = \mathbf{F}_{ext} + (\mathbf{u} - \mathbf{v})\frac{dM}{dt} \qquad (9\text{-}21a)$$

or

$$M\frac{d\mathbf{v}}{dt} = \mathbf{F}_{ext} + \mathbf{v}_{rel}\frac{dM}{dt}. \qquad (9\text{-}21b)$$

The last term in Eq. 9-21b, $[\mathbf{v}_{rel}\,(dM/dt)]$, is the rate at which momentum is being transferred into (or out of) the system by the mass that the system has ejected (or collected). It can be interpreted as the force exerted *on* the system by the mass that leaves it (or joins it). For a rocket, this term is called the *thrust* and it is the rocket designer's aim to make it as large as possible. Inspection of Eq. 9-21b shows that this requires that the rocket eject as much mass per unit time as possible and that the speed of the ejected mass relative to the rocket be as high as possible. We can rewrite Eq. 9-21b as

$$M\frac{d\mathbf{v}}{dt} = \mathbf{F}_{ext} + \mathbf{F}_{reaction}$$

in which $\mathbf{F}_{reaction}\ (= \mathbf{v}_{rel}\,dM/dt)$ is the reaction force exerted on the system by the mass that leaves it.

EXAMPLE 8

A machine gun is mounted on a car that can roll with negligible friction on a horizontal surface as in Fig. 9-11a. The mass of the system (car + gun) at a particular instant is M. At that same instant the gun is firing bullets of mass m whose velocity, in the reference frame shown, is $\mathbf{u}$. The velocity of the car in this same frame is $\mathbf{v}$ and the velocity of the bullets *with respect to the car* is $\mathbf{u} - \mathbf{v}$. The number of bullets fired per unit time is n. What is the acceleration of the car?

We select the car and gun as our system. Because its mass M is variable, we apply Newton's second law in the form given in Eq. 9-21b. Since no net external force acts on the system, we have $\mathbf{F}_{ext} = 0$ in that equation, yielding

$$M\frac{d\mathbf{v}}{dt} = \mathbf{v}_{rel}\frac{dM}{dt}.$$

Now $d\mathbf{v}/dt$ is $\mathbf{a}$, the acceleration of the system; $\mathbf{v}_{rel}$ is $\mathbf{u} - \mathbf{v}$, pointing to the left in Fig. 9-11a, and dM/dt is $-mn$. Inserting these in the equation above yields

$$\mathbf{a} = \frac{d\mathbf{v}}{dt} = -\frac{\mathbf{v}_{rel}(mn)}{M}.$$

This shows that $\mathbf{a}$ points in the direction opposite to $\mathbf{v}_{rel}$, that is, $\mathbf{a}$ points to the right in Fig. 9-11a. If $v_{rel} = 500$ m/s, $m = 10$ g, $n = 10/$s, and $M = 200$ kg at some instant, then at that instant

$$a = \frac{(500\text{ m/s})(10^{-2}\text{ kg})(10/\text{sec})}{200\text{ kg}} = 0.25\text{ m/s}^2.$$

The magnitude of the average "thrust" of the ejected bullets on the system (car + gun) at this instant is given by

$$F = v_{rel}nm = (500\text{ m/s})(10/\text{s})(10^{-2}\text{ kg})$$

$$= 50\text{ N}.$$

figure 9-11

(a) Example 8. A machine gun is fixed to a car that rolls with negligible friction. The gun fires bullets of mass m at a rate (number per unit time) n, the velocity of the bullets with respect to the gun being $\mathbf{u} - \mathbf{v}$. At the instant shown some bullets have already left the system. The velocities indicated for the car and the bullets are those that would be measured by an observer in a reference frame fixed to the rails as shown. The reaction force on the system is $\mathbf{F} = -mn\mathbf{v}_{rel} = (dM/dt)\mathbf{v}_{rel}$. *(b)* A rocket moves through space with negligible external forces. Gas particles are ejected from the exhaust, the particles having a velocity $\mathbf{u} - \mathbf{v}$ with respect to the rocket. The rate at which mass is expelled at the exhaust is $-dM/dt$. The reaction force on the rocket is $\mathbf{F} = (dM/dt)\mathbf{v}_{rel}$. The velocities indicated for the rocket and exhaust gases are relative to the ground.

In Figure 9-11b we show the analogous situation for a rocket. It is instructive to view this problem from the point of view of Newton's third law and the momentum principle. Choose a fixed-mass system (rocket + gas) and attach a reference frame to its center of mass. The rocket forces a jet of hot gases from its exhaust; this is the action force. The jet of hot gases exerts a force on the rocket, propelling it forward. This is the reaction force. These forces are internal forces in the system (rocket + gas). In the absence of external forces the total momentum of the system is constant (the center of mass, initially at rest, remains at rest). The individual parts of the system (rocket and gases) may change their momentum, however; with respect to the center of mass frame, the hot gases acquire momentum in the backward direction and the rocket acquires an equal amount of momentum in the forward direction.

You can analyze the system (bullets + car and gun) in a similar way.

EXAMPLE 9

A rocket weighs 30,000 lb when fueled up on the launching pad. It is fired vertically upward and, at burnout, weighs 10,000 lb. Gases are exhausted at the rate of 10 slugs/s with a velocity of 5000 ft/s, relative to the rocket (exhaust velocity), both quantities being assumed to be constant while the fuel is burning.

(a) What is the thrust? The thrust $\mathbf{F}$ is the last term in Eq. 9-21b, or

$$F = v_{rel}\frac{dM}{dt} = (5000 \text{ ft/s})(10 \text{ slugs/s}) = 50,000 \text{ lb}.$$

Note that initially, when the fuel tanks are full, the net upward force acting on the rocket (neglecting air resistance) is the thrust (50,000 lb) minus the initial weight (30,000 lb) or 20,000 lb. Just before burnout the net upward force is 50,000 lb minus 10,000 lb or 40,000 lb.

(b) If we could neglect *all external forces*, including gravity and air resistance, what would be the speed of the rocket at burnout?

If we put $\mathbf{F}_{ext} = 0$ in Eq. 9-21b, we have

$$M\frac{d\mathbf{v}}{dt} = \mathbf{v}_{rel}\frac{dM}{dt} \quad \text{or} \quad d\mathbf{v} = \mathbf{v}_{rel}\frac{dM}{M}.$$

Integrating this expression (see Appendix I) from the instant the velocity is $\mathbf{v}_0$ and the mass of the rocket is M_0 to the instant when the velocity is $\mathbf{v}$ and the mass of the rocket is M, we obtain

$$\int_{\mathbf{v}_0}^{\mathbf{v}} d\mathbf{v} = \mathbf{v}_{rel}\int_{M_0}^{M}\frac{dM}{M},$$

the exhaust velocity being assumed constant during this time. This yields

$$\mathbf{v} - \mathbf{v}_0 = -\mathbf{v}_{rel}\ln\left(M_0/M\right) = -\mathbf{v}_{rel}\ln\left(1 + \frac{M_0 - M}{M}\right).$$

Hence the change in velocity of the rocket in any interval of time depends only on the exhaust velocity (being opposite in direction from it) and on the fraction of mass exhausted during that time interval.

In our example, $v_0 = 0$ and $M_0/M = (30,000/10,000) = 3.0$, so that the speed of the rocket at burnout is

$$v = v_{rel}\ln\left(M_0/M\right) = (5000 \text{ ft/s})\ln 3.0 = 3800 \text{ mi/h}.$$

If the external forces of gravity and air resistance were taken into account, the final speed would be smaller.*

Assuming that the rocket starts from rest $(v_0 = 0)$ with an initial mass M_0 and reaches a final velocity v_f at burnout when its mass is M_f, we can write the rocket equation above as

$$\frac{M_f}{M_0} = e^{-v_f/v_{rel}}$$

in which v_{rel} is the exhaust velocity.

The classical rocket (or variable mass) equations imply that the speed of the rocket can increase to any value provided only that the rocket expels enough propellant so that the final remaining mass is sufficiently small. However, we know from relativistic mechanics that a rocket cannot be accelerated to a speed equal to or greater than the speed of light. Once the rocket's speed approaches the relativistic range the classical equations are no longer applicable. One must take into account the variation of inertial mass of a particle with speed and the relativistic velocity formula. The resulting equations apply to a relativistic rocket.†

EXAMPLE 10

Sand drops from a stationary hopper at a rate dM/dt onto a conveyor belt moving with velocity $\mathbf{v}$ in the reference frame of the laboratory, as in Fig. 9-12. What force is required to keep the belt moving at a speed v?

This is a clear-cut example of a force associated with change of mass alone, the velocity being constant. We take as our system the belt of varying mass so that Eq. 9-21b applies. We must put $d\mathbf{v}/dt = 0$ in that equation because the velocity of the belt is constant. Furthermore, to an observer at rest on the belt, the falling sand (and the hopper) would appear to have a horizontal motion with speed v in a direction opposite to that shown for the belt in the laboratory. Therefore $\mathbf{v}_{rel} = -\mathbf{v}$ in Eqs. 9-21. More formally, $\mathbf{v}_{rel} = \mathbf{u} - \mathbf{v}$; but $\mathbf{u} = 0$, so that $\mathbf{v}_{rel} = -\mathbf{v}$. Making these substitutions yields

---

*For an exact solution of the classical rocket problem see "Variable-Mass Dynamics" by J. L. Meriam, *Journal of Engineering Education*, December 1960.

† See "The Equation of Motion for Relativistic Particles and Systems with a Variable Rest Mass," by Kalman B. Pomeranz, *American Journal of Physics*, December 1964.

$$0 = \mathbf{F}_{\text{ext}} - \mathbf{v}\frac{dM}{dt}$$

or

$$\mathbf{F}_{\text{ext}} = \mathbf{v}\frac{dM}{dt}.$$

In this example, dM/dt is positive because the system is gaining mass with time. Hence, as expected, the necessary external force must point in the direction in which the belt moves. Note that, in the absence of friction, the mass of the belt itself does not enter the problem.

The power supplied by the external force is

$$P = \mathbf{F} \cdot \mathbf{v} = \mathbf{v} \cdot \mathbf{F} = \mathbf{v} \cdot (\mathbf{v}\, dM/dt) = v^2(dM/dt).$$

Since $v = $ a constant, we can write this as

$$P = \frac{d(Mv^2)}{dt} = 2\frac{d}{dt}\left(\frac{1}{2}Mv^2\right) = 2\frac{dK}{dt}.$$

This tells us that the power required to keep the belt moving is *twice* the rate at which the kinetic energy of the system is increasing; note that we need not consider the kinetic energy of the belt itself because — its speed being constant — its kinetic energy does not change. It is clear that mechanical energy is not conserved in this case. Where is the other half of the power going? In which of the previous examples did we have conservation of momentum without conservation of mechanical energy?

The student should be able to solve Example 10 alternatively by choosing a fixed-mass system and applying the momentum principle.*

questions

1. Must there necessarily be any mass at the center of mass of a system?
2. Does the center of mass of a solid body necessarily lie within the body? If not, give examples.
3. How is the center of mass concept related to the concept of geographic center of the country? To the population center of the country? What can you conclude from the fact that the geographic center differs from the population center?
4. An amateur sculptor decides to portray a bird (Fig. 9-13). Luckily the final model is actually able to stand upright. The model is formed of a single sheet of metal of uniform thickness. Of the points shown, which is most likely to be the center of mass?

figure 9-13
Question 4

* See "Force, Momentum Change, and Motion" by Martin S. Tiersten, *American Journal of Physics*, January 1969, for an excellent general reference on systems of fixed and variable mass.

5. The location of the center of mass of a group of particles with respect to those particles does not depend on the reference frame used to describe the system. Is that so? Can you choose a reference frame whose origin is actually at the center of mass?

6. If only an external force can change the state of motion of the center of mass of a body, how does it happen that the internal force of the brakes can bring a car to rest?

7. Can a body have energy without having momentum? Explain. Can a body have momentum without having energy? Explain.

8. A light and a heavy body have equal kinetic energies of translation. Which one has the larger momentum?

9. A bird is in a wire cage hanging from a spring balance. Is the reading of the balance when the bird is flying about greater than, less than, or the same as that when the bird sits in the cage?

10. Can a sailboat be propelled by air blown at the sails from a fan attached to the boat?

11. A canoeist in a still pond can reach shore by jerking sharply on the rope attached to the bow of the canoe. How do you explain this? (yes she can! — its true).

12. How might a person standing at rest on a frictionless horizontal surface get altogether off of it?

13. A man stands still on a large sheet of slick ice; in his hand he holds a lighted firecracker. He throws the firecracker into the air. Describe briefly, but as exactly as you can, the motion of the center of mass of the firecracker and the motion of the center of mass of the system consisting of man and firecracker. It will be most convenient to describe each motion during each of the following periods: (a) after he throws the firecracker, but before it explodes; (b) between the explosion and the first piece of firecracker hitting the ice; (c) between the first fragment hitting the ice and the last fragment landing; (d) during the time when all fragments have landed but none has reached the edge of the ice.

14. As stated in the text one cannot use the equation $\mathbf{F}_{ext} = d(M\mathbf{v})/dt$ for a system of variable mass. To show this (a) put the equation in the equivalent form $\left(\mathbf{F}_{ext} - M\dfrac{d\mathbf{v}}{dt}\right)/(dM/dt) = \mathbf{v}$ and (b) show that one side of this equation has the same value in all inertial frames, whereas the other side does not. Hence the equation cannot be generally valid. (c) Show that Eq. 9-20 leads to no such contradiction.

15. You throw an ice cube with velocity $\mathbf{v}$ into a hot gravity-free, evacuated space. The cube gradually melts to liquid water and then boils to water vapor. (a) Is it a system of particles all the time? (b) If so, is it the same system of particles? (c) Does the motion of the center of mass undergo any abrupt changes? (d) Does the total linear momentum change? (e) Would your answers change if the space were not gravity free?

16. In 1920 a prominent newspaper editorialized as follows about the pioneering rocket experiments of Robert H. Goddard, dismissing the notion that a rocket could operate in a vacuum: "That Professor Goddard, with his 'chair' in Clark College and the countenancing of the Smithsonian Institution, does not know the relation of action to reaction, and of the need to have something better than a vacuum against which to react — to say that would be absurd. Of course, he seems only to lack the knowledge ladled out daily in high schools." What is wrong with this argument?

17. The final velocity of the final stage of a multistage rocket is much greater than the final velocity of a single-stage rocket of the same total weight and fuel supply. Explain this fact.

18. As a rocket expels burned fuel the location of the center of mass of the rocket (in a frame attached to the rocket) changes. Must one take this into account in an exact solution of the rocket problem?

19. Explain clearly the distinction between the origin of the varying mass of a classical system and that of a relativistic system.

20. Can you think of variable mass systems other than the examples given in the text?

problems

SECTION 9-1

1. Show that the ratio of the distances of two particles from their center of mass is the inverse ratio of their masses.

2. Experiments using the diffraction of electrons show that the distance between the centers of the carbon (C) and oxygen (O) atoms in the carbon monoxide gas molecule is 1.130×10^{-10} m. Locate the center of mass of a CO molecule relative to the carbon atom.

3. The mass of the moon is about 0.013 times the mass of the earth, and the distance from the center of the moon to the center of the earth is about 60 times the radius of the earth. How far is the center of mass of the earth-moon system from the center of the earth? Take the earth's radius to be 6400 km. *Answer:* 4900 km.

4. The masses and coordinates of four particles are as follows: 5.0 kg, $x = y = 0.0$ cm; 3.0 kg, $x = y = 8.0$ cm; 2.0 kg, $x = 3.0$ cm, $y = 0.0$ cm; 6.0 kg, $x = -2.0$ cm, $y = -6.0$ cm. Find the coordinates of the center of mass of this collection of particles.

5. In the ammonia (NH_3) molecule, the three hydrogen (H) atoms form an equilateral triangle, the distance between centers of the atoms being 1.628×10^{-10} m, so that the center of the triangle is 9.39×10^{-11} m from each hydrogen atom. The nitrogen (N) atom is at the apex of a pyramid, the three hydrogens constituting the base (see Fig. 9-14). The hydrogen-nitrogen distance is 1.014×10^{-10} m. Locate the center of mass relative to the nitrogen atom.
   Answer: 6.74×10^{-12} m toward the plane of the hydrogens, along the axis of symmetry.

6. Find the center of mass of a homogenous semicircular plate. Let a be the radius of the circle.

figure 9-14
Problem 5

SECTION 9-2

7. Two blocks of masses 1.0 kg (weight 2.2 lb) and 3.0 kg (weight 6.6 lb) connected by a spring rest on a frictionless surface. If the two are given velocities such that the first travels at 1.7 m/s (5.6 ft/s) toward the center of mass which remains at rest, what is the velocity of the second?
   Answer: 0.57 m/s (1.9 ft/s), toward center of mass.

8. Two particles P and Q are initially at rest 1.0 m apart. P has a mass of 0.10 kg and Q a mass of 0.30 kg. P and Q attract each other with a constant force of 1.0×10^{-2} N. No external forces act on the system. (a) Describe the motion of the center of mass. (b) At what distance from P's original position do the particles collide?

9. A man of mass m clings to a rope ladder suspended below a balloon of mass M. The balloon is stationary with respect to the ground. (a) If the man begins to climb the ladder at a speed v (with respect to the ladder), in what direction and with what speed (with respect to the earth) will the balloon move? (b) What is the state of motion after the man stops climbing?

   Answer: (a) down, $\dfrac{m}{m + M} v$. (b) Balloon again stationary.

10. A cannon and a supply of cannon balls are inside a sealed railroad car as in Fig. 9-15. The cannon fires to the right; the car recoils to the left. The cannon balls remain in the car after hitting the far wall. Show that no matter

figure 9-15
Problem 10

how the cannon balls are fired the railroad car cannot travel more than its length L, assuming it starts from rest.

11. A dog, weighing 10 lb, is standing on a flatboat so that he is 20 ft from the shore. He walks 8.0 ft on the boat toward shore and then halts. The boat weighs 40 lb, and one can assume there is no friction between it and the water. How far is he from the shore at the end of this time? (Hint: The center of mass of boat + dog does not move. Why?) The shoreline is also to the left in Fig. 9-16.   *Answer:* 14 ft.

figure 9-16
Problem 11

12. A ball of mass m and radius R is placed inside a larger hollow sphere with the same mass and inside radius $2R$. The combination is at rest on a frictionless surface in the position shown in Fig. 9-17. The smaller ball is released, rolls around the inside of the hollow sphere, and finally comes to rest at the bottom. How far will the larger sphere have moved during this process?

figure 9-17
Problem 12

13. Ricardo, mass 80 kg, and Carmelita are enjoying Lake Merced at dusk in a 30-kg canoe. When the canoe is at rest in the placid water they change seats, which are 3.0 m apart and symmetrically located with respect to the canoe's center. Ricardo notices that the canoe moved 0.40 m relative to a submerged log, and calculates Carmelita's mass, which she has declined to tell him. What is it?   *Answer:* 58 kg.

14. An 80-kg man is standing at the rear of a 400-kg iceboat that is moving at 4.0 m/s across ice that may be considered to be frictionless. He decides to walk to the front of the 18 m-long boat and does so at a speed of 2.0 m/s with respect to the boat. How far did the boat move across the ice while he was walking?

SECTION 9-3

15. How fast must an 1800-lb (mass = 816 kg) Volkswagen travel (a) to have the same momentum as a 5850-lb (mass = 2650 kg) Cadillac going 10 mi/h (16 km/h)? (b) To have the same kinetic energy? (c) Make the same calculations using a 10-ton (mass = 9080 kg) truck instead of a Cadillac?
Answer: (a) 33 mi/h (52 km/h). (b) 18 mi/h (29 km/h). (c) 110 mi/h (180 km/h); 33 mi/h (52 km/h).

16. A 50-g ball is thrown into the air with an initial speed of 15 m/s at an angle of 45°. (a) What are the values of the kinetic energy of the ball initially and

just before it hits the ground? (b) Find the corresponding values of the momentum (magnitude and direction). (c) Show that the change in momentum is just equal to the weight of the ball multiplied by the time of flight.

17. A 5.0-kg object with a speed of 30 m/s strikes a steel plate at an angle of 45° and rebounds at the same speed and angle (Fig. 9-18). What is the change (magnitude and direction) of the linear momentum of the object?
    Answer: 210 kg · m/s, perpendicular to the plate.

18. Two bodies, each made up of weights from a set, are connected by a light cord which passes over a light, frictionless pulley with a diameter of 5.0 cm. The two bodies are at the same level. Each originally has a mass of 500 g. (a) Locate their center of mass. (b) Twenty grams are transferred from one body to the other, but the bodies are prevented from moving. Locate the center of mass. (c) The two bodies are now released. Describe the motion of the center of mass and determine its acceleration.

figure 9-18
Problem 17

SECTION 9-4

19. A 200-lb man standing on a surface of negligible friction kicks forward a 0.10-lb stone lying at his feet so that it acquires a speed of 10 ft/s. What velocity does the man acquire as a result?
    Answer: 5.0×10^{-3} ft/s, backward.

20. A pellet gun fires ten 2.0-g pellets per second with a speed of 500 m/s. The pellets are stopped by a rigid wall. (a) What is the momentum of each pellet? (b) What is the kinetic energy of each pellet? (c) What is the average force exerted by the pellets on the wall?

21. A machine gun fires 50-g bullets at a speed of 1000 m/s. The gunner, holding the machine gun in his hands, can exert an average force of 180 N against the gun. Determine the maximum number of bullets he can fire per minute.
    Answer: 220 bullets per minute.

22. A very flexible uniform chain of mass M and length L is suspended from one end so that it hangs vertically, the lower end just touching the surface of a table. The upper end is suddenly released so that the chain falls onto the table and coils up in a small heap, each link coming to rest the instant it strikes the table. Find the force exerted by the table on the chain at any instant, in terms of the weight of chain already on the table at that moment.

SECTION 9-5

23. A body of mass 8.0 kg is traveling at 2.0 m/s under the influence of no external force. At a certain instant an internal explosion occurs, splitting the body into two chunks of 4.0 kg mass each; 16 J of translational kinetic energy are imparted to the two-chunk system by the explosion. Neither chunk leaves the line of the original motion. Determine the speed and direction of motion of each of the chunks after the explosion.
    Answer: One chunk comes to rest. The other moves ahead with a speed of 4.0 m/s.

24. The last stage of a rocket is traveling at a speed of 25,000 ft/s (7600 m/s). This last stage is made up of two parts which are clamped together, namely, a rocket case with a mass of 20 slugs (290 kg) and a payload capsule with a mass of 10 slugs (150 kg). When the clamp is released, a compressed spring causes the two parts to separate with a relative speed of 3000 ft/s (910 m/s). (a) What are the speeds of the two parts after they have separated? Assume that all velocities are along the same line. (b) Find the total kinetic energy of the two parts before and after they separate and account for the difference, if any.

25. A radioactive nucleus, initially at rest, decays by emitting an electron and a neutrino at right angles to one another. The momentum of the electron is 1.2×10^{-22} kg·m/s and that of the neutrino is 6.4×10^{-23} kg·m/s. (a) Find the

direction and magnitude of the momentum of the recoiling nucleus. (b) The mass of the residual nucleus is 5.8×10^{-26} kg. What is its kinetic energy of recoil?

Answer: (a) 1.4×10^{-22} kg · m/s, 150° from the electron track and 120° from the neutrino track. (b) 1.0 eV.

26. Each minute, a special game-warden's machine gun fires 220 10-g rubber bullets with a muzzle velocity of 1200 m/s. How many bullets must be fired at an 85-kg animal charging toward the warden at 4.0 m/s in order to stop the animal in its tracks? (Assume the bullets travel horizontally and drop to the ground after striking the target.)

27. A vessel at rest explodes, breaking into three pieces. Two pieces, having equal mass, fly off perpendicular to one another with the same speed of 30 m/s. The third piece has three times the mass of each other piece. What is the direction and magnitude of its velocity immediately after the explosion? *Answer:* 14 m/s, 135° from either other piece.

28. A shell is fired from a gun with a muzzle velocity of 1500 ft/s, at an angle of 60° with the horizontal. The shell explodes into two fragments of equal mass 50 s after leaving the gun. One fragment, whose speed immediately after the explosion is zero, falls vertically. How far from the gun does the other fragment land, assuming level terrain?

29. A block of mass m rests on a wedge of mass M which, in turn, rests on a horizontal table, as shown in Fig. 9-19. All surfaces are frictionless. If the system starts at rest with point P of the block a distance h above the table, find the velocity of the wedge the instant point P touches the table.

$$\text{Answer: } \sqrt{\frac{2m^2gh\,\cos^2\alpha}{(M+m)(M+m\,\sin^2\alpha)}}.$$

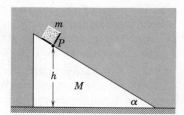

figure 9-19
Problem 29

SECTION 9-7

30. (a) Show that the rocket speed is equal to the exhaust speed when the ratio M_0/M is e (about 2.7). Specify the coordinate system in which this result holds. (b) Show also that the rocket speed is twice the exhaust speed when M_0/M is e^2 (about 7.4).

31. A rocket is moving away from the solar system at a speed of 6.0×10^3 m/s. It fires its rocket engine, which ejects exhaust with a relative velocity of 3.0×10^3 m/s. The mass of the rocket at this time is 4.0×10^4 kg, and it experiences an acceleration of 2.0 m/s². (a) What is the velocity of the exhaust relative to the solar system? (b) At what rate was exhaust ejected during the firing? *Answer:* (a) 3.0×10^3 m/s. (b) 27 kg/s.

32. A widely used rocket fuel is kerosene and liquid oxygen, capable of giving an exhaust velocity v_{rel} of 8000 ft/s (about 1.5 mi/s). (a) Neglect gravity and the weight of fuel tanks, pumps, etc., and find how many pounds of this fuel one needs for each pound of payload in order to get a rocket, starting from rest, to reach a velocity of 7.5 mi/s (the velocity of escape from the earth is 7.0 mi/s). (b) In the Mariner probe to Mars the initial weight was about 200,000 lb and the payload about 500 lb, a "fuel" to payload ratio of 400 to 1. Starting a rocket from rest, what final velocity is achievable under these circumstances? (c) The actual final rocket velocity was about 15 mi/s, much greater than the value found in (b). Explain this, considering the following factors: the external forces and weight neglected in (a) must be taken into account; the rocket uses a number of stages; the initial rocket velocity is that of the earth's surface, in a reference frame attached to the sun.

33. A 6000-kg rocket is set for vertical firing. If the exhaust speed is 1000 m/s, how much gas must be ejected each second to supply the thrust needed (a) to overcome the weight of the rocket, and (b) to give the rocket an initial upward acceleration of 20 m/s²? *Answer:* (a) 59 kg/s. (b) 180 kg/s.

34. Consider a particle acted on by a force having the same direction as its velocity. (a) Using the relativistic relation $F = d(mv)/dt$ for a single particle, show that

$$F \, ds = mv \, dv + v^2 \, dm,$$

where ds is an infinitesimal displacement. (b) Using the relativistic relation $v^2 = (1 - m_0^2/m^2)c^2$, show that

$$mv \, dv = \frac{m_0^2 c^2}{m^2} \, dm.$$

(c) Substitute the relations for $mv \, dv$ and v^2 into result (a) and show that

$$W = \int F \, ds = (m - m_0)c^2.$$

35. A railroad flatcar of weight W can roll without friction along a straight horizontal track as shown. Initially a man of weight w is standing on the car which is moving to the right with speed v_0. What is the change in velocity of the car if the man runs to the left (Fig. 9-20) so that his speed relative to the car is v_{rel} just before he jumps off at the left end?
    Answer: $wv_{rel}/(W + w)$.

figure 9-20
Problem 35

36. Assume that the car in Problem 35 is initially at rest. It holds n men each of weight w. If each man in succession runs with a relative velocity v_{rel} and jumps off the end, do they impart to the car a greater velocity than if they all run and jump at the same time?

37. A toboggan weighing 12 lb and carrying 80 lb of sand slides from rest down an icy slope 300 ft long, inclined 30° below the horizontal. The toboggan has a hole in the bottom, so that the sand leaks out at the rate of 5.0 lb/s. How long does it take the toboggan to reach the bottom of the slope?
    Answer: 6.1 s.

38. Two long barges are floating in the same direction in still water, one with a speed of 10 km/h and the other with a speed of 20 km/h. While they are passing each other, coal is shoveled from the slower to the faster one at a rate of 1000 kg/min. How much additional force must be provided by the driving engines of each barge if neither is to change speed? Assume that the shoveling is always perfectly sideways and that the frictional forces between the barges and the water do not depend on the weight of the barges.

39. A jet airplane is traveling 180 m/s (600 ft/s). The engine takes in 68 m³ (2400 ft³) of air making a mass of 70 kg (4.8 slugs) each second. The air is used to burn 2.9 kg (0.20 slugs) of fuel each second. The energy is used to compress the products of combustion and to eject them at the rear of the plane at 490 m/s (1600 ft/s) relative to the plane. Find (a) the thrust of the jet engine and (b) the delivered power (horsepower).
    Answer: (a) 2.3×10^4 N (5100 lb). (b) 4.1×10^6 W (5600 hp).

40. A freight car, open at the top, weighing 10 tons, is coasting along a level track with negligible friction at 2.0 ft/s when it begins to rain hard. The raindrops fall vertically with respect to the ground. What is the speed of the car when it has collected 0.50 ton of rain? What assumptions, if any, must you make to get your answer?

41. A flexible inextensible string of length l is threaded into a smooth tube, into which it snugly fits. The tube contains a right-angled bend, and is positioned in the vertical plane so that one arm is vertical and the other horizontal. Initially, at $t = 0$, a length y_0 of the string is hanging down in the vertical

arm. The string is released and slides through the tube, so that at any subsequent time t later, it is moving with a speed dy/dt, where $y(t)$ is the length of the string that is then hanging vertically. (a) Show that in terms of the variable mass problem $\mathbf{v}_{\text{rel}} = 0$, so that the equation of motion has the form $m \, d\mathbf{v}/dt = \mathbf{F}_{\text{ext}}$. (b) Show that the specific equation of motion is $(d^2y/dt^2) = gy$. (c) Show that conservation of mechanical energy leads to $(dy/dt)^2 - gy^2 = a$ constant, and that this is consistent with (b). (d) Show that $y = (y_0/2)(e^{\sqrt{g/l}\,t} + e^{-\sqrt{g/l}\,t})$ is a solution to the equation of motion [by substitution into (b)] and discuss the solution.

10
collisions

We learn much about atomic, nuclear, and elementary particles experimentally by observing collisions between them. On a larger scale we can interpret such things as the properties of gases in terms of particle collisions. In this chapter we apply the principles of conservation of energy and conservation of momentum to the collisions of particles.

In a collision a relatively large force acts on each colliding particle for a relatively short time. The basic idea of a "collision" is that the motion of the colliding particles (or of at least one of them) changes rather abruptly and that we can make a relatively clean separation of times that are "before the collision" and those that are "after the collision."

When a bat strikes a baseball for example, the beginning and the end of the collision can be determined fairly precisely. The bat is in contact with the ball for an interval that is quite short in comparison to the time during which we are watching the ball. During the collision the bat exerts a large force on the ball (Fig. 10-1). This force varies with time in a complex way that we can measure only with difficulty. Both the ball and the bat are deformed during the collision.* Forces that act for a time that is short compared to the time of observation of the system are called *impulsive* forces.

When an alpha particle (He[4]) "collides" with a nucleus of gold (Au[197]), the force acting between them may be the well-known repulsive electrostatic force associated with the charges on the particles. The particles may not "touch," but we still may speak of a "collision" because a relatively strong force, acting for a time that is short in comparison to the

10-1
WHAT IS A COLLISION?

figure 10-1
A high-speed flash photograph of a bat striking a baseball. Notice the deformation of the ball, indicating the enormous magnitude of the impulsive force at this instance. (Courtesy Harold E. Edgerton, Massachusetts Institute of Technology, Cambridge, Mass.)

* See "Batting the Ball" by P. Kirkpatrick, *American Journal of Physics*, August 1963.

time that the alpha particle is under observation, has a marked effect on the motion of the alpha particle.

When a proton (H^1 or p) with energy of, say, 25 MeV, "collides" with a nucleus of, say, a silver isotope (perhaps Ag^{107}), the particles may actually "touch," the predominant force then acting between them being, not the electrostatic repulsive force, but the strong, short-range, attractive nuclear force (see page 106). The proton may enter the silver nucleus, forming a compound structure. A short time later—the "collision interval" may be 10^{-18} s—this compound structure may break up into *two different* particles, according to a scheme such as

$$p + Ag^{107} \rightarrow \alpha + Pd^{104},$$

in which α $(= He^4)$ is an alpha particle. Thus we may broaden the concept of collision to include events (usually called *reactions*) in which the identities of the interacting particles change during the event. The conservation principles are applicable to all these examples.

We may, if we wish, broaden our definition of "collision" even further to include the spontaneous decay of a single particle into two or more other particles. An example is the decay of the elementary particle called the *sigma particle* into two other particles, the *pion* and the *neutron* (see Appendix I) or

$$\Sigma^- \rightarrow \pi^- + n.$$

Although two bodies do not come in contact in this process (unless we consider it in reverse), it has many features in common with collisions: (1) there is a clean distinction between "before the event" and "after the event," and (2) the laws of conservation of momentum and energy permit us to learn much about such processes by studying the "before" and "after" situations, even though we may know little about the force laws that operate during the "event" itself.

In studying collisions in this chapter our aim will be this: given the initial motions of the colliding particles, what can we learn about their final motions from the principles of conservation of momentum and energy, assuming that we know nothing about the forces acting during the collision?

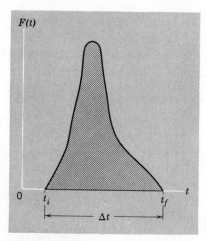

figure 10-2
How an impulsive force $F(t)$ might vary with time during a collision starting at time t_i and ending at t_f.

10-2
IMPULSE AND MOMENTUM

Let us assume that Fig. 10-2 shows the magnitude of the force exerted on a body during a collision. We assume that the force has a constant direction. The collision begins at time t_i and ends at time t_f, the force being zero before and after collision. From Eq. 9-12 we can write the change in momentum $d\mathbf{p}$ of a body in a time dt during which a force $\mathbf{F}$ acts on it as

$$d\mathbf{p} = \mathbf{F} \, dt. \tag{10-1}$$

We can find the change in momentum of the body during a collision by integrating over the time of collision, that is,

$$\mathbf{p}_f - \mathbf{p}_i = \int_{\mathbf{p}_i}^{\mathbf{p}_f} d\mathbf{p} = \int_{t_i}^{t_f} \mathbf{F} \, dt \tag{10-2}$$

in which the subscripts i $(= initial)$ and f $(= final)$ refer to the times before and after the collision, respectively. The integral of a force over the time interval during which the force acts is called the *impulse* $\mathbf{J}$ of the force. Hence the change in momentum of a body acted on by an impulsive force is equal to the impulse. Both impulse and momentum are vectors and both have the same units and dimensions.

The impulsive force represented in Fig. 10-2 is assumed to have a constant direction. The impulse of this force, $\int_{t_i}^{t_f} \mathbf{F}\, dt$, is represented in magnitude by the area under the force-time curve.*

Consider now a collision between two particles, such as those of masses m_1 and m_2, shown in Fig. 10-3. During the brief collision these particles exert large forces on one another. At any instant $\mathbf{F}_1$ is the force exerted on particle 1 by particle 2 and $\mathbf{F}_2$ is the force exerted on particle 2 by particle 1. By Newton's third law these forces at any instant are equal in magnitude but oppositely directed.

The change in momentum of particle 1 resulting from the collision is

$$\Delta\mathbf{p}_1 = \int_{t_i}^{t_f} \mathbf{F}_1\, dt = \overline{\mathbf{F}}_1\, \Delta t$$

in which $\overline{\mathbf{F}}_1$ is the average value of the force $\mathbf{F}_1$ during the time interval of the collision $\Delta t = t_f - t_i$.

The change in momentum of particle 2 resulting from the collision is

$$\Delta\mathbf{p}_2 = \int_{t_i}^{t_f} \mathbf{F}_2\, dt = \overline{\mathbf{F}}_2\, \Delta t$$

in which $\overline{\mathbf{F}}_2$ is the average value of the force $\mathbf{F}_2$ during the time interval of the collision $\Delta t = t_f - t_i$.

If no other forces act on the particles, then $\Delta\mathbf{p}_1$ and $\Delta\mathbf{p}_2$ give the total change in momentum for each particle. But we have seen that at each instant $\mathbf{F}_1 = -\mathbf{F}_2$, so that $\overline{\mathbf{F}}_1 = -\overline{\mathbf{F}}_2$, and therefore

$$\Delta\mathbf{p}_1 = -\Delta\mathbf{p}_2.$$

If we consider the two particles as an isolated system, the total momentum of the system is

$$\mathbf{P} = \mathbf{p}_1 + \mathbf{p}_2,$$

and the total *change* in momentum of the system as a result of the collision is zero, that is,

$$\Delta\mathbf{P} = \Delta\mathbf{p}_1 + \Delta\mathbf{p}_2 = 0.$$

Hence, *if there are no external forces* the total momentum of the system is not changed by the collision. The impulsive forces acting during the collision are internal forces which have no effect on the total momentum of the system.

We have defined a collision as an interaction which occurs in a time Δt that is negligible compared to the time during which we are observing the system. We can also characterize a collision as an event in which the external forces that may act on the system are negligible compared to the impulsive collision forces. When a bat strikes a baseball, a golf club strikes a golf ball, or one billiard ball strikes another,

10-3
CONSERVATION OF MOMENTUM DURING COLLISIONS

figure 10-3
Two "particles" m_1 and m_2, in collision, experience equal and opposite forces along their line of centers, according to Newton's third law; $\mathbf{F}_2(t) = -\mathbf{F}_1(t)$.

---

* The impulse J, defined from Eq. 10-2, does not depend critically on the precise values of t_i and t_f as long as these times are far enough apart to include the crosshatched area of Fig. 10-2. For reasons that will appear later we usually choose t_i and t_f with a separation that is *just large enough* to make a clean distinction between the "collision" and the "before and after intervals."

external forces act on the system. Gravity or friction exerts forces on these bodies, for example; these external forces may not be the same on each colliding body nor are they necessarily canceled by other external forces. Even so it is quite safe to neglect these external forces during the collision and to assume momentum conservation provided, as is almost always true, that the external forces are negligible compared to the impulsive forces of collision. As a result the change in momentum of a particle during a collision arising from an external force is negligible compared to the change in momentum of that particle arising from the impulsive collisional force (Fig. 10-4).

For example, when a bat strikes a baseball, the collision lasts only a small fraction of a second. Because the change in momentum is large and the time of collision is small, it follows from

$$\Delta \mathbf{p} = \overline{\mathbf{F}} \, \Delta t$$

that the average impulsive force $\overline{\mathbf{F}}$ is relatively large. Compared to this force, the external force of gravity is negligible. *During the collision* we can safely ignore this external force in determining the change in motion of the ball; the shorter the duration of the collision the more likely this is to be true.

In practice, therefore, *we can apply the principle of momentum conservation during collisions if the time of collision is small enough.* We can then say that the momentum of a system of particles just before the particles collide is equal to the momentum of the system just after the particles collide.

We can always calculate the motions of bodies after collision from their motions before collision if we know the forces that act during the collision, and if we can solve the equations of motion. Often we do not know these forces. However, the principle of conservation of momentum must hold during the collision. We already know that the principle of conservation of total energy holds. Although we may not know the details of the interaction, we can use these principles in many cases to predict the results of the collision.

Collisions are usually classified according to whether or not *kinetic energy* is conserved in the collision. When kinetic energy is conserved, the collision is said to be *elastic.* Otherwise, the collision is said to be *inelastic.* Collisions between atomic, nuclear, and fundamental particles are *sometimes* (but not always) elastic. These are, in fact, the only truly elastic collisions known. Collisions between gross bodies are always inelastic to some extent. We can often treat such collisions as approximately elastic, however, as, for example, collisions between ivory or glass balls. When two bodies stick together after collision, the collision is said to be *completely inelastic.* For example, the collision between a bullet and a block of wood into which it is fired is completely inelastic when the bullet remains embedded in the block. The term completely inelastic does *not* mean that all the initial kinetic energy is lost; as we shall see, it means rather that the loss is as great as is consistent with momentum conservation.

Even if the forces of collision are not known, we can find the motions of the particles after collision from the motions before collision, provided the collision is completely inelastic, or, if the collision is elastic, provided the collision is a one-dimensional one. For a one-dimensional collision the relative motion after collision is along the

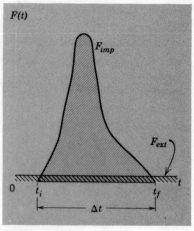

figure 10-4
During a collision, the impulsive force F_{imp} is generally much greater than any external forces F_{ext}, which may act on the system.

10-4
COLLISIONS IN ONE DIMENSION

same line as the relative motion before collision. We restrict ourselves to one-dimensional motion for the present.

Consider first an *elastic* one-dimensional collision. We can imagine two smooth nonrotating spheres moving initially along the line joining their centers, then colliding head-on and moving along the same straight line without rotation after collision (see Fig. 10-5). These bodies exert forces on each other during the collision that are along the initial line of motion, so that the final motion is also along this same line.

figure 10-5
Two spheres before and after an elastic collision. The velocity, $\mathbf{v}_{1i} - \mathbf{v}_{2i}$, of m_1 relative to m_2 before collision is equal to the velocity, $\mathbf{v}_{2f} - \mathbf{v}_{1f}$, of m_2 relative to m_1 after collision.

The masses of the spheres are m_1 and m_2, the (scalar) velocity components being v_{1i} and v_{2i} before collision and v_{1f} and v_{2f} after collision.* We take the positive direction of the momentum and velocity to be to the right. We assume, unless we specify otherwise, that the speeds of the colliding particles are low enough so that we need not use the relativistic expressions for momentum and kinetic energy. Then from conservation of momentum we obtain

$$m_1 v_{1i} + m_2 v_{2i} = m_1 v_{1f} + m_2 v_{2f}.$$

Because we are considering an elastic collision the kinetic energy is conserved by definition and we obtain

$$\tfrac{1}{2}m_1 v_{1i}^2 + \tfrac{1}{2}m_2 v_{2i}^2 = \tfrac{1}{2}m_1 v_{1f}^2 + \tfrac{1}{2}m_2 v_{2f}^2.$$

It is clear at once that, if we know the masses and the initial velocities, we can calculate the two final velocities v_{1f} and v_{2f} from these two equations.

The momentum equation can be written as

$$m_1(v_{1i} - v_{1f}) = m_2(v_{2f} - v_{2i}), \tag{10-3}$$

and the energy equation can be written as

$$m_1(v_{1i}^2 - v_{1f}^2) = m_2(v_{2f}^2 - v_{2i}^2). \tag{10-4}$$

Dividing Eq. 10-4 by Eq. 10-3, and assuming $v_{2f} \neq v_{2i}$ and $v_{1f} \neq v_{1i}$ (see Question 7), we obtain

$$v_{1i} + v_{1f} = v_{2f} + v_{2i}$$

and, after rearrangement,

$$v_{1i} - v_{2i} = v_{2f} - v_{1f} \tag{10-5}$$

This tells us that in an elastic one-dimensional collision, the relative velocity of approach before collision is equal to the relative velocity of separation after collision.

---

*The notation used is easy to interpret and to remember and reveals much information in a simple compact way. The number subscripts, such as 1 and 2, specify the particle and the letter subscripts, i and f, indicate initial value (before the collision) and final value (after the collision), respectively.

To find the velocity components v_{1f} and v_{2f} after collision from the velocity components v_{1i} and v_{2i} before collision, we can use any two of the three previous numbered equations. Thus from Eq. 10-5

$$v_{2f} = v_{1i} + v_{1f} - v_{2i}.$$

Inserting this into Eq. 10-3 and solving for v_{1f}, we find that

$$v_{1f} = \left(\frac{m_1 - m_2}{m_1 + m_2}\right)v_{1i} + \left(\frac{2m_2}{m_1 + m_2}\right)v_{2i}.$$

Likewise, inserting $v_{1f} = v_{2f} + v_{2i} - v_{1i}$ (from Eq. 10-5) into Eq. 10-3 and solving for v_{2f}, we obtain

$$v_{2f} = \left(\frac{2m_1}{m_1 + m_2}\right)v_{1i} + \left(\frac{m_2 - m_1}{m_1 + m_2}\right)v_{2i}.$$

There are several cases of special interest. For example, when the colliding particles have the same mass, m_1 equals m_2 so that the two previous equations become simply

$$v_{1f} = v_{2i} \quad \text{and} \quad v_{2f} = v_{1i}.$$

That is, in a one-dimensional elastic collision of two particles of equal mass, the particles simply exchange velocities during collision.

Another case of interest is that in which one particle m_2 is initially at rest. Then v_{2i} equals zero and

$$v_{1f} = \left(\frac{m_1 - m_2}{m_1 + m_2}\right)v_{1i}, \qquad v_{2f} = \left(\frac{2m_1}{m_1 + m_2}\right)v_{1i}.$$

Of course, if $m_1 = m_2$ also, then $v_{1f} = 0$ and $v_{2f} = v_{1i}$ as we expect. The first particle is "stopped cold" and the second one "takes off" with the velocity the first one originally had. If, however, m_2 is very much greater than m_1, we obtain

$$v_{1f} \cong -v_{1i} \quad \text{and} \quad v_{2f} \cong 0.$$

That is, when a light particle collides with a very much more massive particle at rest, the velocity of the light particle is approximately reversed and the massive particle remains approximately at rest. For example, suppose that we drop a ball vertically onto a horizontal surface attached to the earth. This is in effect a collision between the ball and the earth. If the collision is elastic, the ball will rebound with a reversed velocity and will reach the same height from which it fell.

If, finally, m_2 is very much smaller than m_1, we obtain

$$v_{1f} \cong v_{1i} \qquad v_{2f} \cong 2v_{1i}.$$

This means that the velocity of the massive incident particle is virtually unchanged by the collision with the light stationary particle, but that the light particle rebounds with approximately twice the velocity of the incident particle. The motion of a bowling ball is hardly affected by collision with an inflated beach ball of the same size, but the beach ball bounces away quickly.

Neutrons produced in a reactor from the fission of uranium atoms move very fast and must be slowed down if they are to produce more fissions. Assuming that they make elastic collisions with the nuclei at rest, what material should be picked to moderate (that is, to slow down) the neutrons in the reactor? We can answer this from the considerations just discussed. If the stationary targets were massive nuclei, like lead, the neutrons would simply bounce back with practically the same speed they had initially. If the stationary targets were very much lighter than the neutrons, like electrons, the neutrons would move on with practically the same velocity they had initially. However, if the stationary targets are particles of nearly the same mass, the neutrons will be brought almost to rest in a (head-on) collision with them. Hence, hydrogen, whose nucleus (proton) has nearly the same mass as a neutron, should be most effective. Other considerations affect the choice of a moderator for neutrons, but momentum and energy considerations alone limit the choice to the lighter elements.

If a collision is *inelastic* then, by definition, the kinetic energy is not conserved. The final kinetic energy may be less than the initial value, the difference being ultimately converted to heat energy or to potential energy of deformation in the collision, for example; or the final kinetic energy may exceed the initial value, as when potential energy is released in the collision. In any case, the conservation of momentum still holds, as does the conservation of *total* energy.

Let us consider finally a *conpletely inelastic* collision. The two particles stick together after collision, so that there will be a final common velocity $\mathbf{v}_f$. It is not necessary to restrict the discussion to one-dimensional motion. Using only the conservation of momentum principle, we find

$$m_1\mathbf{v}_{1i} + m_2\mathbf{v}_{2i} = (m_1 + m_2)\mathbf{v}_f. \qquad (10\text{-}6)$$

This determines $\mathbf{v}_f$ when $\mathbf{v}_{1i}$ and $\mathbf{v}_{2i}$ are known.

EXAMPLE 1

A baseball weighing 0.35 lb is struck by a bat while it is in horizontal flight with a speed of 90 ft/s. After leaving the bat the ball travels with a speed of 110 ft/s in a direction opposite to its original motion. Determine the impulse of the collision.

We cannot calculate the impulse from the definition $\mathbf{J} = \int \mathbf{F}\,dt$ because we do not know the force exerted on the ball as a function of time. However, we have seen (Eq. 10-2) that the change in momentum of a particle acted on by an impulsive force is equal to the impulse. Hence

$$\mathbf{J} = \text{change in momentum} = \mathbf{p}_f - \mathbf{p}_i$$

$$= m\mathbf{v}_f - m\mathbf{v}_i = \left(\frac{W}{g}\right)(\mathbf{v}_f - \mathbf{v}_i).$$

Assuming arbitrarily that the direction of $\mathbf{v}_i$ is positive, the impulse is then

$$J = \left(\frac{0.35\text{ lb}}{32\text{ ft/s}^2}\right)(-110\text{ ft/s} - 90\text{ ft/s}) = -2.2\text{ lb·s}$$

The minus sign shows that the direction of the impulse acting on the ball is opposite that of the original velocity of the ball.

We cannot determine the force of the collision from the data we are given. Actually, any force whose impulse is -2.2 lb·s will produce the same change in momentum. For example, if the bat and ball were in contact for 0.0010 s, the average force during this time would be

$$\bar{F} = \frac{\Delta p}{\Delta t} = \frac{-2.2\text{ lb·s}}{0.0010\text{ s}} = -2200\text{ lb}.$$

For a shorter contact time the average force would be greater. The actual force would have a maximum value greater than this average value.

How far would gravity cause the baseball to fall during its collision time?

EXAMPLE 2

(a) By what fraction is the kinetic energy of a neutron (mass m_1) decreased in a head-on elastic collision with an atomic nucleus (mass m_2) initially at rest?

The initial kinetic energy of the neutron K_i is $\frac{1}{2}m_1v_{1i}^2$. Its final kinetic energy K_f is $\frac{1}{2}m_1v_{1f}^2$. The fractional decrease in kinetic energy is

$$\frac{K_i - K_f}{K_i} = \frac{v_{1i}^2 - v_{1f}^2}{v_{1i}^2} = 1 - \frac{v_{1f}^2}{v_{1i}^2}.$$

But, for such a collision,
$$v_{1f} = \left(\frac{m_1 - m_2}{m_1 + m_2}\right)v_{1i},$$

so that
$$\frac{K_i - K_f}{K_i} = 1 - \left(\frac{m_1 - m_2}{m_1 + m_2}\right)^2 = \frac{4m_1m_2}{(m_1 + m_2)^2}.$$

(b) Find the fractional decrease in the kinetic energy of a neutron when it collides in this way with a lead nucleus, a carbon nucleus, and a hydrogen nucleus. The ratio of nuclear mass to neutron mass $(= m_2/m_1)$ is 206 for lead, 12 for carbon, and 1 for hydrogen.

For lead, $m_2 = 206m_1$,

$$\frac{K_i - K_f}{K_i} = \frac{4 \times 206}{(207)^2} = 0.02 \quad \text{or} \quad 2\%.$$

For carbon, $m_2 = 12m_1$,

$$\frac{K_i - K_f}{K_i} = \frac{4 \times 12}{(13)^2} = 0.28 \quad \text{or} \quad 28\%.$$

For hydrogen, $m_2 = m_1$,

$$\frac{K_i - K_f}{K_i} = \frac{4 \times 1}{(2)^2} = 1 \quad \text{or} \quad 100\%.$$

These results explain why paraffin, which is rich in hydrogen, is far more effective in slowing down neutrons than is lead.

EXAMPLE 3

The Ballistic Pendulum. The ballistic pendulum is used to measure bullet speeds. The pendulum is a large wooden block of mass M hanging vertically by two cords. A bullet of mass m, traveling with a horizontal speed v_i, strikes the pendulum and remains embedded in it (Fig. 10-6). If the collision time (the time required for the bullet to come to rest with respect to the block) is very small compared to the time of swing of the pendulum, the supporting cords remain approximately vertical during the collision. Therefore, no external horizontal force acts on the system (bullet + pendulum) during collision, and the horizontal component of momentum is conserved. The speed of the system after collision v_f is much less than that of the bullet before collision. This final speed can be easily determined, so that the original speed of the bullet can be calculated from momentum conservation.

The initial momentum of the system is that of the bullet mv_i, and the momentum of the system just after collision is $(m + M)v_f$, so that

$$mv_i = (m + M)v_f.$$

After the collision is over, the pendulum and bullet swing up to a maximum height y, where the kinetic energy left after impact is converted into gravitational potential energy. Then, using the conservation of mechanical energy for this part of the motion, we obtain

$$\tfrac{1}{2}(m + M)v_f^2 = (m + M)gy.$$

Solving these two equations for v_i, we obtain

$$v_i = \frac{m + M}{m}\sqrt{2gy}.$$

Hence, we can find the initial speed of the bullet by measuring m, M, and y.

figure 10-6
Example 3. A ballistic pendulum consisting of a large wooden block of mass M suspended by cords. When a bullet of mass m and velocity $\mathbf{v}_i$ is fired into it, the block swings, rising a maximum distance y.

The kinetic energy of the bullet initially is $\frac{1}{2}mv_i^2$ and the kinetic energy of the system (bullet + pendulum) just after collision is $\frac{1}{2}(m + M)v_f^2$. The ratio is

$$\frac{\frac{1}{2}(m + M)v_f^2}{\frac{1}{2}mv_i^2} = \frac{m}{m + M}.$$

For example, if the bullet has a mass $m = 5$ g and the block has a mass $M = 2000$ g, only about one-fourth of 1% of the original kinetic energy remains; over 99% is converted to other forms of energy, such as heat energy.

The velocity of the center of mass of two particles is not changed by their collision, for the collision, whether elastic or inelastic, does not change the total momentum of the system of two particles, only the distribution of momentum between the two particles. The momentum of the system can be written (Eq. 9-15) as $\mathbf{P} = (m_1 + m_2)\mathbf{v}_{cm}$. If no external forces act on the system, then $\mathbf{P}$ is constant before and after the collision, and the center of mass moves with uniform velocity throughout.

If we choose a reference frame attached to the center of mass, then in this center-of-mass reference frame, $\mathbf{v}_{cm} = 0$ and $\mathbf{P} = 0$. There is a great simplicity and symmetry in describing collisions with respect to the center of mass, and it is customary to do so in nuclear physics. For whether collisions are elastic or inelastic, momentum is conserved, and in the center of mass reference frame the total momentum is zero. These results hold in two and three dimensions as well as in one because momentum is a vector quantity.

As an example, consider a head-on elastic collision between two particles m_1 and m_2. Let m_2 equal $3m_1$ and let m_2 be at rest, so that v_{2i} equals zero in the laboratory reference frame. The total momentum of the two particles is just that of the incident particle m_1v_{1i}, so that

$$m_1v_{1i} = (m_1 + m_2)v_{cm}$$

or

$$v_{cm} = \left(\frac{m_1}{m_1 + m_2}\right)v_{1i} = \frac{1}{4}v_{1i}.$$

After the collision, m_1 has a velocity $v_{1f} = -\frac{1}{2}v_{1i}$ and m_2 has a velocity $v_{2f} = \frac{1}{2}v_{1i}$. The total momentum of the two particles $(m_1v_{1f} + m_2v_{2f})$ is the same as before the collision, and the motion of the center of mass is unchanged (check this). In Fig. 10-7a we show a series of "snapshots" of the collision taken at equal time

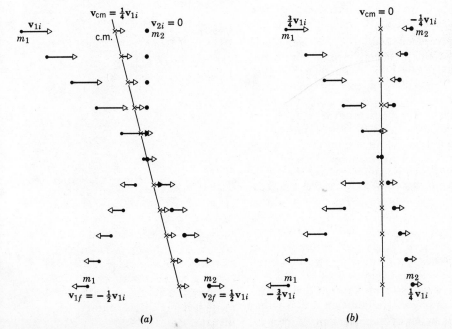

figure 10-7

(a) An elastic collision in the laboratory reference frame. (b) The same elastic collision in the center-of-mass reference frame.

intervals as seen in the laboratory reference frame. In Fig. 10-7b we show the same situation as seen in the center-of-mass reference frame, where v_{cm} is zero. Notice the symmetry of the particles' motions when described in this way. The particle coming from the left has a speed $\frac{3}{4}v_{1i}$ with respect to the center of mass (where v_{1i} is the speed of m_1 in the laboratory frame) and recedes with this same speed. The particle coming from the right has a speed $\frac{1}{4}v_{1i}$ with respect to the center of mass and recedes with this same speed.

If the collision is completely inelastic, the motion after collision is simply that of two particles moving along together at the center of mass. In Figs. 10-8a and 10-8b we show how the collision of Fig. 10-7, now assumed to be completely inelastic, would be described in the laboratory and the center-of-mass reference frames, respectively. These figures are exactly like the previous ones until the collision occurs; after the collision, however, the motion of the center of mass describes that of the entire system.

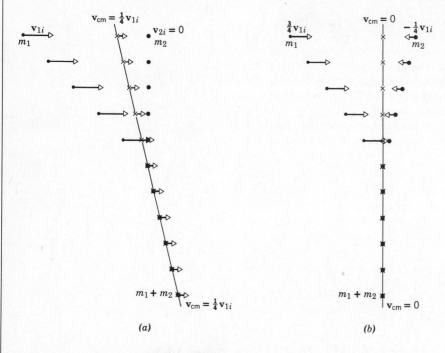

(a)   (b)

figure 10-8
(a) A completely inelastic collision in the laboratory reference frame. (b) The same completely inelastic collision in the center-of-mass reference frame. In each case the motion before collision is the same as that of Fig. 10-7.

The distinction between kinetic energy and momentum and the relationship of these concepts to force were not clearly understood until late in the eighteenth century. Scientists argued whether kinetic energy or momentum was the "true" measure of the effect of a force on a body. Descartes argued that when bodies interact, all that can happen is the transfer of momentum from one body to another, for the total momentum of the universe remains constant; hence the only "true" measure of a force is the change in momentum it produces in a given time. Leibnitz attacked this view and argued that the "true" measure of a force is the change it produces in kinetic energy (called by him *vis viva* or living force, taken to be twice what we now call kinetic energy).

In his treatise on mechanics (1743), D'Alembert dismissed the argument as being pointless and arising from a confusion of terminology. The cumulative effect of a force can be measured by its integrated effect over *time*, $\int F\,dt$, which produces a change in momentum, *or* by its integrated effect over *space*, $\int F\,dx$, which produces a change in kinetic energy. Both concepts are useful and valid, although different. Which one we use depends on what we are interested in or what is more convenient. As our present study of collisions illustrates, we frequently use both concepts in the same problem (see Question 23).

A more modern view is to look for quantities of the motion that are invariant, rather than focusing on the concept of force. The question as to whether

10-5
THE "TRUE" MEASURE OF A FORCE

the energy or the momentum is the "real" quantity of motion becomes pointless for there is no unique "quantity of motion." Instead, *both* energy *and* momentum may be regarded as invariant quantities of the motion in that for an isolated system the total of each of these quantities, summed up over all parts of the system, remains constant with time. There may be an exchange of energy, and of momentum, between different parts of an isolated system, but the total of each quantity is conserved.

In two or three dimensions (except for a completely inelastic collision) the conservation laws alone cannot tell us the motion of particles after a collision if we know the motion before the collision. For example, for a two-dimensional elastic collision, which is the simplest case, we have four unknowns, namely the two components of velocity for each of two particles after collision; but we have only three known relations between them, one for the conservation of kinetic energy and a conservation of momentum relation for each of the two dimensions. Hence we need more information than just the initial conditions. When we do not know the actual forces of interaction, as is often the case, the additional information must be obtained from experiment. It is simplest to specify the angle of recoil of one of the colliding particles.

Let us consider what happens when one particle is projected at a target particle which is at rest. This case is not as restrictive as it may seem, for we can always pick our reference frame to be one in which the target particle is at rest before the collision. Much experimental work in nuclear physics involves projecting nuclear particles at a target which is stationary in the laboratory reference frame. In such collisions, because of momentum conservation, the motion is in a plane determined by the lines of recoil of the colliding particles. The initial motion need not be along the line joining the centers of the two particles. The force of interaction may be electromagnetic (in which we include "contact" forces; see page 106), gravitational, or nuclear. The particles need not "touch"; strong forces, which act at relatively close distances of approach and for a time short compared to the observation time, deflect the particles from their initial courses.

A typical situation is shown in Fig. 10-9. The distance b between the initial line of motion and a line parallel to it through the center of the

10-6
COLLISIONS IN TWO AND THREE DIMENSIONS

figure 10-9
Two particles, m_1 and m_2, undergoing a collision. The open circles indicate their positions before collision, the shaded ones after collision. Initially m_2 is at rest. The impact parameter b is the distance by which the collision misses being head-on.

target particle is called the *impact parameter*. This is a measure of the directness of the collision, $b = 0$, corresponding to a head-on collision. The direction of motion of the incident particle m_1 after collision makes an angle θ_1 with the initial direction, and the target projectile m_2, initially at rest, moves in a direction after collision making an angle θ_2 with the initial direction of the incident projectile. Applying the conservation of momentum, which is a vector relation, we obtain two scalar equations; for the *x*-component of motion we have

$$m_1 v_{1i} = m_1 v_{1f} \cos \theta_1 + m_2 v_{2f} \cos \theta_2,$$

and for the *y*-component

$$0 = m_1 v_{1f} \sin \theta_1 - m_2 v_{2f} \sin \theta_2.$$

Let us now assume that the collision is *elastic*. Here the conservation of kinetic energy applies and we obtain a third equation,

$$\tfrac{1}{2} m_1 v_{1i}^2 = \tfrac{1}{2} m_1 v_{1f}^2 + \tfrac{1}{2} m_2 v_{2f}^2.$$

If we know the initial conditions (m_1, m_2, and v_{1i}), we are left with four unknowns (v_{1f}, v_{2f}, θ_1, and θ_2) but only three equations relating them. We can determine the motion after collision only if we specify a value for one of these quantities, such as θ_1.

Two skaters collide and embrace, as Fig. 10-10 suggests. One, whose mass m_1 is 70 kg ($W_1 = m_1 g = 150$ lb), is initially moving east at a speed v_1 of 6.0 km/h (3.7 mi/h). The other, whose mass m_2 is 50 kg ($W_2 = m_2 g = 110$ lb), is initially moving north at a speed v_2 of 8.0 km/h (5.0 mi/h). (*a*) What is the final velocity of the couple? (*b*) What fraction of the initial kinetic energy of the skaters is lost because of the collision?

EXAMPLE 4

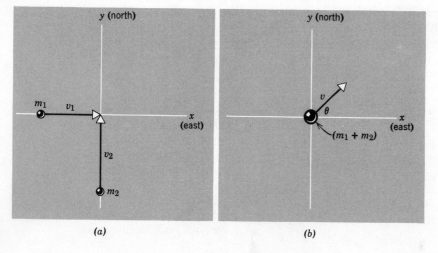

(*a*)　　　　　　　　　(*b*)

figure 10-10
Example 4. (*a*) initial situation.
(*b*) final situation.

(*a*) Figure 10-10 shows the initial and final situations. Because $\mathbf{P} = \mathbf{P}_f$ (no external forces act) we can write for the *x*-component of momentum

$$m_1 v_1 = (m_1 + m_2) v \cos \theta,$$

and for the *y*-component of momentum

$$m_2 v_2 = (m_1 + m_2) v \sin \theta.$$

Dividing the second equation by the first, we get

$$\tan \theta = \frac{m_2 v_2}{m_1 v_1} = \frac{(50 \text{ kg})(8.0 \text{ km/h})}{(70 \text{ kg})(6.0 \text{ km/h})}$$

$$= 0.95 \quad \text{or} \quad \theta = 43°,$$

which gives the direction of the final velocity.

Then, from the y-component equation we have

$$v = \frac{m_2 v_2}{(m_1 + m_2) \sin \theta}$$

$$= \frac{(50 \text{ kg})(8.0 \text{ km/h})}{(70 \text{ kg} + 50 \text{ kg}) \sin 43°}$$

$$= 4.9 \text{ km/h} \; (3.0 \text{ mi/h}),$$

which gives the magnitude of the final velocity.

(b) The initial kinetic energy of the skaters is

$$K_i = \tfrac{1}{2} m_1 v_1^2 + \tfrac{1}{2} m_2 v_2^2$$

$$= \tfrac{1}{2}(70 \text{ kg})(6.0 \text{ km/h})^2 + \tfrac{1}{2}(50 \text{ kg})(8.0 \text{ km/h})^2$$

$$= \left(2860 \frac{\text{kg km}^2}{\text{h}^2}\right)\left(\frac{1 \text{ h}}{3600 \text{ s}}\right)^2\left(\frac{10^3 \text{ m}}{1 \text{ km}}\right)^2$$

$$= 220 \text{ J}.$$

The final kinetic energy of the couple is

$$K_f = \tfrac{1}{2}(m_1 + m_2) v^2$$

$$= \tfrac{1}{2}(70 \text{ kg} + 50 \text{ kg})(4.9 \text{ km/h})^2$$

$$= \left(1440 \frac{\text{kg km}^2}{\text{h}^2}\right)\left(\frac{1 \text{ h}}{3600 \text{ s}}\right)^2\left(\frac{10^3 \text{ m}}{1 \text{ km}}\right)^2$$

$$= 110 \text{ J}.$$

Hence,

$$\frac{K_i - K_f}{K_i} = \frac{220 \text{ J} - 110 \text{ J}}{220 \text{ J}} = \frac{1}{2}$$

so that 50% of the initial kinetic energy is lost in the collision.

EXAMPLE 5

A gas molecule having a speed of 300 m/s collides elastically with another molecule of the same mass which is initially at rest. After the collision the first molecule moves at an angle of 30° to its initial direction. Find the speed of each molecule after collision and the angle made with the incident direction by the recoiling target molecule.

This example corresponds exactly to the situation discussed before the last example, with $m_1 = m_2$, $v_{1i} = 300$ m/s, and $\theta_1 = 30°$. Setting m_1 equal to m_2, we have the relations

$$v_{1i} = v_{1f} \cos \theta_1 + v_{2f} \cos \theta_2,$$

$$v_{1f} \sin \theta_1 = v_{2f} \sin \theta_2,$$

and

$$v_{1i}^2 = v_{1f}^2 + v_{2f}^2.$$

We must solve for v_{1f}, v_{2f}, and θ_2. To do this we square the first equation (rewriting it as $v_{1i} - v_{1f} \cos \theta_1 = v_{2f} \cos \theta_2$), and add this to the square of the second equation (noting that $\sin^2 \theta + \cos^2 \theta = 1$); we obtain

$$v_{1i}^2 + v_{1f}^2 - 2 v_{1i} v_{1f} \cos \theta_1 = v_{2f}^2.$$

Photographs of trajectories of particles undergoing collisions in a cloud chamber, a device that makes these paths visible. The chamber contains saturated water vapor. If the vapor is slightly compressed and then allowed to expand quickly, the water vapor will condense in droplets along the trajectory. The incident particle in all four cases is a helium nucleus (He⁴, or α). In (a) the target is a hydrogen nucleus (H¹ or p). The other tracks are similar, except that in (b) the target is another He⁴ nucleus, whereas in (c) and (d) the targets are fluorine and chlorine nuclei respectively. In general, the particles do not lie in the plane of the photograph. Stereoscopic photos are required for a complete analysis.

figure 10-12
(a) Four proton-proton collisions in a 10″-diameter bubble chamber. The original high-energy proton entered from the upper right. The spiral tracks are low energy electrons. The other tracks passing through the chamber are mesons of various kinds. Stereoscopic viewing shows that the angle between the outgoing tracks in each case is 90°. This is not apparent in the figure because the tracks do not lie in the plane of the figure. (b) A schematic representation of the proton tracks in (a). (Photo courtesy Laurence Radiation Laboratory.)

Combining this with the third equation, we obtain

$$2v_{1f}^2 = 2v_{1i}v_{1f}\cos\theta_1$$

or (because $v_{1f} \neq 0$)

$$v_{1f} = v_{1i}\cos\theta_1 = (300\text{ m/s})(\cos 30°)$$

or

$$v_{1f} = 260\text{ m/s}.$$

From the third equation

$$v_{2f}^2 = v_{1i}^2 - v_{1f}^2 = (300\text{ m/s})^2 - (260\text{ m/s})^2,$$

or

$$v_{2f} = 150\text{ m/s}.$$

Finally, from the second equation

$$\sin\theta_2 = (v_{1f}/v_{2f})\sin\theta_1$$

$$= (260/150)(\sin 30°) = 0.866$$

or

$$\theta_2 = 60°.$$

The two molecules move apart at right angles ($\theta_1 + \theta_2 = 90°$ in Fig. 10-9).

You should be able to show that in an elastic collision between particles of equal mass, one of which is initially at rest, the recoiling particles always move off at right angles to one another.

In Fig. 10-11, we show photographs of four elastic nuclear collisions that take place in a Wilson cloud chamber.* The tracks of the particles are made visible by the trail of droplets left in their wake. In each case the incident particle is an α-particle (He⁴) and the target nucleus is essentially at rest before collision. Notice that as the target mass increases, the angle between the recoiling particles increases (see Problem 42). In case (b) where the target is also an α-particle, stereoscopic photos show that the recoiling particles move off at right angles; the angle is not quite a right angle in the figure because the particles do not lie in the plane of the figure.

Figure 10-12 shows a series of four successive elastic collisions between protons caused when a high energy proton enters a bubble chamber† filled with liquid hydrogen, which supplies the target protons. The tracks of the particles are made visible in this case by the trail of bubbles left in their wake. Since the interacting particles are of equal mass and the collisions are elastic, the particles recoil at right angles to each other; this is apparent when the tracks of Fig. 10-12a are viewed stereoscopically.

10-7
CROSS SECTION

Although we have introduced the concept of the impact parameter b to describe collisions (see Fig. 10-9), it must be clear that, when we deal with particles of atomic or subatomic dimensions, we cannot define the track of the incident particle or the location of the target particle pre-

---

*In 1927, the English physicist, C. T. R. Wilson, received the Nobel prize for inventing the cloud chamber; his investigations started along an entirely different line, namely, an attempt to produce in the laboratory a certain atmospheric phenomenon observed on Ben Nevis, a mountain in Scotland.

†In 1960, the American physicist, Donald Glaser, received the Nobel prize for inventing the bubble chamber; it is said that the concept occurred to him while watching bubbles form in a glass of beer.

cisely enough. In practice, as when we bombard a thin target foil with a beam of α-particles from a cyclotron, we must deal in a statistical way with a large number of independent collisions between the α-particles and the nuclei in the target; the impact parameters for individual collisions cannot be determined.

The situation is much the same as if we were firing a machine gun at random (in the dark, say) at the side of a distant barn of area A on which someone had hung a number of small dinner plates, each of area σ, in random (but not overlapping) positions. If the number of plates is q and if the rate at which bullets strike the barn is R_0, what is the rate R at which plates are broken? It is, on the basis of the random character of the events,

$$R = R_0(\sigma q/A), \qquad (10\text{-}7a)$$

where σq is the total area of all the plates. We could, in fact, use this relation to measure σ, the geometrical area of a single plate. Solving for σ yields

$$\sigma = RA/R_0 q \qquad (10\text{-}7b)$$

which permits us to find σ from measured values of R, A, R_0, and q. We may call σ the *cross section* for the event consisting of the impact of a bullet on a plate.

Let us now consider a more restricted class of event, namely the impact of a bullet on a plate causing the plate to break into (say) exactly five pieces. The rate R_5 at which events of this kind occur is much less than the rate R at which the events described above occur. We may assign a *cross section* σ_5 to these restricted events and may measure it, by analogy with Eq. 10-7, from

$$\sigma_5 = R_5 A/R_0 q. \qquad (10\text{-}8)$$

We can consider other ways of breaking plates such as breaking into thirteen pieces, breaking so that one fragment has an area equal to half the plate or more, breaking so that one fragment flies vertically upward, and so on. Each of these events can be assigned its own cross section σ_x by measuring the rate R_x at which the events occur. *None of these cross sections necessarily has anything to do with the geometrical area of the plate; all are measures of the probability of occurrence of the events to which they are assigned.* Cross sections are important because they are identified with single events and are independent of the details of particular experimental setups. In Eq. 10-8, for example, we would find the same value for σ_5 no matter how large the barn (A), how many plates (q), or how rapid the rate of fire of the machine gun (R_0); the measured value of R_5 would always be such as to yield the same measured value for σ_5.

Similarly, in nuclear physics we often bombard targets with nuclear projectiles, measure the rate at which events of a selected type occur, and assign a cross section to those events. For example, let us bombard a thin gold foil (Au^{197}) with deuterons (H^2, or d) whose energy is, say, 30 Mev. Many events can occur, among them (1) elastic scattering of the deuteron into the forward hemisphere, (2) elastic scattering of the deuteron into the backward hemisphere, (3) inelastic scattering of the deuteron between the angles of 30° and 60° with the direction of the incident beam, (4) the nuclear reaction $d + Au^{197} \rightarrow p + Au^{198}$, and (5) the nuclear reaction $d + Au^{197} \rightarrow n + Hg^{198}$, in which n represents a neutron. Each of these events (and many others that could be written down) has its own cross section σ_x which allows us to calculate the rate R_x at which these events occur if we know the details of the experimental arrangement. The ultimate goal of all the experiments is to understand the nature of nuclear forces.

Let the area of the foil exposed to the beam be A and the thickness of the foil be x. If there are n target particles per unit volume in the foil, the total number of available target particles is nAx. If the *effective area* (i.e., the cross section) for the event we are concerned with is σ_x, the total *effective area* of all the nuclei is $nAx\sigma_x$. If R_0 is the rate at which projectiles strike the target and R_x is the rate at which the events in which we are interested occur, we have, because of the random nature of the events (see Eq. 10-7a),

$$\frac{R_x}{R_0} = \frac{(nAx\sigma_x)}{A}$$

or

$$R_x = R_0 nx\sigma_x. \tag{10-9}$$

Thus we can measure σ_x for the event by measuring R_x, R_0, n, and x and substituting into Eq. 10-9. Cross sections are commonly expressed in *barns* or submultiplies thereof; one barn $= 10^{-28}$ m^2.

Cross sections almost always depend on the energy of the incident particle, often exhibiting sharp peaks as the energy is varied. This simply means that at certain rather precisely defined characteristic energies the reaction is much more likely to "go" than at others. In much the same way a steel plate will vibrate with large amplitude at a series of rather precisely defined characteristic frequencies. We shall study such "resonances" in Chapter 20 and elsewhere.

Figure 10-13 shows one of the thousands of cross section curves that have been measured for various atomic and nuclear process. Let a highly collimated beam of neutrons of kinetic energy K fall on a thin foil of cadmium. The process to which the cross section σ in Fig. 10-13 refers is any process (absorption, elastic scattering, or inelastic scattering) that results in the removal of a neutron from the collimated beam. The numbers above the various peaks show the particular isotope responsible for

figure 10-13
The cross section as a function of energy for processes that remove neutrons from a collimated beam as they pass through cadmium. Note that both scales are logarithmic.

that peak; this can be learned from other experiments using foils made of the separated isotopes. The strong peak labeled "113" that occurs at 0.17 electron volts is caused by the reaction

$$Cd^{113} + n \rightarrow Cd^{114} + \gamma$$

in which γ represents a gamma ray. This reaction, which has a peak cross section of 7600 barns, is responsible for the very large absorbing power of cadmium for slow neutrons. Note that both scales in Fig. 10-13 are logarithmic.

EXAMPLE 6

(a) About 1910, Geiger and Marsden, working under Ernest Rutherford at the University of Manchester, performed a series of classic experiments that established the fact that atoms consisted of a small nucleus surrounded by a cloud of electrons rather than a sphere of distributed positive and negative charges, as Thomson had suggested earlier.

This experiment was in essence that shown very schematically in Fig. 10-14. Here α-particles from a polonium source are allowed to strike a gold foil 4.0×10^{-7} m thick. It is found that although most of the α-particles pass through the foil (forward scattering), about 1 in 6.17×10^4 are scattered backward, that is, are deflected through an angle greater than 90°. The number of gold atoms per unit volume in the foil is $5.9 \times 10^{28}/m^3$. What is the scattering cross section in barns for backward scattering (1 barn = 10^{-28} m²)?

From

$$n x \sigma = \text{fraction scattered backward}$$

we have

$$(5.9 \times 10^{28}/m^3)(4.0 \times 10^{-7} \text{ m})\sigma = 1/(6.17 \times 10^4)$$

or

$$\sigma = 6.9 \times 10^{-28} \text{ m}^2 = 6.9 \text{ barns.}$$

This is the cross section for backward scattering.

(b) Rutherford reasoned that the backward scattering could not be caused by electrons in the atom; the α-particles are so much more massive than the electrons that they would hardly be deflected at all by them, let alone be scattered backward. He then suggested the nuclear model of the atom, attributing the scattering to collisions between α-particles and the massive positive core of the atom, the nucleus.

Assuming that the cross section for backward scattering is approximately equal to the area offered by a gold nucleus for direct collisions, estimate the *effective* size of a gold nucleus.

If the *effective* radius of the gold nucleus is taken to be r, we have

$$\sigma = \pi r^2,$$

$$r^2 = \sigma/\pi = 6.9 \times 10^{-28} \text{ m}^2/\pi,$$

or

$$r = 1.5 \times 10^{-14} \text{ m.}$$

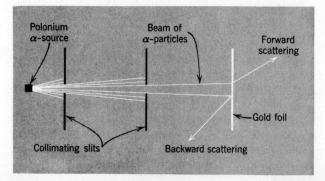

figure 10-14
Example 6. α-particles stream from a polonium source and a beam is formed by collimating slits. Some of the α-particles are scattered backward by the gold foil target; the rest pass through the foil.

This is the approximate radius of a gold *nucleus* which compares with the value of about 1.5×10^{-10} m for the gold *atom*. Hence the massive nucleus is concentrated in a very small region of the atom (about 1 part in 10^{12} by volume).

10-8
REACTIONS AND DECAY PROCESSES

We stated in Section 10-1 that reactions and radioactive decay processes, for atoms, nuclei, and elementary particles, can be treated by the same methods used in collision studies, namely: We can apply the principles of conservation of linear momentum and energy to the (well-defined) periods "before the event" and "after the event." For these processes we must use the conservation of *total* energy because kinetic energy is *not* conserved. In this section we only consider examples in which the speeds of the particles are negligible with respect to the speed of light. This means that we may use the classical expressions for momentum and energy and need not use the relativistic expressions.

EXAMPLE 7

Nuclear Reactions. A thin film containing a fluorine (F^{19}) compound is bombarded by a beam of protons (p) which has been accelerated to an energy of 1.85 MeV (million electron volts; 1 MeV $= 1.60 \times 10^{-13}$ J) in a Van de Graaff accelerator. Some of the protons interact with the fluorine nuclei to produce the following nuclear reaction:

$$F^{19} + p \rightarrow O^{16} + \alpha$$

It is observed that the α-particles (which are helium nuclei) that emerge at *right angles* to the incident proton beam (see Fig. 10-15) have speeds of 1.95×10^7 m/s. What can you learn about the reaction by applying the laws of conservation of linear momentum and of total energy? The masses involved are, to a precision good enough for our purposes,

$$m_p = 1.01 \text{ u} \qquad m_O = 16.0 \text{ u}$$

$$m_F = 19.0 \text{ u} \qquad m_\alpha = 4.00 \text{ u},$$

in which 1 u (*unified atomic mass unit*) $= 1.66 \times 10^{-27}$ kg.

figure 10-15
Example 7. The nuclear reaction $p + F^{19} \rightarrow \alpha + O^{16}$, showing the situation before and after the event, in the laboratory reference frame.

The x- and y-components of linear momentum are conserved, which means that they have the same values before and after the reaction. In the laboratory reference frame of Fig. 10-15, then

$$m_p v_p = m_O v_O \cos \theta \qquad (x\text{-component}) \qquad (10\text{-}10)$$

and

$$0 = m_\alpha v_\alpha - m_O v_O \sin \theta \qquad (y\text{-component}) \qquad (10\text{-}11)$$

For the conservation of total energy we write

$$Q + \tfrac{1}{2}m_p v_p^2 = \tfrac{1}{2}m_0 v_0^2 + \tfrac{1}{2}m_\alpha v_\alpha^2 \tag{10-12}$$

in which it is clear that Q is the amount by which the kinetic energy of the system after the reaction exceeds the kinetic energy of the system before the reaction. Note that we have assumed that the particles are moving slowly enough so that we may use the classical expression for kinetic energy ($\tfrac{1}{2}mv^2$) rather than the relativistic one [$m_0 c^2(1/\sqrt{1 - v^2/c^2} - 1)$]. If Q is positive, kinetic energy must be generated by the reaction.

The energy represented by Q can only come from differences in the rest energies of the particles before or after the reaction, according to Einstein's well-known relation $E = \Delta mc^2$ (see Section 8-9). Thus (if Q is positive), we expect that the rest mass of the system after the reaction would be slightly less than its rest mass before the reaction and that Q would indeed be given by the Einstein relation

$$Q = \Delta mc^2$$

$$= [(m_p + m_F) - (m_\alpha + m_0)]c^2. \tag{10-13}$$

Note that Eqs. 10-12 and 10-13 are independent relations for Q, being connected through Einstein's mass-energy relation.

The three conservation equations contain just three unknowns, v_0, θ, and Q. To find Q from them let us first eliminate θ between the first two equations by squaring and adding (recalling that $\cos^2 \theta + \sin^2 \theta = 1$). We obtain

$$m_p^2 v_p^2 + m_\alpha^2 v_\alpha^2 = m_0^2 v_0^2.$$

We can now eliminate v_0 between this relation and Eq. 10-12. You can show that, after a little rearrangement, we obtain

$$Q = K_\alpha(1 + m_\alpha/m_0) - K_p(1 - m_p/m_0). \tag{10-14}$$

From the data given we know that $K_p(= \tfrac{1}{2}m_p v_p^2) = 1.85$ MeV and

$$K_\alpha = \tfrac{1}{2}m_\alpha v_\alpha^2$$

$$= \tfrac{1}{2}(4.00 \text{ u})(1.66 \times 10^{-27} \text{ kg/u})(1.95 \times 10^7 \text{ m/s})^2$$

$$= (1.26 \times 10^{-12} \text{ J})(1 \text{ MeV}/1.60 \times 10^{-13} \text{ J})$$

$$= 7.88 \text{ MeV}.$$

We may now calculate Q from Eq. 10-14 as

$$Q = (7.88 \text{ MeV})(1 + 4.00/16.0) - (1.85 \text{ MeV})(1 - 1.01/16.0) = 8.13 \text{ MeV}.$$

Thus, by using the principles of conservation of linear momentum and total energy, we can calculate Q for the reaction without making any observations on the recoiling O^{16} nucleus. If we want to know v_0 and θ for this nucleus we can easily calculate them from Eqs. 10-10 and 10-11.

The result $Q = 8.13$ MeV is an important bit of information about the reaction. From Eq. 10-13, which is a relation for Q independent of Eq. 10-14, we can now calculate that the decrease in rest mass during the reaction is given by

$$\Delta m = Q/c^2$$

$$= (8.13 \text{ MeV} \times 1.60 \times 10^{-13} \text{ J/MeV})/(3.00 \times 10^8 \text{ m/s})^2$$

$$= (1.44 \times 10^{-29} \text{ kg})(1 \text{ u}/1.66 \times 10^{-27} \text{ kg})$$

$$= 0.00873 \text{ u}.$$

We can verify this result by calculating $\Delta m [= (m_p + m_F) - (m_\alpha + m_0)]$ from very precise measurements of the four separate masses made in a mass spectrometer (see Problem 47). The excellent agreement that we get shows once again the essential validity of Einstein's mass-energy relationship.

1. Explain how conservation of momentum applies to a handball bouncing off a wall.

2. How can you reconcile the sailing of a sailboat into the wind with the conservation of momentum principle?

3. Is it true that the acceleration of a baseball after it has been hit does not depend on who hit it?

4. Many features on cars, such as collapsible steering wheels and padded dashboards, are meant to transfer momentum more safely for passengers. Explain, using the impulse concept.

5. C. R. Daish (see "At Impact, Clubhead Travels 100 Mph," *Museum*, December 1973) states that, for professional golfers, the initial speed of the ball of the clubhead is about 140 mi/h. He also says: (a) ". . . if the Empire State Building could be swung at the ball at the same speed as the clubhead, the initial ball velocity would only be increased by about 2% . . ."; and (b) that, once the golfer has started his downswing, camera clicking, sneezing, etc., can have no effect on the motion of the ball. Can you give qualitative arguments to support these two statements?

6. The blades of a turbine are usually curved rather than flat in shape so that the fluid striking them follows a path resembling a u-turn. Convince yourself about the fluid's motion and explain the advantage of the curved shape over the flat one.

7. It is obvious from inspection of Eqs. 10-3 and 10-4 that a valid solution to the problem of finding the final velocities of two particles in a one-dimensional elastic collision is $v_{1f} = v_{1i}$ and $v_{2f} = v_{2i}$. What does this mean physically?

8. Consider a one-dimensional elastic collision between a given incoming body A and a body B initially at rest. How would you choose the mass of B, in comparison to the mass of A, in order that B should recoil with (a) the greatest speed, (b) the greatest momentum, and (c) the greatest kinetic energy?

9. A football player, momentarily at rest on the field, is about to catch a football when he is tackled by a running player on the other team. This is certainly a collision (inelastic!) and momentum must be conserved. In the reference frame of the football field there is momentum before the collision but there seems to be none after the collision. Is linear momentum really conserved?

10. Steel is more elastic than rubber. Explain what this means.

11. Two clay balls of equal mass and speed strike each other head-on, stick together, and come to rest. Kinetic energy is certainly not conserved. What happened to it? Is momentum conserved?

12. Discuss the possibility that, if only we could take into account internal motions of atoms and such in bodies, *all* collisions are elastic.

13. In commenting on the fact that kinetic energy is not conserved in a totally inelastic collision, a student observed that kinetic energy clearly is not conserved in an explosion and that a totally inelastic collision is merely the reverse of an explosion. Is this a useful or valid observation?

14. A sand glass is being weighed on a sensitive balance, first when sand is dropping in a steady stream from the upper to the lower part and then again after the upper part is empty. Are the two weights the same or not? Explain your answer.

15. Give a plausible explanation for the breaking of wooden boards or of bricks by a karate punch (see "Karate Strikes" by Jearl D. Walker, *American Journal of Physics*, October 1975).

16. In which one of the following cases is the linear momentum of the italicized objects most nearly conserved? (a) a *ball* falling freely in vacuum; (b) an *automobile* making a turn at constant speed; (c) a *rubber ball* as it bounces from the floor; (d) *two balls* as they collide at right angles; (e) a *bullet* and the *gun* from which it is fired by a man holding the gun.

17. If (only) two particles collide, are we ever forced to resort to a three-dimensional description to describe the event? Explain.

18. In a two body collision *in the center-of-mass reference frame* the momenta of the particles are equal and opposite to one another both before and after the collision. Is the line of relative motion necessarily the same after collision as before? Under what conditions would the magnitudes of the velocities of the bodies increase? decrease? remain the same as a result of the collision?

19. When dealing with atoms, nuclei, or elementary particles, what does it mean to say that two such bodies "touch" during a collision?

20. When the forces of interaction between two particles have an infinite range, such as the mutual gravitational attraction between two bodies, can the cross section for "collision" be finite? Is it at all useful to regard this interaction as a collision?

21. Why does the computation of the radius of the gold nucleus in Example 5 give only an approximate answer?

22. Could we determine in principle the cross section for a collision by using only one bombarding particle and one target particle? In practice?

23. We have seen that the conservation of momentum may apply whether kinetic energy is conserved or not. What about the reverse, that is, does the conservation of kinetic energy imply the conservation of momentum in classical physics? (See "Connection between Conservation of Energy and Conservation of Momentum" by Carl G. Adler, *Am. J. Phys.*, May 1976.)

SECTION 10-2

1. A ball of mass m and speed v strikes a wall perpendicularly and rebounds with undiminished speed. If the time of collision is t, what is the average force exerted by the ball on the wall?   *Answer:* $2mv/t$.

2. A stream of water impinges on a stationary "dished" turbine blade, as shown in Fig. 10-16. The speed of the water is u, both before and after it strikes the curved surface of the blade, and the mass of water striking the blade per unit time is constant at the value μ. Find the force exerted by the water on the blade.

3. A 150-g (0.01 slug) ball is moving at a speed of 40 m/s (130 ft/s) when it is struck by a bat that reverses its direction and gives it a speed of 60 m/s (200 ft/s). What average force was exerted by the bat if it was in contact with the ball for 5.0 ms?   *Answer:* 3000 N (660 lb).

4. A 1.0-kg ball drops vertically onto the floor with a speed of 25 m/s. It rebounds with an initial speed of 10 m/s. (a) What impulse acts on the ball during contact? (b) If the ball is in contact for 0.020 s, what is the average force exerted on the floor?

5. A cue strikes a billiard ball, exerting an average force of 50 N over a time of 10 ms. If the ball has mass 0.20 kg, what speed does it have after impact? *Answer:* 2.5 m/s.

6. A croquet ball (mass 0.50 kg) is struck by a mallet, receiving the impulse shown in the graph (Fig. 10-17). What is the ball's velocity just after the force has become zero?

7. The force on a 10-kg (0.69 slug) object increases uniformly from zero to 50 N (11 lb) in 4.0 s. What is the object's final speed if it started from rest? *Answer:* 10 m/s (32 ft/s).

8. A golfer hits a golf ball imparting to it an initial velocity of magnitude 5.0×10^3 cm/s directed 30° above the horizontal. Assuming that the mass of the ball is 25 g and the club and ball are in contact for 0.010 s find (a) the impulse imparted to the ball; (b) the impulse imparted to the club; (c) the average force exerted on the ball by the club; (d) the work done on the ball.

9. A stream of water from a hose is sprayed on a wall. If the speed of the water is 5.0 m/s (16 ft/s) and the hose sprays 300 cm³/s (0.011 ft³/s), what is the

problems

figure 10-16
Problem 2

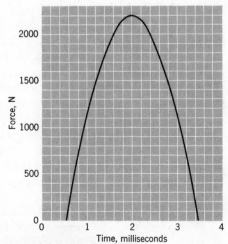

figure 10-17
Problem 6

average force exerted on the wall by the stream of water? Assume that the water does not spatter back appreciably. The density of water is 1.0 g/cm³ (1.9 slug/ft³).   *Answer:* 1.5 N (0.33 lb).

10. Two spacecraft are separated by exploding a small charge placed between them. If the masses of the crafts are 1200 kg and 1800 kg and the impulse of the force of the explosion is 600 N·s, what is the relative speed of recession of the two craft?

SECTION 10-4

11. A bullet of mass 10 g strikes a ballistic pendulum of mass 2.0 kg. The center of mass of the pendulum rises a vertical distance of 12 cm. Assuming the bullet remains embedded in the pendulum, calculate its initial speed.
    Answer: 310 m/s.

12. A 6.0-kg box sled is traveling across the ice at a speed of 9.0 m/s when a 12-kg package is dropped into it vertically. Describe the subsequent motion of the sled.

13. (a) Show that in a one-dimensional elastic collision the speed of the center of mass of two particles, m_1 moving with initial speed v_{1i} and m_2 moving with initial speed v_{2i} is

$$v_{\text{cm}} = \left(\frac{m_1}{m_1 + m_2}\right)v_{1i} + \left(\frac{m_2}{m_1 + m_2}\right)v_{2i}.$$

(b) Use the expressions obtained for v_{1f} and v_{2f}, the particles' speeds after collision, to derive the same result for v_{cm} *after* the collision.

14. A body of 2.0 kg mass makes an elastic collision with another body at rest and continues to move in the original direction but with one-fourth of its original speed. What is the mass of the struck body?

15. In a breech-loading automatic firearm of early vintage the reloading mechanism at the rear of the bore is activated when the breech-block, which recoils after the bullet is fired, compresses a spring by a predetermined amount d. (a) Show that the speed v of the bullet of mass m must be at least $d\sqrt{kM}/m$ on firing, for automatic loading, where k is the force constant of the spring and M is the mass of the breech-block. (b) In what sense, if any, can this process be regarded as a collision?

16. A steel ball weighing 1.0 lb is fastened to a cord 27 in. long and is released when the cord is horizontal. At the bottom of its path the ball strikes a 5.0-lb steel block initially at rest on a frictionless surface (Fig. 10-18). The collision is elastic. Find (a) the speed of the ball and (b) the speed of the block just after the collision.

figure 10-18
Problem 16

17. A bullet weighing 1.0×10^{-2} lb (mass = 4.5×10^{-3} kg) is fired horizontally into a 4.0-lb (mass = 1.8 kg) wooden block at rest on a horizontal surface. The coefficient of kinetic friction between block and surface is 0.20. The bullet comes to rest in the block which moves 6.0 ft (1.8 m). Find the speed of the bullet.   *Answer:* 3500 ft/s (1100 m/s).

18. Two pendulums each of length l are initially situated as in Fig. 10-19. The first pendulum is released and strikes the second. Assume that the collision is completely inelastic and neglect the mass of the strings and any frictional effects. How high does the center of mass rise after the collision?

figure 10-19
Problem 18

19. Two particles, one having twice the mass of the other, are held together with a compressed spring between them. The energy stored in the spring is 60 J. How much kinetic energy does each particle have after they are released?
    Answer: 20 J for the heavy particle; 40 J for the light particle.

20. A railroad freight car weighing 32 tons and traveling 5.0 ft/s overtakes one weighing 24 tons traveling 3.0 ft/s in the same direction. (a) Find the speed of the cars after collision and the loss of kinetic energy during collision if the cars couple together. (b) If the collision is elastic, the freight cars do not couple but separate after collision. What are their speeds?

21. An electron collides elastically with a hydrogen atom initially at rest. The initial and final motions are along the same straight line. What fraction of the electron's initial kinetic energy is transferred to the hydrogen atom? The mass of the hydrogen atom is 1840 times the mass of the electron. *Answer: 0.22%.*

22. A block of mass $m_1 = 100$ kg is at rest on a very long frictionless table, one end of which is terminated in a wall. Another block of mass m_2 is placed between the first block and the wall and set in motion to the left with constant speed v_{2i}, as in Fig. 10-20. Assuming that all collisions are completely elastic, find the value of m_2 for which both blocks move with the same velocity after m_2 has collided once with m_1 and once with the wall. The wall has infinite mass effectively.

figure 10-20
Problem 22

23. An electron, mass m, collides head-on with an atom, mass M, initially at rest. As a result of the collision a characteristic amount of energy E is stored internally in the atom. What is the minimum initial speed v_0 that the electron must have? (Hint: Conservation principles lead to a quadratic equation for the final electron velocity v and a quadratic equation for the final atom velocity V. The minimum value v_0 follows from the requirement that the radical in the solutions for v and V be real.)

Answer: $v_0 = \left(2E\dfrac{M+m}{Mm}\right)^{1/2}$.

24. A ball of mass m is projected with speed v_i into the barrel of a spring-gun of mass M initially at rest on a frictionless surface; see Fig. 10-21. The mass m sticks in the barrel at the point of maximum compression of the spring. No energy is lost in friction. What fraction of the initial kinetic energy of the ball is stored in the spring?

figure 10-21
Problem 24

25. A box is put on a scale that is adjusted to read zero when the box is empty. A stream of pebbles is then poured into the box from a height h above its bottom at a rate of μ (pebbles per second). Each pebble has a mass m. If the collisions between the pebbles and the box are completely inelastic, find the scale reading at time t after the pebbles begin to fill the box. Determine a numerical answer when $\mu = 100$ s^{-1}, $h = 25$ ft, $mg = 0.010$ lb, and $t = 10$ s. *Answer: 11 lb.*

26. A scale is adjusted to read zero. Particles fall from a height of 9.0 ft (2.7 m) colliding with the balance pan of the scale; the collisions are elastic, that is, the particles rebound upward with the same speed. If each particle has a mass of 1/128 slug (110 g) and collisions occur at the rate of 32 s^{-1}, what is the scale reading in pounds (kg)?

27. Mass m_1 collides head-on with m_2, initially at rest, in a completely inelastic collision. (a) What is the kinetic energy of the system before collision? (b) What is the kinetic energy of the system after collision? (c) What fraction of the original kinetic energy was converted into heat energy? (d) Let $\mathbf{v}_{cm}$ be the velocity of the center of mass of the system. View the collision from a primed reference frame moving with the center of mass so that $v_{1i}' = v_{1i} - v_{cm}$, $v_{2i}' = -v_{cm}$. Repeat parts (a), (b), and (c), as seen by an observer in this reference frame. Is the mechanical energy converted to heat energy the same in each case? Explain.
Answer: (a) $m_1 v_{1i}^2/2$. (b) $m_1^2 v_{1i}^2/2(m_1 + m_2)$. (c) $m_2/(m_1 + m_2)$.
(d) $m_1 m_2 v_{1i}^2/2(m_1 + m_2)$; zero; one; yes.

28. An elevator is moving up at 6.0 ft/s in a shaft. At the instant the elevator is 60 ft from the top, a ball is dropped from the top of the shaft. The ball rebounds elastically from the elevator roof. (a) To what height can it rise relative to the top of the shaft? (b) Do the same problem assuming the elevator is moving down at 6.0 ft/s.

29. A block of mass $m_1 = 2.0$ kg slides along a frictionless table with a speed of 10 m/s. Directly in front of it, and moving in the same direction, is a block of mass $m_2 = 5.0$ kg moving at 3.0 m/s. A massless spring with a spring constant $k = 1120$ N/m is attached to the backside of m_2, as shown in Fig. 10-22.

figure 10-22
Problem 29

When the blocks collide, what is the maximum compression of the spring?
Assume that the spring does not bend and always obeys Hooke's law.
Answer: 0.25 m.

30. The two masses on the right of Fig. 10-23 are slightly separated and initially
at rest; the left mass is incident with speed v_0. Assuming head-on elastic
collisions, (a) if $M \leq m$, show that there are two collisions and find all final
velocities; (b) if $M > m$, show that there are three collisions and find all
final velocities.

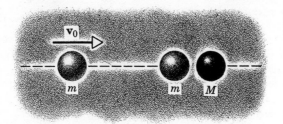

figure 10-23
Problem 30

31. Consider a situation such as that in the previous problem (10-30) but in
which the collisions now may be either all elastic, all inelastic, or some
elastic and some inelastic; also, the masses are now m, m', and M. Show
that to transfer the maximum kinetic energy from m to M the intermediate
body should have a mass $m' = \sqrt{mM}$, that is, the geometric mean of the
adjacent masses. (It is interesting to note that this same relation exists between masses of successive layers of air in the exponential horn in acoustics. See "Energy Transfer in One-Dimensional Collisions of Many Objects" by John B. Hart and Robert B. Herrmann, *American Journal of
Physics*, January, 1968.)

SECTION 10-6

32. Two vehicles A and B are traveling west and south, respectively, toward
the same intersection where they collide and lock together. Before the collision A (total weight, 900 lb) is moving with a speed of 40 mi/h and B
(total weight, 1200 lb) has a speed of 60 mi/h. Find the magnitude and direction of the velocity of the (interlocked) vehicles immediately after the
collision.

33. Two balls A and B, having different but unknown masses, collide. A is initially at rest when B has a speed v. After collision B has a speed $v/2$ and
moves at right angles to its original motion. (a) Find the direction in which
ball A moves after collision. (b) Can you determine the speed of A from the
information given? Explain.
Answer: (a) 117° from the final direction of B. (b) No.

34. A billiard ball moving at a speed of 2.2 m/s strikes an identical stationary
ball a glancing blow. After the collision one ball is found to be moving at a
speed of 1.1 m/s in a direction making a 60° angle with the original line of
motion. (a) Find the velocity of the other ball. (b) Can the collision be inelastic, given these data?

35. An α-particle collides with an oxygen nucleus, initially at rest. The α-par-
ticle is scattered at an angle of 64° from its initial direction of motion and
the oxygen nucleus recoils at an angle of 51° on the other side of this initial
direction. What is the ratio, α-particle to nucleus, of the final speeds of
these particles? The mass of the oxygen nucleus is four times that of the
α-particle.   *Answer:* 3.46.

36. A deuteron is a nuclear particle made up of one proton and one neutron. Its
mass is about 3.4×10^{-24} g. A deuteron, accelerated by a cyclotron to a speed
of 10^9 cm/s, collides with another deuteron at rest. (a) If the two particles
stick together head-on to form a helium nucleus, find the speed of the resulting nucleus. (b) The helium nucleus then breaks up into a neutron with

a mass of about 1.7×10^{-24} g and a helium isotope of mass 5.1×10^{-24} g. If the neutron is given off at right angles to the direction of the original velocity with a speed of 5.0×10^8 cm/s, find the magnitude and direction of the velocity of the helium isotope.

37. A certain nucleus, at rest, disintegrates into three particles. Two of them are detected, with masses and velocities as shown in Fig. 10-24. (a) What is the momentum of the third particle, which is known to have a mass of 12×10^{-27} kg? (b) How much energy was involved in the disintegration process.   *Answer:* (a) $(-1.0\,\mathbf{i} + 0.64\,\mathbf{j}) \times 10^{-19}$ N·s. (b) 1.1×10^{-12} J.

38. In 1932 Chadwick, in England, demonstrated the existence and properties of the neutron (one of the fundamental particles making up the atom) with the device shown in Fig. 10-25. In an evacuated chamber, a sample of radioactive polonium decays to yield α-rays (helium nuclei). These nuclei impinge on a block of beryllium inducing a process whereby neutrons are emitted. (In the reaction He and Be combine to form radioactive carbon, which decays to stable carbon + neutrons.) The neutrons strike a film of paraffin (CH_4), releasing hydrogen nuclei which are detected in an ionization chamber. In other words, an elastic collision occurs in which the momentum of the neutron is partially transferred to the hydrogen nucleus.

figure 10-24
Problem 37

figure 10-25
Problem 38

(a) Find an expression for the maximum speed V_H that the hydrogen nucleus (mass m_H) can achieve. Let the incoming neutrons have mass m_n and speed v_n. (Hint: Will more energy be transferred in a head-on collision or in a glancing collision?)

(b) One of Chadwick's goals was to find the mass of his new particle. Inspection of expression (a) which contains this parameter, however, shows that *two* unknowns are present, v_n and m_n (v_H is known; it can be measured with the ionization chamber). To eliminate the unknown v_n, he substituted a paracyanogen (CN) block for the paraffin. The neutrons then underwent elastic collisions with nitrogen nuclei instead of hydrogen nuclei. Of course, expression (a) still holds if v_N is written for v_H and m_N for m_H. Therefore if v_H and v_N are measured in separate experiments, v_N can be eliminated between the two expressions for hydrogen and nitrogen to yield a value for m_n. Chadwick's values were

$$v_H = 3.3 \times 10^9 \text{ cm/s,}$$

$$v_N = 0.47 \times 10^9 \text{ cm/s.}$$

What is his value for m_n? How does this compare with the established value $m_n = 1.00867$ u? (Take $m_H = 1.0$ u, $m_N = 14$ u.)

39. A ball with an initial speed of 10 m/s collides elastically with two identical balls whose centers are on a line perpendicular to the initial velocity and which are initially in contact with each other (Fig. 10-26). The first ball is aimed directly at the contact point and all the balls are frictionless. Find the velocities of all three balls after the collision. (Hint: The directions of the

two originally stationary balls can be found by considering the direction of the impulse they receive during the collision.)

Answer: $\mathbf{v}_2$ and $\mathbf{v}_3$ will be at 30° to $\mathbf{v}_0$ and will have a magnitude of 6.9 m/s. $\mathbf{v}_1$ will be in the opposite direction to $\mathbf{v}_0$ and will have magnitude 2.0 m/s.

40. After a totally inelastic collision, two objects of the same mass and initial speed are found to move away together at half their initial speed. Find the angle between the initial velocities of the objects.

41. Show that a slow neutron that is scattered through 90° in an elastic collision with a deuteron, initially at rest in a tank of heavy water, loses two-thirds of its initial kinetic energy to the struck deuteron.

42. Show that, in the case of an elastic collision between a particle of mass m_1 with a particle of mass m_2 initially at rest, (a) the maximum angle θ_m through which m_1 can be deflected by the collision is given by $\cos^2 \theta_m = 1 - m_2^2/m_1^2$, so that $0 \leq \theta_m \leq \pi/2$, when $m_1 > m_2$; (b) $\theta_1 + \theta_2 = \pi/2$, when $m_1 = m_2$; (c) θ_1 can take on all values between 0 and π, when $m_1 < m_2$.

figure 10-26
Problem 39

SECTION 10-7

43. A sphere of radius r_1 impinges on a sphere of radius r_2. What is the cross section for a contact collision?   *Answer:* $\pi(r_1 + r_2)^2$.

44. A beam of slow neutrons strikes an aluminum foil 1.0×10^{-5} m thick. Some neutrons are captured by the aluminum that becomes radioactive and decays by emitting an electron (β^-) forming silicon:

$$n + \text{Al}^{27} \rightarrow \text{Al}^{28} \rightarrow \text{Si}^{28} + \beta^-.$$

Suppose the neutron flux is 3.0×10^{16} $m^{-2} \cdot s^{-1}$ and the neutron capture cross section is 0.23 b. How many transmutations per unit area will occur each second?

45. A beam of fast neutrons impinges on a 5.0-mg sample of Cu^{65}, a stable isotope of copper. A possibility exists that the copper nucleus may capture a neutron to form Cu^{66}, which is radioactive and decays to Zn^{66}, which is again stable. If a study of the electron emission of the copper sample implies that 4.6×10^{11} neutron captures occur each second, what is the neutron capture cross section in barns for this process? The intensity of the neutron beam is 1.1×10^{18} neutrons $m^{-2} \cdot s^{-1}$.   *Answer:* 90 barns.

46. In a thick foil there are a great many layers of target particles so that the number of projectile particles reaching a layer will depend on how many have been scattered out by previous layers. Let the number of particles reaching a layer at a depth s be N and the number lost by scattering from that layer by $-dN$; then show that

$$-\frac{dN}{N} = n\sigma ds$$

and

$$N = N_0 e^{-n\sigma s}$$

where N_0 is the number of particles incident on the face of the foil $(s = 0)$ of unit area and n is the number of scatterers per unit volume.

SECTION 10-8

47. The precise masses in the reaction

$$p + F^{19} \rightarrow \alpha + O^{16}$$

have been determined by mass spectrometer measurements and are

$$m_p = 1.00783 \text{ u} \qquad m_\alpha = 4.00260 \text{ u}$$
$$m_F = 18.99840 \text{ u} \qquad m_0 = 15.99491 \text{ u}$$

Calculate the Q of the reaction from these data and compare with the Q calculated in Example 7 from reaction studies.   *Answer:* 8.14 MeV.

48. An elementary particle called Σ^-, at rest in a certain reference frame, decays spontaneously into two other particles according to

$$\Sigma^- \to \pi^- + n.$$

The masses are

$$m_\Sigma = 2340.5 \; m_e$$
$$m_\pi = 273.2 \; m_e$$
$$m_n = 1838.65 \; m_e$$

where m_e is the electron mass. (a) How much kinetic energy is generated in this process? (b) Which of the decay products (π^- and n) gets the larger share of this kinetic energy? Of the momentum?

49. The Q of the reaction in which a U^{236} nucleus at rest splits into just two fragments of masses 132 u and 98 u is 192 MeV. (a) How much energy was lost through radiation? (b) What is the speed of each fragment? (c) What is the kinetic energy of each fragment?

   Answer: (a) 5400 MeV. (b) $v_{132} = 1.09 \times 10^7$ m/s; $v_{98} = 1.47 \times 10^7$ m/s.
   (c) $K_{132} = 81.7$ MeV; $K_{98} = 110$ MeV.

11
rotational kinematics

So far we have dealt mostly with the translational motion of single particles or of rigid bodies, that is, of bodies whose parts all have a fixed relationship to each other. No real body is truly rigid, but many bodies, such as molecules, steel beams, and planets, are rigid enough so that, in many problems, we can ignore the fact that they warp, bend, or vibrate. As Fig. 3-1 suggests, we say that a rigid body moves in pure *translation* if each particle of the body undergoes the same displacement as every other particle in any given time interval.

In this chapter, however, we are interested in *rotation* rather than translation. For the time being we again restrict ourselves to single particles and to rigid bodies, which means that we shall not consider such rotational motions as those of the solar system or of water in a spinning beaker. *We shall also deal only with rotation about axes that remain fixed in the reference frame in which we observe the rotation.*

Figure 11-1 shows the rotational motion of a rigid body about a fixed axis, in this case the *z*-axis of our reference frame. Let *P* represent a particle in the rigid body, arbitrarily selected and described by the position vector **r**. We then say that: *A rigid body moves in pure rotation if every particle of the body* (such as *P* in Fig. 11-1) *moves in a circle, the centers of which are on a straight line called the axis of rotation* (the *z*-axis in Fig. 11-1). If we draw a perpendicular from any point in the body to the axis, each such line will sweep through the same angle in any given time interval as another such line. Thus we can describe the pure rotation of a rigid body by considering the motion of any one of the particles (such as *P*) that make it up. (We must rule out, however, particles that are on the axis of rotation. Why?)

The general motion of a rigid body is a combination of translation

11-1
ROTATIONAL MOTION

figure 11-1
A rigid body rotating about the *z*-axis. Each point in the body, such as *P*, describes a circle about this axis.

and rotation however, rather than one of pure rotation. We can locate a rigid body that is moving in pure translation by giving the three coordinates x, y, z of any point in it (its center of mass, say) in a particular reference frame. For a body that rotates as it moves translationally we need, in the most general case, three more coordinates, such as angles, to specify the orientation of the body with respect to the reference frame. Figure 11-2 (see also Fig. 9-1) shows a special case of rigid body motion combining translation and rotation. The figure is an extension of Fig. 3-1 in which the body now rotates as it moves translationally. To locate this body we must not only locate point O in the body in the xy reference frame but we must also say how the $x'y'$ reference frame, which is fixed in the body, is oriented with respect to the xy frame.

figure 11-2
A rigid body moving in combined translational and rotational motion as seen from reference frame x, y. Notice that the reference frame fixed on the body (x', y') changes its orientation with respect to x, y as the motion proceeds. Compare with Figs. 3-1 and 9-1. This figure represents a special case in that the translational motion occurs in two dimensions only (the xy plane) and the rotational motion occurs about an axis that maintains a fixed direction (the z'-axis).

As we saw in Chapter 9 we can describe the translational motion of any system of particles—whether rigid or not—whether rotating or not—by imagining that all of the mass M of the body is concentrated at the center of mass and that $\mathbf{F}_{ext}$, the resultant of the external forces acting on the body, acts at this point. The acceleration of the center of mass is then given by Eq. 9-10 or $\mathbf{F}_{ext} = M\mathbf{a}_{cm}$. It is very helpful to be able to represent the translational motion of a rigid body by the motion of a single point—its center of mass; all that is left is to determine its rotational motion. We shall discuss such combined translational and rotational motions in the next chapter. This will be simpler to do after we have studied pure rotation about a fixed axis.

We now return, therefore, to the pure rotation of a rigid body about a fixed axis (Fig. 11-1). First, we must *describe* the rotational motion. We call this description rotational kinematics; we must define the variables of angular motion and relate them to each other, just as in particle kinematics (see Chapter 4) we defined the variables of translational motion and related them to each other. The next part of our program is to relate the rotational motion of a body to the properties of the body and of its environment. This is rotational dynamics. In this chapter we study the kinematics of rotation. We develop the dynamics of rotation in the next chapter.

In Fig. 11-1 let us pass a plane through P at right angles to the axis of rotation. This plane, which cuts through the rotating body, contains the circle in which particle P moves. Figure 11-3 shows this plane, as we look downward on it from above, along the z-axis in Fig. 11-1.

We can tell exactly where the entire rotating body is in our reference frame if we know the location of any single particle (P) of the body in this frame. Thus, for the kinematics of this problem, we need only consider the (two-dimensional) motion of a particle in a circle.

The angle θ in Fig. 11-3 is the angular position of particle P with respect to the reference position. We arbitrarily choose the positive sense of rotation in Fig. 11-3 to be counterclockwise, so that θ increases for counterclockwise rotation and decreases for clockwise rotation.

It is convenient to measure θ in radians* rather than in degrees. By definition θ is given in radians by the relation

$$\theta = s/r,$$

in which s is the arc length shown in Fig. 11-3.

Let the body of Fig. 11-3 be rotating counterclockwise. At time t_1 the angular position of P is θ_1 and at a later time t_2 its angular position is θ_2. This is shown in Fig. 11-4, which gives the positions of P and of the position vector $\mathbf{r}$ at these times; the outline of the body itself has been omitted in that figure for simplicity. The *angular displacement* of P will be $\theta_2 - \theta_1 = \Delta\theta$ during the time interval $t_2 - t_1 = \Delta t$. We define the *average angular speed* $\overline{\omega}$ of particle P in this time interval as

$$\overline{\omega} = \frac{\theta_2 - \theta_1}{t_2 - t_1} = \frac{\Delta\theta}{\Delta t}.$$

We define the *instantaneous angular speed* ω as the limit approached by this ratio as Δt approaches zero:

$$\omega = \lim_{\Delta t \to 0} \frac{\Delta\theta}{\Delta t} = \frac{d\theta}{dt}. \tag{11-1}$$

For a rigid body all radial lines fixed in it perpendicular to the axis of rotation rotate through the same angle in the same time, so that the angular speed ω about this axis is the same for each particle in the body. Thus ω is characteristic of the body as a whole. Angular speed has the dimensions of an inverse time (T^{-1}); its units are commonly taken to be radians/second (rad/s) or revolutions/second (rev/s).

If the angular speed of P is not constant, then the particle has an angular acceleration. Let ω_1 and ω_2 be the instantaneous angular speeds at the times t_1 and t_2 respectively; then the *average angular acceleration* $\overline{\alpha}$ of the particle P is defined as

$$\overline{\alpha} = \frac{\omega_2 - \omega_1}{t_2 - t_1} = \frac{\Delta\omega}{\Delta t}.$$

The *instantaneous angular acceleration* is the limit of this ratio as Δt

11-2
ROTATIONAL KINEMATICS— THE VARIABLES

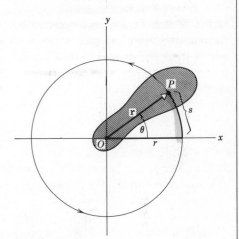

figure 11-3
A cross sectional view of the rigid body of Fig. 11-1, showing point P and vector $\mathbf{r}$ of that figure. Point P, which is fixed in the rotating body, rotates counterclockwise about the origin in a circle of radius r.

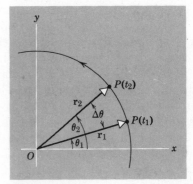

figure 11-4
The reference line r $(= OP)$, fixed in the body of Figs. 11-1 and 11-3, is displaced through angle $\Delta\theta$ $(= \theta_2 - \theta_1)$ in time Δt $(= t_2 - t_1)$.

---

* The radian is a purely geometrical unit having no physical dimension because it is the ratio of two lengths. Since the circumference of a circle of radius r is $2\pi r$, there are 2π rad in a complete circle, that is, $\theta = 2\pi r/r = 2\pi$. Therefore 2π rad $= 360°$, π rad $= 180°$, and 1 rad $\cong 57.3°$.

approaches zero, or

$$\alpha = \lim_{\Delta t \to 0} \frac{\Delta \omega}{\Delta t} = \frac{d\omega}{dt}. \qquad (11\text{-}2)$$

Because ω is the same for all particles in the rigid body, it follows from Eq. 11-2 that α must be the same for each particle and thus α, like ω, is a characteristic of the body as a whole. Angular acceleration has the dimensions of an inverse time squared (T^{-2}); its units are commonly taken to be radians/second^2 (rad/s^2) or revolutions/second^2 (rev/s^2).

The rotation of a particle (or a rigid body) *about a fixed axis* has a formal correspondence to the translational motion of a particle (or a rigid body) *along a fixed direction*. The kinematical variables are θ, ω, and α in the first case and x, v, and a in the second. These quantities correspond in pairs: θ to x, ω to v, and α to a. Note that the angular quantities differ dimensionally from the corresponding linear quantities by a length factor. Note, too, that all six quantities may be treated as scalars in this special case. For example, a particle at any instant can be moving in one direction or the other along its straight-line motion, corresponding to a positive or a negative value for v; similarly a particle at any instant can be rotating in one direction or another about its fixed axis, corresponding to a positive or a negative value for ω.

When, in translational motion, we remove the restriction that the motion be along a straight line and consider the general case of motion in three dimensions along a curved path, the linear variables x, v, and a reveal themselves as the scalar components of the kinematic vectors **r**, **v**, and **a**. In Section 11-4, we shall see to what extent the rotational kinematic variables reveal themselves as vectors when we remove the restriction of a fixed axis of rotation.

11-3 ROTATION WITH CONSTANT ANGULAR ACCELERATION

For translational motion of a particle or a rigid body along a fixed direction, such as the x-axis, we have seen (in Chapter 3) that the simplest type of motion is that in which the acceleration a is zero. The next simplest type corresponds to a = a constant (other than zero); for this motion we derived the equations of Table 3-1, which connect the kinematic variables x, v, a, and t in all possible combinations.

For the rotational motion of a particle or a rigid body around a fixed axis the simplest type of motion is that in which the angular acceleration α is zero (such as uniform circular motion). The next simplest type of motion, in which α = a constant (other than zero), corresponds exactly to linear motion with a = a constant (other than zero). As before, we can derive four equations linking the four kinematic variables θ, ω, α, and t in all possible combinations. You can either derive these angular equations by the methods used to derive the linear equations (see Example 2) or you may write them down at once by substituting corresponding angular quantities for the linear quantities in the linear equations.

We list both sets of equations in Table 11-1, having chosen $x_0 = 0$ and $\theta_0 = 0$ in these relations for simplicity. Here ω_0 is the angular speed at the time $t = 0$. You should check these equations dimensionally before verifying them. Both sets of equations hold not only for particles but also for rigid bodies.

For the angular quantities, we arbitrarily select one of the two possible directions of rotation about the fixed axis as the direction in which θ is increasing. From Eq. 11-1 ($\omega = d\theta/dt$) we see that if θ is increasing with

Table 11-1
Motion with constant linear or angular acceleration

	Translational Motion (Fixed Direction)	Rotational Motion (Fixed Axis)	
(3-12)	$v = v_0 + at$	$\omega = \omega_0 + \alpha t$	(11-3)
(3-14)	$x = \dfrac{v_0 + v}{2} t$	$\theta = \dfrac{\omega_0 + \omega}{2} t$	(11-4)
(3-15)	$x = v_0 t + \frac{1}{2} a t^2$	$\theta = \omega_0 t + \frac{1}{2} \alpha t^2$	(11-5)
(3-16)	$v^2 = v_0^2 + 2ax$	$\omega^2 = \omega_0^2 + 2\alpha\theta$	(11-6)

time, ω is positive. Similarly, from Eq. 11-2 $(\alpha = d\omega/dt)$, we see that if ω is increasing with time, α is positive. There are corresponding sign conventions for the linear quantities.

A grindstone has a constant angular acceleration α of 3.0 rad/s². Starting from rest a line, such as OP in Fig. 11-5, is horizontal. Find (a) the angular displacement of the line OP (and hence of the grindstone) and (b) the angular speed of the grindstone 2.0 s later.

(a) α and t are given; we wish to find θ. Hence, we use Eq. 11-5,

$$\theta = \omega_0 t + \tfrac{1}{2}\alpha t^2.$$

At $t = 0$, we have $\omega = \omega_0 = 0$ and $\alpha = 3.0$ rad/s². Therefore, after 2.0 s,

$$\theta = (0)(2.0 \text{ s}) + \tfrac{1}{2}(3.0 \text{ rad/s}^2)(2.0 \text{ s})^2 = 6.0 \text{ rad} = 0.96 \text{ rev}.$$

(b) α and t are given; we wish to find ω. Hence, we use Eq. 11-3

$$\omega = \omega_0 + \alpha t,$$

and

$$\omega = 0 + (3.0 \text{ rad/s}^2)(2.0 \text{ s}) = 6.0 \text{ rad/s}.$$

Using Eq. 11-6 as a check, we have

$$\omega^2 = \omega_0^2 + 2\alpha\theta,$$

$$\omega^2 = 0 + (2)(3.0 \text{ rad/s}^2)(6.0 \text{ rad}) = 36 \text{ rad}^2/\text{s}^2,$$

$$\omega = 6.0 \text{ rad/s}.$$

Derive the equation $\omega = \omega_0 + \alpha t$ for constant angular acceleration.

(a) Starting from the definition of angular acceleration,

$$\alpha = \frac{d\omega}{dt},$$

we have

$$\alpha \, dt = d\omega$$

or

$$\int \alpha \, dt = \int d\omega.$$

But α is a constant, so that

$$\alpha \int dt = \int d\omega.$$

If at $t = 0$ we call the angular speed ω_0, then

$$\alpha \int_0^t dt = \int_{\omega_0}^{\omega} d\omega$$

EXAMPLE 1

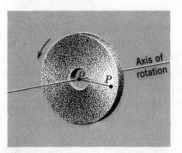

figure 11-5
Example 1. The line OP is attached to a grindstone rotating as shown about an axis through O that is fixed in the reference frame of the observer.

EXAMPLE 2

or

$$\alpha t = \omega - \omega_0$$

and

$$\omega = \omega_0 + \alpha t.$$

(b) We can also derive the result by making use of the fact that the average acceleration equals the instantaneous acceleration when the acceleration is constant. The average acceleration is

$$\bar{\alpha} = \frac{\omega - \omega_0}{t - t_0}.$$

For constant acceleration we have $\alpha = \bar{\alpha}$. Letting $t_0 = 0$, we obtain

$$\alpha = \frac{\omega - \omega_0}{t}$$

or

$$\omega = \omega_0 + \alpha t.$$

Compare this derivation with that of the corresponding linear relation $v = v_0 + at$ in Section 3-8.

The linear displacement, velocity, and acceleration are vectors. The corresponding angular quantities *may* be vectors also, for in addition to a magnitude we must also specify a direction for them, namely, the direction of the axis of rotation in space. Because we considered rotation only about a fixed axis, we were able to treat θ, ω, and α as scalar quantities. If the direction of the axis changes, however, we can no longer avoid the question "are rotational quantities vectors?" We can find out only by seeing whether or not they obey the laws of vector addition.

Let us discuss first the angular displacement θ. The magnitude of the angular displacement of a body is the angle through which the body turns. Angular displacements, however, are *not* vectors because they do *not* add like vectors. For example, give two successive rotations θ_1 and θ_2 to a book which initially lies in a horizontal plane. Let rotation θ_1 be a 90° clockwise turn about a vertical axis through the center of the book as we view it from above. Let θ_2 be a 90° clockwise turn about a north-south axis through the center of the book as we view it looking north. In one case, apply operation θ_1 first and then θ_2. In the other case, apply operation θ_2 first and then θ_1. You should try this for yourself. Now, if angular displacements are vector quantities, they must add like vectors. In particular, they must obey the law of vector addition $\boldsymbol{\theta}_1 + \boldsymbol{\theta}_2 = \boldsymbol{\theta}_2 + \boldsymbol{\theta}_1$, which tells us that the order in which we add vectors does not affect their sum. This law fails for finite angular displacements (see exercise above and also Fig. 11-6a). Hence finite angular displacements are *not* vector quantities.

Suppose that instead of 90° rotations we had made 3° rotations. The result of $\boldsymbol{\theta}_1 + \boldsymbol{\theta}_2$ would still differ from the result of $\boldsymbol{\theta}_2 + \boldsymbol{\theta}_1$, but the difference would be much smaller. In fact, as the two angular displacements are made smaller, the difference between the two sums rapidly disappears (Fig. 11-6b,c). If the angular displacements are made infinitesimal, the order of addition no longer affects the result. Hence *infinitesimal angular displacements are vectors*.

Quantities defined in terms of infinitesimal angular displacements may themselves be vectors. For example, the angular velocity is $\boldsymbol{\omega} = d\boldsymbol{\theta}/dt$. Since $d\boldsymbol{\theta}$ is a vector and dt a scalar, the quotient is a vector.

11-4
ROTATIONAL QUANTITIES AS VECTORS

figure 11-6
(a) A book rotated θ_1 (90° as shown about an axis at right angles to the page) and then θ_2 (90° as shown about a north-south axis) has a different final orientation than if rotated first through θ_2 and then θ_1. This property is called the noncommutivity of finite angles under addition: $\theta_1 + \theta_2 \neq \theta_2 + \theta_1$. (b) The middle group is the same except that the angular displacements are smaller, being 45°. Although the final orientations still differ, they are much nearer each other. (c) The lower group repeats the experiment for 20° displacements. We see here that $\theta_1 + \theta_2 \cong \theta_2 + \theta_1$. As $\theta_1, \theta_2 \to 0$, the final positions approach each other. Finite angles under addition tend to commute as the angles become very small. Infinitesimal angles *do* commute under addition, making it possible to treat them as vectors.

Therefore the angular velocity is a vector. In Fig. 11-7a, for example, we represent the angular velocity ω of the rotating rigid body by an arrow drawn along the axis of rotation; in Fig. 11-7b we represent the rotation of a particle (such as P in Fig. 11-7a) about a fixed axis in just the same way. The length of the arrow is made proportional to the magnitude of the angular velocity. The sense of the rotation determines the direction

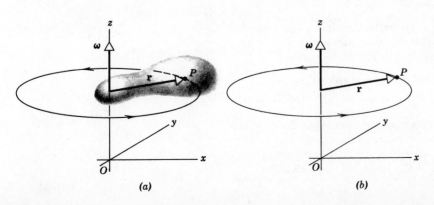

figure 11-7
The angular velocity ω of (a) a rotating rigid body and (b) a rotating particle, about a fixed axis.

in which the arrow points along the axis. By convention, if the fingers of the *right hand* curl around the axis in the direction of rotation of the body, the extended thumb points along the direction of the angular velocity vector. For the rigid body of Fig. 11-1, therefore, the angular velocity vector will be in the positive *z*-direction. In Fig. 11-3, $\boldsymbol{\omega}$ will be perpendicular to the page pointing up out of the page, corresponding to the counter-clockwise rotation. The angular velocity of the turntable of a phonograph is a vector pointing down. Notice that nothing moves in the direction of the angular velocity vector. The vector represents the angular velocity of the rotational motion taking place in the plane perpendicular to it.

Angular acceleration is also a vector quantity. This follows from the definition $\boldsymbol{\alpha} = d\boldsymbol{\omega}/dt$, in which $d\boldsymbol{\omega}$ is a vector and dt a scalar. Later we shall encounter other rotational quantities that are vectors, such as torque and angular momentum.

A disk spins on a horizontal shaft mounted in bearings, with an angular speed ω_1 of 100 rad/s as in Fig. 11-8*a*. The entire disk and shaft assembly are placed on a turntable rotating about a vertical axis at $\omega_2 = 30.0$ rad/s, counterclockwise as we view it from above. Describe the rotation of the disk as seen by an observer in the room.

EXAMPLE 3

(*a*)

(*b*)

figure 11-8
Example 3. (*a*) A spinning disc on a rotating turntable. (*b*) The angular velocities add like vectors.

The disk is subject to two angular velocities simultaneously; we can describe its resultant motion by the vector sum of these vectors. The angular velocity $\boldsymbol{\omega}_1$ associated with the shaft rotation has a magnitude of 100 rad/s and occurs about an axis that is not fixed but, as seen by an observer in the room, rotates in a horizontal plane at 30 rad/s. The angular velocity $\boldsymbol{\omega}_2$ associated with the turntable is fixed vertically and has a magnitude of 30 rad/s.

The resultant angular velocity of the disk $\boldsymbol{\omega}$ is the vector sum of $\boldsymbol{\omega}_1$ and $\boldsymbol{\omega}_2$. The magnitude of $\boldsymbol{\omega}$ is

$$\omega = \sqrt{\omega_1^2 + \omega_2^2} = \sqrt{(100 \text{ rad/s})^2 + (30.0 \text{ rad/s})^2}$$

$$= 104 \text{ rad/s}.$$

The direction of $\boldsymbol{\omega}$ is not fixed in our observer's reference frame but rotates at the same angular rate as the turntable. The vector $\boldsymbol{\omega}$ does not lie in the horizontal plane but points above it by an angle θ (see Fig. 11-8*b*), where

$$\theta = \tan^{-1} \omega_2/\omega_1 = \tan^{-1} (30.0 \text{ rad/s})/(100 \text{ rad/s})$$

$$= \tan^{-1} 0.300 = 16.7°.$$

We can describe the motion of the disk as a simple rotation about this new axis (whose direction in our observer's reference frame is changing with time as described above) at an angular rate of 104 rad/s. How would the situation change if the direction of rotation of the disk, or of the turntable, were changed?

In Sections 4-4 and 4-5 we discussed the linear velocity and acceleration of a particle moving in a circle. When a rigid body rotates about a fixed axis, every particle in the body moves in a circle. Hence we can describe the motion of such a particle either in linear variables or in angular variables. The relation between the linear and angular variables enables us to pass back and forth from one description to another and is very useful.

Consider a particle at P in the rigid body, a distance r from the axis through O. This particle moves in a circle of radius r as the body rotates, as in Fig. 11-9a. The reference position is Ox. The particle moves through a distance s along the arc when the body rotates through an angle θ, such that

$$s = \theta r \qquad (11\text{-}7)$$

where θ is in radians.

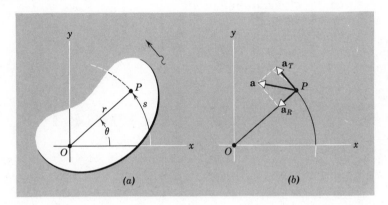

(a)   (b)

figure 11-9
(a) A rigid body rotates about a fixed axis through O perpendicular to the page. The point P sweeps out an arc s which subtends an angle θ. *(b)* The acceleration **a** of point P has components $\mathbf{a}_T$ (tangential) where $a_T = \alpha r$ and $\mathbf{a}_R$ (radial) where $a_R = v^2/r = \omega^2 r$ ($\omega =$ angular speed).

Differentiating both sides of this equation with respect to the time, and noting that r is constant, we obtain

$$\frac{ds}{dt} = \frac{d\theta}{dt} r.$$

But ds/dt is the linear speed of the particle at P and $d\theta/dt$ is the angular speed ω of the rotating body so that

$$v = \omega r. \qquad (11\text{-}8)$$

This is a relation between the *magnitudes* of the linear velocity and the angular velocity; the linear speed of a particle in circular motion is the product of the angular speed and the distance r of the particle from the axis of rotation.

Differentiating Eq. 11-8 with respect to the time, we have

$$\frac{dv}{dt} = \frac{d\omega}{dt} r.$$

But dv/dt is the magnitude of the *tangential* component of acceleration of the particle (see Section 4-5) and $d\omega/dt$ is the magnitude of the angular acceleration of the rotating body, so that

$$a_T = \alpha r. \qquad (11\text{-}9)$$

Hence the magnitude of the tangential component of the linear acceleration of a particle in circular motion is the product of the magnitude of the angular acceleration and the distance r of the particle from the axis of rotation.

We have seen that the *radial* component of acceleration is v^2/r for a particle moving in a circle. This can be expressed in terms of angular speed by use of Eq. 11-8. We have

$$a_R = \frac{v^2}{r} = \omega^2 r. \qquad (11\text{-}10)$$

The resultant acceleration of point P is shown in Fig. 11-9b.

Equations 11-7 through 11-10 enable us to describe the motion of one point on a rigid body rotating about a fixed axis *either* in angular variables *or* in linear variables. We might ask why we need the angular variables when we are already familiar with the equivalent linear variables. The answer is that the angular description offers a distinct advantage over the linear description when various points on the same rotating body must be considered. Different points on the same rotating body do not have the same linear displacement, speed, or acceleration, but *all* points on a rigid body rotating about a fixed axis do have the same *angular* displacement, speed, or acceleration at any instant. By the use of angular variables we can describe the motion of the body as a whole in a simple way.

EXAMPLE 4

If the radius of the grindstone of Example 1 is 0.50 m, calculate (a) the linear or tangential speed of a particle on the rim, (b) the tangential acceleration of a particle on the rim, and (c) the centripetal acceleration of a particle on the rim, at the end of 2.0 s.

We have $\alpha = 3.0$ rad/s^2, $\omega = 6.0$ rad/s after 2.0 s, and $r = 0.50$ m. Then,

(a)
$$v = \omega r$$
$$= (6.0 \text{ rad/s})(0.50 \text{ m})$$
$$= 3.0 \text{ m/s} \qquad \text{(linear speed)};$$

(b)
$$a_T = \alpha r$$
$$= (3.0 \text{ rad/s}^2)(0.50 \text{ m})$$
$$= 1.5 \text{ m/s}^2 \qquad \text{(tangential acceleration)};$$

(c)
$$a_R = v^2/r = \omega^2 r$$
$$= (6.0 \text{ rad/s})^2(0.50 \text{ m})$$
$$= 18 \text{ m/s}^2 \qquad \text{(centripetal acceleration)}.$$

(d) Are the results the same for a particle halfway in from the rim, that is, at $r = 0.25$ m?

The *angular* variables are the same for this point as for a point on the rim. That is, once again

$$\alpha = 3.0 \text{ rad/s}^2, \qquad \omega = 6.0 \text{ rad/s}.$$

But now $r = 0.25$ m, so that for this particle

$$v = 1.5 \text{ m/s}, \qquad a_T = 0.75 \text{ m/s}^2, \qquad a_R = 9.0 \text{ m/s}^2.$$

11-6
RELATION BETWEEN LINEAR AND ANGULAR KINEMATICS FOR A PARTICLE IN CIRCULAR MOTION — VECTOR FORM

Notice that the relations deduced in the previous section are relations between *scalar* quantities, both the linear and angular variables being expressed in scalar form. Let us now use vector methods, making an analysis essentially like that of Section 4-5 except that we now introduce the angular variables. This will illustrate, for a familiar special case, the more general approach and prepare the way for situations in which vector methods are essential. We continue to restrict ourselves to rotation about a fixed axis.

Figure 11-10a shows a particle P, rotating about a fixed axis through the origin, at times t and $t + \Delta t$. The particle moves in a circle of constant radius r; beyond this there are no restrictions on its motion and in general $\boldsymbol{\omega}$ and $\boldsymbol{\alpha}$ may have values that vary as the particle moves. We can express the restriction to a constant radius by

$$\mathbf{r} = \mathbf{u}_r r, \tag{11-11}$$

in which $\mathbf{u}_r$ is a unit vector in the direction of $\mathbf{r}$.

(a)   (b)   (c)

figure 11-10
(a) The particle P rotates through an angle $\Delta\theta$ in time Δt. The unit vectors, in polar coordinates, are shown at each point. (b) The change in $\mathbf{u}_r$; note that $\Delta\mathbf{u}_r$, as $\Delta\theta \to 0$, points in the direction of $\mathbf{u}_\theta$. (c) The change in $\mathbf{u}_\theta$; note that $\Delta\mathbf{u}_\theta$, as $\Delta\theta \to 0$, points in the direction of $-\mathbf{u}_r$.

Differentiating Eq. 11-11, remembering that r (but not $\mathbf{r}$ or $\mathbf{u}_r$, since their directions change) is a constant, we have

$$\frac{d\mathbf{r}}{dt} = \frac{d\mathbf{u}_r}{dt} r. \tag{11-12}$$

Now $d\mathbf{r}/dt$ is $\mathbf{v}$, the linear velocity of the particle. To evaluate $d\mathbf{u}_r/dt$, consider Fig. 11-10b, which shows the unit vector $\mathbf{u}_r$ for two different positions of P, corresponding to a rotation through a (small) angle $\Delta\theta$. Using the definition of angular measure in radians we obtain the *magnitude* of the (vector) change $\Delta\mathbf{u}_r$ in $\mathbf{u}_r$ from

$$\Delta u_r = (1)\,\Delta\theta,$$

in which the factor (1) reminds us that the two unit vectors in Fig. 11-10b have unit length. The above equation will be correct if $\Delta\theta$ is small enough so that we can neglect the difference between the chord and the arc in the small triangle in Fig. 11-10b. The change in $\mathbf{u}_r$ is a vector, $\Delta\mathbf{u}_r$, whose magnitude is given by the above equation; its *direction*, again assuming that $\Delta\theta$ is small enough, is given by the unit vector $\mathbf{u}_\theta$. This follows because, if $\Delta\mathbf{u}_r$ in Fig. 11-10b is translated to point P in Fig. 11-10a, we see that, as $\Delta\theta \to 0$, it points in the direction of $\mathbf{u}_\theta$. Thus we find

$$\Delta\mathbf{u}_r \cong \mathbf{u}_\theta\,\Delta\theta.$$

Dividing by Δt and allowing Δt to approach zero, we have

$$\frac{d\mathbf{u}_r}{dt} = \mathbf{u}_\theta \frac{d\theta}{dt} = \mathbf{u}_\theta \omega. \tag{11-13}$$

Substituting these results into Eq. 11-12 yields, then,

$$\mathbf{v} = \mathbf{u}_\theta \omega r. \tag{11-14a}$$

The scalar relationship that corresponds to this is

$$v = \omega r \tag{11-14b}$$

and is one of the relationships, obtained before, connecting the linear speed v of a particle in circular motion with its angular speed ω.

To find the relation between linear and angular acceleration we differentiate Eq. 11-14a, remembering that r is a constant although $\mathbf{u}_\theta$ and ω vary. We have

$$\frac{d\mathbf{v}}{dt} = \mathbf{u}_\theta \frac{d\omega}{dt} r + \omega \frac{d\mathbf{u}_\theta}{dt} r. \tag{11-15}$$

Now $dv/dt = a$, the linear acceleration of the particle and $d\omega/dt = \alpha$, its angular acceleration. From Fig. 11-10c, guided by the derivation of Eq. 11-13, you should be able to prove that

$$\frac{d\mathbf{u}_\theta}{dt} = -\mathbf{u}_r\omega. \qquad (11\text{-}16)$$

The minus sign comes in because when we translate $\Delta\mathbf{u}_\theta$ in Fig. 11-10c to point P, we see that, as $\Delta\theta \to 0$, it points radially inward, in the direction opposite to $\mathbf{u}_r$.

Making these substitutions into Eq. 11-15 yields

$$\mathbf{a} = \mathbf{u}_\theta\alpha r - \mathbf{u}_r\omega^2 r. \qquad (11\text{-}17)$$

Thus, as we know from Section 4-5, $\mathbf{a}$ has a radial (or centripetal) component $\mathbf{a}_R$ and a tangential component $\mathbf{a}_T$. Their magnitudes, from Eq. 11-17, are

$$a_T = \alpha r \qquad (11\text{-}18a)$$

and (using Eq. 11-14b)

$$a_R = \omega^2 r = v^2/r. \qquad (11\text{-}18b)$$

The last is the familiar result derived in Section 4-4. In Supplementary Topic I we derive the relations between the linear and angular kinematic variables for a particle free to move in a plane but not restricted to circular motion. Equations 11-14a and 11-17 will prove to be special cases of the more general relationships derived there.

Equations 11-14a and 11-17 are relations between the linear kinematic variables in vector form and the angular kinematic variables in scalar form. We should be able to derive relationships in which *each* set of variables is expressed in vector form. Let us do so now. This will be especially useful in cases where the axis of rotation is not fixed.

Figure 11-11 shows the vectors $\mathbf{r}$, $\mathbf{v}$, $\mathbf{a}_T$, $\mathbf{a}_R$, $\boldsymbol{\omega}$, and $\boldsymbol{\alpha}$ for the rotating particle of Fig. 11-7b. The angular quantities are on the axis of rotation, pointing in the direction given by the right-hand rule of page 22. We declare—and shall prove—that the relationships we seek are

$$\mathbf{v} = \boldsymbol{\omega} \times \mathbf{r} \qquad (11\text{-}19)$$

and

$$\mathbf{a} = \mathbf{a}_T + \mathbf{a}_R, \qquad (11\text{-}20a)$$

in which

$$\mathbf{a}_T = \boldsymbol{\alpha} \times \mathbf{r} \quad \text{and} \quad \mathbf{a}_R = \boldsymbol{\omega} \times \mathbf{v}. \qquad (11\text{-}20b)$$

In Section 2-4 (which you may wish to reread) we defined the vector product of two vectors. If $\mathbf{c} = \mathbf{a} \times \mathbf{b}$, then the *magnitude of* $\mathbf{c}$ is $ab \sin \phi$, where ϕ is the angle between $\mathbf{a}$ and $\mathbf{b}$. In applying this part of the definition to Eqs. 11-19 and 11-20 we note (see Fig. 11-11) that $\boldsymbol{\omega}$ and $\mathbf{r}$, $\boldsymbol{\omega}$ and $\mathbf{v}$, and $\boldsymbol{\alpha}$ and $\mathbf{r}$ are each mutually perpendicular; thus the angle ϕ for each of these three pairs of vectors is 90°. In Eq. 11-19 we have, for magnitudes

$$v = \omega r \sin 90° = \omega r,$$

which is exactly Eq. 11-14b. In Eqs. 11-20b we have, for magnitudes

$$a_R = \omega v = \omega(\omega r) = \omega^2 r$$

and

$$a_T = \alpha r.$$

These relations agree with Eqs. 11-18b and a exactly.

It remains to be seen whether *directions* are correctly given by Eqs. 11-19 and 11-20b. For the vector product $\mathbf{c} = \mathbf{a} \times \mathbf{b}$, Fig. 2-12 shows that the direction of $\mathbf{c}$ is found by sweeping $\mathbf{a}$ into $\mathbf{b}$ through the (smaller) angle between them with the fingers of the right hand; the extended right thumb then points in the direc-

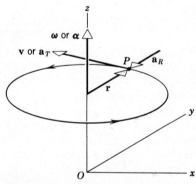

figure 11-11
The directions of the vectors $\mathbf{r}$, $\mathbf{v}$, $\mathbf{a}_T$, $\mathbf{a}_R$, ω and α for a particle rotating in a circle about the z-axis.

tion of **c**. You can readily check that, in Fig. 11-11, the directions of the vectors **v**, $\mathbf{a}_T$ and $\mathbf{a}_R$ are indeed correctly given by Eqs. 11-19 and 11-20b.

questions

1. In Section 11-1 we stated that, in general, six variables are required to locate a rigid body with respect to a particular reference frame. How many variables are required to locate the body of Fig. 11-2 with respect to the x-y frame shown in that figure? If this number is not six, account for the difference.

2. An irregular body is free to rotate about its center of mass which is placed at the origin of a reference frame. How would you specify its orientation?

3. Could the angular quantities θ, ω, and α be expressed in terms of degrees instead of radians in the kinematical equations?

4. Explain why the radian measure of angle is equally satisfactory for all systems of units. Is the same true for degrees?

5. If a car's speedometer is set to read at a speed proportional to the rotational speed of its rear wheels, is it necessary to correct the reading when snow tires replace regular ones?

6. How could you express simply the relationship between the angular velocities of a pair of gears which are coupled?

7. A wheel is rotating about an axis through its center perpendicular to the plane of the wheel. Consider a point on the rim. When the wheel rotates with *constant angular velocity*, does the point have a radial acceleration? A tangential acceleration? When the wheel rotates with *constant angular acceleration*, does the point have a radial acceleration? A tangential acceleration? Do the magnitudes of these accelerations change?

8. Suppose you were asked to determine the equivalent distance traveled by a phonograph needle in playing, say, a 12-in., $33\frac{1}{3}$ rpm record. What information do you need? Discuss from the points of view of reference frames (a) fixed in the room, (b) fixed on the rotating record, and (c) fixed on the record arm.

9. (a) Describe the vector that would represent the angular velocity of the earth rotating about its axis. (b) Describe the vector that would represent the angular velocity of the earth rotating about the sun.

10. It is convenient to picture rotational vectors as lying along the axis of rotation. Is there any reason why they could not be pictured as merely parallel to the axis, but located anywhere? Recall that we are free to slide a displacement vector along its own direction or translate it sideways without changing its value.

11. In a centrifuge particles will tend to separate from the fluid in which they are suspended if their density (mass/volume) differs from that of the fluid. Discuss the dynamical principles upon which the operation of a centrifuge depends. View the situation from both an inertial (laboratory) frame and a noninertial (rotating) frame.

12.* A marksman stands at the center of a merry-go-round firing at a target fixed to a post on its outer perimeter. How, if at all, must the man take into account the (constant) angular velocity of the merry-go-round if he is to hit the target? What if the positions of marksman and target were reversed?

13.* A man on a merry-go-round rotating at constant angular velocity ω releases a cake of ice that he had been holding fixed to the merry-go-round at a radial distance r_0 from the center. Describe the motion of the ice in the reference frame of (a) a ground observer and (b) the man on the merry-go-round. Neglect frictional forces but describe all other forces.

14.* A man on a rotating merry-go-round kicks a cake of ice outward along a radial line. What is its subsequent motion as seen by an observer (a) on the

* See Supplementary Topic I.

merry-go-round and (b) on the ground? Assume that frictional forces may be neglected.

SECTION 11-2

1. What is the angular speed of (a) the second hand of a watch; (b) of the minute hand?   *Answer:* (a) 0.10 rad/s. (b) 1.7×10^{-3} rad/s.

2. A phonograph record on a turntable rotates at 33 rev/min. What is the linear speed of a point on the record at the needle at (a) the beginning and (b) the end of the recording? The distances of the needle from the turntable axis are 5.9 and 2.9 in., respectively, at these two positions.

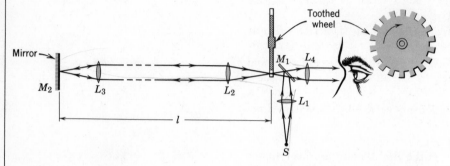

figure 11-12
Problem 3

3. One method of measuring the speed of light makes use of a rotating toothed wheel. A beam of light passes through a slot at the outside edge of the wheel, as in Fig. 11-12, travels to a distant mirror, and returns to the wheel just in time to pass through the next slot in the wheel. One such toothed wheel has a radius of 5.0 cm and 500 teeth at its edge. Measurements taken when the mirror was 500 m from the wheel indicated a speed of light of 3.0×10^5 km/s. (a) What was the (constant) angular speed of the wheel? (b) What was the linear speed of a point on its edge?
Answer: (a) 3.8×10^3 rad/s. (b) 190 m/s.

4. If an airplane propeller of radius 5.0 ft (1.5 m) rotates at 2000 rev/min and the airplane is propelled at a ground speed of 300 mi/h (480 km/h), what is the speed of a point on the tip of the propeller, as seen by (a) the pilot and (b) an observer on the ground?

5. The angular position of a point on the rim of a rotating wheel is described by $\theta = 4.0t - 3.0t^2 + t^3$, where θ is in radians and t in seconds. What is the average acceleration for the time interval which begins at $t = 2.0$ s and ends at $t = 4.0$ s?   *Answer:* 12 rad/s².

6. The angle turned through by the flywheel of a generator during a time interval t is given by

$$\theta = at + bt^3 - ct^4,$$

where a, b, and c are constants. What is the expression for its angular acceleration?

7. A wheel rotates with an angular acceleration α given by

$$\alpha = 4at^3 - 3bt^2,$$

where t is the time and a and b are constants. If the wheel has an initial angular speed ω_0, write the equations for (a) the angular speed and (b) the angle turned through as functions of time.
Answer: (a) $\omega_0 + at^4 - bt^3$. (b) $\theta_0 + \omega_0 t + at^5/5 - bt^4/4$.

8. A planet P revolves around the sun S in a circular orbit, with the sun at the center, which is coplanar with and concentric to, the circular orbit of the earth E around the sun. P and E revolve in the same direction. The times required for the revolution of P and E around the sun are T_P and T_E. Let T_S

be the time required for P to make one revolution around the sun relative to E: show that $1/T_S = 1/T_E - 1/T_P$. Assume $T_P > T_E$.

9. A solar day is the time interval between two successive appearances of the sun overhead at a given longitude, that is, the time for one complete rotation of the earth relative to the sun. A sidereal day is the time for one complete rotation of the earth relative to the fixed stars, that is, the time interval between two successive overhead observations of a fixed direction in the heavens called the vernal equinox. (a) Show that there is exactly one less (mean) solar day in a year than there are (mean) sidereal days in a year. (b) If the (mean) solar day is exactly 24 hours, how long is a (mean) sidereal day?   Answer: (b) 23 h 56 min.

SECTION 11-3

10. The angular speed of an automobile engine is increased from 1200 rev/min to 3000 rev/min in 12 s. (a) What is its angular acceleration, assuming it to be uniform? (b) How many revolutions does the engine make during this time?

11. A phonograph turntable rotating at 78 rev/min slows down and stops in 30 s after the motor is turned off. (a) Find its (uniform) angular acceleration. (b) How many revolutions did it make in this time?
Answer: (a) −0.27 rad/s². (b) 20.

12. A heavy flywheel rotating on its axis is slowing down because of friction in its bearings. At the end of the first minute its angular velocity is 0.90 of its angular velocity ω_0 at the start. Assuming constant frictional forces, find its angular velocity at the end of the second minute.

13. While waiting to board a helicopter, you notice that the rotor's motion changed from 300 rev/min to 225 rev/min in one minute. (a) Find the average angular acceleration during the interval. (b) Assuming that this acceleration remains constant, calculate how long it will take for the rotor to stop. (c) How many revolutions will the rotor make after your second observation?   Answer: (a) −0.13 rad/s². (b) 4.0 min. (c) 340.

14. A wheel has a constant angular acceleration of 3.0 rad/s². In a 4.0-s interval, it turns through an angle of 120 rad. Assuming the wheel started from rest, how long had it been in motion at the start of this 4.0-s interval?

15. A uniform disk rotates about a fixed axis starting from rest and accelerating with constant angular acceleration. At one time it is rotating at 10 rev/s. After completing 60 more complete revolutions its angular speed is 15 rev/s. Calculate (a) the angular acceleration, (b) the time required to complete the 60 revolutions mentioned, (c) the time required to attain the 10 rev/s angular speed, and (d) the number of revolutions from rest until the time the disk attained the 10 rev/s angular speed.
Answer: (a) 1.04 rev/s². (b) 4.8 s. (c) 9.6 s. (d) 48.

16. A flywheel completes 40 revolutions as it slows from an angular speed of 1.5 rad/s to a complete stop. Assuming uniform acceleration, (a) what is the time required for it to come to rest? (b) What is the angular acceleration? (c) How much time is required for it to complete the first one-half of the 40 revolutions?

17. An automobile traveling 60 mi/h (97 km/h) has wheels of 30 in. (76 cm) diameter. (a) What is the angular speed of the wheels about the axle? (b) If the car is brought to a stop uniformly in 30 turns, what is the angular acceleration? (c) How far does the car advance during this braking period?
Answer: (a) 70 rad/s (71 rad/s). (b) −13 rad/s² (−13 rad/s²). (c) 240 ft (72 m).

18. A body moves in the x–y plane such that $x = R \cos \omega t$ and $y = R \sin \omega t$. Here x and y are the coordinates of the body, t is the time, and R and ω are constants. (a) Eliminate t between these equations to find the equation of the curve in which the body moves. What is this curve? What is the meaning of the constant ω? (b) Differentiate the equations for x and y with respect to the time to find the x and y components of the velocity of the body, v_x and

v_y. Combine v_x and v_y to find the magnitude and direction of **v**. Describe the motion of the body. (c) Differentiate v_x and v_y with respect to the time to obtain the magnitude and direction of the resultant acceleration.

19. Wheel A of radius $r_A = 10$ cm is coupled by a belt B to wheel C of radius $r_C = 25$ cm, as shown in Fig. 11-13. Wheel A increases its angular speed from rest at a uniform rate of $\pi/2$ rad/s². Determine the time for wheel C to reach a rotational speed of 100 rev/min, assuming the belt does not slip.
Answer: 17 s.

figure 11-13
Problem 19

SECTION 11-5

20. (a) What is the angular speed about the polar axis of a point on the earth's surface at a latitude of 45°N? (b) What is the linear speed? (c) How do these compare with the similar values for a point at the equator?

21. The earth's orbit about the sun is almost a circle. (a) What is the angular velocity of the earth (regarded as a particle) about the sun and (b) its average linear speed in its orbit? (c) What is the acceleration of the earth with respect to the sun?
Answer: (a) 2.0×10^{-7} rad/s. (b) 3.0×10^4 m/s. (c) 6.0×10^{-3} m/s².

22. What is the angular speed of a car rounding a circular turn of radius 360 ft (110 m) at 30 mi/h (48 km/h)?

23. What is the acceleration of a point on the rim of a 12-in. (30 cm) diameter record rotating at 33.3 rev/min?   *Answer:* 6.1 ft/s² (1.8 m/s²).

24. What is the ratio of the acceleration, associated with the earth's rotation, of a point on the equator, to the acceleration of the earth itself, associated with its motion around the sun? Assume a circular orbit.

25. The flywheel of a steam engine runs with a constant angular speed of 150 rev/min. When steam is shut off, the friction of the bearings and of the air brings the wheel to rest in 2.2 h. (a) What is the average angular acceleration of the wheel? (b) How many rotations will the wheel make before coming to rest? (c) What is the tangential linear acceleration of a particle distant 50 cm from the axis of rotation when the flywheel is turning at 75 rev/min? (d) What is the magnitude of the total linear acceleration of the particle in part (c)?
Answer: (a) -2.0×10^{-3} rad/s². (b) 10^4 rev. (c) -1.0 mm/s². (d) 31 m/s².

26. A rigid body, starting at rest, rotates about a fixed axis with constant angular acceleration α. Consider a particle a distance r from the axis. Express (a) the radial acceleration and (b) the tangential acceleration of this particle in the body in terms of α, r and the time t. (c) If the resultant acceleration of the particle at some instant makes an angle of 60° with the tangential acceleration, what total angle has the body turned through to that instant?

SECTION 11-6

27. Derive Eq. 11-20 by differentiation of Eq. 11-19.

28.* An insect of mass 8.0×10^{-2} g walks out with a constant speed of 1.6 cm/s along a radial line marked on a phonograph turntable rotating at a constant angular velocity of $33\frac{1}{3}$ rev/min. Find (a) the velocity and (b) the acceleration of the insect as seen by the ground observer when the insect is 12 cm from the axis of rotation. (c) What must the minimum coefficient of friction be to allow the insect to get all the way to the edge of the turntable (radius = 16 cm) without slipping?

29.* A virus particle, mass 1.0×10^{-7} g, in solution in a centrifuge is, at a particular moment, at a distance of 6.5 cm from the axis of rotation and moving radially outward at a relatively constant speed of 2.0 mm/s. The centrifuge is rotating at 55,000 rev/min. Discuss the motion quantitatively, giving the magnitude of all forces and accelerations as viewed from a reference frame (a) rotating with the centrifuge and (b) fixed in the laboratory.

* See Supplementary Topic I.

12
rotational dynamics I

In Chapter 11 we considered the kinematics of rotation. In this chapter, following the pattern of our study of translational motion, we study the causes of rotation, a subject called *rotational dynamics*. Rotating systems are made up of particles and we have already learned how to apply the laws of classical mechanics to the motion of particles. For this reason rotational dynamics should contain no features that are fundamentally new. In the same way rotational kinematics contained no basic new features, the rotational parameters θ, ω, and α being related to corresponding translational parameters x, v, and a for the particles that make up the rotating system. As in Chapter 11, however, it is very useful to recast the concepts of translational motion into a new form, especially chosen for its convenience in describing rotating systems.

We restricted our kinematical studies in Chapter 11 to a single but important special case, the rotation of a rigid body about an axis that is fixed in the reference frame in which we make our measurements. In studying rotational dynamics we start from a more fundamental point of view, that of a single particle viewed from an inertial reference frame. Later we shall generalize to systems of many particles, including the special case of a rigid body rotating about a fixed axis. In Chapter 13 we shall discuss the rotation of rigid bodies about axes that are *not* fixed in an inertial reference frame.

In translational motion we associate a *force* with the *linear acceleration* of a body. In rotational motion, what quantity shall we associate with the *angular acceleration* of a body? It cannot be simply force because, as experiment with a heavy revolving door teaches us, a given

12-1
INTRODUCTION

$\tau = r \times F$

12-2
TORQUE ACTING ON A PARTICLE

force (vector) can produce various angular accelerations of the door depending on where the force is applied and how it is directed; a force applied to the hinge line cannot produce any angular acceleration, whereas a force of given magnitude applied at right angles to the door at its outer edge produces a maximum acceleration.

We shall call the rotational analogue of force *torque* and shall now define it for the special case of a single particle observed from an inertial reference frame. Later we shall extend the torque concept to systems of particles (including rigid bodies) and shall show that torque is intimately associated with angular acceleration.

If a force $\mathbf{F}$ acts on a single particle at a point P whose position with respect to the origin O of the inertial reference frame is given by the displacement vector $\mathbf{r}$ (Fig. 12-1), the *torque* $\boldsymbol{\tau}$ acting on the particle *with respect to the origin O* is defined as

$$\boldsymbol{\tau} = \mathbf{r} \times \mathbf{F}. \qquad (12\text{-}1)$$

Torque is a vector quantity. Its magnitude is given by

$$\tau = rF \sin \theta, \qquad (12\text{-}2a)$$

where θ is the angle between $\mathbf{r}$ and $\mathbf{F}$; its direction is normal to the plane formed by $\mathbf{r}$ and $\mathbf{F}$. The sense is given by the right-hand rule for the vector product of two vectors, namely, one swings $\mathbf{r}$ into $\mathbf{F}$ through the smaller angle between them with the curled fingers of the right hand; the direction of the extended thumb then gives the direction of $\boldsymbol{\tau}$.

Torque has the same dimensions as force times distance, or in terms of our assumed fundamental dimensions, M, L, T, it has the dimensions ML^2T^{-2}. These are the same as the dimensions of work. However, torque and work are very different physical quantities. Torque is a vector and work is a scalar, for example. The unit of torque may be the newton-meter $(\text{N} \cdot \text{m})$ or pound-foot $(\text{lb} \cdot \text{ft})$, among other possibilities.

Notice (Eq. 12-1) that the torque produced by a force depends not only on the magnitude and on the direction of the force but also on the point of application of the force relative to the origin, that is, on the vector $\mathbf{r}$. In particular, when particle P is at the origin, so that the line of action of $\mathbf{F}$ passes through the origin, $\mathbf{r}$ is zero and the torque $\boldsymbol{\tau}$ about the origin is zero.

We can also write the magnitude of $\boldsymbol{\tau}$ (Eq. 12-2a) either as

$$\tau = (r \sin \theta)\, F = Fr_{\perp}, \qquad (12\text{-}2b)$$

or as

$$\tau = r(F \sin \theta) = rF_{\perp}, \qquad (12\text{-}2c)$$

in which, as Fig. 12-2a shows, $r_{\perp}\ (= r \sin \theta)$ is the component of $\mathbf{r}$ at right angles to the line of action of $\mathbf{F}$, and $F_{\perp}\ (= F \sin \theta)$ is the component of $\mathbf{F}$ at right angles to $\mathbf{r}$. Torque is often called the *moment of force* and $r_{\perp}$ in Eq. 12-2b is called the *moment arm*. Equation 12-2c shows that only the component of $\mathbf{F}$ perpendicular to $\mathbf{r}$ contributes to the torque. In particular, when θ equals 0 or 180°, there is no perpendicular component $(F_{\perp} = F \sin \theta = 0)$; then the line of action of the force passes through the origin and the moment arm $r_{\perp}$ about the origin is also zero. In this case both Eq. 12-2b and Eq. 12-2c show that the torque $\boldsymbol{\tau}$ is zero.

If, as in Fig. 12-2b, we reverse the direction of $\mathbf{F}$, the magnitude of $\boldsymbol{\tau}$ remains unchanged but the direction of $\boldsymbol{\tau}$ is reversed. Similarly, if, as in Fig. 12-2c, we reverse $\mathbf{r}$, thereby changing the point of application of $\mathbf{F}$, the magnitude of $\boldsymbol{\tau}$ remains unchanged but the direction of $\boldsymbol{\tau}$ is again reversed.

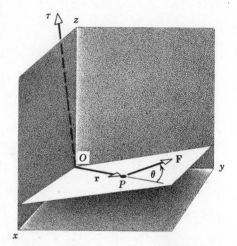

figure 12-1
A force is applied to a particle P, displaced $\mathbf{r}$ relative to the origin. The force vector makes an angle θ with the radius vector $\mathbf{r}$. The torque $\boldsymbol{\tau}$ about O is shown. Its direction is perpendicular to the plane formed by $\mathbf{r}$ and $\mathbf{F}$ with the sense given by the right-hand rule.

figure 12-2
The plane shown is that defined by **r** and **F** in Fig. 12-1. *(a)* The magnitude of τ is given by $Fr_\perp$ (Eq. 12-2b) or by $rF_\perp$ (Eq. 12-2c). *(b)* Reversing **F** reverses the direction of τ but leaves its magnitude unchanged. *(c)* Reversing **r** also reverses the direction of τ but leaves its magnitude unchanged. *(d)* Reversing **F** and **r** leaves both the direction and magnitude of τ unchanged. The directions of τ are represented by $\odot$ (perpendicularly out of the figure, the symbol representing the tip of an arrow) and by $\otimes$ (perpendicularly into the figure, the symbol representing the tail of an arrow).

If, as in Fig. 12-2d, we reverse *both* **r** and **F,** then both the magnitude and the direction of τ remain unchanged. These results follow formally from the facts that: (1) $\sin\theta = \sin(180° - \theta)$, so that Eq. 12-2a for the magnitude of τ is unaffected; (2) reversing the direction of *one* vector in a vector product (either **r** *or* **F**) reverses the direction of the product; and (3) reversing the directions of *both* vectors in a vector product (both **r** *and* **F**) leaves the direction of the product unchanged. You should verify the directions of τ shown in Fig. 12-2, using the right-hand rule.

We have found *linear momentum* to be useful in dealing with the translational motion of single particles or of systems of particles (including rigid bodies). For example, linear momentum is conserved in collisions. For a single particle the linear momentum is $\mathbf{p} = m\mathbf{v}$ (Eq. 9-11); for a system of particles it is $\mathbf{P} = M\mathbf{v}_{cm}$ (Eq. 9-15) in which M is the total system mass and $\mathbf{v}_{cm}$ is the velocity of the center of mass. In rotational motion, what is the analog of linear momentum? We call it *angular momentum* and we define it below for the special case of a single particle. Later we shall broaden the definition to include systems of particles and shall show that angular momentum, as we define it, is as useful a concept in rotational motion as linear momentum is in translational motion.

Consider a particle of mass m and linear momentum **p** at a position **r** relative to the origin O of an inertial reference frame (Fig. 12-3). We define the *angular momentum* **l** of the particle *with respect to the origin O* to be

$$\mathbf{l} = \mathbf{r} \times \mathbf{p}. \tag{12-3}$$

Note that we must specify the origin O in order to define the position vector **r** in the definition of angular momentum.

Angular momentum is a vector. Its magnitude is given by

$$l = rp\sin\theta, \tag{12-4a}$$

12-3
ANGULAR MOMENTUM OF A PARTICLE

$l = r \times p$

$p = mv$

$\therefore l = m(r \times p)$

where θ is the angle between $\mathbf{r}$ and $\mathbf{p}$; its direction is normal to the plane formed by $\mathbf{r}$ and $\mathbf{p}$. The sense is given by the right-hand rule, namely, one swings $\mathbf{r}$ into $\mathbf{p}$, through the smaller angle between them, with the curled fingers of the right hand; the extended right thumb then points in the direction of $\mathbf{l}$.

We can also write the magnitude of $\mathbf{l}$ either as

$$l = (r \sin \theta) \, p = pr_\perp, \qquad (12\text{-}4b)$$

or as

$$l = r(p \sin \theta) = rp_\perp, \qquad (12\text{-}4c)$$

in which $r_\perp \, (= r \sin \theta)$ is the component of $\mathbf{r}$ at right angles to the line of action of $\mathbf{p}$ and $p_\perp \, (= p \sin \theta)$ is the component of $\mathbf{p}$ at right angles to $\mathbf{r}$. Angular momentum is often called *moment of* (linear) *momentum* and $r_\perp$ in Eq. 12-4b is often called the *moment arm*. Equation 12-4c shows that only the component of $\mathbf{p}$ perpendicular to $\mathbf{r}$ contributes to the angular momentum. When the angle θ between $\mathbf{r}$ and $\mathbf{p}$ is 0 or 180°, there is no perpendicular component $(p_\perp = p \sin \theta = 0)$; then the line of action of $\mathbf{p}$ passes through the origin and $r_\perp$ is also zero. In this case both Eqs. 12-4b and 12-4c show that the angular momentum l is zero.

We now derive an important relation between torque and angular momentum. We have seen that $\mathbf{F} = d(m\mathbf{v})/dt = d\mathbf{p}/dt$ for a particle. Let us take the vector product of $\mathbf{r}$ with both sides of this equation, obtaining

$$\mathbf{r} \times \mathbf{F} = \mathbf{r} \times \frac{d\mathbf{p}}{dt}.$$

But $\mathbf{r} \times \mathbf{F}$ is the torque, or moment of a force, about O. We can then write

$$\boldsymbol{\tau} = \mathbf{r} \times \frac{d\mathbf{p}}{dt}. \qquad (12\text{-}5)$$

Next we differentiate Eq. 12-3 and obtain

$$\frac{d\mathbf{l}}{dt} = \frac{d}{dt} (\mathbf{r} \times \mathbf{p}).$$

Now the derivative of a vector product is taken in the same way as the derivative of an ordinary product, except that we must not change the order of the terms. We have

$$\frac{d\mathbf{l}}{dt} = \frac{d\mathbf{r}}{dt} \times \mathbf{p} + \mathbf{r} \times \frac{d\mathbf{p}}{dt}.$$

But $d\mathbf{r}$ is the vector displacement of the particle in the time dt so that $d\mathbf{r}/dt$ is the instantaneous velocity $\mathbf{v}$ of the particle. Also, $\mathbf{p}$ equals $m\mathbf{v}$, so that we can rewrite the equation as

$$\frac{d\mathbf{l}}{dt} = (\mathbf{v} \times m\mathbf{v}) + \mathbf{r} \times \frac{d\mathbf{p}}{dt}.$$

Now $\mathbf{v} \times m\mathbf{v} = 0$, because the vector product of two parallel vectors is zero. Therefore,

$$\frac{d\mathbf{l}}{dt} = \mathbf{r} \times \frac{d\mathbf{p}}{dt}. \qquad (12\text{-}6)$$

Inspection of Eqs. 12-5 and 12-6 shows that

$$\boldsymbol{\tau} = d\mathbf{l}/dt, \qquad (12\text{-}7)$$

which states that *the time rate of change of the angular momentum of*

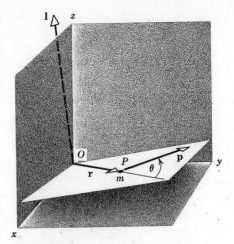

figure 12-3
A particle of mass m is at point P displaced $\mathbf{r}$ relative to the origin, and has linear momentum $\mathbf{p}$. The vector $\mathbf{p}$ makes an angle θ with the radius vector $\mathbf{r}$. The angular momentum $\mathbf{l}$ of the particle with respect to origin O is shown. Its direction is perpendicular to the plane formed by $\mathbf{r}$ and $\mathbf{p}$ with the sense given by the right-hand rule.

a particle is equal to the torque acting on it. This result is the rotational analog of Eq. 9-12, which stated that the time rate of change of the linear momentum of a particle is equal to the force acting on it, that is, that $\mathbf{F} = d\mathbf{p}/dt$.

Equation 12-7, like all vector equations, is equivalent to three scalar equations, namely,

$$\tau_x = (dl_x/dt)_x, \ \tau_y = (dl_y/dt)_y, \ \tau_z = (dl_z/dt)_z. \qquad (12\text{-}8)$$

Hence, the *x*-component of the applied torque is given by the *x*-component of the change with time of the angular momentum. Similar results hold for the *y*- and *z*-directions.

A particle of mass *m* is released from rest at point *a* in Fig. 12-4, falling parallel to the (vertical) *y*-axis. (*a*) Find the torque acting on *m* at any time *t*, with respect to origin *O*. (*b*) Find the angular momentum of *m* at any time *t*, with respect to this same origin. (*c*) Show that the relation $\tau = d\mathbf{l}/dt$ (Eq. 12-7) yields a correct result when applied to this familiar problem.

(*a*) The torque is given by Eq. 12-1 or $\tau = \mathbf{r} \times \mathbf{F}$, its magnitude being given by

$$\tau = rF \sin \theta.$$

In this example $r \sin \theta = b$ and $F = mg$ so that

$$\tau = mgb = \text{a constant}.$$

Note that the torque is simply the product of the force (*mg*) times the moment arm (*b*). The right-hand rule shows that τ is directed perpendicularly into the figure.

(*b*) The angular momentum is given by Eq. 12-3 or $\mathbf{l} = \mathbf{r} \times \mathbf{p}$, its magnitude being given by

$$l = rp \sin \theta.$$

In this example $r \sin \theta = b$ and $p = mv = m(gt)$ so that

$$l = mgbt.$$

The right-hand rule shows that $\mathbf{l}$ is directed perpendicularly into the figure, which means that $\mathbf{l}$ and τ are parallel vectors. The vector $\mathbf{l}$ changes with time in magnitude only, its direction always remaining the same in this case.

(*c*) Since $d\mathbf{l}$, the change in $\mathbf{l}$, and τ are parallel, we can replace the vector relation $\tau = d\mathbf{l}/dt$ by the scalar relation

$$\tau = dl/dt.$$

Using the expressions for τ and l from (*a*) and (*b*) above we have

$$mgb = \frac{d}{dt}(mgbt) = mgb,$$

which is an identity. Thus the relation $\tau = d\mathbf{l}/dt$ yields correct results in this simple case. Indeed, if we cancel the constant *b* out of the first two terms above and if we substitute for *gt* the equivalent quantity *v*, we have

$$mg = \frac{d}{dt}(mv).$$

Since $mg = F$ and $mv = p$, this is the familiar result $F = dp/dt$. Thus, as we indicated earlier, relations such as $\tau = d\mathbf{l}/dt$, though often vastly useful, are not new basic postulates of classical mechanics but are rather the reformulation of the Newtonian laws for rotational motion.

Note that the values of τ and *l* depend on our choice of origin, that is, on *b*. In particular, if $b = 0$, then $\tau = 0$ and $l = 0$.

EXAMPLE 1

figure 12-4
Example 1. A particle of mass *m* drops vertically from point *a*. The torque and the angular momentum about *O* are directed perpendicularly into the figure, as shown by the symbol $\otimes$ at *O*.

So far we have talked only about single particles. Let us now consider a system of many particles. To calculate the total angular momentum **L** of a system of particles about a given point, we must add vectorially the angular momenta of all the individual particles of the system about this same point. For a system containing n particles we have, then,

$$\mathbf{L} = \mathbf{l}_1 + \mathbf{l}_2 + \cdots + \mathbf{l}_n = \sum_{i=1}^{i=n} \mathbf{l}_i,$$

in which the (vector) sum is taken over all particles in the system.

As time goes on, the total angular momentum **L** of the system about a fixed reference point (which we choose, as in our basic definition of **l** in Eq. 12-3, to be the origin of an inertial reference frame) may change. This change, $d\mathbf{L}/dt$, can arise from two sources: (1) torques exerted on the particles of the system by internal forces between the particles and (2) torques exerted on the particles of the system by external forces.

If Newton's third law holds in its so-called strong form, that is, if the forces between any two particles not only are equal and opposite but are also directed along the line joining the two particles, then the total internal torque is zero because the torque resulting from each internal action-reaction force pair is zero.

Hence the first source contributes nothing. For our reference point, therefore, only the second source remains, and we can write

$$\boldsymbol{\tau}_{\text{ext}} = d\mathbf{L}/dt, \tag{12-9}$$

where $\boldsymbol{\tau}_{\text{ext}}$ stands for the *sum* of *all* the *external* torques acting on the system. In words, *the time rate of change of the total angular momentum of a system of particles about the origin of an inertial reference frame is equal to the sum of the* external *torques acting on the system*. Later, for convenience, in situations in which no confusion is likely to arise, we shall drop the subscript on $\boldsymbol{\tau}_{\text{ext}}$.

Equation 12-9 is the generalization of Eq. 12-7 to many particles. When we have only one particle, there are no internal forces or torques. This relation (Eq. 12-9) holds whether the particles that make up the system are in motion relative to each other or whether they have fixed spatial relationships, as in a rigid body.

Equation 12-9 is the rotational analog of Eq. 9-17

$$\mathbf{F}_{\text{ext}} = d\mathbf{P}/dt \tag{9-17}$$

which tells us that for a system of particles (rigid body or not) the resultant external *force* acting on the system equals the time rate of change of the *linear momentum* of the system.

As we have derived it, Eq. 12-9 holds when $\boldsymbol{\tau}$ and **L** are measured with respect to the origin of an inertial reference frame. We may well ask whether it still holds if we measure these two vectors with respect to an arbitrary point (a particular particle, say) in the moving system. In general, such a point would move in a complicated way as the body or system of particles translated, tumbled, and changed its configuration and Eq. 12-9 would *not* apply to such a reference point. However, if the reference point is chosen to be the center of mass of the system, even though this point is not fixed in our reference frame, then Eq. 12-9 *does* hold.* This is another remarkable property of the center of mass. Thus we can separate the general motion of a system of particles into the

---

* See Problem 10 of this chapter and K. R. Symon, *Mechanics*, 3d ed., Addison-Wesley Publishing Co., 1972, Section 4.2.

translational motion of its center of mass (Eq. 9-17) and rotational motion about its center of mass (Eq. 12-9).

We shall now confine our attention to an important special case of a system of particles, *a rigid body*. In a rigid body the particles in the system always maintain the same positions with respect to one another. In studying the rotation of a rigid body we shall consider first the special case, often encountered, in which the axis of rotation is fixed* in an inertial reference frame. Later we shall investigate more general systems and motions.

Let us now imagine a rigid body rotating with angular speed ω about an axis that is fixed in a particular inertial frame, as in Fig. 11-1. Each particle in such a rotating body has a certain amount of kinetic energy. A particle of mass m at a distance r from the axis of rotation moves in a circle of radius r with an angular speed ω about this axis and has a linear speed $v = \omega r$. Its kinetic energy therefore is $\frac{1}{2}mv^2 = \frac{1}{2}mr^2\omega^2$. The total kinetic energy of the body is the sum of the kinetic energies of its particles.

If the body is rigid, as we assume in this section, ω is the same for all particles. The radius r may be different for different particles. Hence the total kinetic energy K of the rotating body can be written as

$$K = \tfrac{1}{2}(m_1 r_1^2 + m_2 r_2^2 + \cdots)\omega^2 = \tfrac{1}{2}(\Sigma\, m_i r_i^2)\omega^2.$$

The term $\Sigma\, m_i r_i^2$ is the sum of the products of the masses of the particles by the squares of their respective distances from the axis of rotation. If we denote this quantity by I, then

$$I = \Sigma\, m_i r_i^2 \qquad (12\text{-}10)$$

is called the *rotational inertia*, or moment of inertia,† of the body with respect to the particular axis of rotation.

Note that *the rotational inertia of a body depends on the particular axis about which it is rotating* as well as on the shape of the body and the manner in which its mass is distributed. Rotational inertia has the dimensions ML^2 and is usually expressed in $\text{kg} \cdot \text{m}^2$ or $\text{slug} \cdot \text{ft}^2$.

In terms of rotational inertia we can now write the kinetic energy of the rotating rigid body as

$$K = \tfrac{1}{2}I\omega^2. \qquad (12\text{-}11)$$

This is analogous to the expression for the kinetic energy of translation of a body, $K = \frac{1}{2}Mv^2$. We have already seen that the angular speed ω is analogous to the linear speed v. Now we see that the rotational inertia I is analogous to the mass, or the translational inertia M. Although the

---

* As stated in Section 12-4, we can separate the general motion of a system of particles into translational motion of its center of mass and rotational motion about its center of mass. Hence the considerations of this chapter apply also to rotations about an axis that is *not* fixed in an inertial reference frame, provided (1) the axis passes through the center of mass and (2) the moving axis always has the same direction in space, that is, the axis at one instant is parallel to the axis at any other instant. Although we shall often refer to a "fixed axis" in what follows we shall always mean to include this special case of a moving axis.

† The term *moment of inertia* is widely used for this second moment of mass even though there are first, third, and other moments of mass. We choose to emphasize the term *rotational inertia*, however, chiefly because I (the rotational inertia) plays the same role, we shall see, in rotational motion as M (the mass, or the translational inertia) plays in translational motion.

figure 12-5
An experiment to show that
$I_a < I_b < I_c$. The three lead bodies
have the same mass M but the mass
is distributed differently about the
axis of rotation.

mass of a body does not depend on its location, the rotational inertia of a body does depend on the axis about which it is rotating.

We should understand that the rotational kinetic energy given by Eq. 12-11 is simply the sum of the ordinary translational kinetic energy of all the parts of the body and not a new kind of energy. Rotational kinetic energy is simply a convenient way of expressing the kinetic energy for a rotating rigid body.

Equations 12-10 and 12-11 show that the rotational energy of a body, for a given angular speed ω, depends not only on the mass of the body but also on the way that mass is distributed around the axis of rotation. The experiment suggested in Fig. 12-5 makes this convincing. The figure shows three identical aluminum shafts, to each of which is attached a body of mass M, made of lead. In (a) the mass is very close to the shaft so that the quantities r_i in Eq. 12-10 for the particles that make up the body are relatively small, in (b) the particles are, on the average, farther from the shaft and in (c), in which the body is a flywheel, they are still farther, corresponding to still larger values of r_i.

Now let us twist each handle until each shaft, starting from rest, is spinning at the same measured angular speed ω. We know from experience that we shall need to do relatively little work on shaft (a), somewhat more work on shaft (b), and still more on shaft (c). In fact, if we were not certain which body was attached to which shaft we could label the shafts with confidence using this technique. Since the work done on each shaft is equal to the kinetic energy $\frac{1}{2}I\omega^2$ imparted to each shaft, the experimental result, that $K_a < K_b < K_c$ when each shaft has the same angular speed ω, leads to the conclusion that $I_a < I_b < I_c$. This is just what we expect from the defining equation for I (Eq. 12-10). We shall see in Section 12-6 that just as the mass M, which we may call the translational inertia, is a measure of the resistance a body offers to a change in its translational motion, so I, the rotational inertia, is a measure of the resistance a body offers to a change in its rotational motion about a given axis.

Consider a body consisting of two spherical masses of 5.0 kg each connected by a light rigid rod 1.0 m long (Fig. 12-6). Treat the spheres as point particles and neglect the mass of the rod. Determine the rotational inertia (or moment of

EXAMPLE 2

inertia) of the body (a) about an axis normal to it through its center at C, and (b) about an axis normal to it through one sphere.

(a) If the axis is normal to the page through C, we have

$$I_C = \Sigma\ m_i r_i^2 = m_a r_a^2 + m_b r_b^2$$

$$= (5.0\ kg)(0.50\ m)^2 + (5.0\ kg)(0.50\ m)^2 = 2.5\ kg \cdot m^2.$$

(b) If the axis is normal to the page through A or B, we have

$$I_A = m_a r_a^2 + m_b r_b^2 = (5.0\ kg)(0\ m)^2 + (5.0\ kg)(1.0\ m)^2 = 5.0\ kg \cdot m^2,$$

$$I_B = m_a r_a^2 + m_b r_b^2 = (5.0\ kg)(1.0\ m)^2 + (5.0\ kg)(0\ m)^2 = 5.0\ kg \cdot m^2.$$

Hence the rotational inertia of this rigid dumbell model is twice as great about an axis through an end as it is about an axis through the center.

figure 12-6
Example 2. Calculating the rotational inertia of a dumbell.

For a body that is not composed of discrete point masses but is instead a continuous distribution of matter, the summation in $I = \Sigma\ m_i r_i^2$ becomes an integration. We imagine the body to be subdivided into infinitesimal elements, each of mass dm. We let r be the distance from such an element to the axis of rotation. Then the rotational inertia is obtained from

$$I = \int r^2\ dm, \tag{12-12}$$

where the integral is taken over the whole body. The procedure by which the summation Σ of a discrete distribution is replaced by the integral $\int$ for a continuous distribution is the same as that discussed for the center of mass in Section 9-1.

For bodies of irregular shape the integrals may be hard to evaluate. For bodies of simple geometrical shape the integrals are relatively easy when an axis of symmetry is chosen as the axis of rotation.

Let us illustrate the procedure for an annular cylinder (or ring) about the cylinder axis (Fig. 12-7). The most convenient mass element is an infinitesimally thin cylinder shell of radius r, thickness dr, and length L. If the density of the material, that is, the mass per unit volume, is called ρ, then

$$dm = \rho\ dV,$$

where dV is the volume of the cylindrical shell of mass dm. We have

$$dV = (2\pi r\ dr)L,$$

so that

$$dm = 2\pi L \rho r\ dr.$$

Then the rotational inertia about the cylinder axis is

$$I = \int r^2\ dm = 2\pi L \int_{R_1}^{R_2} \rho r^3\ dr.$$

Here R_1 is the radius of the inner cylindrical wall and R_2 is the radius of the outer cylindrical wall.

If this body did not have a uniform constant density, we would have to know how ρ depends on r before we could carry out the integration. Let us assume for simplicity that the density is uniform. Then

$$I = 2\pi L \rho \int_{R_1}^{R_2} r^3\ dr = 2\pi L \rho\ \frac{R_2^4 - R_1^4}{4} = \rho \pi (R_2^2 - R_1^2)L\ \frac{R_2^2 + R_1^2}{2}.$$

figure 12-7
Calculating the rotational inertia of an annular cylinder.

The mass M of the annular cylinder is the product of its density ρ by its volume $\pi(R_2^2 - R_1^2)L$, or

$$M = \rho\pi(R_2^2 - R_1^2)L.$$

The rotational inertia of the *annular cylinder* (or ring) of mass M, inner radius R_1 and outer radius R_2, is therefore

$$I = \tfrac{1}{2}M(R_1^2 + R_2^2)$$

about the cylinder axis.

If the inner radius is zero, R_1 equals zero, and we have a *solid cylinder* (or disk). Then

$$I = \tfrac{1}{2}MR^2$$

about the cylinder axis, where R is the radius of the solid cylinder of mass M.

A *hoop* can be thought of as a very thin-walled hollow cylinder. In that case

$$R_1 \cong R_2 \cong R,$$

and

$$I = MR^2$$

is the rotational inertia of a hoop of mass M and radius R about the cylinder axis.

This result for the thin hoop is obvious since every mass point in the hoop is the same distance R from the central axis. For the solid cylinder (or disk) having the *same mass* as the hoop, the rotational inertia (or moment of inertia) would naturally be less than that of the hoop, because most of the cylinder (or disk) is less than a distance R from the axis.

The rotational inertias about certain axes of some common solids (of uniform density) are listed in Table 12-1. Each of these results can be derived by integration in a manner similar to that of our illustration. The total mass of the body is denoted by M in each equation.

There is a simple and very useful relation between the rotational inertia I of a body about any axis and its rotational inertia I_{cm} with respect to a parallel axis *through the center of mass*. If M is the total mass of the body and h the distance between the two axes, the relation is

$$I = I_{cm} + Mh^2. \qquad (12\text{-}13)$$

The proof of this relation (parallel-axis theorem) follows. Let C be the center of mass of the arbitrarily shaped body whose cross section is shown in Fig. 12-8.

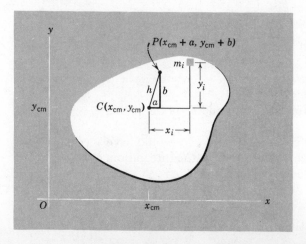

figure 12-8
Derivation of the parallel-axis theorem. Knowing the rotational inertia about an axis through C, we can find its value about a parallel axis through P.

Table 12-1

Hoop about cylinder axis $\qquad$ $I = MR^2$ $\quad a$	Annular cylinder (or ring) about cylinder axis $\qquad$ $I = \frac{M}{2}(R_1{}^2 + R_2{}^2)$ $\quad b$
Solid cylinder about cylinder axis $\qquad$ $I = \frac{MR^2}{2}$ $\quad c$	Solid cylinder (or disk) about a central diameter $\qquad$ $I = \frac{MR^2}{4} + \frac{Ml^2}{12}$ $\quad d$
Thin rod about axis through center $\perp$ to length $\qquad$ $I = \frac{Ml^2}{12}$ $\quad e$	Thin rod about axis through one end $\perp$ to length $\qquad$ $I = \frac{Ml^2}{3}$ $\quad f$
Solid sphere about any diameter $\qquad$ $I = \frac{2MR^2}{5}$ $\quad g$	Thin spherical shell about any diameter $\qquad$ $I = \frac{2MR^2}{3}$ $\quad h$
Hoop about any diameter $\qquad$ $I = \frac{MR^2}{2}$ $\quad i$	Hoop about any tangent line $\qquad$ $I = \frac{3MR^2}{2}$ $\quad j$

The center of mass has coordinates x_{cm} and y_{cm}. We choose the x-y plane to include C, so that z_{cm} equals zero. Consider an axis through C at right angles to the plane of the paper and another axis parallel to it through P at $(x_{cm} + a)$ and $(y_{cm} + b)$. The distance between the axes is $h = \sqrt{a^2 + b^2}$. Then the square of the distance of a particle from the axis through C is $x_i^2 + y_i^2$, where x_i and y_i measure the coordinates of a mass element m_i relative to the axis through C. The square of its distance from an axis through P is $(x_i - a)^2 + (y_i - b)^2$. Hence the rotational inertia about an axis through P is

$$I = \Sigma \, m_i[(x_i - a)^2 + (y_i - b)^2]$$
$$= \Sigma \, m_i(x_i^2 + y_i^2) - 2a \, \Sigma \, m_i x_i - 2b \, \Sigma \, m_i y_i + (a^2 + b^2) \, \Sigma \, m_i.$$

From the definition of center of mass,

$$\Sigma \, m_i x_i = \Sigma \, m_i y_i = 0,$$

so that the two middle terms are zero. The first term is simply the rotational inertia about an axis through the center of mass I_{cm} and the last term is Mh^2. Hence it follows that $I = I_{cm} + Mh^2$.

With the aid of this formula several of the results of Table 12-1 can be deduced from previous results. For example, (f) follows from (e), and (j) follows from (i) with the aid of Eq. 12-13. The formula will prove to be especially useful in problems that combine rotational and translational motion.

12-6 ROTATIONAL DYNAMICS OF A RIGID BODY

In this section we continue to study the special case of a rigid body confined to rotate about an axis that is fixed* in an inertial reference frame. First we shall review the concept of torque as applied to such a rigid body; then we shall show how the torque is related to the angular acceleration of the body about this axis.

Suppose that we apply a torque τ to one of the particles in a rigid body. Since all the particles of a truly rigid body maintain a fixed spatial relationship to all the other particles that make up the body, the torque may be said to act on the rigid body as a whole. In general, the vector τ will not lie along the axis around which the body is free to rotate. We are not concerned in this section with the actual torques applied to the body but only with the components of these torques that lie along the axis.† Only these components can cause the body to rotate about this axis. Torque components perpendicular to the axis tend to turn the axis from its fixed position. We have specifically assumed, however, that the axis maintains a fixed direction. The body may, for example, be attached to a shaft that is held in a fixed position by bearings at each end; if an applied torque has a component at right angles to the shaft, tending to turn it, the bearings will automatically apply an equal and opposite counter-torque to the shaft, canceling the effect of this component.

In Fig. 12-9 (compare Fig. 11-3) we show a section through a rigid body that is free to rotate about the z-axis of an inertial reference frame. A force $\mathbf{F}$, taken for convenience to be in the x-y plane of the section, acts on a particle at point P in the body, the position of P with respect to the rotational axis (the z-axis) being defined by the vector $\mathbf{r}$. The torque acting on the particle at P may be said to act on the rigid body as a whole and is given by Eq. 12-1, or

$$\boldsymbol{\tau} = \mathbf{r} \times \mathbf{F}.$$

Because we have chosen $\mathbf{r}$ and $\mathbf{F}$ to lie in the x-y plane, the torque τ will point along the z-axis. The right-hand rule shows that it points perpendicularly *out of* the plane of Fig. 12-9. If $\mathbf{r}$ and $\mathbf{F}$ did *not* lie in the plane of the figure, τ would not be parallel to the z-axis and we would concern ourselves here only with the component of τ along this axis. The magnitude of τ is given by Eq. 12-2 or

$$\tau = rF \sin \theta$$

which, as we have seen, can also be written as $\tau = rF_\perp$ or $\tau = Fr_\perp$.

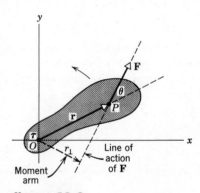

figure 12-9
A force $\mathbf{F}$ acts on the particle P in a rigid body, exerting a torque $\tau = \mathbf{r} \times \mathbf{F}$ on the body, with respect to an axis through O at right angles to the plane of the figure. The moment arm $r_\perp$ is also shown, as is the torque τ, which is a vector emerging perpendicularly from the page.

---

* See the footnote on page 237.
† As for any other vector, we can speak of the vector component of a torque in any given direction, such as a given axis. For torque—and for other angular quantities—we also often speak of the component *around* a given direction or axis. The meaning is the same.

A wagon wheel is free to rotate about a horizontal axis through O. A force of 10 lb is applied to a spoke at the point P, 1.0 ft from the center. OP makes an angle of 30° with the horizontal (x-axis) and the force is in the plane of the wheel making an angle of 45° with the horizontal (x-axis). What is the torque on the wheel?

The angle between the displacement vector $\mathbf{r}$ from O to P and the applied force $\mathbf{F}$ (Fig. 12-10) is θ, where

$$\theta = 45° - 30° = 15°.$$

Then the magnitude of the torque is

$$\tau = rF \sin \theta$$

$$= (1.0 \text{ ft})(10 \text{ lb})(\sin 15°) = 2.6 \text{ lb} \cdot \text{ft}.$$

It is clear that we can obtain this same result from $\tau = rF_\perp$ or $\tau = Fr_\perp$ as well (see Eqs. 12-2). The torque $(\boldsymbol{\tau} = \mathbf{r} \times \mathbf{F})$ is a vector pointing out $\odot$ along the axis through O having a magnitude 2.6 lb · ft.

We now investigate the relationship between the torque applied to the rigid body of Fig. 12-9 and the angular acceleration of this body. Let us observe the rigid body for an infinitesimal time dt, during which it will rotate through an infinitesimal angle $d\theta$. We have seen earlier that we can describe the rotation of a rigid body about a fixed axis by examining the motion of any single point fixed in the body, such as P in Fig. 12-9. For convenience, then, we ignore the body itself in Fig. 12-11 and focus our attention on the representative point P and on the vector $\mathbf{r}$ which locates point P with respect to the axis of rotation.

During the time dt, the point P will move an infinitesimal distance ds along a circular path of radius r as the body rotates through an infinitesimal angle $d\theta$, where

$$ds = r\, d\theta.$$

The work dW done by this force during this infinitesimal rotation is

$$dW = \mathbf{F} \cdot d\mathbf{s} = F \cos \phi\, ds = (F \cos \phi)(r\, d\theta),$$

where $F \cos \phi$ is the component of $\mathbf{F}$ in the direction of $d\mathbf{s}$.

The term $(F \cos \phi)r$, however, is the magnitude of the instantaneous torque exerted by $\mathbf{F}$ on the rigid body about the axis perpendicular to the page through O, so that

$$dW = \tau\, d\theta. \tag{12-14}$$

This differential expression for the work done in rotation (about a fixed axis) is equivalent to the expression $dW = F\, dx$ for the work done in translation (along a straight line).

To obtain the rate at which work is done in rotational motion (about a fixed axis), we divide both sides of Eq. 12-14 by the infinitesimal time interval dt during which the body is displaced through $d\theta$, obtaining

$$\frac{dW}{dt} = \tau \frac{d\theta}{dt}$$

or

$$P = \tau\omega,$$

giving the instantaneous power P. This last expression is the rotational analog of $P = Fv$ for translational motion (along a straight line).

If now a number of forces $\mathbf{F}_1$, $\mathbf{F}_2$, etc., are applied to the body in the

EXAMPLE 3

figure 12-10
Example 3.

figure 12-11
In time dt point P in the rigid body of Fig. 12-9 moves a distance ds along the arc of a circle of radius r. The rigid body (not shown) and the vector $\mathbf{r}$ that locates point P in it each rotate through an angle $d\theta$ during this interval.

plane normal to its axis of rotation, the work done by these forces on the body in a rotation $d\theta$ will be

$$dW = F_1 \cos \phi_1 r_1 \, d\theta + F_2 \cos \phi_2 r_2 \, d\theta + \cdots,$$

$$= (\tau_1 + \tau_2 + \cdots) \, d\theta = \tau \, d\theta,$$

where $r_1 \, d\theta$ equals ds_1, the displacement of the point at which $\mathbf{F}_1$ is applied, and ϕ_1 is the angle between $\mathbf{F}_1$ and $d\mathbf{s}_1$, etc., and where τ is now the magnitude of the *component* of the *resultant* torque along the axis through O. In computing this sum each torque is considered positive or negative according to the sense in which it alone would tend to rotate the body about its axis. We can arbitrarily call the torque associated with a force positive if the effect of the force, acting alone, is to produce a counterclockwise rotation; then the torque is negative if the effect is to produce a clockwise rotation.

There is no internal motion of particles within a truly rigid body. The particles always maintain a fixed position relative to one another and move only with the body as a whole. Hence there can be no dissipation of energy within a truly rigid body. We can therefore equate the rate at which work is being done on the body to the rate at which its kinetic energy is increasing. The rate at which work is being done on the rigid body is

$$\frac{dW}{dt} = \tau \frac{d\theta}{dt} = \tau\omega. \qquad (12\text{-}15)$$

The rate at which the kinetic energy of the rigid body is increasing is

$$\frac{d}{dt}\left(\tfrac{1}{2}I\omega^2\right).$$

But I is constant because the body is rigid and the axis is fixed. Hence

$$\frac{d}{dt}\left(\tfrac{1}{2}I\omega^2\right) = \tfrac{1}{2}I\frac{d}{dt}(\omega^2) = I\omega\frac{d\omega}{dt} = I\omega\alpha. \qquad (12\text{-}16)$$

Equating the right-hand members of Eqs. 12-15 and 12-16, we obtain

$$\tau\omega = I\alpha\omega,$$

or

$$\tau = I\alpha. \qquad (12\text{-}17)$$

Equation 12-17 refers to the rotational motion of a rigid body about a fixed axis. The torque τ, the angular velocity $\boldsymbol{\omega}$, and the angular acceleration $\boldsymbol{\alpha}$ are all constrained to point along this axis, in one direction or the other. The equivalent translational case is that in which the force $\mathbf{F}$ acting on a body, its velocity $\mathbf{v}$, and its acceleration $\mathbf{a}$ all point along a given straight line, in one direction or the other.

The above six quantities are vectors, but when they are directed along a fixed line, they can have only two directions. By taking one of these directions as $+$ and the other as $-$, we can treat these vectors algebraically and deal with their magnitudes only. Thus, in deriving Eq. 12-17 $(\tau = I\alpha)$, we have simply transformed Newton's second law $(F = Ma)$, written in scalar form suitable to describe rectilinear motion, into rotational terms. This suggests that just as we associate a force with the linear acceleration of a body, so we may associate a torque with the angular acceleration of a body about a given axis. The rotational inertia I is a measure of the resistance a body offers to having its rotational motion changed by a given torque just as the translational inertia, or mass, M is a measure of the resistance a body offers to having its translational motion changed by a given force.

In Table 12-2 we compare the translational motion of a rigid body along a straight line with the rotational motion of a rigid body about a fixed axis.

Table 12-2

Rectilinear Motion		Rotation about a Fixed Axis	
Displacement	x	Angular displacement	θ
Velocity	$v = \dfrac{dx}{dt}$	Angular velocity	$\omega = \dfrac{d\theta}{dt}$
Acceleration	$a = \dfrac{dv}{dt}$	Angular acceleration	$\alpha = \dfrac{d\omega}{dt}$
Mass (translational inertia)	M	Rotational inertia	I
Force	$F = Ma$	Torque	$\tau = I\alpha$
Work	$W = \int F\,dx$	Work	$W = \int \tau\,d\theta$
Kinetic energy	$\frac{1}{2}Mv^2$	Kinetic energy	$\frac{1}{2}I\omega^2$
Power	$P = Fv$	Power	$P = \tau\omega$
Linear momentum	Mv	Angular momentum	$I\omega$

figure 12-12
Example 4. A steady downward force **T** produces rotation of the disk. Example 5. Here **T** is supplied by the falling mass m.

The rotation of a rigid body about a fixed axis (to which $\tau = I\alpha$ applies) is not the most general kind of rotary motion in that the body may not be rigid and the axis may not be fixed in an inertial reference frame. In this general case Eq. 12-9, or $\tau_{ext} = d\mathbf{L}/dt$, applies. As we have already pointed out, this is equivalent to Newton's second law for the general translational motion of a system of particles, namely, Eq. 9-17, or $\mathbf{F}_{ext} = d\mathbf{P}/dt$.

In the rest of this chapter we confine ourselves to the rotations of rigid bodies about fixed axes. In Chapter 13 we shall consider some more general kinds of rotary motion.

EXAMPLE 4

A uniform disk of radius R and mass M is mounted on an axle supported in fixed frictionless bearings, as in Fig. 12-12. A light cord is wrapped around the rim of the wheel and a steady downward pull **T** is exerted on the cord. Find the angular acceleration of the wheel and the tangential acceleration of a point on the rim.

The torque about the central axis is $\tau = TR$, and the rotational inertia of the disk about its central axis is $I = \frac{1}{2}MR^2$. From

$$\tau = I\alpha,$$

we have

$$TR = (\tfrac{1}{2}MR^2)\alpha,$$

or

$$\alpha = \frac{2T}{MR}.$$

If the mass of the disk is taken to be $M = 2.50$ kg, its radius $R = 0.20$ m, and the force $T = 5.0$ N, then

$$\alpha = \frac{(2)(5.0 \text{ N})}{(2.50 \text{ kg})(0.20 \text{ m})} = 20 \text{ rad/s}^2.$$

The tangential acceleration of a point on the rim is given by

$$a = R\alpha = (20 \text{ rad/s}^2)(0.20 \text{ m}) = 4.0 \text{ m/s}^2.$$

EXAMPLE 5

Suppose that we hang a body of mass m from the cord in the previous problem. Find the angular acceleration of the disk and the tangential acceleration of a point on the rim in this case.

Now, let T be the tension in the cord. Since the suspended body will accelerate downward, the magnitude of the downward pull of gravity on it, mg, must exceed the magnitude of the upward pull of the cord on it, T. The acceleration a of the suspended body is the same as the tangential acceleration of a point on the rim of the disk. From Newton's second law

$$mg - T = ma.$$

The resultant torque on the disk is TR and its rotational inertia is $\frac{1}{2}MR^2$, so that from

$$\tau = I\alpha$$

we obtain

$$TR = \tfrac{1}{2}MR^2\alpha.$$

Using the relation $a = R\alpha$, we can write this last equation as

$$2T = Ma.$$

Solving the first and last equations simultaneously leads to

$$a = \left(\frac{2m}{M + 2m}\right)g,$$

and

$$T = \left(\frac{Mm}{M + 2m}\right)g.$$

If now we let the disk have a mass $M = 2.50$ kg and a radius $R = 0.20$ m as before, and we let the suspended body weigh 5.0 N, we obtain

$$a = \frac{2mg}{M + 2m} = \frac{(2)(5.0 \text{ N})}{(2.50 \text{ kg}) + 2(5/9.8) \text{ kg}} = 2.85 \text{ m/s}^2,$$

$$\alpha = \frac{a}{R} = \frac{(2.85 \text{ m/s}^2)}{0.20 \text{ m}} = 14.3 \text{ rad/s}^2.$$

Notice that the accelerations are less for a suspended 5.0 N body than they were for a steady 5.0 N pull on the string (Example 4). This corresponds to the fact that the tension in the string supplying the torque is now less than 5.0 N, namely

$$T = \frac{Mmg}{M + 2m} = \frac{(2.50 \text{ kg})(5.0 \text{ N})}{(2.50 + 1.0) \text{ kg}} = 3.6 \text{ N}.$$

The tension in the string must be less than the weight of the suspended body if the body is to accelerate downward.

EXAMPLE 6

Assuming that the disk of Example 5 starts from rest, compute the work done by the applied torque on the disk in 2.0 s. Compute also the increase in rotational kinetic energy of the disk.

Since the applied torque is constant, the resulting angular acceleration is constant. The total angular displacement in constant angular acceleration is obtained from Eq. 11-5,

$$\theta = \omega_0 t + \tfrac{1}{2}\alpha t^2,$$

in which

$$\omega_0 = 0, \qquad \alpha = 14.3 \text{ rad/s}^2, \qquad t = 2.0 \text{ s},$$

so that

$$\theta = 0 + (\tfrac{1}{2})(14.3 \text{ rad/s}^2)(2.0 \text{ s})^2 = 28.6 \text{ rad}.$$

For constant torque the work done in a finite angular displacement is

$$W = \tau(\theta_2 - \theta_1),$$

in which

$$\tau = TR = (3.6 \text{ N})(0.20 \text{ m}) = 0.72 \text{ N} \cdot \text{m},$$

and

$$\theta_2 - \theta_1 = \theta = 28.6 \text{ rad}.$$

Therefore

$$W = (0.72 \text{ N} \cdot \text{m})(28.6 \text{ rad}) = 20.5 \text{ J}.$$

This work must result in an increase in rotational kinetic energy of the disk. Starting from rest the disk acquires an angular speed ω. The rotational energy is $\frac{1}{2}I\omega^2 = \frac{1}{2}(\frac{1}{2}MR^2)\omega^2$. To obtain ω we use Eq. 11-3,

$$\omega = \omega_0 + \alpha t,$$

in which

$$\omega_0 = 0, \qquad t = 2.0 \text{ s}, \qquad \alpha = 14.3 \text{ rad/s}^2,$$

so that

$$\omega = 0 + (14.3 \text{ rad/s}^2)(2.0 \text{ s}) = 28.6 \text{ rad/s}.$$

Then

$$\tfrac{1}{2}I\omega^2 = (\tfrac{1}{4})(2.50 \text{ kg})(0.20 \text{ m})^2(28.6 \text{ rad/s})^2 = 20.5 \text{ J},$$

as before. Hence the increase in kinetic energy of the disk is equal to the work done by the resultant force on the disk, as it must be.

EXAMPLE 7

Show that the conservation of mechanical energy holds for the system of Example 5.

The resultant force acting on the system is the force of gravity on the suspended body. This is a conservative force. Viewing the system as a whole, we see that the suspended body loses potential energy U as it descends,

$$U = mgy,$$

where y is the vertical distance through which the block descends. At the same time the suspended body gains kinetic energy of translation and the disk gains kinetic energy of rotation. The total gain in kinetic energy is

$$\tfrac{1}{2}mv^2 + \tfrac{1}{2}I\omega^2,$$

where v is the linear speed of the suspended mass. We must show then that

$$mgy = \tfrac{1}{2}mv^2 + \tfrac{1}{2}I\omega^2.$$

For the linear motion starting from rest we have $v^2 = 2ay$. From Example 5, we obtained $a = 2\ mg/(M + 2m)$. Hence

$$mgy = \frac{mgv^2}{2a} = \tfrac{1}{2}\ mv^2\left(\frac{g}{a}\right) = \tfrac{1}{2}\ mv^2\left(\frac{M + 2m}{2m}\right) = \tfrac{1}{4}\ (M + 2m)v^2.$$

We also know that $\omega = v/R$ and $I = \frac{1}{2}MR^2$. Substituting these relations into the right-hand side of the conservation equation, we obtain

$$\tfrac{1}{2}mv^2 + \tfrac{1}{2}I\omega^2 = \tfrac{1}{2}mv^2 + \tfrac{1}{2}(\tfrac{1}{2}MR^2)(v^2/R^2) = \tfrac{1}{4}(M + 2m)v^2.$$

The mechanical energy is therefore conserved.

EXAMPLE 8

Derive the relation $L = I\omega$, shown in Table 12-2, for the angular momentum of a rigid body confined to rotate about a fixed axis.

Starting from the scalar relation $\tau = I\alpha$ and the definition of $\alpha\ (= d\omega/dt)$, we may write

$$\tau = I\alpha = I(d\omega/dt) = d(I\omega)/dt,$$

in which the last step is justified by the fact that I is a constant for a given rigid body and a specified (fixed) axis of rotation.

Next we use the vector relation $\tau_{\text{ext}} = d\mathbf{L}/dt$ (Eq. 12-9) and write the corresponding relation for the *scalar components*, τ and dL, of τ_{ext} and $d\mathbf{L}$ along the fixed axis of rotation, obtaining

$$\tau = dL/dt.$$

Simply by comparing the two equations above we obtain the relation sought, namely

$$L = I\omega. \qquad (12\text{-}18)$$

Like Eq. 12-17 ($\tau = I\alpha$), this is a scalar relation holding for the rotation of a rigid body about a fixed axis. L is the component along that axis of the vector angular momentum $\mathbf{L}$ of the rigid body and I must refer to that same axis.

Equation 12-18 is the rotational analog of the expression $P = Mv$ for the *linear* momentum of a rigid body of mass M in pure translational motion with linear speed v. It gives the *angular* momentum about a fixed axis of a rigid body having rotational inertia I and angular speed ω about that same axis.

Up until now we have considered only bodies rotating about some fixed axis. If a body is rolling, however, it is rotating about an axis and also moving translationally. Therefore it would seem that the motion of rolling bodies must be treated as a combination of translational and rotational motion. It is also possible, however, to treat a rolling body as though its motion is one of pure rotation. We wish to illustrate the equivalence of the two approaches.

Consider, for example, a cylinder rolling along a level surface, as in Fig. 12-13. At any instant the bottom of the cylinder is at rest on the surface, since it does not slide. The axis normal to the diagram through the point of contact P is called the *instantaneous axis of rotation*. At that instant the linear velocity of every particle of the cylinder is directed at right angles to the line joining the particle and P and its magnitude is proportional to this distance. This is the same as saying that the cylinder is rotating about a fixed axis through P with a certain angular speed ω, *at that instant*. Hence, at a given instant the motion of the body is equivalent to a pure rotation. The total kinetic energy can, therefore, be written as

$$K = \tfrac{1}{2}I_P\omega^2, \qquad (12\text{-}19)$$

where I_P is the rotational inertia about the axis through P.

Let us now use the parallel axis theorem, which tells us that

$$I_P = I_{\text{cm}} + MR^2,$$

where I_{cm} is the rotational inertia of the cylinder of mass M and radius R about a parallel axis through the center of mass. Equation 12-19 now becomes

$$K = \tfrac{1}{2}I_{\text{cm}}\omega^2 + \tfrac{1}{2}MR^2\omega^2. \qquad (12\text{-}20)$$

The quantity $R\omega$ is the speed with which the center of mass of the cylinder is moving with respect to the fixed point P. Let $R\omega = v_{\text{cm}}$. Equation 12-20 then becomes

$$K = \tfrac{1}{2}I_{\text{cm}}\omega^2 + \tfrac{1}{2}Mv_{\text{cm}}^2. \qquad (12\text{-}21)$$

Now notice that the speed of the center of mass with respect to P is the same as the speed of P with respect to the center of mass. Hence, the angular speed ω of the center of mass about P as seen by someone at P is the same as the angular speed of a particle at P about C as seen

12-7
THE COMBINED TRANSLATIONAL AND ROTATIONAL MOTION OF A RIGID BODY

O MIT !

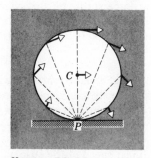

figure 12-13
A rolling body may at any instant be thought of as rotating about a perpendicular axis through its point of contact P.

by someone at C (moving along with the cylinder). This is equivalent to saying that any reference line in the cylinder turns through the same angle in a given time whether it is observed from a reference frame fixed with respect to the surface on which the cylinder is rolling or from a frame moving translationally with respect to this fixed frame. We can therefore interpret Eq. 12-21, which was derived on the basis of a pure rotational motion, in another way; that is, the first term, $\frac{1}{2}I_{cm}\omega^2$, is the kinetic energy the cylinder would have if it were merely rotating about an axis through its center of mass, without translational motion; and the second term, $\frac{1}{2}Mv_{cm}{}^2$, is the kinetic energy the cylinder would have if it were moving translationally with the speed of its center of mass, without rotating. Notice that there is now no reference at all to the instantaneous axis of rotation. In fact, Eq. 12-21 applies to any body that is moving and rotating about an axis perpendicular to its motion whether or not it is rolling on a surface.

The combined effects of translation of the center of mass and rotation about an axis through the center of mass are equivalent to a pure rotation with the same angular speed about an axis through the point of contact of a rolling body.

To illustrate this result simply, let us consider the instantaneous speed of various points on the rolling cylinder. If the speed of the center of mass (as seen by an observer fixed with respect to the surface) is v_{cm}, the instantaneous angular speed about an axis through P is $\omega = v_{cm}/R$. A point Q at the top of the cylinder will therefore have a speed $\omega 2R = 2v_{cm}$ at that instant. The point of contact P is instantaneously at rest. Hence, from the point of view of pure rotation about P, the situation is as shown in Fig. 12-14.

Now let us regard the rolling as a combination of translation of the center of mass and rotation about the cylinder axis through C. If we consider translation only, all points on the cylinder have the same speed v_{cm}, the speed of the center of mass. This is shown in Fig. 12-15a. If we consider the rotation only, the center is at rest, whereas the point Q at the top has a speed ωR in the x-direction and the point P at the bottom of the cylinder has a speed ωR in the $-x$-direction. This is shown in Fig. 12-15b. Now let us combine these two results. Recalling that $\omega = v_{cm}/R$, we obtain

for the point Q $\qquad v = v_{cm} + \omega R = v_{cm} + \dfrac{v_{cm}}{R}R = 2v_{cm},$

for the point C $\qquad v = v_{cm} + 0 = v_{cm},$

for the point P $\qquad v = v_{cm} - \omega R = v_{cm} - \dfrac{v_{cm}}{R}R = 0.$

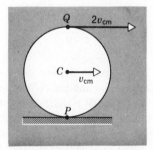

figure 12-14
Since Q and C have the same angular velocity about P, therefore Q, being twice as far from P, moves with twice the linear velocity of C.

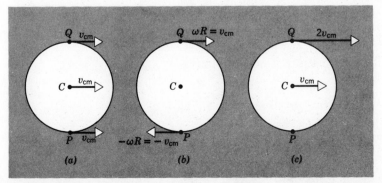

figure 12-15
(a) For pure translation, all points move with the same velocity. *(b)* For pure rotation about C, opposite points move with opposite velocities. *(c)* Combined rotation and translation is obtained by adding together corresponding vectors in *(a)* and *(b)*.

This result, shown in Fig. 12-15c is exactly the same as that obtained from the purely rotational point of view, Fig. 12-14.

EXAMPLE 9

Consider a solid cylinder of mass M and radius R rolling down an inclined plane without slipping. Find the speed of its center of mass when the cylinder reaches the bottom.

The situation is illustrated in Fig. 12-16. We can use the conservation of energy to solve this problem. The cylinder is initially at rest. In rolling down the incline the cylinder loses potential energy of an amount Mgh, where h is the height of the incline. It gains kinetic energy equal to

$$\tfrac{1}{2}I_{cm}\omega^2 + \tfrac{1}{2}Mv^2,$$

where v is the linear speed of the center of mass and ω is the angular speed about the center of mass at the bottom.

We have then the relation

$$Mgh = \tfrac{1}{2}I_{cm}\omega^2 + \tfrac{1}{2}Mv^2,$$

in which

$$I_{cm} = \tfrac{1}{2}MR^2 \quad \text{and} \quad \omega = \frac{v}{R}.$$

Hence

$$Mgh = \tfrac{1}{2}(\tfrac{1}{2}MR^2)\left(\frac{v}{R}\right)^2 + \tfrac{1}{2}Mv^2 = (\tfrac{1}{4} + \tfrac{1}{2})Mv^2,$$

$$v^2 = \tfrac{4}{3}gh \quad \text{or} \quad v = \sqrt{\tfrac{4}{3}gh}.$$

The speed of the center of mass would have been $v = \sqrt{2gh}$ if the cylinder had *slid* down a *frictionless* incline. The speed of the rolling cylinder is, therefore, less than the speed of the sliding cylinder, because for the rolling cylinder, part of the lost potential energy has been transformed into rotational kinetic energy, leaving less available for the translational part of the kinetic energy. Although the rolling cylinder arrives later at the bottom of the incline than an identical sliding cylinder started at the same time down a frictionless, but otherwise identical, incline, both arrive at the bottom with the same amount of energy; the rolling cylinder happens to be rotating as it moves, whereas the sliding one does not rotate as it moves.

figure 12-16
Example 9. A cylinder rolling down an incline.

Notice that static friction is needed to cause the cylinder to rotate. Remembering that friction is a dissipative force, how can you justify using the conservation of mechanical energy in this problem?

EXAMPLE 10

The previous result was derived by use of energy methods. Solve the same problem using only dynamical methods.

The force diagram is shown in Fig. 12-17. Mg is the weight of the cylinder

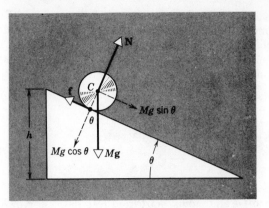

figure 12-17
Example 10. Dynamic solution of the motion of a cylinder rolling down an incline.

acting vertically down through the center of mass,* **N** is the normal force exerted by the incline on the cylinder, and **f** is the force of static friction acting along the incline at the point of contact.

The *translational* motion of a body is obtained by assuming that all the external forces act at its center of mass. Using Newton's second law, we obtain

$$N - Mg \cos \theta = 0 \qquad \text{for motion normal to the incline,}$$

and

$$Mg \sin \theta - f = Ma \qquad \text{for motion along the incline.}$$

The *rotational* motion about the center of mass follows from

$$\tau = I_{cm}\alpha.$$

Neither **N** nor **Mg** can cause rotation about C because their lines of action pass through C, and they have zero moment arms. The force of friction has a moment arm R about C, so that

$$fR = I_{cm}\alpha.$$

But

$$I_{cm} = \tfrac{1}{2}MR^2 \qquad \text{and} \qquad \alpha = \frac{a}{R}$$

so that

$$f = I_{cm}\alpha/R = Ma/2.$$

Substituting this into the second translational equation, we find

$$a = \tfrac{2}{3}g \sin \theta.$$

That is, the acceleration of the center of mass for the rolling cylinder ($\tfrac{2}{3}g \sin \theta$) is less than the acceleration of the center of mass for the cylinder sliding down the incline ($g \sin \theta$).

This result holds at any instant, regardless of the position of the cylinder along the incline. The center of mass moves with constant linear acceleration. To obtain the speed of the center of mass, starting from rest, we use the relation

$$v^2 = 2as,$$

so that

$$v^2 = 2(\tfrac{2}{3}g \sin \theta)s = \tfrac{4}{3}g \frac{h}{s} s = \tfrac{4}{3}gh$$

or

$$v = \sqrt{\tfrac{4}{3}gh}.$$

---

* In drawing the vector diagram for this problem we tacitly assume that the total weight of the body can be thought of as acting at the center of mass. We saw in Section 9-2 that this is justified for analyzing the translational motion. However, later in the problem we use this result in analyzing the rotational motion as well. We shall justify this procedure in Section 14-3, where it is shown that the *weight* of a body can be considered to act at its center of mass for both translational *and* rotational motion.

This result is the same as that obtained before by the energy method. The energy method is certainly simpler and more direct. However, if we are interested in knowing what the forces are, such as **N** and **f**, we must use a dynamical method.

This method determines the minimum force of static friction needed for rolling:

$$f = Ma/2 = (M/2)(\tfrac{2}{3}g \sin \theta) = \tfrac{1}{3}Mg \sin \theta.$$

What would happen if the force of static friction between the surfaces were less than this value?

A sphere and a cylinder, having the same mass and radius, start from rest and roll down the same incline. Which body gets to the bottom first?

For a sphere I_{cm} equals $\tfrac{2}{5}MR^2$. Using the dynamical method we obtain

$$Mg \sin \theta - f = Ma, \qquad \text{translation of cm,}$$

$$fR = I_{cm}\alpha = (\tfrac{2}{5}MR^2)(a/R), \qquad \text{rotation about cm,}$$

or

$$f = \tfrac{2}{5}Ma \quad \text{and} \quad a = \tfrac{5}{7}g \sin \theta, \quad \text{sphere.}$$

For the cylinder (Example 10)

$$a = \tfrac{2}{3}g \sin \theta, \quad \text{cylinder.}$$

Hence the acceleration of the center of mass of the sphere is at all times greater than the acceleration of the center of mass of the cylinder. Since both bodies start from rest at the same instant, the sphere will reach the bottom first.

Which body has the greater rotational energy at the bottom? Which body has the greater translational energy at the bottom?

Note carefully that neither the mass nor the radius of the rolling object enters the previous results. How then would we expect the behavior of cylinders of different mass and radii to compare? How would we expect the behavior of spheres of different mass and radii to compare? How would the behavior of a cylinder and sphere having different masses and radii compare?

EXAMPLE 11

A uniform solid cylinder of radius r and mass m is given an initial angular velocity ω_0 and then dropped on a flat horizontal surface. The coefficient of kinetic friction between the surface and the cylinder is μ_k. Initially the cylinder slips but after a time t pure rolling begins. (a) What is the velocity V the center of mass at the time t? (b) What is the value of t?

(a) Figure 12-18 shows the forces that act on the cylinder.

The acceleration a of the center of mass is constant, since all the forces are constant, so that for the translational motion we can write

$$F = ma = m\left(\frac{V_f - V_i}{t - 0}\right).$$

Here, $V_i = 0$ and $V_f = V$, the velocity at t when pure rolling begins. Also, the resultant force F is $\mu_k mg$, so that

$$\mu_k mg = mV/t. \qquad (12\text{-}22)$$

The angular acceleration α about an axis through the center of mass is also constant (why?), so that for the rotational motion we can write

$$\tau = I\alpha = I\left(\frac{\omega_f - \omega_i}{t - 0}\right).$$

Here, $\omega_f = \omega = V/r$, the angular velocity at time t, and $\omega_i = \omega_0$. Also, the magni-

EXAMPLE 12

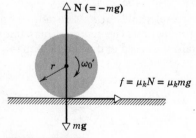

figure 12-18
Example 12

tude of the resultant torque τ is $\mu_k\ mg\ r$. The torque causes an angular deceleration so that

$$\mu_k mg\ r = (\tfrac{1}{2}mr^2)\left(\frac{\omega_0 - V/r}{t}\right).$$ (12-23)

If we eliminate t from our two equations (e.g., divide Eq. 12-23 by Eq. 12-22)
and solve for V (please do the algebra), we obtain

$$V = \tfrac{1}{3}\omega_0 r.$$

Note that V does not depend on the value of m, g, or μ_k. What, however, if
any of these quantities were zero?

(b) By eliminating V from Eqs. 12-22 and 12-23 we can solve for t (please do
the algebra) and find

$$t = \frac{\omega_0 r}{3\mu_k g}.$$

It is worth noting that in this problem neither mechanical energy, linear
momentum, nor angular momentum are conserved, but the *changes* in momentum and angular momentum are directly related because the force of friction is
responsible for both.

___

questions

1. What are the dimensions of angular momentum? Can you find any significance in the fact that they are the same as those of energy multiplied by
   time?
2. Is the vector product of two vectors necessarily an axial vector?*
3. Can the mass of a body be considered as concentrated at its center of mass
   for purposes of computing its rotational inertia?
4. About what axis would a uniform cube have its minimum rotational
   inertia?
5. If two circular disks of the same weight and thickness are made from metals
   having different densities, which disk, if either, will have the larger rotational inertia about its central axis?
6. The rotational inertia of a body of rather complicated shape is to be determined. The shape makes a mathematical calculation from $\int r^2\ dm$ exceedingly difficult. Suggest ways in which the rotational inertia could be
   measured experimentally.

figure 12-19
Question 7

Ring   Cube   Cylinder   Prism   Sphere

7. Five solids are shown in cross section (Fig. 12-19). The cross sections have
   equal heights and maximum widths. The axes of rotation are perpendicular
   to the sections through the points shown. The solids have equal masses.
   Which one has the largest rotational inertia about a perpendicular axis
   through the center of mass? Which the smallest?

* See Supplementary Topic II.

figure 12-20
Question 8

8. In Fig. 12-20a a meter stick, half of which is wood — the other half steel — is pivoted at the wooden end at O and a force is applied to the steel end at a. In Fig. 12-20b the stick is pivoted at the steel end at O' and the same force is applied at the wooden end at a'. Does one get the same angular acceleration in each case? Explain.

9. A person can distinguish between a raw egg and a hard-boiled one by spinning each one on the table. Explain how. Also, if you stop a spinning raw egg with your fingers and release it very quickly, it will resume spinning. Why?

10. Torque has the same dimensions as work or energy. Is torque, therefore, work or energy?

11. Comment on each of these assertions about skiing. (a) In downhill racing one wants skis that do not turn easily. (b) In slalom racing, one wants skis that turn easily. (c) Therefore, the rotational inertia of downhill skis should be larger than that of slalom skis. (See "The Physics of Ski Turns" by J. I. Shonie and D. L. Nordick in *The Physics Teacher*, December 1972.)

12. Considering that there is low friction between skis and snow and that the skier's center of mass is about over the center of the skis, how does a skier exert torques to turn or to stop a turn? (See "The Physics of Ski Turns" by J. I. Shonie and D. L. Nordick in *The Physics Teacher*, December 1972.)

13. Do the expressions for a and T in Example 5 give reasonable results for the special cases in which $g = 0$, $M = 0$, $M \to \infty$, $m = 0$, and $m \to \infty$?

14. The total momentum of a system of particles does not depend on the motions of the particles relative to the center of mass of the system. Can a similar statement be made about the total kinetic energy of a system of particles?

15. A cylindrical drum, pushed along by a board from an initial position shown in Fig. 12-21, rolls forward on the ground a distance $l/2$, equal to half the length of the board. There is no slipping at any contact. Where is the board then? How far has the man walked?

figure 12-21
Question 15

16. For storing wind energy or solar energy, flywheels have been suggested. The amount of energy that can be stored in a flywheel depends on the density and tensile strength of the material making up the flywheel and for a given weight one wants the lowest density strong material available. (See "Flywheels" by R. F. Post and S. F. Post, *Scientific American*, December 1973.) Can you make this plausible?

17. A solid wooden sphere rolls down two different inclined planes of the same height but different angles of inclines. Will it reach the bottom with the same speed in each case? Will it take longer to roll down one incline than the other? If so, which one and why?

18. Two heavy disks are connected by a short rod of much smaller radius. The system is placed on an inclined plane so that the disks hang over the sides and the system rolls down on the rod without slipping (Fig. 12-22). Near the bottom of the incline the disks touch the horizontal table top and the system takes off with greatly increased translational speed. Explain carefully.

19. When a logger cuts down a tree he makes a cut on the side facing the direction in which he wants it to fall. Explain why. Would it be safe to stand directly behind the tree on the opposite side of the fall?

20. Consider a straight stick standing on end on (frictionless) ice. What would be the path of its center of mass if it falls?

figure 12-22
Question 18

21. A yo-yo is resting on a horizontal table and is free to roll (Fig. 12-23). If the string is pulled by a horizontal force such as $\mathbf{F}_1$, which way will the yo-yo roll? What happens when the force $\mathbf{F}_2$ is applied (its line of action passes through the point of contact of the yo-yo and table)? If the string is pulled vertically with the force $\mathbf{F}_3$, what happens?

22. You are looking at the wheel of an automobile traveling at constant speed. Some one says to you: "The top of the wheel is moving twice as fast as the axle but the bottom is not moving at all." Can you accept this statement? Discuss.

23. State Newton's three laws of motion in words suitable for rotating bodies.

figure 12-23
Question 21

problems

SECTION 12-2

1. (a) Given that $\mathbf{r} = \mathbf{i}x + \mathbf{j}y + \mathbf{k}z$ and $\mathbf{F} = \mathbf{i}F_x + \mathbf{j}F_y + \mathbf{k}F_z$, find the torque $\tau = \mathbf{r} \times \mathbf{F}$. (b) Show that if $\mathbf{r}$ and $\mathbf{F}$ lie in a given plane, then τ has no component in that plane.

   Answer: (a) $\mathbf{i}(yF_z - zF_y) + \mathbf{j}(zF_x - xF_z) + \mathbf{k}(xF_y - yF_x)$.

2. Show that the angular momentum about any point of a single particle moving with constant velocity remains constant throughout the motion.

SECTION 12-3

3. A particle P with mass 2.0 kg has position $\mathbf{r}$ and velocity $\mathbf{v}$ as shown in Fig. 12-24. It is acted on by the force $\mathbf{F}$. All three vectors lie in a common plane. Presume that $r = 3.0$ m, $v = 4.0$ m/s, and $F = 2.0$ N. Compute (a) the angular momentum of the particle and (b) the torque acting on the particle. What are the directions of these two vectors?

   Answer: (a) 12 kg · m²/s; out of page. (b) 3.0 N · m; out of page.

4. If we are given r, p, and θ, we can calculate the angular momentum of a particle from Eq. 12-4a. Sometimes, however, we are given the components (x, y, z) of $\mathbf{r}$ and (p_x, p_y, p_z) of $\mathbf{p}$ instead. (a) Show that the components of l along the x-, y-, and z-axes are then given by

$$l_x = yp_z - zp_y,$$
$$l_y = zp_x - xp_z,$$
$$l_z = xp_y - yp_x.$$

   (b) Show that if the particle moves only in the x-y plane, the resultant angular momentum vector has only a z-component.

5. (a) In Example 1, express $\mathbf{F}$ and $\mathbf{r}$ in terms of unit vectors and compute τ. Do the same in Example 3. (b) In example 1, express $\mathbf{p}$ and $\mathbf{r}$ in unit vectors and compute $\mathbf{l}$.

   Answer: (a) $\tau = +\mathbf{k}mgb$; 2.6 $\mathbf{k}$, lb · ft. (b) $\mathbf{l} = +\mathbf{k}mgbt$.

SECTION 12-4

6. In Fig. 12-25 are shown the lines of action and the moment arms of two forces about the origin O. Imagine these forces to be acting on a rigid body pivoted at O, all vectors shown being in the plane of the figure, and find the magnitude and the direction of the resultant torque on the body.

7. Two particles, each of mass m and speed v, travel in opposite directions along parallel lines separated by a distance d. Show that the vector angular momentum of this system of particles is the same no matter what point is taken as the origin.

8. Three particles, each of mass m, are fastened to each other and to a rotation axis by three light strings each with length l as shown in Fig. 12-26. The combination rotates around the rotational axis with angular velocity ω in such a way that the particles remain in a straight line. (a) Calculate the rotational inertia of the combination about O. (b) What is the angular mo-

figure 12-24
Problem 3

figure 12-25
Problem 6

figure 12-26
Problems 8, 13

mentum of the middle particle? (c) What is the total angular momentum of the three particles? Express your answers in terms of m, l, and ω.

9. Starting from Newton's third law, prove that the resultant internal torque on a system of particles is zero.

10. *Relation between the Resultant External Torque and the Angular Momentum of a System of Particles about the Center of Mass of the System.* Let $\mathbf{r}_{cm}$ be the position vector of the center of mass C of a system of particles with respect to the origin O of an inertial reference frame, and let $\mathbf{r}_i'$ be the position vector of the ith particle, of mass m_i, with respect to the center of mass C. Hence $\mathbf{r}_i = \mathbf{r}_{cm} + \mathbf{r}_i'$ (see Fig. 12-27). Now define the total angular momentum of the system of particles relative to the center of mass C to be $\mathbf{L}' = \sum_i \mathbf{r}_i' \times \mathbf{p}_i'$, where $\mathbf{p}_i' = m_i \, d\mathbf{r}_i'/dt$.

(a) Show that $\mathbf{p}_i' = m_i d\mathbf{r}_i/dt - m_i d\mathbf{r}_{cm}/dt = \mathbf{p}_i - m_i \mathbf{v}_{cm}$. (b) Show next that $d\mathbf{L}'/dt = \sum_i \mathbf{r}_i' \times d\mathbf{p}_i'/dt$. (c) Combine the results of (a) and (b) and, using the definition of center of mass and Newton's third law, show that $\tau'_{ext} = d\mathbf{L}'/dt$, where τ'_{ext} is the sum of all the external torques acting on the system about its center of mass.

figure 12-27
Problem 10

SECTION 12-5

11. Assume the earth to be a sphere of uniform density. (a) What is its rotational kinetic energy? Take the radius of the earth to be 6.4×10^3 km and the mass of the earth to be 6.0×10^{24} kg. (b) Suppose this energy could be harnessed for our use. For how long could the earth supply 1.0 kW of power to each of the 4.2×10^9 persons on earth?
Answer: (a) 2.6×10^{29} J. (b) 2.0×10^9 yr.

12. The oxygen molecule has a total mass of 5.30×10^{-26} kg and a rotational inertia of 1.94×10^{-46} kg $\cdot$ m^2 about an axis through the center perpendicular to the line joining the atoms. Suppose that such a molecule in a gas has a mean speed of 500 m/s and that its rotational kinetic energy is two-thirds of its translational kinetic energy. Find its average angular velocity.

13. Presume that the strings in Problem 8 are all replaced with uniform rods, each of mass M. (a) What is the total rotational inertia of the system about O? (b) What is the rotational kinetic energy of the system?
Answer: (a) $14\,ml^2 + 9\,Ml^2$. (b) $(7\,m + 9\,M/2)l^2\omega^2$.

14. (a) Show that a solid cylinder of mass M and radius R is equivalent to a thin hoop of mass M and radius $R/\sqrt{2}$, for rotation about a central axis. (b) The radial distance from a given axis at which the mass of a body could be concentrated without altering the rotational inertia of the body about that axis is called the *radius of gyration*. Let k represent radius of gyration and show that

$$k = \sqrt{I/M}.$$

This gives the radius of the "equivalent hoop" in the general case.

15. A thin rod of length l and mass m is suspended freely from one end. It is pulled aside and swung about a horizontal axis, passing through its lowest position with an angular speed ω. How high does its center of mass rise above its lowest position? Neglect friction and air resistance.
Answer: $l^2\omega^2/6\,g$.

16. (a) Prove that the rotational inertia of a thin rod of length l about an axis through its center perpendicular to its length is $I = \frac{1}{12}Ml^2$. (See Table 12-1.) (b) Use the parallel-axis theorem to show that $I = \frac{1}{3}Ml^2$ when the axis of rotation is through one end perpendicular to the length of the rod.

17. (a) Show that the sum of the rotational inertias of a plane laminar body about any two perpendicular axes in the plane of the body is equal to the rotational inertia of the body about an axis through their point of intersection perpendicular to the plane. (b) Apply this to a circular disk to find its rotational inertia about a diameter as axis.   *Answer:* (b) $MR^2/4$.

18. Show that the rotational inertia of a rectangular plate of sides a and b about an axis perpendicular to the plate through its center is $\frac{1}{12}M(a^2 + b^2)$.

19. A meter stick is held vertically with one end on the floor and is then allowed to fall. Find the speed of the other end when it hits the floor, assuming that the end on the floor does not slip. *Answer:* 5.4 m/s.

20. A tall chimney cracks near its base and falls over. Express (a) the radial and (b) the tangential linear acceleration of the top of the chimney as a function of the angle θ made by the chimney with the vertical. (c) Can the resultant linear acceleration exceed g? (d) The chimney cracks up during the fall. Explain how this can happen. (See 'More on the Falling Chimney' by Albert A. Bartlett, in *The Physics Teacher*, September, 1976.)

SECTION 12-6

21. An automobile engine develops 100 hp $(7.5 \times 10^4$ W) when rotating at a speed of 1800 rev/min. What torque does it deliver?
Answer: 290 ft · lb (400 N · m).

22. Calculate (a) the torque, (b) the energy, and (c) the average power required to accelerate the earth from rest to its present angular speed about its axis in one day.

23. A pulley having a rotational inertia of 1.0×10^4 g · cm² and a radius of 10 cm is acted upon by a force, applied tangentially at its rim, that varies in time as $F = 0.50\,t + 0.30\,t^2$, where F is in newtons and t is in seconds. If the pulley was initially at rest, find its angular velocity after 3.0 seconds.
Answer: 5.0×10^2 rad/s.

24. A wheel of mass M and radius of gyration k (see Problem 14) spins on a fixed horizontal axle passing through its hub. The hub rubs the axle of radius a at only the topmost point, the coefficient of kinetic friction being μ_k. The wheel is given an initial angular velocity ω_0. Assume uniform deceleration and find (a) the elapsed time and (b) the number of revolutions before the wheel comes to a stop.

25. A uniform steel rod of length 1.20 m and mass 6.40 kg has attached to each end a small ball of mass 1.06 kg. The rod is constrained to rotate in a horizontal plane about a vertical axis through its midpoint. At a certain instant it is observed to be making 39.0 rev/s. Because of axle friction it comes to rest 32.0 s later. Compute, assuming a constant frictional torque, (a) the angular acceleration, (b) the retarding torque exerted by axle friction, (c) the total work done by the axle friction, and (d) the number of revolutions executed during the 32.0 s. (e) Suppose, however, that the frictional torque is known not to be constant. Which, if any, of the quantities (a), (b), (c), or (d) can still be computed without requiring any additional information? If such exists, give its value.
Answer: (a) −7.67 rad/s². (b) −11.7 N · m. (c) 4.58 × 10⁴ J. (d) 624 rev. (e) The total work done; 4.58 × 10⁴ J.

26. The angular momentum of a flywheel having a rotational inertia of 0.125 kg · m² $(9.22 \times 10^{-2}$ slug · ft²) decreases from 3.0 (2.21) to 2.0 (1.48) kg · m²/s (slug · ft²/s) in a period of 1.5 s. (a) What is the average torque acting on the flywheel during this period? (b) Assuming a uniform angular acceleration, through how many revolutions will the flywheel have turned? (c) How much work was done? (d) What was the average power supplied by the flywheel?

27. In an Atwood's machine (Fig. 5-9) one block has a mass of 500 g and the other a mass of 460 g. The pulley, which is mounted in horizontal frictionless bearings, has a radius of 5.0 cm. When released from rest the heavier block is observed to fall 75 cm in 5.0 s. What is the rotational inertia of the pulley? *Answer:* 1.4×10^{-2} kg · m².

28. A uniform spherical shell rotates about a vertical axis on frictionless bearings (Fig. 12-28) A light cord passes around the equator of the shell, over a

pulley, and is attached to a small object that is otherwise free to fall under the influence of gravity. What is the speed of the object after it has fallen a distance h from rest?

29. A 6.0-lb block is put on a plane inclined 30° to the horizontal and is attached by a cord parallel to the plane over a pulley at the top to a hanging block weighing 18 lb. The pulley weighs 2.0 lb and has a radius of 0.33 ft. The coefficient of kinetic friction between block and plane is 0.10. Find (a) the acceleration of the hanging block and (b) the tension in the cord on each side of the pulley. Assume the pulley to be a uniform disk.
Answer: (a) 19 ft/s². (b) $T_{18} = 7.6$ lb; $T_6 = 7.0$ lb.

figure 12-28
Problem 28

SECTION 12-7

30. A hoop of radius 10 ft (3.0 m) weighs 320 lb (mass = 150 kg). It rolls along a horizontal floor so that its center of mass has a speed of 0.50 ft/s (0.15 m/s). How much work has to be done to stop it?

31. An automobile has a total mass of 1700 kg. It accelerates from rest to 40 km/h in 10 s. Each wheel has a mass of 32 kg and a radius of gyration (see Problem 14) of 0.30 m. Find, for the end of the 10-s interval, (a) the rotational kinetic energy of each wheel about its axle, (b) the total kinetic energy of each wheel; (c) the total kinetic energy of the automobile.
Answer: (a) 990 J. (b) 3000 J. (c) 1.1×10^5 J.

32. Show that a cylinder will slip on an inclined plane of inclination angle θ if the coefficient of static friction between plane and cylinder is less than $\frac{1}{3} \tan \theta$.

33. A 10-ft-long ladder rests against a wall and makes an angle of 60° with the horizontal floor. If it starts to slip, where is the instantaneous axis of rotation?
Answer: 5.0 ft horizontally from the wall and $5\sqrt{3}$ ft vertically above the ground.

34. A sphere rolls up an inclined plane of inclination angle 30°. At the bottom of the incline the center of mass of the sphere has a translational speed of 16 ft/s. (a) How far does the sphere travel up the plane? (b) How long does it take to return to the bottom?

35. A body of radius R and mass m is rolling horizontally without slipping with speed v. It then rolls up a hill to a maximum height h. If $h = 3\,v^2/4\,g$, (a) what is the body's rotational inertia? (b) What might the body be?
Answer: (a) $\frac{1}{2}mR^2$. (b) A solid circular cylinder.

36. A small sphere rolls without slipping on the inside of a large hemisphere whose axis of symmetry is vertical. It starts at the top from rest. (a) What is its kinetic energy at the bottom? What fraction is rotational? What translational? (b) What normal force does the small sphere exert on the hemisphere at the bottom? Take the radius of the small sphere to be r, that of the hemisphere to be R, and let m be the mass of the sphere.

37. A uniform disk, of mass M and radius R, lies on one side initially at rest on a frictionless horizontal surface. A constant force **F** is then applied tangentially at its perimeter by means of a string wrapped around its edge. Describe the subsequent (rotational and translational) motion of the disk.
Answer: $\alpha = 2\,F/MR$; $a = F/M$.

38. A tape of negligible mass is wrapped around a cylinder of mass M, radius R. The tape is pulled vertically upward at a speed that just prevents the center of mass from falling as the cylinder unwinds the tape. (a) What is the tension in the tape? (b) How much work has been done on the cylinder once it has reached an angular velocity ω? (c) What is the length of tape unwound in this time?

39. A cylinder of length L and radius R has a weight W. Two cords are wrapped around the cylinder, one near each end, and the cord ends are attached to hooks on the ceiling. The cylinder is held horizontally with the two cords exactly vertical and is then released (Fig. 12-29). Find (a) the tension in each

figure 12-29
Problem 39

cord as they unwind and (b) the linear acceleration of the cylinder as it falls.
Answer: (a) W/6. (b) 2 g/3.

40. A homogeneous sphere starts from rest at the upper end of the track shown in Fig. 12-30 and rolls without slipping until it rolls off the right-hand end. If $H = 204$ ft and $h = 64$ ft and the track is horizontal at the right-hand end, determine the distance to the right of point A at which the ball strikes the horizontal base line.

41. A length l of flexible tape is tightly wound. It is then allowed to unwind as it rolls down a steep incline that makes an angle θ with the horizontal, the upper end of the tape being tacked down (Fig. 12-31). Show that the tape unwinds completely in a time $T = \sqrt{3\, l/g}\,\sin\theta$.

42. A small solid marble of mass m and radius r rolls without slipping along the loop-the-loop track shown in Fig. 12-32, having been released from rest somewhere on the straight section of track. (a) From what minimum height above the bottom of the track must the marble be released in order that it just stay on the track at the top of the loop? (The radius of the loop-the-loop is R; assume $R \gg r$.) (b) If the marble is released from height $6R$ above the bottom of the track, what is the horizontal component of the force acting on it at point Q?

43. A yo-yo is made from two uniform disks of radius R and combined mass m. The short shaft connecting the disks has a very small radius r. A string of length $(L + R)$ is wrapped around the shaft several times by an expert yo-yo operator, who releases it with zero speed. Assume the string is vertical at all times. (a) What is the tension in the string during the descent and subsequent ascent of the yo-yo? (b) How long does it take the yo-yo to return to the operator's hands?

Answer: (a) $mgR^2/(R^2 + 2r^2)$, ascent and descent. (b) $\dfrac{2}{r}\sqrt{L(2\,r^2 + R^2)/g}$.

44. A uniform solid cylinder of radius R is given an angular velocity ω_0 about its axis and is then dropped vertically onto a flat horizontal table. The table is not frictionless, so the cylinder begins to move as it slips. What is the velocity of the center of mass of the cylinder when pure rolling sets it?

45. A solid cylinder of weight 50 lb (mass = 23 kg) and radius 3.0 in (7.6 cm) has a light thin tape wound around it. The tape passes over a light, smooth fixed pulley to a 10-lb (mass = 4.5 kg) body, hanging vertically (Fig. 12-33). If the plane on which the cylinder moves is inclined 30° to the horizontal, find (a) the linear acceleration of the cylinder down the incline and (b) the tension in the tape, assuming no slipping.
Answer: (a) 1.4 ft/s² (0.47 m/s²). (b) 11 lb (48 N).

46. A billiard ball is struck by a cue as in Fig. 12-34. The line of action of the applied impulse is horizontal and passes through the center of the ball. The initial velocity $\mathbf{v}_0$ of the ball, its radius R, its mass M, and the coefficient of friction μ between the ball and the table are all known. How far will the ball move before it ceases to slip on the table?

figure 12-30
Problem 40

figure 12-31
Problem 41

figure 12-32
Problem 42

figure 12-33
Problem 45

figure 12-34
Problem 46

13
rotational dynamics II and the conservation of angular momentum

In Chapter 12 we discussed the dynamics of the rotational motion of a rigid body about an axis that was fixed in an inertial reference frame. We saw that the scalar relation $\tau = I\alpha$ (Eq. 12-17), in which only torque components along the axis of rotation were considered, was sufficient to solve dynamical problems in this special case.

In this chapter we shall first consider the rotation of a rigid body about an axis that is *not* fixed in an inertial reference frame. To solve dynamical problems in this more general case we shall use the general (vector) relation for rotational motion, namely,

$$\boldsymbol{\tau} = d\mathbf{L}/dt \qquad (12\text{-}9)$$

in which we have dropped the subscript on τ_{ext} for convenience.

Later we shall consider once more the rotation of particles and rigid bodies about fixed axes. This time, however, we specifically examine the action of torques which have components at right angles to the axis. Our point of departure here will not be Eq. 12-17 ($\tau = I\alpha$) but again the more general Eq. 12-9 ($\boldsymbol{\tau} = d\mathbf{L}/dt$).

Finally we shall consider systems on which no external torques act and shall introduce the important principle of *conservation of angular momentum*.

Figure 13-1*a* shows a top spinning about its axis of symmetry, the point of the top being fixed at the origin O of an inertial reference frame. We know from experience that the axis of such a rapidly spinning top will move around the vertical axis, sweeping out a cone. This motion is called *precession*. Let us see if we can predict this motion from the prin-

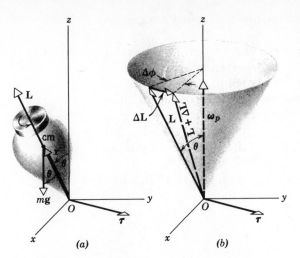

figure 13-1
(a) A precessing top, showing the angular momentum **L**, the weight m**g** and the vector **r** which locates the center of mass. (b) The cone swept out by the precessing axis of the top. The angular velocity of precession $\boldsymbol{\omega}_p$ is shown pointing vertically upward.

ciples of classical mechanics and, in particular, if we can calculate ω_p, the angular speed of the precessional motion.

At the instant shown in Fig. 13-1a the top has an angular velocity $\boldsymbol{\omega}$ about its own axis. It also has an angular momentum **L** about this same axis,* the axis making an angle θ with the vertical.

Two forces act on the top, an upward force on the pivot at O and the pull of gravity, or weight, which acts downward at the center of mass. The upward force passes through O and thus can exert no torque about that point because its moment arm is zero. The weight m**g**, however, exerts a torque about O given by

$$\boldsymbol{\tau} = \mathbf{r} \times \mathbf{F} = \mathbf{r} \times m\mathbf{g},$$

where **r** locates the center of mass with respect to the pivot. This equation requires that $\boldsymbol{\tau}$ be perpendicular to the plane formed by **r** and m**g**; application of the right-hand rule shows that its direction is as shown in Fig. 13-1a. Note that $\boldsymbol{\tau}$, as well as **L** and **r**, rotates about the axis at angular speed ω_p as the top precesses.

When a torque acts on a rigid body it changes the angular momentum of that body according to the fundamental relation (Eq. 12-9)

$$\boldsymbol{\tau} = d\mathbf{L}/dt. \qquad \text{(Eq. 12-9)}$$

Being a vector, **L** can change in magnitude, in direction, or in both. Equation 12-9 shows that the *change* in **L** (that is, $d\mathbf{L}$) must point in the direction of $\boldsymbol{\tau}$. Figure 13-1a shows us that $\boldsymbol{\tau}$ is *at right angles* to **L**; thus the change in **L** brought about by the action of the torque must also be at right angles to **L**.

To examine the matter quantitatively let us observe the top for a time Δt. During this interval a change in **L** of

$$\Delta \mathbf{L} = \boldsymbol{\tau}\, \Delta t \qquad (13\text{-}1)$$

is predicted by Eq. 12-9 (if Δt is small enough). This change $\Delta \mathbf{L}$, which, like $\boldsymbol{\tau}$, is at right angles to **L**, is displayed in Fig. 13-1b where we see the

---

* The vector $\boldsymbol{\omega}$ always points along the (fixed) axis of rotation of a spinning body but, in general, the vector **L** does not (see Section 13-3). For bodies with symmetry about the rotational axis, however, both $\boldsymbol{\omega}$ and **L** point along this axis, assuming that the axis is fixed. We can assume that $\boldsymbol{\omega}$ and **L** are coaxial for the spinning top of Fig. 13-1a if $\omega \gg \omega_p$, that is, if the precession rate is relatively slow so that the axis, although not fixed, changes direction only slowly.

cone swept out by the precessing axis of the top; the top itself is omitted here for clarity.

The angular momentum of the top at the end of the time interval Δt is the vector sum of **L** and Δ**L**. Since Δ**L** is perpendicular to **L** and is assumed to be very small in magnitude compared to it, the new angular momentum vector has the same *magnitude* as the old one but a different *direction*. Hence the head of the angular momentum vector swings around in a horizontal circle as time goes on (Fig. 13-1*b*). Since this vector always lies along the axis of rotation of the top, we have qualitatively accounted for the precession of the top.

The angular speed of precession ω_p follows from Fig. 13-1*b* in which

$$\omega_p = \Delta\phi/\Delta t.$$

But, since $\Delta L \ll L$, we have (see Eq. 13-1),

$$\Delta\phi \cong \Delta L/L \sin\theta = \tau\,\Delta t/L \sin\theta$$

or

$$\omega_p = \Delta\phi/\Delta t = \tau/L \sin\theta. \tag{13-2a}$$

Since (Fig. 13-1*a*)

$$\tau = rmg \sin(180° - \theta) = rmg \sin\theta$$

we have finally

$$\omega_p = mgr/L. \tag{13-2b}$$

Notice that the precessional angular velocity is independent of θ and varies inversely as the magnitude of the angular momentum. If the angular momentum is large, the precessional angular velocity will be small.

We can express Eq. 13-2*b* in vector form. We start by rewriting Eq. 13-2*a* as

$$\tau = \omega_p L \sin\theta.$$

Now $\boldsymbol{\omega}_p$ is a vector pointing vertically upward in Fig. 13-1*b*, and θ in that figure is the angle between $\boldsymbol{\omega}_p$ and **L**. We recognize the right side of the above equation as the magnitude of the vector product $\boldsymbol{\omega}_p \times$ **L** and we see that this equation gives the magnitude of $\boldsymbol{\tau}$ in the vector relation

$$\boldsymbol{\tau} = \boldsymbol{\omega}_p \times \mathbf{L}. \tag{13-3}$$

This is the general vector expression relating the precessional angular velocity to $\boldsymbol{\tau}$ and to **L**; you should show that Eq. 13-2*b* may be derived readily from it. Application of right-hand rule to Fig. 13-1*b* shows that the order of factors on the right side of Eq. 13-3 is correct, that is, $\boldsymbol{\omega}_p \times$ **L** gives the correct direction as well as the correct magnitude for $\boldsymbol{\tau}$.

EXAMPLE 1

A student holds a rim-loaded bicycle wheel, rotating at a relatively high angular speed ω, with its shaft horizontal as in Fig. 13-2*a*. Her physics instructor now asks her to turn the shaft rapidly (for a time Δt) so that the shaft points at a small angle $\Delta\theta$ above the horizontal as in Fig. 13-2*b*. The instructor also asks the student to keep the shaft in a vertical plane at all times. What torques must the student exert on the shaft if she is to follow these instructions?

The student will be well aware from the strain in her wrists that she must exert a torque on the shaft simply to hold it in a horizontal position. This torque, which is needed to counteract the turning effect of the force of gravity that acts at the center of mass, is directed along a horizontal axis and emerges perpendicularly out of the plane of Fig. 13-2. The student must supply this torque whether or not the wheel is rotating.

figure 13-2
(a) A student holds a heavy, rim-loaded, rapidly spinning bicycle wheel by the shaft and (b) tilts the shaft upward from the horizontal through a small angle.

If now the student turns the shaft of the spinning wheel upward, she will find that the wheel will swerve around to her right, perhaps rather violently, so that she will have failed to keep the shaft in a vertical plane. If she is to keep the shaft in this plane while she is tilting it upward, she must exert a torque on the shaft (about an almost vertical axis) tending to turn it to her left to counteract this effect. Let us see why this is so.

In tilting the shaft one changes its angular momentum $\mathbf{L}$, in time Δt, by an amount $\Delta \mathbf{L}$, as Fig. 13-2b shows. During this interval, then, the student must exert an average torque on the wheel given from Eq. 12-9 as

$$\bar{\tau} = \Delta \mathbf{L}/\Delta t;$$

the magnitude of $\bar{\tau}$ is given by

$$\bar{\tau} = \Delta L/\Delta t = L \sin \Delta\theta/\Delta t.$$

This average torque $\bar{\tau}$ has the same direction as $\Delta \mathbf{L}$, that is, it is approximately vertically upwards if the angle $\Delta\theta$ in Fig. 13-2b is not too large. We can see that such a torque would tend to turn the shaft to the left if the wheel was not rotating. This torque *must* be supplied by the student as she is tilting the shaft of the spinning wheel upward; if she fails to do so, the shaft will not remain in a vertical plane.

You should experiment with such a spinning wheel yourself, working out the relationships between the vectors $\mathbf{L}$, $\Delta \mathbf{L}$, and $\boldsymbol{\tau}$. If one is not available you can experiment with a toy gyroscope, although this fails to give the kinesthetic appreciation of $\boldsymbol{\tau} = d\mathbf{L}/dt$ that is provided by a rim-loaded, rapidly spinning wheel.

There is an analogy between the experiment of Fig. 13-2 and another experiment in which the student is asked to swing a heavy weight (attached to a stout cord) around in a horizontal circle at constant speed. In this latter experiment the student, during a time Δt, must change the direction of the *linear momentum* $\mathbf{P}$ of the weight, leaving its magnitude unchanged. To do so, she must supply a *force* that points at right angles to $\mathbf{P}$ (in the direction of $\Delta \mathbf{P}$), that is, radially inward. In the experiment of Fig. 13-2 the student must, during a time Δt, change the direction of the *angular momentum* $\mathbf{L}$ of the wheel, leaving its magnitude unchanged. To do so she must supply a *torque* that points at right angles to $\mathbf{L}$ (in the direction of $\Delta \mathbf{L}$), that is, vertically upward.*

13-3
ANGULAR MOMENTUM AND ANGULAR VELOCITY

In this section it is our purpose to examine the relationship between the angular momentum and the angular velocity for particles and rigid bodies rotating about an axis fixed in an inertial reference frame.

First we consider a single particle of mass m moving with speed v in a circle about the z-axis of an inertial reference frame as in Fig. 13-3. Its angular velocity $\boldsymbol{\omega}$ points upward and can be taken to lie along the z-axis. Its angular momentum $\mathbf{l}$ with respect to the origin O of the reference frame is given by Eq. 12-3, or

$$\mathbf{l} = \mathbf{r} \times \mathbf{p},$$

*This analogy is explored by A. E. Benfield, *American Journal of Physics*, September 1958. See also Problem 5.

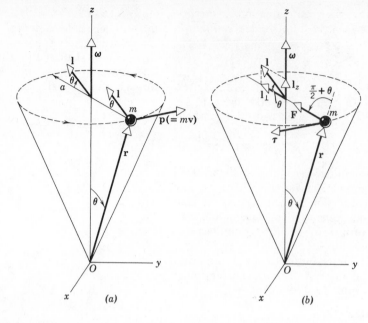

figure 13-3
(a) A particle of mass m rotating with speed v in a circle of radius a about the z-axis of an inertial reference frame. The angular momentum about O, $\mathbf{l}$ $(= \mathbf{r} \times \mathbf{p})$, is shown; for convenience, this vector is also shown translated to the center of the circle. *(b)* The same configuration, showing $\mathbf{l}$ and its components and also the centripetal force $\mathbf{F}$ and the torque τ about O.

where $\mathbf{r}$ and $\mathbf{p}$ $(= m\mathbf{v})$ are shown in the figure. The vector $\mathbf{l}$ is perpendicular to the plane formed by $\mathbf{r}$ and $\mathbf{p}$, which means that $\mathbf{l}$ is *not* parallel to $\boldsymbol{\omega}$. Note that $\mathbf{l}$ has a (vector) component $\mathbf{l}_z$ which is parallel to $\boldsymbol{\omega}$, but it has another (vector) component $\mathbf{l}_\perp$ which is perpendicular to $\boldsymbol{\omega}$. Note too, that if we choose our origin to lie in the plane of the circulating particle, then $\mathbf{l}$ *is* parallel to $\boldsymbol{\omega}$.

The perhaps unexpected result that $\mathbf{l}$ and $\boldsymbol{\omega}$ are not parallel in this simple case may cause some concern. However, this result is quite in accord with the general relationship $\boldsymbol{\tau} = d\mathbf{l}/dt$ for a torque acting on a single particle. The vector $\mathbf{l}$ is changing with time as the particle motion proceeds, the change being entirely in direction and not in magnitude, just as it was for the precessing top in the preceding section. Since the right side of the preceding relationship $(= d\mathbf{l}/dt)$ has a nonzero value, the left side $(= \boldsymbol{\tau})$ must also have a nonzero value; that is, a torque must act on the particle with respect to origin O.

There is indeed such a torque. For if the particle moves in a circle, a centripetal force $\mathbf{F}$ must act on it, as in Fig. 13-3b. We may imagine that $\mathbf{F}$ is provided by the tension in a light cord that ties the rotating particle to the z-axis. The torque about O is provided by $\mathbf{F}$ and is given by Eq. 12-1.

$$\boldsymbol{\tau} = \mathbf{r} \times \mathbf{F}.$$

The torque $\boldsymbol{\tau}$ is tangent to the circle (perpendicular to the plane formed by $\mathbf{r}$ and $\mathbf{F}$) and in the direction shown in Fig. 13-3b, as you may verify from the right-hand rule.

Show that the moving particle of Fig. 13-3 satisfies the relation $\boldsymbol{\tau} = d\mathbf{l}/dt$ quantitatively.

EXAMPLE 2

The proof is along the same lines as that of Section 13-2 for the spinning top because, from a vector point of view, the two problems are identical. In each case we have the precession of an angular momentum vector ($\mathbf{L}$ for the top and $\mathbf{l}$ for the particle of Fig. 13-3) about a vertical axis, at a rate which we called ω_p for the top and which we call ω for the particle. In each case we have a torque which is always at right angles to the plane formed by $\mathbf{L}$ (or $\mathbf{l}$) and $\boldsymbol{\omega}_p$ (or $\boldsymbol{\omega}$).

Thus, since the two problems are formally identical, it suffices to inquire whether the rotating particle of Fig. 13-3 obeys the vector equation for precession ($\tau = \boldsymbol{\omega}_p \times \mathbf{L}$; Eq. 13-3). This equation was derived for the precessing top directly from—and is directly equivalent to—the relation $\tau = d\mathbf{L}/dt$ (Eq. 12-9). We can write the relation $\tau = \boldsymbol{\omega}_p \times \mathbf{L}$ in terms of magnitudes as

$$\tau = \omega l \sin (90° - \theta) = \omega l \cos \theta, \qquad (13\text{-}3)$$

in which we have substituted ω for ω_p and l for L and have noted in Fig. 13-3a that the angle between $\boldsymbol{\omega}$ and $\mathbf{l}$ is $90° - \theta$. For τ and l, again using the notation of Fig. 13-3a, we can write

$$\tau = Fr \sin (90° + \theta) = [m\omega^2(r \sin \theta)](r)(\cos \theta)$$

and

$$l = rp \sin 90° = r(mv) = (r)(m)[\omega(r \sin \theta)],$$

in which $r \sin \theta$ is the radius a of the circle in which the particle moves, $(90° + \theta)$ is the angle between $\mathbf{r}$ and $\mathbf{F}$, and $90°$ is the angle between $\mathbf{r}$ and $\mathbf{p}$. Substituting these two expressions into Eq. 13-3 yields

$$m\omega^2 r^2 \sin \theta \cos \theta = \omega(m\omega r^2 \sin \theta) \cos \theta,$$

which is an identity. In terms of *magnitudes* we have proved our point. Refer to Fig. 13-3 and make certain that the *direction of* τ is that of $d\mathbf{l}/dt$ (Eq. 12-7), or alternatively of $\boldsymbol{\omega} \times \mathbf{l}$ (Eq. 13-2b).

Let us now investigate the relationship between l_z and $\boldsymbol{\omega}$ for the particle of Fig. 13-3. From Example 2 we have

$$l = mr^2\omega \sin \theta.$$

From Fig. 13-3b we see that

$$l_z = l \sin \theta = m\omega r^2 \sin^2 \theta.$$

Now $r \sin \theta = a$, the radius of the circle in which the particle moves. This leads to

$$l_z = ma^2\omega, \qquad (13\text{-}4)$$

in which ma^2 is the rotational inertia I of the particle with respect to the z-axis. Thus

$$l_z = I\omega, \qquad (13\text{-}5)$$

which is to be compared with Eq. 12-18 ($L = I\omega$) for the rotation of a rigid body about a fixed axis. Note that the vector relation $\mathbf{l} = I\boldsymbol{\omega}$ is *not* correct in this case because $\mathbf{l}$ and $\boldsymbol{\omega}$ do *not* point in the same direction. However, $\mathbf{l}_z$ and $\boldsymbol{\omega}$ do, so that we could write Eq. 13-5 in vector form as $\mathbf{l}_z = I\boldsymbol{\omega}$.

Now let us add another particle of mass m to the system of Fig. 13-3. In particular, let us add this particle in the same orbit, moving with the same speed, but always at a diametrically opposite point, on the other side of the axis of rotation. The angular momentum $\mathbf{l}_2$ with respect to O for this second particle will have the same magnitude as that of $\mathbf{l}_1$ for the first particle and it will make the same angle ($90° - \theta$) with the z-axis, but it will have a different orientation around that axis. As Fig. 13-4a shows, $\mathbf{l}_2$ will lie in a plane formed by $\boldsymbol{\omega}$ and by $\mathbf{l}_1$ but will be on the opposite side of the z-axis from $\mathbf{l}_1$. The vectors $\mathbf{l}_1$ and $\mathbf{l}_2$ include an angle of $180° - 2\theta$.

The total angular momentum $\mathbf{L}$ of the system of two particles is the vector sum of the angular momenta of the separate particles, that is, $\mathbf{L} = \mathbf{l}_1 + \mathbf{l}_2$. The resultant vector $\mathbf{L}$, as Fig. 13-4b shows, points along the z-axis (in the direction of $\boldsymbol{\omega}$) and is constant in magnitude. Note that

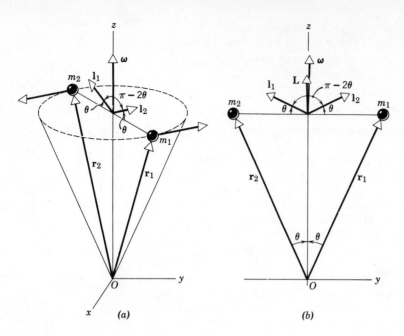

figure 13-4

(a) Two particles of mass m rotating as in Fig. 13-3 but maintaining diametrically opposite positions. (b) A cross section through the two particles, showing that the total angular momentum $\mathbf{L}$ ($= \mathbf{l}_1 + \mathbf{l}_2$) for the two-particle system points along the axis of rotation, in the same direction as $\boldsymbol{\omega}$.

this statement is true no matter where the origin O is located along the axis of rotation.

The fact that $\mathbf{L} = $ a constant (in both magnitude and direction), for this two-particle system, means that $d\mathbf{L}/dt = 0$, which in turn (Eq. 12-9) means that $\boldsymbol{\tau} = 0$ for this system. Convince yourself (Fig. 13-3b will be helpful) that this is the case, the torques for the two particles about O being equal in magnitude but oppositely directed so that the torque acting on the two-particle system is zero.

The fact that $\boldsymbol{\omega}$ and $\mathbf{L}$ point in the same direction in this problem but did not for the case of a single particle can be traced to the fact that, in the two-particle system, the particles have the same mass and are in diametrically opposite positions at the same distance from the rotation axis.

We can now extend our system to a rigid body, made up of many particles. If the body is symmetric about the axis of rotation, by which we mean that for every mass element in the body there must be an identical mass element diametrically opposite the first element and at the same distance from the axis of rotation, then the body can be regarded as made up of sets of particle pairs of the kind we have been discussing. Since $\mathbf{L}$ and $\boldsymbol{\omega}$ are parallel for all such pairs, they are also parallel for rigid bodies that possess this kind of symmetry. Note that, in Table 12-1, all the systems except f and j meet this criterion.

For such symmetrical rigid bodies $\mathbf{L}$ and $\boldsymbol{\omega}$ are parallel and we can write Eq. 12-18 ($L = I\omega$) in vector form as

$$\mathbf{L} = I\boldsymbol{\omega}. \tag{13-6}$$

Do not forget, however, that, if $\mathbf{L}$ stands for the *total* angular momentum, then Eq. 13-6 applies *only* to bodies that have symmetry* about

---

*We have oversimplified the symmetry requirement. Every rigid body, no matter how irregular its shape, has three perpendicular axes through its center of mass, about each of which $\mathbf{L}$ and $\boldsymbol{\omega}$ have the same direction, being related by $L = I\omega$. These axes are called the *principal axes*. The axis of a figure of revolution is always a principal axis, as are axes at right angles to it through the center of mass. In general, however, $\mathbf{L}$ and $\boldsymbol{\omega}$ point in different directions for axes that are not principal axes. See Arnold Sommerfeld, *Mechanics*, Chapter IV, Academic Press, New York, (1964 paperback edition).

the (fixed) rotational axis. If **L** stands for the vector component of angular momentum along the rotational axis (that is, for $\mathbf{L}_z$), then Eq. 13-6 is equivalent to Eq. 12-8 and holds for *any* rigid body, symmetrical or not, that is rotating about a fixed axis.

EXAMPLE 3

In Example 5, Chapter 12, find the acceleration of the falling mass by direct application of Eq. 12-9 ($\tau = dL/dt$).

The system of Fig. 12-12, consisting of the wheel M and the mass m is acted on by two external forces, the downward pull of gravity mg acting on mass m and the upward force exerted by the bearings of the shaft of the cylinder, which we take as our origin. The tension in the cord is an internal force and does not act from the outside on the system (wheel + weight). Only the first of these external forces exerts a torque about the origin and its magnitude is $(mg)R$.

The angular momentum of the system about the origin at any instant is

$$L = I\omega + (mv)R,$$

in which $I\omega$ is the angular momentum of the (symmetrical) disk and $(mv)R$ is the angular momentum (= linear momentum × moment arm) of the falling mass about the origin. Both these contributions to L point in the same direction, namely, perpendicularly out of the plane of Fig. 12-12.

Applying $\tau = dL/dt$ (in scalar form) yields

$$(mg)R = \frac{d}{dt}(I\omega + mvR)$$

$$= I(d\omega/dt) + mR(dv/dt)$$

$$= I\alpha + mRa.$$

Since $a = \alpha R$ and $I = \frac{1}{2}MR^2$, this reduces to

$$mgR = (\tfrac{1}{2}MR^2)(a/R) + mRa$$

or

$$a = \frac{2\,mg}{M + 2m}.$$

EXAMPLE 4

A simple example of an unsymmetrical rotating rigid body is a dumbbell-type rod whose bar makes an angle θ with the fixed axis of rotation passing through its center of mass. The rod rotates at a constant angular velocity ω about this axis, the vector $\boldsymbol{\omega}$ thus pointing along this axis, as shown in Fig. 13-5. Experience tells us that such a system is "unbalanced" or "lop-sided," and if it were not securely fastened to the vertical shaft near C, it would break away from the shaft at high angular velocities. It would tend to move until the angle θ becomes 90°, in which limiting position the system would then be symmetrical about the shaft.

(*a*) Show qualitatively that in the unsymmetrical case, shown in Fig. 13-5, **L** and $\boldsymbol{\omega}$ are not parallel.

Each particle of mass m has an angular momentum with respect to C given by $\mathbf{r} \times \mathbf{p}$ for that particle. At the instant shown the upper particle is moving into the page at right angles to it, and the lower particle is moving out of the page at right angles to it. The momentum vectors of the two masses are therefore equal but opposite, and so are their position vectors with respect to C. Hence, by application of the right-hand rule in $\mathbf{r} \times \mathbf{p}$, we find that **l** is the same for each particle and that their sum, the total angular momentum vector **L** of the dumbbell, is as shown in the figure, at right angles to the bar in the plane of the page. Hence **L** and $\boldsymbol{\omega}$ are *not* parallel at this instant. It is clear that as the dumbbell itself rotates, the angular momentum vector, while constant in magnitude, rotates around the fixed axis of rotation.

figure 13-5
Example 4.

(b) The fact that **L** and **ω** do not point in the same direction is perfectly consistent with the fundamental relation $\tau = d\mathbf{L}/dt$. We have seen twice before (see Section 13-2 and Example 1) that an angular momentum vector of constant magnitude that rotates around a fixed axis must have associated with it a torque τ that is at right angles to the plane formed by **L** and **ω**. At the instant shown in Fig. 13-5 this plane is the plane of the figure. Is there such a torque in this problem and if so, where does it come from?

There is indeed such a torque and it arises from the unbalanced sideways forces exerted by the bearings on the shaft and transmitted by the shaft to the dumbbell bar. At the instant shown in the figure the upper end of the dumbbell would tend to move outward to the right. The shaft would be pulled to the right against the upper bearing, which in turn exerts a force **F** on the shaft that points to the left. Similarly, the lower end of the dumbbell tends to move outwards to the left. The shaft would be pulled to the left against the lower bearing, which in turn exerts a force $-\mathbf{F}$ on the shaft that points to the right. The torque τ about C as a result of these forces points perpendicularly out of the page, at right angles to the plane formed by **L** and **ω**, and in the right direction to account for the rotary motion of **L** (you should check this).

The forces **F** and $-\mathbf{F}$ lie in the plane of Fig. 13-5 at the instant shown. As the dumbbell rotates, these forces, and therefore the torque τ, rotate with it, so that τ always remains at right angles to the plane formed by **ω** and **L** (compare with Fig. 13-1) The rotating forces **F** and $-\mathbf{F}$ cause a "wobble" in the upper and lower bearings. The bearings and their supports must be made strong enough to provide these forces. For a symmetrical rotating body there is no bearing wobble and the shaft rotates smoothly.

Bearing wobble and internal strains can cause serious practical problems when objects, such as turbine rotors, are made to rotate at high speeds. Although designed to be symmetrical, such rotors, because of small errors of blade placement, etc., may be slightly unsymmetrical. They may be restored to symmetry by the addition or removal of metal at appropriate places; this is done by spinning the wheel in a special device such that bearing wobble can be measured quantitatively and the appropriate corrective measure computed and indicated automatically. We are all familiar with lead weights placed at strategic points on automobile tire rims to reduce wobble at high speeds due to unbalance.

13-4
CONSERVATION OF ANGULAR MOMENTUM

In Chapter 12, we found that the time rate of change of the total angular momentum of a system of particles about a point fixed in an inertial reference frame (or about the center of mass) is equal to the sum of the *external* torques acting on the system, that is,

$$\tau_{\text{ext}} = d\mathbf{L}/dt. \qquad (12\text{-}9)$$

Suppose now that $\tau_{\text{ext}} = 0$; then $d\mathbf{L}/dt = 0$ so that **L** = a constant.

When the resultant external torque acting on a system is zero, the total vector angular momentum of the system remains constant. This is the *principle of the conservation of angular momentum.*

For a system of n particles, the total angular momentum **L** about some point is

$$\mathbf{L} = \mathbf{l}_1 + \mathbf{l}_2 + \cdots + \mathbf{l}_n.$$

When the resultant external torque on the system is zero, we have

$$\mathbf{L} = \text{a constant} = \mathbf{L}_0, \qquad (13\text{-}7)$$

where $\mathbf{L}_0$ is the constant total angular momentum vector. The angular momenta of the individual particles may change, but their vector sum $\mathbf{L}_0$ remains constant in the absence of a net external torque.

Angular momentum is a vector quantity so that Eq. 13-7 is equivalent to three scalar equations, one for each coordinate direction through the reference point. The conservation of angular momentum therefore supplies us with three conditions on the motion of a system to which it applies.

For a system consisting of a rigid body rotating about an axis (the z-axis, say) that is fixed in an inertial reference frame, we have

$$\mathbf{L}_z = I\omega, \tag{13-6}$$

where $\mathbf{L}_z$ is the component of the angular momentum along the rotation axis and I is the rotational inertia for this same axis. It is possible for the rotational inertia I of a rotating body to change by rearrangement of its parts. If no net external torque acts, then $\mathbf{L}_z$ must remain constant and, if I does change, there must be a compensating change in ω. The principle of conservation of angular momentum in this case is expressed as

$$I\omega = I_0\omega_0 = \text{a constant.} \tag{13-8}$$

Equation 13-8 holds not only for rotation about a fixed axis but also about an axis through the center of mass of the system that moves so that it always remains parallel to itself (see footnote, p. 237).

Acrobats, divers, ballet dancers, ice skaters, and others often use this principle. Because I depends on the square of the distance of the parts of the body from the axis of rotation, a large variation is possible by extending or pulling in the limbs. Consider the diver* in Fig. 13-6. Let us assume that as he leaves the diving board he has a certain angular speed ω_0 about a horizontal axis through the center of mass, such that he would rotate through half a turn before he strikes water. If he wishes to make a one and one-half turn somersault instead, in the same time, he must triple his angular speed. Now there are no external forces acting

figure 13-6
A diver leaves the diving board with arms and legs outstretched and with some initial angular velocity. Since no torques are exerted on him about his center of mass, L $(= I\omega)$ is constant while he is in the air. When he pulls his arms and legs in, since L decreases, ω increases. When he again extends his limbs, his angular velocity drops back to its initial value. Notice the parabolic motion of his center of mass, common to all two-dimensional motion under the influence of gravity.

* See "The Mechanics of Swimming and Diving" by R. L. Page in *The Physics Teacher*, February 1976, for an interesting biomechanical analysis.

on him except gravity, and gravity exerts no torque about his center of mass. His angular momentum therefore remains constant, and $I_0\omega_0 = I\omega$. Since $\omega = 3\omega_0$, the diver must change his rotational inertia about the horizontal axis through the center of mass from the initial value I_0 to a value I, such that I equals $\frac{1}{3}I_0$. This he does by pulling in his arms and legs toward the center of his body. The greater his initial angular speed and the more he can reduce his rotational inertia, the greater the number of revolutions he can make in a given time.

We should notice that the rotational kinetic energy of the diver is not constant. In fact, in our example, since

$$I\omega = I_0\omega_0$$

and

$$I < I_0,$$

it follows that

$$\tfrac{1}{2}I\omega^2 > \tfrac{1}{2}I_0\omega_0^2,$$

and the diver's rotational kinetic energy *increases*. This increase in energy is supplied by the diver, who does work when he pulls the parts of his body together.

In a similar way the ice skater and ballet dancer can increase or decrease the angular speed of a spin about a vertical axis. A cat manages to land on its feet after a fall by using the same principles, the tail serving as a useful, but unessential, extra appendage.

EXAMPLE 5

A small object of mass m is attached to a light string which passes through a hollow tube. The tube is held by one hand and the string by the other. The object is set into rotation in a circle of radius r_1 with a speed v_1. The string is then pulled down, shortening the radius of the path to r_2 (Fig. 13-7). Find the new linear speed v_2 and the new angular speed ω_2 of the object in terms of the initial values v_1 and ω_1 and the two radii.

The downward pull on the string is transmitted as a radial force on the object. Such a force exerts a zero torque on the object about the center of rotation. Since no torque acts on the object about its axis of rotation, its angular momentum in that direction is constant. Hence

initial angular momentum = final angular momentum,

$$mv_1r_1 = mv_2r_2,$$

and

$$v_2 = v_1\left(\frac{r_1}{r_2}\right).$$

Since $r_1 > r_2$, the object speeds up on being pulled in.

In terms of angular speed, since v_1 equals $\omega_1 r_1$ and v_2 equals $\omega_2 r_2$,

$$mr_1^2\omega_1 = mr_2^2\omega_2$$

and

$$\omega_2 = \left(\frac{r_1}{r_2}\right)^2 \omega_1,$$

so that there is an even greater increase in angular speed over the initial value (see Problem 31). What effect does the force of gravity (the object's weight) have on this analysis?

figure 13-7
Example 5. A mass at the end of a cord moves in a circle of radius r_1 with angular speed ω_1. The cord passes down through a tube. **F** supplies the centripetal force.

EXAMPLE 6

A student sits on a stool that is free to rotate about a vertical axis. He holds his arms extended horizontally with an 8.0-lb (= 35.6 N) weight in each hand. The instructor sets him rotating with an angular speed of 0.50 rev/s. Assume that friction is negligible and exerts no torque about the vertical axis of rotation.

Assume that the rotational inertia of the student remains constant at 4.0 slug $\cdot$ ft² (= 5.4 kg $\cdot$ m²) as he pulls his hands to his sides and that the change in rotational inertia is due only to pulling the weights in. Take the original distance of the weights from the axis of rotation to be 3.0 ft (= 0.91 m) and their final distance 0.50 ft (= 0.15 m). Find the final angular speed of the student.

The only external force is gravity acting through the center of mass, and that exerts no torque about the axis of rotation. Hence the angular momentum is conserved about this axis and

$$\text{initial angular momentum} = \text{final angular momentum},$$

$$I_0\omega_0 = I\omega.$$

We have

$$I = I_{\text{student}} + I_{\text{weights}},$$

$$I_0 = 4.0 + 2\left(\frac{8.0}{32}\right)(3.0)^2 = 8.5 \text{ slug} \cdot \text{ft}^2 \ (= 11.6 \text{ kg} \cdot \text{m}^2),$$

$$I = 4.0 + 2\left(\frac{8.0}{32}\right)\left(\frac{1}{2}\right)^2 = 4.1 \text{ slug} \cdot \text{ft}^2 \ (= 5.6 \text{ kg} \cdot \text{m}^2),$$

$$\omega_0 = 0.50 \text{ rev/s} = \pi \text{ rad/s}.$$

Therefore

$$\omega = \frac{I_0}{I}\omega_0 = \frac{8.5}{4.1}\pi \text{ rad/s} = 2.1\pi \text{ rad/s} \cong 1.0 \text{ rev/s}.$$

The final angular speed is approximately doubled.

If we had allowed for the decrease in I caused by the arms being pulled in, the final angular speed would have been much greater.

What change would friction make? Is kinetic energy conserved as the student pulls in his arms and then puts them out again, assuming there is no friction? Explain.

EXAMPLE 7

A classroom demonstration that illustrates the vector nature of the law of conservation of angular momentum is worth considering.

A student stands on a platform that can rotate only about a vertical axis. In his hand he holds the axle of a rim-loaded bicycle wheel with its axis vertical. The wheel is spinning about this vertical axis with an angular speed ω_0, but the student and platform are at rest. The student tries to change the direction of rotation of the wheel. What happens?

Let us choose as the system the student plus platform plus wheel. The initial total angular momentum of this system is $I_0\boldsymbol{\omega}_0$, arising from the spinning wheel, I_0 being the rotational inertia of the wheel about its axis and $\boldsymbol{\omega}_0$ pointing vertically upward. Figure 13-8a shows the initial condition.

The student next turns the axis of the wheel through an angle θ from the vertical (to do this he must supply a torque; see Example 1. This torque, however, is *internal* to the system as we have defined it). Since there is no *external* component of torque on the system about the vertical axis, the vertical component of angular momentum of the system must be conserved. The wheel, however, is now spinning about an axis making an angle θ with the vertical so that it contributes a vertical component of angular momentum of only $I_0\omega_0 \cos\theta$ to the system. Hence the student and platform must supply the additional angular momentum about the vertical axes, and they begin to rotate about a vertical axis. This extra vertical angular momentum $I_p\omega_p$ when added to $I_0\omega_0 \cos\theta$ must equal the initial vertical angular momentum of the system $I_0\omega_0$. That is,

$$I_p\omega_p = I_0\omega_0(1 - \cos\theta).$$

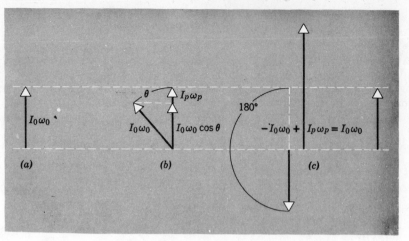

$I_0\omega_0$

θ $I_p\omega_p$

$I_0\omega_0$ $I_0\omega_0\cos\theta$

$180°$

$-I_0\omega_0 + I_p\omega_p = I_0\omega_0$

(a)   (b)   (c)

figure 13-8
Example 7. (a) The initial angular momentum of the system is shown. In (b), the wheel has been tilted an angle θ. Since no external torque in the vertical direction has been exerted on the system, the angular momentum in that direction must be conserved. The deficit, $(1 - \cos\theta)I_0\omega_0$, is made up by rotation of the student and platform. In (c), the wheel has been tilted 180°. The deficit is now $2I_0\omega_0$, which is now, as before, made up by the student and platform.

This is shown in Fig. 13-8b. I_p is the rotational inertia of student and platform with respect to the vertical axis, and ω_p is their angular speed about this axis.

If the student turns the wheel through an angle $\theta = 180°$, the student and platform acquire a vertical angular momentum of $2I_0\omega_0$. The total vertical angular momentum of the system is still being conserved at the initial value $I_0\omega_0$, as shown in Fig. 13-8c.

Consider now the angular momentum of the wheel alone. As the student turns the axis of the wheel through an angle θ he exerts a torque on it which lasts for the time Δt that it takes to reorient the shaft. The vertical component of the reaction to this "torque-impulse" acts on the student and accounts for the vertical angular momentum acquired by him and the platform.

The wheel, held with its shaft fixed at an angle θ with the vertical axis, precesses about this axis just like the top of Fig. 13-1. As for the top, a horizontal torque, which always remains at right angles to the plane defined by the vertical axis and the axis of the wheel, must be provided, in this case by the student.

The precise analysis of the motion of this system depends only on the application of the equation $\boldsymbol{\tau} = d\mathbf{L}/dt$ and the vector nature of the quantities involved. You might want to work this out in more detail, as an exercise.

13-5
SOME OTHER ASPECTS OF THE CONSERVATION OF ANGULAR MOMENTUM

The conservation of angular momentum principle holds in atomic and nuclear physics as well as in celestial and macroscopic regions. Because Newtonian mechanics does not hold in the atomic and nuclear domain, this conservation law must be more fundamental than Newtonian principles. In our derivation of this principle we must have made more rigid assumptions than we needed to. This is true even in the framework of classical mechanics. Note the key role played by Newton's third law in our deduction of this conservation principle. This law was used to justify the assumption that the sum of the internal torques was zero. It was necessary to assert not only that the action and reaction forces were equal and opposite (the "weak" form of the third law) but also that these forces were directed along the line joining the two particles (the "strong" form of the third law). The strong form is known to be violated in some electromagnetic interactions. However, the assumption that the sum of the internal torques in a system of particles is zero can be proven on the basis of a much less stringent requirement than that the third law should hold.*

The law of conservation of angular momentum, as we have formulated it, holds for a system of bodies whenever the bodies can be treated as particles, that is, whenever effects due to the rotation of the individual bodies can be neglected. When the individual bodies have rotation, the conservation of angular momentum principle is still valid, providing we include the angular momentum asso-

* See E. Gerjuoy, *American Journal of Physics*, Vol. 17, 477 (1949).

ciated with this rotation. However, the bodies then are no longer simple particles whose motion can be described by particle dynamics.

In atomic and nuclear physics we find that the "elementary particles" such as electrons, protons, mesons, neutrons, etc. (see Appendix I) have angular momentum associated with an intrinsic spinning motion, as well as with orbital motion about some external point. When we use the law of conservation of total angular momentum we must include this *spin* angular momentum in the total. A fundamental aspect of atomic, molecular, and nuclear systems is that their angular momenta can take on only definite discrete values, rather than a continuum of values. Angular momentum is said to be *quantized*. Hence, angular momentum plays a central role in the description of the behavior of such systems (see Problems 9 and 10). These ideas will be developed in later chapters.

If we were to regard the sun, planets, and satellites as particles having no intrinsic spinning motion, the angular momentum of the solar system would turn out not to be constant. But these bodies do have intrinsic rotations; in fact, tidal forces convert some of the intrinsic spinning angular momentum

Table 13-1
Summary of equations for rotary motion

Eq. No.	Equation	Remarks
	I. Defining Equations	
12-1	$\boldsymbol{\tau} = \mathbf{r} \times \mathbf{F}$	Torque on a particle about a point O, due to a resultant force $\mathbf{F}$
	$\boldsymbol{\tau}_{\text{ext}} = \Sigma\boldsymbol{\tau}_i = \Sigma(\mathbf{r}_i \times \mathbf{F}_i)$	Resultant external torque on a system of particles about a point O
12-3	$\mathbf{l} = \mathbf{r} \times \mathbf{p}$	Angular momentum of a particle about a point O
	$\mathbf{L} = \Sigma\mathbf{l}_i = \Sigma(\mathbf{r}_i \times \mathbf{p}_i)$	Resultant angular momentum of a system of particles about a point O
	II. General Relations	
12-7	$\boldsymbol{\tau} = d\mathbf{l}/dt$	The law of motion for a single particle acted on by a torque. It is the rotational analog of $\mathbf{F} = d\mathbf{p}/dt$ (Eq. 9-12). Equation 12-7 holds only if $\boldsymbol{\tau}$ and $\mathbf{l}$ are measured with respect to any point O fixed in an inertial reference frame
12-9	$\boldsymbol{\tau}_{\text{ext}} = d\mathbf{L}/dt$	The law of motion for a system of particles acted on by a resultant external torque $\boldsymbol{\tau}_{\text{ext}}$. It is the rotational analog of $\mathbf{F} = d\mathbf{P}/dt$ (Eq. 9-7). Equation 12-9 holds only if $\boldsymbol{\tau}_{\text{ext}}$ and $\mathbf{L}$ are measured with respect to (*a*) any point O fixed in an inertial reference frame *or* (*b*) the center of mass of the system
	III. Special Case of a Rigid Body Rotating about an Axis Fixed in an Inertial Reference Frame (see footnote, p. 237)	
12-17	$\tau = I\alpha$	$\boldsymbol{\alpha}$ is constrained to lie along the axis; I must also refer to this axis and τ must be the scalar component of $\boldsymbol{\tau}_{\text{ext}}$ directed along this same axis. It is the rotational analog of $F = Ma$ for rectilinear motion
12-18	$L = I\omega$	$\boldsymbol{\omega}$ is constrained to lie along the axis; I must also refer to this axis and L must be the scalar component along this axis of the total angular momentum. If the rotation axis has special symmetry (that is, if it is a principal axis; see footnote, p. 266), then $\mathbf{L}$ and $\boldsymbol{\omega}$ are both axially directed. It is the rotational analog of $P = Mv$ for rectilinear motion

into orbital angular momentum of the planets and satellites. When we use the law of conservation of the angular momentum, we must include this spin angular momentum in the total. The conservation of angular momentum plays a key role in the evaluation of theories of the origin of the solar system, the contraction of giant stars, and other problems in astronomy. We will consider some astronomical applications in Chapter 16.

The basis for this rather simple way of analyzing the total angular momentum of atomic or astronomical systems is a theorem (see Problem 15) that the *total* angular momentum **L** of any system with respect to the origin of an inertial reference frame may be computed by adding the angular momentum with respect to its center of mass (*spin* angular momentum) to the angular momentum arising from the motion of the center of mass with respect to the origin (*orbital* angular momentum).

The conservation laws of total energy, of linear momentum, and of angular momentum are fundamental to physics, being valid in all modern physical theories. We shall have occasion to use them many times in later chapters.

The subject of the rotary motions of particles and rigid bodies is reasonably complicated, so much so in fact that a completely general treatment is beyond our scope here. It seems advisable to collect in one place all equations dealing with rotational dynamics and to comment on the conditions under which they can be used. We do this in Table 13-1 (see previous page.)

13-6
ROTATIONAL DYNAMICS —A REVIEW

questions

1. We have encountered many vector quantities so far, including position, displacement, velocity, acceleration, force, momentum, and angular momentum. Which of these are defined independent of the choice of the origin in the reference frame?

2. (a) In Example 1, why would merely turning the shaft up send the wheel to the student's right? (b) If the student is anchored to the floor of a large spaceship that is drifting in a region free from gravity, in what way, if any, would this affect her performance of the experiment?

3. If the top of Fig. 13-1 were not spinning, it would tip over. If its spin angular momentum is large compared to the change caused by the applied torque, the top precesses. What happens in between when the top spins slowly?

4. A Tippy-Top, having a section of a spherical surface of large radius on one end and a stem for spinning it on the opposite end, will rest on its spherical surface with no spin but flips over when spun, so as to stand on its stem. Explain. (See "The Tippy-Top" by George D. Freier, *The Physics Teacher*, January 1967.) If you can't find a Tippy-Top, use a hard-boiled egg; the "stand-on-end" behavior of the spinning egg is most easily followed if you put an ink mark on the "pointed" end of the egg.

5. A famous physicist (R. W. Wood), who was fond of practical jokes, mounted a rapidly spinning flywheel in a suitcase which he gave to a porter with instructions to follow him. What happens when the porter is led quickly around a corner? Explain in terms of $\tau = d\mathbf{L}/dt$.

6. A single-engine airplane must be "trimmed" to fly level. (Trimming consists of raising one aileron and lowering the opposite one.) Why is this necessary? Is this necessary on a twin-engine plane under normal circumstances?

7. The propeller of an aircraft rotates clockwise as seen from the rear. When the pilot pulls upward out of a steep dive, he finds it necessary to apply left rudder at the bottom of the dive if he is to maintain his heading. Explain.

8. Why does a long bar help a tightrope walker to keep her balance?

9. You are walking along a narrow rail and you start to lose your balance. If you start falling to the right, which way do you turn your body to regain balance? Explain.

10. Describe, in terms of $\tau = d\mathbf{L}/dt$, the rotational dynamics of the wheels on a fast train going around a curve.

11. Can you suggest a simple theory to explain the stability of a moving bicycle? (See "The Stability of the Bicycle" by David E. H. Jones, *Physics Today*, April 1970.)

12. Explain, in terms of angular momentum and rotational inertia, exactly how one "pumps up" a swing. (See "Pumping on a Swing" by P. L. Tea and H. Falk, *American Journal of Physics*, December 1968; "The Child's Swing" by B. F. Gore, *American Journal of Physics*, March 1970; "On Initiating the Motion in a Swing" by J. T. McMullan, *American Journal of Physics*, May 1972 and "How Children Swing" by S. M. Curry, *American Journal of Physics*, October 1976.)

13. In order to get a billiard ball to roll without sliding from the start the cue must hit the ball not at the center (that is, a height above the table equal to the ball's radius R) but exactly at a height $\frac{2}{5} R$ *above* the center. Explain. (See Arnold Sommerfeld, *Mechanics, Volume I of Lectures on Theoretical Physics*, Academic Press, New York (1964 paperback edition), pp. 158-161, for a supplement on the mechanics of billiards. See also "Some Pitfalls in Demonstrating Conservation of Momentum" by H. L. Armstrong, *American Journal of Physics*, January 1968.)

14. There are points on a bat where, if the ball is hit there, your hands will sting and the bat might break. Explain. (See "Batting the Ball" by P. Kirkpatrick, *American Journal of Physics*, August 1963.)

15. Assume that a uniform rod rests in a vertical position on a surface of negligible friction. The rod is then given a horizontal blow at its lower end. Describe the motion of the center of mass of the rod and of its upper endpoint.

16. A cylinder rotates with angular speed ω about an axis through one end, as in Fig. 13-9. Choose an appropriate origin and show qualitatively the vectors $\mathbf{L}$ and $\boldsymbol{\omega}$. Are these vectors parallel? Do symmetry considerations enter here?

Bearing — Fixed axis

figure 13-9
Question 16

17. Consider the motion of a football tumbling through the air. Is the angular momentum with respect to the center of mass of the football conserved in flight? Does the magnitude or the direction of the angular velocity change with respect either to axes fixed in space or in the body?

18. In Chapter 1 the melting of the polar icecaps was cited as a possible cause of the variation in the earth's time of rotation. Explain.

19. Many great rivers flow toward the equator. What effect does the sediment they carry to the sea have on the rotation of the earth?

20. A man turns on a rotating table with an angular speed ω. He is holding two equal masses at arm's length. Without moving his arms, he drops the two masses. What change, if any, is there in his angular speed? Is the angular momentum conserved? Explain.

21. In Example 5, if the string is released suddenly back to where the object can travel in a circle of radius r_1, will the object return to its original speed? What happens if one repeatedly pulls down on and suddenly releases the string? Explain the behavior in terms of work-energy and torque-angular momentum considerations.

22. A circular turntable rotates at constant angular velocity about a vertical axis. There is no friction and no driving torque. A circular pan rests on the turntable and rotates with it: see Fig. 13-10. The bottom of the pan is covered with a layer of ice of uniform thickness, which is, of course, also rotating with the pan. The ice melts but none of the water escapes from the pan. Is the angular velocity now greater than, the same as, or less than the original velocity? Give reasons for your answer.

Pan
Ice
Turntable
ω

figure 13-10
Question 22

SECTION 13-1

1. The time integral of a torque is called the *angular impulse.* Starting from $\tau = dL/dt$, show that the resultant angular impulse equals the change in angular momentum. This is the rotational analog of the linear impulse-momentum theorem.

SECTION 13-2

2. A top is spinning at 30 Hz (cycles/s) about an axis making an angle of 30° with the vertical. Its mass is 0.50 kg (3.4×10^{-2} slug) and its rotational inertia is 5.0×10^{-4} kg · m² (3.7×10^{-4} slug · ft²). The center of mass is 4.0 cm (1.6 in.) from the pivot point. If the spin is clockwise as seen from above, what is the magnitude and direction of the angular velocity of precession?

3. A gyroscope consists of a rotating disc of 0.05-m radius suitably mounted at the center of a 0.12-m axle so that it can spin and precess freely. Find the rate of precession (in rev/min) if the axle is supported at one end and is horizontal, and the gyroscope's spin rate is 1000 rev/min. *Answer:* 43 rev/min.

4. The gyroscope of Problem 3 is modified by attaching a small weight to the distant end of the axle. Find the new rate of precession (in rev/min) as a function of the ratio r = (mass of added weight)/(mass of gyroscope disc).

SECTION 13-3

5. Start from Eq. 11-20*b*, $\mathbf{a}_R = \boldsymbol{\omega} \times \mathbf{v}$, for a particle in circular motion and show that the force required for uniform circular motion is $\mathbf{F} = \boldsymbol{\omega} \times \mathbf{p}$. Compare this to Eq. 13-2*b*, $\boldsymbol{\tau} = \boldsymbol{\omega}_p \times \mathbf{L}$, and explain how the precessing spinning top can be regarded as a rotational analog to uniform circular motion.

6. Two wheels, *A* and *B*, are arranged by a belt system as in Fig. 13-11. The radius of *B* is three times the radius of *A*. What would be the ratio of the rotational inertias I_A/I_B if (*a*) both wheels have the same angular momenta? (*b*) both wheels have the same rotational kinetic energy?

7. Show that $\mathbf{L} = I\boldsymbol{\omega}$ for the two-particle system of Fig. 13-4.

8. Figure 13-12 shows a symmetrical rigid body rotating about a fixed axis. The origin of coordinates is fixed for convenience at the center of mass. Prove, by summing over the contributions made to the angular momentum by all of the mass elements m_i into which the body is divided, that $\mathbf{L} = I\boldsymbol{\omega}$, where $\mathbf{L}$ is the *total* angular momentum.

9. (*a*) Assume that the electron moves in a circular orbit about the proton in a hydrogen atom. If the centripetal force on the electron is supplied by an electrical force $e^2/4\pi\epsilon_0 r^2$, where e is the magnitude of the charge of an electron and of a proton, r is the orbit radius, and ϵ_0 is a constant, show that the radius of the orbit is

$$r = \frac{e^2}{4\pi\epsilon_0 m v^2},$$

where m is the mass of the electron and v is its speed.

(*b*) Assume now that the angular momentum of the electron about the nucleus can only have values that are integral multiples n of $h/2\pi$, where h is a constant called *Planck's constant.* Show that the only electronic orbits possible are those with a radius

$$r = \frac{nh}{2\pi m v}.$$

(*c*) Combine these results to eliminate v and show that the only orbits consistent with both requirements have radii

$$r = \frac{n^2\epsilon_0 h^2}{\pi m e^2}.$$

Hence the allowed radii are proportional to the square of the integers $n = 1, 2, 3$, etc. When $n = 1$, r is smallest and has the value 0.528×10^{-10} meter.

problems

figure 13-11
Problem 6

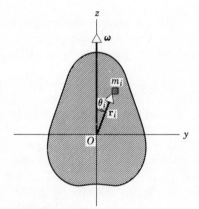

figure 13-12
Problem 8

10. In 1913, Niels Bohr postulated that any mechanical rotating system with rotational inertia I can have an angular momentum whose values can take on only integral multiples of a particular number $h/2\pi = 1.054 \times 10^{-34}$ J · s. In other words,

$$L = I\omega = n(h/2\pi),$$

where n is any positive integer or zero. We say that L is quantized because it is no longer allowed to have any value whatsoever. (a) Show that this postulate restricts the kinetic energy the rotating system can have to a set of discrete values, that is, that the energy is quantized. (b) Consider the so-called *rigid rotator*, consisting of a mass m constrained to rotate in a circle of radius R. With what angular speeds could the mass rotate if the postulate were correct? What values of kinetic energy may it assume? (c) Draw an energy-level diagram of some sort indicating how the spacing between the energy levels varies as n increases. It might look something like Fig. 13-13. Certain low-energy diatomic molecules behave like a rigid rotator.

figure 13-13
Problem 10

11. Using data in the appendix, find (a) the angular momentum of the earth's spin about its own axis, (b) the angular momentum of the earth's orbital motion about the sun.
   Answer: (a) 7.1×10^{33} kg · m²/s. (b) 2.7×10^{40} kg · m²/s.

12. A stick has a mass 0.30 slug (4.4 kg) and a length 4.0 ft (1.2 m). It is initially at rest on a frictionless horizontal plane and is struck perpendicularly by a horizontal impulsive force of impulse 3.0 lb · s (13 N · s) at a distance $l = 1.5$ feet (0.46 m) from the center. Determine the subsequent motion.

13. The moon revolves about the earth in such a way that we always see the same face of the moon. (a) How are the spin and orbital parts of the angular momentum of the moon related? (b) By how much would its spin angular momentum have to change if we were to be able to see all the moon's surface during the course of a month?
   Answer: (a) $L_{spin}/L_{orbital} = \frac{2}{5}(R_m/R_{e-m})^2$ in which R_m is the lunar radius and R_{e-m} is the earth-moon distance. (b) Increase or decrease by $\frac{1}{2}$ of present value.

14. A cylinder rolls down an inclined plane of angle θ. Show, by direct application of Eq. 12-9 ($\tau = dL/dt$), that the acceleration of its center of mass is $\frac{2}{3}g \sin \theta$. Compare this method with that used in Example 10 of Chapter 12.

15. *Relation between the Total Angular Momentum of a System of Particles and the Orbital and Spin Angular Momenta.* The total angular momentum of a system of particles relative to the origin O of an inertial reference frame is given by $\mathbf{L} = \Sigma \mathbf{r}_i \times \mathbf{p}_i$, where $\mathbf{r}_i$ and $\mathbf{p}_i$ are measured with respect to O.

   (a) Use the relations $\mathbf{r}_i = \mathbf{r}_{cm} + \mathbf{r}_i'$ and $\mathbf{p}_i = m_i\mathbf{v}_{cm} + \mathbf{p}_i'$ of Problem 10 of Chapter 12 to express $\mathbf{L}$ in terms of the positions $\mathbf{r}_i'$ and momenta $\mathbf{p}_i'$ relative to the center of mass C. (b) Use the definition of center of mass and the definition of angular momentum $\mathbf{L}'$ with respect to the center of mass (problem 10 of Chapter 12) to obtain $\mathbf{L} = \mathbf{L}' + \mathbf{r}_{cm} \times M\mathbf{v}_{cm}$. (c) Show how this result can be interpreted as regarding the total angular momentum to be the sum of spin angular momentum (angular momentum relative to the center of mass) and orbital angular momentum (angular momentum of the motion of the center of mass C with respect to O if all the system's mass were concentrated at C).

16. A thin rectangular sheet, of length a and width b, rotates about one of its diagonals with constant angular speed ω, the axis being fixed in an inertial reference frame. Find the direction and the magnitude of the angular momentum $\mathbf{L}$ with respect to an origin at the center of mass.

17. The axis of the cylinder in Fig. 13-14 is fixed. The cylinder is initially at rest. The block of mass M is initially moving to the right without friction with speed v_1. It passes over the cylinder to the dotted position. When it first makes contact with the cylinder, it slips on the cylinder, but the friction is large enough so that slipping ceases before M loses contact with the

figure 13-14
Problem 17

cylinder. The cylinder has a radius R and a rotational inertia I. Find the final speed v_2 in terms of v_1, M, I, and R. This can be done most easily by using the relation between impulse and change in momentum.
Answer: $v_1/(1 + I/MR^2)$.

18. A stick, length l, lies on a frictionless horizontal table. It has a mass M and is free to move in any way on the table. A hockey puck m, moving as shown in Fig. 13-15, with speed v collides elastically with the stick. (a) What quantities are conserved in the collision? (b) What must be the mass m of the puck so that it remains at rest immediately after the collision?

figure 13-15
Problem 18

19. At what point below the suspension at one end of a uniform rod of length $2L$, hanging vertically, should you strike it to start its oscillatory motion without imparting an initial horizontal reaction force at the point of suspension?   *Answer:* $4L/3$.

20. Two cylinders having radii R_1 and R_2 and rotational inertias I_1 and I_2, respectively, are supported by fixed axes perpendicular to the plane of Fig. 13-16. The large cylinder is initially rotating with angular velocity ω_0. The small cylinder is moved to the right until it touches the large cylinder and is caused to rotate by the frictional force between the two. Eventually, slipping ceases, and the two cylinders rotate at constant rates in opposite directions. (a) Find the final angular velocity ω_2 of the small cylinder in terms of I_1, I_2, R_1, R_2, and ω_0. (b) Is total angular momentum conserved in this case?

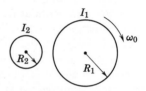

figure 13-16
Problem 20

21. A billiard ball, initially at rest, is given a sharp impulse by a cue. The cue is held horizontally a distance h above the centerline as in Fig. 13-17. The ball leaves the cue with a speed v_0 and, because of its "forward english," eventually acquires a final speed of $\frac{9}{7}v_0$. Show that

$$h = \tfrac{4}{5}R,$$

where R is the radius of the ball.

22. In Problem 21, imagine **F** to be applied below the centerline. (a) Show that it is impossible, with this "reverse english," to reduce the forward speed to zero, without rolling having set in, unless $h = R$. (b) Show that it is impossible to give the ball a backward velocity unless **F** has a downward, vertical component.

figure 13-17
Problem 21

SECTION 13-4
23. A man stands on a frictionless rotating platform which is rotating with an angular speed of 1.0 rev/s (Hz); his arms are outstretched and he holds a weight in each hand. With his hands in this position the total rotational inertia of the man, the weights, and the platform is 6.0 kg · m². If by drawing in the weights the man decreases the rotational inertia to 2.0 kg · m², (a) what is the resulting angular speed of the platform? (b) By how much is the kinetic energy increased?   *Answer:* (a) 3.0 Hz. (b) By a factor of 3.

24. Two skaters, each of mass 50 kg, approach each other along parallel paths separated by 3.0 m. They have equal and opposite velocities of 10 m/s. The first skater carries a long light pole, 3.0 m long, and the second skater grabs the end of it as he passes. (Assume frictionless ice.) (a) Describe quantitatively the motion of the skaters after they are connected by the pole. (b) By pulling on the pole, the skaters reduce their distance apart to 1.0 m. What is their motion then? (c) Compare the kinetic energy of the system in parts (a) and (b). Where does the change come from?

25. A wheel is rotating with an angular speed of 800 rev/min on a shaft whose rotational inertia is negligible. A second wheel, initially at rest and with twice the rotational inertia of the first, is suddenly coupled to the same shaft. (a) What is the angular speed of the resultant combination of the shaft and two wheels? (b) Account for any changes in rotational kinetic energy experienced by this system.
    Answer: (a) 267 rev/min. (b) The system loses two-thirds of its kinetic energy.

26. With center and spokes of negligible mass, a certain bicycle wheel has a thin rim of radius 1.14 ft and weight 8.36 lb; it can turn on its axle with negligible friction. A man holds the wheel above his head with the axis vertical while he stands on a turntable free to rotate without friction; the wheel rotates clockwise, as seen from above, with an angular speed 57.7 rad/s, and the turntable is initially at rest. The rotational inertia of wheel-plus-man-plus-turntable about the common axis of rotation is 1.54 slug · ft². (a) The man's hand suddenly stops the rotation of the wheel (relative to the turntable). Determine the resulting angular velocity (magnitude and direction) of the system. (b) The experiment is repeated with noticeable friction introduced into the axle of the wheel, which, starting from the same initial angular speed (57.7 rad/s) gradually comes to rest (relative to the turntable) while the man holds the wheel as described above. (The turntable is still free to rotate without friction.) Describe what happens to the system, giving as much quantitative information as the data permit.

27. The rotor of an electric motor has a rotational inertia $I_m = 2 \times 10^{-3}$ kg · m² about its central axis. The motor is mounted parallel to the axis of a space probe having a rotational inertia $I_p = 12$ kg · m² about its axis. Calculate the number of revolutions required to turn the probe through 30° about its axis.
    Answer: 500 rev.

28. In a lecture demonstration, a toy train track is mounted on a large wheel that is free to turn with negligible friction about a vertical axis. A toy train of mass m is placed on the track and, with the system initially at rest, the electrical power is turned on. The train reaches a steady speed v with respect to the track. What is the angular velocity ω of the wheel, if its mass is M and its radius R? (Neglect the mass of the spokes of the wheel.)

29. A girl (mass M) stands on the edge of a frictionless merry-go-round (mass $10 M$, radius R, rotational inertia I) that is not moving. She throws a rock (mass m) in a horizontal direction that is tangent to the outer edge of the merry-go-round. The speed of the rock, relative to the ground, is v. What are (a) the angular speed of the merry-go-round and (b) the linear speed of the girl after the rock is thrown?
    Answer: (a) $mvR/(I + MR^2)$. (b) $vmR^2/(I + mR^2)$.

30. In a playground there is a small merry-go-round of radius 4.0 ft (1.2 m) and mass 12.0 slugs (180 kg). The radius of gyration (see Problem 14 of Chapter 12) is 3.0 ft (0.91 m). A child of mass 3.0 slugs (44 kg) runs at a speed of 10 ft/s (3.0 m/s) tangent to the rim of the merry-go-round when it is at rest and then jumps on. Neglect friction between the bearings and the shaft of the merry-go-round and find the angular velocity of the merry-go-round and child.

31. A uniform flat disk of mass M and radius R rotates about a horizontal axis through its center with angular speed ω_0. (a) What is its kinetic energy? Its angular momentum? (b) A chip of mass m breaks off the edge of the disk at an instant such that the chip rises vertically above the point at which it broke off (Fig. 13-18). How high above the point does it rise before starting to fall? (c) What is the final angular speed of the broken disk? The final angular momentum and energy?
    Answer: (a) $MR^2\omega_0^2/4$; $MR^2\omega_0/2$. (b) $R^2\omega_0^2/2$ g. (c) ω_0; $(M/2 - m)R^2\omega_0$; $(M/2 - m)R^2\omega_0^2/2$.

figure 13-18
Problem 31

32. A cockroach, mass m, runs counterclockwise around the rim of a lazy Susan (a circular dish mounted on a vertical axle) of radius R and rotational

inertia I with frictionless bearings. The cockroach's speed (relative to the earth) is v, whereas the lazy Susan turns clockwise with angular speed ω_0. The cockroach finds a bread crumb on the rim and of course, stops. (a) What is the angular speed of the lazy Susan after the cockroach stops? (b) Is energy conserved?

33. In Example 5 compare the kinetic energies of the object in two different orbits. Use the work-energy theorem to explain the difference quantitatively.

34. A particle is projected horizontally along the interior of a smooth hemispherical bowl of radius r which is kept at rest (Fig. 13-19). We wish to find the initial speed v_0 required for the particle to just reach the top of the bowl. Find v_0 as a function of θ_0, the initial angular position of the particle. (*Hint:* Use conservation principles.)

35. On a large horizontal frictionless circular track, radius R, lie two small masses m and M, free to slide on the track. Between the two masses is squeezed a spring which, however, is not fastened to m and M. The two masses are held together by a string. (a) If the string breaks, the compressed spring (assumed massless) shoots off the two masses in opposite directions; the spring itself is left behind. The balls collide when they again meet on the track (Fig. 13-20). Where does this collision take place? (You might find it convenient to express the answer in terms of the angle M travels through.) (b) If the potential energy initially stored in the spring was U_0, what is the time it takes after the string breaks for the collision to take place? (c) Assuming the collision to be perfectly elastic and head-on, where will the balls again collide after the first collision?

   Answer: (a) $2\pi m/(m+M)$ rad. (b) $[2\pi^2 mMR^2/(m+M)U_0]^{1/2}$. (c) At the point of origin.

figure 13-19
Problem 34

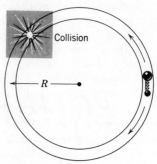

figure 13-20
Problem 35

14
equilibrium of rigid bodies

The towers supporting a suspension bridge must be strong enough so that they do not collapse under the weight of the bridge and its traffic load; the landing gear of an aircraft must not collapse if the pilot makes a poor landing; the tines of a fork must not bend when we cut a tough steak. In all such problems the engineer is concerned that these presumed rigid structures do indeed remain rigid under the forces, and the associated torques, that act on them.

In such problems the engineer must ask two questions. (1) What forces and torques act on the presumed rigid body? (2) Considering its design and the materials used, will the body remain rigid under the action of these forces and torques? In this chapter we are concerned only with the first of these questions; the engineering student will deal at length with the second question in later courses.

We note that the presumed rigid bodies of the preceding section (that is, the bridge towers, the landing gear, and the fork) are in *mechanical equilibrium*. A rigid body is in mechanical equilibrium if, as viewed from an inertial reference frame, (1) the linear acceleration $\mathbf{a}_{cm}$ of its center of mass is zero and (2) its angular acceleration $\boldsymbol{\alpha}$ about any axis fixed in this reference frame is zero.

This definition does not require the body to be at rest with respect to the observer but only to be unaccelerated. Its center of mass, for example, may be moving with constant velocity $\mathbf{v}_{cm}$ and the body may be rotating about a fixed axis with constant angular velocity $\boldsymbol{\omega}$. If the body is actually at rest (so that $\mathbf{v}_{cm} = 0$ and $\boldsymbol{\omega} = 0$), we often speak of *static equilibrium*, the central subject of this chapter. However, as we shall

14-1
RIGID BODIES

14-2
THE EQUILIBRIUM OF A RIGID BODY

see, the restrictions imposed on the forces and torques are the same whether or not the equilibrium is static. Furthermore, we can transform any case of (nonstatic) equilibrium to one of static equilibrium by choosing an appropriate new reference frame.

The translational motion of a rigid body of mass M is governed by Eq. 9-10, or

$$\mathbf{F}_{ext} = M\mathbf{a}_{cm,}$$

in which $\mathbf{F}_{ext}$ is the vector sum of all the external forces acting on the body. Because $\mathbf{a}_{cm}$ must be zero for equilibrium, the first condition of equilibrium (static or otherwise) is: *The vector sum of all the external forces acting on a body in equilibrium must be zero.*

We can write condition (1) as

$$\mathbf{F} = \mathbf{F}_1 + \mathbf{F}_2 + \cdots = 0, \qquad (14\text{-}1)$$

in which we have dropped the subscript on $\mathbf{F}_{ext}$ for convenience. This vector equation leads to three scalar equations.

$$F_x = F_{1x} + F_{2x} + \cdots = 0,$$
$$F_y = F_{1y} + F_{2y} + \cdots = 0, \qquad (14\text{-}2)$$
$$F_z = F_{1z} + F_{2z} + \cdots = 0,$$

which state that the sum of the components of the forces along each of any three mutually perpendicular directions is zero.

The second requirement for equilibrium is that $\boldsymbol{\alpha} = 0$ for any axis. Since the angular acceleration of a rigid body is associated with torque— recall that $\tau = I\alpha$ for a fixed axis—we can state this second condition of equilibrium (static or otherwise) as: *The vector sum of all the external torques acting on a body in equilibrium must be zero.*

We can write condition (2) as

$$\boldsymbol{\tau} = \boldsymbol{\tau}_1 + \boldsymbol{\tau}_2 + \cdots = 0. \qquad (14\text{-}3)$$

This vector equation leads to three scalar equations

$$\tau_x = \tau_{1x} + \tau_{2x} + \cdots = 0,$$
$$\tau_y = \tau_{1y} + \tau_{2y} + \cdots = 0, \qquad (14\text{-}4)$$
$$\tau_z = \tau_{1z} + \tau_{2z} + \cdots = 0,$$

which state that, at equilibrium, the sum of the components of the torques acting on a body, along each of any three mutually perpendicular directions, is zero.

The resultant torque $\boldsymbol{\tau}$ in Eq. 14-3, which must be zero for mechanical equilibrium, is defined with respect to a particular origin O. The quantities τ_x, τ_y, and τ_z in Eq. 14-4 are the scalar components of $\boldsymbol{\tau}$ and refer to any set of three mutually perpendicular axes whose origin is at O, no matter how these axes are oriented in space. This follows because, if a vector is zero, its scalar components must be zero no matter how we orient the axes of the reference frame. You may wonder whether the choice of an origin is essential. The answer—as we shall show below—is that it is not, because (for a body in translational equilibrium), if $\boldsymbol{\tau} = 0$ for any single origin O it is also zero for any other origin in the reference frame. The substance of this paragraph then is that condition 2 is satisfied for a body in translational equilibrium if we can show either that (a) $\boldsymbol{\tau} = 0$ with respect to *any* one point (Eq. 14-3) or that (b) the torque components along *any* three mutually perpendicular axes are zero (Eq. 14-4).

Let us now assume that we have a rigid body in translational equilibrium, so that $\mathbf{F} = \mathbf{F}_1 + \mathbf{F}_2 + \cdots = 0$ (Eq. 14-1). We now wish to show that the torque about *any* point (such as P in Fig. 14-1) will be zero if the torque about one particular point (such as O in Fig. 14-1) is zero. The figure shows three of the n forces, $\mathbf{F}_1, \mathbf{F}_2 \cdots \mathbf{F}_n$, applied at various points on a rigid body and pointing in various directions. The points of application with respect to O are identified by displacement vectors, of which $\mathbf{r}_1$ is an example. The arbitrary point P is identified by displacement vector $\mathbf{r}_P$; the vector $\mathbf{r}_1 - \mathbf{r}_P$ locates the point of application of $\mathbf{F}_1$ with respect to point P.

We can write for the resultant torque about O (see Eq. 12-1)

$$\boldsymbol{\tau}_O = \mathbf{r}_1 \times \mathbf{F}_1 + \mathbf{r}_2 \times \mathbf{F}_2 + \cdots + \mathbf{r}_n \times \mathbf{F}_n$$

and for the torque about P,

$$\boldsymbol{\tau}_P = (\mathbf{r}_1 - \mathbf{r}_P) \times \mathbf{F}_1 + (\mathbf{r}_2 - \mathbf{r}_P) \times \mathbf{F}_2 + \cdots + (\mathbf{r}_n - \mathbf{r}_P) \times \mathbf{F}_n.$$

We can expand the latter equation as

$$\boldsymbol{\tau}_P = [\mathbf{r}_1 \times \mathbf{F}_1 + \mathbf{r}_2 \times \mathbf{F}_2 + \cdots + \mathbf{r}_n \times \mathbf{F}_n] - [\mathbf{r}_P \times (\mathbf{F}_1 + \mathbf{F}_2 + \cdots + \mathbf{F}_n)].$$

Now if, as we have assumed, the first condition of equilibrium is satisfied, then $\mathbf{F}_1 + \mathbf{F}_2 + \cdots + \mathbf{F}_n = 0$ and the second term above in the square brackets vanishes. The first term in the square brackets is simply $\boldsymbol{\tau}_O$ (see above) so that, under these conditions

$$\boldsymbol{\tau}_P = \boldsymbol{\tau}_O.$$

Thus, for a body in translational equilibrium, if $\boldsymbol{\tau}_O = 0$, then $\boldsymbol{\tau}_P = 0$, where P is an arbitrary point.

Hence we have *six independent conditions* on our forces for a body to be in equilibrium. These conditions are the six algebraic relations of Eqs. 14-2 and 14-4. These six conditions are a condition on each of the six degrees of freedom of a rigid body, three translational and three rotational.

Often we deal with problems in which all the forces lie in a plane. Then we have only three conditions on our forces: The sum of their components must be zero for each of any two perpendicular directions in the plane, and the sum of their torques about any one axis perpendicular to the plane must be zero. These conditions correspond to the three degrees of freedom for motion in a plane, two of translation and one of rotation.

We shall limit ourselves henceforth mostly to planar problems in order to simplify the calculations. This does not impose any fundamental restriction on the general principles. Also, as a matter of convenience, we shall consider only the case of static equilibrium, in which bodies are actually at rest in our chosen inertial reference frame.

figure 14-1
We display three of the n forces, $\mathbf{F}_1, \mathbf{F}_2, \mathbf{F}_3 \dots \mathbf{F}_n$, that act on a rigid body, not shown. In the text we show that if $\tau = 0$ for point O it also vanishes for any point such as P, assuming that the body is in translational equilibrium.

14-3
CENTER OF GRAVITY

One of the forces encountered in rigid-body motions is the force of gravity. Actually, for an extended body, this is not just one force but the resultant of a great many forces. Each particle in the body is acted on by a gravitational force. If the body of mass M is imagined to be divided into a large number of particles, say n, the gravitational force exerted by the earth on the ith particle of mass m_i is $m_i\mathbf{g}$. This force is directed down toward the earth. If the acceleration due to gravity $\mathbf{g}$ is the same everywhere in a region, we say that a uniform gravitational field exists there; that is, $\mathbf{g}$ has the same magnitude and direction everywhere in that region. For a rigid body in a uniform gravitational field, $\mathbf{g}$ must be the same for each particle in the body and the weight forces on the particles must be parallel to one another. If we assume that the earth's gravitational field is uniform, we can show that all the individual weight forces acting on a body can be replaced by a single force $M\mathbf{g}$ acting down at the center of mass of the body. This is equivalent to showing that the individual weight forces, acting downward, can be counteracted in their accelerating effects by a single force $\mathbf{F}$ $(= -M\mathbf{g})$ acting upward, *provided this force $\mathbf{F}$ is applied at the center of mass of the body.*

Figure 14-2 shows two typical particles or mass elements m_1 and m_2, selected from the n such elements into which the rigid body has been divided. An upward force $\mathbf{F}$ $(= -M\mathbf{g})$ is applied at a certain point O. It remains to show that the body is in mechanical equilibrium *if* (and only if) point O is the center of mass. Condition 1 for equilibrium (Eq. 14-1) has already been satisfied by our choice of the magnitude and direction of $\mathbf{F}$. That is,

$$\mathbf{F} + m_1\mathbf{g} + m_2\mathbf{g} + \cdots + m_n\mathbf{g} = 0,$$

or

$$\mathbf{F} = -(m_1 + m_2 + \cdots + m_n)\mathbf{g} = -M\mathbf{g},$$

which corresponds to our assumption.

It remains to prove that $\tau = 0$ for any single point in the body, such as O. This is the second condition for equilibrium. By choosing O as our origin we insure that the torque of $\mathbf{F}$ about this point is zero, because the moment arm of $\mathbf{F}$ is zero for this point. The torque about O due to the gravitational pull on the mass elements is

$$\boldsymbol{\tau} = \mathbf{r}_1 \times m_1\mathbf{g} + \mathbf{r}_2 \times m_2\mathbf{g} + \cdots + \mathbf{r}_n \times m_n\mathbf{g}$$

which (because m_1, m_2, etc., are scalars) we can write as

$$\boldsymbol{\tau} = m_1\mathbf{r}_1 \times \mathbf{g} + m_2\mathbf{r}_2 \times \mathbf{g} + \cdots + m_n\mathbf{r}_n \times \mathbf{g}.$$

Because $\mathbf{g}$ is the same in each term, we can factor it out, obtaining

$$\boldsymbol{\tau} = (m_1\mathbf{r}_1 + m_2\mathbf{r}_2 + \cdots + m_n\mathbf{r}_n) \times \mathbf{g}$$

$$= \left(\sum_{1}^{n} m_i\mathbf{r}_i \right) \times \mathbf{g},$$

in which the sum is taken over all the mass elements that make up the body.

Now *if* point O is the center of mass of the body, the sum above is zero. This follows from the definition of the center of mass (see Eq. 9-5b and the discussion following it). We conclude then that *if* point O is the center of mass, then $\tau = 0$ and the second condition for mechanical equilibrium is satisfied.*

Thus the gravitational forces acting on the individual mass elements that make up a rigid body are equivalent in their translational and rotational effects to a single force equal to $M\mathbf{g}$, the total weight of the body, acting at the center of mass. The point of application of the equivalent resultant gravitational force is often called the *center of gravity.*

The fact that the center of gravity and the center of mass coincided came about only because we assumed that the earth's gravitational field $\mathbf{g}$ was the same for all parts of the rigid body. Actually this assumption is not strictly true,

figure 14-2
An irregular body is divided into n mass elements of which two typical elements m_1 and m_2 are shown. In the text we prove that, if the gravitational field is uniform, the body can be held in translational and rotational equilibrium by a single force $\mathbf{F} = (-M\mathbf{g})$ directed upward and applied at the center of mass of the body.

*A body can be suspended in rotational equilibrium from *any* point; see Fig. 14-4. However, the center of mass is the *only* point that gives = 0 for all rotational orientations of the body.

for the magnitude of **g** changes with distance from the center of the earth and furthermore the direction of **g** is radially in toward the center of the earth from any point (Chapter 16). To see the effect this has, let us consider a very long uniform stick inclined to the vertical in the earth's gravitational field, as in Fig. 14-3. The center of gravity of a body is the point at which the equivalent resultant gravitational force on it acts. This point must be the same as the point at which a single oppositely directed force is applied for the body to be kept in equilibrium. If the field were uniform, a single upward force of magnitude Mg at the center of mass would keep the stick in translational and rotational equilibrium. But the field is not uniform, and the value of g at m_1 is less than the value of g at m_n. The point at which a single force must be applied to keep the body in equilibrium is therefore at a point P some distance below the center of mass. Furthermore, if the orientation of the body is changed, the position of the point P, required for application of an equilibrium force, changes. Hence center of gravity really has little usefulness in such a case. Not only does it not coincide with the center of mass, but its position changes with respect to the body as the body is moved.

Because almost all problems in mechanics involve objects having dimensions small compared to the distances over which **g** changes appreciably, we can assume that **g** is uniform over the body. The center of mass and the center of gravity can then be taken as the same point. In fact, we can use this coincidence to determine experimentally the center of mass in irregularly shaped objects. For example, let us locate the center of mass of a thin plate of irregular shape, as in Fig. 14-4. We suspend the body by a cord from some point A on its edge. When the body is at rest, the center of gravity must lie directly under the point of support somewhere on the line Aa, for only then can the torque resulting from the cord and the weight add to zero. We next suspend the body from another point B on its edge. Again, the center of gravity must lie somewhere on Bb. The only point common to the lines Aa and Bb is O, the point of intersection, so that this point must be the center of gravity. If now we suspend the body from any other point on its edge, as C, the vertical line Cc will pass through O. Since we have assumed a uniform field, the center of gravity coincides with the center of mass, which is therefore located at O.

figure 14-3
The center of mass C and center of gravity P in reality do not coincide, since the earth's gravitational field is not uniform.

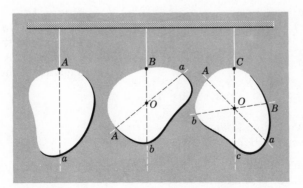

figure 14-4
Since the center of mass O always hangs directly below the point of suspension, hanging a plate from two different points determines O.

In applying the conditions for equilibrium (zero resultant force and zero resultant torque about any axis), we can clarify and simplify the procedure by proceeding as follows.

First, we draw an imaginary boundary around the system under consideration. This assures that we see clearly just what body or system of

14-4
EXAMPLES OF EQUILIBRIUM

bodies it is to which we are applying the laws of equilibrium. This process is called *isolating the system.*

Second, we draw vectors representing the magnitude, direction, and point of application of all *external* forces. An external force is one that acts from outside the boundary which was drawn earlier. Examples of external forces often encountered are gravitational forces and forces transmitted by strings, wires, rods, and beams which cross the boundary. A question sometimes arises about the direction of a force. In this case make an imaginary cut through the member transmitting the force at the point where it crosses the boundary. If the ends of this cut tend to pull apart, the force acts outward. If you are in doubt, choose the direction arbitrarily. A negative value for a force in the solution means that the force acts in the direction opposite to that assumed. Note that only external forces acting on the system need be considered; all internal forces cancel one another in pairs.

Third, we choose a convenient coordinate system along whose axes we resolve the external forces before applying the first condition of equilibrium (Eq. 14-2). The object here is to simplify the calculations. The preferable coordinate system is usually obvious.

Fourth, we choose a convenient coordinate system along whose axes we resolve the external torques before applying the second condition of equilibrium (Eq. 14-4). The object again is to simplify calculations and we may use different coordinate systems in applying the two conditions for static equilibrium if this proves to be convenient. Suppose that an axis passes through the point at which two forces concur and is at right angles to the plane formed by these forces; these forces will automatically have no torque component along (or about) this axis. The torque components resulting from all external forces must be zero about any axis for equilibrium. Internal torques will cancel in pairs and need not be considered.

(a) A uniform steel meter bar rests on two scales at its ends (Fig. 14-5). The bar weighs 4.0 lb. Find the readings on the scales.

EXAMPLE 1

(a)   (b)

figure 14-5
(a) Example 1a. A uniform steel bar rests on two spring scales. (b) Example 1b. A weight is suspended a quarter of the way from one end.

Our system is the bar. The forces acting on the bar are **W,** the gravitational force acting down at the center of gravity, and $\mathbf{F}_1$ and $\mathbf{F}_2$, the forces exerted upward on the bar at its ends by the scales. These are shown in Fig. 14-5a. By Newton's third law, the force exerted by a scale on the bar is equal and opposite to that exerted by the bar on the scale. Therefore, to obtain the readings on the scales, we must determine the magnitudes of $\mathbf{F}_1$ and $\mathbf{F}_2$.

For translational equilibrium (Eq. 14-1) the condition is

$$\mathbf{F}_1 + \mathbf{F}_2 + \mathbf{W} = 0.$$

All forces act vertically, so that if we choose the y-axis to be vertical, no other axes need be considered. Then we get the scalar equation

$$F_1 + F_2 - 4.0 \text{ lb} = 0.$$

For rotational equilibrium, the component of the resultant torque on the bar must be zero about *any* axis. We have seen that it is enough to show that the torque components are zero for any set of three mutually perpendicular axes. These components are certainly zero for any two perpendicular axes that lie in the plane of Fig. 14-5a (Why?). It remains to require that the resultant torque is zero about any one axis at right angles to the plane of the figure. Let us choose an axis through the center of gravity. Then, taking clockwise rotation as positive and counterclockwise rotation as negative, the condition for rotational equilibrium (Eq. 14-4) is

$$F_1\left(\frac{l}{2}\right) - F_2\left(\frac{l}{2}\right) + W(0) = 0,$$

or

$$F_1 - F_2 = 0.$$

Combining the two equations, we obtain

$$F_1 + F_2 = 2F_1 = 2F_2 = 4.0 \text{ lb},$$

$$F_1 = F_2 = 2.0 \text{ lb}.$$

Each scale reads 2.0 lb, as we might have expected.

If we had chosen an axis through one end of the bar, we would have obtained the same result. For example, taking torques about an axis through the right end, we obtain

$$F_1(l) - W\left(\frac{l}{2}\right) + F_2(0) = 0,$$

or

$$F_1 = \frac{W}{2} = \frac{4.0 \text{ lb}}{2} = 2.0 \text{ lb}.$$

Combining this with $F_1 + F_2 = 4.0$ lb, we obtain $F_2 = 2.0$ lb, as before.

(b) Suppose that a 6.0-lb block is placed at the 25-cm mark on the meter bar. What do the scales read now?

The external forces acting on the bar are shown in Fig. 14-5b, where **w** is the force exerted on the bar by the block. The first condition for equilibrium is

$$F_1 + F_2 - W - w = 0.$$

With $W = 4.0$ lb and $w = 6.0$ lb, we obtain

$$F_1 + F_2 = 10 \text{ lb}.$$

If we take an axis through the left end of the bar, the second condition for equilibrium is

$$w\left(\frac{l}{4}\right) + W\left(\frac{l}{2}\right) - F_2(l) = 0.$$

With $W = 4.0$ lb and $w = 6.0$ lb, we obtain

$$F_2 = 3.5 \text{ lb}.$$

Putting this result into the first equation, we obtain

$$F_1 + 3.5 \text{ lb} = 10 \text{ lb},$$

$$F_1 = 6.5 \text{ lb}.$$

The left-hand scale reads 6.5 lb and the right-hand scale reads 3.5 lb at equilibrium.

Why do we obtain only two conditions on the forces in this problem rather than the three conditions expected for problems in which all forces lie in the same plane?

(a) A 60-ft ladder weighing 100 lb rests against a wall at a point 48 ft above the ground. The center of gravity of the ladder is one-third the way up. A 160-lb man climbs halfway up the ladder. Assuming that the wall (but not the ground) is frictionless, find the forces exerted by the system on the ground and on the wall.

EXAMPLE 2

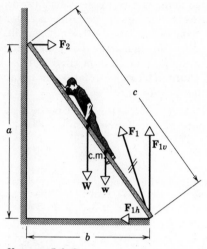

The forces acting on the ladder are shown in Fig. 14-6. $\mathbf{W}$ is the weight of the man standing on the ladder and $\mathbf{w}$ is the weight of the ladder itself. A force $\mathbf{F}_1$ is exerted by the ground on the ladder. $\mathbf{F}_{1v}$ is the vertical component and $\mathbf{F}_{1h}$ is the horizontal component of this force (due to friction). The wall, being frictionless, can exert only a force normal to its surface, called $\mathbf{F}_2$. We are given the following data:

$$W = 160 \text{ lb}, \qquad w = 100 \text{ lb},$$
$$a = 48 \text{ ft}, \qquad c = 60 \text{ ft}.$$

From the geometry we conclude that $b = 36$ ft. The line of action of $\mathbf{W}$ intersects the ground at a distance $b/2$ from the wall and the line of action of $\mathbf{w}$ intersects the ground at a distance $2b/3$ from the wall.

We choose the x-axis to be along the ground and the y-axis along the wall. Then, the conditions on the forces for translational equilibrium (Eq. 14-2) are

$$F_2 - F_{1h} = 0,$$

$$F_{1v} - W - w = 0.$$

figure 14-6
Example 2.

For rotational equilibrium (Eq. 14-4) choose an axis through the point of contact with the ground and obtain

$$F_2(a) - W\left(\frac{b}{2}\right) - w\left(\frac{b}{3}\right) = 0.$$

Using the data given, we obtain

$$F_2(48 \text{ ft}) - (160 \text{ lb})(18 \text{ ft}) - (100 \text{ lb})(12 \text{ ft}) = 0,$$

$$F_2 = 85 \text{ lb},$$

$$F_{1h} = F_2 = 85 \text{ lb},$$

$$F_{1v} = 160 \text{ lb} + 100 \text{ lb} = 260 \text{ lb}.$$

By Newton's third law the forces exerted by the ground and the wall on the ladder are equal but opposite to the forces exerted by the ladder on the ground and the wall, respectively. Therefore, the normal force on the wall is 85 lb, and the force on the ground has components of 260 lb down and 85 lb to the right.

(b) If the coefficient of static friction between the ground and the ladder is $\mu_s = 0.40$, how high up the ladder can the man go before it starts to slip?

Let x be the fraction of the total length of the ladder the man can climb before slipping begins. Then our equilibrium conditions are

$$F_2 - F_{1h} = 0,$$
$$F_{1v} - W - w = 0,$$

and

$$F_2 a - Wbx - w\left(\frac{b}{3}\right) = 0.$$

Now we obtain

$$F_2(48 \text{ ft}) = (160 \text{ lb})(36 \text{ ft})x + (100 \text{ lb})(12 \text{ ft}),$$

$$F_2 = (120x + 25) \text{ lb}.$$

Hence

$$F_{1h} = (120x + 25) \text{ lb},$$

and, as before,

$$F_{1v} = 260 \text{ lb}.$$

The maximum force of static friction is given by

$$F_{1h} = \mu_s F_{1v} = (0.40)(260 \text{ lb}) = 104 \text{ lb}.$$

Therefore,

$$F_{1h} = (120x + 25) \text{ lb} = 104 \text{ lb}$$

and

$$x = \tfrac{79}{120},$$

so that the man can climb up the ladder

$$60x \text{ ft} = 39.5 \text{ ft}$$

before slipping begins.

In this example the ladder is treated as a one-dimensional object, with only one point of contact at the wall and ground. You should reflect on how this limits consideration of the less artificial case of two contact points at each end.

The reason for assuming that the wall is frictionless is discussed later in this section. Can you guess what it is?

EXAMPLE 3

A uniform beam is hinged at the wall. A wire connected to the wall a distance d above the hinge is attached to the other end of the beam. The beam makes an angle of 30° with the horizontal when a weight w is hung from a string fastened to the end of the beam. If the beam has a weight W and a length l, find the tension in the wire and the forces exerted by the hinge on the beam.

The situation is depicted in Fig. 14-7, in which all the forces acting on the beam are shown. The wire pulling on the beam makes some angle α with the horizontal so that the tension $\mathbf{T}$ in the wire has horizontal and vertical components T_h and T_v, respectively, as shown. The force $\mathbf{F}$ exerted by the hinge on the beam also has horizontal and vertical components F_h and F_v, respectively. $\mathbf{W}$ is the weight of the beam, acting at its center of gravity, and $\mathbf{w}$ is the tension in the string that transmits the weight of the suspended body to the beam.

Choosing our axes to be horizontal and vertical, we obtain for translational equilibrium

$$F_v + T_v - W - w = 0,$$

and

$$F_h - T_h = 0.$$

Choosing an axis through the point of intersection of $\mathbf{T}$ and $\mathbf{w}$ (Why?), we obtain for rotational equilibrium

$$F_v(l \cos 30°) - F_h(l \sin 30°) - \frac{W(l \cos 30°)}{2} = 0.$$

Our unknowns are T_h, T_v, F_h, and F_v. Let us assign the following values to the other quantities:

$$W = 60 \text{ N}, \quad w = 40 \text{ N}, \quad l = 3.0 \text{ m}, \quad d = 2.0 \text{ m}.$$

figure 14-7
Example 3.

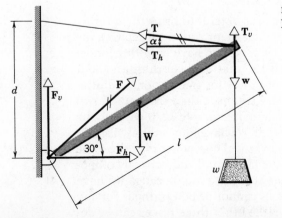

Therefore

(1) $$F_v + T_v = 100 \text{ N},$$

(2) $$F_h = T_h,$$

and

$$F_v(3)(0.866) = F_h(1.5) + (60)(1.5)(0.866),$$

or

(3) $$F_v = F_h(5.0/8.66) + 30 \text{ N}.$$

Recall that we have four unknowns, namely F_v, F_h, T_v, and T_h. We need another relation between these quantities if we are to solve the problem. This relation follows from the fact that $\mathbf{T}_v$ and $\mathbf{T}_h$ must add to give a resultant vector $\mathbf{T}$ directed along the wire. The wire cannot supply or support a force transverse to its orientation. (Notice that this is not true for the beam, however.) Hence our fourth relation is

$$T_v = T_h \tan \alpha,$$

where $\tan \alpha = (d - l \sin 30°)/l \cos 30° = 1.0/5.2$, so that

(4) $$T_v = T_h/5.2.$$

Combining (1) and (4) we obtain

$$F_v = 100 \text{ N} - T_h/5.2.$$

Combining (2) and (3), we obtain

$$F_v = T_h(5.0/8.66) + 30 \text{ N}.$$

Solving these equations simultaneously, we obtain

$$T_h = 91.0 \text{ N},$$

and

$$F_v = 82.5 \text{ N}.$$

From (2) we obtain

$$F_h = 91.0 \text{ N}.$$

From (1) we obtain

$$T_v = 17.5 \text{ N}.$$

The tension in the wire will then be

$$T = \sqrt{T_h{}^2 + T_v{}^2} = 92.7 \text{ N},$$

and the hinge will exert a horizontal force of 91.0 N and a vertical force of 82.5 N on the beam.

In the preceding examples we have been careful to limit the number of unknown forces to the number of independent equations relating the forces. When all the forces act in a plane, we can have only three independent equations of equilibrium, one for rotational equilibrium about any axis normal to the plane and two others for translational equilibrium in the plane. However, we often have more than three unknown forces. For example, in the ladder problem of Example 2a, if we drop the artificial assumption of a frictionless wall, we have four unknown scalar quantities, namely, the horizontal and vertical components of the force acting on the ladder at the wall and the horizontal and vertical components of the force acting on the ladder at the ground. Because we have only three scalar equations, these forces cannot be determined. For any value assigned to one unknown force, the other three forces can be determined. But if we have no basis for assigning any particular value to an unknown force, there are an infinite number of solutions mathematically possible. We must therefore find another independent relation between the unknown forces if we hope to solve the problem uniquely.

Another simple example of such underdetermined structures is the automobile. In this case we wish to determine the forces exerted by the ground on each of the four tires when the car is at rest on a horizontal surface. If we assume that these forces are normal to the ground, we have four unknown scalar quantities. All other forces, such as the weight of the car and passengers, act normal to the ground. Therefore, we have only three independent equations giving the equilibrium conditions, one for translational equilibrium in the single direction of all the forces and two for rotational equilibrium about the two axes perpendicular to each other in a horizontal plane. Again the solution of the problem is indeterminate, mathematically. A four-legged table with all its legs in contact with the floor is a similar example.

Of course, since there is actually a unique solution to any real physical problem, we must find a physical basis for the additional independent relation between the forces that enable us to solve the problem. The difficulty is removed when we realize that structures are never perfectly rigid, as we have tacitly assumed throughout. Actually our structures will be somewhat deformed. For example, the automobile tires and the ground will be deformed, as will the ladder and wall. The laws of elasticity and the elastic properties of the structure determine the nature of the deformation and will provide the necessary additional relation between the four forces. A complete analysis therefore requires not only the laws of rigid body mechanics but also the laws of elasticity. In courses of civil and mechanical engineering, many such problems are encountered and analyzed in this way. We shall not consider the matter further here.

14-5
STABLE, UNSTABLE, AND NEUTRAL EQUILIBRIUM OF RIGID BODIES IN A GRAVITATIONAL FIELD

In Chapter 8 we saw that the gravitational force is a conservative force. For conservative forces we can define a potential energy function $U(x,y,z)$, where U is related to $\mathbf{F}$ by

$$F_x = -\partial U/\partial x, \qquad F_y = -\partial U/\partial y, \qquad F_z = -\partial U/\partial z.$$

At points where $\partial U/\partial x$ is zero, a particle subject to this conservative force will be in translational equilibrium in the x-direction, for then F_x equals zero. Likewise, at points where $\partial U/\partial y$ or $\partial U/\partial z$ are zero, a particle will be in translational equilibrium in the y- and z-directions, respectively. The derivative of U at a point will be zero when U has an extreme value (maximum or minimum) at that point or when U is constant with respect to the variable coordinate.

When U is a minimum, the particle is in *stable* equilibrium; any displacement from this position will result in a restoring force tending to return the particle to the equilibrium position. Another way of stating this is to say that if a body is in stable equilibrium, work must be done on it by an external agent to change its position. This results in an increase in its potential energy.

When U is a maximum, the particle is in *unstable* equilibrium; any displacement from this position will result in a force tending to push the particle farther from the equilibrium position. In this case no work must be done on the particle by an external agent to change its position; the work done in displacing the body is supplied internally by the conservative force, resulting in a decrease in potential energy.

When U is constant, the particle is in *neutral* equilibrium. In this case a par-

figure 14-8

A gravitational potential surface, which may be thought of as a real surface. A particle placed at *A*, *B* or *C* remains at rest; a plane tangent to any of these points is horizontal. We say that a particle here is in equilibrium. At *A*, a particle, if slightly displaced, tends to return to *A*. *A* represents a point of *stable equilibrium*. At *B*, a particle, if slightly displaced, tends to increase its displacement. Thus *B* represents a point of *unstable equilibrium*. At *C*, the particle, if slightly displaced in the direction *aa'*, will tend to return to *C*, but if it is displaced in direction *bb'*, it will tend to increase its displacement. *C* is called a *saddle point* since a saddle has somewhat this shape. Neutral equilibrium, experienced by a particle anywhere on a plane horizontal surface is not illustrated.

ticle can be displaced slightly without experiencing either a repelling or restoring force.

Notice that a particle can be in equilibrium with respect to one coordinate without necessarily being in equilibrium with respect to another coordinate, as for example a freely falling ball. Furthermore, a particle may be in stable equilibrium with respect to one coordinate and in unstable equilibrium with respect to another coordinate, as for example a particle at a saddle point (Fig. 14-8).

All these remarks apply to particles, that is, to translational motion. Suppose now we treat a rigid body. We must consider rotational equilibrium as well as translational equilibrium. The problem of a rigid body in a gravitational field is particularly simple, however, because *all the gravitational forces on the particles of the rigid body can be considered to act at one point, both for translational and rotational purposes.* We can replace this entire rigid body, for purposes of equilibrium under gravitational forces, by a single particle having the equivalent mass at the center of gravity.

For example, consider a cube at rest on one side on a horizontal table. The center of gravity is shown at the center of the central cross section of the cube in Fig. 14-9a. Let us supply a force to the cube so as to rotate it without its slipping about an axis along an edge. Notice that the center of gravity is raised and that work is done on the cube, which increases its potential energy. If the force is removed, the cube tends to return to its original position, its increased potential energy being converted into kinetic energy as it falls back. This initial position is, therefore, one of *stable* equilibrium. In terms of a particle of equivalent mass at the center of gravity, this process is described by the dotted line which indicates the path taken by the center of gravity during this motion. The particle is seen to have a minimum potential energy in the position of stable equilibrium, as required. We can conclude that the rigid body will be in stable equilibrium if the application of any force can raise the center of gravity of the body but not lower it.

If the cube is rotated until it balances on an edge, as in Fig. 14-9b, then once again the cube is in equilibrium. This equilibrium position is seen to be un-

figure 14-9

Equilibrium of an extended body. *(a)* A cube resting on one side is in *stable equilibrium* since its center of gravity *C* is raised if the cube is tipped by a horizontal force **F**. *(b)* A cube resting on one edge is in *unstable equilibrium* since *C* falls if the cube is tipped by **F**. *(c)* A circular cylinder is in *neutral equilibrium* since *C* neither rises nor falls when **F** is applied. Compare these criteria for equilibrium with those given in Fig. 14-8. How are the criteria in the two figures related?

stable. The application of even the slightest horizontal force will cause the cube to fall away from this position with a decrease of potential energy. The particle of equivalent mass at the center of gravity follows the dotted path shown. At the position of unstable equilibrium this particle has a maximum potential energy, as required. We can conclude that the rigid body will be in unstable equilibrium if the application of any horizontal force tends to lower the center of gravity of the body.

The neutral equilibrium of a rigid body is illustrated by a cylinder on a horizontal table (Fig. 14-9c). If the cylinder is subjected to any horizontal force, the center of gravity is neither raised nor lowered but moves along the horizontal dotted line. The potential energy of the cylinder is constant during the displacement, as is that of the particle of equivalent mass at the center of gravity. The system has no tendency to move in any direction when the applied force is removed. A rigid body will be in neutral equilibrium if the application of any horizontal force neither raises nor lowers the center of gravity of the body.

Under what circumstances would a *suspended* rigid body be in stable equilibrium? When would a *suspended* rigid body be in unstable equilibrium, and when would it be in neutral equilibrium?

questions

1. Are Eqs. 14-1 and 14-3 both necessary and sufficient conditions for mechanical equilibrium? For static equilibrium?
2. A wheel rotating at constant angular velocity ω about a fixed axis is in mechanical equilibrium because no net external force or torque acts on it. However, the particles that make up the wheel undergo a centripetal acceleration **a** directed toward the axis. Since $\mathbf{a} \neq 0$ how can the wheel be said to be in equilibrium?
3. Give several examples of a body which is not in equilibrium, even though the resultant of all the forces acting on it is zero.
4. If a body is not in translational equilibrium, will the torque about any point be zero if the torque about some particular point is zero?
5. Which is more likely to break in use, a hammock stretched tightly between two trees or one that sags quite a bit? Prove your answer.
6. A ladder is at rest with its upper end against a wall and the lower end on the ground. Is it more likely to slip when a man stands on it at the bottom or at the top? Explain.
7. In Example 2, if the wall were rough, would the empirical laws of friction supply us with the extra condition needed to determine the extra (vertical) force exerted by the wall on the ladder?
8. In Example 3, why isn't it necessary to consider friction at the hinge?
9. A picture hangs from a wall by two wires. What orientation should the wires have to be under minimum tension? Explain how equilibrium is possible

with any number of orientations and tensions, even though the picture has a definite mass.

10. Show how to use a spring balance to weigh objects well beyond the maximum reading of the balance.

11. Do the center of mass and the center of gravity coincide for a building? For a lake? Under what conditions does the difference between the center of mass and the center of gravity of a body become significant?

12. If a rigid body is thrown into the air without spinning, it does not spin during its flight, provided air resistance can be neglected. What does this simple result imply about the location of the center of gravity?

13. Explain, using forces and torques, how a tree can maintain equilibrium in a high wind.

14. Is there such a thing as a truly rigid body?

15. You are sitting in the driver's seat of a parked automobile. You are told that the forces exerted upward by the ground on each of the four tires are different. Discuss the factors that enter into a consideration of whether this statement is true or false.

16. A uniform block, in the shape of a rectangular parallelepiped of sides in the ratio 1:2:3, lies on a horizontal surface. In which position, that is, on which of its three different faces, can it be said to be most stable, if any?

17. A virus particle in a rotating liquid-filled centrifuge tube is in uniform circular motion (that is, in *accelerated* motion) as viewed by an observer in the laboratory. An observer rotating with the centrifuge, however, would declare the particle to be *unaccelerated*. Explain how the virus particle can be in equilibrium for this second observer but not for the first.

18. In Chapter 5 we *defined* force in terms of acceleration from the relation **F** = **ma**. For a body in equilibrium, however, there are no accelerations. How, then, can we give meaning to the forces acting on such a body?

SECTION 14-2

1. Prove that when only three forces act on a body in equilibrium, they must be coplanar and their lines of action must meet at a point or at infinity.

2. A uniform sphere of weight w and radius r is being held by a rope attached to a frictionless wall a distance L above the center of the sphere, as in Fig. 14-10. Find (a) the tension in the rope and (b) the force exerted on the sphere by the wall.

3. A uniform sphere of weight w lies at rest wedged between two inclined planes of inclination angles θ_1 and θ_2 (Fig. 14-11). (a) Assume that no friction is involved and determine the forces (directions and magnitude) that the planes exert on the sphere. (b) What change would it make, in principle, if friction were taken into account?

   Answer: (a) $F_1 = w \sin \theta_2/\sin (\theta_2 - \theta_1)$; $F_2 = w \sin \theta_1/\sin (\theta_2 - \theta_1)$; normal to the planes.

problems

figure 14-10
Problem 2

figure 14-11
Problem 3

4. Two identical uniform smooth spheres, each of weight W, rest as shown in Fig. 14-12 at the bottom of a fixed, rectangular container. Find, in terms of W, the forces acting on the spheres by (a) the container surfaces and (b) by one another, if the line of centers of the spheres makes an angle of 45° with the horizontal.

5. A flexible chain of weight W hangs between two fixed points, A and B, at the same level, as shown in Fig. 14-13. Find (a) the vector force exerted by the chain on each end point and (b) the tension in the chain at the lowest point.

Answer: (a) $\dfrac{W}{2 \sin \theta}$, tangent to chain. (b) $\dfrac{1}{2} W \cot \theta$.

figure 14-12
Problem 4

SECTION 14-3

6. A nonuniform bar of weight W is suspended at rest in a horizontal position by two light cords as shown in Fig. 14-14; the angle one cord makes with the vertical is $\theta = 36.9°$; the other makes the angle $\phi = 53.1°$ with the vertical. If the length l of the bar is 6.1 m, compute the distance x from the left-hand end to the center of gravity.

7. A circular section of radius r is cut out of a uniform disk of radius R, the center of the hole being $R/2$ from the center of the original disk. Locate the center of gravity of the resulting body.

Answer: Along a line from the center of the hole through the center of the disk, beyond the latter point by a distance $Rr^2/2(R^2 - r^2)$.

figure 14-13
Problem 5

figure 14-14
Problem 6

SECTION 14-4

8. A beam is carried by three men, one man at one end and the other two supporting the beam between them on a crosspiece so placed that the load is equally divided among the three men. Find where the crosspiece is placed. Neglect the mass of the crosspiece.

9. In Fig. 14-15 a man is trying to get his car out of the mud on the shoulder of a road. He ties one end of a rope tightly around the front bumper and the other end tightly around a telephone pole 60 ft away. He then pushes sideways on the rope at its midpoint with a force of 125 lb, displacing the center of the rope 1.0 ft from its previous position and the car almost moves. What force does the rope exert on the car? (The rope stretches somewhat under the tension.)   *Answer:* 1900 lb.

figure 14-15
Problem 9

10. Forces $\mathbf{F}_1$, $\mathbf{F}_2$, and $\mathbf{F}_3$ act on the structure of Fig. 14-16 as shown. We wish to put the structure in equilibrium by applying a force, at a point such as P, whose vector components are $\mathbf{F}_h$ and $\mathbf{F}_v$. We are given that $a = 2.0$ m, $b = 3.0$ m, $c = 1.0$ m, $F_1 = 20$ N, $F_2 = 10$ N, and $F_3 = 5.0$ N. Find (a) F_h, (b) F_v, and (c) d.

figure 14-16
Problem 10

11. What force *F* applied horizontally at the axle of the wheel is necessary to raise the wheel over an obstacle of height *h*? Take *r* as the radius of the wheel and *W* as its weight (Fig. 14-17).

    Answer: $W\sqrt{h(2r-h)}/(r-h)$.

12. A trap door in a ceiling is 3.0 ft (0.91 m) square, weighs 25 lb (mass = 11 kg), and is hinged along one side with a catch at the opposite side. If the center of gravity of the door is 4.0 in. (10 cm) from the door's center and closer to the hinged side, what forces must (a) the catch and (b) the hinge sustain?

13. A meter stick balances on a knife edge at the 50.0-cm mark. When two nickels are stacked over the 12.0-cm mark, the loaded stick is found to balance at the 45.5-cm mark. A nickel has a mass of 5.0 g. What is the mass of the meter stick? Try this technique and check your answer experimentally.   *Answer:* 74.4 g.

14. A balance is made up of a rigid rod free to rotate about a point not at the center of the rod. It is balanced by unequal weights placed in the pans at each end of the rod. When an unknown mass *m* is placed in the left-hand pan, it is balanced by a mass m_1 placed in the right-hand pan, and similarly when the mass *m* is placed in the right-hand pan, it is balanced by a mass m_2 in the left-hand pan. Show that

$$m = \sqrt{m_1 m_2}.$$

15. An automobile weighing 3000 lb (mass = 1360 kg) has a wheel base of 120 in. (305 cm). Its center of gravity is located 70.0 in. (178 cm) behind the front axle. Determine (a) the force exerted on each of the front wheels (assumed the same) and (b) the force exerted on each of the back wheels (assumed the same) by the level ground.
    Answer: (a) 625 lb (2780 N). (b) 875 lb (3890 N).

16. A crate in the form of a 4.0-ft cube contains a piece of machinery whose design is such that the center of gravity of the crate and its contents is located 1.0 ft above its geometrical center. (a) If the crate is to be slid down a ramp without tipping over, what is the maximum angle which the ramp may make with the horizontal? (b) What is the maximum value for the coefficient of static friction between the crate and the ramp in this case such that the crate will just begin to slide?

17. A door 7.0 ft (2.1 m) high and 3.0 ft (0.91 m) wide weighs 60 lb (mass = 27 kg). A hinge 1.0 ft (0.30 m) from the top and another 1.0 ft (0.30 m) from the bottom each support half the door's weight. Assume that the center of gravity is at the geometrical center of the door and determine the horizontal and vertical force components exerted by each hinge on the door. *Answer:* 30 lb (130 N) vertical; 18 lb (80 N) horizontal, oppositely directed.

18. Four bricks, each of length *l*, are put on top of one another (see Fig. 14-18) in such a way that part of each extends beyond the one beneath. Show that the largest equilibrium extensions are (a) top brick overhanging the one below by *l*/2, (b) second brick from top overhanging the one below by *l*/4, and (c) third brick from top overhanging the bottom one by *l*/6.

19. The system shown in Fig. 14-19 is in equilibrium. The mass hanging from the end of the strut *S* weighs 500 lb (mass = 230 kg), and the strut itself weighs 100 lb (mass = 45 kg). Find (a) the tension *T* in the cable and (b) the force exerted on the strut by the pivot *P*.
    Answer: (a) 1500 lb (6800 N). (b) $F_h = 1300$ lb (5900 N); $F_v = 1350$ lb (6100 N).

20. A 100-lb plank, of length *l* = 20 ft, rests on the ground and on a frictionless roller (not shown) at the top of a wall of height *h* = 10 ft (see Fig. 14-20). The center of gravity of the plank is at its center. The plank remains in equilibrium for any value of $\theta \geq 70°$, but slips if $\theta < 70°$. (a) Draw a diagram showing all forces acting on the plank. (b) Find the coefficient of static friction between the plank and the ground.

21. A thin horizontal bar *AB* of negligible weight and length *l* is pinned to a vertical wall at *A* and supported at *B* by a thin wire *BC* that makes an angle θ

figure 14-17
Problem 11

figure 14-18
Problem 18

figure 14-19
Problem 19

figure 14-20
Problem 20

with the horizontal. A weight W can be moved anywhere along the bar as defined by the distance x from the wall (Fig. 14-21). (a) Find the tension T in the thin wire as a function of x. Find (b) the horizontal and (c) the vertical components of the force exerted on the bar by the pin at A.

Answer: (a) $Wx/(l \sin \theta)$. (b) $Wx/(l \tan \theta)$. (c) $W(1 - x/l)$.

22. A homogeneous sphere of radius r and weight **W** slides along the floor under the action of a constant horizontal force **P** applied to a string, as shown in Fig. 14-22. (a) Show that if μ is the coefficient of kinetic friction between sphere and floor, the height h is given by $h = r(l - \mu W/P)$. (b) Show that the sphere is not in translational equilibrium under these circumstances. Is there any point about which the sphere is in rotational equilibrium? (c) Can one get the sphere to be in both rotational *and* translational equilibrium by a different choice of h? By a different direction for **P**? Explain.

figure 14-21
Problem 21

figure 14-22
Problem 22

23. In the stepladder shown in Fig. 14-23 AC and CE are 8.0 ft long and hinged at C. BD is a tie rod 2.5 ft long, halfway up. A man weighing 192 lb climbs 6.0 ft along the ladder. Assuming that the floor is frictionless and neglecting the weight of the ladder, find (a) the tension in the tie rod and (b) the forces exerted on the ladder by the floor. (*Hint:* It will help to isolate parts of the ladder in applying the equilibrium conditions.)

Answer: (a) 47 lb. (b) $F_A = 120$ lb; $F_E = 72$ lb.

24. A cubical box is filled with sand and weighs 200 lb (890 N). It is desired to "roll" the box by pushing horizontally on one of the upper edges. (a) What minimum force is required? (b) What minimum coefficient of static friction is required? (c) Is there a more efficient way to roll the box? If so, find the lowest possible force that would be required to be applied directly to the box.

25. By means of a turnbuckle G, a tension force **T** is produced in bar AB of the square frame $ABCD$ in Fig. 14-24. Determine the forces produced in the other bars. The diagonals AC and BD pass each other freely at E. Symmetry considerations can lead to considerable simplification in this and similar problems.

Answer: Bars AD, BC, and DC are in tension (force T); diagonals AC and BD are in compression (force $\sqrt{2}\, T$).

26. A well-known problem is the following (see, for example, *Scientific American*, November 1964, p. 128): Uniform bricks are placed one upon another in such a manner as to have the maximum offset. This is accomplished by having the center of gravity of the top brick directly above the edge of the brick below it, the center of gravity of the two top bricks combined directly above the edge of the third brick from the top, etc. (a) Justify this criterion for maximum offset. (b) Show that, if the process is continued downward, one can obtain as large an offset as he wants. (Martin Gardner, in the article referred to above, states: "With 52 playing cards, the first placed so that its end is flush with a table edge, the maximum overhang is a little more than $2\frac{1}{4}$ card lengths. . . .") (c) Suppose now, instead, one piles up uniform bricks so that the end of one brick is offset from the one below it by a constant fraction, $1/n$, of a brick length l. How many bricks, N, can one use in this process before the pile will fall over? Check the plausibility of your answer for $n = 1$, $n = 2$, $n = \infty$.

figure 14-23
Problem 23

figure 14-24
Problem 25

SECTION 14-5

27. A bowl having a radius of curvature r rests on a rough horizontal table. Show that the bowl will be in stable equilibrium about the center point at its bottom only if the center of mass of the material piled up in the bowl is not as high as r above the center of the bowl.

28. A cube of uniform density and edge a is balanced on a cylindrical surface of radius r as shown in Fig. 14-25. Show that the criterion for stable equilibrium of the cube, assuming that friction is sufficient to prevent slipping, is $r > a/2$.

figure 14-25
Problem 28

15
oscillations

Any motion that repeats itself in equal intervals of time is called *periodic motion*. As we shall see, the displacement of a particle in periodic motion can always be expressed in terms of sines and cosines. Because the term *harmonic* is applied to expressions containing these functions, periodic motion is often called *harmonic motion*.

If a particle in periodic motion moves back and forth over the same path, we call the motion *oscillatory* or *vibratory*. The world is full of oscillatory motions. Some examples are the oscillations of the balance wheel of a watch, a violin string, a mass attached to a spring, atoms in molecules or in a solid lattice, and air molecules as a sound wave passes by.

Many oscillating bodies do not move back and forth between precisely fixed limits because frictional forces dissipate the energy of motion. Thus a violin string eventually stops vibrating and a pendulum stops swinging. We call such motions *damped* harmonic motions. Although we cannot eliminate friction from the periodic motions of gross objects, we can often cancel out its damping effect by feeding energy into the oscillating system so as to compensate for the energy dissipated by friction. The main spring of a watch and the falling weight in a pendulum clock supply external energy in this way, so that the oscillating system, that is, the balance wheel or the pendulum, moves as if it were undamped.

Not only mechanical systems can oscillate. Radio waves, microwaves, and visible light are oscillating magnetic and electric field vectors. Thus a tuned circuit in a radio and a closed metal cavity in which microwave energy is introduced can oscillate electromagnetically. The analogy is close, being based on the fact that *mechanical and electro-*

magnetic oscillations are described by the same basic mathematical equations. We will make the most of this analogy in later chapters.

The *period* T of a harmonic motion is the time required to complete one round trip of the motion, that is, one complete oscillation or *cycle*. The *frequency* of the motion ν is the number of oscillations (or cycles) per unit of time. The frequency is therefore the reciprocal of the period, or

$$\nu = 1/T. \qquad (15\text{-}1)$$

The SI unit of frequency is the cycle per second, or *hertz* (Hz).* The position at which no net force acts on the oscillating particle is called its *equilibrium* position. The *displacement* (linear or angular) is the distance (linear or angular) of the oscillating particle from its equilibrium position at any instant.

Let us focus attention on a particle oscillating back and forth along a straight line between fixed limits. Its displacement **x** changes periodically in both magnitude and direction. Its velocity **v** and acceleration **a** also vary periodically in magnitude and direction and, in view of the relation $\mathbf{F} = m\mathbf{a}$, so does the force **F** acting on the particle.

Forces associated with harmonic motion are the most general kinds of forces that we have discussed so far. In the earlier chapters we dealt only with constant forces (and accelerations). Later, when we considered forces that are not constant but instead vary with time, we examined a force (and thus an acceleration) that varied in direction although its magnitude was constant (the centripetal force of Section 6-3), and a force (and thus an acceleration) which varied in magnitude although its direction was constant (the impulsive force of Section 10-1). Here, in harmonic motion, the force, and the acceleration, vary both in direction *and* magnitude.

In terms of energy, we can say that a particle undergoing harmonic motion passes back and forth through a point (its equilibrium position) at which its potential energy is a minimum. A swinging pendulum is a good example, its potential energy being a minimum at the bottom of the swing, that is, at the equilibrium position. Figure 15-1a shows the generalized case of a particle oscillating between the limits x_1 and x_2, O being the equilibrium position. Figure 15-1b shows the corresponding potential energy curve, which has a minimum value at that position. The force acting on the particle at any position is derivable from the potential energy function; it is given by Eq. 8-7,

$$F = -dU/dx, \qquad (8\text{-}7)$$

and is displayed in Fig. 15-1c. The force is zero at the equilibrium position O, points to the right (that is, has a positive value) when the particle is to the left of O, and points to the left (that is, has a negative value) when the particle is to the right of O. The force is *a restoring force* because it always acts to accelerate the particle in the direction of its equilibrium position. Hence in harmonic motion the position of equilibrium is always one of *stable* equilibrium.

The total mechanical energy E for an oscillating particle is the sum of its kinetic energy and its potential energy, or

$$E = K + U \qquad (15\text{-}2)$$

figure 15-1
(a) A particle of mass m oscillates harmonically between points x_1 and x_2, O being the equilibrium position. *(b)* The potential energy of the particle as a function of position. The force acting on the particle at position x is given by $F = -dU/dx$. *(c)* The force acting on the particle as a function of position x; note that the force is directed toward the equilibrium position.

---

* This frequency unit is named after Heinrich Hertz (1857–1894) whose research provided the experimental confirmation of the electromagnetic waves predicted theoretically by James Clerk Maxwell (1831–1879).

in which E remains constant if no nonconservative forces, such as the force of friction, are acting. Figure 15-2 shows E for the motion of Fig. 15-1. Note how Eq. 15-2 is satisfied for the particle in the typical position shown. The particle cannot move outside the limits x_1 and x_2 because, in these regions, U exceeds E. This, as Eq. 15-2 shows, would require the kinetic energy to be negative, an impossibility.

For a given environment, that is, for a given function $U(x)$, an oscillating particle can have various total energies, depending on how we set it into motion initially. Thus the total energy may be E', rather than E, in which case the limits of oscillation would be x_1' and x_2', as Fig. 15-2 shows, rather than x_1 and x_2.

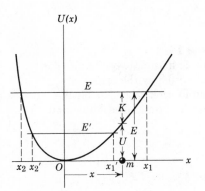

figure 15-2
The total mechanical energy E for the motion of Fig. 15-1 is shown. If the total mechanical energy of the oscillating particle is reduced to E', the limits of oscillation are reduced to x_1' and x_2' respectively.

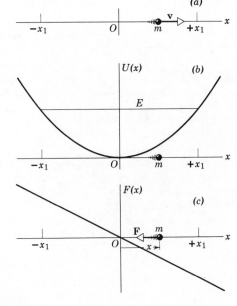

figure 15-3
(a) A particle of mass m oscillates with simple harmonic motion between points $+x_1$ and $-x_1$, O being the equilibrium position. (b) The potential energy $U(x)$ and the total mechanical energy E. (c) The force acting on the particle. Compare this figure carefully with Fig. 15-1, which illustrates the general case of harmonic motion.

15-2
THE SIMPLE HARMONIC OSCILLATOR

Let us consider an oscillating particle (Fig. 15-3a) moving back and forth about an equilibrium position through a potential that varies as

$$U(x) = \tfrac{1}{2}kx^2 \qquad (15\text{-}3)$$

in which k is a constant; see Fig. 15-3b. The force acting on the particle is given by Eq. 8-7, or

$$F(x) = -dU/dx = -d(\tfrac{1}{2}kx^2)/dx = -kx; \qquad (15\text{-}4)$$

DERIVATIVE OF $-\frac{1}{2}Kx^2 = Kx$

see Fig. 15-3c. Such an oscillating particle is called a *simple harmonic oscillator* and its motion is called *simple harmonic motion*. In such motion, as Eq. 15-3 shows, the potential energy curve varies as the square of the displacement, and, as Eq. 15-4 shows, the force acting on the particle is proportional to the displacement but is opposite to it in direction. In simple harmonic motion the limits of oscillation are equally spaced about the equilibrium position. This is not true for the more general motion of Fig. 15-1 which, although harmonic, is not simple harmonic. The magnitude of the maximum displacement, that is, the quantity x_1 in Fig. 15-3, always taken as positive, is called the *amplitude* of the simple harmonic motion.

You will have recognized Eq. 15-3 $[U(x) = \tfrac{1}{2}kx^2]$ as the expression for the potential energy of an "ideal" spring, compressed or extended by a distance x; see Section 8-4. In this same section an ideal spring was defined as one in which the force exerted by the stretched or compressed spring is given by $F(x) = -kx$ (see Eq. 15-4), k being called the *force constant*.

Hence, a body of mass m attached to an ideal spring of force constant k and free to move over a frictionless horizontal surface is an example of a simple harmonic oscillator (see Fig. 15-4). Note that there is a position (the equilibrium position; see Fig. 15-4b) in which the spring exerts no force on the body. If the body is displaced to the right (as in Fig. 15-4a), the force exerted by the spring on the body points to the left and is given by $F = -kx$. If the body is displaced to the left (as in Fig. 15-4c), the force points to the right and is also given by $F = -kx$. In each case the force is a *restoring* force. The motion of the oscillating mass is *simple harmonic motion*.

figure 15-4
A simple harmonic oscillator. The force exerted by the spring is shown in each case. The block slides on a frictionless horizontal table.

Let us apply Newton's second law, $F = ma$, to the motion of Fig. 15-4. For F we substitute $-kx$ (from Eq. 15-4) and for the acceleration a we put in d^2x/dt^2 ($= dv/dt$). This gives us

$$-kx = m\frac{d^2x}{dt^2}$$

or

$$\frac{d^2x}{dt^2} + \frac{k}{m}x = 0. \tag{15-5}$$

This equation involves derivatives and is, therefore, called a *differential equation*. To solve this equation means to determine how the displacement x of the particle must depend on the time t in order that the equation be satisfied. When we know how x depends on time, we know the motion of the particle; thus, Eq. 15-5 is called the *equation of motion* of a simple harmonic oscillator. We shall solve this equation and describe the motion in detail in the next section.

The simple harmonic oscillator problem is important for two reasons: First, most problems involving mechanical vibrations reduce to that of the simple harmonic oscillator at small amplitudes of vibration, or to a combination of such vibrations. This is equivalent to saying that if we consider a small enough portion of the restoring force curve of Fig. 15-1c (around the origin), it becomes arbitrarily close to a straight line which, as Fig. 15-3c shows, is characteristic of simple harmonic motion. Or, in other words, the potential energy curve of Fig. 15-1b for general oscillatory motion reduces to that of Fig. 15-3b for simple harmonic oscillation when the amplitude of vibration is made sufficiently small about the equilibrium position O.

Second, as we have indicated, equations like Eq. 15-5 occur in many physical problems in acoustics, in optics, in mechanics, in electrical circuits, and even in atomic physics. The simple harmonic oscillator exhibits features common to many physical systems.

Equation 15-4 $(F = -kx)$ is an empirical relation known as *Hooke's law*. It is a special case of a more general relation, dealing with the deformation of elastic bodies, discovered by Robert Hooke (1635–1703).* It is obeyed by springs and other elastic bodies provided the deformation is not too great. If the solid is deformed beyond a certain point, called its *elastic limit*, it will not even return to its original shape when the applied force is removed (Fig. 15-5). It turns out that Hooke's law holds almost up to the elastic limit for many common materials. The range of applied forces over which Hooke's law is valid is called the "proportional region." Beyond the elastic limit, the force can no longer be specified by a potential energy function, because the force then depends on many factors including the speed of deformation and the previous history of the solid.

Notice that the restoring force and potential energy function of the simple harmonic oscillator are the same as that of a solid deformed in one dimension in the proportional region. If the deformed solid is released, it will vibrate, just as the simple harmonic oscillator does. Therefore, as long as the amplitude of the vibration is small enough, that is, as long as the deformation remains in the proportional region, mechanical vibrations behave exactly like simple harmonic oscillators. It is easy to generalize this discussion to show that any problem involving mechanical vibrations of small amplitude in three dimensions reduces to a combination of simple harmonic oscillators.

The vibrating string or membrane, sound vibrations, the vibrations of atoms in solids, and electrical or acoustical oscillations in a cavity can be described in a form which is mathematically identical to a system of harmonic oscillators. The analogy enables us to solve problems in one area by using the techniques developed in other areas.

Let us now solve the equation of motion of the simple harmonic oscillator,

$$\frac{d^2x}{dt^2} + \frac{k}{m}x = 0. \tag{15-6}$$

Recall that any system of mass m upon which a force $F = -kx$ acts will be governed by this equation. In the case of a spring, the proportionality constant k is the force constant of the spring, which is a measure of its stiffness. In other oscillating systems the proportionality constant k may be related to other physical features of the system, as we shall see later. We can use the oscillating mass-spring system as our prototype.

*Hooke first expressed his law in 1676 as a Latin cryptogram *ceiiinosssttuv*. Two years later he deciphered this as *ut tensio sic vis*, which we may translate as: *the (ex)tension is proportional to the force*.

figure 15-5
Typical graph of applied force F versus resulting elongation of an aluminum bar under tension. The sample was a foot long and a square inch in cross section. Notice that we may write $F = kx$ only for the portion Oa, since beyond this point the slope is no longer constant but varies in a complicated way with x. At some point such as b (the *elastic limit*) the sample does not return to its original length when the applied force is removed. Between b and b' the elongation increases, even though the force is held constant; the material flows like a viscous fluid. At c, the sample can be stretched no farther; any increase in elongation results in the sample's breaking in two. The applied force is equal in magnitude to the restoring force so that no minus sign appears in the relation $F = kx$.

15-3
SIMPLE HARMONIC MOTION

Equation 15-6 is a differential equation. It gives a relation between a function of the time $x(t)$ and its second time derivative d^2x/dt^2. To find the position of the particle as a function of the time, we must find a function $x(t)$ which satisfies this relation.

We can rewrite Eq. 15-6 as

$$\frac{d^2x}{dt^2} = -\left(\frac{k}{m}\right)x. \tag{15-7}$$

Equation 15-7 then requires that $x(t)$ be some function whose second derivative is the negative of the function itself, except for a constant factor k/m. We know from the calculus, however, that the sine function or the cosine function has this property.* For example,

$$\frac{d}{dt}\cos t = -\sin t \quad \text{and} \quad \frac{d^2}{dt^2}\cos t = -\frac{d}{dt}\sin t = -\cos t.$$

This property is not affected if we multiply the cosine function by a constant A. We can allow for the fact that the sine function will do as well, and for the fact that Eq. 15-7 contains a constant factor, by writing as a tentative solution of Eq. 15-7,

$$x = A \cos(\omega t + \phi). \tag{15-8}$$

Here since

$$\cos(\omega t + \phi) = \cos\phi \cos\omega t - \sin\phi \sin\omega t = a\cos\omega t + b\sin\omega t,$$

the constant ϕ allows for any combination of sine and cosine solutions. Hence, with the (as yet) unknown constants A, ω, and ϕ, we have written as general a solution to Eq. 15-7 as we can. In order to determine these constants such that Eq. 15-8 is actually the solution of Eq. 15-7, we differentiate Eq. 15-8 twice with respect to the time. We have

$$\frac{dx}{dt} = -\omega A \sin(\omega t + \phi)$$

and

$$\frac{d^2x}{dt^2} = -\omega^2 A \cos(\omega t + \phi).$$

Putting this into Eq. 15-7, we obtain

$$-\omega^2 A \cos(\omega t + \phi) = -\frac{k}{m} A \cos(\omega t + \phi).$$

Therefore, if we choose the constant ω such that

$$\omega^2 = \frac{k}{m}, \tag{15-9}$$

then

$$x = A \cos(\omega t + \phi)$$

is in fact a solution of the equation of a simple harmonic oscillator.

The constants A and ϕ are still undetermined and, therefore, still completely arbitrary. This means that any choice of A and ϕ whatsoever will satisfy Eq. 15-7, so that a large variety of motions is possible for the oscillator. Actually, this is characteristic of a differential equation of motion, for such an equation does not describe just one single motion but a group or family of possible motions which have some features in common but differ in other ways. In this case ω is common to all the allowed motions, but A and ϕ may differ among them. We

---

* Harmonic motion is not only periodic but also bounded. Only the sine and cosine functions (or combinations of them) have both these properties.

shall see later that A and ϕ are determined for a particular harmonic motion by how the motion starts.

Let us find the *physical* significance of the constant ω. If we increase the time t in Eq. 15-8 by $2\pi/\omega$, the function becomes

$$x = A \cos\,[\omega(t + 2\pi/\omega) + \phi],$$
$$= A \cos\,(\omega t + 2\pi + \phi),$$
$$= A \cos\,(\omega t + \phi).$$

That is, the function merely repeats itself after a time $2\pi/\omega$. Therefore, $2\pi/\omega$ is the *period* of the motion T. Since $\omega^2 = k/m$, we have

$$T = \frac{2\pi}{\omega} = 2\pi\sqrt{\frac{m}{k}}. \qquad (15\text{-}10)$$

Hence, all motions given by Eq. 15-7 have the same period of oscillation, and this is determined only by the mass m of the oscillating particle and the force constant k of the spring. The *frequency ν* of the oscillator is the number of complete vibrations per unit time and is given by

$$\nu = \frac{1}{T} = \frac{\omega}{2\pi} = \frac{1}{2\pi}\sqrt{\frac{k}{m}}. \qquad (15\text{-}11)$$

Hence,
$$\omega = 2\pi\nu = \frac{2\pi}{T}. \qquad (15\text{-}12)$$

The quantity ω is called the *angular frequency*; it differs from the frequency ν by a factor 2π. It has the dimension of reciprocal time (the same as angular speed) and its unit is the radian/second. In Section 15-6 we shall give a geometric meaning to this angular frequency.

The constant A has a simple physical meaning. The cosine function takes on values from -1 to 1. The *displacement* x from the central equilibrium position $x = 0$, therefore, has a maximum value of A; see Eq. 15-8. We call A ($= x_{\text{max}}$) the *amplitude* of the motion. Because A is not fixed by our differential equation, motions of various amplitudes are possible, but all have the same frequency and period. *The frequency of a simple harmonic motion is independent of the amplitude of the motion.*

The quantity $(\omega t + \phi)$ is called the *phase* of the motion. The constant ϕ is called the *phase constant*. Two motions may have the same amplitude and frequency but differ in phase. If $\phi = -\pi/2$, for example,

$$x = A \cos\,(\omega t + \phi) = A \cos\,(\omega t - 90°)$$
$$= A \sin \omega t$$

so that the displacement is zero at the time $t = 0$. When $\phi = 0$, the displacement $x = A \cos \omega t$ is a maximum at the time $t = 0$. Other initial displacements correspond to other phase constants.

The amplitude A and the phase constant ϕ of the oscillation are determined by the initial position and speed of the particle. These two initial conditions will specify A and ϕ exactly.* Once the motion has started, however, the particle will continue to oscillate with a constant amplitude and phase constant at a fixed frequency, unless other forces disturb the system.

---

*A phase constant may be increased by any integral multiple of 2π, or of 360°, and it will still describe the motion equally well.

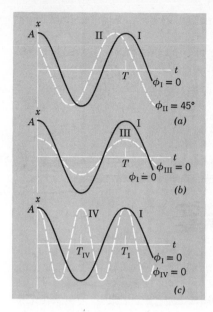

figure 15-6
Several solutions of the simple
harmonic oscillator equation. *(a)*
Both solutions have the same
amplitude and period but differ in
phase by 45°. *(b)* Both have the
same period and phase constant but
differ in amplitude by a factor of 2.
(c) Both have the same phase
constant and amplitude but differ
in period by a factor of 2.

PHASE – ϕ
FREQUENCY – # OF HUMPS
AMPLITUDE – HIEGHTH OF HUMPS

In Fig. 15-6 we plot the displacement x versus the time t for several
simple harmonic motions described by Eq. 15-8. Three comparisons are
made. In Fig. 15-6a, I and II have the same amplitude and frequency but
differ in phase by $\phi = \pi/4$ or 45°. In Fig. 15-6b, I and III have the same
frequency and phase constant but differ in amplitude by a factor of 2.
In Fig. 15-6c, I and IV have the same amplitude and phase constant but
differ in frequency by a factor $\frac{1}{2}$ or in period by a factor 2. Study these
curves carefully to become familiar with the terminology used in simple harmonic motion.

Another distinctive feature of simple harmonic motion is the relation between the displacement, the velocity, and the acceleration of the
oscillating particle. Let us compare these quantities for curve I of Fig.
15-6, which is typical. In Fig. 15-7 we plot separately the displacement
x versus the time t, the velocity $v = dx/dt$ versus the time t, and the
acceleration $a = dv/dt = d^2x/dt^2$ versus the time t. The equations of
these curves are

$$x = A \cos (\omega t + \phi),$$

$$v = \frac{dx}{dt} = -\omega A \sin (\omega t + \phi), \qquad (15\text{-}13)$$

$$a = \frac{dv}{dt} = -\omega^2 A \cos (\omega t + \phi).$$

For the case plotted we have taken $\phi = 0$. The units and scale of displacement, velocity, and acceleration are omitted for simplicity of comparison. Notice that (see Eq. 15-13) the maximum displacement is A,
the maximum speed is ωA, and the maximum acceleration is $\omega^2 A$.

When the displacement is a maximum in either direction, the speed
is zero because the velocity must now change its direction. The acceleration at this instant, like the restoring force, has a maximum value
but is directed opposite to the displacement. When the displacement is
zero, the speed of the particle is a maximum and the acceleration is
zero, corresponding to a zero restoring force. The speed increases as the
particle moves toward the equilibrium position and then decreases as
it moves out to the maximum displacement, just as for a pendulum bob.

figure 15-7
The relations between displacement,
velocity, and acceleration in simple
harmonic motion. The phase
constant ϕ is zero in this particular
case since the displacement is
maximum at $t = 0$; see Eq. 15-8.

figure 15-8
The force acting on, and the acceleration, velocity and displacement of a mass m undergoing simple harmonic motion. Compare carefully with Fig. 15-7.

In Fig. 15-8 we show the instantaneous values of $\mathbf{x}$, $\mathbf{v}$, and $\mathbf{a}$ at four instants in the motion of a particle oscillating at the end of a spring.

Equation 15-2 tells us that for harmonic motion, including simple harmonic motion, in which no dissipative forces act, the total mechanical energy $E\ (= K + U)$ is conserved (remains constant). We can now study this in more detail for the special case of simple harmonic motion, for which the displacement is given by

15-4
ENERGY CONSIDERATIONS IN SIMPLE HARMONIC MOTION

$$x = A \cos (\omega t + \phi). \tag{15-8}$$

The potential energy U at any instant is given by

$$U = \tfrac{1}{2}kx^2$$

$$= \tfrac{1}{2}kA^2 \cos^2 (\omega t + \phi). \tag{15-14}$$

$\}$ *Know!*

The potential energy has a maximum value of $\tfrac{1}{2}kA^2$. During the motion the potential energy varies between zero and this maximum value, as the curves in Fig. 15-9a and 15-9b show.

The kinetic energy K at any instant is $\tfrac{1}{2}mv^2$. Using the relations

$$v = dx/dt = -\omega A \sin (\omega t + \phi)$$

and

$$\omega^2 = k/m,$$

we obtain

$$K = \tfrac{1}{2}mv^2,$$

$$= \tfrac{1}{2}m\omega^2 A^2 \sin^2 (\omega t + \phi),$$

$$= \tfrac{1}{2}kA^2 \sin^2 (\omega t + \phi). \tag{15-15}$$

The kinetic energy, therefore, has a maximum value of $\tfrac{1}{2}kA^2$ or $\tfrac{1}{2}m(\omega A)^2$, in agreement with the maximum speed ωA noted earlier. During the

(a)

$K + U = K_{max} = U_{max} = \frac{1}{2}kA^2 = E$

$U = U_{max} \cos^2 \omega t$

$K = K_{max} \sin^2 \omega t$

(b)

$K_{max} = U_{max} = \frac{1}{2}kA^2 = E$

$K(x) = \frac{1}{2}mv^2$

$U(x) = \frac{1}{2}kx^2$

figure 15-9

Energies of a simple harmonic oscillator. *(a)* Potential energy $(- - - -)$, kinetic energy $(- \cdot -)$, and total energy $(— —)$ plotted as a function of time. *(b)* Potential, kinetic, and total energy plotted as a function of displacement from the equilibrium position. Compare with Fig. 8-4.

motion the kinetic energy varies between zero and this maximum value, as shown by the curves in Fig. 15-9a and 15-9b.

The total mechanical energy is the sum of the kinetic and the potential energy. Using Eqs. 15-14 and 15-15 we obtain

$$E = K + U = \frac{1}{2}kA^2 \sin^2 (\omega t + \phi) + \frac{1}{2}kA^2 \cos^2 (\omega t + \phi) = \frac{1}{2}kA^2. \quad (15\text{-}16)$$

We see that the total mechanical energy is constant, as we expect, and has the value $\frac{1}{2}kA^2$. At the maximum displacement the kinetic energy is zero, but the potential energy has the value $\frac{1}{2}kA^2$. At the equilibrium position the potential energy is zero, but the kinetic energy has the value $\frac{1}{2}kA^2$. At other positions the kinetic and potential energies each contribute energy whose sum is always $\frac{1}{2}kA^2$. This constant total energy is shown in Fig. 15-9a and 15-9b. *The total energy of a particle executing simple harmonic motion is proportional to the square of the amplitude of the motion.* It is clear from Fig. 15-9a that the *average* kinetic energy for the motion during one period is exactly equal to the *average* potential energy and that each of these average quantities is $\frac{1}{4}kA^2$.

Equation 15-16 can be written quite generally as

$$K + U = \frac{1}{2}mv^2 + \frac{1}{2}kx^2 = \frac{1}{2}kA^2. \quad (15\text{-}17)$$

From this relation we obtain $v^2 = (k/m)(A^2 - x^2)$ or

$$v = \frac{dx}{dt} = \pm \sqrt{\frac{k}{m}(A^2 - x^2)}. \quad (15\text{-}18)$$

This relation shows clearly that the speed is a maximum at the equilibrium position $x = 0$ and zero at the maximum displacement $x = A$. In fact, we can start from the conservation of energy principle, Eq. 15-17 (in which $\frac{1}{2}kA^2 = E$), and by integration of Eq. 15-18 obtain the displacement as a function of time. The result is identical with Eq. 15-8 which we deduced from the differential equation of the motion, Eq. 15-6. (See Problem 29.)

The effect of dissipative forces will be discussed in Section 15-9.

EXAMPLE 1

The horizontal spring of Fig. 15-4 is found to be stretched 3.0 in. from its equilibrium position when a force of 0.75 lb acts on it. Then a 1.5-lb body is attached to the end of the spring and is pulled 4.0 in. along a horizontal frictionless table from the equilibrium position. The body is then released and executes simple harmonic motion.

(a) What is the force constant of the spring?

A force of 0.75 lb *on* the spring produces a displacement of 0.25 ft. Hence,

$$k = F/x = 0.75 \text{ lb}/0.25 \text{ ft} = 3.0 \text{ lb/ft}.$$

Why didn't we use $k = -F/x$ here?

(b) What is the force exerted by the spring on the 1.5-lb body just before it is released?

The spring is stretched 4.0 in. or $\frac{1}{3}$ ft. Hence, the force exerted *by* the spring is

$$F = -kx = -(3.0 \text{ lb/ft})(\tfrac{1}{3} \text{ ft}) = -1.0 \text{ lb}.$$

The minus sign indicates that the force is directed opposite to the displacement.

(c) What is the period of oscillation after release?

$$T = 2\pi \sqrt{\frac{m}{k}} = 2\pi \sqrt{\frac{1.5/32}{3.0}} \text{ s} = \frac{\pi}{4} \text{s} = 0.79 \text{ s}.$$

This corresponds to a frequency $\nu (= 1/T)$ of 1.3 Hz and to an angular frequency $\omega (= 2\pi\nu)$ of 8.0 rad/s.

(d) What is the amplitude of the motion?

The maximum displacement corresponds to zero kinetic energy and a maximum potential energy. This is the initial condition before release, so that the amplitude is the initial displacement of 4.0 in. Hence, $A = \frac{1}{3}$ ft.

(e) What is the maximum speed of the vibrating body?

From Eq. 15-13, $v_{max} = \omega A = (2\pi/T)A$,

$$v_{max} = \left(\frac{2\pi}{\pi/4} \text{ s}^{-1}\right)(\tfrac{1}{3} \text{ ft}) = 2.7 \text{ ft/s}.$$

The maximum speed occurs at the equilibrium position, where $x = 0$. This value is achieved twice in each period, the velocity being -2.7 ft/s when the body passes through $x = 0$ after release and $+2.7$ ft/s when the body passes through $x = 0$ on the return trip.

(f) What is the maximum acceleration of the body?

From Eq. 15-13, $a_{max} = \omega^2 A = (k/m)A$,

$$a_{max} = \left(\frac{3.0}{1.5/32}\right)\left(\frac{1}{3}\right) \text{ ft/s}^2 = 21 \text{ ft/s}^2.$$

The maximum acceleration occurs at the ends of the path where $x = \pm A$ and $v = 0$. Hence, $a = -21$ ft/s² at $x = +A$ and $a = +21$ ft/s² at $x = -A$, the acceleration and displacement being oppositely directed.

(g) Compute the velocity, the acceleration, and the kinetic and potential energies of the body when it has moved in halfway from its initial position toward the center of motion.

At this point,
$$x = \frac{A}{2} = \tfrac{1}{6} \text{ ft},$$

so that from Eq. 15-18,

$$v = -\frac{2\pi}{T} \sqrt{A^2 - x^2}$$

$$= -\frac{2\pi}{\pi/4} \sqrt{\left(\frac{1}{3}\right)^2 - \left(\frac{1}{6}\right)^2} \text{ ft/s} = -\frac{4}{\sqrt{3}} \text{ ft/s} = -2.3 \text{ ft/s},$$

$$a = -\frac{k}{m}x = \frac{-3.0}{1.5/32}\left(\frac{1}{6}\right) \text{ ft/s}^2 = -11 \text{ ft/s}^2,$$

$$K = \tfrac{1}{2}mv^2 = \left(\frac{1}{2}\right)\left(\frac{1.5}{32}\right)\left(\frac{4}{\sqrt{3}}\right)^2 \text{ ft} \cdot \text{lb} = \tfrac{1}{8} \text{ ft} \cdot \text{lb},$$

$$U = \tfrac{1}{2}kx^2 = \left(\frac{1}{2}\right)(3)\left(\frac{1}{6}\right)^2 \text{ ft} \cdot \text{lb} = \tfrac{1}{24} \text{ ft} \cdot \text{lb}.$$

(h) Compute the total energy of the oscillating system.

Since the total energy is conserved, we can compute it at any stage of the motion. Using previous results, we obtain

$$E = K + U = (\tfrac{1}{8} + \tfrac{1}{24}) \text{ ft} \cdot \text{lb} = \tfrac{1}{6} \text{ ft} \cdot \text{lb}, \qquad \text{(particle at } x = A/2)$$

$$E = U_{max} = \tfrac{1}{2}kx_{max}^2 = \left(\frac{1}{2}\right)(3)\left(\frac{1}{3}\right)^2 \text{ ft} \cdot \text{lb} = \tfrac{1}{6} \text{ ft} \cdot \text{lb}, \qquad \text{(particle at } x = A)$$

$$E = K_{max} = \tfrac{1}{2}mv_{max}^2 = \left(\frac{1}{2}\right)\left(\frac{1.5}{32}\right)\left(\frac{8}{3}\right)^2 \text{ ft} \cdot \text{lb} = \tfrac{1}{6} \text{ ft} \cdot \text{lb}. \qquad \text{(particle at } x = 0)$$

(i) What is the displacement of the body as a function of time?

In general, we have

$$x = A \cos (\omega t + \phi).$$

We have already found that $A = \tfrac{1}{3}$ ft. We must now determine ω and ϕ. We obtain

$$\omega = \frac{2\pi}{T} = \frac{2\pi}{\pi/4} = 8 \text{ rad/s},$$

so that, with our particular units,

$$x = \tfrac{1}{3} \cos (8t + \phi).$$

At the time $t = 0$, $x = \tfrac{1}{3}$ ft, so that at that instant

$$x = \tfrac{1}{3} \cos \phi = \tfrac{1}{3}$$

or $\phi = 0$ rad.

Therefore, with $A = \tfrac{1}{3}$ ft, $\omega = 8$ rad/s, and $\phi = 0$ rad, we obtain

$$x = \tfrac{1}{3} \cos 8t.$$

This describes the motion of the body, where x is in feet, t is in seconds, and the angle $8t$ is in radians.

A few physical systems that move with simple harmonic motion are considered here. We will discuss others from time to time throughout the text.

15-5
APPLICATIONS OF SIMPLE HARMONIC MOTION*

The Simple Pendulum. A simple pendulum is an idealized body consisting of a point mass, suspended by a light inextensible cord. When pulled to one side of its equilibrium position and released, the pendulum swings in a vertical plane under the influence of gravity. The motion is periodic and oscillatory. We wish to determine the period of the motion.

Figure 15-10 shows a pendulum of length l, particle mass m, making an angle θ with the vertical. The forces acting on m are mg, the gravita-

---

* See "A Repertoire of S.H.M." by Eli Maor, *The Physics Teacher*, October 1972, for a full discussion of 16 physical systems that exhibit simple harmonic motion.

tional force, and **T,** the tension in the cord. Choose axes tangent to the circle of motion and along the radius. Resolve mg into a radial component of magnitude $mg \cos \theta$, and a tangential component of magnitude $mg \sin \theta$. The radial components of the forces supply the necessary centripetal acceleration to keep the particle moving on a circular arc. The tangential component is the restoring force acting on m tending to return it to the equilibrium position. Hence, the restoring force is

$$F = -mg \sin \theta.$$ FOR A SIMPLE PENDULUM.

Notice that the restoring force is not proportional to the angular displacement θ but to $\sin \theta$ instead. The resulting motion is, therefore, not simple harmonic. However, *if the angle θ is small,* $\sin \theta$ is very nearly equal to θ in radians.* The displacement along the arc is $x = l\theta$, and for small angles this is nearly straight-line motion. Hence, assuming

$$\sin \theta \cong \theta,$$ (SMALL ANGLE APPROXIMATION)

we obtain

$$F = -mg\theta = -mg\frac{x}{l} = -\left(\frac{mg}{l}\right)x.$$

For *small displacements,* therefore, the restoring force is proportional to the displacement and is oppositely directed. This is exactly the criterion for simple harmonic motion. The constant mg/l represents the constant k in $F = -kx$. Check the dimensions of k and mg/l. The period of a simple pendulum when its amplitude is small is

$$T = 2\pi \sqrt{\frac{m}{k}} = 2\pi \sqrt{\frac{m}{mg/l}} \qquad \text{or} \qquad T = 2\pi \sqrt{\frac{l}{g}}. \qquad (15\text{-}19)$$

Notice that the period is independent of the mass of the suspended particle.

When the amplitude of the oscillation is not small, the general equation for the period can be shown to be

$$T = 2\pi \sqrt{\frac{l}{g}}\left(1 + \frac{1}{2^2} \cdot \sin^2\frac{\theta_m}{2} + \frac{1}{2^2} \cdot \frac{3^2}{4^2} \cdot \sin^4\frac{\theta_m}{2} + \cdots\right). \qquad (15\text{-}20)$$

Here θ_m is the maximum angular displacement and the succeeding terms become smaller and smaller. The period can then be computed to any desired degree of accuracy by taking enough terms in the infinite series. When $\theta_m = 15°$, corresponding to a total to-and-fro angular displacement of $30°$, the true period differs from that given by Eq. 15-19 by less than 0.5%.

Because the period of a simple pendulum is practically independent of the amplitude, the pendulum is useful as a timekeeper. As damping forces reduce the amplitude of swing, the period remains very nearly unchanged. In a pendulum clock energy is supplied automatically by an escapement mechanism to compensate for frictional loss. The pendulum clock with escapement was invented by Christian Huygens (1629–1695).

The simple pendulum also provides a convenient method for measuring the value of g, the acceleration due to gravity. We need not perform a free-fall experiment here, but instead we merely measure l and T.

figure 15-10
The forces acting on a simple pendulum are the tension **T** in the string and the weight mg of mass m. The magnitudes of the radial and tangential components of mg are labeled.

* For example,.

θ	$\sin \theta$	Difference, %
$0° = 0.00000$ rad	0.00000	0.00
$2° = 0.03491$ rad	0.03490	0.03
$5° = 0.08727$ rad	0.08716	0.24
$10° = 0.17453$ rad	0.17365	0.50
$15° = 0.26180$ rad	0.25882	1.14

The Torsional Pendulum. In Fig. 15-11 we show a disk suspended by a wire attached to the center of mass of the disk. The wire is securely fixed to a solid support and to the disk. At the equilibrium position of the disk a radial line is drawn from its center to P, as shown. If the disk is rotated in a horizontal plane to the radial position Q, the wire will be twisted. The twisted wire will exert a torque on the disk tending to return it to the position P. This is a restoring torque. For small twists the restoring torque is found to be proportional to the amount of twist, or the angular displacement (Hooke's law), so that $F = -Kx$

$$\tau = -\kappa\theta.$$ FOR TORSIONAL PENDULUM   (15-21)

Here κ is a constant that depends on the properties of the wire and is called the *torsional* constant. The minus sign shows that the torque is directed opposite to the angular displacement θ. Equation 15-21 is the condition for *angular simple harmonic motion.*

The equation of motion for such a system is

$$\tau = I\alpha = I\frac{d\omega}{dt} = I\frac{d^2\theta}{dt^2},$$

so that, on using Eq. 15-21, we obtain

$$-\kappa\theta = I\frac{d^2\theta}{dt^2}$$

or

$$\frac{d^2\theta}{dt^2} = -\left(\frac{\kappa}{I}\right)\theta.$$ (15-22)

Notice the similarity between Eq. 15-22 for simple angular harmonic motion and Eq. 15-7 for simple linear harmonic motion. In fact, the equations are mathematically identical. We have simply substituted angular displacement θ for linear displacement x, rotational inertia I for mass m, and torsional constant κ for force constant k. By substituting these correspondences, we find the solution of Eq. 15-22, therefore, to be a simple harmonic oscillation in the angle coordinate θ, namely

$$\theta = \theta_m \cos(\omega t + \phi).$$ (15-23)

Here, θ_m is the maximum angular displacement, that is, the amplitude of the angular oscillation. In Fig. 15-11 the disk oscillates about the equilibrium position $\theta = 0$ (line OP), the total angular range being $2\theta_m$ (from OQ to OR).

The period of the oscillation by analogy with Eq. 15-10 is

$$T = 2\pi\sqrt{\frac{I}{\kappa}}.$$ (15-24)

If κ is known and T is measured, the rotational inertia I about the axis of rotation of any oscillating rigid body can be determined. If I is known and T is measured, the torsional constant κ of any sample of wire can be determined.

Many laboratory instruments involve torsional oscillations, notably the galvanometer. The Cavendish balance is a torsional pendulum (Chapter 16). The balance wheel of a watch is another example of angular harmonic motion, the restoring torque here being supplied by a spiral hairspring.

figure 15-11
The torsional pendulum. The line drawn from the center to P oscillates between Q and R, sweeping out an angle $2\theta_m$ where θ_m is the (angular) amplitude of the motion.

LINEAR HARMONIC MOTION

$-\frac{K}{m}x$

PERIOD OF OSCILLATION FOR LINEAR HARMONIC MOTION

EXAMPLE 2

A thin rod of mass 0.10 kg and length 0.10 m is suspended by a wire which passes through its center and is perpendicular to its length. The wire is twisted and the rod set oscillating. The period is found to be 2.0 s. When a flat body in the shape of an equilateral triangle is suspended similarly through its center of mass, the period is found to be 6.0 s. Find the rotational inertia of the triangle about this axis.

The rotational inertia of the rod is $Ml^2/12$ (see Table 12-1). Hence

$$I_{rod} = \frac{(0.10 \text{ kg})(0.10 \text{ m})^2}{12} = 8.3 \times 10^{-5} \text{ kg} \cdot \text{m}^2.$$

From Eq. 15-24,

$$\frac{T_{rod}}{T_{triangle}} = \left(\frac{I_{rod}}{I_{triangle}}\right)^{1/2} \quad \text{or} \quad I_{triangle} = I_{rod}\left(\frac{T_t}{T_r}\right)^2,$$

so that

$$I_{triangle} = (8.3 \times 10^{-5} \text{ kg} \cdot \text{m}^2)\left(\frac{6.0 \text{ s}}{2.0 \text{ s}}\right)^2 = 7.5 \times 10^{-4} \text{ kg} \cdot \text{m}^2.$$

Does the amplitude of the oscillation affect the period in these cases?

The Physical Pendulum. Any rigid body mounted so that it can swing in a vertical plane about some axis passing through it is called a physical pendulum. This is a generalization of the simple pendulum in which a weightless cord holds a single particle. Actually all real pendulums are physical pendulums.

For convenience we choose our pendulum to be a laminar body, such as may be cut out from a sheet of plywood with a jigsaw, and we choose the axis of oscillation to be at right angles to the plane of this body. We lose nothing essential by this restriction.

In Fig. 15-12 a body of irregular shape is pivoted about a horizontal frictionless axis through P and displaced from the equilibrium position by an angle θ. The equilibrium position is that in which the center of mass of the body, C, lies vertically below P. The distance from pivot to center of mass is d, the rotational inertia of the body about an axis through the pivot is I, and the mass of the body is M. The restoring torque for an angular displacement θ is

$$\boxed{\tau = -Mgd \sin \theta} \quad \text{FOR A PHYSICAL PENDULUM}$$

and is due to the tangential component of the force of gravity. Since τ is proportional to $\sin \theta$, rather than θ, the condition for simple angular harmonic motion does not, in general, hold here. For small angular displacements, however, the relation $\sin \theta \cong \theta$ is, as before, an excellent approximation, so that for *small amplitudes*,

$$\tau = -Mg \, d \, \theta$$

or

$$\tau = -\kappa\theta,$$

where

$$\kappa = Mgd.$$

But

$$\tau = I\frac{d^2\theta}{dt^2} = I\alpha,$$

so that

$$\frac{d^2\theta}{dt^2} = \frac{\tau}{I} = -\frac{\kappa}{I}\theta.$$

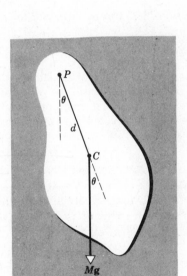

figure 15-12
A laminar physical pendulum, with center of mass C, is pivoted at P and displaced an angle θ from its equilibrium position (when C hangs directly below P). Its weight Mg supplies a restoring torque.

Hence, the period of a physical pendulum oscillating with small amplitude is

$$T = 2\pi\sqrt{\frac{I}{\kappa}} = 2\pi\sqrt{\frac{I \cdot}{Mgd}}. \qquad (15\text{-}25)$$

At larger amplitudes the physical pendulum still has a harmonic motion, but not a simple harmonic one.

Notice that this treatment applies to a laminar object of *any* shape and that the pivot can be located *anywhere*. As a special case consider a point mass m suspended at the end of a weightless string of length l. Here

$$I = ml^2, \qquad M = m, \qquad d = l,$$

so that

$$T = 2\pi\sqrt{\frac{I}{Mgd}} = 2\pi\sqrt{\frac{l}{g}},$$

RELATIONSHIP BETWEEN A SIMPLE PENDULUM AND A PHYSICAL PENDULUM.

which is the period of a simple pendulum with small amplitude. The physical pendulum is often used for accurate determinations of g.

Equation 15-25 can be solved for the rotational inertia I, giving

$$I = \frac{T^2 Mgd}{4\pi^2}. \qquad (15\text{-}26)$$

The quantities on the right are all directly measurable. The center of mass can be determined by suspension as was shown in Fig. 14-4. Hence, the rotational inertia about an axis of rotation (other than through the center of mass) of a body of any shape can be determined by suspending the body as a physical pendulum from that axis.

Find the length of a simple pendulum whose period is equal to that of a particular physical pendulum.

Equating the period of a simple pendulum to that of a physical pendulum, we obtain

$$T = 2\pi\sqrt{\frac{l}{g}} = 2\pi\sqrt{\frac{I}{Mgd}}$$

or

$$l = \frac{I}{Md}. \qquad (15\text{-}27)$$

Hence, as far as its period of oscillation is concerned, the mass of a physical pendulum may be considered to be concentrated at a point whose distance from the pivot is $l = I/Md$. This point is called the *center of oscillation* of the physical pendulum. Notice that it depends on the location of the pivot for any given body.

EXAMPLE 3

A disk is pivoted at its rim (Fig. 15-13). Find its period for small oscillations and the length of the equivalent simple pendulum.

The rotational inertia of a disk about an axis through its center is $\frac{1}{2}Mr^2$, where r is the radius and M is the mass of the disk. The rotational inertia about the pivot at the rim is

$$I = \tfrac{1}{2}Mr^2 + Mr^2 = \tfrac{3}{2}Mr^2.$$

The period then, with $d = r$, is

EXAMPLE 4

$$T = 2\pi\sqrt{\frac{I}{Mgr}} = 2\pi\sqrt{\frac{3}{2}\frac{Mr^2}{Mgr}} = 2\pi\sqrt{\frac{3}{2}\frac{r}{g}},$$

independent of the mass of the disk.

The simple pendulum having the same period has a length

$$l = \frac{I}{Mr} = \tfrac{3}{2}r$$

or three-fourths the diameter of the disk. The center of oscillation of the disk pivoted at P is, therefore, at O, a distance $\tfrac{3}{2}r$ below the point of support. Is any particular mass required of the equivalent simple pendulum?

If we pivot the disk at a point midway between the rim and the center, as at O, we find that $I = \tfrac{3}{4}Mr^2$ and $d = \tfrac{1}{2}r$. The period T is

$$T = 2\pi\sqrt{\tfrac{3}{2}r/g}$$

just as before. This illustrates a general property of the center of oscillation O and the point of support P, namely, if the pendulum is pivoted about a new axis through O, its period is unchanged and P becomes the new center of oscillation.

If the disk were pivoted at the center, what would be its period of oscillation?

EXAMPLE 5

The center of oscillation of a physical pendulum has another interesting property. If an impulsive force (assumed horizontal and in the plane of oscillation) acts at the center of oscillation, no reaction is felt at the point of support. Prove this for an impulsive force $\mathbf{F}$ acting toward the left at point O in Fig. 15-13. Assume the pendulum to be initially at rest.

This is a case of combined translation and rotation (see Sec. 12-7). The translation effect, acting alone, would make P in Fig. 15-13 move to the left with an acceleration

$$\mathbf{a}_{\text{left}} = F/M.$$

The rotational effect, acting alone, would produce a clockwise angular acceleration about C of

$$\alpha = \tau/I$$
$$= (F)(\tfrac{1}{2}r)/(\tfrac{1}{2}Mr^2)$$
$$= F/Mr.$$

Because of this angular acceleration P would move to the right with an acceleration

$$\mathbf{a}_{\text{right}} = \alpha r$$
$$= (F/Mr)(r) = F/M.$$

Thus $\mathbf{a}_{\text{left}} = -\mathbf{a}_{\text{right}}$ and there is no movement at point P.

When viewed from this point of view the center of oscillation is often called the *center of percussion*. Baseball players know that unless the ball hits the bat at just the right spot (center of percussion) the impact will sting their hands. The "sting" has a different direction depending on whether the ball strikes on one side or the other of this spot.

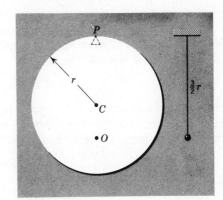

figure 15-13
Example 4. A physical pendulum consisting of a disk pivoted at the edge *(P)*, along with a simple pendulum having the same period. *O* is the center of oscillation.

EXAMPLE 6

The period of a disk of radius 10.2 cm executing small oscillations about a pivot at its rim is measured to be 0.784 s. Find the value of g, the acceleration due to gravity at that location.

From $T = 2\pi\sqrt{\tfrac{3}{2}r/g}$, we obtain

$$g = \frac{6\pi^2 r}{T^2}.$$

With $T = 0.784$ s and $r = 0.102$ m, we obtain

$$g = \frac{6\pi^2 \cdot 0.102}{(0.784)^2} \text{ m/s}^2 = 9.82 \text{ m/s}^2.$$

Let us consider the relation between simple harmonic motion along a straight line and uniform circular motion. This relation is useful in describing many features of simple harmonic motion. It also gives a simple geometric meaning to the angular frequency ω and the phase constant ϕ. Uniform circular motion is also an example of a combination of simple harmonic motions, a phenomenon we deal with rather often in wave motion.

In Fig. 15-14 Q is the point moving around a circle of radius A with a constant angular speed of ω, expressed, say, in radians/second. P is the perpendicular projection of Q on the horizontal diameter, along the x-axis. Let us call Q the *reference point* and the circle on which it moves the *reference circle*. As the reference point revolves, the projected point P moves back and forth along the horizontal diameter. The x-component of Q's displacement is always the same as the displacement of P; the x-component of the velocity of Q is always the same as the velocity of P; and the x-component of the acceleration of Q is always the same as the acceleration of P.

Let the angle between the radius OQ and the x-axis at the time $t = 0$ be called ϕ. At any later time t, the angle between OQ and the x-axis is $(\omega t + \phi)$, the point Q moving with constant angular speed. The x-coordinate of Q at any time is, therefore,

$$x = A \cos (\omega t + \phi). \qquad (15\text{-}28)$$

Hence, the projected point P moves with simple harmonic motion along the x-axis. Therefore, *simple harmonic motion can be described as the projection along a diameter of uniform circular motion.*

15-6
RELATION BETWEEN SIMPLE HARMONIC MOTION AND UNIFORM CIRCULAR MOTION

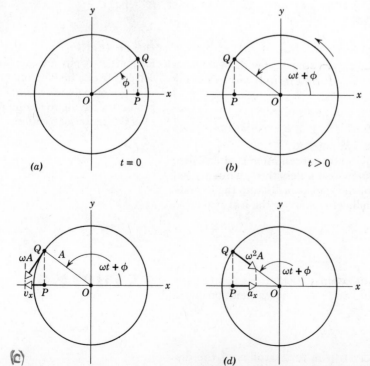

(a) $t = 0$

(b) $t > 0$

(c)

(d)

figure 15-14
The relation of simple harmonic motion to uniform circular motion. Q moves in uniform circular motion and P in simple harmonic motion. Q has angular speed ω, P angular frequency ω. (a, b) The x-component of Q's displacement is always equal to P's displacement. (c) The x-component of Q's velocity is always equal to P's velocity. (d) The x-component of Q's acceleration is always equal to P's acceleration.

The angular frequency ω of simple harmonic motion is the same as the angular speed of the reference point. The frequency of the simple harmonic motion is the same as the number of revolutions per unit time of the reference point. Hence, $\nu = \omega/2\pi$ or $\omega = 2\pi\nu$. The time for a complete revolution of the reference point is the same as the period T of the simple harmonic motion. Hence, $T = 2\pi/\omega$ or $\omega = 2\pi/T$. The phase of the simple harmonic motion, $\omega t + \phi$, is the angle that OQ makes with the x-axis at any time t (Fig. 15-14b,c,d). The angle that OQ makes with the x-axis at the time $t = 0$ (Fig. 15-14a) is ϕ, the phase constant or initial phase of the motion. The amplitude of the simple harmonic motion is the same as the radius of the reference circle.

The tangential velocity of the reference point Q has a magnitude of ωA. Hence, the x-component of this velocity (Fig. 15-14c) is

$$v_x = -\omega A \sin(\omega t + \phi).$$

[margin handwritten: $V_x = -\omega A \sin(\omega t + \phi);$]

[margin handwritten: RELATIONSHIP BETWEEN SIMPLE HARMONIC MOTION AND UNIFORM CIRCULAR MOTION. FOR ANGULAR VELOCITY.]

This relation gives a negative v_x when Q and P are moving to the left and a positive v_x when Q and P are moving to the right. Notice that v_x is zero at the end points of the simple harmonic motion, where $\omega t + \phi$ is zero and π, as required.

The acceleration of point Q in uniform circular motion is directed radially inward and has a magnitude of $\omega^2 A$. The acceleration of the projected point P is the x-component of the acceleration of the reference point Q (Fig. 15-14d). Hence,

$$a_x = -\omega^2 A \cos(\omega t + \phi)$$

[margin handwritten: $a_x = -\omega^2 A \cos(\omega t + \phi)$]

[margin handwritten: RELATIONSHIP FOR ANGULAR ACCELERATION IN SIMPLE HARMONIC MOTION.]

gives the acceleration of the point executing simple harmonic motion. Notice that a_x is zero at the midpoints of the simple harmonic motion, where $\omega t + \phi = \pi/2$ or $3\pi/2$, as required.

These results are all identical with those of simple harmonic motion along the x-axis; see Eqs. 15-13.

If we had taken the perpendicular projection of the reference point onto the y-axis, instead, we would have obtained for the motion of the y-projected point

$$y = A \sin(\omega t + \phi). \tag{15-29}$$

This is again a simple harmonic motion. It differs only in phase from Eq. 15-28, for if we replace ϕ by $\phi - \pi/2$, then $\cos(\omega t + \phi)$ becomes $\sin(\omega t + \phi)$. It is clear that the projection of uniform circular motion along *any* diameter gives a simple harmonic motion.

Conversely, uniform circular motion can be described as a combination of two simple harmonic motions. It is that combination of two simple harmonic motions, occuring along perpendicular lines, which have the same amplitude and frequency but differ in phase by 90°. When one component is at the maximum displacement, the other component is at the equilibrium point. If we combine these components (Eqs. 15-28 and 15-29), we obtain at once the relation

$$r = \sqrt{x^2 + y^2} = A.$$

By writing the relations for v_y and a_y (you should do this) and combining corresponding quantities, we obtain also the relations

$$v = \sqrt{v_x^2 + v_y^2} = \omega A,$$
$$a = \sqrt{a_x^2 + a_y^2} = \omega^2 A.$$

[margin handwritten: $v_x = -\omega A \sin(\omega t + \phi) + v_y = -\omega A \cos(\omega t + \phi)$ $-\omega A \sin^2(\omega t + \phi) - \omega A \cos(\omega t + \phi)$]

These relations correspond respectively to the magnitudes of the displacement, the velocity, and the acceleration in uniform circular motion.

It will be possible for us to analyze many complicated motions as combinations of individual simple harmonic motions. Circular motion is a particularly simple combination. In the next section we shall consider other combinations of simple harmonic motions.

In example 1 we considered a body executing a horizontal simple harmonic motion. The equation of that motion (units?) was

$$x = \tfrac{1}{3} \cos 8t.$$

This motion can also be represented as the projection of uniform circular motion along a horizontal diameter.

(a) Give the properties of the corresponding uniform circular motion.

The x-component of the circular motion is given by

$$x = A \cos (\omega t + \phi).$$

Therefore, the reference circle must have a radius $A = \tfrac{1}{3}$ ft, the initial phase or phase constant must be $\phi = 0$, and the angular velocity must be $\omega = 8$ rad/s, in order to obtain the equation $x = \tfrac{1}{3} \cos 8t$ for the horizontal projection.

(b) From the motion of the reference point determine the time required for the body to come halfway in toward the center of motion from its initial position.

As the body moves halfway in, the reference point moves through an angle of $\omega t = 60°$ (Fig. 15-15). The angular velocity is constant at 8 rad/s so that the time required to move through 60° is

$$t = \frac{60°}{\omega} = \frac{\pi/3 \text{ rad}}{8 \text{ rad/s}} = \frac{\pi}{24}\text{ s} = 0.13 \text{ s}.$$

The time may also be computed directly from the equation of motion. Thus,

$$x = \tfrac{1}{3} \cos 8t \quad \text{and} \quad x = \frac{A}{2} = \tfrac{1}{6}.$$

Hence

$$\tfrac{1}{6} = \tfrac{1}{3} \cos 8t \quad \text{or} \quad 8t = \cos^{-1}(\tfrac{1}{2}) = \frac{\pi}{3}.$$

Therefore,

$$t = \frac{\pi}{24}\text{ s} = 0.13 \text{ s}.$$

Often two linear simple harmonic motions *at right angles* are combined. The resulting motion is the sum of two independent oscillations. Consider first the case in which the frequencies of the vibrations are the same, such as

$$x = A_x \cos (\omega t + \phi_x),$$
$$y = A_y \cos (\omega t + \phi_y). \tag{15-30}$$

The x- and y-motions have different amplitudes and different phase constants, however.

If the phase constants are the same so that $\phi_x = \phi_y = \phi$, the resulting motion is a straight line. This can be shown analytically, for when we eliminate t from the equations

$$x = A_x \cos (\omega t + \phi) \quad y = A_y \cos (\omega t + \phi),$$

we obtain $\qquad\qquad y = (A_y/A_x)x.$

EXAMPLE 7

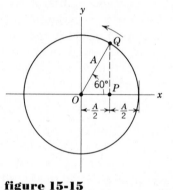

figure 15-15
Example 7. The particles Q and P of Fig. 15-14 are shown for $\omega t = 60°$. Since ω is known, t may be found.

15-7
COMBINATIONS OF HARMONIC MOTIONS

This is the equation of a straight line, whose slope is A_y/A_x. In Fig. 15-16a and b we show the resultant motion for two cases, $A_y/A_x = 1$ and $A_y/A_x = 2$. In these cases both the x- and y-displacements reach a maximum at the same time and reach a minimum at the same time. They are in phase.

If the phase constants are different, the resulting motion will not be a straight line. For example, if the phase constants differ by $\pi/2$, the maximum x-displacement occurs when the y-displacement is zero and vice versa. When the amplitudes are equal, the resulting motion is circular; when the amplitudes are unequal, the resulting motion is elliptical. Two cases, $A_y/A_x = 1$ and $A_y/A_x = 2$, are shown in Fig. 15-16c and d, for $\phi_x = \phi_y + \pi/2$. The cases $A_y/A_x = 1$ and $A_y/A_x = 2$, for $\phi_x = \phi_y - \pi/4$, are shown in Fig. 15-16e and f.

All possible combinations of two simple harmonic motions at right angles having the same frequency correspond to *elliptical* paths, the circle and straight line being special cases of an ellipse. This can be shown analytically by combining Eqs. 15-30 and eliminating the time; you can show that the resulting equation is that of an ellipse. The shape of the ellipse depends only on the ratio of the amplitudes, A_y/A_x, and the *difference* in phase between the two oscillations, $\phi_x - \phi_y$. The actual motion can be either clockwise or counterclockwise, depending on which component leads in phase.

A simple way to produce such patterns is by means of an oscillo-

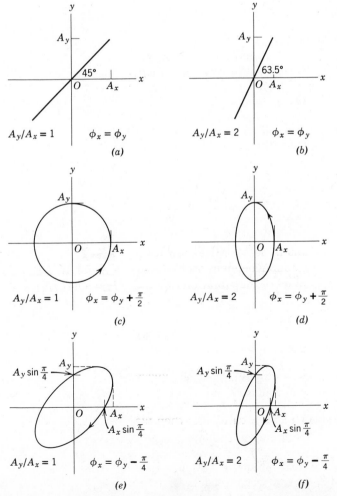

figure 15-16
Simple harmonic motions in two dimensions. *(a)* The amplitudes of x and y (namely A_x and A_y) are the same, as are their phase constants. *(b)* y's amplitude is twice x's but their phase constants are the same. *(c)* Their amplitudes are equal, but x leads y in phase by 90°. *(d)* Same as *(c)* except that y's amplitude is twice x's. *(e)* Equal amplitudes, but x lags y in phase by 45°. *(f)* Same as *(e)* except that y's amplitude is twice x's.

scope. In this, electrons are deflected by each of two electric fields at right angles to one another. The field strengths alternate sinusoidally with the same frequency, but their phases and amplitudes can be varied. In this way the electrons can be made to trace out the various patterns discussed above on a fluorescent screen. We can also produce these patterns mechanically by means of a pendulum swinging with small amplitude but not confined to one vertical plane. Such combinations of two simple harmonic motions at right angles having the same frequency are particularly important in the study of polarized light and alternating current circuits.

Combinations of simple harmonic motions of the same frequency in the *same direction*, but with different amplitudes and phases, are of special interest in the study of diffraction and interference of light, sound, and electromagnetic radiation. This will be discussed later in the text.

If two oscillations of *different frequencies* are combined at right angles, the resulting motion is more complicated. The motion is not even periodic unless the two component frequencies ω_1 and ω_2 are the ratio of two integers (see Problem 49). Oscillations of different frequencies in the same direction may also be combined. The treatment of this motion is particularly important in the case of sound vibrations and will be discussed in Chapter 20.

15-8
TWO-BODY OSCILLATIONS

The simple harmonic oscillator of Fig. 15-4 is a mass m coupled by a spring of force constant k to a solid wall. The wall is rigidly connected to the earth, so that this system is really a two-body system, connected by a spring, one of the bodies being effectively of infinite mass. This solid support remains at rest in an inertial reference frame so that the change in length of the spring is equal to the displacement of the mass m; the other end of the spring does not move. In this case we defined the potential energy $U(x)$ of the oscillating system of Fig. 15-4 to be a function of the displacement x of the mass m alone (see Figs. 15-3, 9). This again is equivalent to assuming that one end of the spring is connected to an infinite mass so that the extension of the spring is determined by the motion of mass m alone.

Often in nature we find two-body oscillating systems in which we *cannot* take the mass of one of the bodies to be infinite and we must consider the motions of both bodies in an appropriate inertial reference frame. Examples are diatomic molecules such as H_2, CO, HCl, etc., which can oscillate along their axis of symmetry. The coupling between the atoms that make up these molecules is electromagnetic, but we may imagine them, for our purpose, to be connected by a tiny, massless spring.

The surprising thing about two-body oscillators is that, by redefining terms slightly and by introducing a new concept (that of *reduced mass*), we can describe the oscillations by exactly the same equations that we have already derived for the (effectively) one-body system of Fig. 15-4. Let us prove this.

Figure 15-17a shows two bodies m_1 and m_2 connected by a (massless) spring of force constant k; the system is free to oscillate on a frictionless horizontal surface. We locate the ends of the spring by the coordinates $x_1(t)$ and $x_2(t)$, as shown. The length of the spring at any instant is $x_1 - x_2$. If its normal, unstressed length is l, then the *change* in length of the spring, $x(t)$, is given by

$$x = (x_1 - x_2) - l. \qquad (15-31)$$

If x is positive, the spring is stretched, if $x = 0$, the spring has its normal length, and if x is negative, it is compressed.

In Fig. 15-17a we assume, for concreteness, that the spring is stretched, so that $x > 0$. We show also the force **F** exerted by the spring on m_2 and the force $-\mathbf{F}$

figure 15-17
(a) Two bodies of masses m_1 and m_2 connected by a (massless) spring whose unstressed length is l. (b) A single body of mass μ (the reduced mass) connected by an identical spring to a rigid wall.

exerted on m_1. These two forces are equal and opposite, as the figure shows, and have the common magnitude $F = kx$.

If we apply Newton's second law, $F = ma$, to masses m_1 and m_2, we obtain

$$m_1 \frac{d^2 x_1}{dt^2} = -kx$$

and

$$m_2 \frac{d^2 x_2}{dt^2} = +kx.$$

Let us now multiply the first equation by m_2 and the second equation by m_1 and subtract. We obtain

$$m_1 m_2 \frac{d^2 x_1}{dt^2} - m_1 m_2 \frac{d^2 x_2}{dt^2} = -m_2 kx - m_1 kx,$$

which we can write as

$$\frac{m_1 m_2}{m_1 + m_2} \frac{d^2}{dt^2}(x_1 - x_2) = -kx. \tag{15-32}$$

Let us call the quantity $m_1 m_2 / (m_1 + m_2)$, which has the dimensions of mass, the *reduced mass* of the system and give it the symbol μ; that is,

$$\mu = \frac{m_1 m_2}{m_1 + m_2}. \tag{15-33}$$

Because l is a constant, $d^2(x_1 - x_2)/dt^2 = d^2x/dt^2$ (see Eq. 15-31) and Eq. 15-32 now can be written as

$$\frac{d^2 x}{dt^2} + \frac{k}{\mu} x = 0. \tag{15-34}$$

This is identical in form to Eq. 15-5 which we developed for the single-body oscillation of Fig. 15-4. The differences are that (1) x in Eq. 15-34 is the *relative* displacement of the two blocks from their equilibrium positions (see Eq. 15-31) rather than the displacement of a single block from its equilibrium position, and (2) μ is the *reduced mass* of the pair of blocks rather than the mass of a single block.

Note from Eq. 15-33, which we can write either as

$$\mu = m_1 \frac{m_2}{m_1 + m_2} = m_2 \frac{m_1}{m_1 + m_2}$$

or as

$$\frac{1}{\mu} = \frac{1}{m_1} + \frac{1}{m_2},$$

that (for finite masses) μ is always *smaller* than m_1 or m_2; hence the name *reduced* mass. Equation 15-34 leads, by way of the derivation that follows Eq. 15-6, to

$$\nu = \frac{1}{2\pi} \sqrt{\frac{k}{\mu}} \quad \text{or} \quad T = 2\pi \sqrt{\frac{\mu}{k}} \tag{15-35}$$

for the frequency and period of oscillation of the system of Fig. 15-17a. It is clear that this system has the same frequency and period as a single block of mass μ, connected by a similar spring to a rigid wall, as in Fig. 15-17b. Hence, the two-body oscillation of Fig. 15-17a is equivalent to the one-body oscillation of Fig. 15-17b. One particle moves relative to the other particle as though the other particle were fixed and the mass of the moving one were reduced to μ. The reduced mass concept is applied widely in physics.

We can solve Eq. 15-34, as in Section 15-3, to yield these relations:

$$x = A \cos (\omega t + \phi),$$

$$v = dx/dt = -\omega A \sin (\omega t + \phi),$$

and

$$a = dv/dt = -\omega^2 A \cos (\omega t + \phi).$$

They are identical with Eqs. 15-13 except that here x, v, and a are the *relative* displacement, velocity, and acceleration, respectively, of the two blocks. Thus

$$x = (x_1 - x_2) - l,$$

$$v = dx/dt = v_1 - v_2, \qquad (15\text{-}36)$$

and

$$a = dv/dt = a_1 - a_2,$$

in which the subscripts refer to the two blocks.

The potential energy of a two-body, simple harmonic oscillator is given by $U(x) = \frac{1}{2}kx^2$ which shows clearly, because x depends on the positions of both blocks (see Eq. 15-36), that the potential energy is a characteristic of the system as a whole.

Many actual two-body oscillators, although harmonic, are not simple harmonic; their potential energy curves, like that of Fig. 8-7a which refers to a diatomic molecule, are not parabolic. Even such oscillators, however, behave like simple harmonic oscillators for small enough amplitudes of oscillation about the equilibrium position. Note, too, that x in Fig. 8-7a has a different meaning than we have assigned to it in this chapter; it is the actual separation, rather than (see Eq. 15-36) the difference between the actual separation and the equilibrium separation. Thus in Fig. 8-7a the stable equilibrium position corresponds, not to $x = 0$ as in Fig. 15-2, but to $x = \sqrt[6]{2a/b}$. This change is only a change in the origin of the x-axis of the potential energy curve and has no fundamental significance.

figure 15-18
A damped harmonic oscillator. A disk is attached to the mass and immersed in a fluid which exerts a damping force $-b\, dx/dt$. The elastic restoring force is $-kx$.

15-9
DAMPED HARMONIC MOTION

Up to this point we have assumed that no frictional forces act on the oscillator. If this assumption held strictly, a pendulum or a weight on a spring would oscillate indefinitely. Actually, the amplitude of the oscillation gradually decreases to zero as a result of friction. The motion is said to be damped by friction and is called *damped harmonic motion*. Often the friction arises from air resistance or internal forces. The magnitude of the frictional force usually depends on the speed. In most cases of interest the frictional force is proportional to the velocity of the body but directed opposite to it. An example of a damped oscillator is shown in Fig. 15-18.

The equation of motion of the damped simple harmonic oscillator is given by the second law of motion, $F = ma$, in which F is the sum of the restoring force $-kx$ and the damping force $-b\, dx/dt$. Here b is a positive constant. We obtain

$$F = ma,$$

or

$$-kx - b\frac{dx}{dt} = m\frac{d^2x}{dt^2}$$

or

$$m\frac{d^2x}{dt^2} + b\frac{dx}{dt} + kx = 0. \qquad (15\text{-}37)$$

If b is small, the solution of this differential equation (given without proof)* is

$$x = Ae^{-bt/2m} \cos (\omega' t + \phi), \tag{15-38}$$

where

$$\omega' = 2\pi\nu' = \sqrt{\frac{k}{m} - \left(\frac{b}{2m}\right)^2}. \tag{15-39}$$

In Fig. 15-19 we plot the displacement x as a function of the time t for oscillatory motion with small damping.

We can interpret the solution as follows. First, the frequency is smaller and the period is longer when friction is present. Friction slows down the motion, as might be expected. If no friction were present, b would equal zero and ω' would equal $\sqrt{k/m}$ or ω, which is the angular frequency of undamped motion. When friction is present, ω' is less than ω, as shown by Eq. 15-39. Second, the amplitude of the motion gradually decreases to zero. The time interval τ during which the amplitude drops to $1/e$ of its initial value is called the *mean lifetime* of the oscillation. The amplitude factor is $Ae^{-bt/2m}$, so that $\tau = 2m/b$. Once again, if there were no friction present, b would equal zero and the amplitude would have the constant value A as time went on; the lifetime would be infinite.

If the force of friction is great enough, b becomes so large that Eq. 15-38 is no longer a valid solution of the equation of motion.* Then the motion will not be periodic at all. The body merely returns to its equilibrium position when released from its initial displacement A.

In damped harmonic motion the energy of the oscillator is gradually dissipated by friction and falls to zero in time.

Thus far we have discussed only the natural oscillations of a body, that is, the oscillations that occur when the body is displaced and then released. For a mass attached to a spring the natural frequency is

$$\omega = 2\pi\nu = \sqrt{\frac{k}{m}}$$

in the absence of friction and

$$\omega' = 2\pi\nu' = \sqrt{\frac{k}{m} - \left(\frac{b}{2m}\right)^2},$$

in the presence of a small frictional force bv.

A different situation arises, however, when the body is subject to an oscillatory external force. As examples, a bridge vibrates under the influence of marching soldiers, the housing of a motor vibrates owing to periodic impulses from an irregularity in the shaft, and a tuning fork vibrates when exposed to the periodic force of a sound wave. The oscillations that result are called *forced oscillations*. These forced oscillations have the frequency of the *external force* and not the natural frequency of the body. However, the response of the body depends on the relation between the forced and the natural frequency. A succession of small impulses applied at the proper frequency can produce an oscillation of large amplitude. A child using a swing learns that by pumping at proper time intervals he can make the swing move with a large amplitude. The problem of forced oscillations is a very general one. Its solution is useful in acoustic systems, alternating current circuits, and atomic physics as well as in mechanics.

The equation of motion of a forced oscillator follows from the second law of motion. In addition to the restoring force $-kx$ and the damping force $-b\,dx/dt$, we have also the applied oscillating external force. For simplicity let this external force be given by $F_m \cos \omega'' t$. Here F_m is the maximum value of the external force and ω'' $(= 2\pi\nu'')$ is its angular frequency. We can imagine such a force applied directly to the suspended mass of Fig. 15-18, if we wish, for concreteness.

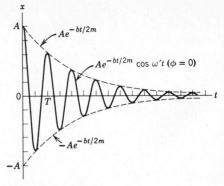

figure 15-19
Damped harmonic motion plotted versus time. The motion is oscillatory with everdecreasing amplitude. The amplitude $(- - -)$ is seen to start with value A and decay exponentially to zero as $t \to \infty$.

15-10
FORCED OSCILLATIONS AND RESONANCE

---

* See, for example, K. R. Symon, *Mechanics*, third edition, Addison-Wesley Publishing Company, 1971, Section 2.9.

From $\qquad\qquad\qquad\qquad\qquad F = ma,$

we obtain $\qquad\qquad -kx - b\dfrac{dx}{dt} + F_m \cos \omega''t = m\dfrac{d^2x}{dt^2}$

or $\qquad\qquad m\dfrac{d^2x}{dt^2} + b\dfrac{dx}{dt} + kx = F_m \cos \omega''t. \qquad\qquad (15\text{-}40)$

The solution of this equation (given without proof)* is

$$x = \frac{F_m}{G} \sin (\omega''t - \phi), \qquad\qquad (15\text{-}41)$$

where $\qquad\qquad G = \sqrt{m^2(\omega''^2 - \omega^2)^2 + b^2\omega''^2}, \qquad\qquad (15\text{-}42)$

and $\qquad\qquad\qquad \phi = \cos^{-1}\dfrac{b\omega''}{G}. \qquad\qquad (15\text{-}43)$

Let us consider the resulting motion in a qualitative way.

Notice (Eq. 15-41) that the system vibrates with the frequency of the driving force, ω'', rather than with its natural frequency ω, and that the motion is undamped harmonic motion.

The simplest case is that in which there is no damping, which means that $b = 0$ in Eq. 15-42. The factor G, which has the value $|m(\omega''^2 - \omega^2)|$ for $b = 0$, is large when the frequency of the driving force ω'' is very different from the natural undamped frequency of the system ω. This means that the amplitude of the resultant motion, F_m/G, is small. As the driving frequency approaches the natural frequency, that is, as $\omega'' \to \omega$, we see that $G \to 0$ and the amplitude $F_m/G \to \infty$. Actually some damping is always present so that the amplitude of oscillation, although it may become large, remains finite in practice.

For actual, damped oscillators (for which $b \neq 0$ in Eq. 15-42), there is a characteristic value of the driving frequency ω'' at which the amplitude of oscillation is a maximum. This condition is called *resonance*† and the value of ω'' at which resonance occurs is called the *resonant frequency*. The smaller the damping in a given system the closer is the resonant frequency to the natural undamped frequency ω. Frequently the damping is small enough so that the resonant frequency can be taken to equal the natural undamped frequency ω with small error. Likewise, for small damping, the natural undamped frequency $\omega (= \sqrt{k/m})$ can be taken to equal the natural damped frequency ω' (see Eq. 15-39) with small error.

In Fig. 15-20 we have drawn five curves giving the amplitude of the forced vibrations as a function of the ratio of the driving frequency ω'' to the undamped natural frequency ω. Each of the five curves corresponds to a different value of the damping constant b. Curve (a) shows the amplitude when $b = 0$, that is, when there is no damping. In this case, as we have seen, the amplitude becomes infinite at $\omega'' = \omega$ because energy is being fed into the system continuously by the applied force and none of it is dissipated. In practice, some friction is always present, so the amplitude reaches a large, but finite, value. Of course, when the amplitude gets so large that Hooke's law no longer holds and the elastic limit is exceeded, the system is no longer governed by Eq. 15-40. Often the system breaks, as in the Tacoma Bridge disaster (Fig. 15-21). Curves (b) and (c) give the amplitude of forced vibration for two cases of increasing damping.

The displacement caused by a constant force F_m applied to a system with a force constant k is simply F_m/k. Notice (Fig. 15-20) that the amplitude of the

---

* *Ibid.*, Section 2.10.

† Resonance, defined here to occur at the frequency at which the forced oscillations have their maximum amplitude, may be defined in other ways as, for example, at the frequency at which maximum power is transferred from the driving unit to the oscillating system or at which the speed of the oscillating mass is a maximum. The definitions are not equivalent; we will discuss the matter further when we deal with forced electrical oscillations; see Problem 55.

figure 15-21
On July 1, 1940, the Tacoma Narrows Bridge at Puget Sound, Washington, was completed and opened to traffic. Just four months later a mild gale set the bridge oscillating until the main span broke up, ripping loose from the cables and crashing into the water below. The wind produced a fluctuating resultant force in resonance with a natural frequency of the structure. This caused a steady increase in amplitude until the bridge was destroyed. Many other bridges were later redesigned to make them aerodynamically stable.

forced vibrations is rather large compared to this static displacement. A column of soldiers marching in step across a bridge can set it vibrating with a destructively large amplitude if the frequency of their steps happens to be some natural frequency of the bridge. This is the reason why soldiers break step when crossing a bridge. Resonance considerations are very important in many electrical, acoustic, and atomic devices, as we shall see later.

questions

1. Give some examples of motions that are approximately simple harmonic. Why are motions that are exactly simple harmonic rare?

2. A typical screen-door spring is tension-stressed in its normal state, that is, adjacent turns cling to each other and resist separation. Does such a spring obey Hooke's law?

3. Is Hooke's law obeyed, even approximately, by a diving board? A trampoline? A coiled spring made of lead wire?

4. A spring has a force constant k, and a mass m is suspended from it. The spring is cut in half and the same mass is suspended from one of the halves. How are the frequencies of oscillation, before and after the spring is cut, related?

5. An unstressed spring has a force constant k. It is stretched by a weight hung from it to an equilibrium length well within the elastic limit. Does the spring have the same force constant k for displacements from this new equilibrium position?

6. Suppose we have a block of unknown mass and a spring of unknown force constant. Show how we can predict the period of oscillation of this block-spring system simply by measuring the extension of the spring produced by attaching the block to it.

7. Any real spring has mass. If this mass is taken into account, explain qualitatively how this will change our expressions for the period of oscillation of a spring-and-mass system (see Problem 31).

8. Can one have an oscillator which even for smaller amplitudes is not simple harmonic? That is, can one have a nonlinear restoring force in an oscillator even at arbitrarily small amplitudes?

9. How are each of the following properties of a simple harmonic oscillator affected by doubling the amplitude: period, force constant, total mechanical energy, maximum velocity, maximum acceleration?

10. What changes could you make in a harmonic oscillator that would double the maximum speed of the oscillating mass?

11. We think of energy exchange for a mass-spring system as a transfer between U and K, their sum E remaining constant; see Fig. 15-9. Suppose a mass is oscillating between two stretched springs, as in Fig. 15-23. A student says: "Consider the mass instantaneously at rest at one end of its limit of oscillation. Here $K = 0$. However, when the mass starts to move toward its equilibrium position K increases. Also because $U = \frac{1}{2}kx^2$ (Eq. 8-11) both springs increase their potential energy because the sign of x (compression or extension) does not matter. Therefore K and U both increase. How can their sum $(= E)$ be constant?"
What is wrong with this argument?

12. A person stands on a bathroom-type scale which rests on a platform suspended by a large spring. The whole system executes simple harmonic motion in a vertical direction. Describe the variation in scale reading during a period of motion.

13. Could we ever construct a simple pendulum?

14. Could standards of mass, length, and time be based on properties of a pendulum? Explain.

15. Show that as the amplitude θ_m in Eq. 15-20 approaches 180° the period approaches infinity. Is this reasonable?

16. Predict by qualitative arguments whether a pendulum oscillating with large amplitude will have a period longer or shorter than the period for oscillations with small amplitude. (Consider extreme cases.)

17. What happens to the frequency of a swing as its oscillations die down from large amplitude to small?

18. How is the period of a pendulum affected when its point of suspension is (a) moved horizontally with acceleration a; (b) moved vertically upward with acceleration a; (c) moved vertically downward with acceleration $a < g$. Which case, if any, applies to a pendulum mounted on a cart rolling down an inclined plane?

19. Why was an axis through the center of mass excluded in using Eq. 15-26 to determine I? Does this equation apply to such an axis? How can you determine I for such an axis using physical pendulum methods?

20. A hollow sphere is filled with water through a small hole in it. It is hung by a long thread and, as the water slowly flows out of the hole at the bottom,

one finds that the period of oscillation first increases and then decreases. Explain.

21. (a) The effect of the mass, m, of the cord attached to the bob, of mass M, of a pendulum is to increase the period over that for a simple pendulum in which $m = 0$. Make this plausible. (b) Although the effect of the mass of the cord on the pendulum is to increase its period, a cord of length l swinging without anything on the end ($M = 0$) has a period less than that of a simple pendulum of length l. Make that plausible. (See "Effect of the Mass of the Cord on the Period of a Simple Pendulum," by H. L. Armstrong, *American Journal of Physics*, June, 1976.)

22. Two pendula, each consisting of a disk attached to a light bar, are identical except for the coupling between disk and bar. In one the bar is rigidly mounted to the disk; in the other ball-bearings are used so that the disk would be free to spin about the end of the bar, for example. Both pendula are hung, pulled aside to the same height, and released. Which has the greater period? Explain.

23. Will the frequency of oscillation of a torsional pendulum change if it is taken to the moon? a simple pendulum? a mass-spring oscillator? a physical pendulum?

24. How can a pendulum be used so as to trace out a sinusoidal curve?

25. What component simple harmonic motions would give a figure 8 as the resultant motion?

26. Is there a connection between the F vs. x relation at the molecular level and the macroscopic relation between F and x in a spring?

27. (a) Under what circumstances would the reduced mass of a two-body system be equal to the mass of one body? Explain. (b) What is the reduced mass if the bodies have equal mass? (c) Do cases (a) and (b) give the extreme values of the reduced mass?

28. Why are damping devices often used on machinery? Give an example.

29. Give some examples of common phenomena in which resonance plays an important role.

30. The lunar ocean tide is much more important than the solar ocean tide (see Question 18 of Chapter 16, for example). The opposite is true for tides in the earth's atmosphere, however. Explain this, using resonance ideas, given the fact that the atmosphere has a natural period of oscillation of nearly 12 hours.

problems

SECTION 15-3

1. A 4.0-kg block extends a spring 16 cm from its unstretched position. The block is removed and a 0.50-kg body is hung from the same spring. If the spring is then stretched and released, what is its period of oscillation? *Answer:* 0.28 s.

2. A 2.0-kg mass hangs from a spring. A 300-g body hung below the mass stretches the spring 2.0 cm farther. If the 300-g body is removed and the mass is set into oscillation, find the period of motion.

3. The scale of a spring balance reading from 0 to 32 lb is 4.0 in. long. A package suspended from the balance is found to oscillate vertically with a frequency of 2.0 Hz. How much does the package weigh? *Answer:* 19 lb.

4. An automobile can be considered to be mounted on a spring as far as vertical oscillations are concerned. The springs of a certain car are adjusted so that the vibrations have a frequency of 3.0 Hz. (a) What is the spring's force constant if the car weighs 3200 lb? (b) What will the vibration frequency be if five passengers, averaging 160 lb each, ride in the car?

328 is in the left margin along with CHAP. 15 OSCILLATIONS

5. (a) Show that the general relations for the period and frequency of any simple harmonic motion are

$$T = 2\pi\sqrt{-\frac{x}{a}} \quad \text{and} \quad \nu = \frac{1}{2\pi}\sqrt{-\frac{a}{x}}.$$

(b) Show that the general relations for the period and frequency of any simple angular harmonic motion are

$$T = 2\pi\sqrt{-\frac{\theta}{\alpha}} \quad \text{and} \quad \nu = \frac{1}{2\pi}\sqrt{-\frac{\alpha}{\theta}}.$$

6. The endpoint of a spring vibrates with a period of 2.0 s when a mass m is attached to it. When this mass is increased by 2.0 kg the period is found to be 3.0 s. Find the value of m.

7. A particle executes linear harmonic motion about the point $x = 0$. At $t = 0$ it has displacement $x = 0.37$ cm and zero velocity. The frequency of the motion is 0.25 Hz. Determine (a) the period, (b) the angular frequency, (c) the amplitude, (d) the displacement at time t, (e) the velocity at time t, (f) the maximum speed, (g) the maximum acceleration, (h) the displacement at $t = 3.0$ s, and (i) the speed at $t = 3.0$ s.
  Answer: (a) 4.0 s. (b) $\pi/2$ rad/s. (c) 0.37 cm. (d) 0.37 cos $(\pi t/2)$, in centimeters. (e) -0.58 sin $(\pi t/2)$, in centimeters per second. (f) 0.58 cm/s. (g) 0.91 cm/s². (h) Zero. (i) 0.58 cm/s.

8. A small body of mass 0.10 kg ($W = mg = 0.22$ lb) is undergoing simple harmonic motion of amplitude 1.0 m (3.3 ft) and period 0.20 s. (a) What is the maximum value of the force acting on it? (b) If the oscillations are produced by a spring, what is the force constant of the spring?

9. The vibration frequencies of atoms in solids at normal temperatures are of the order 10^{13} Hz. Imagine the atoms to be connected to one another by "springs." Suppose that a single silver atom vibrates with this frequency and that all the other atoms are at rest. Compute the force constant of a single spring. One mole of silver has a mass of 108 g and contains 6.02 $\times 10^{23}$ atoms. Assume that the atom interacts only with its nearest neighbor.
  Answer: 710 N/m.

10. A block is on a piston which is moving vertically with a simple harmonic motion of period 1.0 s. (a) At what amplitude of motion will the block and piston separate? (b) If the piston has an amplitude of 5.0 cm, what is the maximum frequency for which the block and piston will be in contact continuously?

11. A block is on a horizontal surface which is moving horizontally with a simple harmonic motion of frequency 2.0 Hz. The coefficient of static friction between block and plane is 0.50. How great can the amplitude be if the block does not slip along the surface? *Answer:* 3.1 cm.

12. The end of one of the prongs of a tuning fork that executes simple harmonic motion of frequency 1000 Hz has an amplitude of 0.40 mm. Find (a) the maximum acceleration and maximum speed of the end of the prong and (b) the acceleration and the speed of the end of the prong when it has a displacement 0.20 mm. (c) Express the end's position as a function of time if it is at equilibrium when $t = 0$.

13. A body oscillates with simple harmonic motion according to the equation

$$x = 6.0 \cos (3\pi t + \pi/3)$$

where x is in meters, t is in seconds, and the numbers in the parentheses are in radians. What is (a) the displacement, (b) the velocity, (c) the acceleration, and (d) the phase at the time $t = 2.0$ s. Find also (e) the frequency ν, and (f) the period of the motion.
  Answer: (a) 3.0 m. (b) -49 m/s. (c) -270 m/s². (d) 20 rad. (e) 1.5 Hz. (f) 0.67 s.

14. A loudspeaker produces a musical sound by the oscillation of a diaphragm. If the amplitude of oscillation is limited to 1.0×10^{-3} mm, what frequencies will result in the acceleration of the diaphragm exceeding g?

15. Two particles execute simple harmonic motion of the same amplitude and frequency along the same straight line. They pass one another when going in opposite directions each time their displacement is half their amplitude. What is the phase difference between them? *Answer*: 120°.

16. Two particles oscillate in simple harmonic motion along a common straight line segment of length A. Each particle has a period of 1.5 s but they differ in phase by 30°. (a) How far apart are they (in terms of A) 0.50 s after the lagging particle leaves one end of the path? (b) Are they moving in the same direction, toward each other, or away from each other at this time?

17. A massless spring of force constant 7.0 N/m is cut into halves. (a) What is the force constant of each half? (b) The two halves, suspended separately, support a block of mass M (see Fig. 15-22). If the system vibrates at a frequency of 3.0 Hz, what is the value of the mass M?
Answer: (a) 14 N/m. (b) 79 g.

figure 15-22
Problem 17

18. A uniform spring whose unstressed length is l has a force constant k. The spring is cut into two pieces of unstressed lengths l_1 and l_2, where $l_1 = nl_2$ and n is an integer. What are the corresponding force constants k_1 and k_2 in terms of n and k? Check your result for $n = 1$ and $n = \infty$.

19. Two springs are attached to a mass m and to fixed supports as shown in Fig. 15-23. Show that the frequency of oscillation in this case is

$$\nu = \frac{1}{2\pi}\sqrt{\frac{k_1 + k_2}{m}}.$$

(The electrical analog of this system is a series combination of two capacitors.)

figure 15-23
Problem 19

figure 15-24
Problem 20

20. Two equal masses m and three identical springs of force constant k are arranged as shown in Fig. 15-24. (a) Let x_1, x_2 represent the displacement of each mass from its equilibrium position and show that

$$m\frac{d^2x_1}{dt^2} = k(x_2 - 2x_1)$$

and

$$m\frac{d^2x_2}{dt^2} = k(x_1 - 2x_2).$$

(b) Find the frequencies of vibration for the system by assuming a solution of the form $x_1 = A_1 \sin \omega t$ and $x_2 = A_2 \sin \omega t$.

21. Two springs are joined and connected to a mass m as shown in Fig. 15-25. The surfaces are frictionless. If the springs separately have force constants k_1 and k_2, show that the frequency of oscillation of m is

$$\nu = \frac{1}{2\pi}\sqrt{\frac{k_1 k_2}{(k_1 + k_2)m}}.$$

(The electrical analog of this system is a parallel connection of two capacitors.)

figure 15-25
Problem 21

22. The force of interaction between two atoms in certain diatomic molecules can be represented by $F = -a/r^2 + b/r^3$, in which a and b are positive constants and r is the separation distance of the atoms. Make a graph of F vs. r. Then (a) show that the separation at equilibrium is b/a; (b) show that for small oscillations about this equilibrium separation the force constant is a^4/b^3; (c) find the period of this motion.

23. A massless spring of force constant 19 N/m (1.3 lb/ft) hangs vertically. A body of mass 0.20 kg (W = mg = 0.44 lb) is attached to its free end and then released. Assume that the spring was unstretched before the body was released. Find (a) how far below the initial position the body descends, (b) the frequency, and (c) the amplitude of the resulting motion, assumed to be simple harmonic.

    Answer: (a) 0.21 m (0.68 ft). (b) 1.6 Hz (1.5 Hz). (c) 0.11 m (0.34 ft).

24. An oscillating mass-spring system has a mechanical energy of 1.0 J (0.74 ft · lb), amplitude of 0.10 m (0.33 ft), and maximum speed of 1.0 m/s (3.3 ft/s). Find (a) the force constant of the spring, (b) the mass, and (c) the frequency of oscillation.

25. (a) When the displacement is one-half the amplitude A, what fraction of the total energy is kinetic and what fraction is potential in simple harmonic motion? (b) At what displacement is the energy half kinetic and half potential?

    Answer: (a) $\frac{3}{4}$: $\frac{1}{4}$. (b) $A/\sqrt{2}$.

26. (a) Prove that in simple harmonic motion the average potential energy equals the average kinetic energy when the average is taken with respect to time over one period of the motion, and that each average equals $\frac{1}{4}kA^2$. (See Fig. 15-9a.) (b) Prove that when the average is taken with respect to position over one cycle, the average potential energy equals $\frac{1}{6}kA^2$ and the average kinetic energy equals $\frac{1}{3}kA^2$. (See Fig. 15-9b.) (c) Explain physically why the two results above (a and b) are different.

27. *Vertical Spring in a Uniform Gravitational Field.* Consider a massless spring of force constant k in a uniform gravitational field. Attach a mass m to the spring. (a) Show that if $x = 0$ marks the slack position of the spring, the static equilibrium position is given by $x = mg/k$ (see Fig. 15-26). (b) Show that the equation of motion of the mass-spring system is

$$ m\frac{d^2x}{dt^2} + kx = mg $$

figure 15-26
Problem 27

and that the solution for the displacement as a function of time is $x = A \cos (\omega t + \phi) + mg/k$, where $\omega = \sqrt{k/m}$ as before. (c) Show, therefore, that the system has the same ω, v, a, ν, and T in a uniform gravitational field as in the absence of such a field, with the one change that the equilibrium position has been displaced by mg/k. (d) Now consider the energy of the system, $\frac{1}{2}mv^2 + \frac{1}{2}kx^2 + mg(h - x) =$ constant, and show that time differentiation leads to the equation of motion of part (b). (e) Show that when the mass falls from $x = 0$ to the static equilibrium position, $x = mg/k$, the loss in gravitational potential energy goes half into a gain in elastic potential energy and half into a gain in kinetic energy. (f) Finally, consider the system in motion at the static equilibrium position. Compute separately the change in gravitational potential energy and in elastic potential energy when the mass moves *up* through a displacement A, and when the mass moves *down* through a displacement A. Show that the *total* change in potential energy is the same in each case, namely $\frac{1}{2}kA^2$.

    In view of the results (c) and (f), one can simply ignore the uniform gravitational field in the analysis merely by shifting the reference position from $x = 0$ to $x_0 = x - mg/k = 0$. The new potential energy curve $[U(x_0) = \frac{1}{2}kx_0^2 +$ constant] has the same parabolic shape as the potential energy curve in the absence of a gravitational field $[U(x) = \frac{1}{2}kx^2]$.

28. An 8.0-lb block is suspended from a spring with a force constant of 3.0 lb/in. A bullet weighing 0.10 lb is fired into the block from below with a speed of 500 ft/s and comes to rest in the block. (a) Find the amplitude of the resulting simple harmonic motion. (b) What fraction of the original kinetic energy of the bullet is stored in the harmonic oscillator? Is energy lost in this process? Explain your answer.

29. Start from Eq. 15-17 for the conservation of energy (with $\frac{1}{2}kA^2 = E$) and

obtain the displacement as a function of the time by integration of Eq. 15-18. Compare with Eq. 15-8.

30. Attach a solid cylinder to a horizontal massless spring so that it can *roll without slipping* along a horizontal surface, as in Fig. 15-27. The force constant k of the spring is 3.0 N/m. If the system is released from rest at a position in which the spring is stretched by 0.25 m, find (a) the translational kinetic energy and (b) the rotational kinetic energy of the cylinder as it passes through the equilibrium position. (c) Show that under these conditions the center of mass of the cylinder executes simple harmonic motion with a period

$$T = 2\pi \sqrt{3M/2k},$$

where M is the mass of the cylinder.

31. If the mass of a spring m_s is not negligible but is small compared to the mass m of the object suspended from it, the period of motion is $T = 2\pi \sqrt{(m + m_s/3)/k}$. Derive this result. (*Hint:* The condition $m_s \ll m$ is equivalent to the assumption that the spring stretches proportionally along its length.) (See H. L. Armstrong, *American Journal of Physics*, 37, 447 (1969) for a complete solution of the general case.)

figure 15-27
Problem 30

SECTION 15-5

32. What is the length of a simple pendulum whose period is 1.00 s at a point where $g = 32.2$ ft/s²?

33. A simple pendulum of length 1.00 m (3.28 ft) makes 100 complete oscillations in 204 s at a certain location. What is the acceleration due to gravity at this point?   *Answer:* 9.49 m/s² (31.1 ft/s²).

34. A solid sphere of mass 2.0 kg ($W = mg = 4.4$ lb) and diameter 0.30 m (0.98 ft) is suspended on a wire. Find the period of angular oscillation for small displacements if the torque constant of the wire is 6.0×10^{-3} N · m/rad (4.4×10^{-3} lb · ft/rad).

35. A circular hoop of radius 2.0 ft and weight 8.0 lb is suspended on a horizontal nail. (a) What is its frequency of oscillation for small displacements from equilibrium? (b) What is the length of the equivalent simple pendulum?   *Answer:* (a) 0.45 Hz. (b) 4.0 ft.

36. Determine the largest amplitude of a simple pendulum such that Eq. 15-19 for the period is correct to within 1.0%.

37. A long uniform rod of length l and mass m is free to rotate in a horizontal plane about a vertical axis through its center. A spring with force constant k is connected horizontally between the end of the rod and a fixed wall as shown in Fig. 15-28. What is the period of the small oscillations that result when the rod is pushed slightly to one side and released?
Answer: $2\pi \sqrt{m/3k}$.

figure 15-28
Problem 37

38. The balance wheel of a watch vibrates with an angular amplitude of π radians and a period of 0.50 s. Find (a) the maximum angular speed of the wheel, (b) the angular speed of the wheel when its displacement is $\pi/2$ rads, and (c) the angular acceleration of the wheel when its displacement is $\pi/4$ radians.

39. (a) What is the frequency of a simple pendulum 2.0 m long? (b) Assuming small amplitudes, what would its frequency be in an elevator accelerating upward at a rate of 2.0 m/s²? (c) What would its frequency be in free fall?
Answer: (a) 0.35 Hz. (b) 0.39 Hz. (c) Zero.

40. A simple pendulum of length l and mass m is suspended in a car that is traveling with a constant speed v around a circle of radius R. If the pendulum undergoes small oscillations in a radial direction about its equilibrium position, what will its frequency of oscillation be?

41. Prove, for the generalized physical pendulum of Fig. 15-12, that the centers of oscillation and percussion coincide. See Examples 4 and 5 for a special case.

42. A pendulum is formed by pivoting a long thin rod of length l and mass m about a point on the rod which is a distance d above the center of the rod. (a) Find the small amplitude period of this pendulum in terms of d, l, m, and g. (b) Show that the period has a *minimum* value when $d = l/\sqrt{12} = 0.289l$.

43. A disk 1.0 m in diameter is cut from a metal sheet. The disk is made to swing as a pendulum by drilling a small hole in it and mounting it on a nail driven into a wall. Let l be the distance from the nail to the center of the plate. (a) For what value or values of l will the period be 1.7 s? (b) Suppose you want the period to be as small as possible. What value of l would you use?   *Answer:* (a) 0.30 m; 0.42 m. (b) 0.35 m.

44. (a) Show that the maximum tension in the string of a simple pendulum, when the amplitude θ_m is small, is $mg(1 + \theta_m{}^2)$. (b) At what position of the pendulum is the tension a maximum?

SECTION 15-7

45. Electrons in an oscilloscope are deflected by two mutually perpendicular electric fields in such a way that at any time t the displacement is given by

$$x = A \cos \omega t, \qquad y = A \cos (\omega t + \phi_y).$$

(a) Describe the path of the electrons and determine its equation when $\phi_y = 0°$. (b) When $\phi_y = 30°$. (c) When $\phi_y = 90°$.
Answer: (a) Straight line, $y = \pm x$. (b) Ellipse, $y^2 - \sqrt{3}xy + x^2 = A^2/4$. (c) Circle, $x^2 + y^2 = A^2$.

46. Sketch the path of a particle which moves in the x-y plane according to the equations $x = A \cos (\omega t - \pi/2)$, $y = 2A \cos (\omega t)$, in which x and y are in meters and t is in seconds.

47. The figure shown in Fig. 15-29 is the result of combining the two simple harmonic motions $x = A_x \cos \omega_x t$ and $y = A_y \cos (\omega_y t + \phi_y)$. (a) What is the value of A_x/A_y? (b) What is the value of ω_x/ω_y? (c) What is the value of ϕ_y? *Answer:* (a) 1.0, (b) 0.50, (c) $\pm \frac{1}{2}\pi$.

48. A particle, mass m, moves in a fixed plane along the trajectory $\mathbf{r} = \mathbf{i}\, A \cos \omega t + \mathbf{j}\, A \cos 3\, \omega t$. (a) Sketch the trajectory of the particle. (b) Find the particle's angular momentum as a function of time. (c) Find the force acting on the particle. Also find (d) its potential energy and (e) its total energy as functions of time. (f) Is the motion periodic? If so, what is the period?

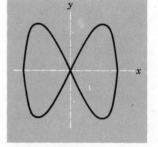

figure 15-29
Problem 47

49. *Lissajous Figures.* When oscillations at right angles are combined, the frequencies for the motion of the particle in the x- and y-directions need not be equal, so that in the general case Eqs. 15-30 become

$$x = A_x \cos (\omega_x t + \phi_x) \qquad \text{and} \qquad y = A_y \cos (\omega_y t + \phi_y).$$

The path of the particle is no longer an ellipse but is called a *Lissajous curve*, after Jules Antoine Lissajous who first demonstrated such curves in 1857. (a) If ω_x/ω_y is a rational number, so that the angular frequencies ω_x and ω_y are "commensurable," then the curve is closed and the motion repeats itself at regular intervals of time. Assume $A_x = A_y$ and $\phi_x = \phi_y$ and draw the Lissajous curve for $\omega_x/\omega_y = \frac{1}{2}$, $\frac{1}{3}$, and $\frac{2}{3}$. (b) Let ω_x/ω_y be a rational number, either $\frac{1}{2}$, $\frac{1}{3}$, or $\frac{2}{3}$ say, and show that the shape of the Lissajous curve depends upon the phase difference $\phi_x - \phi_y$. Draw curves for $\phi_x - \phi_y = 0$, $\pi/4$, and $\pi/2$ rad. (c) If ω_x/ω_y is not a rational number, then the curve is "open." Convince yourself that after a long time the curve will have passed through every point lying in the rectangle bounded by $x = \pm A_x$ and $y = \pm A_y$, the particle never passing twice through a given point with the same velocity. For definiteness, assume $\phi_x = 0$ throughout.

SECTION 15-8

50. (a) What is the reduced mass of each of the following diatomic molecules: O_2, HCl, and CO? Express your answers in unified atomic mass units, the mass of a hydrogen atom being approximately 1.00 u. (b) An HCl molecule is known to vibrate at a fundamental frequency of $\nu = 8.7 \times 10^{13}$ Hz.

What is the effective "force constant" k for the coupling forces between the atoms? In terms of your experience with ordinary springs, would you say that this "molecular spring" is relatively stiff or not?

51. (a) Show that when $m_2 \to \infty$ in Eq. 15-33, $\mu \to m_1$. (b) Show that the effect of a noninfinite wall $(m_2 < \infty)$ on the oscillations of a mass m_1 at the end of a spring attached to the wall is to reduce the period, or increase the frequency, of oscillation compared to (a). (c) Show that when $m_2 = m_1$ the effect is as though the spring were cut in half, each mass oscillating independently about the center of mass at the middle.

52. The spring in Fig. 15-17a has a force constant $k = 250$ N/m (17 lb/ft). Let $m_1 = 1.0$ kg $(W_1 = m_1 g = 2.2$ lb) and $m_2 = 3.0$ kg $(W_2 = m_2 g = 6.6$ lb). (a) What is the oscillation frequency of the two-body system? (b) What is the ratio K_1/K_2 of the kinetic energies of the bodies?

53. Show that the kinetic energy of the two-body oscillator of Fig. 15-17a is given by $K = \frac{1}{2}\mu v^2$, where μ is the reduced mass and v $(= v_1 - v_2)$ is the relative velocity. It may help to note that linear momentum is conserved while the system oscillates.

SECTION 15-9

54. For the system shown in Fig. 15-18, the block has a mass of 1.5 kg and the spring constant $k = 8.0$ N/m. Suppose the block is pulled down a distance of 12 cm and released. If the friction force is given by $-b\, dx/dt$, where $b = 0.23$ kg/s, find the number of oscillations made by the block during the time interval required for the amplitude to fall to one-third of its initial value.

SECTION 15-10

55. Starting from Eq. 15-41, find the velocity v $(= dx/dt)$, in forced oscillatory motion. Show that the velocity amplitude is $v_m = F_m/[(m\omega'' - k/\omega'')^2 + b^2]^{1/2}$.

The equations of Section 15-10 are identical in form with those representing an electrical circuit containing a resistance R, an inductance L, and a capacitance C in series with an alternating emf $V = V_m \cos \omega'' t$. Hence, b, m, k, and F_m are analogous to R, L, $1/C$, and V_m, respectively, and x and v are analogous to electric charge q and current i, respectively. In the electrical case the current amplitude i_m, analogous to the velocity amplitude v_m above, is used to describe the quality of the resonance.

16 *gravitation*

From at least the time of the Greeks two problems were the subjects of searching inquiry: (1) the tendency of objects such as stones to fall to earth when released, and (2) the motions of the planets, including the sun and the moon, which were classified with the planets in those times. In early days these problems were thought of as completely separate. It is one of Newton's achievements that, building on the work of his predecessors, he saw them clearly as aspects of a single problem and subject to the same laws.

In 1665 the 23-year-old Newton was driven from Cambridge to Lincolnshire when the college was dismissed because of the plague. About 50 years later he wrote, ". . . in the same year (1665) I began to think of gravity extending to the orb of the Moon . . . and having thereby compared the force requisite to keep the Moon in her orb with the force of gravity at the surface of the earth, and found them to answer pretty nearly."

Newton's young friend William Stukeley wrote of having tea with Newton under some apple trees when Newton said that the setting was the same as when he got the idea of gravitation. "It was occasion'd by the fall of an apple,† as he sat in a contemplative mood . . . and thus by degrees he began to apply this property of gravitation to the motion of the earth and the heavenly bodys . . ." (See Fig. 16-1.)

We can compute the acceleration of the moon toward the earth from its period of revolution and the radius of its orbit. We obtain 0.0089 ft/s²

---

* See "A Background to Newtonian Gravitation" by V. V. Raman in *The Physics Teacher*, November 1972.
† There is little basis for the belief that the apple hit Newton on the head!

(see Example 4, Chapter 4). This value is about 3600 times smaller than g, the acceleration due to gravity at the surface of the earth. Newton, guided as he says by Kepler's third law (see below and see Problem 25), sought to account for this difference by assuming that the acceleration of a falling body is inversely proportional to the square of its distance from the earth.

The question of what we mean by "distance from the earth" immediately arises. Newton eventually came to regard every particle of the earth as contributing to the gravitational attraction it had on other bodies. He made the daring assumption that the mass of the earth could be treated as if it were all concentrated at its center. (See Section 16-6.)

We can treat the earth as a particle with respect to the sun, for example. It is not obvious, however, that we can treat the earth as a particle with respect to an apple located only a few feet above its surface. If we make this assumption, however, a falling body near the earth's surface is a distance of one earth radius from the effective center of attraction of the earth, or 4000 mi. The moon is about 240,000 mi away. The inverse square of the ratio of these distances is $(4000/240,000)^2 = 1/3600$, in agreement with the ratio of the accelerations of the moon and the apple. In Newton's words, quoted above, it does indeed "answer pretty nearly."

Newton did not publish his conclusions in full until 1678, some 22 years after he had conceived the basic ideas. He did so then in his master-work, the *Principia*. Quite apart from the apple-earth problem which we mentioned above, there was a real uncertainty about the radius of the earth, a needed parameter in the calculations. Finally, there was Newton's general reluctance to publish anything; he was a shy and introspective man and abhorred controversy. Bertrand Russell wrote of him: "If he had encountered the sort of opposition with which Galileo had to contend, it is probable that he would never have published a line." Edmund Halley, of Halley's comet fame, virtually forced Newton to publish the *Principia*. The mathematician Augustus DeMorgan wrote of Halley: ". . . but for him, in all human probability, that work would not have been thought of, nor when thought of written, nor when written printed."

In the *Principia* Newton went beyond the apple-earth and the moon-earth problems and extended his law of gravitation to *all* bodies, in a way that we will discuss in the next section.

There are three overlapping realms in which we can discuss gravitation. (1) The gravitational attraction between two bowling balls, for example, although measurable by sensitive techniques, is too weak to fall within our ordinary sense perceptions. (2) The attraction of ourselves and objects around us by the earth is a controlling feature of our lives from which we can escape only by extreme measures. The designers of our space program have the gravitational force constantly in mind as a central and controlling factor. (3) On a cosmic scale, that is, in the realm of the solar system and of the formation and interaction of stars and galaxies, gravitation is by far the dominant force.

The earliest serious attempts to explain the kinematics of the solar system were made by the Greeks. Ptolemy (Claudius Ptolemaus, second century A.D.) developed a geocentric (Ptolemaic) scheme for the solar system in which, as the name implies, the earth remains stationary at the center whereas the planets, including the sun and the moon, revolve around it. This should not be a surprising deduction. The earth seems to us to be a substantial body. Shakespeare referred to it as ". . .

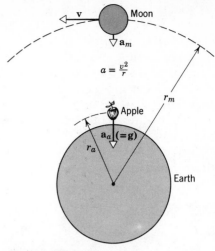

figure 16-1
Both the moon and the apple are accelerated toward the center of the earth. The difference in their motions arises because the moon has a tangential velocity v whereas the apple does not.

this goodly frame, the earth. . . ." Even today, in teaching navigational astronomy, we use a geocentric reference frame and in ordinary conversation we use terms such as "sunrise," which imply such a frame.

Simple circular orbits cannot account for the complicated motions of the planets so that Ptolemy had to use the concept of epicycles, in which a planet moves around a circle whose center moves around another circle centered on the earth. (See Fig. 16-2b.) He also had to resort to several other geometrical arrangements, each of which preserved the supposed sanctity of the circle as a central feature of planetary motions. We now know that it is not a circle that is fundamental but an ellipse, with the sun at one focus (see below).

figure 16-2
(a) The Copernican view of the solar system. The sun is at the center and the planets move around it. *(b)* The Ptolemaic view of the solar system. The earth is at the center and the planets move around it. Both investigators introduced geometrical complexities to explain the complex motion of the planets. In *(b)*, for example, Mars travels about a (circular) epicycle whose center travels about a (circular) deferent. The arrangements of Copernicus, essentially equally complex, are not shown. The basic difference is whether or not the sun or the earth is to be at the center of the planetary motions. (See *The Crime of Galileo*, by Giorgio De Santillana, Chicago: University of Chicago Press, 1955. See also *The Copernican Revolution*, by Thomas S. Kuhn, Cambridge, Mass.: Harvard University Press, 1957.)

In the sixteenth century Copernicus (1473–1543) proposed a heliocentric (Copernican) scheme, in which the sun was at the center of the solar system, the earth moving about it as one of its planets: see Fig. 16-2a. It is often thought that the Copernicus scheme is so much simpler than that of Ptolemy that it should have been adopted at once. This is not true. Copernicus still believed in the sanctity of circles and his use of epicycles and other arrangements was about as great as that of Ptolemy; these are not shown in Fig. 16-2a. Copernicus however, by putting the sun at the center of things, gave a much simpler description and more natural explanation of certain features of planetary motion. Above all he laid the indispensable groundwork from which our modern view of the solar system developed.

The growing controversy over the two theories stimulated astronomers to obtain more accurate observational data. Such data were compiled by Tycho Brahe* (1546–1601), who was the last great astronomer to make observations without the use of a telescope.† His data on planetary motions were analyzed and interpreted for about twenty years by Johannes Kepler (1571–1630), who had been Brahe's assistant. Kepler found important regularities in the motion of the planets. These regularities are known as *Kepler's three laws of planetary motion.*

1. All planets move in elliptical orbits having the sun as one focus (the law of orbits).
2. A line joining any planet to the sun sweeps out equal areas in equal times (the law of areas).
3. The square of the period of any planet about the sun is proportional to the cube of the planet's mean distance from the sun (the law of periods).

Kepler's laws lent strong support to the Copernican theory. They showed the great simplicity with which planetary motions could be described when the sun was taken as the reference body. However, these laws were empirical; they simply described the observed motion of the planets without any theoretical interpretation. Kepler had no concept of force as a cause of such regularities.** In fact, the concept of force was not yet clearly formulated. It was, therefore, a great triumph for Newton's ideas that he could *derive* Kepler's laws from his laws of motion and his law of gravitation. Newton's law of gravitation in this case required each planet to be attracted toward the sun with a force proportional to the mass of the planet and inversely proportional to the square of its distance from the sun.

In this way Newton was able to account for the motion of the planets in the solar system and of bodies falling near the surface of the earth with one common concept. He thereby synthesized into one theory the previously separate sciences of terrestrial mechanics and celestial mechanics. The real scientific significance of Copernicus' work lies in the fact that the heliocentric theory opened the way for this synthesis.‡ Subsequently, on the assumption that the earth rotates and revolves about the sun, it became possible to explain such diverse phenomena as the daily and the annual apparent motion of the stars, the flattening of the earth from a spherical shape, the behavior of the tradewinds, and many other things that could not have been tied together so simply in a geocentric theory.

It is instructive to review the development of our understanding of the motions of the bodies in the solar system in terms of the program of classical mechanics that we outlined in Chapter 5; see page 73. Historically, there were four "breakthroughs."

* See "Copernicus and Tycho" by Owen Gingerich, in *Scientific American*, December 1973.
† The first scientifically useful telescope was built in 1609 by Galileo. With it he discovered the inner moons of Jupiter and the phases of Venus. Galileo was a strong advocate of the Copernican theory and used his observations to argue in its behalf. Newton, incidentally, invented a telescope, the reflecting type.
** See 'How Did Kepler Discover His First Two Laws' by Curtis Wilson in *Scientific American*, March 1972.
‡ Newton's work certainly built on or was influenced by the work of others. Among them we must include Galileo, Kepler, Halley, and Hooke.

1. Copernicus pointed out that the sun and not the earth is the central body of the solar system. In today's language he gave us a reference frame (the sun) much more suitable than the one previously used (the earth) for describing the motions of the solar system. Among other advantages, the Copernican frame, fixed with respect to the sun but not rotating with it, is essentially an *inertial* reference frame; the reference frame fixed to the revolving earth on which we live cannot be so considered for problems involving planetary motions.

2. Brahe made accurate measurements of the motions of the planets as viewed from the earth. He provided the necessary observational data that made further progress possible.

3. Kepler, studying Brahe's data, deduced from it the three simple empirical laws of planetary motion that we have discussed above. Adopting Copernicus' reference frame, he displayed the kinematic information about planetary motions in simple form.

4. Newton discovered the laws of motion for mechanical systems in general as well as the particular force law that applies to the motions of the planets, namely the law of universal gravitation.

Thus, over a span of about 200 years, we see emerging (1) the appropriate reference frame, (2) accurate kinematical information, (3) the empirical laws of planetary motion, and (4) the general laws of classical mechanics and the force law appropriate to planetary motion.

The force between any two particles having masses m_1 and m_2 separated by a distance r is an attraction acting along the line joining the particles and has the magnitude

$$F = G \frac{m_1 m_2}{r^2},$$

(16-1)

where G is a universal constant having the same value for all pairs of particles.

16-2
THE LAW OF UNIVERSAL GRAVITATION

This is Newton's law of universal gravitation. It is important to stress at once many features of this law in order that we understand it clearly.

First, the gravitational forces between two particles are an action-reaction pair. The first particle exerts a force on the second particle that is directed toward the first particle along the line joining the two. Likewise, the second particle exerts a force on the first particle that is directed toward the second particle along the line joining the two. These forces are equal in magnitude but oppositely directed.

The universal constant G must not be confused with the **g** which is the acceleration of a body arising from the earth's gravitational pull on it. The constant G has the dimensions L^3/MT^2 and is a scalar; **g** has the dimensions L/T^2, is a vector, and is neither universal nor constant.

Notice that Newton's law of universal gravitation is not a defining equation for any of the physical quantities (force, mass, or length) contained in it. According to our program for classical mechanics in Chapter 5, force is defined from Newton's second law, $\mathbf{F} = m\mathbf{a}$. The essence of this law, however, is the assumption that the force on a particle, so defined, can be related in a simple way to measurable properties of the particle and of its environment, that is, the existence of simple force laws is assumed. The law of universal gravitation is such a simple law. The constant G must be found from experiment. Once G is determined for a given pair of bodies, we can use that value in the law of gravitation to determine the gravitational forces between any other pair of bodies.

Notice also that Eq. 16-1 expresses the force between mass *particles*. If we want to determine the force between extended bodies, as for example the earth and the moon, we must regard each body as decomposed into particles. Then the interaction between all particles must be computed. Integral calculus makes such a calculation possible. Newton's motive in developing the calculus arose in part from a desire to solve such problems. In general, it is incorrect to assume that all the mass of a body can be concentrated at its center of mass for gravitational purposes. This assumption is correct for uniform spheres, however, a result that we shall use often and shall prove in Section 16-6.

Implicit in the law of universal gravitation is the idea that the gravitational force between two particles is independent of the presence of other bodies or the properties of the intervening space. The correctness of this idea depends on the correctness of the deductions using it and has so far been borne out. This fact has been used by some to rule out the possible existence of "gravity screens."

We can express the law of universal gravitation in vector form. Let the displacement vector $\mathbf{r}_{12}$ point *from* the particle of mass m_1 *to* the particle of mass m_2, as Fig. 16-3a shows. The gravitational force $\mathbf{F}_{21}$, exerted *on* m_2 *by* m_1, is given in direction and magnitude by the vector relation

$$\mathbf{F}_{21} = -G \frac{m_1 m_2}{r_{12}{}^3} \mathbf{r}_{12} \qquad (16\text{-}2a)$$

in which r_{12} is the magnitude of $\mathbf{r}_{12}$. The minus sign in Eq. 16-2a shows that $\mathbf{F}_{21}$ points in a direction opposite to $\mathbf{r}_{12}$; that is, the gravitational force is *attractive*, m_2 feeling a force directed toward m_1 (see Fig. 16-3). That Eq. 16-2a is indeed an inverse square law can be seen by writing it as $\mathbf{F}_{21} = -(Gm_1m_2/r_{12}{}^2)(\mathbf{r}_{12}/r_{12})$; here the displacement vector divided by its own magnitude, $\mathbf{r}_{12}/r_{12}$, is simply a unit vector $\mathbf{u}_r$ in the direction of the displacement. If we express the relation in scalar form by equating the magnitudes of each side, a factor r_{12} in the numerator cancels one of the factors of $r_{12}{}^3$ in the denominator and the inverse square relation of Eq. 16-1 results.

The force exerted *on* m_1 *by* m_2 is clearly

$$\mathbf{F}_{12} = -G \frac{m_2 m_1}{r_{21}{}^3} \mathbf{r}_{21}. \qquad (16\text{-}2b)$$

Note, in Eqs. 16-2, that $\mathbf{r}_{21} = -\mathbf{r}_{12}$ (see Fig. 16-3a,b) so that, as we expect, $\mathbf{F}_{12} = -\mathbf{F}_{21}$ (see Fig. 16-3c); that is, the gravitational forces acting on the two bodies form an action-reaction pair.

figure 16-3
The force exerted on m_2 (by m_1), $\mathbf{F}_{21}$, is directed opposite to the displacement, $\mathbf{r}_{12}$, of m_2 from m_1. The force exerted on m_1 (by m_2), $\mathbf{F}_{12}$, is directed opposite to the displacement, $\mathbf{r}_{21}$, of m_1 from m_2. $\mathbf{F}_{21} = -\mathbf{F}_{12}$, the forces being an action-reaction pair.

16-3
THE CONSTANT OF UNIVERSAL GRAVITATION, G

To determine the value of G it is necessary to measure the force of attraction between two known masses. The first accurate measurement was made by Lord Cavendish in 1798. Significant improvements were made by Poynting and Boys in the nineteenth century. The present accepted value of G is*

$$G = 6.6720 \times 10^{-11} \text{ N} \cdot \text{m}^2/\text{kg}^2,$$

accurate to about 0.0006×10^{-11} N $\cdot$ m²/kg². In the British engineering system this value is 3.436×10^{-8} lb $\cdot$ ft²/slug².

The constant G can be determined by the maximum deflection

* See "A New Determination of the Constant of Gravitation" by A. H. Cook, in *Contemporary Physics*, May 1968, for a good review of the principles and methods used.

figure 16-4
The Cavendish balance, used for
experimental verification of
Newton's law of universal
gravitation. Masses m,m are
suspended from a fiber. Masses
M,M can rotate on a stationary
support. An image of the lamp
filament is reflected by the mirror
attached to m,m onto the scale so
that any rotation of m,m can
be measured.

method illustrated in Fig. 16-4. Two small balls, each of mass m, are
attached to the ends of a light rod. This rigid "dumbbell" is suspended,
with its axis horizontal, by a fine vertical fiber. Two large balls each
of mass M are placed near the ends of the dumbbell on opposite sides.
When the large masses are in the positions A and B, the small masses
are attracted, by the law of gravitation, and a torque is exerted on the
dumbbell rotating it counterclockwise, as viewed from above. When
the large masses are in the positions A' and B', the dumbbell rotates
clockwise. The fiber opposes these torques as it is twisted. The angle
θ through which the fiber is twisted when the balls are moved from one
position to the other is measured by observing the deflection of a beam
of light reflected from the small mirror attached to it. If the masses
and their distances of separation and the torsional constant of the fiber
are known, we can calculate G from the measured angle of twist. The
force of attraction is very small so that the fiber must have an extremely small torsion constant if we are to obtain a detectable twist.
In Example 1 at the end of this section some data are given from which
G can be calculated.

The masses in the Cavendish balance of Fig. 16-4 are, of course, not
particles but extended objects. Since they are uniform spheres, however, they act gravitationally as though all their mass were concentrated at their centers (Section 16-6).

Because G is so small, the gravitational forces between bodies on the
earth's surface are extremely small and can be neglected for ordinary
purposes. For example, two spherical objects each having a mass of
100 kg (about 220-lb weight) and separated by 1.0 m at their centers
attract each other with a force

$$F = \frac{Gm_1m_2}{r^2} = \frac{(6.67 \times 10^{-11}) \times (100) \times (100)}{(1.0)^2}\,\text{N}$$

$$= 6.7 \times 10^{-7}\,\text{N}$$

or about 1.5×10^{-7} lb! The Cavendish experiment must be a very deli-
cate one indeed. Even so, it is often performed as an experiment in an
introductory physics laboratory.

The large gravitational force which the earth exerts on all bodies near its surface is due to the extremely large mass of the earth. In fact, we can determine the mass of the earth from the law of universal gravitation and the value of G calculated from the Cavendish experiment. For this reason Cavendish is said to have been the first person to "weigh" the earth. Consider the earth, mass M_e, and an object on its surface of mass m. The force of attraction is given both by

$$F = mg$$

and

$$F = \frac{GmM_e}{R_e^2}.$$

Here R_e is the radius of the earth, which is the separation of the two bodies, and g is the acceleration due to gravity at the earth's surface. Combining these equations we obtain

$$M_e = \frac{g\,R_e^2}{G} = \frac{(9.80 \text{ m/s}^2)(6.37 \times 10^6 \text{ m})^2}{6.67 \times 10^{-11} \text{ N} \cdot \text{m}^2/\text{kg}^2} = 5.97 \times 10^{24} \text{ kg}$$

or

$$6.6 \times 10^{21} \text{ tons} \qquad \text{"weight."}$$

If we were to divide the total mass of the earth by its total volume, we would obtain the average density of the earth. This turns out to be 5.5 g/cm³, or about 5.5 times the density of water. The average density of the rock on the earth's surface is much less than this value. We conclude that the interior of the earth contains material of density greater than 5.5 g/cm³. From the Cavendish experiment we have obtained information about the nature of the earth's core. (See Question 7 and Problem 16.)

Let the small spheres of Fig. 16-4 each have a mass of 10.0 g and let the light rod be 50.0 cm long. The period of torsional oscillation of this system is found to be 769 s. Then two large fixed spheres each of mass 10.0 kg are placed near each suspended sphere so as to produce the maximum torsion. The angular deflection of the suspended rod is then 3.96×10^{-3} rad and the distance between centers of the large and small spheres is 10.0 cm. Calculate the universal constant of gravitation G from these data.

The period of torsional oscillation is given by Eq. 15-24,

$$T = 2\pi \sqrt{\frac{I}{\kappa}}.$$

For the rigid dumbbell, if we neglect the contribution of the light rod,

$$I = \Sigma\, mr^2 = (10.0 \text{ g})(25.0 \text{ cm})^2 + (10.0 \text{ g})(25.0 \text{ cm})^2$$

or

$$I = 1.25 \times 10^{-3} \text{ kg} \cdot \text{m}^2.$$

With $T = 769$ s, we can obtain the torsional constant κ as

$$\kappa = \frac{4\pi^2 I}{T^2} = \frac{(4\pi^2)(1.25 \times 10^{-3} \text{ kg} \cdot \text{m}^2)}{(769 \text{ s})^2} = 8.34 \times 10^{-8} \text{ kg} \cdot \text{m}^2/\text{s}^2.$$

The relation between the applied torque and the angle of twist is $\tau = \kappa\theta$. We now know κ and the value of θ at maximum deflection.

The torque arises from the gravitational forces exerted by the large spheres on the small ones. This torque will be a maximum for a given separation when the line joining the centers of these spheres is at right angles to the rod. The force on *each* small sphere is

$$F = \frac{GMm}{r^2},$$

and the moment arm for each force is half the length of the rod ($l/2$). Then,

$$\text{torque} = \text{force} \times \text{moment arm}$$

or

$$\tau = 2\frac{GMm}{r^2}\frac{l}{2}.$$

Combining this with

$$\tau = \kappa\theta,$$

we obtain

$$G = \frac{\kappa\theta r^2}{Mml} = \frac{(8.34 \times 10^{-8}\ \text{kg} \cdot \text{m}^2/\text{s}^2)(3.96 \times 10^{-3}\ \text{rad})(0.100\ \text{m})^2}{(10.0\ \text{kg})(0.0100\ \text{kg})(0.500\ \text{m})}$$

$$= 6.63 \times 10^{-11}\ \text{N} \cdot \text{m}^2/\text{kg}^2.$$

Notice that this result is about 1% lower than the accepted value. What have we neglected in this calculation that might account for this difference?

The gravitational force on a body is proportional to its mass, as Eq. 16-1 shows. This proportionality between gravitational force and mass is the reason we ordinarily consider the theory of gravitation to be a branch of mechanics, whereas theories of other kinds of force (electromagnetic, nuclear, etc.) may not be.

An important consequence of this proportionality is that we can measure a mass by measuring the gravitational force on it. This can be done by using a spring balance or by comparing the gravitational force on one mass with that on a standard mass, as in a balance; in other words, we can determine the mass of a body by weighing it. This gives us a more practical and more convenient method of measuring mass than is given by our original definition of mass (Section 5-4).

The question arises whether these two methods really measure the same property. The word *mass* has been used in two quite different experimental situations. For example, if we try to push a block that is at rest on a horizontal frictionless surface, we notice that it requires some effort to move it. The block seems to be inert and tends to stay at rest, or if it is moving, it tends to keep moving. Gravity does not enter here at all. It would take the same effort to accelerate the block in gravity-free space. It is the mass of the block which makes it necessary to exert a force to change its motion. This is the mass occurring in $\mathbf{F} = m\mathbf{a}$ in our original experiments in dynamics. We call this mass m the *inertial mass*. Now there is a different situation which involves the mass of the block. For example, it requires effort just to hold the block up in the air at rest above the earth. If we do not support it, the block will fall to the earth with accelerated motion. The force required to hold up the block is equal in magnitude to the force of gravitational attraction between it and the earth. Here inertia plays no role whatever; the property of material bodies, that they are attracted to other objects such as the earth, does play a role. The force is given by

$$F = G\frac{m'M_e}{R_e^2},$$

where m' is the *gravitational mass* of the block. *Are the gravitational mass m' and the inertial mass m of the block really the same?* Let us look more carefully at this.

Consider two particles A and B of gravitational masses m_A' and m_B' acted on by a third particle C of gravitational mass m_C'. Let the third particle be an equal distance r from the other two. Then, the gravitational force exerted on A by C is

$$F_{AC} = G\frac{m_A'm_C'}{r^2},$$

16-4
INERTIAL AND GRAVITATIONAL MASS

and the gravitational force exerted on B by C is

$$F_{BC} = G\frac{m_B'm_C'}{r^2}.$$

The ratio of the gravitational forces on A and B is the ratio of their gravitational masses; that is,

$$\frac{F_{AC}}{F_{BC}} = \frac{m_A'}{m_B'}.$$

Now suppose that the third body C is the earth. Then F_{AC} and F_{BC} are what we have called the *weights* of bodies A and B. Hence,

$$\frac{W_A}{W_B} = \frac{m_A'}{m_B'}.$$

Therefore, the law of universal gravitation contains within it the result that the weights of various bodies, at the same place on the earth, are exactly proportional to their *gravitational masses*.

Now suppose we measure the inertial masses m_A and m_B of the particles A and B by dynamical experiments, perhaps using a spring as in Section 5-4. Having done this, we then let these particles fall to the earth from a given place and measure their accelerations. We find experimentally that objects of different *inertial masses* all fall with the same acceleration g arising from the earth's gravitational pull. But the earth's gravitational pulls on these bodies are their weights, so that using the second law of motion we obtain

$$W_A = m_A g,$$

$$W_B = m_B g,$$

or

$$\frac{W_A}{W_B} = \frac{m_A}{m_B}.$$

In other words, the weights of bodies at the same place on the earth are exactly proportional to their *inertial masses* as well. Hence, inertial mass and gravitational mass are at least proportional to one another. In fact, they appear to be identical.

Newton devised an experiment to test directly the apparent equivalence of inertial and gravitational mass. If we go back (Section 15-5) and look up the derivation of the period of a simple pendulum, we find that the period (for small angles) was given by

$$T = 2\pi\sqrt{\frac{ml}{m'g}},$$

where m in the numerator refers to the inertial mass of the pendulum bob and m' in the denominator is the gravitational mass of the pendulum bob, such that $m'g$ gives the gravitational pull on the bob. Only if we assume that m equals m', as we did there implicitly, do we obtain the expression

$$T = 2\pi\sqrt{\frac{l}{g}}$$

for the period. Newton made a pendulum bob in the form of a thin shell. Into this hollow bob he put different substances, being careful always to have the same *weight* of substance as determined by a balance. Hence, in all cases the force on the pendulum was the same at the same angle. Because the external shape of the bob was always the same, even the air resistance on the moving pendulum was the same. As one substance replaced another inside the bob, any difference in acceleration could only be due to a difference in the *inertial* mass. Such a difference would show up by a change in the period of the pendulum. But in all cases Newton found the period of the pendulum to be the same, always given by $T = 2\pi\sqrt{l/g}$. Hence, he concluded that $m = m'$ and that inertial and gravitational masses are equivalent.

In 1909, Eötvös devised an apparatus which could detect a difference of 5 parts in 10^9 in gravitational force. He found that equal inertial masses always experienced equal gravitational forces within the accuracy of his apparatus. A refined version of the Eötvös experiment was reported in 1964 by R. H. Dicke and his collaborators, who improved the accuracy of the original experiment by a factor of several hundred.*

In classical physics the equivalence of gravitational and inertial mass was looked upon as a remarkable accident having no deep significance. But in modern physics this equivalence is regarded as a clue leading to a deeper understanding of gravitation (see Section 16-13). This was, in fact, an important clue leading to the development of the general theory of relativity.

Up to this point we have taken the acceleration due to gravity g as a constant. From Newton's law of gravitation, however, it is apparent that g will vary with altitude, that is, with distance from the center of the earth. We have already pointed this out specifically in the moon-apple discussion. Let us compute the change in g that occurs as we proceed outward from the earth's surface. From Eq. 16-1,

16-5
VARIATIONS IN ACCELERATION DUE TO GRAVITY

$$F = G\frac{m_1 m_2}{r^2},$$

we obtain, on differentiating with respect to r,

$$dF = -2\frac{Gm_1 m_2}{r^3}\, dr.$$

Combining these two equations, we obtain

$$\frac{dF}{F} = -2\frac{dr}{r}.$$

Therefore, the fractional change in F is twice the fractional change in r. The minus sign indicates that the force decreases as the separation distance increases. If we let m_1 be the earth's mass and m_2 the object's mass, the gravitational force on the object attributable to the earth is

$$F = m_2 g$$

directed toward the earth. If we differentiate this expression, we obtain

$$dF = m_2 dg,$$

and on dividing this equation by the previous one we find that

$$\frac{dF}{F} = \frac{dg}{g} = -2\frac{dr}{r}. \tag{16-3}$$

For example, in going up 16 km from the earth's surface, r changes from about 6400 km to 6416 km, a relative increase of 1/400. Therefore,† g must change by about $-1/200$ over this distance, or from about 980 cm/s^2 to about 975 cm/s^2. Hence, g is really very nearly constant near the earth's surface at a given latitude. At higher altitudes, such as those for a typical satellite orbit or for the moon's orbit, g drops appreciably, as Table 16-1 shows.

---

* See "The Eötvös Experiment," by R. H. Dicke, *Scientific American*, December 1961, for an elegant review of this subject.

† Equation 16-3 is a differential expression and is exact. The corresponding expression obtained when dr is replaced by a finite change Δr is a good approximation, provided that $\Delta r/r$ is very small.

Table 16-1
Variation of g **with altitude at 45° latitude**

Altitude, meters	g, meters/second²	Altitude, meters	g, meters/second²
0	9.806	32,000	9.71
1,000	9.803	100,000	9.60
4,000	9.794	500,000	8.53
8,000	9.782	1,000,000[1]	7.41
16,000	9.757	380,000,000[2]	0.00271

[1] Typical satellite orbit altitude (= 620 mi).
[2] Radius of moon's orbit (= 240,000 mi).

Measurements of **g** are an essential source of information about the shape of the earth. To define the problem more closely we usually consider not the earth itself but an imaginary closed surface called the *geoid*. Over the oceans the geoid is defined to coincide with mean sea level, whereas over the continents it is defined as a continuation of this level; in principle the position of the geoid can be found by digging small sea-level canals across the continents and noting the mean water level. The geoid is a surface of constant gravitational potential; at any point the direction of a plumb line is at right angles to it.

The ancient Greeks believed the earth to be round and one of them, Eratosthenes (c 276–194 B.C.), measured the radius of the earth on the assumption that it is a sphere. He obtained a value of 7400 km, which is to be compared with the modern value of 6371 km. Later it was learned by measurement that, to a good second approximation, the geoid is not a sphere but is an ellipsoid of revolution, flattened along the earth's rotational axis and bulging at the equator. The equatorial radius, in fact, exceeds the polar radius by 21 km. This flattening is caused by centrifugal effects in the rotating, plastic earth. The geoidic surface is not exactly ellipsoidal, lying outside the ellipsoid of closest fit under mountain masses and inside it over the oceans.

The fact that the equator is farther from the center of the earth than are the poles means that there should be a steady increase in the measured value of g as one goes from the equator (latitude 0°) to either pole (latitude 90°). This is shown in Table 16-2. As Example 2 shows, however, about half of this variation can be accounted for by another effect, namely, the change in the effective value of g caused by the earth's rotation. If the earth were rotating fast enough, for example, objects on its surface at the equator would seem to be weightless, which means that the effective value of $g = (W/m)$ would be zero. For all rotational speeds less than this critical value, g has a definite nonzero value which is, however, less than the value it would have at the same point on a nonrotating earth.

Table 16-2
Variation of g **with latitude at sea level**

Latitude	g, meters/second²	Latitude	g, meters/second²
0°	9.78039	50°	9.81071
10°	9.78195	60°	9.81918
20°	9.78641	70°	9.82608
30°	9.79329	80°	9.83059
40°	9.80171	90°	9.83217

In 1959, it was observed that the orbit of the Vanguard artificial earth satellite, calculated using values of g based on an ellipsoidal geoid, did not agree exactly with the observed orbit. It was concluded that the geoid is best approximated not by an ellipsoid of revolution but by a slightly pear-shaped figure, the small end of the "pear" being in the northern hemisphere and extending about 15 m above the reference ellipsoid. The motion of a satellite is governed at all times by the value of **g** at its position. Thus an artificial earth satellite forms a useful probe to explore the values of g near the surface of the earth and from this to deduce information about the shape of the geoid.*

Effect on g of the rotation of the earth. Figure 16-5 is a schematic view of the earth looking down on the north pole. In it we show an enlarged view of a body of mass m hanging from a spring balance at the equator. The forces on this body are the upward pull of the spring balance w, which is the apparent weight of the body, and the downward pull of the earth's gravitational attraction $F = GmM_e/R_e^2$. This body is not in equilibrium because it experiences a centripetal acceleration a_R as it rotates with the earth. There must, therefore, be a net force acting on the body toward the center of the earth. Consequently, the force F of gravitational attraction (the true weight of the body) must exceed the upward pull of the balance w (the apparent weight of the body).

From the second law of motion we obtain

$$F - w = ma_R,$$

$$\frac{GM_em}{R_e^2} - mg = ma_R,$$

$$g = \frac{GM_e}{R_e^2} - a_R \qquad \text{at the equator.}$$

At the poles $a_R = 0$ so that

$$g = \frac{GM_e}{R_e^2} \qquad \text{at the poles.}$$

This is the value of g we would obtain anywhere (assuming a spherical earth) were the rotation of the earth to be neglected.

Actually the centripetal acceleration is not directed in toward the center of the earth other than at the equator. It is directed perpendicularly in toward the earth's axis of rotation at any given latitude. The detailed analysis is, therefore, really a two-dimensional one. However, the extreme case is at the equator. There

$$a_R = \omega^2\, R_e = \left(\frac{2\pi}{T}\right)^2 R_e = \frac{4\pi^2 R_e}{T^2},$$

in which ω is the angular speed of the earth's rotation, T is the period, and R_e is the radius of the earth. Using the values

$$R_e = 6.37 \times 10^6 \text{ m,}$$

$$T = 8.64 \times 10^4 \text{ s,}$$

we obtain

$$a_R = 0.0336 \text{ m/s}^2.$$

Referring to Table 16-2, we see that this effect is enough to account for more than half the difference between the observed values of g at low and high latitudes.

* See "Satellite Orbits and Their Geophysical Implications," by D. G. King-Hele, *Contemporary Physics*, April 1961, "Refining the Shape of the Earth" by D. G. King-Hele and G. E. Cook, *Nature*, 246, 86 (1973), and "The Shape of the Earth" by D. G. King-Hele, *Science*, June 25, 1976.

EXAMPLE 2

figure 16-5
Example 2. Effect of the earth's rotation on the weight of a body as measured by a spring balance.

We have already used the fact that a large sphere attracts particles outside it, just as though the mass of the sphere were concentrated at its center. Let us now prove this result.

Consider a uniformly dense spherical shell whose thickness t is small compared to its radius r (Fig. 16-6). We seek the gravitational force it exerts on an external particle P of mass m.

We assume that each particle of the shell exerts on P a force which is proportional to the mass of the small part, inversely proportional to the square of the distance between that part of the shell and P, and directed along the line joining them. We must then obtain the resultant force on P attributable to all parts of the spherical shell.

The small part of the shell at A attracts m with a force $\mathbf{F}_1$. A small part of equal mass at B, equally far from m but diametrically opposite A, attracts m with a force $\mathbf{F}_2$. The resultant of these two forces on m is $\mathbf{F}_1 + \mathbf{F}_2$. Notice, however, that the vertical components of these two forces cancel one another and that the horizontal components, $F_1 \cos \alpha$ and $F_2 \cos \alpha$, are equal. By dividing the spherical shell into pairs of particles like these, we can see at once that all transverse forces on m cancel in pairs. A small mass in the upper hemisphere exerts a force having an upward component on m that will annul the downward component of force exerted on m by an equal symmetrically located mass in the lower hemisphere of the shell. To find the resultant force on m arising from the shell, we need consider only horizontal components.

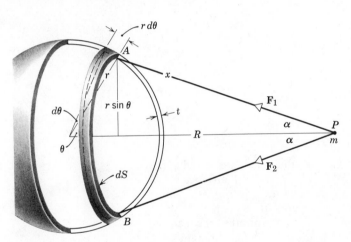

figure 16-6
Gravitational attraction of a section dS of a spherical shell of matter on a particle of mass m.

Let us take as our element of mass of the shell a circular strip labeled dS in the figure. Its length is $2\pi(r \sin \theta)$, its width is $r\, d\theta$, and its thickness is t. Hence, it has a volume

$$dV = 2\pi t r^2 \sin \theta\, d\theta.$$

Let us call the density ρ, so that the mass within the strip is

$$dM = \rho\, dV = 2\pi t \rho r^2 \sin \theta\, d\theta.$$

The force exerted by dM on the particle of mass m at P is horizontal and has the value

$$dF = G\frac{m\, dM}{x^2} \cos \alpha$$

$$= 2\pi G t \rho m r^2 \frac{\sin \theta\, d\theta}{x^2} \cos \alpha. \qquad (16\text{-}4)$$

The variables x, α, and θ are related. From the figure we see that

$$\cos \alpha = \frac{R - r \cos \theta}{x}. \qquad (16\text{-}5)$$

Since, by the law of cosines,

$$x^2 = R^2 + r^2 - 2Rr \cos \theta, \qquad (16\text{-}6)$$

we have

$$r \cos \theta = \frac{R^2 + r^2 - x^2}{2R}. \qquad (16\text{-}7)$$

On differentiating Eq. 16-6, we obtain

$$2x \, dx = 2Rr \sin \theta \, d\theta$$

or

$$\sin \theta \, d\theta = \frac{x}{Rr} \, dx. \qquad (16\text{-}8)$$

We now put Eq. 16-7 into Eq. 16-5 and then put Eqs. 16-5 and 16-8 into Eq. 16-4. As a result we eliminate θ and α and obtain

$$dF = \frac{\pi G t \rho m r}{R^2} \left(\frac{R^2 - r^2}{x^2} + 1 \right) dx.$$

This is the force exerted by the circular strip dS on the particle m.

We must now consider every element of mass in the shell and sum up over all the circular strips in the entire shell. This operation is an integration over the shell with respect to the variable x. But x ranges from a minimum value of $R - r$ to a maximum value $R + r$.

Since

$$\int_{R-r}^{R+r} \left(\frac{R^2 - r^2}{x^2} + 1 \right) dx = 4r,$$

we obtain the resultant force

$$F = \int_{R-r}^{R+r} dF = G \frac{(4\pi r^2 \rho t) m}{R^2} = G \frac{Mm}{R^2}, \qquad (16\text{-}9)$$

where

$$M = (4\pi r^2 t \rho)$$

is the total mass of the shell. This is exactly the same result we would obtain for the force between *particles* of mass M and m separated a distance R. We have proved, therefore, that *a uniformly dense spherical shell attracts an external mass point as if all its mass were concentrated at its center.*

A solid sphere can be regarded as composed of a large number of concentric shells. If each spherical shell has a uniform density, even though different shells may have different densities, the same result applies to the solid sphere. Hence, a body such as the earth, the moon, or the sun, to the extent that they are such spheres, may be regarded gravitationally as point particles to bodies outside them.

Notice that our proof applies only to spheres and then only when the density is constant over the sphere or a function of radius alone.

An interesting result of some significance is the force exerted by a spherical shell on a particle *inside* it. This force is *zero*. To prove this we refer to Fig. 16-7, where m is shown inside the shell. Notice that R is now smaller than r. The limits of our integration over x are now $r - R$ to $R + r$. But

$$\int_{r-R}^{R+r} \left(\frac{R^2 - r^2}{x^2} + 1 \right) dx = 0,$$

so that $F = 0$.

This last result, although not obvious, is plausible because the mass elements of the shell on opposite sides of m now exert forces of opposite directions on m inside. The total annulment depends on the fact that the force varies precisely as an inverse square of the separation distance of two particles. (See Problem 18.) Important consequences of this result will be discussed in the chapters on electricity. There we shall see that the electrical force between charged par-

figure 16-7
Gravitational attraction of a section dS of a spherical shell of matter on a particle of mass m. Here the particle is inside the shell.

ticles also depends inversely on the square of the distance between them. A consequence of interest in gravitation is that the gravitational force exerted by the earth on a particle decreases as the particle goes deeper into the earth, *assuming a constant density for the earth*, for the portions of matter in shells external to the position of the particle exert no force on it, the force becoming zero at the center of the earth. Hence, g would be a maximum at the earth's surface and decrease both outward and inward from that point *if the earth had constant density*. Can you imagine a spherically symmetric distribution of the earth's mass* which would not give this result? (See Problem 16.)

Suppose a tunnel could be dug through the earth from one side to the other along a diameter, as shown in Fig. 16-8.

(a) Show that the motion of a particle dropped into the tunnel is simple harmonic motion. Neglect all frictional forces and assume that the earth has a uniform density.

The gravitational attraction of the earth for the particle at a distance r from the center of the earth arises entirely from that portion of matter of the earth in shells internal to the position of the particle. The external shells exert no force on the particle. Let us assume that the earth's density is uniform at the value ρ. Then the mass inside a sphere of radius r is

$$M' = \rho V' = \rho \frac{4\pi r^3}{3}.$$

This mass can be treated as though it were concentrated at the center of the earth for gravitational purposes. Hence, the force on the particle of mass m is

$$F = \frac{-GM'm}{r^2}.$$

The minus sign is used to indicate that the force is attractive and directed toward the center of the earth.

Substituting for M', we obtain

$$F = -G \frac{(\rho 4\pi r^3)m}{3r^2} = -\left(G\rho \frac{4\pi m}{3}\right) r = -kr.$$

Here $G\rho 4\pi m/3$ is a constant, which we have called k. The force is, therefore, proportional to the displacement r but oppositely directed. This is exactly the criterion for simple harmonic motion.

(b) If mail were delivered through this chute, how much time would elapse between deposit at one end and delivery at the other end?

The period of this simple harmonic motion is

$$T = 2\pi \sqrt{\frac{m}{k}} = 2\pi \sqrt{\frac{3m}{G\rho 4\pi m}} = \sqrt{\frac{3\pi}{G\rho}}.$$

Let us take $\rho = 5.51 \times 10^3$ kg/m³ and $G = 6.67 \times 10^{-11}$ N · m²/kg². This gives

$$T = \sqrt{\frac{3\pi}{G\rho}} = \sqrt{\frac{3\pi}{(6.67 \times 10^{-11})(5.51 \times 10^3)}} \text{ s} = 5050 \text{ s} = 84.2 \text{ min.}$$

The time for delivery is one-half period, or about 42 min. Notice that this time is independent of the mass of the mail.

The earth does not really have a uniform density. Suppose ρ were some function of r, rather than a constant. What effect would this have on our problem?

EXAMPLE 3

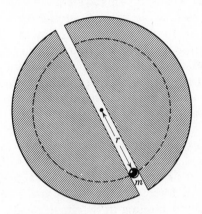

figure 16-8
Example 3. Particle moving in a tunnel through the earth.

---

* See, in this connection, "Comment on 'The Radial Variation of g in a Spherically Symmetric Mass with Nonuniform Density'" by K. Sundaralingam, in *American Journal of Physics*, September 1974.

The motions of bodies in the solar system can be deduced from the laws of motion and the law of universal gravitation. As Kepler pointed out (see page 337), all planets move in elliptical orbits, the sun being at one focus. We can learn a lot about planetary motion by considering the special case of circular orbits. We shall neglect the forces between planets, considering only the interaction between the sun and a given planet. These considerations apply equally well to the motion of a satellite (natural or artificial) about a planet.

Consider two spherical bodies of masses M and m moving in circular orbits under the influence of each other's gravitational attraction. The center of mass of this system of two bodies lies along the line joining them at a point C such that $mr = MR$ (Fig. 16-9). If there are no external forces acting on this system, the center of mass has no acceleration. In this case we choose C to be the origin of our reference frame. The large body of mass M moves in an orbit of constant radius R and the small body of mass m in an orbit of constant radius r, both having the same angular velocity ω. In order for this to happen, the gravitational force acting on each body must provide the necessary centripetal acceleration. Because these gravitational forces are simply an action-reaction pair, the centripetal forces must be equal but oppositely directed. That is, $m\omega^2 r$ (the magnitude of the centripetal force exerted by M on m) must equal $M\omega^2 R$ (the magnitude of the centripetal force exerted by m on M). That this is so follows at once, for $mr = MR$ so that $m\omega^2 r = M\omega^2 R$. The specific requirement, then, is that the gravitational force on either body must equal the centripetal force needed to keep it moving in its circular orbit, that is,

$$\frac{GMm}{(R + r)^2} = m\omega^2 r. \qquad (16\text{-}10)$$

If one body has a much greater mass than the other, as in the case of the sun and a planet, its distance from the center of mass is much smaller than that of the other body. Let us therefore assume that R is negligible compared to r. Equation 16-10 then becomes

$$GM_s = \omega^2 r^3,$$

where M_s is the mass of the sun. If we express the angular velocity in terms of the period of the revolution, $\omega = 2\pi/T$, we obtain

$$GM_s = \frac{4\pi^2 r^3}{T^2}. \qquad (16\text{-}11)$$

This is a basic equation of planetary motion; it holds also for elliptical orbits if we define r to be the semi-major axis of the ellipse. Let us consider some of its consequences.

One immediate consequence of Eq. 16-11 is that it predicts Kepler's third law of planetary motion in the special case of circular orbits. For we can express Eq. 16-11 as

$$T^2 = \frac{4\pi^2}{GM_s} r^3.$$

Notice that the mass of the planet is not involved in this expression. Here, $4\pi^2/GM_s$ is a constant, the same for all planets.

When the period T and radius r of revolution are known for any planet, Eq. 16-11 can be used to determine the mass of the sun. For example, the earth's period is

$$T = 365 \text{ days} = 3.15 \times 10^7 \text{ s},$$

16-7
THE MOTIONS OF PLANETS AND SATELLITES

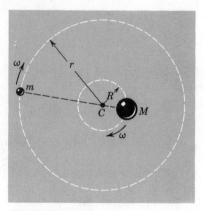

figure 16-9
Two bodies moving in circular orbits under the influence of each other's gravitational attraction. They both have the same angular velocity ω.

and its orbital radius is

$$r = 93 \times 10^6 \text{ mi} = 1.5 \times 10^{11} \text{ m}.$$

Hence,

$$M_s = \frac{4\pi^2 r^3}{GT^2} = \frac{(4\pi^2)(1.5 \times 10^{11} \text{ m})^3}{(6.67 \times 10^{-11} \text{ N} \cdot \text{m}^2/\text{kg}^2)(3.15 \times 10^7 \text{ s})^2} \cong 2.0 \times 10^{30} \text{ kg}.$$

The mass of the sun is thus about 300,000 times the mass of the earth. The error made in neglecting R compared to r is seen to be trivial, for

$$R = \frac{m}{M} r = \frac{1}{300,000} r \cong 300 \text{ mi}; \qquad \frac{R}{r} 100\% \cong \frac{1}{3000} \text{ of } 1\%.$$

In a similar manner we can determine the mass of the earth from the period and radius of the moon's orbit about the earth. (See Problem 22.)

If we know the mass of the sun M_s and the period of revolution T of any planet about it, we can determine the radius of the planet's orbit r from Eq. 16-11. Since the period is easily obtained from astronomical observations, this method of determining a planet's distance from the sun is fairly reliable.

Equation 16-11 holds also for the motion of artificial satellites about the earth; we need only substitute the mass of the earth M_e for M_s in that equation.

Kepler's second law of planetary motion (see page 337) must, of course, hold for circular orbits. In such orbits both ω and r are constant so that equal areas are swept out in equal times by the line joining a planet and the sun. For the exact elliptical orbits, however, or for any orbit in general, both r and ω will vary. Let us consider this case.

Figure 16-10 shows a particle revolving about C along some arbitrary path. The area swept out by the radius vector in a very short time interval Δt is shown shaded in the figure. This area, neglecting the small triangular region at the end, is one-half the base times the altitude or approximately $\frac{1}{2}(r\omega \, \Delta t) \cdot r$. This expression becomes more exact in the limit as $\Delta t \to 0$, the small triangle going to zero more rapidly than the large one. The rate at which area is being swept out instantaneously is therefore

$$\lim_{\Delta t \to 0} \frac{\frac{1}{2}(r\omega \, \Delta t)(r)}{\Delta t} = \tfrac{1}{2}\omega r^2.$$

But $m\omega r^2$ is simply the angular momentum of the particle about C. Hence, Kepler's second law, which requires that the rate of sweeping

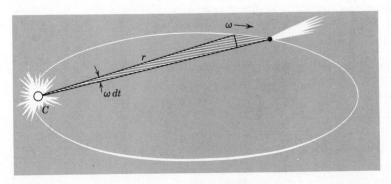

figure 16-10
A comet* moving along an elliptical path with the sun C at the focus of the ellipse. In time dt the comet (or a planet) sweeps out an angle $d\theta = \omega dt$.

* See "The Nature of Comets" by Fred L. Whipple, in *Scientific American*, February 1974, for a fascinating discussion of the properties and possible origin of comets.

out of area $\frac{1}{2}\omega r^2$ be constant, is entirely equivalent to the statement that *the angular momentum of any planet about the sun remains constant.* The angular momentum of the particle about C cannot be changed by a force on it directed toward C. Kepler's second law would, therefore, be valid for any *central force*, that is, any force directed toward the sun. The exact nature of this force—how it depends on distance of separation or other properties of the bodies—is not revealed by this law.

It is Kepler's first law that requires the gravitational force to depend exactly on the inverse square of the distance between two bodies, that is, on $1/r^2$. Only such a force, it turns out, can yield planetary orbits which are elliptical with the sun at one focus.

EXAMPLE 4

A planet revolves around the sun in a elliptical orbit of eccentricity e. Find the ratio of the time spent by the planet between the ends of the minor axis close to the sun to the period of revolution.

By Kepler's first law, the sun is at one focus of the ellipse. (In Fig. 16-11, the ellipse shown has a much larger eccentricity than does the orbit of any planet in the solar system.) The major axis (length $2a$) and the minor axis (length $2b$) intersect at the center C of the ellipse and the distance CF from the center of the ellipse to the focus F is ea by the definition of eccentricity. Notice that for a circular orbit the eccentricity would be zero.

Let the period of revolution be T, and the time required for the planet to travel from B to D, on that part of the ellipse which is close to the sun, be t. Then, if A = area of the ellipse and A' = shaded area, we have, by the conservation of angular momentum (or by the equivalent statement that the rate of sweeping out of area is constant),

$$\frac{A}{T} = \frac{A'}{t}.$$

But, $A' = \frac{1}{2}A - A''$, where A'' = area of triangle BDF. Therefore,

$$\frac{t}{T} = \frac{A'}{A} = \frac{\frac{1}{2}A - A''}{A} = \frac{1}{2} - \frac{A''}{A} = \frac{1}{2} - \frac{\frac{1}{2}(2b)(ae)}{\pi ab}$$

and

$$\frac{t}{T} = \frac{1}{2} - \frac{e}{\pi}.$$

Notice that this reduces to $\frac{1}{2}$ for a circular orbit $(e = 0)$. Why is the ratio less than one-half for elliptical orbits?

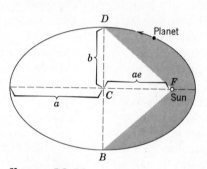

figure 16-11
Example 4. A planet revolves around the sun in an elliptical orbit.

Newton's laws of motion and his law of universal gravitation are in almost complete agreement with astronomical observations.* In our calculation we considered the motion of a planet about the sun as a "two-body" problem. However, we saw that the motion of the sun could be neglected while retaining a high degree of accuracy because of the large ratio of solar mass to planetary mass. This reduced the problem to that of motion of one body about a center of force. If we had required greater accuracy, we would have had to include the sun's motion in our problem (see Problem 28). In fact, for an exact treatment we would have to take into account the effect of other planets and satellites on the motion of sun and planet. This "many-body" problem is quite formidable, but it can be solved by approximation methods to a high degree of accuracy. The results of such calculations are in excellent agreement with astronomical observations.

---

* The major axis of the elliptical orbit of Mercury rotates slightly more than that predicted by Newtonian mechanics when the perturbing influence of other planets is included. This effect is accounted for by the general theory of relativity.

A basic fact of gravitation is that two masses exert forces on one another. We can think of this as a direct interaction between the two mass particles, if we wish. This point of view is called *action-at-a-distance*, the particles interacting even though they are not in contact. Another point of view is the *field* concept, which regards a mass particle as modifying the space around it in some way and setting up a *gravitational field*. This field then acts on any other mass particle in it, exerting the force of gravitational attraction on it. The field, therefore, plays an intermediate role in our thinking about the forces between mass particles. According to this view we have two separate parts to our problem. First, we must determine the field established by a given distribution of mass particles; and secondly, we must calculate the force that this field exerts on another mass particle placed in it.

For example, consider the earth as an isolated mass. If a body is now brought in the vicinity of the earth, a force is exerted on it. This force has a definite direction and magnitude at each point in space. The direction is radially in toward the center of the earth and the magnitude is *mg*. We can, therefore, associate with each point near the earth a vector **g** which is the acceleration that a body would experience if it were released at this point. We call **g** the *gravitational field strength* at the point in question. Since

$$\mathbf{g} = \frac{\mathbf{F}}{m},\qquad\qquad (16\text{-}12)$$

we may define gravitational field strength at any point as the gravitational force per unit mass at that point.* We calculate the force from the field simply by multiplying **g** by the mass *m* of the particle placed at any point.

The gravitational field is an example of a *vector field*, each point in this field having a vector associated with it. There are also scalar fields, such as the temperature field in a heat-conducting solid. The gravitational field arising from a fixed distribution of matter is also an example of a *stationary field*, because the value of the field at a given point does not change with time.

The field concept is particularly useful for understanding electromagnetic forces between moving electric charges. It has distinct advantages, both conceptually and in practice, over the action-at-a-distance concept. The field concept was not used in Newton's day. It was developed much later by Faraday for electromagnetism and only then applied to gravitation. Subsequently, this point of view was adopted for gravitation in the general theory of relativity. The chief purpose of introducing it here is to give the student an early familiarity with a concept that proves to be important in the development of physical theory.

---

* In Eq. 16-12 *g* is defined as the gravitational force per unit mass; at a point *P* a distance *R* from the center of a spherically symmetric mass *M*, it is given by GM/R^2. This *g* differs from the *g* whose magnitude is displayed in Tables 16-1 and 16-2 in that, as Example 2 shows, the centripetal acceleration of a body moving around the earth is already taken into account so that what is described in these tables is an *effective g*. For example, the effective *g* in an orbiting earth satellite is zero, as we have all seen on television transmissions from such satellites. This is because GM/R^2 in Example 2 is exactly equal to a_R in that example. However, the gravitational field at the site of the orbiting satellite, which is given just by GM/R^2, is *not* zero.

EXAMPLE 5

In Chapter 15 we derived the formula for the period of a simple pendulum, $T = 2\pi \sqrt{l/g}$. Keeping in mind that the earth's gravitational field is not uniform over large distances, as was assumed for small distances, what is the longest period a simple pendulum could have in the vicinity of the earth's surface?

The formula $T = 2\pi \sqrt{l/g}$, although not applicable when g varies over the pendulum's path, suggests that we increase the length of the pendulum. Let us make the length infinite. The pendulum bob would then travel along the arc of a circle of infinite radius, that is, along a straight line, as shown in Fig. 16-12. The direction of the earth's gravitational field is everywhere radially in toward the center of the earth, so that its direction changes along the arc. Let us assume that the bob of mass m has an amplitude that is small compared to the radius of the earth. Then the bob is always a distance R_e, the earth's radius, from the center of the earth, to a good approximation. Then the force F on m is

$$F = \frac{GM_e m}{R_e^2} = mg,$$

where M_e is the mass of the earth. This force is directed toward the earth's center as shown. The component of this vector force along x, the line of motion of the bob, is

$$F_x = F \cos \theta = -F \frac{x}{R_e} = -\frac{GM_e m}{R_e^3} x,$$

where the minus sign indicates that the force is directed opposite to the displacement from $x = 0$. We can write this as

$$F_x = -kx,$$

where $k = GM_e m/R_e^3$, a constant.

The formula for the period of a simple harmonic oscillator is $T = 2\pi \sqrt{m/k}$. Hence,

$$T = 2\pi \sqrt{\frac{m}{k}} = 2\pi \sqrt{\frac{m}{GM_e m/R_e^3}} = 2\pi \sqrt{\frac{R_e}{GM_e/R_e^2}} = 2\pi \sqrt{\frac{R_e}{g}},$$

because g at the earth's surface equals GM_e/R_e^2. Putting in $R_e = 6.37 \times 10^6$ m and $g = 9.80$ m/s², we obtain $T = 84.3$ min as the longest period of a simple pendulum in the vicinity of the earth's surface. (See Question 37.)

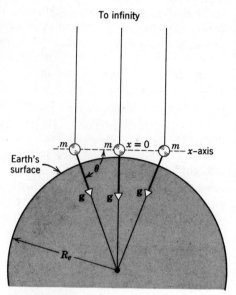

figure 16-12
Example 5. A simple pendulum suspended at infinity.

16-9 GRAVITATIONAL POTENTIAL ENERGY

In Chapter 8 we discussed the gravitational potential energy of a particle (mass m) and the earth (mass M). We considered only the special case in which the particle remains close to the earth so that we could assume the gravitational force acting on the particle to be constant for all positions of the particle. In this section we remove that restriction and consider particle-earth separations that may be appreciably greater than the earth's radius.

Equation 8-5b, which we may write as,

$$\Delta U = U_b - U_a = -W_{ab}, \qquad (8\text{-}5b)$$

defines the change ΔU in the potential energy of any system, in which a conservative force (gravity, say) acts, as the system changes from configuration a to configuration b. W_{ab} is the work done by that conservative force as the system changes.

The potential energy of the system in any arbitrary configuration b is (see Eq. 8-5b)

$$U_b = -W_{ab} + U_a. \qquad (16\text{-}13)$$

To give a value to U_b we must (arbitrarily) choose configuration a to be some agreed-upon reference configuration and we must assign to U_a some (arbitrarily) agreed-upon value, usually zero.

In Chapter 8 we chose as a reference configuration for the earth-particle system that in which the particle is resting on the surface of the earth and we assigned to this configuration the potential energy $U_a = 0$. When the particle is at a height y above the surface of the earth, the potential energy U ($= U_b$) is given from Eq. 16-13 as

$$U = -W_{ab} + 0 = -(-mg)(y) = mgy.$$

The conservative force in question, gravity, points down and has the value $(-mg)$; the displacement of the particle $(+y)$ points up from the reference level; hence the difference in sign for these quantities.

For the more general case, in which the restriction $y \ll R$ (in which R is the radius of the earth) is not imposed, we find it convenient to select a different reference configuration, namely that in which the particle and the earth are infinitely far apart. We assign the value zero to the potential energy of the system in this configuration. Thus the zero-potential-energy configuration is also the zero-force configuration. We made a similar choice when we defined the zero-energy configuration of a spring to be its normal unstressed state, for which the restoring force is zero.

When the particle of mass m is a distance r from the center of the earth, the system potential energy is given by Eq. 16-13 as

$$U(r) = -W_{\infty r} + 0 \qquad (16\text{-}14)$$

in which $W_{\infty r}$ is the work done by the conservative force (gravity) on the particle as the particle moves in from infinity to a distance r from the center of the earth. For simplicity we assume for the present that the particle moves toward the earth along a radial line. The gravitational force $F(r)$ acting on the particle (assuming $r \geqslant R$) will then be $-GMm/r^2$, the minus sign indicating an attractive force, that is, a force that pulls the particle toward the earth. We may then find $U(r)$ from Eq. 16-14 as

$$U(r) = -W_{\infty r}$$

$$= -\int_{\infty}^{r} F(r)\,dr$$

$$= -\int_{\infty}^{r} \left(-\frac{GMm}{r^2}\right)dr = -\frac{GMm}{r}\Big|_{\infty}^{r}$$

$$= -\frac{GMm}{r}. \qquad (16\text{-}15)$$

The minus sign indicates that the potential energy is negative at any finite distance; that is, the potential energy is zero at infinity and decreases as the separation distance decreases. This corresponds to the fact that the gravitational force exerted on the particle by the earth is attractive. As the particle moves in from infinity, the work $W_{\infty r}$ done by this force on the particle is positive, which means, from Eq. 16-14, that $U(r)$ is negative.

Equation 16-15 holds no matter what path is followed by the particle in moving in from infinity to radius r. We can show this by breaking up any arbitrary path into infinitesimal steplike portions, which are drawn alternately along the radius and perpendicular to it (Fig. 16-13). No work is done along perpendicular segments, such as AB, because along them the force is perpendicular to the displacement. But the work done along the radial parts of the path, such as BC, adds up to the work done in going directly along a radial path, such as AE. The work done in moving the particle between two points in a gravitational field is, therefore, independent of the actual path connecting these points. Hence, the gravitational force is a *conservative* force.

figure 16-13
Work done in taking a mass from A to E is independent of the path.

Equation 16-15 shows that the potential energy of the particles M and m is a characteristic of the system $M + m$. The potential energy is a property of the *system* of bodies, rather than of either body alone. The potential energy changes whether M or m is displaced; each is in the gravitational field of the other. Nor does it make any sense to assign part of the potential energy to M and part of it to m. Often, however, we *do* speak of the potential energy of a body m (planet or stone, say) in the gravitational field of a much more massive body M (sun or earth, respectively). The justification for speaking as though the potential energy belongs to the planet or to the stone alone is this: When the potential energy of a system of two bodies changes into kinetic energy, the lighter body gets most of the kinetic energy. The sun is so much more massive than a planet that the sun receives hardly any of the kinetic energy; and the same is true for the earth in the earth-stone system.

We can derive the gravitational force from the potential energy. The relation for spherically symmetric potential energy functions is $F = -dU/dr$; see Eq. 8-7. This relation is the converse of Eq. 16-15. From it we obtain

$$F = -\frac{dU}{dr} = -\frac{d}{dr}\left(-\frac{GMm}{r}\right) = -\frac{GMm}{r^2}. \qquad (16\text{-}16)$$

The minus sign here shows that the force is an attractive one, directed inward along a radius opposite to the radial displacement vector.

We can, if we wish, associate a scalar field with gravitation. We first define, quite generally, the *gravitational potential V* as the *gravitational potential energy per unit mass* of a body in a gravitational field. Then, for the spherically symmetrical body of mass M,

$$V = \frac{U(r)}{m} = -\frac{GM}{r}. \qquad (16\text{-}17)$$

Associated with every point in the space around a mass M we then have a number, the gravitational potential. This gives us a *scalar field*, potential being a scalar quantity. To determine the force exerted by this field on a mass particle m placed in it, we simply compute $-dV/dr$ at the point in question and multiply by m. The force has a magnitude $-m\,dV/dr$ and a direction radially in toward the center of force M.

EXAMPLE 6

Escape velocity. We can readily find the gravitational potential energy of a particle of mass m at the surface of the earth as (Eq. 16-15) $U(R) = -GM_em/R_e$. The amount of work required to move a body from the surface of the earth to infinity is GM_em/R_e, or about 6.0×10^7 J/kg. If we could give a projectile more than this energy at the surface of the earth, then, neglecting the resistance of the earth's atmosphere, it would escape from the earth never to return. As it proceeds outward its kinetic energy decreases and its potential energy increases, but its speed is never reduced to zero. The critical initial speed, called the escape speed v_0, such that the projectile does not return, is given by

$$\tfrac{1}{2}mv_0{}^2 = \frac{GM_em}{R_e}$$

or

$$v_0 = \sqrt{2\frac{GM_e}{R_e}} = 7.0 \text{ mi/s } (25{,}000 \text{ mi/h}) = 11.2 \text{ km/s}.$$

Should a projectile be given this initial speed, it would escape from the earth. For initial speeds less than this the projectile will return. Its kinetic energy becomes zero at some finite distance from the earth and the projectile falls back to earth.*

The lighter molecules in the earth's upper atmosphere can attain high enough speeds by thermal agitation to escape into outer space. Hydrogen gas, which must have been present in the earth's atmosphere a long time ago, has now disappeared from it. Helium gas escapes at a steady rate, much of it resupplied by radioactive decay from the earth's crust. The escape speed for the sun is much too great to allow hydrogen to escape from its atmosphere. On the other hand, the speed of escape on the moon is so small that it can hardly keep any atmosphere at all. (See Problem 30.)

16-10
POTENTIAL ENERGY FOR MANY-PARTICLE SYSTEMS

If two particles are separated by a distance r, their potential energy is given from Eq. 16-14 as

$$U(r) = -W_{\infty r} \qquad (16\text{-}14)$$

in which $W_{\infty r}$ is the work done *by the gravitational force* as the particles move from an infinite separation to separation r. We now give another interpretation to $U(r)$.

Let us balance out the gravitational force by an *external force* applied by some external agent and let us arrange it so that, at all times, this external force is equal and opposite to the gravitational force for each particle. The work done by the *external* force as the particles move from an infinite separation to separation r is not $W_{\infty r}$ but $-W_{\infty r}$; this follows because the displacements are the same but the forces are equal and opposite. Thus we may interpret Eq. 16-14 as follows: *The potential energy of a system of particles is equal to the work that must be done by an external agent to assemble the system, starting from the standard reference configuration*

Thus, if you lift a stone of mass m a distance of y above the earth's surface, you are the external agent (separating earth and stone) and the work you do in "assembling the system" is $+mgy$, which is also the potential energy. Similarly, the work done *by an external agent* as a body of mass m moves in from infinity to a distance r from the earth

* We have ignored the forces exerted on the projectile by bodies other than the earth. At sufficiently great distances from the earth we must take into account the gravitational forces arising from the moon, the planets, the sun etc., so that we can no longer use the simple "two-body" result. A projectile can escape from the earth by being "captured" by another astronomical body, for example, in such "many-body" cases.

is *negative* because the agent must exert a restraining force on the body; this is in agreement with Eq. 16-14.

These considerations hold for systems that contain more than two particles. Consider three bodies of masses m_1, m_2, and m_3. Let them initially be infinitely far from one another. The problem is to compute the work done by an external agent to bring them into the positions shown in Fig. 16-14. Let us first bring m_2 in toward m_1 from an infinite separation to the separation $\mathbf{r}_{12}$. The work done against the gravitational force exerted by m_1 on m_2 is $-Gm_1m_2/r_{12}$. Now let us bring m_3 in from infinity to the separation $\mathbf{r}_{13}$ from m_1 and $\mathbf{r}_{23}$ from m_2. The work done against the gravitational force exerted by m_1 on m_3 is $-Gm_1m_3/r_{13}$ and that against the gravitation force exerted by m_2 on m_3 is $-Gm_2m_3/r_{23}$. The total work done in assembling this system is the total potential energy of the system

$$-\left(\frac{Gm_1m_2}{r_{12}} + \frac{Gm_1m_3}{r_{13}} + \frac{Gm_2m_3}{r_{23}}\right).$$

Notice that no vector operations are needed in this procedure.

No matter how we assemble the system, that is, regardless of which bodies are moved or which paths are taken, we always find this same amount of work required to bring the bodies into the configuration of Fig. 16-13 from an initial infinite separation. The potential energy must, therefore, be associated with the system rather than with any one or two bodies. If we wanted to separate the system into three isolated masses once again, we would have to supply an amount of energy

$$+\left(\frac{Gm_1m_2}{r_{12}} + \frac{Gm_1m_3}{r_{13}} + \frac{Gm_2m_3}{r_{23}}\right).$$

This energy may be regarded as a sort of binding energy holding the mass particles together in the configuration shown.

Just as we can associate elastic potential energy with the compressed or stretched configuration of a spring holding a mass particle, so we can associate gravitational potential energy with the configuration of a system of mass particles held together by gravitational forces. Similarly, if we want to think of the elastic potential energy of a particle as being stored in the spring, so we can think of the gravitational potential energy as being stored in the gravitational field of the system of particles. A change in either configuration results in a change of potential energy.

These concepts occur again when we meet forces of electric or magnetic origin, or, in fact, of nuclear origin. Their application is rather broad in physics. The advantage of the energy method over the dynamical method is derived from the fact that the energy method uses scalar quantities and scalar operations rather than vector quantities and vector operations. When the actual forces are not known, as is often the case in nuclear physics, the energy method is essential.

figure 16-14
Three masses m_1, m_2, and m_3 brought together from infinity.

What is the binding energy of the earth-sun system? Neglect the presence of other planets or satellites.

For simplicity assume that the earth's orbit about the sun is circular at a radius r_{es}. The work done against the gravitational force to bring the earth and sun from an infinite separation to a separation r_{es} is

$$-G\frac{M_sM_e}{r_{es}} \cong -5.0 \times 10^{33} \text{ J},$$

EXAMPLE 7

where we take $M_s \cong 330{,}000 M_e$, $M_e = 6.0 \times 10^{24}$ kg, $r_{es} = 150 \times 10^9$ m. The minus sign indicates that the force is attractive, so that work is done by the gravitational force. It would take an equivalent amount of work by an outside agent to separate these bodies completely from rest. Because the kinetic energy of the earth in its orbit is half the magnitude of the potential energy of the earth-sun system, only half of this work is needed to break up the system, so that the effective binding energy, assuming that the earth-sun-system is at rest after breakup, is about 2.5×10^{33} J.

What effect does the presence of the moon and other planets have on the energy binding the earth to the solar system?

Consider again the motion of a body of mass m (planet or satellite, say) about a massive body of mass M (sun or earth, say). We shall consider M to be at rest in an inertial reference frame with the body m moving about it in a circular orbit. The potential energy of the system is

$$U(r) = -\frac{GMm}{r},$$

where r is the radius of the circular orbit. The kinetic energy of the system is

$$K = \tfrac{1}{2} m \omega^2 r^2$$

the sun being at rest. From the equation preceding Eq. 16-11 we obtain

$$\omega^2 r^2 = \frac{GM}{r},$$

so that

$$K = \frac{1}{2} \frac{GMm}{r}.$$

The total energy is

$$E = K + U = \frac{1}{2}\frac{GMm}{r} - \frac{GMm}{r} = -\frac{GMm}{2r}. \tag{16-18}$$

This energy is constant and is negative. Now the kinetic energy can never be negative, but from Eq. 16-18 we see that it must go to zero as the separation goes to infinity. The potential energy is always negative, except for its zero value at infinite separation. The meaning of the total negative energy, then, is that the system is a closed one, the planet m always being bound to the attracting solar center M and never escaping from it (Fig. 16-15).

Even when we consider elliptical orbits, in which r and ω vary, the total energy is negative. It is also constant, corresponding to the fact that gravitational forces are conservative. Hence, both the total energy and the total angular momentum are constant in planetary motion. These quantities are often called *constants of the motion*. We obtain the actual orbit of a planet with respect to the sun by starting with these conservation relations and eliminating the time variable by use of the laws of dynamics and gravitation. The result is that planetary orbits are elliptical.

In the earlier theories of the atom, as in the Bohr theory of the hydrogen atom, these identical mechanical relations are used in describing the motion of an electron about an attracting nuclear center. These same relations are used for open orbits (total energy positive) as in the experiments of Rutherford on the scattering of charged nuclear particles. Central forces, and particularly inverse square forces, are often encountered in physical systems.

16-11
ENERGY CONSIDERATIONS IN THE MOTIONS OF PLANETS AND SATELLITES

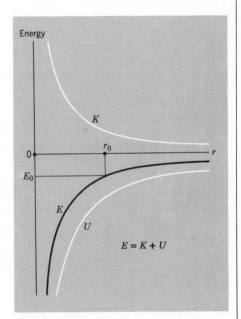

figure 16-15
Kinetic energy K, potential energy U, and total energy $E = U + K$ of a body in circular planetary motion. A planet with total energy $E_0 < 0$ will remain in an orbit of radius r_0. The farther the planet is from the sun, the greater (that is, less negative) its (constant) total energy E. To escape from the center of force and still have kinetic energy at infinity, it would need positive total energy.

In describing the experiments which were fundamental to our definitions of force and mass, we had to assume some reference frame relative to which accelerations could be measured. If the reference frame itself were erratically accelerated, we would not observe any regularity in our measured accelerations. As a matter of fact, our laboratory experiments are performed in a reference frame which is fixed to the earth. We have already discussed the effect that the rotation of the earth about its own axis has on our measurements. What effect does the motion of the earth as a whole about the sun, or some other cosmic body, have?

The acceleration of the earth with respect to the sun is $\omega^2 r$ or about 6×10^{-3} m/s². It would seem at first that this acceleration, small as it is, might prove disturbing in experiments involving small forces. That this is not the case, however, follows from the universality of the law of gravitation. Not only the earth but also the masses we use in our laboratory apparatus are accelerated toward the sun at practically the same rate.

Let us compute the error made in neglecting the earth's orbital acceleration. The acceleration of the earth toward the sun is k/r^2, where r is the distance from the center of the sun and the center of the earth and k is GM_s. Consider now a body on that side of the earth most distant from the sun. We can imagine that we are weighing it on a spring scale, for example. Then, the part of its acceleration toward the sun which is due to the gravitational attraction of the sun itself is

$$\frac{k}{(r + r_0)^2} = \frac{k}{r^2}\left(1 - \frac{2r_0}{r} + \cdots + \text{much smaller terms}\right),$$

where r_0 is the radius of the earth. The *difference* between the acceleration of the earth due to the sun's attraction (that is, k/r^2) and the acceleration of the apparatus due to the sun's attraction (the expression above) would be less than $(k/r^2)(2r_0/r)$. But $2r_0/r$ is about 10^{-4}. The difference, then, would be less than 10^{-4} of the earth's acceleration, or less than 10^{-6} m/s². The relative acceleration of the body and the earth due to the sun's attraction is about one-ten-millionth as strong as the gravitational acceleration of the body due to the earth. The moon has a similar effect of comparable magnitude on the body. Hence, only if we were measuring to one part in a million, would we need to consider seriously the accelerating nature of a reference frame attached to the earth. For almost all practical purposes the earth is good enough as an inertial reference frame.

16-12 THE EARTH AS AN INERTIAL REFERENCE FRAME

16-13 THE PRINCIPLE OF EQUIVALENCE

Consider two reference frames: (1) a nonaccelerating (inertial) reference frame S in which there is a uniform gravitational field and (2) a reference frame S' which is accelerating uniformly with respect to an inertial frame but in which there is no gravitational field. In his general theory of relativity, Albert Einstein showed that two such frames are exactly equivalent physically. That is, experiments carried out under the same conditions in these two frames should give the same results. This is the *principle of equivalence*.

Suppose that a spaceship is at rest in an inertial reference frame S in which there is a uniform gravitational field, say at the surface of the earth. Inside the spaceship objects such as an apple, which are released, will fall with an acceleration, say **g**, in the gravitational field; objects which are at rest—such as an astronaut sitting on the floor or a package on a spring scale attached to the ceiling —will experience a force, exerted by the floor or the spring respectively, opposing their weight.

Now suppose that the rockets are turned on and that the spaceship proceeds to a region of outer space where there is no gravitational field. Let the acceleration of the spaceship, our new frame S', be $\mathbf{a} = -\mathbf{g}$ with respect to the inertial reference frame S; that is, the ship is accelerating away from the earth beyond the region where the earth's field (or any other gravitational field) is appreciable. The conditions in the spaceship will now be similar to those in a spaceship at rest on the surface of the earth. Inside the ship, if the astronaut releases an apple, it will accelerate downward relative to the spaceship with an acceleration **g.** In fact, since all bodies that are free of any forces move with uniform velocity relative to the inertial frame *S*, all such bodies appear to fall with the *same* acceleration **g** with respect to the spaceship, *S'*. Furthermore, objects which are at rest relative to the spaceship — such as an astronaut sitting on the floor or a package on a spring scale attached to the ceiling — will experience forces indistinguishable from the forces which balanced their weight in the case when the spaceship was at rest in a gravitational field in *S*.

Indeed, if the astronaut did not know that rockets were accelerating his ship from *S*, he would be justified in concluding that he was in a gravitational field — a field whose pull accelerated the falling apple in *S'* and whose pull required that a balancing force be applied to the package (the tension in the spring) and to the spaceman (the normal force of the floor) to keep them at rest in *S'*. The astronaut simply could not tell the difference, from observations in his own frame, between a situation in which his ship was accelerating relative to an inertial frame in a region having no gravitational field and a situation in which the spaceship was unaccelerated in an inertial frame in which a uniform gravitational field existed. The two situations are exactly equivalent.

Einstein pointed out that, from the principle of equivalence, it follows that one cannot speak of the absolute acceleration of a reference frame, only a relative one, just as it followed from the special theory of relativity that one cannot speak of the absolute velocity of a reference frame, only a relative one. It also follows from the principle of equivalence that inertial mass and gravitational mass are equal. For all bodies which are free of any forces will move with uniform velocity relative to an inertial reference frame no matter what their inertial masses are, and they should, therefore, all have the same acceleration relative to an accelerated reference frame. Hence, from the principle of equivalence of *S* and *S'*, all bodies should fall with the same acceleration in a homogeneous gravitational field.

From the discussion so far we see that a uniform gravitational field can be imitated by a "field of acceleration." Indeed, a uniform gravitational field can be "transformed away" by transforming to a reference frame accelerating in the direction of the field with an acceleration equal in magnitude to that due to the field. In this new frame a particle whose motion was originally subject to a gravitational field is now a free particle. For example, in an artificial earth satellite an apple released by an astronaut will not fall relative to the satellite and the astronaut himself will be free of the forces which countered the pull of gravity before launching, so that he feels weightless. In general, however, gravitational fields, such as that of the earth, are not uniform through all space, so that one cannot replace the gravitational fields throughout space simply by transforming to a single reference frame accelerating with respect to the source of the field. One would need a different accelerated frame at each point in space to imitate the entire gravitational field.

questions

1. Modern observational astronomy and navigation procedures make use of the geocentric (or Ptolemaic) point of view (by using the rotating "celestial sphere"). Is this wrong? If not, then what criterion determines the system (the Copernican or Ptolemaic) we use? When would we use the heliocentric (or Copernican) system?

2. If the force of gravity acts on all bodies in proportion to their masses, why doesn't a heavy body fall correspondingly faster than a light body?

3. How does the weight of a body vary en route from the earth to the moon? Would its mass change?

4. At the earth's surface a freely suspended object is given a horizontal blow by a hammer. The object is taken to the moon, suspended freely, and given an equal horizontal blow with the same hammer. How is the horizontal speed resulting on the moon related to the horizontal speed on the earth?

5. Would we have more sugar to the pound at the pole or the equator? What about sugar to the kilogram?

6. What approximately is the *gravitational* force of attraction between a normal woman and a typical man 10 m away? When they are dancing? Compare with typical body weights.

7. Does the concentration of the earth's mass near its center change the variation of g with height compared with a homogeneous sphere? How?

8. Because the earth bulges near the equator, the source of the Mississippi River, although high above sea level, is nearer to the center of the earth than is its mouth. How can the river flow "uphill"?

9. The earth is an oblate spheroid because of the "flattening" effect of the earth's rotation. Is a degree of latitude larger or smaller near either pole than near the equator? Why?

10. Why can we learn more about the shape of the earth by studying the motion of an artificial satellite than by studying the motion of the moon?

11. How can one determine the mass of the moon?

12. One clock is based on an oscillating spring, the other on a pendulum. Both are taken to Mars. Will they keep the same time there that they kept on Earth? Will they agree with each other? Explain. Mars has a mass 0.1 that of Earth and a radius half as great.

13. From Kepler's second law and observations of the sun's motion as seen from the earth, we can conclude that the earth is closer to the sun during winter in the Northern hemisphere than during summer. Why isn't it colder in summer than in winter?

14. Does the law of universal gravitation require the planets of our solar system to have the actual orbits observed? Would planets of another star, similar to our Sun, have the same orbits? Suggest factors that might have determined the special orbits observed.

15. How is the orbital speed of a planet related to its (assumed circular) orbital radius?

16. The gravitational force exerted by the sun on the moon is about twice as great as the gravitational force exerted by the earth on the moon. Why then doesn't the moon escape from the earth (during a solar eclipse, for example)?

17. Explain why the following reasoning is wrong. "The sun attracts all bodies on the earth. At midnight, when the sun is directly below, it pulls on an object in the same direction as the pull of the earth on that object; at noon, when the sun is directly above, it pulls on an object in a direction opposite to the pull of the earth. Hence, all objects should be heavier at midnight (or night) than they are at noon (or day)."

18. The gravitational attraction of the sun and the moon on the earth produces tides. The sun's tidal effect is about half as great as that of the moon's. The direct pull of the sun on the earth, however, is about 175 times that of the moon. Why is it then that the moon causes the larger tides?

19. If lunar tides slow down the rotation of the earth (owing to friction), the angular momentum of the earth decreases. What happens to the motion of the moon as a consequence of the conservation of angular momentum? Does the sun (and solar tides play a role here? (See "Tides and the Earth-Moon System" by Peter Goldreich, *Scientific American*, April 1972 and "Tides of the British Seas" by Frank Sandon, in *Physics Education*, June 1975.)

20. Would you expect the total energy of the solar system to be constant? The total angular momentum? Explain your answers.

21. Discuss how the period of a simple pendulum changes if it is in an assembly that a rocket will propel from earth to a stable satellite orbit about the earth.

22. Does a rocket really need the escape speed of 25,000 mi/h initially to escape from the earth?

23. Objects at rest on the earth's surface move in circular paths with a period of 24 h. Are they "in orbit" in the sense that an earth satellite is in orbit?

Why not? What would the length of the "day" have to be to put such objects in true orbit?

24. An artificial satellite of the earth releases a package. Neglecting effects of air resistance, determine whether the package will strike the earth and if so, where—at a point ahead, directly below, or behind the satellite at the instant of impact, or directly under the satellite at release time?

25. Neglecting air friction and technical difficulties, can a satellite be put into an orbit by being fired from a huge cannon at the earth's surface? Explain.

26. Can a satellite move in a stable orbit in a plane not passing through the earth's center? Explain.

27. As measured by an observer on earth would there be any difference in the periods of two satellites, each in a circular orbit near the earth in an equatorial plane, but one moving eastward and the other westward?

28. After Sputnik I was put into orbit we were told that it would not return to earth but would burn up in its *descent*. Considering the fact that it did not burn up in its *ascent*, how is this possible?

29. In which case do astronauts experience greater acceleration, when being launched into orbit or on reentry and return to earth?

30. Show that a satellite may speed down: that is, show that if frictional forces cause a satellite to lose total energy, it will move into an orbit closer to the earth and may have increased kinetic energy.

31. An artificial satellite is in a circular orbit about the earth. How will its orbit change if one of its rockets is momentarily fired (a) toward the earth, (b) away from the earth, (c) in a forward direction, (d) in a backward direction, (e) at right angles to the plane of the orbit?

32. Inside a spaceship what difficulties would you encounter in walking? In jumping? In drinking?

33. We have all seen TV transmissions from orbiting satellites and watched objects floating around in effective zero gravity. Suppose an astronaut, bracing himself against the satellite frame, kicks a floating bowling ball. Will he stub his toe? Explain.

34. If a planet of given density were made larger, its force of attraction for an object on its surface would increase because of the planet's greater mass but would decrease because of the greater distance from the object to the center of the planet. Which effect predominates?

35. Consider a hollow spherical shell. How does the gravitational potential inside compare with that on the surface? What is the gravitational field strength inside?

36. A stone is dropped along the center of a deep vertical mine shaft. Assume no air resistance but consider the earth's rotation. Will the stone continue along the center of the shaft? If not, describe its motion.

37. Use qualitative arguments to explain why the following four periods are equal (all are 84 min, assuming a uniform earth density): (a) time of revolution of a satellite just above the earth's surface; (b) period of oscillation of mail in a tunnel through the earth; (c) period of simple pendulum having a length equal to the earth's radius in a uniform field 9.8 N/kg; (d) period of an infinite simple pendulum in the earth's real gravitational field.

38. The "action-at-a-distance" view of the gravitational force implies that the action is instantaneous. Actually, present physical theory assumes that gravitation propagates with a finite speed and this is taken into account in the modification of classical physics represented by general relativity theory. (See "Gravitational Waves—a Progress Report" by Jonothan L. Logan, in *Physics Today*, March 1973 for a discussion of the ideas and attempts at experimental verification.) What would happen to classical deductions if it were assumed there that the action were not instantaneous? (See also "Infinite Speed of Propagation of Gravitation in Newtonian Physics" by I. J. Good, *American Journal of Physics*, July 1975.)

39. Can one regard gravity as a "fictitious" force arising from the acceleration of one's reference frame relative to an inertial reference frame, rather than a "real" force?

problems

SECTION 16-2

1. How far from the earth must a body be along a line toward the sun so that the sun's gravitational pull balances the earth's? The sun is 9.3×10^7 mi away and its mass is $3.24 \times 10^5 M_e$.   *Answer:* 1.6×10^5 mi.

SECTION 16-3

2. What is the percentage change in the acceleration of the earth toward the sun from a total eclipse of the sun to the point where the moon is on a side of the earth directly opposite the sun?

SECTION 16-5

3. At what altitude above the earth's surface would the acceleration of gravity be 4.9 m/s²? The mass of the earth is 6.0×10^{24} kg and its mean radius is 6.4×10^6 m.   *Answer:* 2.6×10^6 m.

4. (a) What is the period of a "seconds pendulum" (period = 2 s on earth) on the surface of the moon? The moon's mass is 7.35×10^{22} kg and its radius is 1,720 km. (b) Why should a "seconds pendulum" have a period of two seconds rather than one second?

5. Certain neutron stars (extremely dense stars) are believed to be rotating at about one revolution per second. If such a star has a radius of 20 km, what must be its mass so that objects on its surface will be attracted to the star and not "thrown off" by the rapid rotation?   *Answer:* 4.7×10^{24} kg.

6. The fact that g varies from place to place over the earth's surface drew attention when Jean Richer in 1672 took a pendulum clock from Paris to Cayenne, French Guiana, and found that it lost 2.5 min/day. If $g = 9.81$ m/s² in Paris, what is g in Cayenne?

7. If a pendulum has a period of exactly one second at the Equator, what would be its period at the South Pole?   *Answer:* 0.9974 s.

8. Masses m, assumed equal, hang from strings of different lengths on a balance at the surface of the earth, as shown in Fig. 16-16. If the strings have negligible mass and differ in length by h, (a) show that the error in weighing, associated with the fact that W' is closer to the earth than W, is $W' - W = 8\pi G\rho mh/3$ in which ρ is the mean density of the earth (5.5 g/cm³). (b) Find the difference in length which will give an error of one part in a million.

9. A scientist is making a precise measurement of g at a certain point in the Indian Ocean (on the equator) by timing the swings of a pendulum of accurately known construction. To provide a stable base the measurements are conducted in a submerged submarine. It is observed that a slightly different result for g is obtained when the submarine is moving eastward through the point than when it is moving westward, the speed in each case being 16 km/h. Account for this difference and calculate the fractional error $\Delta g/g$ in g.   *Answer:* 6.6×10^{-5}.

10. A body is suspended on a spring balance in a ship sailing along the equator with a speed v. (a) Show that the scale reading will be very close to $W_0(1 \pm 2\omega v/g)$, where ω is the angular speed of the earth and W_0 is the scale reading when the ship is at rest. (b) Explain the plus or minus.

SECTION 16-6

11. Two concentric shells of uniform density of mass M_1 and M_2 are situated as shown in Fig. 16-17. Find the force on a particle of mass m when the particle is located at (a) $r = a$, (b) $r = b$, and (c) $r = c$. The distance r is measured from the center of the shells.
Answer: (a) $G(M_1 + M_2)m/a^2$. (b) GM_1m/b^2. (c) Zero.

figure 16-16
Problem 8

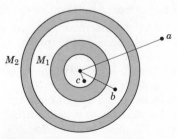

figure 16-17
Problem 11

12. With what speed would mail pass through the center of the earth if it were delivered by the chute of Example 3?

13. The sun, mass 2.0×10^{30} kg, is revolving about the center of the Milky Way galaxy, which is 2.4×10^{20} m away. It completes one revolution every 2.5×10^8 yr. Estimate the number of stars in the Milky Way, assuming a circular orbit.   *Answer:* 6.5×10^{10}.

14. Consider an inertial reference frame whose origin is fixed at the center of mass of the system earth + falling object. (a) Show that the acceleration toward the center of mass of *either* body is independent of the mass of that body. (b) Show that the mutual, or relative, acceleration of the two bodies depends on the sum of the masses of the two bodies. Comment on the meaning, then, of the statement that a body falls toward the earth with an acceleration that is independent of its mass.

15. The following problem is from the 1946 "Olympic" examination of Moscow State University (see Fig. 16-18): A spherical hollow is made in a lead sphere of radius R, such that its surface touches the outside surface of the lead sphere and passes through its center. The mass of the sphere before hollowing was M. With what force, according to the law of universal gravitation, will the lead sphere attract a small sphere of mass m, which lies at a distance d from the center of the lead sphere on the straight line connecting the centers of the spheres and of the hollow?

figure 16-18
Problem 15

$$Answer:\ \frac{GmM}{d^2}\left[1 - \frac{1}{8(1 - R/2d)^2}\right].$$

16. The variation of g in the earth's interior is given in the accompanying table. The earth's radius is 6400 km.

Depth, km	g, m/s²	Depth, km	g, m/s²
0	9.82	1400	9.88
33	9.85	1600	9.86
100	9.89	1800	9.85
200	9.92	2000	9.86
300	9.95	2200	9.90
413	9.98	2400	9.98
600	10.01	2600	10.09
800	9.99	2800	10.26
1000	9.95	2900	10.37
1200	9.91	4000	8.00

Within the earth's central core (below 2900 km) the values of g diminish monotonically (not linearly) from 10.37 m/s² to zero. The actual variation of g below 4000 km is uncertain. (a) Plot qualitatively g versus r (where r *is the distance from the earth's center*) from 0 to 6400 km. (b) Explain carefully how the earth's density must vary as we proceed from its surface to its center in order to account for this variation of g. (c) Take $\rho = 1$ at the surface (its average value is actually 3.0 g/cm³), and plot qualitatively ρ versus r. Assume throughout that ρ and g are spherically symmetrical.

17. (a) Show that in a chute through the earth along a chord line, rather than along a diameter, the motion of an object will be simple harmonic; assume a uniform earth density. (b) Find the period. (c) Will the object attain the same maximum speed along a chord as it does along a diameter?
Answer: (b) 84 min. (c) No.

18. Consider a mass particle at a point P anywhere inside a spherical shell of matter. Assume the shell is of uniform thickness and density. Construct a narrow double cone with apex at P intercepting areas dA_1 and dA_2 on the shell (Fig. 16-19). (a) Show that the resultant gravitational force exerted on the particle at P by the intercepted mass elements is zero. (b) Show then

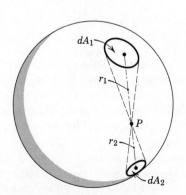

figure 16-19
Problem 18

that the resultant gravitational force of the entire shell on an internal particle is zero everywhere. (This method was devised by Newton.)

SECTION 16-7

19. (a) Can a satellite be sent out to a distance where it will revolve about the earth with an angular velocity equal to that at which the earth rotates, so that it remains always above the same point on the earth? (b) What would be the radius of the orbit of such a so-called synchronous earth satellite? *Answer:* (a) Yes. The plane of the orbit must be equatorial. (b) 4.2×10^4 km.

20. (a) With what horizontal speed must a satellite be projected at 100 mi (160 km) above the surface of the earth so that it will have a circular orbit about the earth? Take the earth's radius as 4000 mi (6400 km). (b) What will be the period of revolution?

21. The mean distance of Mars from the sun is 1.52 times that of Earth from the sun. Find the number of years required for Mars to make one revolution about the sun. *Answer:* 1.87 yr.

22. Determine the mass of the earth from the period T and the radius r of the moon's orbit about the earth: $T = 27.3$ days and $r = 2.39 \times 10^5$ mi (3.85×10^5 km).

23. (a) Satellite A is in a circular earth orbit with radius R and satellite B is in a circular earth orbit with radius $4R$. Calculate the ratio of the periods of revolution, T_A/T_B. (b) A pendulum and a mass-spring system oscillate with approximately the same frequency on the earth's surface. How do their frequencies compare if they are mounted first in satellite A and then in satellite B?
Answer: (a) $T_A/T_B = \frac{1}{8}$. (b) The pendulum frequency is zero; the mass-spring frequency is unchanged.

24. Consider an artificial satellite in a circular orbit about the earth. State how the following properties of the satellite vary with the radius r of its orbit: (a) period; (b) kinetic energy; (c) angular momentum; (d) speed.

25. Show how, guided by Kepler's third law (p. 337), Newton could deduce that the force holding the moon in its orbit, assumed circular, must vary as the inverse square of the distance from the center of the earth.

26. If a satellite in an elliptical orbit about the earth has a perigee (closest distance of approach) of 300 km above the surface of the earth, and an apogee (furthest distance of approach) of 2000 km above the surface of the earth, then what is the ratio of the orbital speed at perigee to that at apogee?

27. Three identical bodies of mass M are located at the vertices of an equilateral triangle with side L. At what speed must they move if they all revolve under the influence of one another's gravity in a circular orbit circumscribing the triangle while still preserving the equilateral triangle?
Answer: $\sqrt{GM/L}$.

28. (a) Show that the two-body problem of Section 16-7 can be simplified to a one-body problem by use of the reduced mass concept of Section 15-8. That is, show that if we use $\mu = mM/(m + M)$ instead of m, we may solve for the motion of m relative to M exactly as though M were the origin of our inertial reference frame. (b) Show that the assumption made in Section 16-7 that R is negligibly small compared to r is equivalent to assuming that the reduced mass μ is equal to m. (c) Compare μ for the earth-sun system with the earth's mass; compare μ for the moon-earth system with the moon's mass. (d) If we were to use the reduced mass μ of the two-body system instead of m, how would this affect the equations of Section 16-7?

SECTION 16-9

29. Mars has a mean diameter of 6900 km, Earth one of 1.3×10^4 km. The mass of Mars is $0.11 M_e$. (a) How does the mean density of Mars compare with that of Earth? (b) What is the value of g on Mars? (c) What is the escape velocity on Mars? *Answer:* (a) $\rho_M = 0.73\ \rho_E$. (b) 3.7 m/s². (c) 5.0 km/s.

30. (a) Show that to escape from the atmosphere of a planet a necessary condition for a molecule is that it have a speed such that $v^2 > 2GM/r$, where M is the mass of the planet and r is the distance of the molecule from the center of the planet. (b) Determine the escape speed from the earth for an atmospheric particle 1000 km above the earth's surface. (c) Do the same for Mars.

31. It is conjectured that a "burned-out" star could collapse to a "gravitational radius," defined as the radius for which the work needed to remove an object of mass m_0 from the star's surface to infinity equals the rest energy m_0c^2 of the object. Show that the gravitational radius of the sun is GM_s/c^2 and determine its value in terms of the sun's present radius. (For a review of this phenomenon see "Black Holes: New Horizons in Gravitational Theory" by Philip C. Peters, in *American Scientist*, Sept.–Oct. 1974.)
 Answer: $2 \times 10^{-6} \, R_s$.

32. Show that the velocity of escape from the sun at the earth's distance from the sun is $\sqrt{2}$ times the speed of the earth in its orbit, assumed to be a circle.

33. A projectile is fired vertically from the earth's surface with an initial speed of 10 km/s. Neglecting atmospheric friction, how far above the surface of the earth would it go? Take the earth's radius as 6400 km.
 Answer: 2.6×10^4 km.

34. A rocket is accelerated to a speed of $v = 2\sqrt{gR_e}$ near the earth's surface and then coasts upward. (a) Show that it will escape from the earth. (b) Show that very far from the earth its speed is $V = \sqrt{2gR_e}$.

35. Physicists have speculated about the possible existence of bodies with negative mass; for such hypothetical bodies it is postulated that m in the formulas of physics should be replaced by $-m$. Suppose that two particles, of mass $+m$ and $-m$ respectively, are placed a distance d apart. Show (a) the force acting on each and (b) the acceleration of each. Describe the expected motion, assuming that both particles are initially at rest, and show that this motion does not violate the laws of conservation of linear momentum or of mechanical energy. Such negative-mass particles have not yet been found.
 Answer: (a) The force, from Newton's law of gravitation, is repulsive. (b) The accelerations, from Newton's second law, point in the same direction, from the negative to the positive mass.

36. A sphere of matter, mass M, radius a, has a concentric cavity, radius b, as shown in cross section in Fig. 16-20. (a) Sketch the gravitational force F exerted by the sphere on a particle of mass m, located a distance r from the center of the sphere, as a function of r in the range $0 \leq r \leq \infty$. Consider points $r = 0$, b, a, and ∞ in particular. (b) Sketch the corresponding curve for the potential energy $U(r)$ of the system. (c) From these graphs, how would you obtain graphs of the gravitational field strength and the gravitational potential due to the sphere?

37. Two particles of mass m and M are initially at rest an infinite distance apart. Show that at any instant their relative velocity of approach attributable to gravitational attraction is $\sqrt{2G(M + m)/d}$, where d is their separation at that instant.

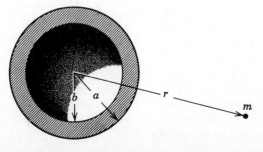

figure 16-20
Problem 36

38. How long will it take a comet, moving in a parabolic path, to move from its point of closest approach to the sun through an angle of 90°, measured at the sun? Let the distance of closest approach to the sun be equal to the radius of the earth's orbit, assumed circular. (*Hint:* See Example 4 and Problem 32.)

SECTION 16-10

39. In a double star, two stars of mass 3×10^{30} kg each rotate about their common center of mass, 10^{11} m away. (*a*) What is their common angular speed? (*b*) Suppose that a meteorite passes through this center of mass moving at right angles to the orbital plane of the stars. What must its speed be if it is to escape from the gravitational field of the double star?
Answer: (*a*) 2×10^{-7} rad/s. (*b*) 9×10^4 m/s.

40. An 800-kg mass and a 600-kg mass are separated by 0.25 m. (*a*) What is the gravitational field strength due to these masses at a point 0.20 m from the 800-kg mass and 0.15 m from the 600-kg mass? (*b*) What is the gravitational potential at this point due to these same masses?

41. Masses of 200 and 800 g are 12 cm apart. (*a*) Find the gravitational force on an object of unit mass situated at a point on the line joining the masses 4.0 cm from the 200-g mass. (*b*) Find the gravitational potential energy per unit mass at that point. (*c*) How much work is needed to move this object to a point 4.0 cm from the 800-g mass along the line of centers?
Answers: (*a*) Zero. (*b*) -10×10^{-5} erg/g. (*c*) -5.0×10^{-6} erg.

42. For interstellar travel, a spaceship must overcome the sun's gravitational field as well as that of the earth. (*a*) What is the total amount of energy required for a 1.0-kiloton (equivalent to 9.1×10^5 kg) spaceship to free itself from the combined earth-sun gravitational field starting from an orbit 300 mi (480 km) above the earth's surface? Neglect all other bodies in the solar system. (*b*) What fraction of this energy is used to overcome the sun's field?

43. (*a*) Write an expression for the potential energy of a body of mass m in the gravitational field of the earth and moon. Let M_e be the earth's mass, M_m the moon's mass, R the distance from the earth's center, and r the distance from the moon's center. (*b*) At what point between the earth and moon will the total gravitational field strength attributable to the earth and moon be zero? (*c*) What will be the gravitational potential and the gravitational field strength on the earth's surface? (*d*) Answer for the moon's surface.
Answer: (*a*) $-Gm(M_e/R + M_m/r)$. (*b*) 3.4×10^8 m from earth. (*c*) -6.3×10^7 J/kg; 9.8 m/s². (*d*) -3.9×10^6 J/kg; 1.6 m/s².

SECTION 16-11

44. Consider two satellites A and B of equal mass m, moving in the same circular orbit of radius r around the earth E but in opposite senses of rotation and therefore on a collision course (see Fig. 16-21). (*a*) In terms of G, M_e, m, and r, find the total mechanical energy $E_A + E_B$ of the two-satellite-plus-earth system before collision. (*b*) If the collision is completely inelastic so that wreckage remains as one piece of tangled material (mass $= 2m$), find the total mechanical energy immediately after collision. (*c*) Describe the subsequent motion of the wreckage.

figure 16-21
Problem 44

45. (*a*) Does it take more energy to get a satellite up to 1000 mi above the earth than to put it in orbit once it is there? (*b*) What about 2000 mi? (*c*) What about 3000 mi? Take the earth's radius to be 4000 miles.
Answer: (*a*) No. (*b*) The same. (*c*) Yes.

46. Two earth satellites, A and B, each of mass m, are to be launched into (nearly) circular orbits about the earth's center. Satellite A is to orbit at an altitude of 4000 mi. Satellite B is to orbit at an altitude of 12,000 mi. The radius of the earth R_e is 4000 mi (Fig. 16-22). (*a*) What is the ratio of the potential energy of satellite B to that of satellite A, in orbit? (Explain the result in terms of the work required to get each satellite from its orbit to infinity.) (*b*) What is the ratio of the kinetic energy of satellite B to that of

satellite *A*, in orbit? (*c*) Which satellite has the greater total energy if each has a mass of 1.0 slug? By how much?

47. A pair of stars rotates about a common center of mass. One of the stars has a mass M which is twice as large as the mass m of the other, that is, $M = 2m$. Their centers are a distance d apart, d being large compared to the size of either star. (*a*) Derive an expression for the period of rotation of the stars about their common center of mass in terms of d, m, and G. (*b*) Compare the angular momenta of the two stars about their common center of mass by calculating the ratio L_m/L_M. (*c*) Compare the kinetic energies of the two stars by calculating the ratio K_m/K_M.
    Answer: (*a*) $2\pi d^{3/2}/\sqrt{3Gm}$. (*b*) 2. (*c*) 2.

figure 16-22
Problem 46

48. A satellite travels initially in an approximately circular orbit 640 km above the surface of the earth; its mass is 220 kg. (*a*) Determine its speed. (*b*) Determine its period. (*c*) For various reasons the satellite loses mechanical energy at the (average) rate of 1.4×10^5 J per complete orbital revolution. Adopting the reasonable approximation that the trajectory is a "circle of slowly diminishing radius," determine the distance from the surface of the earth, the speed, and the period of the satellite at the end of its 1500th orbital revolution. (*d*) What is the magnitude of the average retarding force? (*e*) Is angular momentum conserved?

49. A particle of mass m is subject to an attractive central force of magnitude k/r^2, k being a constant. If at the instant when the particle is at an extreme position in its closed orbit, at a distance a from the center of force, its speed is $\sqrt{k/2ma}$, find (*a*) the other extreme position, and (*b*) the speed of the particle at this position.
    Answer: (*b*) $3\sqrt{k/2ma}$.

SECTION 16-12

50. *Foucault Pendulum.* A pendulum whose upper end is attached so as to allow the pendulum to swing freely in any direction can be used to repeat an experiment first shown publicly by Foucault in Paris in 1851. If the pendulum is set oscillating, the plane of oscillation slowly rotates with respect to a line drawn on the floor, even though the tension in the wire supporting the bob and the gravitational pull of the earth on the bob lie in a vertical plane. (*a*) Show that this is a result of the fact that the earth is not an inertial reference frame. (*b*) Show that for a Foucault pendulum at a latitude angle θ, the period of rotation of the plane is $(24/\sin \theta)$ h. (*c*) Explain in simple terms the result at $\theta = 90°$ (the poles) and $\theta = 0°$ (the equator).

17
fluid statics

It is customary to classify matter, viewed macroscopically, into solids and fluids. A *fluid* is a substance that can *flow*. Hence the term fluid includes liquids and gases. Such classifications are not always clearcut. Some fluids, such as glass or pitch, flow so slowly that they behave like solids for the time intervals that we usually work with them. Plasma, which is highly ionized gas, does not fit easily into any of these categories; it is often called a "fourth state of matter" to distinguish it from the solid, the liquid, and the gaseous state. Even the distinction between a liquid and a gas is not clearcut because, by changing the pressure and temperature properly, it is possible to change a liquid (water, say) into a gas (steam, say) without the appearance of a meniscus and without boiling; the density and viscosity change in a continuous manner throughout the process.* In this text, however, we will define a fluid as it is ordinarily understood, and we will be interested only in those properties of fluids connected with their ability to flow. Therefore, the same basic laws control the static and dynamic behavior of both liquids and gases in spite of the differences between them that we observe at ordinary pressures.

For solids, which have a definite size and shape, we formulated the mechanics of rigid bodies, modified by the laws of elasticity for bodies that cannot be considered perfectly rigid. Since fluids change their shape readily and, in the case of gases, have a volume equal to that of the container in which they are confined, we must develop new techniques for solving problems in fluid mechanics. Our applications of

17-1
FLUIDS

* Pressures higher than the so-called critical point pressure must be employed to do this; for H_2O the critical point pressure is 218 atmospheres.

370

mechanics to continuous media, both solids and fluids, are based on Newton's laws of motion combined with the appropriate force laws. For fluids, as for solids, however, we find it convenient to develop special formulations of these basic laws.

17-2
PRESSURE AND DENSITY

There is a difference in the way a surface force acts on a fluid and on a solid. For a solid there are no restrictions on the direction of such a force, but for a fluid at rest the surface force must always be directed at right angles to the surface. For a fluid at rest cannot sustain a tangential force; the fluid layers would simply slide over one another when subjected to such a force. Indeed, it is the inability of fluids to resist such tangential forces (or shearing stresses) that gives them their characteristic ability to change their shape or to flow.

It is convenient, therefore, to describe the force acting on a fluid by specifying the *pressure p*, which is defined as the magnitude of the *normal* force per unit surface area. Pressure is transmitted to solid boundaries or across arbitrary sections of fluid *at right angles* to these boundaries or sections at every point. Pressure is a scalar quantity. The SI unit of pressure is the *pascal* (abbreviation Pa, 1 Pa = 1 N/m²). This unit is named after the French scientist Blaise Pascal (1623–1662) (see Section 17-4). Other units are bar (1 bar = 10^5 Pa), lb/in.², atmosphere (1 atm = 14.7 lb/in.² = 101,325 Pa), and mm-Hg (760 mm-Hg = 1 atm).

A fluid under pressure exerts a force on any surface in contact with it. Consider a closed surface containing a fluid (Fig. 17-1). An element of the surface can be represented by a vector **S** whose magnitude gives the area of the element and whose direction is taken to be the outward normal to the surface of the element. Then the force **F** exerted by the fluid against this surface element is

$$\mathbf{F} = p\mathbf{S}.$$

Since **F** and **S** have the same direction, the pressure p can be written as

$$p = \frac{F}{S}.$$

We assume that the element of area S is small enough so that the pressure p, defined as above, is independent of the size of the element S. The pressure may actually vary from point to point on the surface.

The density ρ of a homogeneous fluid (its mass divided by its volume) may depend on many factors, such as its temperature and the pressure to which it is subjected. For liquids the density varies very little over wide ranges in pressure and temperature, and we can safely treat it as a constant for our present purposes; see entries under *Water*

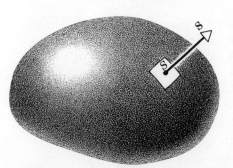

figure 17-1
An element of surface S can be represented by a vector **S**, equal to its area in magnitude and normal to it in direction.

Table 17-1
Densities of some materials and objects in kg/meter³

Interstellar space	$10^{-18} - 10^{-21}$
Best laboratory vacuum	$\sim 10^{-17}$
Hydrogen: at 0°C and 1.0 atm	9.0×10^{-2}
Air: at 0°C and 1.0 atm	1.3
at 100°C and 1.0 atm	0.95
at 0°C and 50 atm	6.5
Styrofoam	$\sim 1 \times 10^{2}$
Ice	0.92×10^{3}
Water: at 0°C and 1.0 atm	1.000×10^{3}
at 100°C and 1.0 atm	0.958×10^{3}
at 0°C and 50 atm	1.002×10^{3}
Aluminum	2.7×10^{3}
Mercury	1.36×10^{4}
Platinum	2.14×10^{4}
The earth: average density	5.52×10^{3}
density of core	9.5×10^{3}
density of crust	2.8×10^{3}
The sun: average density	1.4×10^{3}
density at center	$\sim 1.6 \times 10^{5}$
White dwarf stars (central densities)	$10^{8} - 10^{15}$
A uranium nucleus	$\sim 10^{17}$

in Table 17-1. The density of a gas, however, is very sensitive to changes in temperature and pressure; see entries under *Air* in Table 17-1. This table shows the range of densities that occur in nature. Note that the variation is by a factor of about 10^{38}.

17-3
THE VARIATION OF PRESSURE IN A FLUID AT REST

If a fluid is in equilibrium, every portion of the fluid is in equilibrium. Let us consider a small element of fluid volume submerged within the body of the fluid. Let this element have the shape of a thin disk and be a distance y above some reference level, as shown in Fig. 17-2a. The thickness of the disk is dy and each face has an area A. The mass of this element is $\rho A\,dy$ and its weight is $\rho g A\,dy$. The forces exerted on the element by the surrounding fluid are perpendicular to its surface at each point (Fig. 17-2b).

The resultant horizontal force is zero, for the element has no horizontal acceleration. The horizontal forces are due only to the pressure

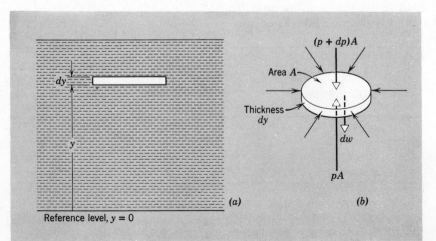

(a)   (b)

figure 17-2
(a) A small volume element of fluid at rest. *(b)* The forces on the element.

of the fluid, and by symmetry the pressure must be the same at all points within a horizontal plane at y.

The fluid element is also unaccelerated in the vertical direction, so that the resultant vertical force on it must be zero. However, the vertical forces are due not only to the pressure of the fluid on its faces but also to the weight of the element. If we let p be the pressure on the lower face and $p + dp$ the pressure on its upper face, the upward force is pA (exerted on the lower face) and the downward force is $(p + dp)A$ (exerted on the upper face) plus the weight of the element dw. Hence, for vertical equilibrium

$$pA = (p + dp)A + dw$$
$$= (p + dp)A + \rho gA\ dy,$$

and

$$\frac{dp}{dy} = -\rho g. \tag{17-1}$$

This equation tells us how the pressure varies with elevation above some reference level in a fluid in static equilibrium. As the elevation increases (dy positive), the pressure decreases (dp negative). The cause of this pressure variation is the weight per unit cross-sectional area of the layers of fluid lying between the points whose pressure difference is being measured.

The quantity ρg is often called the *weight density* of the fluid; it is the weight per unit volume of the fluid. For water, for example, the weight density is 62.4 lb/ft³ (= 9800 N/m³).

If p_1 is the pressure at elevation y_1, and p_2 the pressure at elevation y_2 above some reference level, integration of Eq. 17-1 gives

$$\int_{p_1}^{p_2} dp = -\int_{y_1}^{y_2} \rho g\ dy$$

or

$$p_2 - p_1 = -\int_{y_1}^{y_2} \rho g\ dy. \tag{17-2}$$

For liquids ρ is practically constant because liquids are nearly incompressible, and differences in level are rarely so great that any change in g need be considered. Hence, taking ρ and g as constants, we obtain

$$p_2 - p_1 = -\rho g(y_2 - y_1) \tag{17-3}$$

for a homogeneous liquid.

If a liquid has a free surface, this is the natural level from which to measure distances. To change our reference level to the top surface, we take y_2 to be the elevation of the surface, at which point the pressure p_2 acting on the fluid is usually that exerted by the earth's atmosphere p_0. We take y_1 to be at any level and we represent the pressure there as p. Then,

$$p_0 - p = -\rho g(y_2 - y_1).$$

But $y_2 - y_1$ is the depth h below the surface at which the pressure is p (see Fig. 17-3), so that

$$p = p_0 + \rho gh. \tag{17-4}$$

This shows clearly that the pressure is the same at all points at the same depth.

For gases ρ is comparatively small and the difference in pressure at two points is usually negligible (see Eq. 17-3). Thus, in a vessel con-

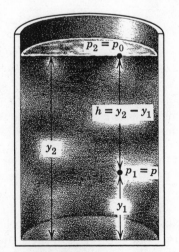

figure 17-3
A liquid whose top surface is open to the atmosphere.

taining a gas the pressure can be taken as the same everywhere. How-ever, this is not the case if $y_2 - y_1$ is very great. The pressure of the air varies greatly as we ascend to great heights in the atmosphere. In fact, in such cases the density ρ varies with altitude and ρ must be known as a function of y before we can integrate Eq. 17-2.

We can get a reasonable idea of the variation of pressure with altitude in the earth's atmosphere if we assume that the density ρ is proportional to the pressure. This would be very nearly true if the temperature of the air remained the same at all altitudes. Using this assumption, and also assuming that the variation of g with altitude is negligible, find the pressure p at an altitude y above sea level.

From Eq. 17-1 we have

EXAMPLE 1

$$\frac{dp}{dy} = -\rho g.$$

Since ρ is proportional to p, we have

$$\frac{\rho}{\rho_0} = \frac{p}{p_0},$$

where ρ_0 and p_0 are the known values of density and pressure at sea level. Then,

$$\frac{dp}{dy} = -g\rho_0 \frac{p}{p_0},$$

so that

$$\frac{dp}{p} = -\frac{g\rho_0}{p_0} dy.$$

Integrating this from the value p_0 at the point $y = 0$ (sea level) to the value p at the point y (above sea level), we obtain

$$\ln \frac{p}{p_0} = -\frac{g\rho_0}{p_0} y$$

or

$$p = p_0 e^{-g(\rho_0/p_0)y}.$$

However,

$$g = 9.80 \text{ m/s}^2, \qquad \rho_0 = 1.20 \text{ kg/m}^3 \text{ (at } 20°\text{C)},$$

$$p_0 = 1.01 \times 10^5 \text{ N/m}^2 = 1.01 \times 10^5 \text{ Pa},$$

so that

$$g\frac{\rho_0}{p_0} = 1.16 \times 10^{-4} \text{ m}^{-1} = 0.116 \text{ km}^{-1}.$$

Hence,

$$p = p_0 e^{-ay}.$$

where $a = 0.116 \text{ km}^{-1}$.

We have seen that because liquids are almost incompressible the lower layers are not noticeably compressed by the weight of the upper layers super-imposed on them and the density ρ is practically constant at all levels. For gases at uniform temperature the density ρ of any layer is proportional to the pressure p at that layer. The variation of pressure with distance above the bottom of the fluid for a gas is different from that for a liquid. Figure 17-4 shows the pressure distribution in water and in air.

Equation 17-3 gives the relation between the pressures at any two points in a fluid, regardless of the shape of the containing vessel. For no matter what the shape of the containing vessel, two points in the fluid can be connected by a path made up of vertical and horizontal steps. For example, consider points A and B in the homogeneous liquid con-

figure 17-4
Example 1. Variation of pressure with altitude in air and with depth in water, assuming p = 1 atm (exactly) at sea level. Note that the pressure scales are different for altitude and depth. The solid line for air is calculated on the assumption that the air has a constant temperature and that g does not change with altitude. The dashed line (the U.S. Standard Atmosphere – 1962) is a more refined calculation in which these assumptions are not made.

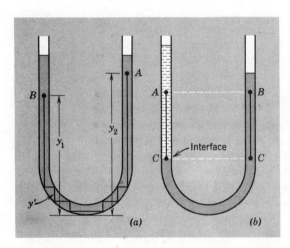

figure 17-5
(a) The difference in pressure between two points A and B in a homogeneous liquid depends only on their difference in elevation $y_2 - y_1$. (b) Two points A and B at the same elevation can be at different pressures if the densities there differ.

tained in the U-tube of Fig. 17-5a. Along the zigzag path from A to B there is a difference in pressure $\rho g y'$ for each vertical segment of length y', whereas along each horizontal segment there is no change in pressure. Hence, the difference in pressure $p_B - p_A$ is ρg times the algebraic sum of the vertical segments from A to B, or $\rho g(y_2 - y_1)$.

If the U-tube contains different imiscible liquids, say a dense liquid in the right tube and a less dense one in the left tube, as shown in Fig. 17-5b, the pressure can be different at the same level on different sides. In the figure the liquid surface is higher in the left tube than in the right. The pressure at A will be greater than at B. The pressure at C is the same on both sides, but the pressure falls less from C to A than from C to B, for a column of liquid of unit cross-sectional area connecting A and C will weigh less than a corresponding column connecting B and C.

figure 17-6
Example 2.

EXAMPLE 2

A U-tube is partly filled with water. Another liquid, which does not mix with water, is poured into one side until it stands a distance d above the water level on the other side, which has meanwhile risen a distance l (Fig. 17-6). Find the density of the liquid relative to that of water.

In Fig. 17-6 points C are at the same pressure. Hence, the pressure drop from C to each surface is the same, for each surface is at atmospheric pressure.

The pressure drop on the water side is $\rho_w g 2l$; the $2l$ comes from the fact that the water column has risen a distance l on one side and fallen a distance l on the other side, from its initial position. The pressure drop on the other side is $\rho g(d + 2l)$, where ρ is the density of the unknown liquid. Hence,

$$\rho_w g 2l = \rho g(d + 2l)$$

and

$$\frac{\rho}{\rho_w} = \frac{2l}{(2l + d)}.$$

The ratio of the density of a substance to the density of water is called the *relative density* (or the *specific gravity*) of that substance.

Figure 17-7 shows a liquid in a cylinder that is fitted with a piston to which we may apply an external pressure p_0. The pressure p at any arbitrary point P a distance h below the upper surface of the liquid is given by Eq. 17-4, or

$$p = p_0 + \rho g h.$$

Let us increase the external pressure by an arbitrary amount Δp_0 (which need not be small compared to p_0). Since liquids are virtually incompressible, the density ρ in the preceding equations remains essentially constant during the process. The equation shows that, to this extent, the change in pressure Δp at the arbitrary point P is equal to Δp_0. This result was stated by Blaise Pascal (see p. 371) and is called *Pascal's principle*. It is usually given as follows: Pressure applied to an enclosed fluid is transmitted undiminished to every portion of the fluid and the walls of the containing vessel. This result is a necessary consequence of the laws of fluid mechanics, rather than an independent principle.

Although we often assume liquids to be incompressible, they are, in fact, slightly compressible. This means that a change of pressure applied to one portion of a liquid propagates through the liquid as a wave at the speed of sound in that liquid. Once the disturbance has died out and equilibrium is established, it is found that Pascal's principle is valid. The principle holds for gases with slight complications of interpretation caused by the large volume changes that may occur when the pressure on a confined gas is changed.

Archimedes' principle is also a necessary consequence of the laws of fluid statics. When a body is wholly or partly immersed in a fluid (either a liquid or a gas) at rest, the fluid exerts pressure on every part of the body's surface in contact with the fluid. The pressure is greater on the parts immersed more deeply. The resultant of all the forces is an upward force called the *buoyancy* of the immersed body. We can determine the magnitude and direction of this resultant force quite simply as follows.

The pressure on each part of the surface of the body certainly does not depend on the material the body is made of. Let us suppose, then, that the body, or as much of it as is immersed, is replaced by fluid like the surroundings. This fluid will experience the pressures that acted on the immersed body (Fig. 17-8) and will be at rest. Hence, the resultant upward force on it will equal its weight and will act vertically upward through its center of gravity. From this follows *Archimedes' principle*, namely, that a body wholly or partly immersed in a fluid is buoyed up with a force equal to the weight of the fluid displaced by the body. We have seen that the force acts vertically up through the center of gravity of the fluid before its displacement. The corresponding point in the immersed body is called its *center of buoyancy*.

17-4
PASCAL'S PRINCIPLE AND ARCHIMEDES' PRINCIPLE

figure 17-7
A fluid in a cylinder fitted with a movable piston. The pressure at any point P is due not only to the weight of the fluid above the level of P but also to the force exerted by the piston.

figure 17-8
Illustrating Archimedes' principle. The fluid exerts a resultant upward force on the immersed body.

EXAMPLE 3

What fraction of the total volume of an iceberg is exposed? The density of ice is $\rho_i = 0.92$ gram/cm³ and that of sea water is $\rho_w = 1.03$ gram/cm³. The weight of the iceberg is

$$W_i = \rho_i V_i g,$$

where V_i is the volume of the iceberg; the weight of the volume V_w of sea water displaced is the buoyant force

$$B = \rho_w V_w g.$$

But B equals W_i, for the iceberg is in equilibrium, so that

$$\rho_w V_w g = \rho_i V_i g,$$

and

$$\frac{V_w}{V_i} = \frac{\rho_i}{\rho_w} = \frac{0.92}{1.03} = 89\%.$$

The volume of water displaced V_w is the volume of the submerged portion of the iceberg, so that 11% of the iceberg is exposed.

17-5
MEASUREMENT OF PRESSURE

Evangelista Torricelli (1608–1647) devised a method for measuring the pressure of the atmosphere by his invention of the mercury barometer in 1643.* The mercury barometer is a long glass tube that has been filled with mercury and then inverted in a dish of mercury, as in Fig. 17-9. The space above the mercury column contains only mercury vapor, whose pressure is so small at ordinary temperatures that it can be neglected. It is easily shown (see Eq. 17-3) that the atmospheric pressure p_0 is

$$p_0 = \rho g h.$$

figure 17-9
The Torricelli barometer.

Most pressure gauges use atmospheric pressure as a reference level and measure the difference between the actual pressure and atmospheric pressure, called the *gauge pressure*. The actual pressure at a point in a fluid is called the *absolute pressure*. Gauge pressure is given either above or below atmospheric pressure.

The pressure of the atmosphere at any point is numerically equal to the weight of a column of air of unit cross-sectional area extending from that point to the top of the atmosphere. The atmospheric pressure at a point, therefore, decreases with altitude. There are variations in atmospheric pressure from day to day because the atmosphere is not static. The mercury column in the barometer will have a height of about 76 cm at sea level, varying with the atmospheric pressure. A pressure equivalent to that exerted by exactly 76 cm of mercury at 0°C under standard gravity, $g = 32.174$ ft/s² $= 980.665$ cm/s², is called *one atmosphere* (1 atm). The density of mercury at this temperature is 13.5950 gram/cm³. Hence, one atmosphere is equivalent to

$$1 \text{ atm} = (13.5950 \text{ gram/cm}^3)(980.665 \text{ cm/s}^2)(76.00 \text{ cm})$$

$$= 1.013 \times 10^5 \text{ N/m}^2 \; (\equiv 1.013 \times 10^5 \text{ Pa})$$

$$= 2116 \text{ lb/ft}^2$$

$$= 14.70 \text{ lb/in.}^2$$

Often pressures are specified by giving the height of mercury column,

* See *The History of the Barometer*, by W. E. K. Middleton, The Johns Hopkins Press (1964) for a fascinating account of the development of the concept of atmospheric pressure and of devices to measure it.

at 0°C under standard gravity, which exerts the same pressure. This is the origin of the expression "centimeters of mercury (cm-Hg)" or "inches of mercury (in-Hg)" pressure. Pressure is the ratio of force to area, however, and not a length.

Torricelli described his experiments with the mercury barometer in letters in 1644 to his friend Michelangelo Ricci in Rome. In them he says that the aim of his investigation was "not simply to produce a vacuum, but to make an instrument which shows the mutations of the air, now heavier and dense, and now lighter and thin." On hearing of the Italian experiments, Blaise Pascal, in France, reasoned that if the mercury column was held up simply by the pressure of the air, the column ought to be shorter at a high altitude. He tried it on a church steeple in Paris, but desiring more decisive results, he wrote to his brother-in-law to try the experiment on the Puy de Dôme, a high mountain in Auvergne. There was a difference of 3 inches in the height of the mercury, "which ravished us with admiration and astonishment." Pascal himself constructed a barometer using red wine and a glass tube 46 feet long.

The chief significance of these experiments at the time was the realization it brought that an evacuated space could be created. Aristotle believed that a vacuum could not exist, and as late a writer as Descartes held the same view. For 2000 years philosophers spoke of the "horror" that nature had for empty space—the *horror vacui*. Because of this nature was said to prevent the formation of a vacuum by laying hold of anything nearby and with it instantly filling up any vacuated space. Hence, the mercury or wine should fill up the inverted tube because "nature abhorred a vacuum." The experiments of Torricelli and Pascal showed that there were limitations to nature's ability to prevent a vacuum. They created a sensation at the time. The goal of producing a vacuum became more of a practical reality through the development of pumps by Otto von Guericke in Germany around 1650 and by Robert Boyle in England around 1660. Even though these pumps were relatively crude, they did provide a tool for experimentation. With a pump and a glass jar, an experimental space could be provided in which to study how the properties of heat, light, sound, and later electricity and magnetism, are affected by an increasingly rarefied atmosphere. Although even today we cannot completely remove every trace of gas from a closed vessel, these seventeenth-century experimenters freed science from the bugaboo of *horror vacui* and spurred efforts to create highly evacuated systems.

Interestingly, within several decades in the seventeenth century no fewer than six important instruments were developed. They are the barometer, air pump, pendulum clock, telescope, microscope, and thermometer. All excited great wonder and curiosity.

The open-tube manometer (Fig. 17-10) measures gauge pressure. It consists of a U-shaped tube containing a liquid, one end of the tube being open to the atmosphere and the other end being connected to the system (tank) whose pressure p we want to measure. From Eq. 17-4 we obtain

$$p - p_0 = \rho g h.$$

Thus the gauge pressure, $p - p_0$, is proportional to the difference in height of the liquid columns in the U-tube. If the vessel contains gas under high pressure, a dense liquid like mercury is used in the tube; water can be used when low gas pressures are involved.

figure 17-10
The open-tube manometer, as used to measure the pressure in a tank.

EXAMPLE 4

An open-tube mercury manometer (Fig. 17-10) is connected to a gas tank. The mercury is 39.0 cm higher on the right side than on the left when a barometer nearby reads 75.0 cm-Hg. What is the absolute pressure of the gas? Express the answer in cm-Hg, atm, Pa, and lb/in.².

The gas pressure is the pressure at the top of the left mercury column. This is the same as the pressure at the same horizontal level in the right column. The pressure at this level is the atmospheric pressure (75.0 cm-Hg) plus the pressure exerted by the extra 39.0-cm column of Hg, or (assuming standard values of mercury density and gravity) a total of 114 cm-Hg. Therefore, the absolute pressure of the gas is

$$114 \text{ cm-Hg} = \tfrac{114}{76} \text{ atm} = 1.50 \text{ atm} = 1.52 \times 10^5 \text{ Pa.} = (1.50)(14.7) \text{ lb/in.}^2$$

$$= 22.1 \text{ lb/in.}^2.$$

What is the gauge pressure of the gas?

questions

1. Make an estimate of the average density of your body. Explain a way in which you could get an accurate value using ideas in this chapter.

2. Persons confined to bed are less likely to develop sores on their bodies if they use a water bed rather than an ordinary mattress. Explain.

3. (a) Two bodies (for example, balls) have the same shape and size but one is denser than the other. Assuming the air resistance to be the same on each, show that when they are released simultaneously from the same height the heavier body will reach the ground first. (b) Two bodies (for example, raindrops) have the same shape and density but one is larger than the other. Assuming the air resistance to be proportional to the body's speed through the air, which body will fall faster?

4. Water is poured to the same level in each of the three vessels shown, all of the same base area (Fig. 17-11). If the pressure is the same at the bottom of each vessel, the force experienced by the base of each vessel is the same. Why then do the three vessels have different weights when put on a scale? This apparently contradictory result is commonly known as the "hydrostatic paradox."

figure 17-11
Question 4.

5. Can a mountain climber climb high enough so that the atmospheric pressure is reduced to one-half of its sea-level value?

6. (a) An ice cube is floating in a glass of water. When the ice melts, will the water level rise? Explain. (b) If the ice cube contains a piece of lead, the water level will fall when the ice melts. Explain.

7. When a slice of lemon is first put into a cup of tea it sinks to the bottom. Later it is found to be floating. What is a likely explanation?

8. Does Archimedes' law hold in a vessel in free fall? In a satellite moving in a circular orbit? Explain.

9. A spherical bob made of cork floats half submerged in a pot of tea at rest on the earth. Will the cork float or sink aboard a spaceship coasting in free space? On the surface of Jupiter?

10. Two hollow bodies of equal weight and volume and having the same shape, except that one has an opening at the bottom and the other is sealed, are immersed to the same depth in water. Is less work required to immerse one than the other? If so, which one and why?

11. A ball floats on the surface of water in a container exposed to the atmosphere. Will the ball remain immersed at its former depth or will it sink or rise somewhat if (a) the container is covered and the air is removed or (b) the container is covered and the air is compressed?

12. Explain why an inflated balloon will rise to a definite height once it starts to rise, whereas a submarine will always sink to the bottom of the ocean once it starts to sink, if no changes are made.

13. Explain how a submarine rises, falls, and maintains a fixed depth. Do fish use the same principles? (See "The Buoyancy of Marine Animals" by Eric Denton in *Scientific American*, July 1960 and "Submarine Physics" by G. P. Harnwell in *American Journal of Physics*, March 1948.)

14. A soft plastic bag weighs the same when empty as when filled with air at atmospheric pressure. Why? Would the weights be the same if measured in a vacuum?

15. A leaky tramp steamer that is barely able to float in the North Sea steams up the Thames estuary toward the London docks. It sinks before it arrives. Why?

16. Is it true that a floating object will only be in stable equilibrium if its center of buoyancy lies above its center of gravity? Illustrate with examples.

17. Why didn't American Indians put seats in their canoes?

18. Very often a sinking ship will turn over as it becomes immersed in water. Why?

19. According to Example 3, 89% of an iceberg is submerged. Yet occasionally icebergs turn over, with possibly disasterous results to nearby shipping. How can this happen considering that so much of their mass is below sea level?

20. A barge filled with scrap iron is in a canal lock. If the iron is thrown overboard, what happens to the water level in the lock?

21. A bucket of water is suspended from a spring balance. Does the balance reading change when a piece of iron suspended from a string is immersed in the water? When a piece of cork is put in the water?

22. Logs dropped upright into a pond do not remain upright, but float "flat" in the water. Explain.

23. Explain why a uniform wooden stick which will float horizontally if it is not loaded will float vertically if enough weight is added to one end. (See Problem 30.)

24. A solid cylinder is placed in a container in contact with the base. When liquid is poured into the container, none of it goes beneath the solid, which remains closely in contact with the base. Is there a buoyant force on the solid? Explain.

25. Estimate with some care the buoyant force exerted by the atmosphere on you.

26. Although there are practical difficulties it is possible in principle to float an ocean liner in a few barrels of water. How would you go about doing this?

27. Is the water at the bottom of the Marianas Trench (11,000 m deep) substantially (a) less or (b) more buoyant than the water on the surface?

28. Mountain climbers use aneroid barometers to estimate their altitude. How can they be useful, considering that the atmospheric pressure at a given location is not constant?

29. What is wrong with this statement? "The atmospheric pressure is 753 mm-Hg when the atmosphere supports, in a barometer, a mercury column 753 mm long."

30. In a barometer, how important is it that the inner diameter of the barometer tube be uniform? That the barometer tube be absolutely vertical?

31. An open-tube manometer has one tube twice the diameter of the other. Explain how this would affect the operation of the manometer. Does it matter which end is connected to the chamber whose pressure is to be measured?

32. Explain how a physician can determine your blood pressure.

33. Liquid containers tend to leak when taken aloft in an airplane. Why? Does it matter whether or not they are right-side up? Does it matter whether or not they are initially completely full?

34. Assuming that a mercury barometer at standard atmospheric pressure on the earth's surface reads 76.0 cm height of mercury column, estimate what would be the height of the mercury column in an artificial earth satellite in orbit about the earth.

35. An open bucket of water is on a smooth plane inclined at an angle α to the

horizontal. Find the equilibrium inclination to the horizontal of the free surface of the water when (a) the bucket is held at rest, $a = 0$ and $v = 0$; (b) the bucket is allowed to slide down at constant speed, $a = 0$, $v = $ constant; (c) the bucket slides down without restraint, $a = $ constant. If the plane is curved so that $a \neq $ constant, what will happen?

36. If a U-tube containing water is rotated about a vertical axis through the center of one limb, the water level will fall in one limb and rise in the other compared to the rest position. Explain carefully. (See Problem 18.)

37. Explain how it can be that pressure is a scalar quantity when forces, which are vectors, can be produced by the action of pressures.

38. A thin-walled pipe will burst more easily if, when there is a pressure differential between inside and outside, the excess pressure is on the outside. Explain.

39. We have considered liquids under compression. Can liquids be put under tension? If so, will they tear under sufficient tension as do solids? (See "The Tensile Strength of Liquids" by Robert E. Apfel in *Scientific American*, December 1972.)

problems

SECTION 17-2

1. Find the pressure increase in the fluid in a syringe when a nurse applies a force of 42 N to the syringe's piston of radius 1.1 cm.
   Answer: 1.1×10^5 Pa.

2. An airtight box having a lid with an area of 12 in.² is partially evacuated. If a force of 108 lb is required to pull the lid off the box, and the outside atmospheric pressure is 15 lb/in.², what was the pressure in the box?

3. In 1654 Otto von Guericke, burgomeister of Magdeburg and inventor of the air pump, gave a demonstration before the Imperial Diet in which two teams of eight horses could not pull apart two evacuated brass hemispheres. (a) Show that the force F required to pull apart the hemispheres is $F = \pi R^2 P$ where R is the (outside) radius of the hemispheres and P is the difference in pressure outside and inside the sphere (Fig. 17-12). (b) Taking R equal to 1.0 ft and the inside pressure as 0.10 atm, what force would the team of horses have had to exert to pull apart the hemispheres? (c) Why were two teams of horses used? Would not one team prove the point just as well?
   Answer: (b) 6000 lb.

figure 17-12
Problem 3

SECTION 17-3

4. Find the total pressure, in lb/in.² (Pa), 500 ft (150 m) below the surface of the ocean. The relative density of sea water is 1.03 and the atmospheric pressure at sea level is 14.7 lb/in.² (1.0×10^5 Pa).

5. Estimate the hydrostatic difference in blood pressure in a person of height 1.83 m (6.00 ft), between the brain and the foot, assuming that the density of blood is 1.06×10^3 kg/m³ (2.06 slug/ft³).
   Answer: 1.90×10^4 Pa (2.75 lb/in.²).

6. The human lungs can operate against a pressure differential of less than one-twentieth a standard atmosphere. If a diver uses a snorkel (long tube) for breathing, how far below water level can he swim?

7. Find the pressure in the atmosphere 16 km (10 mi) above sea level.
   Answer: 1.6×10^4 Pa (2.3 lb/in.²).

8. The height at which the pressure in the atmosphere is just $1/e$ that at sea level is called the *scale height* of the atmosphere at sea level. (a) Show that the scale height H at sea level is also the height of an atmosphere that has the same density everywhere as at sea level and that will exert the same pressure at sea level as the actual infinite atmosphere does. (b) Show that the scale height at sea level is 8.6 km.

9. What would be the height of the atmosphere if the air density (a) were constant and (b) decreased linearly to zero with height? Assume a sea-level density of 1.3 kg/m³. *Answer:* (a) 8.0×10^3 m. (b) 16×10^3 m.

10. A swimming pool has the dimensions 80 ft × 30 ft × 8.0 ft. (a) When it is filled with water, what is the force (due to the water alone) on the bottom? On the ends? On the sides? (b) If you are concerned with whether or not the concrete walls will collapse, is it appropriate to take the atmospheric pressure into account?

11. A simple U-tube contains mercury. When 13.6 cm of water is poured into the right arm, how high does the mercury rise in the left arm from its initial level? *Answer:* 0.50 cm.

12. Water stands at a depth D behind the vertical upstream face of a dam, as shown in Fig. 17-13. Let W be the width of the dam. (a) Find the resultant horizontal force exerted on the dam by the gauge pressure of the water and (b) the net torque due to the gauge pressure of the water exerted about a line through O parallel to the width of the dam. (c) What is the line of action of the equivalent resultant force?

figure 17-13
Problem 12

13. Three liquids that will not mix are poured into a cylindrical container. The amounts and densities of the liquids are 0.50 liter, 2.6 g/cm³; 0.25 liter, 1.0 g/cm³; and 0.40 liter, 0.80 g/cm³. What is the total force acting on the bottom of the container? (Ignore the contribution due to the atmosphere.) *Answer:* 18 N.

14. Two identical cylindrical vessels with their bases at the same level each contain a liquid of density ρ. The area of either base is A, but in one vessel the liquid height is h_1 and in the other h_2. Find the work done by gravity in equalizing the levels when the two vessels are connected.

15. (a) Consider a container of fluid subject to a *vertical upward* acceleration a. Show that the pressure variation with depth in the fluid is given by

$$p = \rho h \, (g + a),$$

where h is the depth and ρ is the density. (b) Show also that if the fluid as a whole undergoes a *vertical downward* acceleration a, the pressure at a depth h is given by

$$p = \rho h (g - a).$$

(c) What is the state of affairs in free fall?

16. (a) Consider the horizontal acceleration of a mass of liquid in an open tank. Acceleration of this kind causes the liquid surface to drop at the front of the tank and to rise at the rear. Show that the liquid surface slopes at an angle θ with the horizontal, where $\tan \theta = a/g$, a being the horizontal acceleration. (b) How does the pressure vary with h, the *vertical* depth below the surface?

17. The surface of contact of two fluids of different densities that are at rest and do not mix is horizontal. Prove this general result (a) from the fact that the potential energy of a system must be a minimum in stable equilibrium; (b) from the fact that at any two points in a horizontal plane in either fluid the pressures are equal.

18. (a) A fluid mass is rotating at constant angular velocity ω about the central vertical axis of a cylindrical container. Show that the variation of pressure in the radial direction is given by

$$\frac{dp}{dr} = \rho \omega^2 r.$$

(b) Take $p = p_c$ at the axis of rotation ($r = 0$) and show that the pressure p at any point r is

$$p = p_c + \tfrac{1}{2}\rho \omega^2 r^2.$$

(c) Show that the liquid surface is of paraboloidal form (Fig. 17-14); that is, a

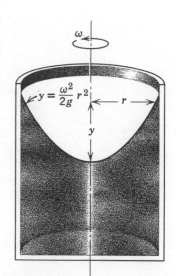

figure 17-14
Problem 18

vertical cross section of the surface is the curve $y = \omega^2 r^2/2g$. (d) Show that the variation of pressure with depth is $dp = \rho g \, dh$.

SECTION 17-4

19. A piston of small cross-sectional area a is used in the hydraulic press to exert a small force f on the enclosed liquid. A connecting pipe leads to a larger piston of cross-sectional area A (Fig. 17-15). (a) What force F will the larger piston sustain? (b) If the small piston has a diameter of 1.5 in. and the large piston one of 21 in., what weight on the small piston will support 2.0 tons on the large piston?   *Answer: (a) fA/a. (b) 20 lb.*

20. A cubical object of dimensions L (2.0 ft) on a side and weight W (1000 lb) in a vacuum is suspended by a rope in an open tank of water of density ρ (2.0 slug/ft³) as in Fig. 17-16. (a) Find the total downward force exerted by the water and the atmosphere on the top of the object of area A (4.0 ft²). (b) Find the total force on the bottom of the object. (c) Find the tension in the rope.

21. (a) What is the minimum area of a block of ice 1.0 ft. (0.3 m) thick floating on water that will hold up an automobile weighing 2500 lb (mass = 1100 kg)? (b) Does it matter where the car is placed on the block of ice? *Answer: (a) 500 ft² (46 m²). (b) Yes.*

22. Three boys each of weight W (80 lb) make a log raft by lashing together logs of diameter D (1.0 ft) and length L (6.0 ft). How many logs will be needed to keep them afloat? Take the relative density of wood to be 0.80.

23. A block of wood floats in water with two-thirds of its volume submerged. In oil it has 0.90 of its volume submerged. Find the density of (a) the wood and (b) the oil.   *Answer: (a) 6.7×10^2 kg/m³. (b) 7.4×10^2 kg/m³.*

24. A block of wood has a mass of 3.67 kg and a relative density of 0.60. It is to be loaded with lead so that it will float in water with 0.90 of its volume immersed. What weight of lead is needed (a) if the lead is on top of the wood? (b) if the lead is attached below the wood? The density of lead is 1.13×10^4 kg/m³.

25. Assume the density of brass weights to be 8.0 g/cm³ and that of air to be 0.0012 g/cm³. What percent error arises from neglecting the buoyancy of air in weighing an object of mass m and density ρ on a beam balance? *Answer: $0.12(1/\rho - 1/8)$, with ρ in g/cm³.*

26. A hollow spherical iron shell floats almost completely submerged in water. If the outer diameter is 2.00 ft and the relative density of iron is 7.80, find the inner diameter.

27. An iron casting containing a number of cavities weighs 60 lb (mass = 27 kg) in air and 40 lb (mass = 18 kg) in water. What is the volume of the cavities in the casting? Assume the relative density of iron to be 7.8. *Answer: 0.20 ft³ (5.5×10^{-3} m³).*

28. A cube floating on mercury has one-fourth of its volume submerged. If enough water is added to cover the cube, (a) what fraction of its volume will remain immersed in mercury? (b) Does the answer depend on the shape of the body?

29. A U-tube is filled with a single homogeneous liquid. The liquid is temporarily depressed in one side by a piston. The piston is removed and the level of the liquid in each side oscillates. Show that the period of oscillation is $\pi\sqrt{2L/g}$ where L is the total length of the liquid in the tube.

30. A cylindrical wooden log is loaded with lead at one end so that it floats upright in water as in Fig. 17-17. The length of the submerged portion is $l = 8.0$ ft. The log is set into vertical oscillation. (a) Show that the oscillation is simple harmonic. (b) Find the period of the oscillation. Neglect the fact that the water has a damping effect on the motion.

31. A long uniform wooden bar with square cross section floats on water either with two opposite surfaces parallel to the water or with all four surfaces at 45° with the water. Which of these positions is assumed for densities of 0.20, 0.50, and 0.80 g/cm³?

figure 17-15
Problem 19

figure 17-16
Problem 20

figure 17-17
Problem 30

32. The tension in a string holding a solid block below the surface of a liquid (of density greater than the solid) is T_0 when the containing vessel (Fig. 17-18) is at rest. Show that the tension T, when the vessel has an upward vertical acceleration a, is given by $T_0(1 + a/g)$.

figure 17-18
Problem 32

18
fluid
dynamics

One way of describing the motion of a fluid is to divide the fluid into infinitesimal volume elements, which we may call fluid particles, and to follow the motion of each of these particles. This is a formidable task. We would give coordinates x, y, z to each such fluid particle and would specify these as functions of the time t. The coordinates x, y, z at the time t of the fluid particle which was at x_0, y_0, z_0 at the time t_0 would be determined by functions $x(x_0,y_0,z_0,t_0,t)$, $y(x_0,y_0,z_0,t_0,t)$, $z(x_0,y_0,z_0,t_0,t)$, which then describe the motion of the fluid. This procedure is a direct generalization of the concepts of particle mechanics and was first developed by Joseph Louis Lagrange (1736–1813).

There is a treatment, developed by Leonhard Euler (1707–1783), which is more convenient for most purposes. In it we give up the attempt to specify the history of each fluid particle and instead specify the density and the velocity of the fluid at each point in space at each instant of time. This is the method we shall follow here. We describe the motion of the fluid by specifying the density $\rho(x,y,z,t)$ and the velocity $\mathbf{v}(x,y,z,t)$ at the point (x,y,z) at the time t. We thus focus our attention on what is happening at a particular point in space at a particular time, rather than on what is happening to a particular fluid particle. Any quantity used in describing the state of the fluid, for example the pressure p, will have a definite value at each point in space and at each instant of time. Although this description of fluid motion focuses attention on a point in space rather than on a fluid particle, we cannot avoid following the fluid particles themselves, at least for short time intervals dt. For it is the particles, after all, and not the space points, to which the laws of mechanics apply. In order to understand the nature of the simplifications we shall make, let us consider first some general characteristics of fluid flow.

18-1
GENERAL CONCEPTS OF FLUID FLOW

1. Fluid flow can be *steady* or *nonsteady*. When the fluid velocity **v** at any given point is constant in time, the fluid motion is said to be steady. That is, at any given point in a steady flow the velocity of each passing fluid particle is always the same. At some other point a particle may travel with a different velocity, but every other particle which passes this second point behaves there just as this particle did when it passed this point. These conditions can be achieved at low flow speeds; a gently flowing stream is an example. In nonsteady flow, as in a tidal bore, the velocities **v** *are* a function of the time. In the case of turbulent flow, such as rapids or a waterfall, the velocities vary erratically from point to point as well as from time to time.

2. Fluid flow can be *rotational* or *irrotational*. If the element of fluid at each point has no net angular velocity about that point, the fluid flow is irrotational. We can imagine a small paddle wheel immersed in the moving fluid (Fig. 18-1). If the wheel moves without rotating, the motion is irrotational; otherwise it is rotational. Rotational flow includes vortex motion, such as whirlpools.

3. Fluid flow can be *compressible* or *incompressible*. Liquids can usually be considered as flowing incompressibly. But even a highly compressible gas may sometimes undergo unimportant changes in density. Its flow is then practically incompressible. In flight at speeds much lower than the speed of sound in air (described by subsonic aerodynamics), the motion of the air relative to the wings is one of nearly incompressible flow. In such cases the density ρ is a constant, independent of x, y, z, and t, and the mathematical treatment of fluid flow is thereby greatly simplified.

4. Finally, fluid flow can be *viscous* or *nonviscous*. Viscosity in fluid motion is the analog of friction in the motion of solids. In many cases, such as in lubrication problems, it is extremely important. Sometimes, however, it is negligible. Viscosity introduces tangential forces between layers of fluid in relative motion and results in dissipation of mechanical energy.

We shall confine our discussion of fluid dynamics for the most part to *steady, irrotational, incompressible, nonviscous* flow. The mathematical simplifications resulting should be obvious. We run the danger, however, of making so many simplifying assumptions that we are no longer talking about a meaningfully real fluid.* Furthermore, it is sometimes difficult to decide whether a given property of a fluid—its viscosity, say—can be neglected in a particular situation. In spite of all this, the restricted analysis that we are going to give has wide application in practice, as we shall see.

figure 18-1
We place a small free-floating paddle wheel in a flowing liquid. If it rotates, we call the flow *rotational*; if not, we call the flow *irrotational*.

18-2 STREAMLINES

In steady flow the velocity **v** at a given point is constant in time. Consider the point P (Fig. 18-2) within the fluid. Since **v** at P does not change in time, every particle arriving at P will pass on with the same speed in the same direction. The same is true about the points Q and R. Hence, if we trace out the path of the particle, as is done in the figure, that curve will be the path of every particle arriving at P. This curve is called a *streamline*. A streamline is parallel to the velocity of the fluid particles at every point. No two streamlines can cross one another, for if

* Richard Feynman has pointed out that John von Neumann called this idealized fluid "dry water."

figure 18-2
A particle passing through points P, Q, and R traces out a streamline, assuming steady flow. Any other particle passing through P must be traveling along the same streamline in steady flow.

they did, an oncoming fluid particle could go either one way or the other, and the flow could not be steady. In steady flow the pattern of streamlines in a flow is stationary with time.*

In principle we can draw a streamline through every point in the fluid. Let us assume steady flow and select a finite number of streamlines to form a bundle, like the streamline pattern of Fig. 18-3. This tubular region is called a *tube of flow*. The boundary of such a tube consists of streamlines and is always parallel to the velocity of the fluid particles. Hence, no fluid can cross the boundaries of a tube of flow and the tube behaves somewhat like a pipe of the same shape. The fluid that enters at one end must leave at the other.

figure 18-3
A tube of flow made up of a bundle of streamlines.

18-3
THE EQUATION OF CONTINUITY

In Fig. 18-4 we have drawn a thin tube of flow. The velocity of the fluid inside, although parallel to the tube at any point, may have different magnitudes at different points. Let the speed be v_1 for fluid particles at P and v_2 for fluid particles at Q. Let A_1 and A_2 be the cross-sectional areas of the tubes perpendicular to the streamlines at the points P and Q, respectively. In the time interval Δt a fluid element travels approximately the distance $v\,\Delta t$. Then the mass of fluid Δm_1 crossing A_1 in the time interval Δt is approximately

$$\Delta m_1 = \rho_1 A_1 v_1\,\Delta t$$

or the *mass flux* $\Delta m_1/\Delta t$ is approximately $\rho_1 A_1 v_1$. We must take Δt small enough so that in this time interval neither v nor A varies appreciably over the distance the fluid travels. In the limit as $\Delta t \to 0$, we obtain the precise definitions:

$$\text{mass flux at } P = \rho_1 A_1 v_1,$$

and

$$\text{mass flux at } Q = \rho_2 A_2 v_2,$$

where ρ_1 and ρ_2 are the fluid densities at P and Q respectively. Because no fluid can leave through the walls of the tube and there are no "sources" or "sinks" wherein fluid can be created or destroyed in the tube, the mass crossing each section of the tube per unit time must be the same. In particular, the mass flux at P must equal that at Q:

$$\rho_1 A_1 v_1 = \rho_2 A_2 v_2,$$

or

$$\rho A v = \text{constant}. \tag{18-1}$$

figure 18-4
A tube of flow used in proving the equation of continuity.

---

* The family of streamlines in a fluid is so drawn that, at any point in the fluid, the direction of the instantaneous velocity **v** for the fluid particle at that point is tangent to the streamline at that point. In nonsteady flow the pattern of streamlines in the fluid changes as time goes on and the path of an individual fluid particle through the fluid does not coincide with a streamline of a given instant. The streamline and the line of motion of the particle touch each other at the point, locating the particle at the instant in question. The path or line of motion and the streamline coincide only for steady flow.

This result (Eq. 18-1) expresses the law of conservation of mass in fluid dynamics.

Would you expect Eq. 18-1 to hold when the flow is (*a*) nonsteady, (*b*) rotational, (*c*) compressible, or (*d*) viscous?

In the more general case in which sources or sinks are present and in which the density varies with time as well as position, mass must still be conserved and we can write (without proof) an *equation of continuity* that expresses this fact. It is

$$\frac{\partial(\rho v_x)}{\partial x} + \frac{\partial(\rho v_y)}{\partial y} + \frac{\partial(\rho v_z)}{\partial z} + \frac{\partial \rho}{\partial t} = S \tag{18-2}$$

in which v_x, v_y, and v_z are the velocity components of the fluid; like the density ρ they vary both with position and time.*

Let us consider a small volume element in such a fluid. It can be shown that:

1. The sum of the first three terms of Eq. 18-2 gives the net outflow, per unit volume, of mass from the volume element.

2. The fourth term gives the rate, per unit volume, at which mass is accumulating within the volume element.

3. The last term, S, gives the rate, per unit volume, at which mass is being introduced into volume element from a "source" (if S is positive) or is disappearing from the volume element into a "sink" (if S is negative).

It is clear that, with these interpretations of its terms, Eq. 18-2 is a statement of the conservation of mass for fluid flow. Is this equation dimensionally correct?

If $S = 0$ in Eq. 18-2, there are no sources or sinks. If the sum of the first three terms is negative, there is a net *inflow* of mass to the volume element. Thus the mass contained in the element must increase with time as fluid "piles up." This is in agreement with Eq. 18-2 because, for the conditions stated, $\partial \rho / \partial t$ must be positive, which means that the density of the fluid (and thus the mass of the fluid) in the volume element is increasing as time goes on.

If the fluid is incompressible, as we shall henceforth assume, then $\rho_1 = \rho_2$ and Eq. 18-1 takes on the simpler form

$$A_1 v_1 = A_2 v_2.$$

or
$$A v = \text{constant}, \tag{18-3}$$

The product Av gives the *volume flux* or flow rate, as it is often called. Its SI units are m³/s. Notice that it predicts that in steady incompressible flow the speed of flow varies inversely with the cross-sectional area, being larger in narrower parts of the tube. The fact that the product Av remains constant along a tube of flow allows us to interpret the streamline picture somewhat. In a narrow part of the tube the streamlines must crowd closer together than in a wide part. Hence, as the distance between streamlines decreases, the fluid speed must increase. Therefore, we conclude that widely spaced streamlines indicate regions of low speed and closely spaced streamlines indicate regions of high speed.

We can obtain another interesting result by applying Newton's second law of motion to the flow of fluid between P and Q (Fig. 18-4). A fluid particle at P with speed v_1 must be decelerated in the forward direction in acquiring the smaller forward speed v_2 at Q. Hence the fluid is decelerated in going from P to Q. The deceleration can come about

---

* Because these four quantities are functions of more than one variable we have written the derivatives in Eq. 18-2 as partial derivatives.

from a difference in pressure acting on the fluid particle flowing from P to Q or from the action of gravity. In a horizontal tube of flow the gravitational force does not change. Hence we can conclude that in steady horizontal flow the pressure is greatest where the speed is least.

Were you ever in a crowd when it started to push its way through a small opened door? Outside in the back of the crowd the cross-sectional area was large, the pressure was great, but the speed of advance rather small. Through the door of small cross section the pressure was relieved and the speed of advance gratifyingly increased. This particular "human fluid" is compressible and viscous and the flow is sometimes turbulent and rotational.

18-4 BERNOULLI'S EQUATION*

Bernoulli's equation is a fundamental relation in fluid mechanics. Like all equations in fluid mechanics it is not a new principle but is derivable from the basic laws of Newtonian mechanics. We will find it convenient to derive it from the work-energy theorem (see Section 7-4), for it is essentially a statement of the work-energy theorem for fluid flow.

Consider the nonviscous, steady, incompressible flow of a fluid through the pipeline or tube of flow in Fig. 18-5. The portion of pipe shown in the figure has a uniform cross section A_1 at the left. It is horizontal there at an elevation y_1 above some reference level. It gradually widens and rises and at the right has a uniform cross section A_2. It is horizontal there at an elevation y_2. Let us concentrate on the portion of fluid represented by both cross-shading and horizontal shading and call this fluid the "system." Consider then the motion of the system from the position shown in (a) to that in (b). At all points in the narrow part of the pipe the pressure is p_1 and the speed v_1; at all points in the wide portion the pressure is p_2 and the speed v_2.

The work-energy theorem (see Eq. 7-14) states: *The work done by the*

figure 18-5
A portion of fluid (cross-shading and horizontal shading) moves through a section of pipeline from the position shown in (a) to that shown in (b).

---

* There are eight Bernoullis listed in the Encyclopedia Brittanica (11th ed.). We refer here to Daniel Bernoulli (1700–1782), perhaps the most renowned member of this famous family.

resultant force acting on a system is equal to the change in kinetic energy of the system. In Fig. 18-5 the forces that do work on the system, assuming that we can neglect viscous forces, are the pressure forces p_1A_1 and p_2A_2 that act on the left- and right-hand ends of the system, respectively, and the force of gravity. As fluid flows through the pipe the net effect, as a comparison of Figs. 18-5a and b shows, is to raise an amount of fluid represented by the cross-shaded area in Fig. 18-5a to the position shown in Fig. 18-5b. The amount of fluid represented by the horizontal shading is unchanged by the flow.

We can find the work W done on the system by the resultant force as follows:

1. The work done on the system by the pressure force p_1A_1 is $p_1A_1 \, \Delta l_1$.
2. The work done on the system by the pressure force p_2A_2 is $-p_2A_2 \, \Delta l_2$. Note that it is negative, which means that positive work is done *by* the system.
3. The work done on the system by gravity is associated with lifting the cross-shaded fluid from height y_1 to height y_2 and is $-mg(y_2 - y_1)$ in which m is the mass of fluid in either cross-shaded area. It too is negative because work is done by the system *against* the gravitational force.

The work W done *on* the system by the *resultant* force is found by adding these three terms, or

$$W = p_1A_1 \, \Delta l_1 - p_2A_2 \, \Delta l_2 - mg(y_2 - y_1).$$

Now $A_1 \, \Delta l_1 (= A_2 \, \Delta l_2)$ is the volume of the cross-shaded fluid element, which we can write as m/ρ, in which ρ is the (constant) fluid density. Recall that the two fluid elements have the same mass, so that in setting $A_1 \, \Delta l_1 = A_2 \, \Delta l_2$ we have assumed the fluid to be incompressible. With this assumption we have

$$W = (p_1 - p_2)(m/\rho) - mg(y_2 - y_1). \qquad (18\text{-}4a)$$

The change in kinetic energy of the fluid element is

$$\Delta K = \tfrac{1}{2}mv_2^2 - \tfrac{1}{2}mv_1^2. \qquad (18\text{-}4b)$$

From the work-energy theorem (Eq. 7-14) we then have

$$W = \Delta K$$

or

$$(p_1 - p_2)(m/\rho) - mg(y_2 - y_1) = \tfrac{1}{2}mv_2^2 - \tfrac{1}{2}mv_1^2, \qquad (18\text{-}5a)$$

which can be rearranged to read

$$p_1 + \tfrac{1}{2}\rho v_1^2 + \rho g y_1 = p_2 + \tfrac{1}{2}\rho v_2^2 + \rho g y_2. \qquad (18\text{-}5b)$$

Since the subscripts 1 and 2 refer to *any* two locations along the pipeline, we can drop the subscripts and write

$$p + \tfrac{1}{2}\rho v^2 + \rho g y = \text{constant.} \qquad (18\text{-}6)$$

Equation 18-6 is called *Bernoulli's equation* for steady, nonviscous, incompressible flow. It was first presented by Daniel Bernoulli in his *Hydrodynamica* in 1738.

Bernoulli's equation is strictly applicable only to steady flow, the quantities involved being evaluated along a streamline. In our figure the streamline used is along the axis of the pipeline. If the flow is irrotational, however, it can be shown (see Problem 25 for a special case) that the constant in Eq. 18-6 is the same for *all* streamlines.

In a nonviscous incompressible fluid we cannot change the temperature of the fluid by mechanical means. Hence, Bernoulli's equation, as we stated it, refers to isothermal (constant temperature) processes. It is possible, however, to change the temperature of a nonviscous *compressible* fluid by mechanical means. We can generalize this equation to include a compressible fluid by adding to the left of Eq. 18-6 a term u, which represents the *internal energy* per unit volume of the fluid. This term (and the pressure p) will have a value that depends on the temperature.

If the flow is *viscous*, forces of a frictional nature act on the fluid so that some of the work done that appeared as a change in kinetic energy in the nonviscous case appears now as heat energy in the fluid. We must then write Eq. 18-5a as

$$(p_1 - p_2)(m/\rho) - mg(y_2 - y_1) = \tfrac{1}{2}mv_2^2 - \tfrac{1}{2}mv_1^2 + Q$$

where Q represents the heat energy generated in the viscous flow from point 1 to point 2. In practice, Bernoulli's equation can be modified accordingly by use of empirical corrections for conversion of mechanical energy to heat energy. However, if the pipe is smooth and the diameter is large compared to the length, and if the fluid flows slowly and has a small viscosity, the heat energy generated is negligible.

Just as the statics of a particle is a special case of particle dynamics, so fluid statics is a special case of fluid dynamics. It should come as no surprise, therefore, that the law of pressure change with height in a fluid at rest is included in Bernoulli's equation as a special case. For let the fluid be at rest; then $v_1 = v_2 = 0$ and Eq. 18-5b becomes

$$p_1 + \rho g y_1 = p_2 + \rho g y_2$$

or
$$p_2 - p_1 = -\rho g(y_2 - y_1),$$

which is the same as Eq. 17-3.

In Eq. 18-6 all terms have the dimension of a pressure (check this). The pressure $p + \rho g h$, which would be present even if there were no flow ($v = 0$), is called the *static pressure*; the term $\tfrac{1}{2}\rho v^2$ is called the *dynamic pressure*.

18-5
APPLICATIONS OF BERNOULLI'S EQUATION AND THE EQUATION OF CONTINUITY

Bernoulli's equation can be used to determine fluid speeds by means of pressure measurements. The principle generally used in such measuring devices is the following: The equation of continuity requires that the speed of the fluid at a constriction increase; Bernoulli's equation then shows that the pressure must fall there. That is, for a horizontal pipe $\tfrac{1}{2}\rho v^2 + p$ equals a constant; if v increases and the fluid is incompressible, p must decrease. This result was also deduced from dynamic considerations in Section 18-3.

1
The Venturi Meter

This (Fig. 18-6) is a gauge put in a flow pipe to measure the flow speed of a liquid. A liquid of density ρ flows through a pipe of cross-sectional area A. At the throat the area is reduced to a and a manometer tube is attached, as shown. Let the manometer liquid, such as mercury, have a density ρ'. By applying Bernoulli's equation and the equation of continuity at points 1 and 2, you can show that the speed of flow at A is

$$v = a \sqrt{\frac{2(\rho' - \rho)gh}{\rho(A^2 - a^2)}}.$$

If we want the volume flux or flow rate R, which is the volume of liquid transported past any point per second, we simply compute

$$R = vA.$$

figure 18-6
The Venturi meter, used to measure the speed of flow of a fluid.

2
The Pitot Tube

This device (Fig. 18-7) is used to measure the flow speed of a gas. Consider the gas, say air, flowing past the openings at a. These openings are parallel to the direction of flow and are set far enough back so that the velocity and pressure outside the openings have the free-stream values. The pressure in the left arm of the manometer, which is connected to these openings, is then the static pressure in the gas stream, p_a. The opening of the right arm of the manometer is at right angles to the stream. The velocity is reduced to zero at b and the gas is stagnant at that point. The pressure at b is the full *ram pressure*, p_b. Applying Bernoulli's equation to points a and b, we obtain

$$p_a + \tfrac{1}{2}\rho v^2 = p_b,$$

where, as shown in the figure, p_b is greater than p_a. If h is the difference in height of the liquid in the manometer arms and ρ' is the density of the manometer liquid, then

$$p_a + \rho' g h = p_b.$$

Comparing these two equations, we find

$$\tfrac{1}{2}\rho v^2 = \rho' g h$$

or

$$v = \sqrt{\frac{2gh\rho'}{\rho}},$$

which gives the gas speed. This device can be calibrated to read v directly and is then known as an air-speed indicator.

figure 18-7
Cross-sectional diagram of a Pitot tube.

Dynamic lift is the force that acts on a body, such as an airplane wing, a hydrofoil, or a helicopter rotor, by virtue of its motion through a fluid. We must distinguish it from *static lift*, which is the buoyant force that acts on a balloon or an iceberg in accord with Archimedes' principle (Section 17-4).

Figure 18-8 shows the streamlines about an airfoil (or wing cross section) attached to an aircraft.* Let us choose the aircraft as our frame of reference, as in a wind tunnel experiment, and let us assume that the air is moving past the wing from right to left.

3
Dynamic Lift

* See "Bernoulli and Newton in Fluid Mechanics," Norman F. Smith, *The Physics Teacher*, November 1972.

See also *The Flettner Ship*, an article by Albert Einstein in his book *Essays in Science*, Philosophical Library, New York. The Flettner ship, like a sailboat, derives its motive power from the wind. Instead of a sail it has a large cylinder that is caused to rotate about a vertical axis by a small motor. The resulting dynamic "lift" (in this case, horizontal) propels the vessel.

figure 18-8
Streamlines about an airfoil. The velocity of the approaching air $\mathbf{v}_a$ is horizontal. That of the leaving air $\mathbf{v}_l$ has a downward component. Thus, because the airfoil has forced the air down, the air, from Newton's third law, must have forced the airfoil up. This is represented by the "lift" $\mathbf{F}$.

The *angle of attack* of the wing causes air to be deflected downward. From Newton's third law the reaction of this downward force of the wing on the air is an upward force $\mathbf{F}$, the lift, exerted by the air on the wing.

The pattern of streamlines is consistent. Above the wing (point 1) the streamlines are closer together than they are below the wing (point 2). Thus $\mathbf{v}_1 > \mathbf{v}_2$ and, from Bernoulli's principle, $p_1 < p_2$, which must be true if there is to be a lift.

As our final example let us compute the thrust on a rocket produced by the escape of its exhaust gases. Consider a chamber (Fig. 18-9) of cross-sectional area A filled with a gas of density ρ at a pressure p. Let there be a small orifice of cross-sectional area A_0 at the bottom of the chamber. We wish to find the speed v_0 with which the gas escapes through the orifice.

4
Thrust on a Rocket

figure 18-9
Fluid streaming out of a chamber.

Let us write Bernoulli's equation (Eq. 18-5b) as

$$p_1 - p_2 = \rho g(y_2 - y_1) + \tfrac{1}{2}\rho(v_2{}^2 - v_1{}^2).$$

For a gas the density is so small that we can neglect the variation in pressure with height in a chamber (see Section 17-3). Hence, if p represents the pressure p_1 in the chamber and p_0 represents the atmospheric pressure p_2 just outside the orifice, we have

$$p - p_0 = \tfrac{1}{2}\rho(v_0{}^2 - v^2)$$

or

$$v_0{}^2 = \frac{2(p - p_0)}{\rho} + v^2, \tag{18-7}$$

where v is the speed of the flowing gas inside the chamber and v_0 is the *speed of efflux* of the gas through the orifice. Although a gas is compressible and the flow may become turbulent, we can treat the flow as steady and incompressible for pressure and efflux speeds that are not too high.

Now let us assume continuity of mass flow (in a rocket engine this is achieved when the mass of escaping gas equals the mass of gas created by burning the fuel), so that (for an assumed constant density)

$$Av = A_0 v_0.$$

If the orifice is very small so that $A_0 \ll A$, then $v_0 \gg v$, and we can neglect v^2 compared to $v_0{}^2$ in Eq. 18-7. Hence, the speed of efflux is

$$v_0 = \sqrt{\frac{2(p - p_0)}{\rho}}. \tag{18-8}$$

If our chamber is the exhaust chamber of a rocket, the thrust on the rocket (Section 9-7) is $v_0 \, dM/dt$. But the mass of gas flowing out in time dt is $dM = \rho A_0 v_0 \, dt$, so that

$$v_0 \frac{dM}{dt} = v_0 \rho A_0 v_0 = \rho A_0 v_0{}^2,$$

and from Eq. 18-8 the thrust is

$$2A_0(p - p_0). \tag{18-9}$$

In Newtonian particle mechanics the derivation of the laws of conservation of linear momentum and angular momentum makes explicit use of Newton's third law of motion. The internal forces and torques in a mechanical system cancel one another because of this third law, leaving only the external forces and torques to contribute to the momenta. In the case of a fluid the internal forces are represented by the pressure within the fluid. But the very concept of pressure itself contains Newton's third law implicitly. The force produced by pressure exerted in one direction across any surface element is equal and opposite to the force exerted in the opposite direction across the same surface element. Also, each of these two forces is applied at the same place, namely at the surface element. Both forces must have the same line of action. Hence, in the equations for the time rate of change of linear momentum or of angular momentum of a fluid, the internal pressures will cancel out. We can conclude then that the time rate of change of the total linear momentum in a volume V of moving fluid is equal to the total *external force* acting on it. Likewise, the time rate of change of the total angular momentum in a volume V of moving fluid is equal to the total *external torque* acting on it. The conservation laws of linear and angular momentum follow.

In the chapter on gravitation we saw how to summarize the physical state of affairs near masses by use of a field. Each point in the field can be regarded as having a vector associated with it, namely **g,** the gravitational force per unit mass at that point. Or, alternately, we can associate a scalar quantity with each point in space, namely the gravitational potential V. We can then draw a surface, called an equipotential surface, through all points that have the same potential. We draw several such surfaces, the potential on one differing by a constant amount from that on the next one, and so on. The gravitational force at any point is then directed along a line passing through this point perpendicular to these surfaces, and its magnitude is determined from the rate of change of potential with distance in this direction, as indicated by the spacing and orientation of the equipotential surfaces. By drawing in lines of force we can picture vividly how space is affected by the presence of mass.

Likewise, in fluid dynamics we can summarize the physical state of affairs within a moving fluid by means of a field of flow. In general, the field of flow is a *vector* field. We associate a vector quantity with each point in space, namely the flow velocity **v** at that point. For a steady flow the field of flow is stationary. Of course, even in this case a particular fluid particle may still have a variable velocity as it moves from point to point in the field. The field gives the properties of the space from which we deduce the behavior of particles in that space. If the flow is irrotational, as well as steady, we call it *potential flow*. Then the flow velocity **v** can be related to a velocity potential ψ, just as in gravitation **g** can be related to the gravitational potential V. If we draw in surfaces of equal velocity potential, as we drew in surfaces of equal gravitational

18-6
CONSERVATION OF MOMENTUM IN FLUID MECHANICS

18-7
FIELDS OF FLOW

figure 18-10
Streamlines (horizontal) and surfaces of equal velocity potential (vertical) for a homogensous field of flow.

potential, we can deduce **v** from the equipotential flow surfaces just as **g** is deduced from the equipotential gravitational surfaces. Hence, a field for potential flow is analogous to a conservative force field.

A flowing fluid mass can always be divided into tubes of flow. When the flow is steady, the tubes remain unchanged in shape and the fluid that is at one instant in a tube remains inside this tube thereafter. We have seen that the flow velocity inside a tube of flow is parallel to the tube and has a magnitude inversely proportional to the area of the cross section (Eq. 18-1). Let us assign such cross sections to the tubes that the constant of proportionality is the same for all of them; if possible we take this constant to be unity. That is, the volume flux is the same for all tubes, namely unit flux. Then the magnitude of the flow velocity can be determined from the areas of the cross sections of the tubes of flow. There is another procedure equivalent to this which consists of setting up a unit area perpendicular to the direction of flow and drawing through it just as many streamlines as the number of units of magnitude of the velocity at that point.

Let us consider some examples of fields of flow. For drawing purposes we consider only *two-dimensional* examples. In these the flow velocity is the same at all points on a line perpendicular to the plane at any point.

In Fig. 18-10 we have drawn a *homogeneous field of flow*. Here all the streamlines are parallel and the flow velocity **v** is the same at all points. We have seen that there are two equivalent ways of deriving the relative magnitudes of the flow velocities from such fields of flow: (a) from the widths of the tubes of flow and (b) from the distances between lines of equal velocity potential. The latter method applies to steady irrotational flow only. For such flows we draw in the lines of equal velocity potential as dashed lines.

In Fig. 18-11 we show the field for a *uniform rotation* (see Problem 18, Chapter 17). Here v is proportional to r. In Fig. 18-12 we draw the field of flow of a *vortex*. In this case v is proportional to $1/r$ (see Problem 29). Notice that both

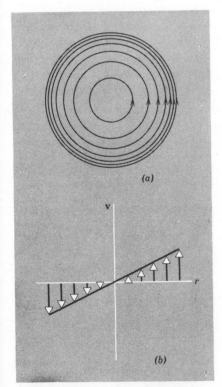

figure 18-11
(a) Uniform rotational field of flow.
(b) Variation of fluid velocity from the center.

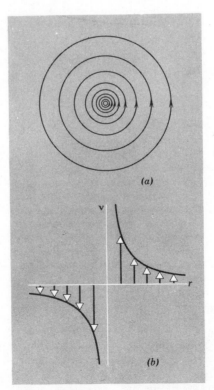

figure 18-12
(a) Vortical field of flow. (b) Variation of fluid velocity from the center.

uniform rotation and vortex motion are represented by circular streamlines but are entirely different kinds of flow. Obviously, the *shapes* of the streamlines give only limited information; their spacing is needed too.

Figure 18-13 represents the field of flow for a *source*. All streamlines are directed radially outward. The source is a line through the center perpendicular to the paper emitting a mass per unit time Q. The field of flow around a linear *sink* is the same as the source except for the sign of the flow, which is directed radially inward.

For a linear source and linear sink which have the same strengths, Q and $-Q$, and are slightly separated, we obtain the combined field called *linear dipole flow*, shown in Fig. 18-14.

As we shall see later the electrostatic field, the magnetic field, and the field of flow for an electric current are also vector fields. In this connection, the homogeneous field (Fig. 18-10) corresponds to the electric field of a plane capacitor, the source field or sink field (Fig. 18-13) correspond to the electric field of a cylindrical capacitor or straight wire of positive or negative charge respectively, and the linear dipole field (Fig. 18-14) corresponds to the electric field of two oppositely charged wires. In all these the field of flow is potential flow and the electric fields are conservative.

The homogeneous field of Fig. 18-10 also represents the magnetic field inside a solenoid. The vortex field of Fig. 18-12 represents the magnetic field around a straight current-carrying wire. This last is an example of a field that is rotational (about the vortex axis).

Because of these analogies between fluid and electromagnetic fields, we can often determine a field of flow, which is difficult to calculate by present mathematical methods, by experimental measurements on appropriate electrical devices.

As we have seen throughout this chapter, the basic field ideas and conservation principles find application in many areas of physics. We shall encounter them many times again.

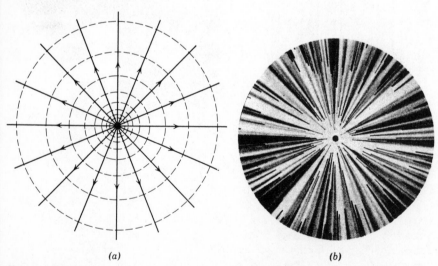

(a)   *(b)*

figure 18-13

(a) Flow from a linear source. *(b)* Fluid flow map of the same. The map in this figure is made by allowing water to flow between a horizontal layer of plate glass and a horizontal layer of plaster. In *(b)* the water comes up through a hole in the center of the plaster and flows out toward the edges. The direction of the flow is made visible by sprinkling the plaster with potassium permanganate crystals which dissolve and color the water a deep purple. (The fluid flow map was made and photographed by Professor A. D. Moore at the University of Michigan, and is taken from *Introduction to Electric Fields,* by W. E. Rogers, McGraw-Hill Book Co., 1954.)

(a)

(b)

figure 18-14
(a) Linear dipole flow. The source is on the left, the sink on the right. (b) A fluid flow map of the same. (The fluid flow map was made and photographed by Professor A. D. Moore at the University of Michigan, and is taken from *Introduction to Electric Fields*, by W. E. Rogers, McGraw-Hill Book Co., 1954.)

questions

1. Briefly describe what is meant by each of the following and illustrate with an example: (a) steady fluid flow; (b) nonsteady fluid flow; (c) rotational fluid flow; (d) irrotational fluid flow; (e) compressible fluid flow; (f) incompressible fluid flow; (g) viscous fluid flow; (h) nonviscous fluid flow.

2. Can you assign a coefficient of static friction between two surfaces, one of which is a fluid surface?

3. It is found that liquid will flow faster and more smoothly from a sealed can when two holes are punctured in the can than when one hole is made. Explain.

4. List all the assumptions made in deriving Bernoulli's equation (Eq. 18-6).

5. Describe the forces acting on an element of fluid as it flows through a pipe of nonuniform cross section.

6. In a lecture demonstration a ping-pong ball is kept in midair by a vertical jet of air. Is the equilibrium stable, unstable, or neutral? Explain.

7. The height of the liquid in the standpipes indicates that the pressure drops along the channel, even though the channel has a uniform cross section and the flowing liquid is incompressible (Fig. 18-15). Explain.

figure 18-15
Question 7

8. The taller the chimney the better the draft taking the smoke out of the fireplace. Explain. Why doesn't the smoke pour into the room containing the fireplace?

9. (a) Explain how a pitcher can make a baseball curve to his right or left? Can we justify applying Bernoulli's equation to such a spinning baseball? (See the Smith reference on p. 392 for an explanation.) (b) Why is it easier to throw a curve with a tennis ball than with a baseball?

10. Not only a ball with a rough surface but also a smooth ball can be made to curve when thrown, but these balls will curve in *opposite* directions. Why? (See "Effect of Spin and Speed on the Curve of a Baseball; and the Magnus Effect for Smooth Spheres" by Lyman J. Briggs, in *American Journal of Physics*, November 1959.)

11. Two rowboats moving parallel to one another in the same direction are pulled toward one another. Two automobiles moving parallel are also pulled together. Explain such phenomena on the basis of Bernoulli's equation.

12. In building "skyscrapers," what forces produced by the movement of air must be counteracted? How is this done? (See "The Wind Bracing of Buildings" by Carl W. Condit in *Scientific American*, February 1974.)

13. Can the action of a parachute in retarding free fall be explained by Bernoulli's equation?

14. Liquid is flowing inside a horizontal pipe which has a constriction along its length. Vertical tube manometers are attached at both the wide portion and the narrow portion of the pipe. If a stopcock at the exit end is closed, will the liquid in the manometer tubes rise or fall? Explain.

15. A stream of water from a faucet becomes narrower as it falls. Explain.

16. Can you explain why water flows in a continuous stream down a vertical pipe, whereas it breaks into drops when falling freely?

17. How does the flush toilet work? Really. (See *Flushed with Pride: The Story of Thomas Crapper*, by W. Reyburn, Englewood Cliffs, N.J.: Prentice-Hall, 1969.)

18. Can you explain why an object falling from a great height reaches a steady terminal speed?

19. Bernoulli's equation (Eq. 18-6) is a statement of energy conservation for fluid motion. In connection with the Venturi meter (p. 391) can you see a formal relationship to energy changes occurring in a roller coaster when it dips down into a valley and climbs up the other side?

20. Sometimes people remove letters from envelopes by cutting a sliver from a narrow end, holding it firmly and blowing toward it. Does Bernoulli's equation play a role in this enterprise? Explain.

21. On takeoff would it be better for an airplane to move into the wind or with the wind? On landing . . . ?

22. Does the difference in pressure between the lower and upper surfaces of an airplane wing depend on the altitude of the moving plane? Explain.

23. The accumulation of ice on an airplane wing may change its shape in such a way that its lift is greatly reduced. Explain.

24. How is an airplane able to fly upside down?

25. An aeronautical engineer claims that he can design a helicopter that will make a "soft" landing without causing a "down draft." Explain whether or not you think this is possible and why.

26. "The characteristic banana-like shape of most returning boomerangs has hardly anything to do with their ability to return. . . . The essential thing is the cross section of the arms, which should be more convex on one side than on the other, like the wing profile of an airplane." (From, "The Aerodynamics of Boomerangs" by Felix Hess, in *Scientific American*, November 1968.) Explain.

27. What powers the flight of soaring birds? (See "The Soaring Flight of Birds" by C. D. Cone, Jr. in *Scientific American*, April 1962.)

28. Why does the factor "2" appear in Eq. 18-9, rather than "1"? One might naively expect that the thrust would simply be the pressure difference times the area, that is, $A_0(p - p_0)$.

29. The destructive effect of a tornado (twister) is greater near the center of the disturbance than near the edge. Explain.

30. When a stopper is pulled from a filled basin, the water drains out while circulating like a small whirlpool. The angular velocity of a fluid element about a vertical axis through the orifice appears to be greatest near the orifice. Explain.

31. Is it true that in bathtubs in the northern hemisphere the water drains out with a counterclockwise rotation and in those in the southern hemisphere with a clockwise rotation? If so, explain and predict what would happen at the equator. (See "Bath-Tub Vortex" by Ascher H. Shapiro in *Nature*, December 15, 1962.)

32. The longer the board and the shallower the water, the farther will a surf board skim across the water. Explain. (See "The Surf Skimmer" by R. D. Edge, in *American Journal of Physics*, July 1968.)

33. When poured from a teapot water has a tendency to run along the underside of the spout. Explain. (See "The Teapot Effect . . . a Problem" by Markus Reiner, in *Physics Today*, September 1956.)

34. Prairie dogs live in large colonies in complex interconnected burrow systems. They face the problem of maintaining a sufficient air supply to their burrows to avoid suffocation. They avoid this by building conical earth mounds about some of their many burrow openings. In terms of Bernoulli's

equation (Eq. 18-6) how does this air conditioning scheme work? Note that because of viscous forces the wind speed over the prairie is less close to ground level than it is even a few inches higher up. (See *New Scientist*, p. 191, 27 January 1972.)

35. Use the criterion of the paddle wheel (Fig. 18-1) to determine which flow fields (Figs. 18-10 through 18-14) are rotational.

36. In steady flow the velocity vector **v** at any point is constant. Can there then be accelerated motion of the fluid particles? Discuss.

SECTION 18-3

problems

1. A garden hose having an internal diameter of 0.75 in. is connected to a lawn sprinkler that consists of an enclosure with 24 holes, each 0.050 in. in diameter. If the water in the hose has a speed of 3.0 ft/s, at what speed does it leave the sprinkler holes?   *Answer:* 28 ft/s.

2. How much work is done by pressure in forcing 50 ft³ (1.4 m³) of water through a 0.50-in. (13-mm) pipe if the difference in pressure at the two ends of the pipe is 15 lb/in.² (1.0×10^5 Pa)?

3. Water flows continuously from the outlet of a faucet of internal diameter d at an initial speed v_0. Determine the diameter of the stream in terms of the distance h below the outlet. (Neglect air resistance and assume droplets are not formed.)

$$Answer: \; d\left[\frac{v_0}{\sqrt{v_0{}^2 + 2gh}}\right]^{1/2}.$$

4. Water is pumped steadily out of a flooded basement at a speed of 5.0 m/s through a uniform hose of radius 1.0 cm. The hose passes out through a window 3.0 m above the water line. How much power is supplied by the pump?

SECTION 18-4

5. A hollow tube has a disc DD attached to its end. When air is blown through the tube, the disc attracts the card CC. Let the area of the card be A and let v be the average airspeed between CC and DD (Fig. 18-16); calculate the resultant upward force on CC. Neglect the card's weight.
Answer: $\frac{1}{2}\rho v^2 A$, where ρ is the density of air.

6. In a horizontal oil pipeline of constant cross-sectional area the pressure decrease between two points 1000 ft apart is 5.0 lb/in.². What is the energy loss per cubic foot of oil per unit distance?

7. Figure 18-17 shows liquid discharging from an orifice in a large tank at a distance h below the water level. (a) Apply Bernoulli's equation to a streamline connecting points 1, 2, and 3, and show that the speed of efflux is

$$v = \sqrt{2gh}.$$

This is known as Torricelli's law. (b) If the orifice were curved directly up-

figure 18-16
Problem 5

figure 18-17
Problem 7

ward, how high would the liquid stream rise? (c) How would viscosity or turbulence affect the analysis?   *Answer:* (b) It would rise to height h.

8. A tank is filled with water to a height H. A hole is punched in one of the walls at a depth h below the water surface (Fig. 18-18). (a) Show that the dis-. tance x from the foot of the wall at which the stream strikes the floor is given by $x = 2\sqrt{h(H-h)}$. (b) Could a hole be punched at another depth so that this second stream would have the same range? If so, at what depth?

figure 18-18
Problem 8

9. The upper surface of water in a standpipe is a height H above level ground. (a) At what depth h should a small hole be put to make the emerging horizontal water stream strike the ground at the maximum distance from the base of the standpipe? (b) What is this maximum distance?
Answer: (a) H/2. (b) H.

10. (a) Consider the stagnant air at the front edge of a wing and the air rushing over the wing surface at a speed v. Assume pressure at the leading edge to be approximately atmospheric and find the greatest value possible for v in streamline flow; assume air is incompressible and use Bernoulli's equation. Take the density of air to be 1.2×10^{-3} g/cm³. (b) How does this compare with the speed of sound of 770 mi/h? Can you explain the difference? Why should there be any connection between these quantities?

11. If a person blows air with a speed of 15 m/s across the top of one side of a U-tube containing water, what will be the difference between the water levels on the two sides? Assume the density of air is 1.2 kg/m³.
Answer: 1.4 cm.

12. A siphon is a device for removing liquid from a container that cannot be tipped. It operates as shown in Fig. 18-19. The tube must initially be filled, but once this has been done the liquid will flow until its level drops below the tube opening at A. The liquid has density ρ and negligible viscosity. (a) With what speed does the liquid emerge from the tube at C? (b) What is the pressure in the liquid at the topmost point B? (c) What is the greatest possible height h_1 that a siphon may lift water?

figure 18-19
Problem 12

SECTION 18-5

13. A Pitot tube is mounted on an airplane wing to determine the speed of the plane relative to the air, which is at a temperature of 0°C. The tube contains alcohol and indicates a level difference of 26 cm. What is the plane's speed relative to the air? The density of alcohol is 0.81×10^3 kg/m³.
Answer: 200 km/h.

14. Models of torpedoes are sometimes tested in a horizontal pipe of flowing water, much as a wind tunnel is used to test model airplanes. Consider a circular pipe of internal diameter 10 in. and a torpedo model, aligned along the axis of the pipe, with a diameter of 2.0 in. The torpedo is to be tested with water flowing past it at 8.0 ft/s. (a) With what speed must the water

flow in the unconstricted part of the pipe? (b) What will the pressure difference be between the constricted and unconstricted parts of the pipe?

15. Water is moving with a speed of 5.0 m/s through a pipe with a cross-sectional area of 4.0 cm². The water gradually descends 10 m as the pipe increases in area to 8.0 cm². (a) What is the speed of flow at the lower level? (b) If the pressure at the upper level is 1.50×10^5 Pa, what is the pressure at the lower level?   *Answer:* (a) 2.5 m/s. (b) 2.6×10^5 Pa.

16. Suppose that two tanks, 1 and 2, each with a large opening at the top, contain different liquids. A small hole is made in the side of each tank at the same depth h below the liquid surface, but the hole in tank 1 has half the cross-sectional area of the hole in tank 2. (a) What is the ratio ρ_1/ρ_2 of the densities of the fluids if it is observed that the mass flux is the same for the two holes? (b) What is the ratio of the flow rates (volume flux) from the two tanks? (c) To what height above the hole in the second tank should fluid be added or drained to equalize the flow rates?

17. A small plane has a wing area (each wing) of 100 ft² (9.3 m²). At a certain air speed, air flows over the upper wing surface at 160 ft/s (49 m/s) and over the lower wing surface at 130 ft/s (40 m/s). What is the weight of the plane? Assume that the plane travels at constant velocity and that the lift effects associated with the fuselage and tail assembly are small. Discuss the lift if the plane, flying at the same air speed, is (a) in level flight, (b) climbing at 15°, and (c) descending at 15°. Take the density of air to be 2.33×10^{-3} slug/ft³ (1.2 kg/m³).
Answer: 2000 lb (8900 N). Lift is the same in all three cases.

18. If the speed of air flow past the lower surface of a wing is 350 ft/s, what speed of flow over the upper surface will give a lift of 20.0 lb/in.²? Take the density of air to be 2.33×10^{-3} slug/ft³.

19. Consider a uniform U-tube with a diaphragm at the bottom and filled with a liquid to different heights in each arm (see Fig. 18-20). Now imagine that the diaphragm is punctured so that the liquid flows from left to right. (a) Show that application of Bernoulli's principle to points 1 and 3 leads to a contradiction. (b) Explain why Bernoulli's principle is not applicable here. (*Hint:* Is the flow steady?)

20. Calculate the speed of efflux of a liquid from an opening in a tank, taking into account the velocity of the top surface of the liquid, as follows. (a) Show, from Bernoulli's equation, that

$$v_0^2 = v^2 + 2gh$$

where v is the speed of the top surface. (b) Then consider the flow as one big tube of flow and obtain v/v_0 from the equation of continuity, so that
$$v_0 = \sqrt{2gh/[1 - (A_0/A)^2]}$$

where A is the tube cross section at the top and A_0 is the tube cross section at the opening. (c) Then show that if the hole is small compared to the area of the surface,

$$v_0 \cong \sqrt{2gh}\,[1 + \tfrac{1}{2}(A_0/A)^2].$$

21. By applying Bernoulli's equation and the equation of continuity to points 1 and 2 of Fig. 18-6, show that the speed of flow at the entrance is

$$v = a\sqrt{\frac{2(\rho' - \rho)gh}{\rho(A^2 - a^2)}}.$$

22. A Venturi meter has a pipe diameter of 10 in. and a throat diameter of 5.0 in. If the water pressure in the pipe is 8.0 lb/in.² and in the throat is 6.0 lb/in.², determine the rate of flow of water in ft³/s (volume flux).

23. Consider the Venturi tube of Fig. 18-6 without the manometer. Let A equal $5a$. Suppose the pressure at A is 2.0 atm. (a) Compute the values of v at A and v' at a that would make the pressure p' at a equal to zero. (b) Compute the corresponding volume flow rate if the diameter at A is 5.0 cm. The

figure 18-20
Problem 19

phenomenon at *a* when p' falls to nearly zero is known as *cavitation*. The water vaporizes into small bubbles.

Answer: (*a*) 20 m/s. (*b*) 8.0×10^{-3} m³/s.

SECTION 18-6

24. (*a*) Consider a stream of fluid of density ρ with speed v_1 passing *abruptly* from a cylindrical pipe of cross-sectional area a_1 into a wider cylindrical pipe of cross-sectional area a_2 (see Fig. 18-21). The jet will mix with the surrounding fluid and, after the mixing, will flow on almost uniformly with an average speed v_2. Without referring to the details of the mixing, use momentum ideas to show that the increase in pressure due to the mixing is approximately

$$p_2 - p_1 = \rho v_2 (v_1 - v_2).$$

(*b*) Show from Bernoulli's principle that in a *gradually* widening pipe we would get

$$p_2 - p_1 = \tfrac{1}{2}\rho(v_1{}^2 - v_2{}^2).$$

(*c*) Find the loss of pressure due to the abrupt enlargement of the pipe. Can you draw an analogy with elastic and inelastic collisions in particle mechanics?

figure 18-21
Problem 24

SECTION 18-7

25. Show that the constant in Bernoulli's equation (Eq. 18-6) is the same for *all* streamlines in the case of the steady, irrotational flow of Fig. 18-10.

26. A force field is conservative if $\oint \mathbf{F} \cdot d\mathbf{s} = 0$. The circle on the integration sign means that the integration is to be taken along a closed curve (a round trip) in the field. A flow is a potential flow (hence irrotational) if $\oint \mathbf{v} \cdot d\mathbf{s} = 0$ for every closed path in the field.

Using this criterion, show that the fields of Figs. (*a*) 18-10 and (*b*) 18-13 are fields of potential flow.

27. The so-called Poiseuille field of flow is shown in Fig. 18-22. The spacing of the streamlines indicates that although the motion is rectilinear, there is a velocity gradient in the transverse direction. Show that such a flow is rotational.

figure 18-22
Problem 27

28. In flows that are sharply curved centrifugal effects are appreciable. Consider an element of fluid which is moving with speed v along a streamline of a curved flow in a horizontal plane (Fig. 18-23).

(*a*) Show that $dp/dr = \rho v^2/r$, so that the pressure increases by an amount $\rho v^2/r$ per unit distance perpendicular to the streamline as we go from the concave to the convex side of the streamline.

(*b*) Then use Bernoulli's equation and this result to show that vr equals a constant, so that speeds increase toward the center of curvature. Hence, streamlines that are uniformly spaced in a straight pipe will be crowded toward the inner wall of a curved passage and widely spaced toward the outer wall. This problem should be compared to Problem 18 of Chapter 17 in which the curved motion is produced by rotating a container. There the speed varied directly with r, but here it varies inversely.

(c) Show that this flow is irrotational.

29. Before Newton proposed his theory of gravitation, a model of planetary motion proposed by René Descartes was widely accepted. In Descartes' model the planets were caught in and dragged along by a whirlpool of ether particles centered around the sun. Newton showed that this vortex scheme contradicted observations, for: (a) The speed of an ether particle in the vortex varies inversely as its distance from the sun. (b) The period of revolution of such a particle varies directly as the square of its distance from the sun. (c) This result contradicts Kepler's third law. Prove (a), (b), and (c).

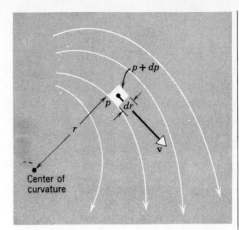

figure 18-23
Problem 28

19

waves in elastic media

Wave motion appears in almost every branch of physics. We are all familiar with water waves. There are also sound waves, as well as light waves, radio waves, and other electromagnetic waves. One formulation of the mechanics of atoms and subatomic particles is called wave mechanics. Clearly the properties and behavior of waves are very important in physics.

In this chapter and the next we confine our attention to waves in deformable or elastic media. These waves, among which ordinary sound waves in air are one example, might be called *mechanical waves*. They originate in the displacement of some portion of an elastic medium from its normal position, causing it to oscillate about an equilibrium position. Because of the elastic properties of the medium, the disturbance is transmitted from one layer to the next. This disturbance, or wave, consequently progresses through the medium. Note that the medium itself does not move as a whole along with the wave motion; the various parts of the medium oscillate only in limited paths. For example, in water waves small floating objects like corks show that the actual motion of various parts of the water is slightly up and down and back and forth. Yet the water waves move steadily along the water. As they reach floating objects they set them in motion, thus transferring energy to them.* Energy can be transmitted over considerable distances by wave motion. The energy in the waves is the kinetic and potential energy of the matter, but the transmission of the energy comes about by its being passed along from one part of the matter to the next, not by any long-range motion of the matter itself. Mechanical waves are charac-

19-1
MECHANICAL WAVES

* See "Ocean Waves," by Willard Bascom, *Scientific American*, August 1959.

404

terized by the transport of energy through matter by the motion of a disturbance in that matter without any corresponding bulk motion of the matter itself.

It is necessary to have a material medium to transmit mechanical waves. We do not need such a medium, however, to transmit electromagnetic waves, light passing freely, for example, through the near vacuum of space from the stars. The properties of the medium that determine the speed of a wave through that medium, as we will see in Section 19-5, are its inertia and its elasticity. All material media, including, say, air, water, and steel, possess these properties and can transmit mechanical waves. It is the elasticity that gives rise to the restoring forces on any part of the medium displaced from its equilibrium position; it is the inertia that tells us how this displaced portion of the medium will respond to these restoring forces. Together these two factors determine the wave speed.

19-2
TYPES OF WAVES

In listing water waves, light waves, and sound waves as examples of wave motion, we are classifying waves according to their broad physical properties. Waves can be classified in other ways.

We can distinguish different kinds of mechanical waves by considering how the motions of the particles of matter are related to the direction of propagation of the waves themselves. If the motions of the matter particles conveying the wave are perpendicular to the direction of propagation of the wave itself, we then have a *transverse* wave. For example, when a vertical string under tension is set oscillating back and forth at one end, a transverse wave travels down the string; the disturbance moves along the string but the string particles vibrate at right angles to the direction of propagation of the disturbance (Fig. 19-1a).

Light waves are not mechanical waves. The disturbance that travels along is not a motion of matter but an electromagnetic field (Chapter 41). But because the electric and magnetic fields are perpendicular to the direction of propagation, light waves are also transverse waves.

If, however, the motion of the particles conveying a mechanical wave is back and forth along the direction of propagation, we then have a *longitudinal wave*. For example, when a vertical spring under tension is set oscillating up and down at one end, a longitudinal wave travels along the spring; the coils vibrate back and forth in the direction in which the disturbance travels along the spring (Fig. 19-1b). Sound waves in a gas are longitudinal waves. We shall discuss them in greater detail in Chapter 20.

Some waves are neither purely longitudinal nor purely transverse. For example, in waves on the surface of water the particles of water move both up and down and back and forth, tracing out elliptical paths as the water waves move by.

Waves can also be classified as one-, two-, and three-dimensional waves, according to the number of dimensions in which they propagate energy. Waves moving along the string or the spring of Fig. 19-1 are one-dimensional. Surface waves or ripples on water, caused by dropping a pebble into a quiet pond, are two-dimensional. Sound waves and light waves which emanate radially from a small source are three-dimensional.

Waves may be classified further according to the behavior of a particle of the matter conveying the wave during the course of time the wave propagates. For example, we can produce a *pulse* traveling down a

(a)

(b)

figure 19-1
(a) In a transverse wave the particles of the medium (stretched string) vibrate at right angles to the direction in which the wave itself is propagated. *(b)* In a longitudinal wave the particles of the medium (stretched spring) vibrate in the same direction as that in which the wave itself is propagated.

stretched string by applying a single sidewise movement at its end. Each particle remains at rest until the pulse reaches it, then it moves during a short time, and then it again remains at rest. If we continue to move the end of the string back and forth (Fig. 19-1a), we produce a *train of waves* traveling along the string. If our motion is periodic, we produce a *periodic train of waves* in which each particle of the string has a periodic motion. The simplest special case of a periodic wave is a *simple harmonic wave* which gives each particle a simple harmonic motion.

Consider a three-dimensional pulse. We can draw a surface through all points undergoing a similar disturbance at a given instant. As time goes on, this surface moves along showing how the pulse propagates. We can draw similar surfaces for subsequent pulses. For a periodic wave we can generalize the idea by drawing in surfaces, all of whose points are in the same phase of motion. These surfaces are called *wavefronts*. If the medium is homogeneous and isotropic, the direction of propagation is always at right angles to the wavefront. A line normal to the wavefronts, indicating the direction of motion of the waves, is called a *ray*.

Wavefronts can have many shapes. If the disturbances are propagated in a single direction, the waves are called *plane waves*. At a given instant conditions are the same everywhere on any plane perpendicular to the direction of propagation. The wavefronts are plane and the rays are parallel straight lines (Fig. 19-2a). Another simple case is that of *spherical waves*. Here the disturbance is propagated out in all directions from a point source of waves. The wavefronts are spheres and the rays are radial lines leaving the point source in all directions (Fig. 19-2b). Far from the source the spherical wavefronts have very small curvature, and over a limited region they can often be regarded as plane. Of course, there are many other possible shapes for wavefronts.

figure 19-2
(a) A plane wave. The planes represent wavefronts spaced a wavelength apart, and the arrows represent rays. (b) A spherical wave. The rays are radial and the wavefronts, spaced a wavelength apart, from spherical shells. Far out from the source, however, small portions of the wavefronts become nearly plane.

We shall refer to all these wave types as we progress through the wave phenomena of physics. In this chapter we often use the transverse wave in a string to illustrate the general properties of waves. In the next chapter we shall see the consequences of these properties for sound, a longitudinal mechanical wave. Later in the text we will discuss the properties of nonmechanical waves such as light waves.

19-3
TRAVELING WAVES

Let us consider a long string stretched in the x-direction along which a transverse wave is traveling. At some instant of time, say $t = 0$, the shape of the string can be represented by

$$y = f(x) \qquad t = 0, \tag{19-1}$$

where y is the transverse displacement of the string at the position x. In Fig. 19-3a we show a possible waveform (a pulse) on the string at $t = 0$. Experiment shows that as time goes on such a wave travels along the string without changing its form, provided internal frictional losses are small enough. At some time t later the wave has traveled a distance vt to the right, where v is magnitude of the wave velocity, assumed constant. The equation of the curve at the time t is therefore

$$y = f(x - vt) \qquad t = t. \tag{19-2}$$

This gives us the same waveform about the point $x = vt$ at time t as we

figure 19-3
(a) The shape of a stretched string (in this case a pulse) at $t = 0$.
(b) At a later time t the pulse has traveled to the right a distance $x = vt$.

had about $x = 0$ at the time $t = 0$ (Fig. 19-3b). Equation 19-2 is the general equation representing a wave of *any* shape *traveling to the right.* To describe a particular shape we must specify exactly what the function f is.*

Let us look more carefully at this equation. If we wish to follow a particular part (or phase) of the wave as times goes on, then in the equation we look at a particular value of y (say, the top of the pulse just described). Mathematically this means we look at how x changes with t when $(x - vt)$ has some particular fixed value. We see at once that as t increases x must increase in order to keep $(x - vt)$ fixed. Hence, Eq. 19-2 does in fact represent a wave traveling to the right (increasing x as time goes on). If we wished to represent a wave *traveling to the left*, we would write

$$y = f(x + vt), \tag{19-3}$$

for here the position x of some fixed phase $(x + vt)$ of the wave decreases as time goes on. The velocity of a particular phase of the wave is easily obtained. For a particular phase of a wave traveling to the right we require that

$$x - vt = \text{constant.}$$

Then differentiation with respect to time gives

$$\frac{dx}{dt} - v = 0 \quad \text{or} \quad \frac{dx}{dt} = v, \tag{19-4}$$

so that v is really the *phase velocity* of the wave. For a wave traveling to the left we obtain $-v$, in the same way, as its phase velocity.†

The general equation of a wave can be interpreted further. Note that for any fixed value of the time t the equation gives y as a function of x. This defines a curve, and this curve represents the actual shape of the string at this chosen time. It gives us a snapshot of the wave at this time. Suppose, on the other hand, we wish to focus our attention on one point of the string, that is, a fixed value of x. Then the equation gives us y as a function of the time t. This describes how the transverse position of this point on the string changes with time.

---

* When we say that "y is a function of $(x - vt)$," we mean that the variables x and t occur only in the combination $x - vt$. For example, $\sin k(x - vt)$, $\log (x - vt)$, and $(x - vt)^3$ are functions of $x - vt$, but $x^2 - vt^2$ is not.

† In disturbances that can be represented as a group of waves, the energy may be transported with a velocity different from the phase velocity of any individual wave. This group velocity will be considered in Chapter 41 in connection with electromagnetic waves. Until then whenever we use the term wave velocity we mean the phase velocity of the wave.

The argument just presented holds for longitudinal waves as well as for transverse waves. The analogous longitudinal example is that of a long straight tube of gas whose axis is taken as the x-axis, and the wave or pulse is a pressure change traveling along the tube. Then the same reasoning leads us to an equation, having the form of Eqs. 19-2 and 19-3, which gives the pressure variations with time at all points of the tube. (See Section 20-3.)

Let us now consider a particular waveform, whose importance will soon become clear. Suppose that at the time $t = 0$ we have a wavetrain along the string given by

$$y = y_m \sin \frac{2\pi}{\lambda} x. \qquad (19\text{-}5)$$

The wave shape is a sine curve (Fig. 19-4). The maximum displacement y_m is the *amplitude* of the sine curve. The value of the transverse displacement y is the same at x as it is at $x + \lambda$, $x + 2\lambda$, etc. The symbol λ is called the *wavelength* of the wavetrain and represents the distance between two adjacent points in the wave having the same phase. As times goes on let the wave travel to the right with a phase velocity v. Hence, the equation of the wave at the time t is

$$y = y_m \sin \frac{2\pi}{\lambda} (x - vt). \qquad (19\text{-}6)$$

Notice that this has the form required for a traveling wave (Eq. 19-2).

figure 19-4
At $t = 0$, the string has a shape $y = y_m \sin 2\pi x/\lambda$ (solid line). At a later time t the sine wave has moved to the right a distance $x = vt$, and the string has a shape given by $y = y_m \sin 2\pi(x - vt)/\lambda$.

The *period* T is the time required for the wave to travel a distance of one wavelength λ, so that

$$\lambda = vT. \qquad (19\text{-}7)$$

Putting this relation into the equation of the wave, we obtain

$$y = y_m \sin 2\pi \left(\frac{x}{\lambda} - \frac{t}{T} \right). \qquad (19\text{-}8)$$

From this form it is clear that y, at any given time, has the same value at $x + \lambda$, $x + 2\lambda$, etc., as it does at x, and that y, at any given position, has the same value at the time $t + T$, $t + 2T$, etc., as it does at the time t.

To reduce Eq. 19-8 to a more compact form, we define two quantities, the *wave number* k and the *angular frequency* ω (see Eq. 15-12). They are given by

$$k = \frac{2\pi}{\lambda} \quad \text{and} \quad \omega = \frac{2\pi}{T}. \qquad (19\text{-}9)$$

In terms of these quantities, the equation of a sine wave traveling to the right (positive x-direction) is

$$y = y_m \sin (kx - \omega t). \qquad (19\text{-}10a)$$

For a sine wave traveling to the left (negative x-direction), we have

$$y = y_m \sin (kx + \omega t). \qquad (19\text{-}10b)$$

Comparing Eqs. 19-7 and 19-9, we see that the phase velocity v of the wave is given by

$$v = \frac{\lambda}{T} = \frac{\omega}{k}. \qquad (19\text{-}11)$$

In the traveling waves of Eqs. 19-10a and 19-10b we have assumed that the displacement y is zero at the position $x = 0$ at the time $t = 0$. This, of course, need not be the case. The general expression for a sinusoidal wavetrain traveling to the right is

$$y = y_m \sin (kx - \omega t - \phi),$$

where ϕ is called the phase constant. For example, if $\phi = -90°$, the displacement y at $x = 0$ and $t = 0$ is y_m. This particular example is

$$y = y_m \cos (kx - \omega t),$$

for the cosine function is displaced by 90° from the sine function.

If we fix our attention on a given point of the string, say $x = \pi/k$, the displacement y at that point can be written* as

$$y = y_m \sin (\omega t + \phi).$$

This is similar to Eq. 15-29 for simple harmonic motion. Hence, any particular element of the string undergoes simple harmonic motion about its equilibrium position as this wavetrain travels along the string.

It is an experimental fact that for many kinds of waves *two or more waves can traverse the same space independently of one another*. The fact that waves act independently of one another means that the displacement of any particle at a given time is simply the sum of the displacements that the individual waves alone would give it. This process of vector addition of the displacements of a particle is called *superposition*. For example, radio waves of many frequencies pass through a radio antenna; the electric currents set up in the antenna by the superposed action of all these waves are very complex. Nevertheless, we can still tune to a particular station, the signal that we receive from it being in principle the same as that which we would receive if all other stations were to stop broadcasting. Likewise, in sound we can listen to notes played by individual instruments in an orchestra, even though the sound wave reaching our ears from the full orchestra is very complex.

For waves in deformable media the superposition principle holds whenever the mathematical relation between the deformation and the restoring force is one of simple proportionality. Such a relation is expressed mathematically by a linear equation. For electromagnetic waves the superposition principle holds because the mathematical relations between the electric and magnetic fields are linear.

The superposition principle seems so obvious that it is worthwhile to point out that it does not always hold. Superposition fails when the equations governing

19-4
THE SUPERPOSITION PRINCIPLE

---

* Using the fact that $\sin (\pi - \theta) = \sin \theta$.

wave motion are not linear. Physically this happens when the wave disturbance is relatively large and the ordinary linear laws of mechanical action no longer hold. For example, beyond the elastic limit Hooke's law no longer holds and the linear relation $F = -kx$ can no longer be used.

As for sound, violent explosions create shock waves. Although shock waves are longitudinal elastic waves in air, they behave differently from ordinary sound waves. The equation governing their propagation is quadratic, and superposition does not hold. With two very loud notes the ear hears something more than just the two individual notes. Those familiar with high-fidelity apparatus will know that "intermodulation distortion" between two tones arises when the system fails to combine the tones linearly, and that this distortion is more apparent when the amplitude of the tones is high. A more obvious physical example is water waves. Ripples cannot travel independently across breakers as they can across gentle swells.

The importance of the superposition principle physically is that, where it holds, it makes it possible to analyze a complicated wave motion as a combination of simple waves. In fact, as was shown by the French mathematician J. Fourier (1768–1830), all that we need to build up the most general form of periodic wave are simple harmonic waves. Fourier showed that any periodic motion of a particle can be represented as a combination of simple harmonic motions. For example, if $y(t)$ represents the motion of a source of waves having a period T, we can analyze $y(t)$ as follows:

$$y(t) = A_0 + A_1 \sin \omega t + A_2 \sin 2\omega t + A_3 \sin 3\omega t + \cdots$$
$$+ B_1 \cos \omega t + B_2 \cos 2\omega t + B_3 \cos 3\omega t + \cdots$$

where $\omega = 2\pi/T$. This expression is called a Fourier series. The A's and B's are constants which have definite values for any particular periodic motion $y(t)$. (See Fig. 19-5, for example.) If the motion is not periodic, as

figure 19-5

(a) The dashed line is a sawtooth wave commonly encountered in electronics. It can be written $y(t) = (\omega/2\pi)t - \frac{1}{2}$ for $0 < t < 2\pi/\omega$, as $y(t) = (\omega/2\pi)t - \frac{3}{2}$ for $2\pi/\omega < t < 4\pi/\omega$, etc. The Fourier series for this function is $y(t) = (1/\pi)[-\sin \omega t - \frac{1}{2} \sin 2\omega t - \frac{1}{3} \sin 3\omega t - \ldots .]$. The solid line is the sum of the first six terms of this series and can be seen to approximate the sawtooth quite closely, except for overshooting near the discontinuities. As more terms of the series are included, the approximation becomes better and better. (b) Here are shown the first six terms of the Fourier series which, when added together, yield the solid curve in (a).

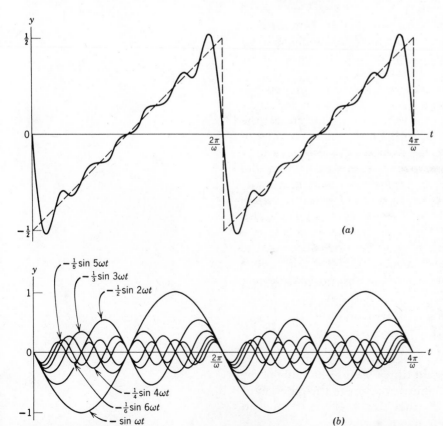

a pulse, the sum is replaced by an integral—the so-called Fourier integral. Hence, any motion of a source of waves can be represented in terms of simple harmonic motions. Because the motion of the source creates the waves, it should come as no surprise that the waves themselves can be analyzed as combinations of simple harmonic waves. Herein lies the importance of simple harmonic motion and simple harmonic waves.

When the elasticity of the medium is such that (for mechanical waves) Hooke's law is not exactly obeyed, then a wave pulse produced at the end of a stretched string may change its shape as it travels along the string. Although each of the component harmonic waves travels without changing its shape, the speed of each component is now different for each frequency (or wavelength). This phenomena is called *dispersion* and the medium is said to be *dispersive* for the wave type in question. As a result the pulse shape can change and the pulse speed may depend on the details of its initial shape. Examples of nondispersive situations are mechanical waves propagated along an ideal (perfectly flexible) stretched string and electromagnetic waves (including light) propagated through a vacuum. Examples of dispersive situations are ocean waves and light waves propagated through a transparent medium such as glass.

Another way in which the wave pulse may change its shape is by loss of mechanical energy to the medium or its surroundings; for example, by air resistance, viscosity, or internal friction. Then the amplitude of the wave decreases with time and the wave is said to be *attenuated*.

For the moment, we will assume that the medium is nondispersive and that there is no dissipation of energy as the wave travels through the medium.

19-5 WAVE SPEED

Given the characteristics of the medium it should be possible to calculate the wave speed from the basic principles of Newtonian mechanics. In this section we continue to focus our attention on transverse waves in a stretched string and in Supplementary Topic III we show how to calculate the speed of such waves in the most general way. Here we consider two other approaches—a treatment based on dimensional analysis and a somewhat less general mechanical analysis in which we compute the speed of a transverse pulse along a stretched string.

We stated in Section 19-1 that the wave speed for a medium depends on the elasticity of the medium and on its inertia. For a stretched string the elasticity is measured by the tension F in the string; the greater the tension the greater will be the elastic restoring force on an element of the string that is pulled sideways. The inertia characteristic is measured by μ, the mass per unit length of the string. Assuming then, that the wave speed v depends only on F and μ, we can use dimensional analysis to find how v depends on these quantities. In terms of mass M, length L, and time T, the dimensions of F are MLT^{-2} and the dimensions of μ are ML^{-1}. The only way these dimensions can be combined to get a velocity (which has the dimensions LT^{-1}) is to take the square root of F/μ. That is, F/μ has the dimensions L^2T^{-2} and $\sqrt{F/\mu}$ has the dimensions LT^{-1} of a velocity. Dimensional analysis cannot account for any dimensionless quantities, so that the result

$$v = \sqrt{\frac{F}{\mu}} \qquad (19\text{-}12)$$

may or may not be complete. The most we can say is that the wave speed is equal to a dimensionless constant times $\sqrt{F/\mu}$. The value of the constant can be obtained from a mechanical analysis of the problem or from experiment. These methods show that the constant is equal to unity and that Eq. 19-12 is correct as it stands.

Now let us *derive* the velocity of a pulse in a stretched string by a mechanical analysis. In Fig. 19-6 we show a wave pulse proceeding from right to left in the string with a speed v. We can imagine the entire string to be moved from left to right with this same speed so that the wave pulse remains fixed in space, whereas the particles composing the string successively pass through the pulse. This simply means that, instead of taking our reference frame to be the walls between which the string is stretched, we choose a reference frame which is in uniform motion with respect to that one. Because Newton's laws involve only accelerations, which are the same in both frames, we can use them in either frame. We just happen to choose a more convenient frame.

We consider a small section of the pulse of length Δl to form an arc of a circle of radius R, as shown in the diagram. If μ is the mass per unit length of the string, the so-called linear density, then $\mu \, \Delta l$, is the mass of this element. The tension F in the string is a tangential pull at each end of this small segment of the string. The horizontal components cancel and the vertical components are each equal to $F \sin \theta$. Hence, the total vertical force is $2F \sin \theta$. Because θ is small, we can take $\sin \theta \cong \theta$ and

$$2F \sin \theta = 2F\theta = 2F \frac{(\Delta l/2)}{R} = F \frac{\Delta l}{R}.$$

figure 19-6
Derivation of wave speed by considering the forces on a section of string of length Δl.

This gives the force supplying the centripetal acceleration of the string particles directed toward O. Now the centripetal force acting on a mass $\mu \, \Delta l$ moving in a circle of radius R with speed v is $\mu \, \Delta l \, v^2/R$; see Section 6-3. Notice that the tangential velocity v of this mass element along the top of the arc is horizontal and is the same as the pulse speed. Combining the equivalent expressions just given we obtain

$$F \frac{\Delta l}{R} = \frac{\mu \, \Delta l \, v^2}{R}$$

or

$$v = \sqrt{\frac{F}{\mu}}.$$

If the amplitude of the pulse were very large compared to the length of the string, we would have been unable to use the approximation $\sin \theta \cong \theta$. Furthermore, the tension F in the string would be changed by the presence of the pulse, whereas we assumed F to be unchanged from the original tension in the stretched string. Therefore, our result, like superposition, holds only for relatively small transverse displacements of the string—which case, however, is widely applicable in practice. Notice also that the wave speed is independent of the shape of the wave, for no particular assumption about the actual shape of the pulse was used in the proof.

The frequency of a wave is naturally determined by the frequency of the source. The speed with which the wave travels through a medium is determined by the properties of the medium, as previously illustrated. Once the frequency ν and speed v of the wave are determined, the wavelength λ is fixed. In fact, from Eq. 19-7 and the relation, $\nu = 1/T$, we have

$$\lambda = \frac{v}{\nu}. \qquad (19\text{-}13)$$

A transverse sinusoidal wave is generated at one end of a long horizontal string by a bar which moves the end up and down through a distance of 0.50 cm. The motion is continuous and is repeated regularly 120 times per second.

(a) If the string has a linear density of 0.25 kg/m and is kept under a tension of 90 N, find the speed, amplitude, frequency, and wavelength of the wave motion.

EXAMPLE 1

The end moves 0.25 cm away from the equilibrium position, first above it, then below it; therefore, the amplitude y_m is 0.25 cm.

The entire motion is repeated 120 times each second so that the frequency is 120 vibrations per second, or 120 Hz.

The wave speed is given by $v = \sqrt{F/\mu}$. But $F = 90$ N and $\mu = 0.25$ kg/m, so that

$$v = \sqrt{\frac{90 \text{ N}}{0.25 \text{ kg/m}}} = 19 \text{ m/s}.$$

The wavelength is given by $\lambda = v/\nu$, so that

$$\lambda = \frac{19 \text{ m/s}}{120 \text{ vib/s}} = 16 \text{ cm}.$$

(b) Assuming the wave moves in the $+x$-direction and that, at $t = 0$, the end of the string described by $x = 0$ is in its equilibrium position $y = 0$, write the equation of the wave.

The general expression for a transverse sinusoidal wave moving in the $+x$-direction is

$$y = y_m \sin (kx - \omega t - \phi).$$

Requiring that $y = 0$ for the conditions $x = 0$ and $t = 0$ yields

$$0 = y_m \sin (-\phi),$$

which means that the phase constant ϕ may be taken to be zero. You should show that integral multiples of π yield the same final results. Hence for this wave

$$y = y_m \sin (kx - \omega t),$$

and with the values just found,

$$y_m = 0.25 \text{ cm},$$

$$\lambda = 16 \text{ cm} \quad \text{or} \quad k = \frac{2\pi}{\lambda} = \frac{2\pi}{16 \text{ cm}} = 0.39 \text{ cm}^{-1},$$

$$v = 19 \text{ m/s} = 1900 \text{ cm/s} \quad \text{or} \quad \omega = vk = (1900 \text{ cm/s})(0.39 \text{ cm}^{-1}) = 740 \text{s}^{-1} = 740 \text{ Hz},$$

we obtain as the equation for the wave

$$y = 0.25 \sin (0.39x - 740t)$$

where x and y are in centimeters and t is in seconds.

As this wave passes along the string, each particle of the string moves up and down at right angles to the direction of the wave motion. Find the velocity and acceleration of a particle 2.0 ft from the end.

EXAMPLE 2

The general form of this wave is

$$y = y_m \sin (kx - \omega t) = y_m \sin k(x - vt).$$

The v in this equation is the constant horizontal velocity of the wavetrain. What we are after now is the velocity of a particle in the string through which this wave moves; this particle velocity is neither horizontal nor constant. In fact, each particle moves vertically, that is, in the y-direction. In order to determine the particle velocity, which we shall designate by the symbol u, let us fix our attention on a particle at a particular position x — that is, x is now a constant in this equation — and ask how the particle displacement y changes with time. With x constant we obtain

$$u = \frac{\partial y}{\partial t} = -y_m \omega \cos (kx - \omega t),$$

in which the *partial derivative* $\partial y/\partial t$ reminds us that although in general y is a function of both x and t, we here assume that x remains constant so that t becomes the only variable. The acceleration a of the particle at this (constant) value of x is

$$a = \frac{\partial^2 y}{\partial t^2} = \frac{\partial u}{\partial t} = -y_m\omega^2 \sin (kx - \omega t) = -\omega^2 y.$$

This shows that for each particle through which this transverse sinusoidal wave passes we have precisely SHM (simple harmonic motion), for the acceleration a is proportional to the displacement y, but oppositely directed.

For a particle at $x = 62$ cm with the wave of Example 1, in which

$$y_m = 0.25 \text{ cm}, \qquad k = 0.39 \text{ cm}^{-1}, \qquad \omega = 740 \text{ s}^{-1},$$

we obtain
$$u = -y_m v \cos (kx - \omega t)$$

or
$$u = -0.25 \ (740) \cos [(0.39)(62) - (740)t] = -185 \cos (24 - 740 \ t)$$

and
$$a = -\omega^2 y$$

or
$$a = -(740)^2 \ 0.25 \sin [(0.39)(62) - (740)t] = -13.7 \times 10^4 \sin (24 - 740 \ t)$$

where t is expressed in seconds u in cm/s and a in cm/s^2.

Can you describe the motion of this particle at the time $t = 4$ s?

19-6
POWER AND INTENSITY IN WAVE MOTION

In Fig. 19-7 we draw an element of the stretched string at some position x and at a particular time t. The *transverse* component of the tension in the string exerted *by* the element to the left of x *on* the element of the right of x is

$$F_{\text{trans}} = -F \frac{\partial y}{\partial x}.$$

$\mathbf{F}$ is the tension in the string; $\partial y/\partial x$ gives the tangent of the angle made by the direction of $\mathbf{F}$ with the horizontal at the time t in question and, because we assume small displacements, this can be taken equal also to the sine of the angle. The transverse force is in the direction of increasing y; in the figure the slope is negative, so the transverse force is positive. The transverse velocity of the particle at x is $\partial y/\partial t$, which may be positive or negative. The power being expended by the force at x, or the energy passing through the position x per unit time in the positive x-direction (see Section 7-6), is

$$P = F_{\text{trans}} \ u = \left(-F \frac{\partial y}{\partial x}\right) \frac{\partial y}{\partial t}.$$

Suppose that the wave on the string is the simple sine wave

$$y = y_m \sin (kx - \omega t).$$

Then the magnitude of the slope at x is

$$\frac{\partial y}{\partial x} = ky_m \cos (kx - \omega t), \qquad [t = \text{constant}]$$

and the transverse force is

$$-F \frac{\partial y}{\partial x} = -Fky_m \cos (kx - \omega t).$$

The transverse velocity of a particle of the string at x is

$$u = \frac{\partial y}{\partial t} = -\omega y_m \cos (kx - \omega t), \qquad [x = \text{constant}].$$

Hence, the power transmitted through x is

$$P = (-Fky_m)(-\omega y_m) \cos^2 (kx - \omega t),$$

$$= y_m{}^2 k\omega F \cos^2 (kx - \omega t).$$

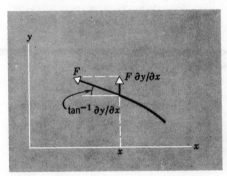

figure 19-7
The transverse component of the tension in the string at each point x is $F \ (\partial y/\partial x)$.

Notice that the power or rate of flow of energy is not constant. The power is not constant because the power input oscillates. As the energy is passed along the string, it is stored in each element of string as a combination of kinetic energy of motion and potential energy of deformation. The situation is much like that in an alternating current circuit; there energy is stored both in the inductor and in the capacitor and the power input also oscillates. For a string the power is absorbed by internal friction and viscous effects and appears as heat energy; in the circuit the power is expended in the resistor and appears as heat energy. The power input to the string or the circuit is often taken to be the *average* over one period of motion. The average power delivered is

$$\overline{P} = \frac{1}{T} \int_t^{t+T} P \, dt,$$

where T is the period. Using the fact that the average value of $\sin^2 \theta$ or $\cos^2 \theta$ over one cycle is $\frac{1}{2}$, we obtain for the string

$$\overline{P} = \tfrac{1}{2} y_m{}^2 k \omega F = 2\pi^2 y_m{}^2 \nu^2 \frac{F}{v},$$

a result which does not depend on x or t. For the string, however, $v = \sqrt{F/\mu}$, so that

$$\overline{P} = 2\pi^2 y_m{}^2 \nu^2 \mu v.$$

The fact that the rate of transfer of energy depends on the square of the wave amplitude and square of the wave frequency is true in general, holding for all types of waves.

Confirm that, if we had picked a wave traveling in the negative x-direction, we would have obtained the negative of this result. That is, the wave delivers power in the direction of wave propagation.

In a three-dimensional wave, such as a light wave or a sound wave from a point source, it is more significant to speak of the *intensity* of the wave. Intensity is defined as the power transmitted across a unit area normal to the direction in which the wave is traveling. Just as with power in the wave in a string, the intensity of a space wave is always proportional to the square of the amplitude.

As a wave progresses through space, its energy may be absorbed. For example, in a viscous medium, such as syrup or lead, mechanical waves would rapidly decay in amplitude and disappear, owing to absorption of energy by internal friction. In most cases of interest to us, however, absorption will be negligible. Throughout this chapter we have assumed that there is no loss of energy in a given wave, no matter how far it travels.

Spherical waves travel from a source of waves whose power output, assumed constant, is P; see Fig. 19-8. Find how the wave intensity depends on the distance from the source. We assume that the medium is isotropic and that the

EXAMPLE 3

figure 19-8
Example 3.

source radiates uniformly in all directions, that is, that its emission is spherically symmetrical.

The intensity of a three-dimensional wave is the power transmitted across a unit area normal to the direction of propagation. As the wavefront expands from a distance r_1 from the source at the center to a distance r_2, its surface area increases from $4\pi r_1^2$ to $4\pi r_2^2$. If there is no absorption of energy, the total energy transported per second by the wave remains constant at the value P, so that

$$P = 4\pi r_1^2 I_1 = 4\pi r_2^2 I_2,$$

$$I = \frac{P}{A}$$

where I_1 and I_2 are the wave intensities at r_1 and r_2 respectively. Hence,

$$\frac{I_1}{I_2} = \frac{r_2^2}{r_1^2}$$

and the wave intensity varies inversely as the square of its distance from the source. Since the intensity is proportional to the square of the amplitude, the amplitude of the wave must vary inversely as the distance from the source.

19-7 INTERFERENCE OF WAVES

Interference refers to the physical effects of superimposing two or more wavetrains. Let us consider two waves of equal frequency and amplitude traveling with the same speed in the same direction $(+x)$ but with a phase difference ϕ between them. The equations of the two waves will be

$$y_1 = y_m \sin (kx - \omega t - \phi) \qquad (19\text{-}14)$$

and

$$y_2 = y_m \sin (kx - \omega t). \qquad (19\text{-}15)$$

— ϕ is not needed.

We can rewrite the first equation in two equivalent forms

$$y_1 = y_m \sin \left[k\left(x - \frac{\phi}{k}\right) - \omega t \right] \qquad (19\text{-}14a)$$

or

$$y_1 = y_m \sin \left[kx - \omega\left(t + \frac{\phi}{\omega}\right) \right]. \qquad (19\text{-}14b)$$

Equations 19-14a and 19-15 suggest that if we take a "snapshot" of the two waves at any time t, we will find them displaced from one another along the x-axis by the constant distance ϕ/k. Equations 19-14b and 19-15 suggest that if we station ourselves at any position x, the two waves will give rise to two simple harmonic motions having a constant time difference ϕ/ω. This gives some insight into the meaning of the phase difference ϕ.

Now let us find the resultant wave, which, on the assumption that superposition occurs, is the sum of Eqs. 19-14 and 19-15 or

$$y = y_1 + y_2 = y_m[\sin (kx - \omega t - \phi) + \sin (kx - \omega t)].$$

From the trigonometric equation for the sum of the sines of two angles

$$\sin B + \sin C = 2 \sin \tfrac{1}{2}(B + C) \cos \tfrac{1}{2}(C - B), \qquad (19\text{-}16)$$

we obtain

$$y = y_m \left[2 \sin\left(kx - \omega t - \frac{\phi}{2}\right)\cos\frac{\phi}{2} \right],$$

$$= \left(2y_m \cos\frac{\phi}{2}\right)\sin\left(kx - \omega t - \frac{\phi}{2}\right). \qquad (19\text{-}17)$$

This resultant wave corresponds to a new wave having the same frequency but with an amplitude $2y_m \cos (\phi/2)$. If ϕ is *very small* (compared to 180°), the resultant amplitude will be nearly $2y_m$. That is, when

ϕ is very small, cos $(\phi/2) \cong$ cos $0° = 1$. When ϕ is *zero*, the two waves have the same phase everywhere. The crest of one corresponds to the crest of the other and likewise for the troughs. The waves are then said to interfere constructively. The resultant amplitude is just twice that of either wave alone. If ϕ is near 180°, on the other hand, the resultant amplitude will be nearly zero. That is, when $\phi \cong$ 180°, cos $(\phi/2) \cong$ cos 90° = 0. When ϕ is *exactly* 180°, the crest of one wave corresponds exactly to the trough of the other. The waves are then said to interfere destructively. The resultant amplitude is zero.

In Fig. 19-9*a* we show the superposition of two wavetrains almost in phase (ϕ small) and in Fig. 19-9*b* the superposition of two wavetrains almost 180° out of phase ($\phi \cong$ 180°). Notice that in these figures the algebraic sum of the ordinates of the thin (component) curves at any value of x equals the ordinate of the thick (resultant) curve. The sum of two waves can, therefore, have different values, depending on their phase relations.

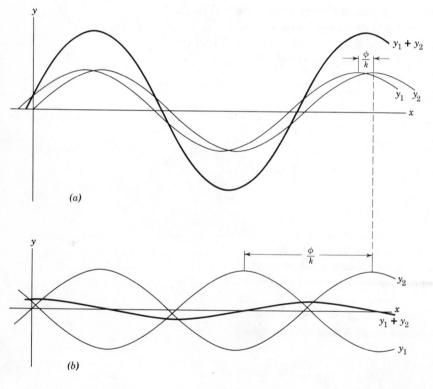

(a)

(b)

figure 19-9

(a) The superposition of two waves of equal frequency and amplitude that are almost in phase results in a wave of almost twice the amplitude of either component. *(b)* The superposition of two waves of equal frequency and amplitude and almost 180° out of phase results in a wave whose amplitude is nearly zero. Note that in both the resultant frequency is unchanged. (The drawings correspond to the instant $t = 0$.)

The resultant wave will be a sine wave, even when the amplitudes of the component sine waves are unequal. Figure 19-10, for example, illustrates the addition of two sine waves of the same frequency and velocity but different amplitudes. The resultant amplitude depends on the phase difference, which is taken as zero in this figure. The result for other phase differences could be obtained by shifting one of the component waves sideways with respect to the other and would give a smaller resultant amplitude. The smallest resultant amplitude would be the difference in the amplitudes of the components, obtained when the phases differ by 180°. However, the resultant is always a sine wave. The addition of any number of sine waves having the same frequency and velocity gives a similar result. The resultant waveform will always have a constant amplitude because the component waves (and their

figure 19-10

The addition of two waves of the same frequency and phase but differing amplitudes (light lines) yields a third wave of the same frequency and phase (heavy line).

resultant) all move with the same velocity and maintain the same relative position. The actual state of affairs can be pictured by having all the waves in Figs. 19-9 and 19-10 move toward the right with the same speed.

In practice, interference effects are obtained from wavetrains which originate in the same source (or in sources having a fixed phase relationship to one another) but which follow different paths to the point of interference. The phase difference ϕ between the waves arriving at a point can be calculated by finding the difference between the paths traversed by them from the source to the point of interference. The path difference is ϕ/k or $(\phi/2\pi)\lambda$. When the path difference is 0, λ, 2λ, 3λ, etc., so that $\phi = 0$, 2π, 4π, etc., the two waves interfere constructively. For path differences of $\frac{1}{2}\lambda$, $\frac{3}{2}\lambda$, $\frac{5}{2}\lambda$, etc., ϕ is π, 3π, 5π, etc., and the waves interfere destructively. We shall return to these matters later in more detail.

19-8
COMPLEX WAVES

The waves we have considered thus far have been of the simple harmonic type, in which the displacements at any time are represented by a sine curve. We have seen that superposition of any number of such waves having the same frequency and velocity, but arbitrary amplitudes and phases, still gives rise to a resultant wave of this simple type. If, however, we superimpose waves that have *different frequencies*, the resulting wave is *complex*. In a complex wave the motion of a particle is no longer simple harmonic motion, and the wave shape is no longer a sine curve. In this section we consider only the qualitative aspects of complex waves. The analytical treatment of such waves will be given when we encounter physical situations described by them. We will look at the results of adding graphically two or more waves traveling with the same speed in the same direction but having various relative frequencies, amplitudes, and phases.

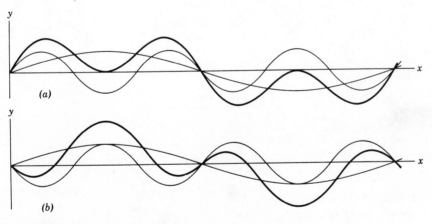

figure 19-11
The addition of two waves with a frequency ratio 3:1 (light lines) yields a wave whose shape (heavy line) depends on the phase relationship of the components. Compare *(a)* and *(b)*.

In Figs. 19-11*a* and 19-11*b* we add two waves having the same amplitude but having frequencies in the ratio 3 to 1; the phase relation is changed from *a* to *b* and we see how changing the phase relation may produce a resultant of very different form. If these represent sound waves, our eardrums will vibrate in a way represented by the resultant in each case, but we will hear and interpret these as the two original frequencies, regardless of their phase relation. If the resultant waves represent visible light, our eyes will receive the same sensation of a mixture of two colors, regardless of the phase relation of the components.

In Fig. 19-12 three waves of different frequencies and amplitudes are added. The resultant complex wave is quite different from a simple periodic wave and, in this respect resembles waveforms normally generated by musical instru-

figure 19-12
The addition of three waves (top)
of differing frequencies yields a
complex waveform (bottom).

figure 19-13
The addition (heavy line) of two
waves of widely differing frequency
(light lines).

ments. In Fig. 19-13 a wave of very high frequency is added to one of very low
frequency. Each component frequency is clearly discernible in the resultant. In
Fig. 19-14 two waves of nearly the same frequency are added. The resultant
wave consists of groups which, in the case of sound, produce the familiar phenomenon of beats (Section 20-6).

figure 19-14
The addition (bottom) of two waves
with nearly the same frequency
(top), illustrating the phenomenon
of beats (see Chapter 20).

In all of these figures the resultant wave is obtained under the assumption
that the principle of superposition holds, by simply adding the displacements
caused by the individual waves at every point. Because all the component
waves travel with the same velocity, the resultant waveform moves with this
same velocity and the wave shape is unchanged.

The cathode-ray oscilloscope (Chapter 27) gives the simplest way of observing how complex waves can be synthesized and analyzed in terms of simple harmonic waves.

19-9 STANDING WAVES

In a one-dimensional body of finite size, such as a taut string held by
two clamps a distance *l* apart, traveling waves in the string are reflected
from the boundaries of the body, that is, from the clamps. Each such
reflection gives rise to a wave traveling in the string in the opposite
direction. The reflected waves add to the incident waves according to
the principle of superposition.

Consider two wavetrains of the same frequency, speed, and amplitude which are traveling in *opposite directions* along a string. Two such

waves may be represented by the equations

$$y_1 = y_m \sin (kx - \omega t),$$

$$y_2 = y_m \sin (kx + \omega t).$$

Hence, the resultant may be written as

$$y = y_1 + y_2 = y_m \sin (kx - \omega t) + y_m \sin (kx + \omega t) \qquad (19\text{-}18a)$$

or, making use of the trigonometric relation of Eq. 19-16, as

$$y = 2y_m \sin kx \cos \omega t. \qquad (19\text{-}18b)$$

Equation 19-18b is the equation of a *standing* wave.* Notice that a particle at any particular point x executes simple harmonic motion as time goes on, and that all particles vibrate with the same frequency. In a traveling wave each particle of the string vibrates with the same amplitude. *Characteristic of a standing wave, however, is the fact that the amplitude is not the same for different particles but varies with the location x of the particle.*† In fact, the amplitude, $2y_m \sin kx$, has a *maximum* value of $2y_m$ at positions where

$$kx = \frac{\pi}{2}, \frac{3\pi}{2}, \frac{5\pi}{2}, \text{ etc.}$$

or

$$x = \frac{\lambda}{4}, \frac{3\lambda}{4}, \frac{5\lambda}{4}, \text{ etc.}$$

These points are called *antinodes* and are spaced one-half wavelength apart. The amplitude has a *minimum* value of zero at positions where

$$kx = \pi, 2\pi, 3\pi, \text{ etc.}$$

or

$$x = \frac{\lambda}{2}, \lambda, \frac{3\lambda}{2}, 2\lambda, \text{ etc.}$$

These points are called *nodes* and are spaced one-half wavelength apart. The separation between a node and an adjacent antinode is one-quarter wavelength.

It is clear that energy is not transported along the string to the right or to the left, for energy cannot flow past the nodal points in the string which are permanently at rest. Hence, the energy remains "standing" in the string, although it alternates between vibrational kinetic energy and elastic potential energy. We call the motion a wave motion because we can think of it as a superposition of waves traveling in opposite directions (Eq. 19-18a). We can equally well regard the motion as an oscillation of the string as a whole (Eq. 19-18b), each particle oscillating with SHM of angular frequency ω and with an amplitude that depends on its location. Each small part of the string has inertia and elasticity, and the string as a whole can be thought of as a collection of coupled oscillators. Hence, the vibrating string is the same in principle‡ as a spring-mass system, except that a spring-mass system has only one natural frequency, and a vibrating string has a large number of natural frequencies (Section 19-10).

---

* Standing waves may also be produced in finite bodies of two or three dimensions; see Chapters 20 and 38 for examples.

† The combining waves moving in opposite directions along the string will still produce standing waves even if their amplitudes are unequal. We consider only the equal-amplitude case here; see Problem 29, however.

‡ For a general discussion see "On the Teaching of 'Standing Waves,'" J. Rekveld, *American Journal of Physics*, March 1958.

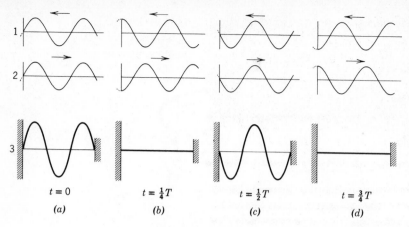

figure 19-15
Standing waves as the superposition of left- and right-going waves; 1 and 2 are the components, 3 the resultant.

In Fig. 19-15, in (a), (b), (c), and (d), we show a standing wave pattern separately at intervals of one-quarter of a period in the lower figures, 3. The traveling waves, one moving in the positive x-direction and the other moving in the negative x-direction, whose superposition can be considered to give rise to the standing wave, are shown for the same quarter-period intervals in the upper figures 2 and 1. Standing waves can also be produced with electromagnetic waves and with sound waves.

In Fig. 19-16 we show how the energy associated with the oscillating string shifts back and forth between kinetic energy of motion K and potential energy of deformation U during one cycle. Compare this with Fig. 8-4, which shows the same thing for a mass-spring oscillator. Oscillating strings often vibrate so rapidly that the eye perceives only a blur whose shape is that of the envelope of the motion; see Fig. 19-17.

figure 19-16
A standing wave in a stretched string, showing one cycle of oscillation. At (a) the string is momentarily at rest and the energy of the system is all potential energy of elastic deformation associated with the transverse displacement of the string. (b) An eighth-cycle later the displacement is reduced and the string is in motion. The two arrows show the velocities of the string particles at the positions shown. K and U have the same value. (c) The string is not displaced, but its particles have their maximum speeds; the energy is all kinetic. The motion continues until the initial condition (a) is reached after which the cycle continues to repeat itself.

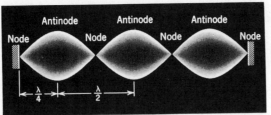

figure 19-17
The envelope of a standing wave, corresponding to a time exposure of the motion, and showing the patterns of nodes and antinodes.

The superposition of an incident wave and a reflected wave, being the sum of two waves traveling in opposite directions, will give rise to a standing wave. We shall now consider the process of reflection of a wave more closely. Suppose a pulse travels down a stretched string which is fixed at one end, as shown in Fig. 19-18a. When the pulse arrives at that end, it exerts an upward force on the support. The support is rigid, however, and does not move. By Newton's third law the support exerts an equal but oppositely directed force on the string. This reaction force generates a pulse at the support, which travels back along the string in a direction opposite to that of the incident pulse. We say that the incident pulse has been *reflected* at the fixed endpoint of the string. Notice that the reflected pulse returns with its transverse displacement reversed. If a wavetrain is incident on the fixed endpoint, a reflected wavetrain is generated at that point in the same way. The displacement of any point along the string is the sum of the displacements caused by the incident and reflected wave. Since the endpoint is fixed, these two waves must always interfere destructively at that point so as to give zero displacement there. Hence, the reflected wave is always 180° out of phase with the incident wave at a fixed boundary. We say that *on reflection from a fixed end a wave undergoes a phase change of 180°.*

Let us now consider the reflection of a pulse at a free end of a stretched string, that is, at an end that is free to move transversely. This can be achieved by attaching the end to a very light ring free to slide without friction along a transverse rod, or (see later) to a long and very much lighter string. When the pulse arrives at the free end, it exerts a force on the element of string there. This element is accelerated and its inertia carries it past the equilibrium point; it "overshoots" and exerts a reaction force on the string. This generates a pulse which travels back along the string in a direction opposite to that of the incident pulse. Once again we get reflection, but now at a free end. The free end will obviously suffer the maximum displacement of the particles on the string; an incident and a reflected wavetrain must interfere constructively at that point if we are to have a maximum there. Hence, the reflected wave is always in phase with the incident wave at that point (see Fig. 19-18b). We say that *at a free end a wave is reflected without change of phase.*

Hence, when we have a standing wave in a string, there will be a node at a fixed end (Fig. 19-18a) and an antinode at a free end (Fig. 19-18b). These ideas will be applied to sound waves and electromagnetic waves in subsequent chapters.

In the treatment just given we have assumed that there is total reflection at the boundary. In general, at a boundary there is partial reflection and partial transmission. For example, suppose that instead of being attached to a rigid wall the string is attached to another string. At the boundary joining the strings the incident wave will be partly reflected and partly transmitted. The amplitude

(a)   (b)

figure 19-18
(a) Reflection of a pulse at the fixed end of a string. The drawings are spaced uniformly in time. The phase is changed by 180° on reflection. (b) Reflection of a pulse at an end free to move in a transverse direction. (The string is attached to a ring which slides vertically without friction.) The phase is unchanged on reflection.

of the reflected wave will be less than that of the incident wave because a transmitted wave continues along the second string and carries away some of the incident energy. If the second string has a greater linear density than the first, the wave reflected back into the first will still suffer a phase shift of 180° on reflection. But because its amplitude is less than the incident wave, the boundary point will not be a node and will move. Thus a net energy transfer occurs along the first string into the second. If the second string has a smaller linear density than the first, partial reflection occurs without change of phase, but once again energy is transmitted to the second string. In practice the best way to realize a "free end" for a string is to attach it to a long and very much lighter string. The energy transmitted is negligible, and the second string serves to maintain the tension in the first one.

It is of interest to note that the transmitted wave travels with a different speed than the incident and reflected waves. The wave speed is determined by the relation $v = \sqrt{F/\mu}$; the tension is the same in both strings, but their densities are different. Hence, the wave travels more slowly in the denser string. The frequency of the transmitted wave is the same as that of the incident and reflected waves. Waves having the same frequency but traveling with different speeds have different wavelengths. Hence, from the relation $\lambda = v/\nu$ we conclude that in the denser string, where v is less, the wavelength is shorter. This phenomenon of change of wavelength as a wave passes from one medium to another will be encountered frequently in our study of light waves.

In general, whenever a system capable of oscillating is acted on by a periodic series of impulses having a frequency equal or nearly equal to one of the natural frequencies of oscillation of the system, the system is set into oscillation with a relatively large amplitude. This phenomenon is called *resonance* (see Section 15-10) and the system is said to resonate with the applied impulses.

Consider a string fixed at both ends. Oscillations or standing waves can be established in the string. The only requirement we have to satisfy is that the endpoints be nodes. There may be any number of nodes in between or none at all, so that the wavelength associated with the standing waves can take on many different values. The distance between adjacent nodes is $\lambda/2$, so that in a string of length l there must be exactly an integral number n of half wavelengths, $\lambda/2$. That is,

$$\frac{n\lambda}{2} = l$$

or
$$\lambda = \frac{2l}{n}, \qquad n = 1, 2, 3, \cdots.$$

But $\lambda = v/\nu$ and $v = \sqrt{F/\mu}$, so that the natural frequencies of oscillation of the system are

$$\nu = \frac{n}{2l} \sqrt{\frac{F}{\mu}}, \qquad n = 1, 2, 3, \cdots. \tag{19-19}$$

If the string is set vibrating and left to itself, the oscillations gradually die out. The motion is damped by dissipation of energy through the elastic supports at the ends and by the resistance of the air to the motion. We can pump energy into the system by applying a driving force. If the driving frequency is near that of any natural frequency of the string, the string will vibrate at that frequency with a large amplitude. Because the string has a large number of natural frequencies, resonance can occur at many different frequencies. A mass-spring system, by contrast, has only one resonant frequency. The difference is associated with

19-10
RESONANCE

the fact that in the mass-spring system the inertia characteristic is concentrated ("lumped") in one part of the system—the mass—and the elastic characteristic is concentrated in a separate part of the system—the spring. We say that this system has *lumped elements*.

A stretched string, on the other hand, is said to have *distributed elements* because every element of the string has both inertia and elastic characteristics. In the mass-spring system, there is only one way to exchange energy between kinetic and potential forms as the system oscillates; energy in kinetic form must be associated with the moving mass and energy in potential form must be associated with the deformed spring. In the stretched string, however, masslike (inertia) and spring-like (elasticity) elements are uniformly distributed along the string. There are many possible ways, rather than a single way, of exchanging energy between kinetic and potential forms as the system oscillates, corresponding to the sequence of allowed values for n in Eq. 19-19.

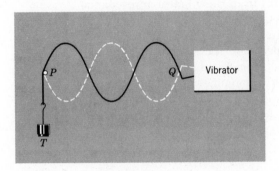

figure 19-19
Standing waves in a driven string when the natural and driving frequencies are very nearly equal.

Resonance in a string is often demonstrated by attaching a string to a fixed end, by means of a weight attached to it over a pulley, and connecting the other end to a vibrator, as shown in Fig. 19-19. The transverse oscillations of the vibrator set up a traveling wave in the string which is reflected back from the fixed end. The frequency of the waves is that of the vibrator, and the wavelength is determined by $\lambda = v/\nu$. The fixed end P is a node, but the end Q vibrates and is not. If we now vary the tension in the string by changing the hanging weight, for example, we can change the wavelength. Changing the tension changes the wave velocity, and the wavelength changes in proportion to the velocity, the frequency being constant. Whenever the wavelength becomes nearly equal to $2l/n$, where l is the length of the string, we obtain standing waves of great amplitude. The string now vibrates in one of its natural modes and resonates with the vibrator. The vibrator does work on the string to maintain these oscillations against the losses due to damping. The amplitude builds up only to the point at which the vibrator expends all its energy input against damping losses. The point Q is almost a node because the amplitude of the vibrator is small compared to that of the string.

Hence, with damping, the resonant frequency is almost, but not quite, a natural frequency of the string. One endpoint is a node, the other almost a node. In between there are points that are almost nodes, points at which the amplitude is very small. These points cannot be true nodes, for energy must flow along the string past them from the vibrator. This situation is analogous to the resonance condition for a damped harmonic oscillator with driving force, discussed in Section 15-10. There, too, the resonant frequency was almost the same as the natural frequency of the system, and the amplitude was large but not infinite.

If no damping were present, the resonant frequency would be exactly a natural frequency. Then the amplitude would build up to infinity as the energy is pumped in. In practice, the system would cease to obey Hooke's law, or the small-oscillations condition, as the amplitude becomes large and the system would break. This happens even with damping, when the damping is small or the driving force is large (as in the Tacoma Bridge disaster, Fig. 15-21).

If the frequency of the vibrator is much different from a natural frequency of the system, as given by Eq. 19-19, the wave reflected at P on returning to Q may be much out of phase with the vibrator, and it can do work on the vibrator. That is, the string can give up some energy to the vibrator as well as receive energy from it. The "standing" wave pattern is not fixed in form but wiggles about. On the average the amplitude is small and not much different from that of the vibrator. This situation is analogous to the erratic motion of a swing being pushed periodically with a frequency other than its natural one. The displacement of the swing is rather small.

Hence, the string absorbs peak energy from the vibrator at resonance. Tuning a radio is an analogous process. By tuning a dial the natural frequency of an alternating current in the receiving circuit is made equal to the frequency of the waves broadcast by the station desired. The circuit resonates with the transmitted signals and absorbs peak energy from the signal. We shall encounter resonance conditions again in sound, in electromagnetism, in optics, and in atomic and nuclear physics. In these areas, as in mechanics, the system will absorb peak energy from the source at resonance and relatively little energy off resonance.

EXAMPLE 4

In a demonstration with the apparatus just described, the vibrator has a frequency $\nu = 20$ Hz and the string has a linear density $\mu = 1.56 \times 10^{-4}$ slug/ft ($= 7.47 \times 10^{-3}$ kg/m) and a length $l = 24$ ft ($= 7.3$ m). The tension F is varied by pulling down on the end of the string over the pulley. If the demonstrator wants to show resonance, starting with one loop and then with two, three, and four loops, what force must he exert on the string?

At resonance,

$$\nu = \frac{n}{2l} \sqrt{\frac{F}{\mu}}.$$

Hence, the tension F is given by

$$F = \frac{4l^2\nu^2\mu}{n^2}.$$

For one loop, $n = 1$, so that

$F_1 = 4l^2\nu^2\mu = 4(24 \text{ ft})^2(20 \text{ s}^{-1})^2(1.56 \times 10^{-4} \text{ slug/ft}) = 144 \text{ lb } (= 640 \text{ N}).$

For two loops, $n = 2$, and

$$F_2 = \frac{4l^2\nu^2\mu}{4} = \frac{F_1}{4} = 36 \text{ lb } (= 160 \text{ N}).$$

Likewise, for three and four loops

$$F_3 = \frac{F_1}{(3)^2} = 16 \text{ lb } (= 71 \text{ N}),$$

$$F_4 = \frac{F_1}{(4)^2} = 9 \text{ lb } (= 40 \text{ N}).$$

Hence, the demonstrator gradually relaxes the tension to obtain resonance with an increasing number of loops. Although the resonant frequency is always the same under these circumstances, the speed of propagation and the wavelength at resonance decrease proportionately.

Taking damping into account, are the tensions given exactly correct?

If the tension were kept fixed, giving a definite wave speed, would we obtain more than one resonance condition by varying the frequency of the vibrator?

questions

1. How could you prove experimentally that energy is associated with a wave?

2. Energy can be transferred by particles as well as by waves. How can we experimentally distinguish between these methods of energy transfer?

3. Can a wave motion be generated in which the particles of the medium vibrate with angular simple harmonic motion? If so, explain how and describe the wave.

4. Are torsional waves transverse or longitudinal? Can they be considered as a superposition of two waves, which are either transverse or longitudinal?

5. How can one create plane waves? Spherical waves?

6. The following functions in which A is a constant are of the form $f(x \pm vt)$:

$$y = A(x - vt), \qquad y = A(x + vt)^2,$$
$$y = A \sqrt{x - vt}, \qquad y = A \ln (x + vt).$$

Explain why these functions are not useful in wave motion.

7. Can one produce on a string a wave form which has a discontinuity in slope at a point, that is, it has a sharp corner? Explain.

8. How do the amplitude and the intensity of surface water waves vary with the distance from the source?

9. The inverse square law does not apply exactly to the decrease in intensity of sounds with distance. Why not?

10. When two waves interfere, does one alter the progress of the other?

11. When waves interfere, is there a loss of energy? Explain your answer.

12. Why don't we observe interference effects between the light beams emitted from two flashlights or between the sound waves emitted by two violins.

13. As Fig. 19-15 shows, twice during a cycle the configuration of standing waves in a stretched string is a straight line, exactly what it would be if the string were not vibrating at all. Discuss from the point of view of energy conservation.

14. If two waves differ only in amplitude and are propagated in opposite directions through a medium, will they produce standing waves? Is energy transported? Are there any nodes? (See Problem 29.)

15. The partial reflection of wave energy by discontinuities in the path of transmission is usually wasteful and can be minimized by insertion of "impedance matching" devices between the sections of the path bordering on the discontinuity. For example, a megaphone helps match the air column of mouth and throat to the air outside the mouth. Give other examples and explain qualitatively how such devices minimize reflection losses (see Problem 29).

16. Consider the standing waves in a string to be a superposition of traveling waves and explain, using superposition ideas, why there are no true nodes in the resonating string of Fig. 19-19, even at the "fixed" end. (*Hint:* Consider damping effects.)

17. Standing waves in a string are demonstrated by an arrangement such as that of Fig. 19-19. The string is illuminated by a fluorescent light and the vibrator is driven by the same electric outlet that powers the light. The string exhibits a curious color variation in the transverse direction. Explain.

18. In the discussion of transverse waves in a string we have dealt only with displacements in a single plane, the x-y plane. If all displacements lie in one plane, the wave is said to be *plane polarized.* Can there be displacements in a plane other than the single plane dealt with? If so, can two differently plane-polarized waves be combined? What appearance would such a combined wave have?

19. A wave transmits energy. Does it transfer momentum? Can it transfer angular momentum? (See Question 18.) (See "Energy and Momentum Transport in String Waves" by D. W. Juenker, *American Journal of Physics,* January 1976.)

SECTION 19-3

1. The speed of electromagnetic waves in vacuum is 3.0×10^8 m/s. (a) Wavelengths in the visible part of the spectrum (light) range from about 4.0×10^{-7} m in the violet to about 7.0×10^{-7} m in the red. What is the range of frequencies of light waves? (b) The range of frequencies for shortwave radio (for example, FM radio and VHF television) is 1.5 MHz (megahertz; see Table 2, Chapter 1) to 300 MHz. What is the corresponding wavelength range? (c) X-rays are also electromagnetic. Their wavelength range extends from about 5.0 nm (nanometer; see Table 2, Chapter 1) to about 1.0×10^{-2} nm. What is the frequency range for X-rays?
Answer: (a) 400 THz (THz = terahertz; see Table 2, Chapter 1) to 800 THz. (b) 1.0 m to 200 m. (c) 6.0×10^4 THz to 3.0×10^7 THz.

2. Show that $y = y_m \sin (kx - \omega t)$ may be written in the alternative forms

$$y = y_m \sin k \, (x - vt), \qquad y = y_m \sin 2\pi \left(\frac{x}{\lambda} - \nu t\right),$$

$$y = y_m \sin \omega \left(\frac{x}{v} - t\right), \qquad y = y_m \sin 2\pi \left(\frac{x}{\lambda} - \frac{t}{T}\right).$$

3. The equation of a transverse wave traveling along a very long string is given by $y = 6.0 \sin (0.020\pi x + 4.0\pi t)$, where x and y are expressed in cm and t in seconds. Calculate (a) the amplitude, (b) the wavelength, (c) the frequency, (d) the speed, (e) the direction of propagation of the wave, and (f) the maximum transverse speed of a particle in the string.
Answer: (a) 6.0 cm. (b) 100 cm. (c) 2.0 Hz. (d) 200 cm/s. (e) negative x-direction. (f) 75 cm/s.

4. A sinusoidal wave travels along a string. If the time for a particular point to move from maximum displacement to zero displacement is 0.17 s, what are (a) the period, and (b) frequency? (c) If the wavelength is 1.4 m what is the speed of the wave?

5. A wave of frequency 500 Hz has a velocity of 350 m/s. (a) How far apart are two points 60° out of phase? (b) What is the phase difference between two displacements at a certain point at times 10^{-3} s apart?
Answer: (a) 12 cm. (b) 180°.

6. Write the equation for a wave traveling in the negative direction along the x-axis and having an amplitude 0.010 m, a frequency 550 Hz, and a speed 330 m/s.

7. (a) A continuous sinusoidal longitudinal wave is sent along a coiled spring from a vibrating source attached to it. The frequency of the source is 25 Hz, and the distance between successive rarefactions in the spring is 24 cm. Find the wave speed. (b) If the maximum longitudinal displacement of a particle in the spring is 0.30 cm and the wave moves in the $-x$-direction, write the equation for the wave. Let the source be at $x = 0$ and the displacement $x = 0$ when $t = 0$ be zero.
Answer: (a) 600 cm/s. (b) $y = 0.30 \sin (0.26x + 160t)$, with x and y in cm and t in seconds.

problems

SECTION 19-5

8. What is the speed of a transverse wave in a rope of length 2.0 m (6.6 ft) and mass 0.060 kg (0.0041 slug) under a tension of 500 N (110 lb)?

9. The linear density of a vibrating string is 1.3×10^{-4} kg/m. A transverse wave is propagating on the string and is described by the equation $y = 0.021 \sin (x + 30t)$, where x and y are measured in meters and t in seconds. What is the tension in the string? *Answer:* 0.12 N.

10. A continuous sinusoidal wave is traveling on a string with velocity 80 cm/s. The displacement of the particles of the string at $x = 10$ cm is found to vary with time according to the equation $y = 5.0 \sin (1.0 - 4.0t)$ in cm. The linear density of the string is 4.0 g/cm. (a) What is the frequency of the wave? (b) What is the wavelength of the wave? (c) Write the general equation giving the transverse displacement of the particles of the string as a function of position and time. (d) Calculate the tension in the string.

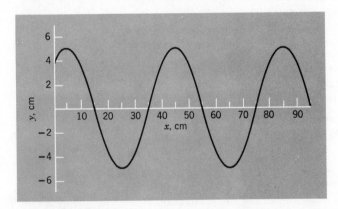

figure 19-20
Problem 11

11. A simple harmonic transverse wave is propagating along a string toward the left (or $-x$) direction. Figure 19-20 shows a plot of the displacement as a function of position at time $t = 0$. The string tension is 3.6 N and its linear density is 25 g/m. Calculate (a) the amplitude, (b) the wavelength, (c) the wave speed, (d) the period, and (e) the maximum speed of a particle in the string. (f) Write an equation describing the traveling wave.
Answer: (a) 5.0 cm. (b) 40 cm. (c) 12 m/s. (d) 0.033 s. (e) 9.4 m/s. (f) 5.0 sin $(0.16x + 190t + 0.93)$, with x and y in cm and t in seconds.

12. Prove that the slope of a string at any point x is numerically equal to the ratio of the particle speed to the wave speed at that point.

13. A uniform circular hoop of string is rotating clockwise in the absence of gravity (see Fig. 19-21). The tangential speed is v_0. Find the speed of waves traveling on this string. (*Remark:* The answer is independent of the radius of the circle and the mass per unit length of the string!) *Answer:* v_0.

14. A uniform rope of mass m and length L hangs from a ceiling. (a) Show that the speed of a transverse wave in the rope is a function of y, the distance from the lower end, and is given by $v = \sqrt{gy}$. (b) Show that the time it takes a transverse wave to travel the length of the rope is given by $t = 2\sqrt{L/g}$. (c) Does the actual mass of the rope affect the results of (a) and (b)?

figure 19-21
Problem 13

SECTION 19-6

15. Spherical waves are emitted from a 1.0-watt source in an isotropic non-absorbing medium. What is the wave intensity 1.0 m from the source? *Answer:* 0.080 W/m².

16. (a) Show that the intensity I (the energy crossing unit area per unit time) is the product of the energy per unit volume u and the speed of propagation v of a wave disturbance. (b) Radio waves travel at a speed of 3.0×10^8 m/s (9.8×10^8 ft/s). Find the energy density in a radio wave 480 km (300 mi) from a 50,000-W (67-hp) source, assuming the waves to be spherical and the propagation to be isotropic.

17. A line source emits a cylindrical expanding wave. Assuming the medium absorbs no energy, find how (a) the intensity and (b) the amplitude of the wave depend on the distance from the source.
Answer: (a) Proportional to r^{-1}. (b) Proportional to $r^{-1/2}$.

18. (a) From Example 2 show that the maximum speed of a particle in a string through which a sinusoidal wave is passing is $u = y_m\omega$. (b) In Example 2 we saw that the particles in the string oscillate with simple harmonic motion. The mechanical energy of each particle is the sum of its potential and kinetic energies and is always equal to the maximum value of its kinetic energy. Consider an element of string of mass $\mu\Delta x$ and show that the energy per unit length of the string is given by

$$E_1 = 2\pi^2\mu v^2 y_m^2.$$

(c) Show finally that the average power or average rate of transfer of energy is the product of the energy per unit length and the wave speed. (d) Do these results hold only for a sinusoidal wave?

19. A wave travels out uniformly in all directions from a point source. (a) Justify the following expression for the displacement y of the medium at any distance r from the source:

$$y = \frac{Y}{r}\,\sin\,k(r - vt).$$

Consider the speed, direction of propagation, periodicity, and intensity of the wave. (b) What are the dimensions of the constant Y?
Answer: (b) L^2.

SECTION 19-7

20. Determine the amplitude of the resultant motion when two sinusoidal motions having the same frequency and traveling in the same direction are combined, if their amplitudes are 3.0 cm and 4.0 cm and they differ in phase by $\pi/2$ rad.

21. A source S and a detector D of high-frequency waves are a distance d apart on the ground. The direct wave from S is found to be in phase at D with the wave from S that is reflected from a horizontal layer at an altitude H (Fig. 19-22). The incident and reflected rays make the same angle with the reflecting layer. When the layer rises a distance h, no signal is detected at D. Neglect absorption in the atmosphere and find the relation between d, h, H, and the wavelength λ of the waves.

Answer: $\lambda = 2\sqrt{4(H + h)^2 + d^2} - 2\sqrt{4H^2 + d^2}$.

figure 19-22
Problem 21

22. Four component sine waves have frequencies in the ratio 1, 2, 3, and 4 and amplitudes in the ratio 1, $\frac{1}{2}$, $\frac{1}{3}$, and $\frac{1}{4}$, respectively. When $t = 0$, at $x = 0$, the first and third components are 180° out of phase with the second and fourth components. Plot the resultant waveform and discuss its nature.

figure 19-23
Problem 23

23. Two pulses are traveling along a string in opposite directions, as shown in Fig. 19-23. (a) If the wave velocity is 2.0 m/s and the pulses are 6.0 cm apart, sketch the patterns after 5.0, 10, 15, 20, 25 ms. (b) What has happened to the energy at $t = 15$ ms?
Answer: (b) Even though the displacement is zero at this instant, the transverse velocities are not. The energy is all kinetic.

24. Three component sinusoidal waves have the same period, but their amplitudes are in the ratio 1, $\frac{1}{2}$, and $\frac{1}{3}$ and their phase angles are 0, $\pi/2$, and π respectively. Plot the resultant waveform and discuss its nature.

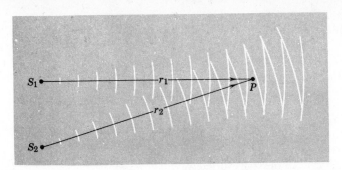

figure 19-24
Problem 25

25. Consider two point sources S_1 and S_2 in Fig. 19-24 which emit waves of the same frequency and amplitude. The waves start in the same phase, and this phase relation at the sources is maintained throughout time. Consider points P at which r_1 is nearly equal to r_2. (a) Show that the superposition of these two waves gives a wave whose amplitude varies with the position P approximately according to

$$\frac{2Y}{r} \cos \frac{k}{2} (r_1 - r_2),$$

in which $r = (r_1 + r_2)/2$. (b) Then show that total annulment occurs when $r_1 - r_2 = (n + \frac{1}{2})\lambda$, n being any integer, and that total re-enforcement occurs when $r_1 - r_2 = n\lambda$.

The locus of points whose difference in distance from two fixed points is a constant is a hyperbola, the fixed points being the foci. Hence each value of n gives a hyperbolic line of constructive interference and a hyperbolic line of destructive interference. At points at which r_1 and r_2 are not approximately equal (as near the sources), the amplitudes of the waves from S_1 and S_2 differ and the annulments are only partial.

SECTION 19-9

26. The equation of a transverse wave traveling in a string is given by

$$y = 10 \cos (0.0079x - 13t - 0.89),$$

in which x and y are expressed in centimeters and t in seconds. Write down the equation of a wave which, when added to the given one, would produce standing waves on the rope.

27. A string vibrates according to the equation

$$y = 0.5 \sin \frac{\pi x}{3} \cos 40\pi t,$$

where x and y are in centimeters and t is in seconds. (a) What are the amplitude and velocity of the component waves whose superposition can give rise to this vibration? (b) What is the distance between nodes? (c) What is the velocity of a particle of the string at the position $x = 1.5$ cm when $t = \frac{9}{8}$ s? *Answer:* (a) 0.25 cm, 120 cm/s. (b) 3.0 cm. (c) Zero.

28. Two transverse sinusoidal waves travel in opposite directions along a string. Each has an amplitude of 0.30 cm and a wavelength of 6.0 cm. The speed of a transverse wave in the string is 1.5 m/s. Plot the shape of the string at each of the following times: $t = 0$ (arbitrary), $t = 5.0$, $t = 10.0$, $t = 15.0$, $t = 20.0$ ms.

29. If an incident traveling wave is only partially reflected from a boundary, the resulting superposition of two waves having different amplitudes and traveling in opposite directions gives a standing wave pattern of waves whose envelope is shown in Fig. 19-25. The standing wave ratio (SWR) is defined as $(A_i + A_r)/(A_i - A_r) = A_{max}/A_{min}$, and the percent reflection is defined as the ratio of the average power in the reflected wave to the average power in the incident wave, times 100. (a) Show that for 100% reflection SWR = ∞ and that for no reflection SWR = 1. (b) Show that a measurement

A_{max} A_{min} A_{max}

figure 19-25
Problem 29

of the SWR just before the boundary reveals the percent reflection occurring at the boundary according to the formula

$$\% \text{ reflection} = [(SWR - 1)^2/(SWR + 1)^2] \times 100.$$

30. Two strings of linear density μ_1 and μ_2 are knotted together at $x = 0$ and stretched to a tension F. A wave $y = A \sin k_1(x - v_1 t)$ in the string of density μ_1 reaches the junction between the two strings, at which it is partly transmitted into the string of density μ_2 and partly reflected. Call these waves $B \sin k_2(x - v_2 t)$ and $C \sin k_1(x + v_1 t)$, respectively. (a) Assuming that $k_2 v_2 = k_1 v_1 = \omega$ and that the displacement of the knot arising from the incident and reflected waves is the same as that arising from the transmitted wave, show that $A = B + C$. (b) If it is assumed that both strings near the knot have the same slope (why?), i.e., that dy/dx in string $1 = dy/dx$ in string 2, show that

$$C = A\frac{(k_2 - k_1)}{(k_2 + k_1)}$$

$$= A\frac{v_1 - v_2}{v_1 + v_2}.$$

Under what conditions is C negative?

31. Consider a standing wave that is the sum of two waves traveling in opposite directions but otherwise identical. Show that the energy in each loop of the standing wave is $2\pi^2 y_m^2 \nu v$.

SECTION 19-10

32. In a laboratory experiment on standing waves a string 3.0 ft (0.9 m) long is attached to the prong of an electrically driven tuning fork which vibrates perpendicular to the length of the string at a frequency of 60 vib/s (60 Hz). The weight of the string is 0.096 lb (mass = 0.044 kg). (a) What tension must the string be under (weights are attached to the other end) if it is to vibrate in four loops? (b) What would happen if the tuning fork is turned so as to vibrate parallel to the length of the string?

33. Vibrations from a 600-cycle/s tuning fork set up standing waves in a string clamped at both ends. The wave speed for the string is 400 m/s. The standing wave has four loops and an amplitude of 2.0 mm. (a) What is the length of the string? (b) Write an equation for the displacement of the string as a function of position and time.
    Answer: (a) 1.3 m. (b) $2.0 \times 10^{-3} \sin 9.4x \cos 3800t$, where x and y are in meters and t in seconds.

34. An aluminum wire of length $l_1 = 60.0$ cm and of cross-sectional area 1.00×10^{-2} cm^2 is connected to a steel wire of the same cross-sectional area. The compound wire, loaded with a block m of mass 10.0 kg, is arranged as shown in Fig. 19-26 so that the distance l_2 from the joint to the supporting pulley is 86.6 cm. Transverse waves are set up in the wire by using an external source of variable frequency. (a) Find the lowest frequency of excitation for which standing waves are observed such that the joint in the wire is a node. (b) What is the total number of nodes observed at this frequency, excluding the two at the ends of the wire? The density of aluminum is 2.60 g/cm^3, and that of steel is 7.80 g/cm^3.

figure 19-26
Problem 34

20
sound
waves

Sound waves are longitudinal mechanical waves. They can be propagated in solids, liquids, and gases. The material particles transmitting such a wave oscillate in the direction of propagation of the wave itself. There is a large range of frequencies within which longitudinal mechanical waves can be generated, sound waves being confined to the frequency range which can stimulate the human ear and brain to the sensation of hearing. This range is from about 20 cycles/sec (or 20 Hz) to about 20,000 Hz and is called the *audible* range. A longitudinal mechanical wave whose frequency is below the audible range is called an *infrasonic* wave, and one whose frequency is above the audible range is called an *ultrasonic* wave.

Infrasonic waves of interest are usually generated by large sources, earthquake waves being an example.* The high frequencies associated with ultrasonic waves† may be produced by elastic vibrations of a quartz crystal induced by resonance with an applied alternating electric field (piezoelectric effect). It is possible to produce ultrasonic frequencies as high as 6×10^8 Hz in this way; the corresponding wavelength in air is about 5×10^{-5} cm, the same as the length of visible light waves.

Audible waves originate in vibrating strings (violin, human vocal cords), vibrating air columns (organ, clarinet), and vibrating plates and membranes (xylophone, loudspeaker, drum). All of these vibrating elements alternately compress the surrounding air on a forward movement and rarefy it on a backward movement. The air transmits these dis-

20-1
AUDIBLE, ULTRASONIC, AND INFRASONIC WAVES

* See "Long Earthquake Waves," by Jack Oliver, *Scientific American*, March 1959.
† See "Applications of Ultrasonics" by Margaret F. Cracknell and Arthur P. Cracknell, *Contemporary Physics*, January 1976.

turbances outward from the source as a wave. Upon entering the ear, these waves produce the sensation of sound. Waveforms which are approximately periodic or consist of a small number of approximately periodic components give rise to a pleasant sensation (if the intensity is not too high), as, for example, musical sounds.* Sound whose waveform is nonperiodic is heard as noise. Noise can be represented as a superposition of periodic waves, but the number of components is very large.

In this chapter we deal with the properties of longitudinal mechanical waves, using sound waves as the prototype.

20-2
PROPAGATION AND SPEED OF LONGITUDINAL WAVES

Sound waves, if unimpeded, will spread out in all directions from a source. It is simpler to deal with one-dimensional propagation, however, than with three-dimensional propagation, so that we consider first the transmission of longitudinal waves in a tube.

Figure 20-1 shows a piston at one end of a long tube filled with a compressible medium. The vertical lines divide the compressional (fluid) medium into thin "slices," each of which contains the same mass of fluid. Where the lines are relatively close together the fluid pressure and density are greater than they are in the normal undisturbed fluid, and conversely. We shall treat the fluid as a continuous medium and ignore for the time being the fact that it is made up of molecules that are in continual random motion.

If we push the piston of Fig. 20-1 forward, the fluid in front of it is compressed, the fluid pressure and density rising above their normal undisturbed values. The compressed fluid moves forward, compressing the fluid layers next to it, and a compressional pulse travels down the tube. If we then withdraw the piston, the fluid in front of it expands, its pressure and density falling below their normal undisturbed values; a pulse of rarefaction travels down the tube. These pulses are similar to transverse pulses traveling along a string, except that the oscillating fluid elements are displaced along the direction of propagation (longitudinal) instead of at right angles to this direction (transverse). If the piston oscillates back and forth, a continuous train of compressions and rarefactions will travel along the tube (Fig. 20-1). As for transverse waves in a string (see Section 19-5) we should be able, using Newton's laws of motion, to express the speed of propagation of this longitudinal wave in terms of an elastic and an inertial property of the medium. We now do so.

figure 20-1
Sound waves generated in a tube by an oscillating piston. The vertical lines divide the compressible medium in the tube into layers of equal mass.

For the moment, let us assume that the tube is very long so that we can ignore reflections from the far end. As for the string of Fig. 19-6, we will consider not an extended wave but a single (compressional) pulse that we might generate by giving the piston in Fig. 20-1 a short, rapid, inward stroke.

Figure 20-2 shows such a pulse (labeled "compressional zone") traveling at speed v along the tube from left to right. For simplicity we have assumed this pulse to have sharply defined leading and trailing edges and to have a uniform fluid pressure and density in its interior. When we analyzed the motion of a transverse pulse in a string, we found it convenient to choose a reference frame in which the pulse remained stationary; we will do this here also. In Fig. 20-2, then, the compressional zone remains stationary in our reference frame while the fluid moves through it from right to left with speed v, as shown.

Let us follow the motion of the element of fluid contained between the ver-

---

* A good general reference on the scientific properties of musical sound is 'The Acoustical Foundations of Music' by John Backus, W. W. Norton & Co., Inc., New York, 1969.

figure 20-2
A compressional pulse travels along a gas-filled tube. In a reference frame in which the undisturbed gas is at rest the pulse moves from left to right with speed v. We view the pulse, however, from a reference frame in which the pulse is stationary; in such a frame the gas outside the pulse streams through the tube from right to left with speed v, as shown. Note that Δv is negative.

tical lines at P in Fig. 20-2. This element moves forward at speed v until it strikes the compressional zone. While it is entering this zone it encounters a difference of pressure Δp between its leading and its trailing edges. The element is compressed and *decelerated*, moving with a lower speed $v + \Delta v$ within the zone, the quantity Δv being negative. The element eventually emerges from the left face of the zone where it expands to its original volume and the pressure differential Δp acts to *accelerate* it to its original speed v. The figure shows the element at point R, having passed through the compressional zone and moving again with speed v, as at P.

Let us apply Newton's laws to the fluid element while it is entering the compressional zone. The resultant force acting during entry points to the right in Fig. 20-2 and has magnitude

$$F = (p + \Delta p)A - pA = \Delta pA$$

in which A is the cross-sectional area of the tube.

The length of the element outside the compressional zone (at P, say) is $v\,\Delta t$, where Δt is the time required for the element to move past any given point. The volume of the element is thus $vA\,\Delta t$ and its mass is $\rho_0 vA\,\Delta t$, where ρ_0 is the density of the fluid outside the compressional zone. The deceleration a experienced by the element as it enters the zone is $-\Delta v/\Delta t$; because Δv is inherently negative, a is positive, which means that, like the force ΔpA in Fig. 20-2, it points to the right. Thus Newton's second law

$$F = ma$$

yields

$$\Delta pA = (\rho_0 vA\,\Delta t)\frac{-\Delta v}{\Delta t},$$

which we may write as

$$\rho_0 v^2 = \frac{-\Delta p}{\Delta v/v}.$$

Now the fluid that would occupy a volume $V = Av\,\Delta t$ at P is compressed by an amount $A(\Delta v)\,\Delta t = \Delta V$ on entering the compressional zone. Hence,

$$\frac{\Delta V}{V} = \frac{A\,\Delta v\,\Delta t}{Av\,\Delta t} = \frac{\Delta v}{v}$$

and we obtain

$$\rho_0 v^2 = \frac{-\Delta p}{\Delta V/V}.$$

The ratio of the change in pressure on a body, Δp, to the fractional change in volume resulting, $-\Delta V/V$, is called the *bulk modulus of elasticity B* of the body. That is, $B = -V\,\Delta p/\Delta V$. B is positive because an increase in pressure causes a decrease in volume. In terms of B, the speed of the longitudinal pulse in the medium of Fig. 20-2 is

$$v = \sqrt{B/\rho_0}. \qquad (20\text{-}1)$$

A more extended analysis than given above shows that Eq. 20-1 applies not only to rectangular pulses of the type displayed in Fig. 20-2 but also to pulses of any shape and to extended wave trains. Notice that the speed of the wave is determined by the properties of the medium through which it propagates, and that an elastic property B and an inertial property ρ_0 are involved. Table 20-1 gives the speed of longitudinal (sound) waves in various media.

Table 20-1
Speed of sound

Medium	Tempera-ture, °C	Speed m/s	Speed ft/s
Air	0	331.3	1,087
Hydrogen	0	1,286	4,220
Oxygen	0	317.2	1,041
Water	15	1,450	4,760
Lead	20	1,230	4,030
Aluminum	20	5,100	16,700
Copper	20	3,560	11,700
Iron	20	5,130	16,800
Extreme values			
Granite		6,000	19,700
Vulcanized rubber	0	54	177

If the medium is a gas, such as air, it is possible to express B in terms of the undisturbed gas pressure p_0. For a sound wave in a gas we obtain

$$v = \sqrt{\gamma p_0/\rho_0},$$

where γ is a constant called the ratio of specific heats for the gas (Chapter 23).

If the medium is a solid, for a thin rod the bulk modulus is replaced by a stretch modulus (called Young's modulus). If the solid is extended, we must allow for the fact that, unlike a fluid, a solid offers elastic resistance to tangential or shearing forces and the speed of longitudinal waves will depend on the shear modulus as well as the bulk modulus.

20-3
TRAVELING LONGITUDINAL WAVES

Consider again the continuous train of compressions and rarefactions traveling down the tube of Fig. 20-1. As the wave advances along the tube, each small volume element of fluid oscillates about its equilibrium position. The displacement is to the right or left along the x-direction of propagation of the wave. For convenience let us represent the displacement of any such volume element (or layers of elements that move in the same way) from its equilibrium position at x by the letter y. It is to be understood that the displacement y is *along the direction of propagation* for a longitudinal wave, whereas for a transverse wave the displacement y is *at right angles to the direction of propagation*. Then the equation of a longitudinal wave traveling to the right may be written as

$$y = f(x - vt).$$

For the particular case of a simple harmonic oscillation we may have

$$y = y_m \cos \frac{2\pi}{\lambda} (x - vt).$$

In this equation v is the speed of the longitudinal wave, y_m is its amplitude, and λ is its wavelength; y gives the displacement of a particle at time t from its equilibrium position at x. As before, we may write this more compactly as

$$y = y_m \cos{(kx - \omega t)}. \qquad (20\text{-}2)$$

It is usually more convenient to deal with pressure variations in a sound wave than with the actual displacements of the particles conveying the wave. Let us therefore write the equation of the wave in terms of the pressure variation rather than in terms of the displacement.

From the relation

$$B = -\frac{\Delta p}{\Delta V/V},$$

we have

$$\Delta p = -B\frac{\Delta V}{V}.$$

Just as we let y represent the displacement from the equilibrium position x, so we now let p represent the *change* from the undisturbed pressure p_0. Then p replaces Δp, and

$$p = -B\frac{\Delta V}{V}.$$

If a layer of fluid at pressure p_0 has a thickness Δx and cross-sectional area A, its volume is $V = A\,\Delta x$. When the pressure changes, its volume will change by $A\,\Delta y$, where Δy is the amount by which the thickness of the layer changes during compression or rarefaction. Hence,

$$p = -B\frac{\Delta V}{V} = -B\frac{A\,\Delta y}{A\,\Delta x}.$$

As we let $\Delta x \rightarrow 0$ so as to shrink the fluid layer to infinitesimal thickness, we obtain

$$p = -B\frac{\partial y}{\partial x}. \qquad (20\text{-}3)$$

We have used partial derivative notation because (see Eq. 20-2) y is a function of both x and t and we take the latter quantity as constant in this discussion. If the particle displacement is simple harmonic, then, from Eq. 20-2, we obtain

$$\frac{\partial y}{\partial x} = -ky_m \sin{(kx - \omega t)},$$

and from Eq. 20-3

$$p = Bky_m \sin{(kx - \omega t)}. \qquad (20\text{-}4)$$

Hence, the pressure variation at each position x is also simple harmonic. Because $v = \sqrt{B/\rho_0}$, we can write Eq. 20-4 more conveniently as

$$p = [k\rho_0 v^2 y_m] \sin{(kx - \omega t)}.$$

Recall that p represents the change from standard pressure p_0. The term in brackets represents the maximum change in pressure and is called the *pressure amplitude*. If we denote this by P, then

$$p = P \sin{(kx - \omega t)}, \qquad (20\text{-}5)$$

where

$$P = k\rho_0 v^2 y_m. \qquad (20\text{-}6)$$

Hence, a sound wave may be considered either as a displacement wave or as a pressure wave. If the former is written as a cosine function, the latter will be a sine function and vice versa. The displacement wave is thus 90° out of phase with the pressure wave. That is, when the displacement from equilibrium at a point is a maximum or a minimum, the excess pressure there is zero; when the displacement at a point is zero, the excess or deficiency of pressure there is a maximum. Equation 20-6 gives the relation between the pressure amplitude (maximum variation of pressure from equilibrium) and the displacement amplitude (maximum variation of position from equilibrium). You should check the dimensions of each side of Eq. 20-6 for consistency. What units may the pressure amplitude have?

The intensity of a wave is proportional to the square of the displacement amplitude of the wave; see Section 19-6. We have just shown that for sound waves the pressure amplitude is proportional to the displacement amplitude. Hence, the intensity of a sound wave is proportional to the square of the pressure amplitude. In fact, when the intensity is expressed in terms of the pressure amplitude, the frequency does not appear explicitly in the expression (see Problem 14). Hence, by measuring pressure changes, the intensities of sounds having *different* frequencies can be compared directly. For this reason instruments that measure pressure changes are preferred to those that measure displacement amplitude. As we shall see in Example 1, the displacement amplitudes would be difficult to measure in any case.

EXAMPLE 1

(a) The maximum pressure variation P that the ear can tolerate in loud sounds is about 28 N/m² (= 28 Pa). Normal atmospheric pressure is about 100,000 Pa. Find the corresponding maximum displacement for a sound wave in air having a frequency of 1000 Hz.

From Eq. 20-6 we have

$$y_m = \frac{P}{k\rho_0 v^2}.$$

From Table 20-1, $v = 331$ m/s so that

$$k = \frac{2\pi}{\lambda} = \frac{2\pi\nu}{v} = \frac{2\pi \times 10^3}{331} \text{ m}^{-1} = 19 \text{ m}^{-1}.$$

The density of air ρ_0 is 1.22 kg/m³. Hence, for $P = 28$ Pa we obtain

$$y_m = \frac{28}{(19)(1.22)(331)^2} \text{ m} = 1.1 \times 10^{-5} \text{ m}.$$

The displacement amplitudes for the *loudest* sounds are about 10^{-5} m, a very small value indeed.

(b) In the faintest sound that can be heard at 1000 Hz the pressure amplitude is about 2.0×10^{-5} Pa. Find the corresponding displacement amplitude.

From $y_m = P/k\rho_0 v^2$, using these values for k, v, and ρ_0, we obtain, with $P = 2.0 \times 10^{-5}$ N/m²,

$$y_m \cong 8 \times 10^{-12} \text{ m} \cong 10^{-11} \text{ m}.$$

This is smaller than the radius of an atom, which is about 10^{-10} m! How can it be that the ear responds to such a small displacement?

In our analysis we have ignored the molecular structure of matter and treated the fluid as a continuous medium. In gases, however, the spaces between molecules are large compared to the diameters of the molecules. The molecules move about at random. The oscillations produced by a sound wave passing through are superimposed on this random thermal motion. An impulse given to one molecule is passed on to another molecule only after the first one has moved through the empty space between them and collided with the second. From this brief discussion, would you ever expect the speed of sound to exceed the average molecular speed in a fluid?

Longitudinal waves traveling along a gas-filled tube are reflected at the ends of the tube, just as transverse waves in a stretched string are reflected at its ends. Interference between the waves traveling in opposite directions gives rise to standing longitudinal waves.

If the end of the tube is closed, the reflected wave is 180° out of phase with the incident wave. This result is a necessary consequence of the fact that the displacement of the small volume elements at a closed end must always be zero. Hence, a closed end is a displacement *node*. If the end of the tube is open, the fluid elements there are free to move. However, the nature of the reflection there depends on whether the tube is wide or narrow compared to the wavelength. If the tube is narrow compared to the wavelength, as in most musical instruments, the reflected wave has nearly the same phase as the incident wave. Then the open end is almost a displacement *antinode*. The exact antinode is usually somewhere near the opening, but the effective length of the air columns of a wind instrument, for example, is not as definite as the length of a string fixed at both ends.

Standing longitudinal waves in a gas column can be dramatically demonstrated by means of the apparatus shown in Fig. 20-3. A source of longitudinal waves, such as the speaker of an audio oscillator at S, sets up vibrations in a flexible diaphragm at one end of the tube. Gas fills the tube from the inlet and passes slowly out through regularly spaced small openings along the top. The escaping gas is lit, giving a series of flames. The frequency of the audio oscillator is varied and when a frequency is found at which the gas column is in resonance, the amplitude of the standing longitudinal waves becomes rather large; then we can see a wavelike variation in the height and width of the gas flames along the tube. The interval between nodes or antinodes is clearly visible. By continuing to vary the frequency we can pass from one resonance condition to another. The natural modes of oscillation of the gas column are determined by the effective length of the column and the wave speed. The wavelength λ at resonance can be taken to be twice the distance between adjacent nodes (or antinodes), and knowing the frequency ν of the source at resonance, we can determine the wave speed in the gas under these conditions from $v = \nu\lambda$. In practice there are more flexible and accurate ways to measure the speed of sound in gases. (See Problem 21 and Example 2.)

In Fig. 20-3 the nodes and antinodes, N and A, refer to the particle *displacements* in the standing wave. At a displacement node, the pressure variations (above and below the average) are a maximum. Hence,

20-4
STANDING LONGITUDINAL WAVES

figure 20-3
Flames show the presence of standing waves in a tube filled with illuminating gas. A and N refer to *displacement* antinodes and nodes, respectively.

a displacement node corresponds to a pressure antinode. At a displacement antinode the pressure remains constant with time. Hence, a displacement antinode corresponds to a pressure node.

This can be understood physically by realizing that two small volume elements of gas on opposite sides of a displacement node are vibrating in *opposite phase*. Hence, when they approach each other, the pressure at this node is increasing, and when they recede from each other, the pressure at this node is decreasing. Two small elements of gas which are on opposite sides of a displacement antinode vibrate *in phase* and therefore give rise to no pressure variations at the antinode.

20-5
VIBRATING SYSTEMS AND SOURCES OF SOUND

If a string fixed at both ends is bowed, transverse vibrations travel along the string; these disturbances are reflected at the fixed ends, and a standing wave pattern is formed. The natural modes of vibration of the string are excited and these vibrations give rise to longitudinal waves in the surrounding air which transmits them to our ears as a musical sound. We have seen (Section 19-10) that a string of length l, fixed at both ends, can resonate at frequencies given by

$$\nu_n = \frac{n}{2l} v = \frac{n}{2l} \sqrt{\frac{F}{\mu}}, \qquad n = 1, 2, 3, \ldots. \qquad (20\text{-}7)$$

Here v is the speed of the transverse waves in the string whose superposition can be thought of as giving rise to the vibrations; the speed $v \ (= \sqrt{F/\mu})$ is the same for all frequencies. At any one of these frequencies the string will contain a whole number n of loops between its ends, and the condition that the ends be nodes is met (Fig. 20-4).

$$\nu_1 = \frac{1}{2}\frac{v}{l} \qquad \lambda_1 = 2l$$
$$\nu_2 = \frac{v}{l} \qquad \lambda_2 = l$$
$$\nu_3 = \frac{3}{2}\frac{v}{l} \qquad \lambda_3 = \frac{2}{3}l$$
$$\nu_4 = 2\frac{v}{l} \qquad \lambda_4 = \frac{1}{2}l$$

figure 20-4
The first four modes of vibration of a string fixed at both ends. Note that $\nu_n\lambda_n = v = \sqrt{F/\mu}$.

The lowest frequency, $\sqrt{F/\mu}/2l$, is called the *fundamental* frequency ν_1 and the others are called *overtones*. Overtones whose frequencies are integral multiples of the fundamental are said to form a harmonic series. The fundamental is the first harmonic. The frequency $2\nu_1$ is the first overtone or the second harmonic, the frequency $3\nu_1$ is the second overtone or the third harmonic, and so on.

If the string is initially distorted so that its shape is the same as *any one* of the possible harmonics, it will vibrate at the frequency of that particular harmonic, when released. The initial conditions usually arise from striking or bowing the string, however, and in such cases not only the fundamental but many of the overtones are present in the resulting vibration. We have a superposition of several natural modes of oscillation. The actual displacement is the sum of the several harmonics with various amplitudes; see Fig. 19-12. The impulses that are sent through the air to the ear and brain give rise to one net effect which is characteristic of the particular stringed instrument. The quality of the sound of a particular note (fundamental frequency) played by an instrument is determined by the number of overtones present and their respective intensities. Figure 20-5 shows the sound spectra and corresponding waveforms for the violin and piano.*

An organ pipe is a simple example of sound originating in a vibrating air column. If both ends of a pipe are open and a stream of air is directed against an edge, standing longitudinal waves can be set up in the tube. The air column will then resonate at its natural frequencies of vibration, given by

$$\nu_n = \frac{n}{2l} v, \qquad n = 1, 2, 3, \ldots.$$

Here v is the speed of the longitudinal waves in the column whose superposition can be thought of as giving rise to the vibrations, and n is the number of half wavelengths in the length l of the column. As with the bowed string, the fundamental and overtones are excited at the same time.

In an open pipe the fundamental frequency corresponds (approximately) to a displacement antinode at each end and a displacement node in the middle, as shown in Fig. 20-6a. The succeeding drawings of Fig. 20-6a show three of the overtones, the second, third, and fourth harmonics. Hence, in an open pipe the fundamental frequency is $v/2l$ and *all* harmonics are present.

In a closed pipe the closed end is a displacement node. Figure 20-6b shows the modes of vibration of a closed pipe. The fundamental frequency is $v/4l$ (approximately), which is one-half that of an open pipe of the same length. The only overtones present are those that give a

figure 20-5
Waveform and sound spectrum for two stringed instruments, the violin and the piano. The fundamental frequency in both cases is 440 cycles/sec (concert A). In each diagram we show only four cycles of the wave. The sound spectrum shows the relative amplitude of the various harmonic components of the wave. Notice the presence of loud higher harmonics (especially the fifth) in the violin spectrum.

(a)

(b)

figure 20-6
(a) The first four modes of an open organ pipe. The distance from the center line of the pipe to the light lines drawn inside the pipe shows the displacement amplitude at each place. *N* and *A* mark the locations of the displacement nodes and antinodes. Note that *both* ends of the pipe are open. *(b)* The first four modes of vibration of a closed organ pipe. Notice that the even-numbered harmonics are absent and the upper end of the pipe is closed.

---

* See "The Physics of the Piano" by E. Donnell Blackham in *Scientific American*, December 1965 and "The Physics of the Violin" by Carleen M. Hutchins in *Scientific American*, November 1962.

displacement node at the closed end and an antinode (approximately) at the open end. Hence, as is shown in Fig. 20-6b, the second, fourth, etc., harmonics are missing. In a closed pipe the fundamental frequency is $v/4l$, and only the *odd* harmonics are present. The quality of the sounds from an open pipe is therefore different from that from a closed pipe.

Vibrating rods, plates, and stretched membranes also give rise to sound waves. Consider a stretched flexible membrane, such as a drumhead. If it is struck a blow, a two-dimensional pulse travels outward from the struck point and is reflected again and again at the boundary of the membrane. If some point of the membrane is forced to vibrate periodically, continuous trains of waves travel out along the membrane. Just as in the one-dimensional case of the string, so here too standing waves can be set up in the two-dimensional membrane. Each of these standing waves has a certain frequency natural to (or characteristic of) the membrane. Again the lowest frequency is called the fundamental and the others are overtones. Generally, a number of overtones are present along with the fundamental when the membrane is vibrating. These vibrations may excite sound waves of the same frequency.

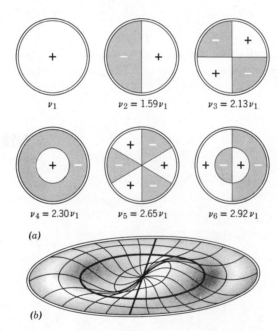

v_1 $v_2 = 1.59v_1$ $v_3 = 2.13v_1$

$v_4 = 2.30v_1$ $v_5 = 2.65v_1$ $v_6 = 2.92v_1$

(a)

(b)

figure 20-7
(a) The first six modes of vibration of a circular drumhead clamped around its periphery. The lines represent nodes, the circumference being a node in every case. The $+$ and $-$ signs represent opposite displacements; at an instant when the $+$ areas are raised, the $-$ areas will be depressed. Note that the frequency of each mode is not an integral multiple of the fundamental v_1 as is the case for strings and tubes. *(b)* A sketch of a drum-head vibrating in mode v_6. The displacement shown here is exaggerated for clarity.

The nodes of a vibrating membrane are lines rather than points (as in a vibrating string) or planes (as in a pipe). Since the boundary of the membrane is fixed, it must be a nodal line. For a circular membrane fixed at its edge, possible modes of vibration together with their nodal lines are shown in Fig. 20-7. The natural frequency of each mode is given in terms of the fundamental v_1. Notice that the frequencies of the overtones are *not* harmonics, that is, they are not integral multiples of v_1. Vibrating rods also have a nonharmonic set of natural frequencies. Rods and plates have limited use as musical instruments for this reason.

In general, we find that all elastic bodies will vibrate freely with a definite set of frequencies for a given set of boundary or end conditions. These frequencies are called proper frequencies, characteristic frequencies, or *eigenfrequencies**

---

* *Eigen* — from the German — meaning *own, individual, characteristic.*

of the system. In general, the eigenfrequencies do *not* form a harmonic series, although some of them may be related as the ratio of whole numbers. In all these cases we have standing waves, and certain regions of the bodies stay at rest all the time. These nodes are curves in two-dimensional bodies and surfaces in three-dimensional bodies.

Recall that for a vibrating string the equation describing a standing wave (see Eq. 19-18b) is of the type

$$y = 2y_m \cos 2\pi\nu t \, \sin \frac{2\pi x}{\lambda}.$$

This holds for a string fixed at both ends ($y = 0$ at $x = 0$ and $x = n\lambda/2$). The picture of the string at any time is determined by the equation

$$y = C \sin \frac{2\pi x}{\lambda} = C \sin \frac{n\pi x}{l} \qquad (t = \text{constant}),$$

where C is a constant "scale factor," whose value varies with time; l is the length of the string, and n is an integer specifying the mode of vibration (the harmonic). This function $\sin 2\pi x/\lambda$ fixes the position of the nodes and is called the proper function, characteristic function, or *eigenfunction* of the string.

Likewise, the nodes of *any* vibrating elastic body are fixed by certain functions of position which are called the eigenfunctions of the problem. In general, these functions are *not* sinusoidal functions but are functions that become zero for certain values of the coordinates. The determination of these functions and the corresponding values of the eigenfrequencies is an important problem in atomic, nuclear, and solid-state physics. They characterize the behavior of such systems. It is in quantum mechanics that the procedure has been successfully worked out for microscopic systems. The results, however, bear a striking analogy to the results of classical vibration and wave theory, as applied to macroscopic systems.

EXAMPLE 2

Figure 20-8 shows a simple apparatus that can be used to measure the speed of sound in air by resonance methods. A vibrating tuning fork of frequency ν is held near the open end of a tube. The tube is partly filled with water. The length of the air column can be varied by changing the water level. It is found that the sound intensity is a maximum when the water level is gradually lowered from the top of the tube a distance a. Thereafter, the intensity reaches a maximum again at distances d, $2d$, $3d$, etc., below the level at a. Find the speed of sound in air.

The sound instensity reaches a maximum when the air column resonates with the tuning fork. The air column acts like a tube closed at one end. The standing wave pattern consists of a node at the water surface and an antinode near the open end. Since the frequency of the source is fixed and the speed of sound in the air column has a definite value, resonance occurs at one specific wavelength,

$$\lambda = \frac{v}{\nu}.$$

The distance d between successive resonance positions is therefore the distance between adjacent nodes. (See Fig. 20-8.) Hence,

$$d = \frac{\lambda}{2} \qquad \text{or} \qquad \lambda = 2d.$$

Combining equations we find

$$2d = \frac{v}{\nu} \qquad \text{or} \qquad v = 2d\nu.$$

figure 20-8
Example 2. Measuring the speed of sound in air. The water level in the tube can be adjusted by raising or lowering the reservoir on the left which is connected to the tube by a rubber hose.

In an experiment with a fork of frequency $\nu = 1080$ cycles/s, d is found to be 15.3 cm. Hence,

$$\lambda = 2d = 30.6 \text{ cm}$$

and $$v = \nu\lambda = (1080)(0.306) \text{ m/s} = 330 \text{ m/s}.$$

What significance does the distance a have? Could gases other than air be used conveniently in this apparatus?

20-6
BEATS

When two wavetrains of the *same frequency* travel along the same line in *opposite directions*, standing waves are formed in accord with the principle of superposition. We may characterize these waves by drawing a plot of the amplitude of oscillation as a function of distance, as in Fig. 20-4. This illustrates a type of interference that we can call *interference in space*.

The same principle of superposition leads us to another type of interference, which we can call *interference in time*. It occurs when two wavetrains of slightly *different frequency* travel in the *same direction*. With sound such a condition exists when, for example, two adjacent piano keys are struck simultaneously.

Consider some one point in space through which the waves are passing. In Fig. 20-9a we plot the displacements produced at such a point by the two waves separately as a function of time. For simplicity we have assumed that the two waves have equal amplitude, although this is not necessary. The resultant vibration at that point as a function of time is the sum of the individual vibrations and is plotted in Fig. 20-9b. We see that the *amplitude* of the resultant wave at the given point is not constant but *varies with time*. In the case of sound the varying amplitude gives rise to variations in loudness which are called *beats*. Two strings may be tuned to the same frequency by tightening one of them while sounding both until the beats disappear.

(a)

(b)

figure 20-9
The beat phenomenon. Two waves of slightly different frequencies, shown in *(a)*, combine in *(b)* to give a wave whose amplitude (dashed line) varies periodically with time. Compare with Fig. 19-14, which shows the same phenomenon displayed as a function of distance.

Let us represent the displacement at the point produced by one wave as

$$y_1 = y_m \cos 2\pi\nu_1 t,$$

and the displacement at the point produced by the other wave of equal amplitude as

$$y_2 = y_m \cos 2\pi\nu_2 t.$$

By the superposition principle, the resultant displacement is

$$y = y_1 + y_2 = y_m(\cos 2\pi\nu_1 t + \cos 2\pi\nu_2 t),$$

and since $\qquad \cos a + \cos b = 2 \cos \dfrac{a-b}{2} \cos \dfrac{a+b}{2},$

this can be written as

$$y = \left[2y_m \cos 2\pi \left(\frac{\nu_1 - \nu_2}{2} \right) t \right] \cos 2\pi \left(\frac{\nu_1 + \nu_2}{2} \right) t. \qquad (20\text{-}8)$$

The resulting vibration may then be considered to have a frequency

$$\bar{\nu} = \frac{\nu_1 + \nu_2}{2},$$

which is the average frequency of the two waves, and an amplitude given by the expression in brackets. Hence, the amplitude itself varies with time with a frequency

$$\nu_{\text{amp}} = \frac{\nu_1 - \nu_2}{2}.$$

If ν_1 and ν_2 are nearly equal, this term is small and the amplitude fluctuates slowly. This phenomenon is a form of amplitude modulation which has a counterpart (side bands) in AM radio receivers.

A beat, that is, a maximum of amplitude, will occur whenever

$$\cos 2\pi \left(\frac{\nu_1 - \nu_2}{2} \right) t$$

equals 1 or −1. Since *each* of these values occurs once in each cycle (see Fig. 19-14), the number of beats per second is *twice* the frequency ν_{amp} or $\nu_1 - \nu_2$. Hence, the number of beats per second equals the difference of the frequencies of the component waves. Beats between two tones can be detected by the ear up to a frequency of about seven per second. At higher frequencies individual beats cannot be distinguished in the sound produced.

20-7 THE DOPPLER EFFECT

When a listener is in motion toward a stationary source of sound, the pitch (frequency) of the sound heard is higher than when he is at rest. If the listener is in motion away from the stationary source, he hears a lower pitch than when he is at rest. We obtain similar results when the source is in motion toward or away from a stationary listener. The pitch of the whistle of the locomotive is higher when the source is approaching the hearer than when it has passed and is receding.

Christian Johann Doppler (1803–1853), an Austrian, in a paper of 1842, called attention to the fact that the color of a luminous body must be changed by relative motion of the body and the observer. This *Doppler effect*, as it is called, applies to waves in general. Doppler himself mentions the application of his principle to sound waves. An experimental test was carried out in Holland in 1845 by Buys Ballot, ". . . using a locomotive drawing an open car with several trumpeters."

We now consider the application of the Doppler effect to sound waves, treating only the special case in which the source and observer move along the line joining them. Let us adopt a reference frame at rest in the medium through which the sound travels. Figure 20-10 shows a source of sound S at rest in this frame and an observer O (note the ear) moving *toward* the source at a speed v_o. The circles represent wavefronts, spaced one wavelength apart, traveling through the medium. If the observer were at rest in the medium, he would receive vt/λ waves in time t, where v is the speed of sound in the medium and λ is the wavelength. Because of his motion toward the source, however, he receives $v_0 t/\lambda$ *additional* waves in this same time t. The frequency ν' that he hears is the number of waves received per unit time or

$$\nu' = \frac{vt/\lambda + v_o t/\lambda}{t} = \frac{v + v_o}{\lambda} = \frac{v + v_o}{v/\nu}.$$

That is,

$$\nu' = \nu \frac{v + v_o}{v} = \nu\left(1 + \frac{v_o}{v}\right). \qquad (20\text{-}9a)$$

The frequency ν' heard by the observer is the ordinary frequency ν heard at rest plus the increase $\nu(v_o/v)$ arising from the motion of the observer. When the observer is in motion *away* from the stationary source, there is a *decrease* in frequency $\nu(v_o/v)$ corresponding to the waves that do not reach the observer each unit of time because of his receding motion. Then

$$\nu' = \nu\left(\frac{v - v_o}{v}\right) = \nu\left(1 - \frac{v_o}{v}\right). \qquad (20\text{-}9b)$$

Hence, the general relation holding when the *source is at rest* with respect to the medium but the *observer is moving* through it is

$$\nu' = \nu\left(\frac{v \pm v_o}{v}\right), \qquad (20\text{-}9)$$

where the *plus* sign holds for motion *toward* the source and the *minus* sign holds for motion *away* from the source. Notice that the cause of the change here is the fact that the observer intercepts more or fewer waves each second because of his motion through the medium.

When the *source* is in motion *toward* a stationary observer, the effect is a shortening of the wavelength (see Fig. 20-11), for the source is following after the approaching waves and the crests therefore come closer together. If the frequency of the source is ν and its speed is v_s, then during each vibration it travels a distance v_s/ν and each wavelength is shortened by this amount. Hence, the wavelength of the sound arriving

figure 20-11
The Doppler effect due to motion of the source. The observer is at rest. Wavefront 1 was emitted by the source when it was at S_1, wavefront 2 was emitted when it was at S_2, etc. At the instant the "snapshot" was taken, the source was at S.

at the observer is not $\lambda = v/\nu$ but $\lambda' = v/\nu - v_s/\nu$. Therefore, the frequency of the sound heard by the observer is *increased*, being

$$\nu' = \frac{v}{\lambda'} = \frac{v}{(v - v_s)/\nu} = \nu\left(\frac{v}{v - v_s}\right). \tag{20-10a}$$

If the source moves *away* from the observer, the wavelength emitted is v_s/ν greater than λ, so that the observer hears a *decreased* frequency, namely

$$\nu' = \frac{v}{(v + v_s)/\nu} = \nu\left(\frac{v}{v + v_s}\right). \tag{20-10b}$$

Hence, the general relation holding when the *observer is at rest* with respect to the medium but the *source is moving* through it is

$$\nu' = \nu\left(\frac{v}{v \mp v_s}\right), \tag{20-10}$$

where the *minus* sign holds for motion *toward* the observer and the *plus* sign holds for motion *away* from the observer. Notice that the cause of the change here is the fact that the motion of the source through the medium shortens or increases the wavelength transmitted through the medium.

If both source *and* observer move through the transmitting medium, the student should be able to show that the observer hears a frequency

$$\nu' = \nu\left(\frac{v \pm v_0}{v \mp v_s}\right), \tag{20-11}$$

where the upper signs (+ numerator, − denominator) correspond to the source and observer moving along the line joining the two in the direction *toward* the other, and the lower signs in the direction *away* from the other. Notice that Eq. 20-11 reduces to Eq. 20-9 when $v_s = 0$ and to Eq. 20-10 when $v_0 = 0$, as it must.

If a vibrating tuning fork on its resonating box is moved rapidly toward a wall, the observer will hear two notes of different frequency. One is the note heard directly from the receding fork and is lowered in pitch by the motion. The other note is due to the waves reflected from the wall, and this is raised in pitch. The superposition of these two wave trains produces beats.

The Doppler effect is important in light. The speed of light is so great that only astronomical or atomic sources, which have high velocities compared to terrestrial macroscopic sources, show pronounced Doppler effects. The astronomical effect consists of a shift in the wavelength observed from light emitted by elements on moving astronomical bodies compared to the wavelength observed from these same elements on earth. (See Chapter 42). An easily observed consequence of the Doppler effect is the broadening (or spread in frequency) of the radiation emitted from hot gases. This broadening results from the fact that the emitting atoms or molecules move in all directions and with varying speeds relative to the observing instruments, so that a spread of frequencies is detected.

There are differences, however, in the Doppler effect formula for light and for sound. In sound it is not just the relative motion of source and observer that determines the frequency change. In fact, as we have seen, even when the relative motion is the same (v_o in Eq. 20-9a equals v_s in Eq. 20-10a), we obtain different quantitative results, depending on whether the source or the observer is moving. This difference occurs because v_o and v_s are relative to the medium in which the sound wave is propagated and because this medium determines the wave speed. Light, however, does not require a material medium for its transmission, and the speed of light relative to the source or the observer is always the same value c, regardless of the motion of these bodies relative to each other. This is a basic postulate of the special theory of relativity (See Supplementary Topic V). Hence, for light only the relative motion of source and observer can lead to physical changes, there being no material medium to use as a reference frame. Although the Doppler formula for light (Chapter 42) differs from that for sound, the effects are qualitatively the same. We can apply Eq. 20-10 to light as a good approximation if v_s is taken to mean the *relative* velocity of source and observer and if v_s is very small compared to the velocity of light.

EXAMPLE 3

Show that Eqs. 20-9 and 20-10 become practically identical when the speed of the sources and the observer are small compared to the speed of sound in the medium.

Let $v_0 = v_s = u$. That is, let u represent the speed of observer *or* source. Then Eq. 20-9 becomes

$$\nu' = \nu\left(1 \pm \frac{u}{v}\right).$$

We must show then that Eq. 20-10,

$$\nu' = \nu\left(\frac{v}{v \mp u}\right),$$

reduces to the previous form when $u/v \ll 1$.

We can rewrite Eq. 20-10 as

$$\nu' = \nu\left(\frac{1}{1 \mp u/v}\right).$$

Now by the binomial expansion

$$\left(\frac{1}{1 \mp u/v}\right) = \left(1 \mp \frac{u}{v}\right)^{-1} = 1 \pm \frac{u}{v} + \left(\frac{u}{v}\right)^2 \pm \cdots .$$

But if u/v is sufficiently small compared to unity that we may neglect $(u/v)^2$ and higher powers, then

$$\left(\frac{1}{1 \mp u/v}\right) \cong 1 \pm \frac{u}{v},$$

and Eq. 20-10 becomes

$$\nu' \cong \nu\left(1 \pm \frac{u}{v}\right),$$

the same as Eq. 20-9.

As a numerical example take $u = 73.0$ mi/h $(= 117.5$ km/h). The speed of sound in air is about 730 mi/h $(= 1175$ km/h). Then if the source has a speed $v_s = u = 73.0$ mi/h toward the stationary observer, the frequency heard by the observer is Eq. 20-10,

$$\nu' = \nu\left(\frac{v}{v - v_s}\right) = \nu\left(\frac{730}{730 - 73.0}\right)$$

or

$$\frac{\nu'}{\nu} = 1.11.$$

If the observer has a speed $v_o = u = 73.0$ mi/h toward the stationary source, the frequency heard by the observer is Eq. 20-9,

$$\nu' = \nu\left(\frac{v + v_o}{v}\right) = \nu\left(\frac{730 + 73.0}{730}\right)$$

or

$$\frac{\nu'}{\nu} = 1.10.$$

Hence, when $u/v = 73.0/730 = 1/10$, the percentage difference in the frequency heard between that for the moving observer and that for the moving source, the relative motion being the same, is only 1%.

When v_o or v_s becomes comparable in magnitude to v, the formulas just given for the Doppler effect usually must be modified. One modification is required because the linear relation between restoring force and displacement assumed

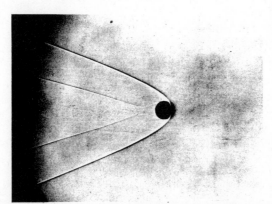

figure 20-12
Top, a group of wavefronts associated with a projectile moving with supersonic speed. The wavefronts are spherical and their envelope is a cone. The student should see the relation between this figure and the previous one. Bottom, a spark photograph of a projectile undergoing this motion. (U.S. Navy Photograph.)

up until now may no longer hold in the medium. The speed of wave propagation is then no longer the normal phase velocity, and the wave shapes change in time. Components of the motion at right angles to the line joining source and observer also contribute to the Doppler effect at these high speeds. When v_0 or v_s exceeds v, the Doppler formula does not apply; for example, if $v_s > v$, the source will get ahead of the wave in one direction; if $v_0 > v$ and the observer moves away from the source, the wave will never catch up with the observer.

There are many instances in which the source moves through a medium at a speed greater than the phase velocity of the wave in that medium. In such cases the wavefront takes the shape of a cone with the moving body at its apex. Some examples are the bow wave from a speedboat on the water and the "shock wave" from an airplane or projectile moving through the air at a speed greater than the velocity of sound in that medium (supersonic speeds). The Cerenkov radiation consists of light waves emitted by charged particles which move through a medium with a speed greater than the phase velocity of light in that medium.*

In Fig. 20-12 we show the present positions of the spherical waves which originated at various positions of the source during its motion. The radius of each sphere at this time is the product of the wave speed v and the time t which has elapsed since the source was at its center. The envelope of these waves is a cone whose surface makes an angle θ with the direction of motion of the source. From the figure we obtain the result

$$\sin \theta = \frac{v}{v_s}.$$

For water waves the cone reduces to a pair of intersecting lines. In aerodynamics the ratio v_s/v is called the *Mach number*.

questions

1. List some sources of infrasonic waves. Of ultrasonic waves.
2. Ultrasound can be used to reveal internal structures of the body. It can, for example, distinguish between liquid and soft human tissues far better than can X-rays. Discuss. (See Problem 4.)
3. What experimental evidence is there for assuming that the speed of sound is the same for all wavelengths?
4. Give a qualitative explanation why the speed of sound in lead is less than that in copper.
5. What quantity, if any, for transverse waves in a string corresponds to the pressure amplitude for longitudinal waves in a tube?
6. A bell is rung for a short time in a school. After a while its sound is inaudible. Trace the sound waves and the energy they transfer from the time of emission until they become inaudible.
7. How can we experimentally locate the positions of nodes and antinodes in a string? In an air column? On a vibrating surface?
8. What physical properties of a sound wave corresponds to the human sensation of pitch, of loudness, and of tone quality?
9. What is the difference between a violin note and the same note sung by a human voice that enables us to distinguish between them?
10. Bells frequently sound much less pleasant than pianos or violins. Why?
11. Does your singing really sound better in a shower? If so, are there physical reasons for this?
12. Discuss the factors that determine the range of frequencies in your voice and the quality of your voice.
13. Explain the origin of the sound in ordinary whistling.

*See "Cerenkov Radiation: its Origin, Properties and Applications," by J. V. Jelley in *Contemporary Physics*, October 1961.

14. What is the common purpose of the valves of a cornet and the slide of a trombone?*

15. The bugle has no valves. How then can we sound different notes on it? To what notes is the bugler limited? Why?*

16. The pitch of the wind instruments rises and that of the string instruments falls as an orchestra "warms up." Explain.*

17. Would a plucked violin string oscillate for a longer or shorter time if the violin had no sounding board? Explain.*

18. Explain how bowing a violin string gets it to vibrate.*

19. Explain the audible tone produced by drawing a wet finger around the rim of a wine glass.

20. Explain how a stringed instrument is "tuned."*

21. A tube can act like an acoustic filter, discriminating against the passage through it of sound of frequencies different from the natural frequencies of the tube. The muffler of an automobile is an example. (a) Explain how such a filter works. (b) How can we determine the cut-off frequency, below which frequency sound is not transmitted?

22. Two ships with steam whistles of the same pitch sound off in the harbor. Would you expect this to produce an interference pattern with regions of high and low intensity?

23. Can sound waves from a *single* tuning fork interfere? How can you explain that the fork is much less audible in certain directions than in others?

24. Two identical tuning forks emit notes of the same frequency. Explain how you might hear beats between them.

25. Suppose that, in the Doppler effect for sound, the source and receiver are at rest in some reference frame but the transmitting medium is moving with respect to this frame. Will there be a change in wavelength, or in frequency, received?

26. Is there a Doppler effect for sound when the observer or the source moves at right angles to the line joining them? How then can we determine the Doppler effect when the motion has a component at right angles to this line?

27. A satellite emits radio waves of constant frequency. These waves are picked up on the ground and made to beat against some standard frequency. The beat frequency is then sent through a loudspeaker and one "hears" the satellite signals. Describe how the sound changes as the satellite approaches, passes overhead, and recedes from the detector on the ground.

28. Discuss factors that improve the acoustics in music halls.†

29. A lightning flash dissipates an enormous amount of energy and is essentially instantaneous. How is that energy transformed into the sound waves of thunder and why is that sound often a spread-out sequence of noises?‡

30. Transverse waves in a string can be polarized (see, for example, Question 18 of Chapter 19). Can sound waves be polarized?

31. Bats can examine the characteristics of objects—such as size, shape, distance, direction, motion—by sensing the way the high-frequency sounds

---

* See the following articles for discussions of the physics of musical instruments: "Acoustics of the Flute" by John W. Coltman, in *Physics Today*, November 1968. "The Physics of Wood Winds" by Arthur H. Benade, in *Scientific American*, October 1960. "The Physics of Brasses" by Arthur H. Benade, in *Scientific American*, July 1973. "The Physics of the Piano" by E. Dornell Blackham, in *Scientific American*, December 1965. "The Physics of Violins" by Carleen M. Hutchins, in *Scientific American*, November 1962. "The Electronic Music Synthesizer and the Physics of Music" by W. M. Hartman, in *American Journal of Physics*, September 1975.

† See "The Development of Architectural Acoustics" by Robert S. Shankland in *American Scientist*, March–April 1972.

‡ See "Thunder" by Arthur A. Few, in *Scientific American*, July 1975.

they emit are reflected off the objects back to the bat. Discuss qualitatively each of these features. (See "Information Content of Bat Sonar Echoes" by J. A. Simmons, D. J. Howell, and N. Suga in *American Scientist*. March–April 1975.

problems

SECTION 20-2

1. The lowest pitch detectable as sound by the average human ear is about 20 Hz and the highest is about 20,000 Hz. What is the wavelength of each in air?   *Answer:* 17 m; 1.7 cm.

2. Bats emit ultrasonic waves. The shortest wavelength emitted in air by a bat is about 0.13 in. (3.3 mm). What is the highest frequency a bat can emit?

3. A sound wave has a frequency of 440 Hz. What is the wavelength of this sound (a) in air and (b) in water?   *Answer:* (a) 75 cm. (b) 3.3 m.

4. Intense ultrasound of frequency 10 MHz is used to modify or destroy tumors in soft tissue. (a) What is the wavelength in air of such a sound wave? (b) If the speed of sound in tissue is 1500 m/s, what is the wavelength of this wave in tissue?

5. (a) A conical loudspeaker has a diameter of 6.0 in. At what frequency will the wavelength of the sound it emits in air be equal to its diameter? Be ten times its diameter? Be one-tenth its diameter? (b) Make the same calculations for a speaker of diameter 12 in. (*Note:* If the wavelength is large compared to the diameter of the speaker, the sound waves spread out almost uniformly in all directions from the speaker, but when the wavelength is small compared to the diameter of the speaker, the wave energy is propagated mostly in the forward direction.)
Answer: (a) 2.2; 0.22; 22 kHz. (b) 1.1; 0.11; 11 kHz.

6. (a) A rule for finding your distance from a lightning flash is to count seconds from the time you see the flash until you hear the thunder and then divide the count by five. The result is supposed to give the distance in miles. Explain this rule and determine the percent error in it at standard conditions. (b) Can you devise a similar rule for the distance in kilometers?

7. The speed of sound in a certain metal is V. One end of a long pipe of that metal of length *l* is struck a hard blow. A listener at the other end hears two sounds, one from the wave that has traveled along the pipe and the other from the wave that has traveled through the air. (a) If *v* is the speed of sound in air, what time interval *t* elapses between the two sounds? (b) Suppose *t* = 1.0 s and the metal is iron. Find the length *l*.
Answer: (a) $l(V - v)/Vv$. (b) 350 m.

8. Two spectators at a soccer game in a large stadium see, and a moment later hear, the ball being kicked on the playing field. If the time delay for one spectator is 0.90 s and for the other 0.60 s, and lines through each spectator and the player kicking the ball meet at an angle of 90°, (a) how far is each spectator from the player? (b) How far are the spectators from each other?

9. A stone is dropped into a well. The sound of the splash is heard 3.0 s later. What is the depth of the well?   *Answer:* 41 m.

SECTION 20-3

10. The pressure in a traveling sound wave is given by the equation

$$p = 1.5 \sin \pi(x - 330t).$$

where *x* is in meters, *t* in seconds, and *p* in pascals. Find (a) the pressure amplitude, (b) the frequency, (c) the wavelength, and (d) the speed of the wave.

11. Two waves give rise to pressure variations at a certain point in space given by

$$p_1 = P \sin 2\pi\nu t,$$

$$p_2 = P \sin 2\pi(\nu t - \phi).$$

What is the pressure amplitude of the resultant wave at this point when $\phi = 0$, $\phi = \frac{1}{4}$, $\phi = \frac{1}{6}$, $\phi = \frac{1}{8}$? All ϕ's are measured in radians.
Answer: 2.00 P; 1.41 P; 1.73 P; 1.85 P.

figure 20-13
Problem 12

12. In Fig. 20-13 we show an acoustic interferometer, used to demonstrate the interference of sound waves. S is a diaphragm that vibrates under the influence of an electromagnet. D is a sound detector, such as the ear or a microphone. Path SBD can be varied in length, but path SAD is fixed.

The interferometer contains air, and it is found that the sound intensity has a minimum value of 100 units at one position of B and continuously climbs to a maximum value of 900 units at a second position 1.65 cm from the first. Find (a) the frequency of the sound emitted from the source, and (b) the relative amplitudes of the waves arriving at the detector for either of the two positions of B. (c) How can it happen that these waves have different amplitudes, considering that they originate at the same source?

figure 20-14
Problem 13

13. A spherical sound source is placed at P_1 near a reflecting wall AB and a microphone is located at point P_2, as shown in Fig. 20-14. The frequency of the sound source P_1 is variable. Find two different frequencies for which the sound intensity, as observed at P_2, will be a maximum. The speed of sound in air is 1100 ft/s. Assume the paths of the interfering waves to be parallel.
Answer: 31 Hz; 94 Hz.

14. Show that the intensity of a sound wave (a) when expressed in terms of the pressure amplitude P, is given by

$$I = \frac{p^2}{2\rho_0 v},$$

where v is the speed of the wave and ρ_0 is the standard density of air, and, (b) when expressed in terms of the displacement amplitude y_m, is given by

$$I = 2\pi^2 \rho_0 v y_m^2 \nu^2,$$

where ν is the frequency of the wave. (c) If two sound waves, one in air and one in water, are equal in intensity, what is the ratio of the pressure amplitude of the wave in water to that of the wave in air? (d) If the pressure amplitudes are equal instead, what is the ratio of the intensities of the waves?

15. A sound wave of frequency 1000 Hz propagating through air has a pressure amplitude of 10 Pa. What are the (a) wavelength, (b) particle displacement amplitude, and (c) maximum particle speed?
Answer: (a) 33 cm. (b) 40 μm. (c) 2.5 cm/s.

16. A note of frequency 300 Hz has an intensity of 1.0 μW/m². What is the amplitude of the air vibrations caused by this sound?

17. A certain loudspeaker produces a sound with a frequency of 2000 Hz and an intensity of 1.2×10^{-7} hp/ft² (9.6×10^{-4} W/m²) at a distance of 20 ft (6.1 m). Presume that there are no reflections and that the loudspeaker emits the same in all directions. (a) What would be the intensity at 100 ft (30 m)? (b) What is the displacement amplitude at 20 ft (6.1 m)? (c) What is the pressure amplitude at 20 ft (6.1 m)?
Answer: (a) 4.8×10^{-9} hp/ft² (4.0×10^{-5} W/m²). (b) 5.7×10^{-7} ft (1.7×10^{-7} m). (c) 1.3×10^{-4} lb/in.² (0.88 Pa).

18. Two sources of sound are separated by a distance of 10 m. They both emit sound at the same amplitude and frequency, 300 Hz, but they are 180° out of phase. At what points along the line between them will the sound intensity be at a relative minimum due to destructive interference?

19. The violin section in some symphony orchestras is divided into two parts,

one placed on each side of the conductor. Consider two violinists 8.0 m apart, symmetrically placed with respect to the conductor and each 5.0 m from him. If the power output from each is 1.0×10^{-4} W, what are (a) the intensity of each playing alone as heard by the conductor, and (b) the combined intensity of both playing together (the same note) as heard by the conductor?   *Answer:* (a) 3.2×10^{-7} W/m². (b) 4.6×10^{-7} W/m².

20. Two loudspeakers, S_1 and S_2, each emit sound of frequency 200 Hz uniformly in all directions (three-dimensional waves). S_1 has an acoustic output of 1.2×10^{-3} W and S_2 one of 1.8×10^{-3} W. The loudspeakers are 7.0 m apart. Consider a point P which is 4.0 m from S_1 and 3.0 m from S_2. (a) How are the phases of the two waves arriving at P related? What is the intensity of sound at P (b) if S_2 is turned off (S_1 on), (c) if S_1 is turned off (S_2 on), and (d) with both S_1 and S_2 on?

SECTION 20-5

21. In Fig. 20-15 a rod R is clamped at its center and a disk D at its end projects into a glass tube, which has cork filings spread over its interior. A plunger P is provided at the other end of the tube. The rod is set into longitudinal vibration and the plunger is moved until the filings form a pattern of nodes and antinodes (the filings form well-defined ridges at the pressure antinodes). If we know the frequency ν of the longitudinal vibrations in the rod, a measurement of the average distance d between successive antinodes determines the speed of sound v in the gas in the tube. Show that

$$v = 2\nu d.$$

This is *Kundt's method* for determining the speed of sound in various gases.

figure 20-15
Problem 21

22. If a violin string is tuned to a certain note, by how much must the tension in the string be increased if it is to emit a note of double the original frequency (that is, a note one octave higher in pitch)?

23. An open organ pipe has a fundamental frequency of 300 Hz. The first overtone of a closed organ pipe has the same frequency as the first overtone of the open pipe. How long is each pipe?   *Answer:* 55 cm; 41 cm.

24. A 3.0-m skip rope is used essentially in its fundamental mode of oscillation. If the rope has a mass of 1.0 kg and the children are pulling back with a force of 10 N, what is the frequency of oscillation?

25. A certain violin string is 50 cm long between its fixed ends and has a mass of 2.0 g. The string sounds an A note (440 Hz) when played without fingering. Where must one put one's finger to play a C (528 Hz)?
Answer: 8.3 cm from one end.

26. The strings of a cello have a length L. (a) By what length l must they be shortened by fingering to change the pitch by a frequency ratio r? (b) Find l, if $L = 0.80$ m and $r = 6/5$, 5/4, 4/3, and 3/2.

27. The water level in a vertical glass tube 1.0 m long can be adjusted to any position in the tube. A tuning fork vibrating at 660 Hz is held just over the open top end of the tube. At what positions of the water level will there be resonance?   *Answer:* Water filled to a height of 7/8, 5/8, 3/8, or 1/8 m.

28. S in Fig. 20-16 is a small loudspeaker driven by an audio oscillator and amplifier, adjustable in frequency from 1000 to 2000 Hz only. D is a piece of cylindrical sheetmetal pipe 18.0 in. long. (a) If the speed of sound in air is 1130 ft/s at the existing temperature, at what frequencies will resonance occur when the frequency emitted by the speaker is varied from 1000 to 2000 Hz? (b) Sketch the displacement nodes for each. Neglect end effects.

29. A well with vertical sides and water at the bottom resonates at 7.0 Hz and at no lower frequency. The air in the well has a density of 1.1 kg/m³ (2.1×10^{-3} slug/ft³), a pressure of 9.5×10^4 Pa (13.8 lb/in.²), and a ratio of specific heats of 7/5. How deep is the well?   *Answer:* 12 m (41 ft).

30. The period of a pulsating variable star may be estimated by considering the star to be executing radial longitudinal pulsations in the fundamental

figure 20-16
Problem 28

standing wave mode; that is, the radius varies periodically with the time, with a displacement antinode at the surface. (a) Would you expect the center of the star to be a displacement node or antinode? (b) By analogy with the open organ pipe, show that the period of pulsation T is given by

$$T = 4R/v_s,$$

where R is the equilibrium radius of the star and v_s is the average sound speed. (c) Typical white dwarf stars have pressures of 10^{22} Pa, densities of 10^{10} kg/m³, ratio of specific heats of 4/3, and radius 0.009 solar radii. What is the approximate pulsation period of a white dwarf? (See "Pulsating Stars" by John R. Percy, in *Scientific American*, June 1975.)

31. A tube 1.0 m (3.3 ft) long is closed at one end. A stretched wire is placed near the open end. The wire is 0.30 m (0.98 ft) long and has a mass of 0.010 kg (6.9×10^{-4} slug). It is fixed at both ends and vibrates in its fundamental mode. It sets the air column in the tube into vibration at its fundamental frequency by resonance. Find (a) the frequency of oscillation of the air column and (b) the tension in the wire.
    Answer: (a) 83 Hz (82 Hz). (b) 82 N (18 lb).

32. A 31.6-cm violin string with linear density 0.65 g/m is placed near a loudspeaker that is fed by an audio-oscillator of variable frequency. It is found that the string is set in oscillation only at the frequencies 880 and 1320 Hz as the frequency of the oscillator is varied continuously over the range 500 to 1500 Hz. What is the tension in the string?

SECTION 20-6

33. Two identical piano wires have a fundamental frequency of 600 Hz when kept under the same tension. What fractional increase in the tension of one wire will lead to the occurrence of six beats per second when both wires vibrate simultaneously?
    Answer: 2.0%.

34. A tuning fork of unknown frequency makes three beats per second with a standard fork of frequency 384 Hz. The beat frequency decreases when a small piece of wax is put on a prong of the first fork. What is the frequency of this fork?

SECTION 20-7

35. A bullet is fired with a speed of 2200 ft/s. Find the angle made by the shock wave with the line of motion of the bullet.
    Answer: 30°.

36. Calculate the speed of the projectile illustrated in the photograph in Fig. 20-12. Assume the speed of sound in the medium through which the projectile is traveling to be 380 m/s.

37. The speed of light in water is about three-fourths the speed of light in vacuum. A beam of high-speed electrons from a betatron emits Cerenkov radiation in water, the wavefront being a cone of angle 120°. Find the speed of the electrons in the water?
    Answer: 2.6×10^8 m/s.

38. A jet plane passes overhead at a height of 5000 m and a speed of Mach 1.5 (that is, 1.5 times the speed of sound). (a) Find the angle made by the shock wave with the line of motion of the jet. (b) How long after the jet has passed directly overhead will the shock wave reach the ground?

39. A whistle of frequency 540 Hz rotates in a circle of radius 2.00 ft an an angular speed of 15.0 rad/s. What is (a) the lowest and (b) the highest frequency heard by a listener a long distance away at rest with respect to the center of the circle?
    Answer: (a) 525 Hz. (b) 555 Hz.

40. (a) Could you go through a red light fast enough to have it appear green? (b) If so, would you get a ticket for speeding? Take $\lambda = 620$ nm (= 620 nanometer = 620×10^{-9} m; see Table 1-2) for red light, $\lambda = 540$ nm for green light, and $c = 3.0 \times 10^8$ m/s as the speed of light.

41. A bat is flittering about in a cave, navigating very effectively by the use of ultrasonic bleeps (short emissions lasting a millisecond or less and repeated several times a second). Assume that the sound emission frequency of the

bat is 39,000 Hz. During one fast swoop directly toward a flat wall surface, the bat is moving at 1/40 of the speed of sound in air. What frequency does he hear reflected off the wall?   *Answer:* 41,000 Hz.

42. A source of sound waves of frequency 1080 Hz moves to the right with a speed of 108 ft/s relative to the ground. To its right is a reflecting surface moving to the left with a speed of 216 ft/s relative to the ground. Take the speed of sound in air to be 1080 ft/s and find (*a*) the wavelength of the sound emitted in air by the source, (*b*) the number of waves per second arriving at the reflecting surface, (*c*) the speed of the reflected waves, (*d*) the wavelength of the reflected waves.

43. A siren emitting a sound of frequency 1000 Hz moves away from you toward a cliff at a speed of 10 m/s. (*a*) What is the frequency of the sound you hear coming directly from the siren? (*b*) What is the frequency of the sound you hear reflected off the cliff? (*c*) Could you hear the beat frequency? Take the speed of sound in air as 330 m/s.
Answer: (*a*) 970 Hz. (*b*) 1030 Hz. (*c*) No. It is too high.

44. Microwaves, which travel with the speed of light, are reflected from a distant airplane approaching the wave source. It is found that when the reflected waves are beat against the waves radiating from the source the beat frequency is 990 Hz. If the microwaves are 0.10 m in wavelength, what is the approach speed of the airplane?

45. Radar measurements in pursuing situations are relatively inaccurate compared to rest situations. (*a*) Consider a radar unit at rest and show that the difference $d\nu$ between the frequency reflected off a car moving at a speed V and the transmitted frequency ν is given approximately by $d\nu/\nu = 2V/c$. (*b*) Now consider the radar unit to be in a pursuing vehicle moving at a speed v and show that $d\nu/\nu = 2(v - V)/c$. Discuss various cases and justify the first sentence: in particular, consider (*c*) the case in which the pursuer (police) is moving at the same speed as the speeder. What Doppler shift is observed in this case?   *Answer:* None.

46. Equation 20-11 for the Doppler effect assumes a reference frame at rest in the medium through which the sound travels. Suppose instead that the reference frame is fixed to the earth and that the medium moves with speed v_m from source to observer. How must you modify Eq. 20-11 in this (more general) case?

47. A girl is sitting near the open window of a train that is moving at a velocity of 10.0 m/s to the east. The girl's uncle stands near the tracks and watches the train move away. The locomotive whistle vibrates at 500 Hz. The air is still. (*a*) What frequency does the uncle hear? (*b*) What frequency does the girl hear? A wind begins to blow from the east at 10 m/s. (*c*) What frequency does the uncle now hear? (*d*) What frequency does the girl now hear?
Answer: (*a*) 485 Hz. (*b*) 500 Hz. (*c*) 486 Hz. (*d*) 500 Hz.

48. A woman standing on the ground beside a highway blows her whistle (pitch 800 Hz) to alert three colleagues waiting 200 m away, one each to the north, east, and south. A fourth confederate is driving west at 40 m/s. A steady wind at 4.0 m/s is blowing from south to north. What is the frequency of the sound heard by each of the waiting women and the driving woman?

21

temperature

In analyzing physical situations we usually focus our attention on some portion of matter which we separate, in our minds, from the environment external to it. We call such a portion the *system*. Everything outside the system which has a direct bearing on its behavior we call the *environment*. We then seek to determine the behavior of the system by finding how it interacts with its environment. For example, a ball can be the system and the environment can be the air and the earth. In free fall we seek to find how the air and the earth affect the motion of the ball. Or the gas in a container can be the system, and a movable piston and a Bunsen burner can be the environment. We seek to find how the behavior of the gas is affected by the action of the piston and burner. In all such cases we must choose suitable observable quantities to describe the behavior of the system. We classify these quantities, which are gross properties of the system measured by laboratory operations, as *macroscopic*. For processes in which heat is involved the laws relating the appropriate macroscopic quantities (which include pressure, volume, temperature, internal energy, and entropy, among others) form the basis for the science of *thermodynamics*. Many of the macroscopic quantities (pressure, volume, and temperature, for example) are directly associated with our sense perceptions. We can also adopt a *microscopic* point of view. Here we consider quantities that describe the atoms and molecules that make up the system, their speeds, energies, masses, angular momenta, behavior during collisions, etc. These quantities, or mathematical formulations based on them, form the basis for the science of *statistical mechanics*. The microscopic properties are not directly associated with our sense perceptions.

For any system the macroscopic and the microscopic quantities must

21-1
MACROSCOPIC AND MICROSCOPIC DESCRIPTIONS

be related because they are simply different ways of describing the same situation. In particular, we should be able to express the former in terms of the latter. The pressure of a gas, viewed macroscopically, is measured operationally using a manometer (Fig. 17-10). Viewed microscopically it is related to the average rate per unit area at which the molecules of the gas deliver momentum to the manometer fluid as they strike its surface. In Section 23-4 we will make this microscopic definition of pressure quantitative. Similarly (see Section 23-5), the temperature of a gas may be related to the average kinetic energy of translation of the molecules.

If the macroscopic quantities can be expressed in terms of the microscopic quantities, we should be able to express the laws of thermodynamics quantitatively in the language of statistical mechanics. We can indeed do this. In the words of R. C. Tolman:

The explanation of the complete science of thermodynamics in terms of the more abstract science of statistical mechanics is one of the greatest achievements of physics. In addition, the more fundamental character of statistical mechanical considerations makes it possible to supplement the ordinary principles of thermodynamics to an important extent.

We begin our examination of heat phenomena in this chapter with a study of temperature. As we progress we shall try to gain a deeper understanding of these phenomena by interweaving the microscopic and the macroscopic description—statistical mechanics and thermodynamics. The interweaving of the microscopic and the macroscopic points of view is characteristic of present-day physics.

21-2
THERMAL EQUILIBRIUM — THE ZEROTH LAW OF THERMODYNAMICS

The sense of touch is the simplest way to distinguish hot bodies from cold bodies. By touch we can arrange bodies in the order of their hotness, deciding that A is hotter than B, B than C, etc. We speak of this as our *temperature* sense. This is a very subjective procedure for determining the temperature of a body and certainly not very useful for purposes of science. A simple experiment, suggested in 1690 by John Locke, shows the unreliability of this method. Let a person immerse his hands, one in hot water, the other in cold. Then let him put both hands in water of intermediate hotness. This will seem cooler to the first hand and warmer to the second hand. Our judgment of temperature can be rather misleading. Further, the range of our temperature sense is limited. What we need is an objective, numerical, measure of temperature.

To begin with, we should try to understand the meaning of temperature. Let an object A which feels cold to the hand and an identical object B which feels hot be placed in contact with each other. After a sufficient length of time, A and B give rise to the same temperature sensation. Then A and B are said to be in *thermal equilibrium* with each other. We can generalize the expression "two bodies are in thermal equilibrium" to mean that the two bodies are in states such that, if the two *were* connected, the combined systems would be in thermal equilibrium. The logical and operational test for thermal equilibrium is to use a third or test body, such as a thermometer. This is summarized in a postulate often called *the zeroth law of thermodynamics: If A and B are each in thermal equilibrium with a third body C (the "thermometer"), then A and B are in thermal equilibrium with each other.*

This discussion expresses the idea that the temperature of a system is a property which eventually attains the same value as that of other systems when all these systems are put in contact. This concept agrees with the everyday idea of temperature as the measure of the hotness or coldness of a system, because as far as our temperature sense can be trusted, the hotness of all objects becomes the same after they have been in contact long enough. The idea contained in the zeroth law, although simple, is not obvious. For example, Jones and Smith each know Green, but they may or may not know each other. Two pieces of iron attract a magnet but they may or may not attract each other.

A more formal, but perhaps more fundamental phrasing of the zeroth law is: *There exists a scalar quantity called* temperature, *which is a property of all thermodynamic systems (in equilibrium states), such that temperature equality is a necessary and sufficient condition for thermal equilibrium.* This statement* justifies our use of temperature as a thermodynamic variable; the formulation given above is the corollary of this new statement. Speaking loosely, the essence of the zeroth law is: *there exists a useful quantity called "temperature."*

21-3 *MEASURING TEMPERATURE*

There are many measurable physical properties that vary as our physiological perception of temperature varies. Among these are the volume of a liquid, the length of a rod, the electrical resistance of a wire, the pressure of a gas kept at constant volume, the volume of a gas kept at constant pressure, and the color of a lamp filament. Any of these properties can be used in the construction of a thermometer — that is, in the setting up of a particular "private" temperature scale. Such a temperature scale is established by choosing a particular thermometric substance and a particular thermometric property of this substance. We then define the temperature scale by an *assumed* continuous monotonic relation between the chosen thermometric property of our substance and the temperature as measured on our (private) scale. For example, the thermometric substance may be a liquid in a glass capillary tube and the thermometric property can be the length of the liquid column; or the thermometric substance may be a gas kept in a container at constant volume and the thermometric property can be the pressure of the gas; and so forth. *We must realize that each choice of thermometric substance and property — along with the assumed relation between property and temperature — leads to an individual temperature scale whose measurements need not necessarily agree with measurements made on any other independently defined temperature scale.*

This apparent chaos in the definition of temperature is removed by universal agreement, within the scientific community, on the use of a particular thermometric substance, a particular thermometric property, and a particular functional relation between measurements of that property and a universally accepted temperature scale. A private temperature scale defined in any other way can then always be calibrated against the universal scale. We describe such a universal scale in Section 21-5 and an equivalent one in Section 25-6.

Suppose that we have chosen a thermometric substance. Let us represent by X the thermometric property that we wish to use in setting up a temperature scale. We arbitrarily choose the following linear function of the property X as the temperature T which the appropriate thermometer, and any system in thermal equilibrium with it, has:

$$T(X) = aX. \tag{21-1}$$

---

* See J. S. Thomsen, "A Restatement of the Zeroth Law of Thermodynamics," *American Journal of Physics*, 30, 294, 1962.

figure 21-1
The National Bureau of Standards triple-point cell. It contains pure water and is sealed after all air has been removed. It is then immersed in a water-ice bath. The system is at the triple point when ice, water, and vapor are all present, and in equilibrium, inside the cell. The thermometer to be calibrated is immersed in the central well.

Water vapor

Water

Ice

Sealed cell

Thermometer well

Water–ice mixture

Vacuum flask

In this expression a is a constant which we must still evaluate. By choosing this linear form for $T(X)$ we have fixed it so that *equal temperature differences*, or temperature intervals, *correspond to equal changes in X*. This means, for example, that every time the mercury column in the mercury-in-glass thermometer changes in length by one unit, the temperature changes by a definite fixed amount, no matter what the starting temperature. It also follows that two temperatures, measured with the same thermometer, are in the same ratio as their corresponding X's, that is,

$$\frac{T(X_1)}{T(X_2)} = \frac{X_1}{X_2}.$$

To determine the constant a, and hence to calibrate the thermometer, we specify a *standard fixed point* at which all thermometers must give the same reading for temperature T. This fixed point is chosen to be that at which ice, liquid water, and water vapor coexist in equilibrium and is called the *triple point of water*. This state can be achieved only at a definite pressure and is unique (Fig. 21-1). The water vapor pressure at the triple point is 4.58 mm-Hg. The temperature at this standard fixed point was arbitrarily set at 273.16 degrees Kelvin and was abbreviated 273.16° K. Later,* the name *kelvin* (symbol K) replaced degree Kelvin (symbol °K) and the unit of thermodynamic temperature was defined as follows: *The kelvin, unit of thermodynamic temperature, is the fraction 1/273.16 of the thermodynamic temperature of the triple point of water.*†

If we indicate values at the triple point by the subscript tr, then, for any thermometer,

$$\frac{T(X)}{T(X_{tr})} = \frac{X}{X_{tr}},$$

---

* Adopted in 1967 at the Tenth General Conference on Weights and Measures.
† The triple point of water was chosen over the freezing point (previously used) because the former is more reproducible, by a factor of ten, than the latter.

where, for *all* thermometers,

$$T(X_{tr}) = 273.16 \text{ K},$$

so that

$$T(X) = 273.16 \text{ K} \frac{X}{X_{tr}}. \tag{21-2}$$

Hence, when the thermometric property has the value X, the temperature T, on the particular private scale selected, is given in K by $T(X)$, when the value of X and X_{tr} are inserted on the right-hand side of this equation.

We can now apply Eq. 21-2 to several thermometers. For a liquid-in glass thermometer X is L, the length of the liquid column, and Eq. 21-2 yields

$$T(L) = 273.16 \text{ K} \frac{L}{L_{tr}}.$$

For a gas at constant pressure, X is V, the volume of the gas, and

$$T(V) = 273.16 \text{ K} \frac{V}{V_{tr}} \quad \text{(constant } P\text{)}.$$

For a gas at constant volume, X is P, the gas pressure, and

$$T(P) = 273.16 \text{ K} \frac{P}{P_{tr}} \quad \text{(constant } V\text{)}.$$

For a platinum resistance thermometer, X is R, the electrical resistance, and

$$T(R) = 273.16 \text{ K} \frac{R}{R_{tr}},$$

and likewise for other thermometric substances and thermometric properties.

EXAMPLE 1

A certain platinum resistance thermometer has a resistance R of 90.35 ohms when its bulb is placed in a triple-point cell like that of Fig. 21-1. What temperature is defined by Eq. 21-2 if the bulb is placed in an environment such that its resistance is 96.28 ohms?

From Eq. 21-2,

$$T(X) = 273.16 \text{ K} \frac{X}{X_{tr}}$$

$$= (273.16 \text{ K})\left(\frac{96.28}{90.35}\right) = 280.6 \text{ K}.$$

Note that this temperature is on a private scale, defined by applying Eq. 21-2 to a particular device, the platinum resistance thermometer.

The question now arises whether the value we obtain for the temperature of a system depends on the choice of the thermometer we use to measure it. We have insured by definition that all the different kinds of thermometers will agree at the standard fixed point, but what happens at other points? We can imagine a series of tests in which the temperature of a given system is measured simultaneously with many different thermometers. Results of such tests show that the thermometers all read differently. Even when different thermometers of the same kind are used, such as constant-volume gas thermometers using different gases, we obtain different temperature readings for a given system in a given state.

Hence, to obtain a definite temperature scale, we must select one particular kind of thermometer as the standard. The choice will be made, not on the basis of experimental convenience, but by inquiring whether the temperature scale defined by a particular thermometer proves to be a useful quantity in the formulation of the laws of physics. The smallest variation in readings is found among different constant-volume gas thermometers, which suggests that we choose a gas as the standard thermometric substance. It turns out that as the amount of gas used in such a thermometer, and therefore its pressure, is reduced, the variation in readings between gas thermometers using different kinds of gas is reduced also. Hence, there seems to be something fundamental about the behavior of a constant-volume thermometer containing a gas at low pressure.

21-4 THE CONSTANT VOLUME GAS THERMOMETER

If the volume of a gas is kept constant, its pressure depends on the temperature and increases steadily with rising temperature. The constant-volume gas thermometer uses the pressure at constant volume as the thermometric property.

The thermometer is shown diagrammatically in Fig. 21-2. It consists of a bulb of glass, porcelain, quartz, platinum or platinum-iridium (depending on the temperature range over which it is to be used), connected by a capillary tube to a mercury manometer. The bulb containing some gas is put into the bath or environment whose temperature is to be measured; by raising or lowering the mercury reservoir the mercury in the left branch of the U-tube can be made to coincide with a fixed reference mark, thus keeping the confined gas at a constant volume. Then we read the height of the mercury in the right branch. The pressure of the confined gas is the difference of the heights of the mercury columns (times ρg) plus the atmospheric pressure, as indicated by the barometer. In practice the apparatus is very elaborate and we must make many corrections, for example, (1) to allow for the small volume change owing to slight contraction or expansion of the bulb and (2) to allow for the fact that not all the confined gas (such as that in the capillary) has been immersed in the bath. Assume that all corrections have been made, and let P be the corrected value of the pressure at the temperature of the bath. Then the temperature is given provisionally (see below) by

$$T(P) = 273.16 \text{ K} \frac{P}{P_{tr}} \qquad (\text{constant } V).$$ (21-3)

The constant-volume thermometer, used as described below, is the thermometer which serves to establish the temperature scale used universally in scientific work today.

figure 21-2
A representation of a constant-volume gas thermometer. As long as the mercury in the left manometer tube remains at the same position on the scale (zero) the volume of the confined gas will be constant. The meniscus can always be brought to the zero position by raising or lowering reservoir R.

21-5 IDEAL GAS TEMPERATURE SCALE

Let a certain amount of gas be put into the bulb of a constant-volume gas thermometer so that when the bulb is surrounded by water at the triple point the pressure P_{tr} is equal to a definite value, say 80 cm-Hg. Now surround the bulb with steam condensing at 1-atm pressure and, with the volume kept constant at its previous value, measure the gas

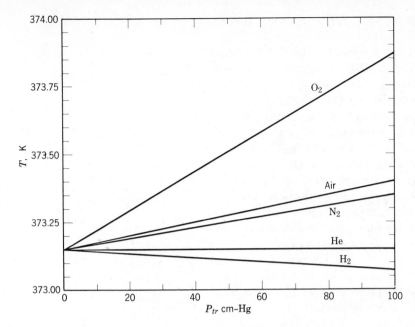

figure 21-3
The readings of a constant-volume gas thermometer for the temperature T of condensing steam as a function of P_{tr}, when different gases are used. As the amount of gas in the thermometer is reduced its pressure P_{tr} at the triple point decreases. Note that at a particular P_{tr} the values of T given by different gas thermometers differ. The discrepancy is small but measurable, being about 0.2% in the most extreme cases shown (O_2 and H_2 at 100 cm-Hg; note that the entire vertical axis covers only 1.00 K). Helium gives nearly the same T at all pressures (the curve is almost horizontal) so that its behavior is the most similar to that of an ideal gas over the entire range shown.

pressure P_s, the pressure at the steam point, in this case, P_{s80}. Then calculate the temperature provisionally from $T(P_{s80}) = 273.16$ K $(P_{s80}/80$ cm-Hg). Next remove some of the gas so that P_{tr} has a smaller value, say 40 cm-Hg. Then measure the new value of P_s and calculate another provisional temperature from $T(P_{s40}) = 273.16$ K $(P_{s40}/40$ cm-Hg). Continue this same procedure, reducing the amount of gas in the bulb again, and at this new lower value of P_{tr} calculating the temperature at the steam point $T(P_s)$. If we plot the values $T(P_s)$ against P_{tr} and have enough data, we can extrapolate the resulting curve to the intersection with the axis where $P_{tr} = 0$.

In Fig. 21-3, we plot curves obtained from such a procedure for constant-volume thermometers of some different gases. These curves show that the temperature readings of a constant-volume gas thermometer depend on the gas used at ordinary values of the reference pressure. However, as the reference pressure is decreased, the temperature readings of constant-volume gas thermometers using different gases approach the same value. Therefore, *the extrapolated value of the temperature depends only on the general properties of gases and not on any particular gas.* We therefore define an *ideal gas temperature scale* by the relation

$$T = 273.16 \text{ K} \lim_{P_{tr} \to 0} \left(\frac{P}{P_{tr}} \right) \qquad \text{(constant } V\text{).} \qquad (21\text{-}4)$$

Our standard thermometer is therefore chosen to be a constant-volume gas thermometer using a temperature scale defined by Eq. 21-4.

Although our temperature scale is independent of the properties of any one particular gas, it does depend on the properties of gases in general (that is, on the properties of a so-called ideal gas). Therefore, to measure a temperature, a gas must be used at that temperature. The lowest temperature that can be measured with any gas thermometer is about 1 K. To obtain this temperature we must use low-pressure helium, for helium becomes a liquid at a temperature lower than any other gas. Therefore we cannot give experimental meaning to temperatures below about 1 K, by means of a gas thermometer.

We would like to define a temperature scale in a way that is *inde-*

pendent of the properties of any particular substance. We will show in Section 25-6 that the *absolute thermodynamic temperature scale,* called the Kelvin scale, is such a scale. We will show also that *the ideal gas scale and the Kelvin scale are identical in the range of temperatures in which a gas thermometer may be used.* For this reason we can write "K" after an ideal gas temperature, as we have already done.

We will also show in Section 25-6 that the Kelvin scale has an *absolute zero* of 0 K and that temperatures below this do not exist. The absolute zero of temperature has defied all attempts to reach it experimentally, although it is possible to come arbitrarily close.* The existence of the absolute zero is inferred by extrapolation. You should not think of absolute zero as a state of zero energy and no motion. The conception that all molecular action would cease at absolute zero is incorrect. This notion assumes that the purely macroscopic concept of temperature is strictly connected to the microscopic concept of molecular motion. When we try to make such a connection, we find in fact that as we approach absolute zero the kinetic energy of the molecules approaches a finite value, the so-called zero-point energy. The molecular energy is a minimum, but not zero, at absolute zero.

Table 21-1
Some temperatures‡ (K)

Carbon thermonuclear reaction	5×10^8
Helium thermonuclear reaction	10^8
Solar interior	10^7
Solar corona	10^6
Shock wave in air at Mach 20	2.5×10^4
Luminous nebulae	10^4
Solar surface	6×10^3
Tungsten melts	3.6×10^3
Lead melts	6.0×10^2
Water freezes	2.7×10^2
Oxygen boils (1 atm)	9.0×10^1
Hydrogen boils (1 atm)	2.0×10^1
Helium (He⁴) boils at 1 atm	4.2
He³ boils at attainable low pressure	3.0×10^{-1}
Adiabatic demagnetization of paramagnetic salts	10^{-3}
Adiabatic demagnetization of nuclei	10^{-6}

‡ See *Scientific American*, September 1954, special issue on heat.

In Table 21-1 we list the temperatures, on the Kelvin scale, of various bodies and processes.

Two temperature scales in common use are the Celsius† and the Fahrenheit scales. These are defined in terms of the Kelvin scale, which is the fundamental temperature scale in science.

The Celsius temperature scale uses the unit "degree Celsius" (sym-

21-6
THE CELSIUS AND FAHRENHEIT SCALES

* It is possible to prepare systems that have *negative Kelvin temperatures.* Surprisingly enough, such temperatures are *not* reached by passing through 0 K but by proceeding through infinite temperatures. That is, negative temperatures are not 'colder' than absolute zero but instead are 'hotter' than infinite temperatures. See *Science by Degrees*, by Castle, Emmerich, Heikes, Miller, and Rayne, published by Walker and Company, New York, 1965. The absolute zero remains experimentally unattainable.
† This scale, based on a scale invented by a Swede named Celsius in 1742, was called the "centigrade" scale until 1948, when the Ninth General Conference on Weights and Measures decided that the name should be changed.

bol °C) equal to the unit "kelvin." If we let T_C represent the Celsius temperature, then

$$T_C = T - 273.15° \qquad (21\text{-}5)$$

relates the Celsius temperature T_C(°C) and the Kelvin temperature T(K). We see that the triple point of water (= 273.16 K by definition) corresponds to 0.01° C. By experiment the temperature at which ice and air-saturated water are in equilibrium at atmospheric pressure—the so-called ice point—proves to be 0.00° C and the temperature at which steam and liquid water are in equilibrium at 1-atm pressure—the so-called steam point—proves to be 100.00° C.

The Fahrenheit scale, still in use in some English-speaking countries (England itself adopted the Celsius scale for commercial and civil use in 1964) is not used in scientific work. The relationship between the Fahrenheit and Celsius scales is defined to be

$$T_F = 32 + \tfrac{9}{5}T_C.$$

From this relation we can conclude that the ice point (0.00° C) equals 32.0° F, that the steam point (100.0° C) equals 212.0° F, and that one Fahrenheit degree is exactly $\tfrac{5}{9}$ as large as one Celsius degree. In Fig. 21-4 we compare the Kelvin, Celsius, and Fahrenheit scales.

figure 21-4
The Kelvin, Celsius, and Fahrenheit temperature scales.

Let us now summarize the ideas of the last few sections. The standard fixed point in thermometry is the triple point of water which is arbitrarily assigned a value of 273.16 K. The constant-volume gas thermometer is the standard thermometer. The extrapolated gas scale is used to define the ideal gas temperature from $T = 273.16 \text{ K} \lim\limits_{P_{tr} \to 0} (P/P_{tr})$.

This scale is identical with the (absolute thermodynamic) Kelvin scale in the range in which a gas thermometer can be used.

By using the standard thermometer in this way, we can experimentally determine other reference points for temperature measurements, called fixed points. We list the basic fixed points adopted for experimental reference in Table 21-2. The temperatures can be expressed on

21-7
THE INTERNATIONAL PRACTICAL TEMPERATURE SCALE

Table 21-2
Fixed points on the international practical temperature scale[a]

Substance	State	Temperature	
		K	°C
Hydrogen	Triple point	13.81	−259.34
Hydrogen	Boiling point[b]	17.042	−256.108
Hydrogen	Boiling point	20.28	−252.87
Neon	Boiling point	27.102	−246.048
Oxygen	Triple point	54.361	−218.789
Oxygen	Boiling point	90.188	−182.962
Water[c]	Triple point	273.16	0.01
Water[c]	Boiling point	375.15	100
Zinc	Freezing point	692.73	419.58
Silver	Freezing point	1235.08	961.93
Gold	Freezing point	1337.58	1064.43

[a] The so-called IPTS-68, adopted in 1968 by the International Committee on Weights and Measures.
[b] This boiling point is for a pressure of 25/76 atm. All other boiling points (and all freezing points) are for a pressure of 1 atm.
[c] The water used should have the isotopic composition of sea water.

the Celsius scale, with the use of Eq. 21-5, once the Kelvin temperature is determined.

Determining ideal gas temperatures is painstaking. It would not make sense to use this procedure to determine temperatures for all work.

Hence, an International Practical Temperature Scale (IPTS) was adopted in 1927 (revised in 1948 and again in 1968) to provide a scale that can be used easily for practical purposes, such as for calibration of industrial or scientific instruments. This scale consists of a set of recipes for providing in practice the best possible approximations to the Kelvin scale. A set of fixed points, the basic points in Table 21-2, is adopted, and a set of instruments is specified to be used in interpolating between these fixed points and in extrapolating beyond the highest fixed point. The IPTS-68 departs from the Kelvin scale at temperatures between the fixed points, but the difference is usually negligible. The IPTS-68 has become the legal standard in nearly all countries.

Common effects of temperature changes are changes in size and changes of state of materials. Let us consider changes of size which occur without changes of state. Consider a simple model of a crystalline solid. The atoms are held together in a regular array by forces of electrical origin. The forces between atoms are like those that would be exerted by a set of springs connecting the atoms, so that we can visualize the solid body as a microscopic bedspring (Fig. 21-5). These "springs" are quite stiff (Problem 9, Chapter 15), and there are about 10^{22} of them per cubic centimeter. At any temperature the atoms of the solid are vibrating. The amplitude of vibration is about 10^{-9} cm, about one-tenth of an atomic diameter, and the frequency about 10^{13} Hz.

When the temperature is increased the average distance between atoms increases, which leads to an expansion of the whole solid body. The change in *any* linear dimension of the solid, such as its length, width, or thickness, is called a linear expansion. If the length of this linear dimension is l, the change in length, arising from a change in temperature ΔT, is Δl. We find from experiment that, if ΔT is small enough, this change in length Δl is proportional to the temperature change ΔT and to the original length l. Hence, we can write

$$\Delta l = \alpha l \, \Delta T, \qquad (21\text{-}6)$$

where α, called the *coefficient of linear expansion*, has different values for different materials. Rewriting this formula we obtain

$$\alpha = \frac{1}{l} \frac{\Delta l}{\Delta T},$$

so that α has the meaning of a fractional change in length per degree temperature change.

Strictly speaking, the value of α depends on the actual temperature and the reference temperature chosen to determine l (see Problem 13). However, its variation is usually negligible compared to the accuracy with which engineering measurements need to be made. We can safely take it as a constant for a given material, independent of the temperature. In Table 21-3 we list the experimental values for the average coefficient of linear expansion of several common solids. For all the substances listed, the change in size consists of an expansion as the

21-8
TEMPERATURE EXPANSION

figure 21-5
A solid behaves in many ways as if it is a microscopic "bedspring" in which the molecules are held together by elastic forces.

Table 21-3
Some values* of $\bar{\alpha}$

Substance	$\bar{\alpha}$ (per C°)	Substance	$\bar{\alpha}$ (per C°)
Aluminum	23×10^{-6}	Hard rubber	80×10^{-6}
Brass	19×10^{-6}	Ice	51×10^{-6}
Copper	17×10^{-6}	Invar	0.7×10^{-6}
Glass (ordinary)	9×10^{-6}	Lead	29×10^{-6}
Glass (pyrex)	3.2×10^{-6}	Steel	11×10^{-6}

* For the range 0° C to 100° C; except $-10°$ C to 0° C for ice.

temperature rises, for $\bar{\alpha}$ is positive. The order of magnitude of the expansion is about 1 millimeter per meter length per 100 Celsius degrees.**

EXAMPLE 2

A steel metric scale is to be ruled so that the millimeter intervals are accurate to within about 5×10^{-5} mm at a certain temperature. What is the maximum temperature variation allowable during the ruling?

From Eq. 21-6,

$$\Delta l = \alpha l \, \Delta T,$$

we have

$$5 \times 10^{-5} \text{ mm} = (11 \times 10^{-6}/\text{C°})(1.0 \text{ mm}) \, \Delta T$$

in which we have used $\bar{\alpha}$ for steel, taken from Table 21-3. This yields $\Delta T \cong$ 5 C°. The temperature maintained during the ruling process must be maintained when the scale is being used and it must be held constant to within about 5 C°.

Note (see Table 21-3) that if the alloy invar is used instead of steel, then for the same required tolerance one can permit a temperature variation of about 75 C°; or for the same temperature variation ($\Delta T = 5$ C°) the tolerance achieved would be more than an order of magnitude better.

On the microscopic level thermal expansion of a solid suggests an increase in the average separation between the atoms in the solid. The potential energy curve for two adjacent atoms in a crystalline solid as a function of their internuclear separation is an asymmetric curve like that of Fig. 21-6. As the atoms move close together, their separation decreasing from the equilibrium value r_0, strong repulsive forces come into play and the potential curve rises steeply $(F = -dU/dr)$; as the atoms move farther apart, their separation increasing from the equilibrium value, somewhat weaker attractive forces take over and the potential curve rises more slowly. At a given vibrational energy the separation of the atoms will change periodically from a minimum to a maximum value, the average separation being greater than the equilibrium separation because of the asymmetric nature of the potential energy curve. At still higher vibrational energy the average separation will be even greater. The effect is enhanced by the fact that in taking a time average of the motion one must allow for the longer time spent at extreme separations (lower vibrational speeds). Because the vibrational energy increases as the temperature rises, the average separation between atoms increases with temperature and the solid as a whole expands.

Note that if the potential energy curve were symmetric about the equilibrium separation, then no matter how large the amplitude of the vibration becomes the average separation would correspond to the equilibrium separation. Hence, thermal expansion is a direct consequence of the deviation from

figure 21-6
Potential energy curve for two adjacent atoms in a crystalline solid as a function of internuclear separation. The equilibrium separation is r_0. Because the curve is asymmetric the average separation (r_1, r_2) increases as the temperature (T_1, T_2), and hence the vibrational energy (E_1, E_2), increases.

** One Celsius degree (1 C°) is a temperature *interval* (ΔT_C) of one unit measured on a Celsius scale. One degree Celsius (1° C) is a specific temperature reading (T_C) on that scale.

symmetry (i.e., the assymetry) of the potential energy curve characteristic of solids.

Some crystalline solids, in certain temperature regions, may contract as the temperature rises. The above analysis remains valid if one assumes that only compressional (i.e., longitudinal) modes of vibration exist or that these modes predominate. However, solids may vibrate in shear-like (i.e., transverse) modes as well and these modes of vibration will allow the solid to contract as the temperature rises, the average separation of the planes of atoms decreasing. For certain types of crystalline structure and in certain temperature regions these transverse modes of vibration may predominate over the longitudinal ones, giving a net negative coefficient of thermal expansion.

It should be emphasized that the microscopic models presented here are oversimplifications of a complex phenomenon which can be treated with greater insight with the use of thermodynamics and quantum theory.

For many solids, called *isotropic*, the percent change in length for a given temperature change is the same for all lines in the solid. The expansion is quite analogous to a photographic enlargement, except that a solid is three-dimensional. Thus, if you have a flat plate with a hole punched in it, $\Delta l/l \; (= \alpha \Delta T)$ for a given ΔT is the same for the length, the thickness, the face diagonal, the body diagonal, and the hole diameter. Every line, whether straight or curved, lengthens in the ratio α per degree temperature rise. If you scratch your name on the plate, the line representing your name has the same fractional change in length as any other line. The analogy to a photographic enlargement is shown in Fig. 21-7.

(a)

(b)

figure 21-7
The same steel rule at two different temperatures. On expansion every dimension is increased by the same proportion: the scale, the numbers, the hole, and the thickness are all increased by the same factor. (The expansion shown, from (a) to (b), is obviously exaggerated, for it would correspond to an imaginary temperature rise of about 100,000 C°!)

With these ideas in mind, you should be able to show (see Problems 14 and 16) that to a high degree of accuracy the fractional change in area A per degree temperature change for an isotropic solid is 2α, that is,

$$\Delta A = 2\alpha A \; \Delta T,$$

and the fractional change in volume V per degree temperature change for an isotropic solid is 3α, that is,

$$\Delta V = 3\alpha V \; \Delta T.$$

Because the shape of a fluid is not definite, only the change in volume with temperature is significant. Gases respond strongly to temperature or pressure changes, whereas the change in volume of liquids with changes in temperature or pressure is very much smaller. If we let β represent the coefficient of volume expansion for a liquid, that is,

$$\beta = \frac{1}{V}\frac{\Delta V}{\Delta T},$$

we find that β is relatively independent of the temperature. Liquids typically expand with increasing temperature, their volume expansion being generally about ten times greater than that of solids.

However, the most common liquid, water, does not behave like other liquids. In Fig. 21-8 we show the expansion curve for water. Notice that above 4° C water expands as the temperature rises, although not linearly. As the temperature is lowered from 4 to 0° C, however, water expands instead of contracting. Such an expansion with decreasing temperature is not observed in any other common liquid; it is observed in rubberlike substances and in certain crystalline solids over limited temperature intervals. The density of water is a maximum at 4° C, where its value* is 1000 kg/m³ or 1.000 g/cm³. At all other temperatures its density is less. This behavior of water is the reason why lakes freeze first at their upper surface.

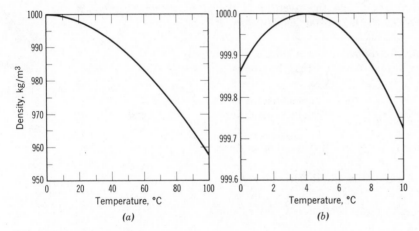

(a)   *(b)*

figure 21-8
(a) The variation with temperature of density of water under atmospheric pressure. *(b)* The variation between 0 and 10° C in more detail.

questions

1. Is temperature a microscopic or macroscopic concept?
2. Are there physical quantities other than temperature that tend to equalize if two different systems are joined?
3. Give a reasonable explanation for this: a piece of ice and a thermometer are suspended in an insulated evacuated enclosure so that they are not in contact and yet the thermometer reading decreases for a time.
4. Can a temperature be assigned to a vacuum?
5. Does our "temperature sense" have a built-in sense of direction; that is, does hotter necessarily mean higher temperature, or is this just an arbitrary convention? Celsius, by the way, originally chose the steam point as 0° C and the ice point as 100° C.
6. Figure 21-1 shows an apparatus by which the triple point of water is realized. How would you modify this apparatus to realize the freezing point of water?
7. How would you suggest measuring the temperature of (a) the sun, (b) the

* It is to this value of *unit* maximum density of water that the relative sizes of the kilogram and meter were originally supposed to correspond. Accurate measurements show, however, that the international standards of mass and length do not correspond exactly to this value. The maximum density of water is actually 999.973 kg/m³ at 3.98° C.

earth's upper atmosphere, (c) an insect, (d) the moon, (e) the ocean floor, and (f) liquid helium?

8. Is one gas any better than another for purposes of a standard constant-volume gas thermometer? What properties are desirable in a gas for such purposes?

9. State some objections to using water-in-glass as a thermometer. Is mercury-in-glass an improvement?

10. Can you explain why the column of mercury first descends and then rises when a mercury-in-glass thermometer is put in a flame?

11. What do the Celsius and Fahrenheit temperature conventions have in common?

12. Considering the Celsius, Fahrenheit, and Kelvin scales, does any one stand out as "Nature's scale"? Discuss.

13. What are the dimensions of α, the coefficient of linear expansion? Does the value of α depend on the unit of length used? When F° are used instead of C° as a unit of temperature change, does the numerical value of α change? If so, how?

14. A metal ball can pass through a metal ring. When the ball is heated, however, it gets stuck in the ring. What would happen if the ring, rather than the ball, were heated?

15. A bimetallic strip, consisting of two different metal strips riveted together, is used as a control element in the common thermostat. Explain how it works.

16. Two strips, one of iron and one of zinc, are riveted together side by side to form a straight bar which curves when heated. Why is it that the iron is on the inside of the curve?

17. Explain how the period of a pendulum clock can be kept constant with temperature by attaching tubes of mercury to the bottom of the pendulum. (See Problem 32.)

18. Explain why some rubberlike substances contract with rising temperature. (See Question 25, Chapter 25.)

19. Explain why the apparent expansion of a liquid in a bulb does not give the true expansion of the liquid.

20. Why do liquids typically have much larger volume coefficients of expansion than solids?

21. Does the change in volume of a body when its temperature is raised depend on whether the body has cavities inside, other things being equal? Consider a solid sphere and a hollow sphere, for example.

22. What difficulties would arise if you defined temperature in terms of the density of water?

23. Explain why lakes freeze first at the surface.

24. What is it that causes water pipes to burst in the winter?

25. What can you conclude about how the melting point of ice depends on pressure from the fact that ice floats on water?

SECTION 21-3

1. A *resistance thermometer* is a thermometer in which the thermometric property is electrical resistance. We are free to define temperatures measured by such a thermometer in Kelvins to be directly proportional to the resistance R, measured in ohms. A certain resistance thermometer is found to have a resistance R of 90.35 ohms when its bulb is placed in water at the triple point temperature (273.16 K). What temperature is indicated by the thermometer if the bulb is placed in an environment such that its resistance is 96.28 ohms? *Answer:* 291.1 K.

problems

2. It is an everyday observation that hot and cold objects cool down or warm up to the temperature of their surroundings. If the temperature ΔT between an object and its surroundings is not too great, the rate of cooling or warming is approximately proportional to the temperature difference between the object and its surroundings; that is,

$$\frac{d\,\Delta T}{dt} = -K\,\Delta T,$$

where K is a constant. The minus sign appears because ΔT decreases with time if ΔT is positive and vice versa. This is known as *Newton's law of cooling*. (a) On what factors does K depend? What are its dimensions? (b) If at some instant $t = 0$ the temperature difference is ΔT_0, show that it is

$$\Delta T = \Delta T_0 e^{-Kt}$$

at a time t later.

3. A mercury-in-glass thermometer is placed in boiling water for a few minutes and then removed. The temperature readings at various times after removal are as follows:

t, s	T, °C	t, s	T, °C	t, s	T, °C	t, s	T, °C
0	98.4	25	65.1	100	50.3	700	26.5
5	76.1	30	63.9	150	43.7	1000	26.1
10	71.1	40	61.6	200	38.8	1400	26.0
15	67.7	50	59.4	300	32.7	2000	26.0
20	66.4	70	55.4	500	27.8	3000	26.0

Plot K as a function of time, assuming Newton's law of cooling to apply [see Problem 2]. To what extent are you justified in assuming that Newton's law of cooling applies here?

SECTION 21-5

4. If the ideal gas temperature at the steam point is 373.15 K, what is the limiting value of the ratio of the pressures of a gas at the steam point and at the triple point of water when the gas is kept at constant volume?

5. Let p_{tr} be the pressure in the bulb of a constant-volume gas thermometer when the bulb is at the triple-point temperature of 273.16 K and p the pressure when the bulb is at room temperature. Given three constant-volume gas thermometers: for No. 1 the gas is oxygen and $p_{tr} = 20$ cm-Hg; for No. 2 the gas is also oxygen but $p_{tr} = 40$ cm-Hg; for No. 3 the gas is hydrogen and $p_{tr} = 30$ cm-Hg. The measured values of p for the three thermometers are p_1, p_2, and p_3. (a) An approximate value of the room temperature T can be obtained with each of the thermometers using

$$T_1 = 273.16 \text{ K} \frac{p_1}{20 \text{ cm-Hg}}; \quad T_2 = 273.16 \text{ K} \frac{p_2}{40 \text{ cm-Hg}};$$

$$T_3 = 273.16 \text{ K} \frac{p_3}{30 \text{ cm-Hg}}.$$

Mark each of the following statements "true" or "false": (1) With the method described, all three thermometers will give the same value of T. (2) The two oxygen thermometers will agree with each other but not with the hydrogen thermometer. (3) Each of the three will give a different value of T. (b) In the event that there is disagreement among the three thermometers, explain how you would change the method of using them to cause all three to give the same value of T.

Answer: (a) (1) False; (2) false; (3) true. (b) Take the limiting value as $p_{tr} \to 0$.

SECTION 21-6

6. (a) The temperature of the surface of the sun is about 6000 K. Express this on the Fahrenheit scale. (b) Express normal human body temperature, 98.6° F, on the Celsius scale. (c) In the continental United States, the highest officially recorded temperature is 134° F at Death Valley, California, and the lowest is −70° F at Rogers Pass, Montana. Express these extremes on the Celsius scale. (d) Express the normal boiling point of oxygen, −183°C, on the Fahrenheit scale. (e) At what Celsius temperature would you find a room to be uncomfortably warm?

7. At what temperature do the following pairs of scales give the same reading? (a) Fahrenheit and Celsius. (b) Fahrenheit and Kelvin. (c) Celsius and Kelvin. *Answer:* (a) −40°. (b) 575°. (c) Not possible.

SECTION 21-7

8. In the interval between 0 and 660° C, a platinum resistance thermometer of definite specifications is used for interpolating temperatures on the International Practical Temperature Scale. The temperature T_C is given by a formula for the variation of resistance with temperature:

$$R = R_0(1 + AT_c + BT_c{}^2).$$

R_0, A, and B are constants determined by measurements at the ice point, the steam point, and the sulphur point. (a) If R equals 10.000 ohms at the ice point, 13.946 ohms at the steam point, and 24.817 ohms at the sulphur point, find R_0, A, and B. (b) Plot R versus T_C in the temperature range from 0 to 660° C.

SECTION 21-8

9. The Pyrex glass mirror in the telescope at the Mount Palomar Observatory has a diameter of 200 in. The temperature ranges from −10° to 50° C on Mount Palomar. Determine the maximum change in the diameter of the mirror. *Answer:* 0.038 in.

10. A circular hole in an aluminum plate is 1.000 in. (2.540 cm) in diameter at 0° C. What is its diameter when the temperature of the plate is raised to 100° C?

11. Steel railroad tracks are laid when the temperature is 0° C. A standard section of rail is then 12.0 m long. What gap should be left between rail sections so that there is no compression when the temperature gets as high as 42° C? *Answer:* 0.55 cm.

12. A steel rod is 3.000 cm in diameter at 25° C. A brass ring has an interior diameter of 2.992 cm at 25° C. At what common temperature will the ring just slide onto the rod?

13. Show that if α is treated as a variable, dependent on the temperature T, then

$$L = L_0\left[1 + \int_{T_0}^{T} \alpha(T)\, dT\right]$$

where L_0 is the length at a reference temperature T_0.

14. The area A of a rectangular plate is ab. Its coefficient of linear expansion is α. After a temperature rise ΔT, side a is longer by Δa and side b is longer by Δb. Show that if we neglect the small quantity $\Delta a\, \Delta b/ab$ (see Fig. 21-9), then $\Delta A = 2\alpha A\, \Delta T$.

15. A glass window is exactly 20 cm (7.9 in.) by 30 cm (11.8 in.) at 10° C. By how much has its area increased when its temperature is 40° C? *Answer:* 0.32 cm² (0.050 in.²).

16. Prove that, if we neglect extremely small quantities, the change in volume of a solid on expansion through a temperature rise ΔT is given by $\Delta V = 3\alpha V\, \Delta T$ where α is the coefficient of linear expansion.

figure 21-9
Problem 14

17. Find the change in volume of an aluminum sphere of 10.0-cm (3.94-in.) radius when it is heated from 0° to 100° C.   *Answer:* 29 cm³ (1.8 in.³).

18. When the temperature of a "copper" penny is raised by 100 C°, its diameter increases by 0.18%. To two significant figures give the percent increase in the (a) area of a face, (b) thickness, (c) volume, and (d) mass of the penny. (e) What is the coefficient of linear expansion?

19. Density is mass per unit volume. If the volume V is temperature dependent, so is the density ρ. Show that the change in density $\Delta\rho$ with change in temperature ΔT is given by

$$\Delta\rho = -\beta\rho\,\Delta T$$

where β is the volume coefficient of expansion. Explain the minus sign.

20. Show that when the temperature of a liquid in a barometer changes by ΔT, and the pressure is constant, the height h changes by $\Delta h = \beta h\,\Delta T$ where β is the coefficient of volume expansion.

21. (a) Show that if the lengths of two rods of different solids are inversely proportional to their respective coefficients of linear expansion at some initial temperature, the difference in length between them will be constant at all temperatures. (b) What should be the lengths of a steel and a brass rod at 0° C so that at all temperatures their difference in length is 0.30 m? *Answer:* (b) Steel, 71 cm; brass, 41 cm.

22. Consider a mercury-in-glass thermometer. Assume that the cross-section of the capillary is constant at A_0, and that V_0 is the volume of the bulb of mercury at 0.00° C. If the mercury just fills the bulb at 0.00° C, show that the length of the mercury column in the capillary at a temperature $t°$ C is

$$l = \frac{V_0}{A_0}\,(\beta - 3\alpha)t,$$

that is, proportional to the temperature, where β is the volume coefficient of expansion of mercury and α is the linear coefficient of expansion of glass.

23. Imagine an aluminum cup of 0.1 liter capacity filled with mercury at 12° C. How much mercury, if any, will spill out of the cup if the temperature is raised to 18° C? (The coefficient of volume expansion of mercury is $1.8 \times 10^{-4}/$C°.)   *Answer:* 70 mm³.

24. A clock pendulum made of Invar has a period of 0.500 s at 20° C. If the clock is used in a climate where the temperature averages 30° C, what correction (approximately) is necessary at the end of 30 days to the time given by the clock?

25. (a) Prove that the change in rotational inertia I with temperature of a solid object is given by $\Delta I = 2\alpha I\,\Delta T$. (b) Prove that the change in period t of a physical pendulum with temperature is given by $\Delta t = \frac{1}{2}\alpha t\,\Delta T$.

26. Consider a uniform solid brass cylinder of mass $M = 0.50$ kg and radius $R = 0.030$ m. The cylinder is placed in frictionless bearings and set to rotate about its cylinder axis with an angular velocity $\omega = 60$ rad/s. (a) What is the angular momentum of the cylinder and how much work is required to reach this rate of rotation, starting from rest? After the cylinder has reached the state of rotation just described we heat it, without mechanical contact, from room temperature (20° C) to 100° C. Take the mean coefficient of linear expansion of brass to be $\bar{\alpha} = 2.0 \times 10^{-5}/$C°. Find the fractional changes, if any, in (b) the angular velocity, (c) the angular momentum, and (d) the kinetic energy of rotation of the cylinder.

27. A 1.0-m long vertical glass tube is half-filled with a liquid at 20° C. How much will the height of the liquid column change when the tube is heated to 30° C? Take $\bar{\alpha}_{glass} = 1.0 \times 10^{-5}/$C° and $\bar{\beta}_{liquid} = 4 \times 10^{-5}/$C°. *Answer:* Increases by 0.10 mm.

28. A solid aluminim cylinder is suspended by a flexible steel belt attached to opposite walls at the same level, as shown in Fig. 21-10. It is required that the axis C of the cylinder not be moved by thermal expansions and contrac-

figure 21-10
Problem 28

tions of the cylinder and belt. The angle $\theta = 50°$ and remains practically unaffected by temperature changes. Find the radius R of the cylinder when $T = 290$ K if $L = 2.5$ m at this temperature. (Neglect the weight of the belt.)

29. Two vertical glass tubes filled with a liquid are connected at their lower ends by a horizontal capillary tube. One tube is surrounded by a bath containing ice and water in equilibrium (0.0° C), the other by a hot-water bath (t). The difference in height of the liquids in the two columns is Δh, and h_0 is the height of the column at 0.0° C. (a) Show how this apparatus (Fig. 21-11), first used in 1816 by Dulong and Petit, can be used to measure the true coefficient of volume expansion β of a liquid (rather than the differential expansion between glass and liquid). (b) Determine β if $t = 16.0°$ C, $h_0 = 126$ cm, and $\Delta h = 1.50$ cm.   *Answer:* (b) $7.44 \times 10^{-4}/C°$.

30. An aluminum cube 20 cm on an edge floats on mercury. How much further will the block sink down when the temperature rises from 270 K to 320 K? (The coefficient of volume expansion of mercury is $1.8 \times 10^{-4}/C°$.)

31. The distance between the towers of the main span of the Golden Gate Bridge at San Francisco is 4200 ft. The sag of the cable halfway between the towers at 50° F is 470 ft. Take $\alpha = 6.5 \times 10^{-6}/F°$ for the cable and compute (a) the change in length of the cable and (b) the change in sag for a temperature change from −20 to 110° F. Assume no bending or separation of the towers and a parabolic shape for the cable.   *Answer:* (a) 3.7 ft. (b) 6.5 ft.

32. A glass tube nearly filled with mercury is attached in tandem to the bottom of an iron pendulum rod 100 cm long. How high must the mercury be in the glass tube so that the center of mass of this pendulum will not rise or fall with changes in temperature? (The cross-sectional area of the tube is equal to that of the iron rod. Neglect the mass of the glass. Iron has a density of 7.87×10^3 kg/m³ and a linear coefficient of expansion equal to $12 \times 10^{-6}/C°$. The coefficient of volume expansion of mercury is $18 \times 10^{-5}/C°$.)

figure 21-11
Problem 29

22
heat and the first law of thermodynamics

When two systems at different temperatures are placed together, the final temperature reached by both systems is somewhere between the two starting temperatures. This is a common observation. Humans have long sought for a deeper understanding of such phenomena. Up to the beginning of the nineteenth century, they were explained by postulating that a material substance, *caloric*, existed in every body. It was believed that a body at high temperature contained more caloric than one at a low temperature. When the two bodies were put together, the body rich in caloric lost some to the other until both bodies reached the same temperature. The caloric theory was able to describe many processes, such as heat conduction or the mixing of substances in a calorimeter, in a satisfactory way. However, the concept of heat as a *substance*, whose total amount remained constant, eventually could not stand the test of experiment. Nevertheless, we still describe many common temperature changes as the transfer of "something" from one body at a higher temperature to one at the lower, and this "something" we call heat. A useful but nonoperational definition, is: *heat is that which is transferred between a system and its surroundings as a result of temperature differences only.*

Eventually it became generally understood that heat is a form of energy rather than a substance. The first conclusive evidence that heat could not be a substance was given by Benjamin Thompson (1753–1814), an American who later became Count Rumford of Bavaria. In a paper read before the Royal Society* in 1798 he wrote:

22-1
HEAT, A FORM OF ENERGY

I . . . am persuaded, that a habit of keeping the eyes open to everything that is going on in the ordinary course of the business of life has oftener led, as it were by accident, or in the playful excursions of the imagination . . . to useful doubts and sensible schemes for investigation and improvement, than all the more intense meditations of philosophers, in the hours expressly set apart for study. It was by accident that I was led to make the Experiments of which I am about to give an account.

Rumford made this discovery while supervising the boring of cannon for the Bavarian government. To prevent overheating, the bore of the cannon was kept full of water. The water was replenished as it boiled away during the boring process. It was accepted that caloric had to be supplied to water to boil it. The continuous production of caloric was explained by assuming that when a substance was more finely subdivided, as in boring, its capacity for retaining caloric became smaller, and that the caloric released in this way was what caused the water to boil. Rumford observed in specific experiments, however, that the water boiled away even when his boring tools became so dull that they were no longer cutting or subdividing matter.

He wrote after ruling out by experiment all possible caloric interpretations,

. . . in reasoning on this subject, we must not forget to consider that most remarkable circumstance, that the source of Heat generated by friction, in these Experiments, appeared evidently to be *inexhaustible* . . . it appears to me to be extremely difficult, it not quite impossible, to form any distinct idea of any thing capable of being excited and communicated in the manner the Heat was excited and communicated in these Experiments, except it be *MOTION*.

Here we have the germ of the idea that the mechanical work expended in the boring process was responsible for the creation of heat. The idea was not clearly put until much later, by others. Instead of the continuous disappearance of mechanical energy and the continuous creation of heat, neither obeying any conservation principle, the whole process is now viewed as a transformation of energy from one form to another, the total energy being conserved.

Although the concept of energy and its conservation seems self-evident today, it was a novel idea as late as the 1850s and had eluded such men as Galileo and Newton. Throughout the subsequent history of physics this conservation idea led to new discoveries. Its early history was remarkable in many ways. Several thinkers arrived at this great concept at about the same time; at first, all of them either met with a cold reception or were ignored. The principle of the conservation of energy was established independently by Julius Mayer (1814–1878) in Germany, James Joule (1818–1889) in England, Hermann von Helmholtz (1821–1894) in Germany, and L. A. Colding (1815–1888) in Denmark.*

It was Joule who showed by experiment that, when a given quantity of mechanical energy is converted to heat, the same quantity of heat

---

* After the posthumous publication of his *Reflections on the Motive Power of Fire* (in 1872, 40 years after his death) it became clear that Sadi Carnot (1796–1832) had arrived at the conservation of energy principle before all the others. It should give some food for thought to realize that the five men who first understood the conservation of energy principle were all young and all had major professional interests outside the field of physics: Mayer (medicine; age 28), Helmholtz (physiology; age 32), Colding (engineering; age 27), Joule (industrial management—he inherited his father's brewery; age 25), and Carnot (engineering; age 34). Rumford, age 45, was an old man by comparison.

is always developed. Thus, the equivalence of heat and mechanical work as two forms of energy was definitely established.

Helmholtz first expressed clearly the idea that not only heat and mechanical energy but all forms of energy are equivalent, and that a given amount of one form cannot disappear without an equal amount appearing in some of the other forms.

22-2
QUANTITY OF HEAT AND SPECIFIC HEAT

The unit of heat Q used to be defined* quantitatively in terms of a specified change produced in a body during a specified process. Thus, if the temperature of one kilogram of water is raised from 14.5 to 15.5° C by heating, we say that one *kilocalorie* (kcal) of heat has been added to the system. The *calorie* (= 10^{-3} kcal) is also used as a heat unit. (Incidentally, the "calorie" used to measure the energy content of foods is actually a kilocalorie.) In the engineering system the unit of heat is the *British thermal unit* (Btu), which is defined as the heat necessary to raise the temperature of one pound of water from 63 to 64° F.

The reference temperatures are stated because, near room temperature, there is a slight variation in the heat needed for a one-degree temperature rise with the temperature interval chosen. We will neglect this variation for most practical purposes. The heat units are related as follows:

$$1.000 \text{ kcal} = 1000 \text{ cal} = 3.968 \text{ Btu.}$$

Substances differ from one another in the quantity of heat needed to produce a given rise of temperature in a given mass. The ratio of the amount of heat energy ΔQ supplied to a body to its corresponding temperature rise ΔT is called the *heat capacity C* of the body; that is,

$$C = \text{heat capacity} = \frac{\Delta Q}{\Delta T}.$$

The word "capacity" may be misleading because it suggests the essentially meaningless statement "the amount of heat a body can hold," whereas what is meant is simply the energy that must be added as heat in order to raise the temperature of the body one degree.

The heat capacity per unit mass of a body, called *specific heat*, is characteristic of the material of which the body is composed:

$$c = \frac{\text{heat capacity}}{\text{mass}} = \frac{\Delta Q}{m \, \Delta T}. \tag{22-1}$$

We properly speak, on the one hand, of the heat capacity of a penny but, on the other, of the specific heat of copper.

Neither the heat capacity of a body nor the specific heat of a material is constant but depends on the location of the temperature interval. The previous equations give only average values for these quantities in the temperature range of ΔT. In the limit, as $\Delta T \to 0$, we can speak of the specific heat at a particular temperature T.

The heat that must be given to a body of mass m, whose material has a specific heat capacity c, to increase its temperature from T_i to T_f, is, assuming $\Delta T \ll T_f - T_i$,

$$Q = \Sigma \, \Delta Q = \sum_{T_i}^{T_f} mc \, \Delta T. \tag{22-2}$$

* We shall see in Section 22-5 how the calorie is now defined.

In the differential limit this becomes

$$Q = m \int_{T_i}^{T_f} c \, dT \qquad (22\text{-}3)$$

where c is a function of the temperature. At ordinary temperatures and over ordinary temperature intervals, specific heats can be considered to be constants. Figure 22-1 shows the variation in the specific heat of water with temperature. Information of this sort is obtained by using an electrical heating coil to supply heat at a rate that can be accurately determined. We see from the graph that the specific heat of water varies less than 1% from its value of 1.000 cal/g·C° at 15° C.

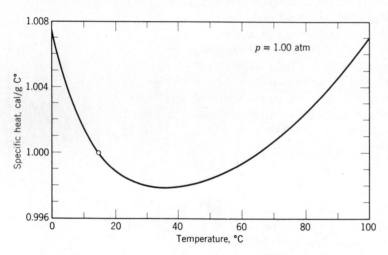

figure 22-1
The variation with temperature of the specific heat of water at a pressure of 1.00 atm. The circle, located at 15° C, suggests the definition of the calorie.

Equation 22-1 does not define specific heat uniquely. We must also specify the conditions under which the heat ΔQ is added to the specimen. We have implied that the condition is that the specimen remain at normal (constant) atmospheric pressure while we add the heat. This is a common condition, but there are many other possibilities, each leading, in general, to a different value for c. To obtain a unique value for c we must specify the conditions, such as specific heat at constant pressure c_p, specific heat at constant volume c_v, etc.

Table 22-1
Values for c_p for some solids
(at room temperature and for $p = 1.0$ atm)

Substance	Specific heat cal/g C°	Specific heat J/g C°	Molecular weight g/mol	Molar heat capacity cal/mol C°	Molar heat capacity J/mol C°
Aluminum	0.215	0.900	27.0	5.82	24.4
Carbon	0.121	0.507	12.0	1.46	6.11
Copper	0.0923	0.386	63.5	5.85	24.5
Lead	0.0305	0.128	207	6.32	26.5
Silver	0.0564	0.236	108	6.09	25.5
Tungsten	0.0321	0.134	184	5.92	24.8

Table 22-1 (second and third columns) shows the specific heats at constant pressure of some solid elements; we will discuss the specific heats of gases later. You should realize from the way the calorie and the Btu are defined that 1 cal/g·C° = 1 kcal/kg C° = 1 Btu/lb F°, exactly. Note that the specific heat of water, equal to 1.00 cal/g·C°, is large compared to that of most substances.

EXAMPLE 1

A 75-gram block of copper, taken from a furnace, is dropped into a 300-gram glass beaker containing 200 grams of water. The temperature of the water rises from 12 to 27° C. What was the temperature of the furnace?

This is an example of two systems originally at different temperatures reaching thermal equilibrium after contact. No mechanical energy is involved, only heat exchange. Hence,

heat from copper = heat to (beaker + water),

$$m_C c_C (T_C - T_e) = (m_G c_G + m_W c_W)(T_e - T_W).$$

The subscript C stands for copper, G for glass, and W for water. The initial copper temperature is T_C, the initial beaker water temperature is T_W, and T_e is the final equilibrium temperature. Substituting the given values, with $c_C = 0.093$ cal/g·C°, $c_G = 0.12$ cal/g·C°, and $c_W = 1.0$ cal/g·C°, we obtain

$$(75 \text{ g})(0.093 \text{ cal/g·C°})(T_C - 27° \text{ C}) = [(300 \text{ g})(0.12 \text{ cal/g·C°})$$
$$+ (200 \text{ g})(1.0 \text{ cal/g·C°})](27° \text{ C} - 12° \text{ C})$$

or, solving for T_C, $\qquad T_C = 530°$ C

What approximations, both experimental and theoretical, were used implicitly to arrive at this answer?

22-3
MOLAR HEAT CAPACITIES OF SOLIDS

From the second column of Table 22-1 we conclude that the specific heats of solids vary widely from one material to another. However quite a different story emerges if we compare samples of materials that contain the same number of molecules rather than samples that have the same mass. We can do this by expressing specific heats (called when so expressed *molar heat capacities*) in cal/mol·C° rather than in cal/g·C°.* In 1819 Dulong and Petit pointed out that the molar heat capacities of all substances, with few exceptions (see carbon in Table 22-1), have values close to 6 cal/mol C°. The molar heat capacity, listed in the fifth and sixth columns of Table 22-1, is found by multiplying the specific heat (second and third columns) by the molecular weight (fourth column). We see that the amount of heat required *per molecule* to raise the temperature of a solid by a given amount seems to be about the same for almost all materials. This is striking evidence for the molecular theory of matter.

Actually molar heat capacities vary with temperature, approaching zero as $T \to 0$ K and approaching the Dulong-Petit value as $T \to \infty$. Since the number of molecules rather than the kind of molecule seems to be important in determining the heat required to increase the temperature of a body by a given amount, we are led to expect that the molar heat capacities of different substances will vary with temperature in much the same way. Figure 22-2 shows that, indeed, the molar heat capacities of various substances can be made to fall on the same curve by a simple, empirical adjustment in the temperature scale. The horizontal scale in Fig. 22-2 is the dimensionless ratio T/T_D, where T is the Kelvin temperature and T_D is a characteristic temperature, called the *Debye temperature*, that has a particular constant value for each material. For lead, T_D has the empirical value of 88 K and for carbon, $T_D = 1860$ K. From these data you can show that a scale value of $T/T_D = 0.600$ corresponds to $T = 53$ K for lead but to $T = 1120$ K for carbon. Alternatively, room temperature (~ 300 K) corresponds to $T/T_D = 3.4$ for lead and to $T/T_D = 0.16$ for carbon. Thus

* A mole (abbr. *mol*) of any substance is the amount of the substance that contains a specified number of elementary entities, namely, 6.02252×10^{23}, called Avogadro's number. This number is the result of the defining relation that one mol of carbon atoms (actually, of the isotope C^{12}) shall have a mass of 12 g, exactly. The *gram molecular weight* M of a substance is the number of grams per mole of that substance. Thus the gram molecular weight of ordinary oxygen molecules is 32.0 g/mol. Although the mole is an amount of substance, we cannot translate it into mass, as grams, until we specify what the elementary entity is; it may be atoms, molecules, ions, electrons, other particles, or specified groups of such particles.

we see from Fig. 22-2 that in the early days, when only room temperature specific heats were available, lead would conform to the Dulong and Petit rule but carbon would seem to be an exception.

The straight line *I* in Fig. 22-2 is the Dulong and Petit value of 1819; it agrees with experiment at high temperature but fails at low temperatures. It corresponds to the assumption that every atom in a solid vibrates independently like a classical oscillator. Curve II is due to Debye (1912). In the Debye theory, a characteristic temperature T_D, which is directly related to a vibrational frequency characteristic of the material, can be obtained independent of specific heat experiments. One then uses quantum principles to analyze the coupled vibrations of the atoms in a solid and obtains a specific heat formula which, in terms of the dimensionless ratio T/T_D, is the same for all substances. The excellent agreement of this formula (curve II) with experiment is a triumph of quantum physics.*

The materials displayed in Fig. 22-2 are "normal" in that they do not melt, boil, change their crystal structure, etc., in the temperature range indicated. Specific heat measurements, which tells us how a solid absorbs energy as its temperature is raised, are a sensitive probe to detect such molecular, atomic, or electronic rearrangements. Figure 22-3, for example, shows the specific heat of tantalum near 4.39 K. Below this transition temperature tantalum loses all its electric resistance—it becomes superconducting. Above this temperature it has the resistance expected of a normal metal.

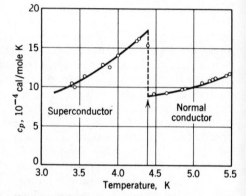

22-4 *HEAT CONDUCTION*

The transfer of energy arising from the temperature difference between adjacent parts of a body is called *heat conduction*. Consider a slab of material of cross-sectional area A and thickness Δx, whose faces are kept at different temperatures. We measure the heat ΔQ that flows perpendicular to the faces in a time Δt. Experiment shows that ΔQ is proportional to Δt and to the cross-sectional area A for a given temperature difference ΔT, and that ΔQ is proportional to $\Delta T/\Delta x$ for a given Δt and A, providing both ΔT and Δx are small. That is,

$$\frac{\Delta Q}{\Delta t} \propto A \frac{\Delta T}{\Delta x} \qquad \text{approximately.} \qquad (22\text{-}4a)$$

---

* The data reported in Fig. 22-2 are values of c_v but those in Table 22-1 are c_p. The former is easier to calculate theoretically because the thermal expansion need not be taken into account, but (for solids) the latter is much easier to measure. The two are related by the simple thermodynamic formula

$$c_p = c_v + T\beta^2/\kappa\rho$$

in which β is the thermal coefficient of volume expansion, κ ($= -\Delta V/V\Delta p$) is the (isothermal) compressibility, and ρ is the density. At room temperature the difference between c_p and c_v for typical solids is about 5%.

In the limit of a slab of infinitesimal thickness dx, across which there is a temperature difference dT, we obtain the fundamental law of heat conduction, in which the heat flow H is given by

$$H = -kA\frac{dT}{dx}. \tag{22-4b}$$

Here H (measured, say, in cal/s; see Eq. 22-4a) is the time rate of heat transfer across the area A, dT/dx is called the *temperature gradient*, and k is a constant of proportionality called the *thermal conductivity*. We choose the direction of heat flow to be the direction in which x increases; since heat flows in the direction of decreasing T, we introduce a minus sign in Eq. 22-4 (that is, we wish H to be positive when dT/dx is negative).

A substance with a large thermal conductivity k is a good heat conductor; one with a small thermal conductivity k is a poor heat conductor, or a good thermal insulator. The value of k depends on the temperature, increasing slightly with increasing temperature, but k can be taken to be practically constant throughout a substance if the temperature difference between its parts is not too great. In Table 22-2 we list values of k for various substances; we see that metals as a group are better heat conductors than nonmetals, and that gases are poor heat conductors.

Let us apply Eq. 22-4b to a rod of length L and constant cross-sectional area A in which a steady state has been reached (Fig. 22-4). In a

Table 22-2
Thermal conductivities
(Gases at 0° C; others at about room temperature)

	kcal/s·m·C°	J/s·m·C°
Metals		
Aluminum	4.9×10^{-2}	20×10^{1}
Brass	2.6×10^{-2}	11×10^{1}
Copper	9.2×10^{-2}	39×10^{1}
Lead	8.3×10^{-3}	35
Silver	9.9×10^{-2}	41×10^{1}
Steel	1.1×10^{-2}	46
Gases		
Air	5.7×10^{-6}	2.4×10^{-2}
Hydrogen	3.3×10^{-5}	1.4×10^{-1}
Oxygen	5.6×10^{-6}	2.3×10^{-2}
Others		
Asbestos	2×10^{-5}	8×10^{-2}
Concrete	2×10^{-4}	8×10^{-1}
Cork	4×10^{-5}	17×10^{-2}
Glass	2×10^{-4}	8×10^{-1}
Ice	4×10^{-4}	17×10^{-1}
Wood	2×10^{-5}	8×10^{-2}

figure 22-4
Conduction of heat through an insulated conducting bar.

steady state the temperature at each point is constant in time. Hence, H is the same at all cross-sections. (Why?) But $H = -kA(dT/dx)$, so that, for a constant k and A, the temperature gradient dT/dx is the same at all cross-sections. Hence, T decreases linearly along the rod so that $-dT/dx = (T_2 - T_1)/L$. Therefore, the time rate of transfer of heat energy is

$$H = kA \frac{T_2 - T_1}{L}. \qquad (22\text{-}5)$$

The phenomenon of heat conduction also shows that the concepts of heat and temperature are distinctly different. Different rods, having the same temperature difference between their ends, may transfer entirely different quantities of heat in the same time.

EXAMPLE 2

Consider a compound slab, consisting of two materials having different thicknesses, L_1 and L_2, and different thermal conductivities, k_1 and k_2. If the temperatures of the outer surfaces are T_2 and T_1, find the rate of heat transfer through the compound slab (Fig. 22-5) in a steady state.

Let T_x be the temperature at the interface between the two materials. Then

$$H_2 = \frac{k_2 A(T_2 - T_x)}{L_2}$$

and

$$H_1 = \frac{k_1 A(T_x - T_1)}{L_1}.$$

In a steady state $H_2 = H_1 = H$, so that

$$\frac{k_2 A(T_2 - T_x)}{L_2} = \frac{k_1 A(T_x - T_1)}{L_1}.$$

Let H be the rate of heat transfer (the same for all sections). Then, solving for T_x and substituting into either of these equations, we obtain

$$H = \frac{A(T_2 - T_1)}{(L_1/k_1) + (L_2/k_2)}.$$

The extension to any number of sections in series is obviously

$$H = \frac{A(T_2 - T_1)}{\Sigma(L_i/k_i)}.$$

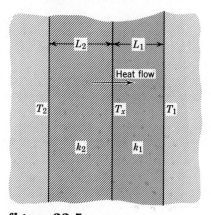

figure 22-5
Example 2. Conduction of heat through two layers of matter with different thermal conductivities.

We have seen earlier that the foot-pound (Section 7-2) was developed as a unit of work and the Btu (together with the calorie; see Section 22-2) as a unit of heat. Work and heat were thought of as separate concepts until Rumford, in 1798, suggested (Section 22-1) that heat had a mechanical aspect, thus proposing a connection between them. This connection was firmly established in the middle of the nineteenth century as the principle of conservation of energy. This principle asserts that heat and work are each forms of energy and that there should be a definite relationship, called the *mechanical equivalent of heat*, between them. It was Joule, in 1850, who first found by experiment how many foot-pounds of work are equivalent to 1 Btu of heat.

Joule used an apparatus in which falling weights rotated a set of paddles in an insulated water container (Fig. 22-6). In one cycle of operation the falling weights do a known amount of work on the water, of mass m, and we note that the temperature rises by ΔT. Now we *could*

22-5
THE MECHANICAL
EQUIVALENT OF HEAT

have produced this same rise in temperature by transfering heat energy Q to the system, given by

$$Q = mc \, \Delta T.$$

Thus, we measure W, observe ΔT, and calculate Q. The results are

$$1 \text{ Btu } (= 252.0 \text{ cal}) = 777.9 \text{ ft} \cdot \text{lb},$$

that is, 777.9 ft · lb of mechanical work will, when converted entirely into heat energy, generate 1 Btu; that is, it will raise the temperature of one pound of water from 63° F to 64° F. We can write this relation in other units as*

$$1 \text{ cal} = 4.186 \text{ J}.$$

It is appropriate that the SI unit of energy is the joule $(= 1 \text{ N} \cdot \text{m} = 1 \text{ kg} \cdot \text{m}^2/\text{s}^2)$. In modern laboratory practice the calorie is not much used or needed. It is, however, deeply embedded in the literature of science. To permit the continued use of this familiar unit — but to recognize the practical importance of the joule — a new calorie, the *thermochemical calorie*, is defined as

$$1 \text{ calorie (thermochemical)} = 4.184 \text{ joule (exactly)}.$$

In ordinary laboratory practice this calorie does not differ significantly from that defined earlier.

Joule also made other experiments (stirring mercury, forcing water through narrow tubes, rubbing together iron rings in a mercury bath, etc.). His conclusions are noteworthy for (1) the skill and ingenuity that he showed, (2) the accuracy of his final results, which differ only by about 1% from present values, and (3) the influence that they had in convincing scientists of the correctness of the concept that heat, like work, is a form of energy.

figure 22-6
Joule's arrangement for measuring the mechanical equivalent of heat. The falling weights turn paddles which stir the water in the container, thus raising its temperature.

22-6
HEAT AND WORK

We have seen that *heat is energy that flows from one body to another because of a temperature difference between them*. The idea that heat is something *in* a body, as the caloric theory assumed, contradicts many experimental facts. It is only as it flows, because of a temperature difference, that the energy is called heat energy. If heat were a substance, or a definite kind of energy that kept its identity while contained in a system, it would not be possible to remove heat indefinitely from a system which does not change. Yet Rumford showed that this was possible. In fact, by continually performing mechanical work in Joule's apparatus, we can obtain an indefinite amount of heat out of the water, by connecting it to a cooler system, for example, without changing the condition of the water.

In the same way work is not something of which a system contains a definite amount. We can put an indefinite amount of work into a system, as Joule's apparatus again illustrates. Work, like heat, involves a transfer of energy. In mechanics, work is involved in energy transfers

* Henry A. Rowland, in 1879, carried out a painstaking determination of the mechanical equivalent of heat which, to this day, remains a model of careful experimentation. His result differs from the accepted value today by only 1 part in 2000. Rowland graduated from Rensselaer Polytechnic Institute in 1870 and in 1876 became the first Professor of Physics at the then newly established Johns Hopkins University, where he conducted this experiment. See "The Education of an American Scientist, Henry A. Rowland," by Samuel Rezneck, *American Journal of Physics*, February, 1960 and 'Rowland's Physics' by John D. Miller, *Physics Today*, July, 1976.

in which temperature played no role. If heat energy is transmitted by temperature differences, we can distinguish heat and work by defining *work as energy that is transmitted from one system to another in such a way that a difference of temperature is not directly involved.* This definition is consistent with our previous use of the term. That is, in the expression $dW = F\,dx$, the force F can arise from electrical, magnetic, gravitational, and other sources. The term *work* includes all these energy transfer processes, but it specifically excludes energy transfer arising from temperature differences.

Consider another simple example, that of rubbing two surfaces together. There is no limit to the amount of heat that can be removed from this system or to the amount of work that can be put into it, so that there is no definite meaning to phrases such as "the heat in the system" or "the work in the system." The quantities Q and W are not characteristic of the (equilibrium) *state* of the system but rather of the *thermodynamic process* by which the system moves from one equilibrium state to another, by interacting with its environment. It is only during such a process that we can give meaning to heat and work; we can then identify Q with the heat transferred to or from the system and W with the work done on or by the system. The study of such processes and of the changes in energy involved in the performance of work and the flow of heat is the subject matter of *thermodynamics*.

In Fig. 22-7 we consider a general thermodynamic process. We must first state definitely what the system is and what the environment is. In the figure we draw a closed surface surrounding the system to define it. In (a) the system is in its *initial state*, in equilibrium with the environment external to it. In (b) the system interacts with its environment through some specific *thermodynamic process*. During this process, energy in the form of heat and/or work may go into or out of the system. Arrows representing the flow of Q or W must pierce the surface enclosing the system. In (c) the system has reached its *final state*, again in equilibrium with the environment external to it.

Figure 22-8 shows a falling weight which turns a generator, which in turn sends an electric current through a resistor immersed in a water container. Let us choose the system to be the generator and the attached electric circuit, the water, and its container. Then the environment is the weight and the earth, which pulls on the weight. The process consists of letting the weight fall a distance h in the earth's gravitational field. During this process the environment (by means of the cord) does work W on the system. There are no temperature differences between the system and its environment and hence $Q = 0$ for this process.

Our choice of a system in thermodynamic problems is arbitrary. Let us now choose the system to be only the water and its container in Fig. 22-8. The environment now is the generator and attached circuit as well as the weight and the earth. For this choice of system there now *is* a temperature difference between the environment (resistor) and the system (water), and heat Q will flow into the system during the process. No forces act through the system boundary to produce displacements, however, and hence $W = 0$ for this process. This example shows that we must first state definitely what the system is and what the environment is before we can decide whether the change in the state of the system is due to the flow of heat or to the performance of work or both. There will be a transfer of heat between system and environment only when a

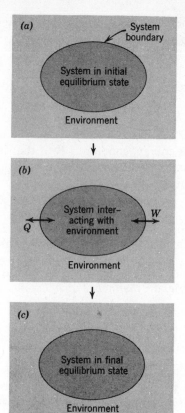

figure 22-7
(a) A system in an initial state, in equilibrium with its surroundings. (b) A thermodynamic process during which the system may exchange heat Q or work W with its environment. (c) A final equilibrium state reached as the result of the process.

figure 22-8
Heat and work. A weight, in falling, does work on an electric generator which sends current through a resistor which heats the water in which it is immersed.

figure 22-9
Work is done by the gas at pressure p as it expands against the piston. Heat may enter or leave the system from the heat reservoir on which the cylinder rests.

temperature difference exists across the system boundary; if no temperature difference exists, the energy transfer involves work.

Let us now compute Q and W for a specific thermodynamic process. Consider a gas in a cylindrical container with a movable piston. Let the gas be the system. Initially it is in equilibrium with the environment external to it (which is the heat reservoir and the piston, shown in Fig. 22-9) and has a pressure p_i and a volume V_i. We can think of the containing walls as the system boundary. Heat can flow into the system or out of it through the bottom of the cylinder and work can be done on the system or by the system by compressing or expanding the gas, respectively, with the piston. Consider a process whereby the system interacts with its environment and reaches a final equilibrium state characterized by a pressure p_f and a volume V_f.

In Fig. 22-9 we show the gas expanding against the piston. The work done by the gas in displacing the piston through an infinitesimal distance ds is

$$dW = \mathbf{F} \cdot d\mathbf{s} = pA\ ds = p\ dV$$

where dV is the differential change in the volume of the gas. In general, the pressure will not be constant during a displacement. To obtain the total work W done on the piston by the gas in a large displacement, we must know how p varies with the displacement. Then we compute the integral

$$W = \int dW = \int_{V_i}^{V_f} p\ dV$$

over the range in volume. This integral can be graphically evaluated as the area under the curve in a $p-V$ diagram, as shown for a special case in Fig. 22-10.

There are many different ways in which the system can be taken

figure 22-10
The work done by a gas is equal to the area under a p-V curve.

from the initial state i to the final state f. For example (Fig. 22-11), the pressure may be kept constant from i to a and then the volume kept constant from a to f. Then the work done by the expanding gas is equal to the area under the line ia. Another possibility is the path ibf, in which case the work done by the gas is the area under the line bf. The continuous curve from i to f is another possible path in which the work done by the gas is still different from the previous two paths. We can see, therefore, that *the work done by a system depends not only on the initial and final states but also on the intermediate states, that is, on the path of the process.*

A similar result follows if we compute the flow of heat during the process. State i is characterized by a temperature T_i and state f by a temperature T_f. The heat flowing into the system, say, depends on how the system is heated. We can heat it at a constant pressure p_i, for example, until we reach the temperature T_f, and then change the pressure at constant temperature to the final value p_f. Or we can first lower the pressure to p_f and then heat it at that pressure to the final temperature T_f. Or we can follow many other paths. Each path gives a different result for the heat flowing into the system. Hence, *the heat lost or gained by a system depends not only on the initial and final states but also on the intermediate states, that is, on the path of the process.* This is an experimental fact. As J. C. Slater has written:

". . . It would be pleasant to be able to say, in a given state of the system, that the system has so and so much heat energy. Starting from the absolute zero of temperature, where we could say that the heat energy was zero, we could heat the body up to the state we were interested in, find $\int dQ$ from absolute zero up to this state, and call that the heat energy. But the stubborn fact remains that we would get different answers if we heated it up in different ways. . . . There is nothing to do about it."

Both heat and work "depend on the path" taken; neither one is independent of the path, and neither one can be conserved alone.

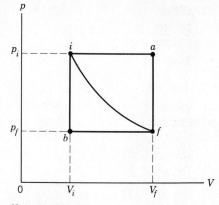

figure 22-11
The work done by a system depends not only on the initial state (i) and the final state (f) but on the intermediate path as well.

22-7
THE FIRST LAW OF THERMODYNAMICS

We can now tie all these ideas together. Let a system change from an initial equilibrium state i to a final equilibrium state f in a definite way, the heat absorbed by the system being Q and the work done by the system being W. Then we compute the $Q - W$. Now we start over and change the system from the same state i to the same state f, but this time in another way by a different path. We do this over and over again, using different paths each time. We find that in every case the quantity $Q - W$ *is the same.* That is, although Q and W separately depend on the path taken, $Q - W$ does *not* depend at all on how we took the system from state i to state f but only on the initial and final (equilibrium) states.

You will recall from mechanics that when an object is moved from an initial point i to a final point f in a gravitational field in the absence of friction, the work done depends only on the positions of the two points and not at all on the path through which the body is moved. From this we concluded that there is a function of the space coordinates of the body whose final value minus its initial value equals the work done in displacing the body. We called it the potential energy function. Now in thermodynamics we find that when a system has its state changed from state i to state f, the quantity $Q - W$ depends *only* on the initial and final coordinates and *not at all* on the path taken between these end points. We conclude that there is a function of the thermodynamic coordinates whose final value minus its initial value equals the change $Q - W$ in the process. We call this function the *internal energy function.*

Now Q is the energy added to the system by the transfer of heat and W is the energy given up by the system in performing work, so that $Q - W$ represents, by definition, *the internal energy change of the system.* Let us represent the internal energy function by the letter U. Then the internal energy of the system in state f, U_f, minus the internal energy of the system in state i, U_i, is simply *the change in internal energy of the system*, and this quantity *has a definite value independent of how the system went from state i to state f.* We have

$$U_f - U_i = \Delta U$$

and

$$\Delta U = Q - W. \tag{22-6}$$

Just as for potential energy, so for internal energy too it is the change that matters. If some arbitrary value is chosen for the internal energy in some standard reference state, its value in any other state can be given a definite value. Equation 22-6 is known as the *first law of thermodynamics.* In applying Eq. 22-6 we must remember that Q is considered positive when heat *enters* the system and W is positive when work is done *by* the system.

If our system undergoes only an infinitesimal change in state, only an infinitesimal amount of heat dQ is absorbed and only an infinitesimal amount of work dW is done, so that the internal energy change dU is also infinitesimal. In such a case, the first law is written in *differential* form as*

$$dU = dQ - dW. \tag{22-7}$$

We may express the first law in words by saying: *Every thermodynamic system in an equilibrium state possesses a state variable called the internal energy U whose change dU in a differential process is given by Eq. 22-7.* Recall that the essential content of the zeroth law of thermodynamics (p. 459) is, speaking loosely: *there exists a useful thermodynamic quantity called "temperature."* The essential content of the first law is: *there exists a useful thermodynamic quantity called "internal energy";* the law also provides, in Eq. 22-6, a recipe for measuring changes in internal energy quantitatively.

The first law of thermodynamics is thought to apply to every process in nature that proceeds between equilibrium states. Note that the *process* may or may not involve equilibrium states. We may apply the first law to the explosion of a firecracker in an insulated steel drum, for example. Because of its generality, the information that the first law gives is far from complete, although exact and correct. There are some very general questions which it cannot answer. For example, although it tells us that energy is conserved in every process, it does not tell us whether any particular process can actually occur. An entirely different generalization, called the second law of thermodynamics, gives us this information, and much of the subject matter of thermodynamics depends on this second law (Chapter 25).

---

* W and Q are not actual functions of the state of a system, that is, they do not depend on the values of the system's coordinates. Hence, dW and dQ are not exact differentials as the term is used in mathematics. All they mean here is a very small quantity. More advanced books write them as $đQ$ and $đW$ to indicate their inexact nature. However, dU is an exact differential, for U is an exact function of the system's coordinates.

We have seen that when a gas expands the work it does on its environment is

$$W = \int p \, dV,$$

where p is the pressure exerted on or by the gas and dV is the differential change in volume of the gas. Consider a special case in which the pressure remains constant while the volume changes by a finite amount, say from V_i to V_f. Then

$$W = \int_{V_i}^{V_f} p \, dV = p \int_{V_i}^{V_f} dV = p(V_f - V_i) \qquad \text{(constant pressure)}.$$

A process taking place at constant pressure is called an *isobaric* process. For example, water is heated in the boiler of a steam engine up to its boiling point and is vaporized to steam; then the steam is superheated, all processes proceeding at a constant pressure.

In Fig. 22-12 we show an isobaric process. The system is H_2O in a cylindrical container. A frictionless airtight piston is loaded with sand to produce the desired pressure on the H_2O and to maintain it automatically. Heat can be transferred from the environment to the system by a Bunsen burner. If the process continues long enough, the water boils and some is converted to steam; we assume that this occurs. The system may expand, very slowly (quasi-statically) but the pressure it exerts on the piston is automatically always the same, for this pressure must be equal to the constant pressure which the piston exerts on the system. If we wedged the piston so that it could not move, or if we added or took away some sand during the heating process, the process would not be isobaric.

Let us consider the boiling process. We know that substances will change their phase from liquid to vapor at a definite combination of values of pressure and temperature. Water will vaporize at 100° C and atmospheric pressure, for example. For a system to undergo a change of phase heat must be added to it, or taken from it, quite apart from the heat necessary to bring its temperature to the required value. Consider the change of phase of a mass m of liquid to a vapor occurring at constant temperature and pressure. Let V_l be the volume of liquid and V_v the volume of vapor. The work done by this substance in expanding from V_l to V_v at constant pressure is

$$W = p(V_v - V_l).$$

Let L represent the heat of vaporization, that is, the heat needed per unit mass to change a substance from liquid to vapor at constant temperature and pressure. Then the heat absorbed by the mass m during the change of state is

$$Q = mL.$$

From the first law of thermodynamics, we have

$$\Delta U = Q - W$$

so that

$$\Delta U = mL - p(V_v - V_l)$$

for this process.

22-8
SOME APPLICATIONS OF THE FIRST LAW OF THERMODYNAMICS

Frictionless and airtight contact

Sand

W

Steam

Liquid water

Q

figure 22-12
Water boiling at constant pressure (isobarically). The pressure is kept constant by the weight of the sand, the piston, and the external atmospheric pressure.

At atmospheric pressure 1.00 g of water, having a volume of 1.00 cm³, becomes 1671 cm³ of steam when boiled. The heat of vaporization of water is 539 cal/g at 1 atm. Hence, if $m = 1.00$ g,

$$Q = mL = 539 \text{ cal}.$$

EXAMPLE 3

This quantity, which represents heat *added* to the system from the environment, is positive.

$$W = p(V_v - V_l) = (1.013 \times 10^5 \text{ N/m}^2)[(1671 - 1) \times 10^{-6} \text{ m}^3]$$

$$= 169.5 \text{ J}.$$

This quantity, which represents work done *by* the system on the environment, is positive.

Since 1 cal equals 4.186 J, $W = 41$ cal. Then,

$$\Delta U = U_v - U_l = mL - p(V_v - V_l) = (539 - 41) \text{ cal}$$

$$= 498 \text{ cal}.$$

This quantity is positive; the internal energy of the system *increases* during this process. Hence, of the 539 cal needed to boil 1 g of water (at 100° C and 1 atm), 41 cal go into external work of expansion and 498 cal go into internal energy added to the system. This energy represents the internal work done in overcoming the strong attraction of H_2O molecules for one another in the liquid state.

How would you expect the 80 cal that are needed to melt 1 g of ice to water (at 0° C and 1 atm) to be shared by the external work and the internal energy?

A process that takes place in such a way that no heat flows into or out of the system is called an *adiabatic process*. Experimentally such processes are achieved either by sealing the system off from its surroundings with heat insulating material or by performing the process quickly. Because the flow of heat is somewhat slow, any process can be made practically adiabatic if it is performed quickly enough.

For an adiabatic process Q equals zero, so that from the first law we obtain

$$\Delta U = U_f - U_i = -W.$$

Hence, the internal energy of a system increases exactly by the amount of work done *on* the system in an adiabatic process. If work is done *by* the system in an adiabatic process, the internal energy of the system decreases by exactly the amount of external work it performs. An increase of internal energy usually raises the system's temperature and conversely, a decrease of internal energy usually lowers the system's temperature. A gas that expands adiabatically does external work and its internal energy decreases; such a process is used to attain low temperatures. The increase of temperature during an adiabatic compression of air is well known from the heating of a bicycle pump.

In Fig. 22-13 we show a simple adiabatic process. The system is a gas inside a cylinder made of heat-insulating material. Heat cannot enter the system from its environment or leave the system to the environment. Again we have a pile of sand on a frictionless airtight piston. The only interaction permitted between system and environment is through the performance of work. Such a process can occur when sand is added or removed from the piston, so that the gas can be compressed or can expand against the piston.

Among the many engineering examples of adiabatic processes are the expansion of steam in the cylinder of a steam engine, the expansion of hot gases in an internal combustion engine, and the compression of air in a Diesel engine or in an air compressor. These processes all occur rapidly enough so that only a very small amount of heat can enter or leave the system through its walls during that short time. The compres-

Frictionless and airtight contact

Sand

Heat-insulating material

W

Gas

figure 22-13
In an adiabatic process there is no flow of heat to or from the system. Here the walls are insulated and, as sand is removed or added, the volume of the gas changes adiabatically.

sions and rarefactions of a sound wave in a gas are adiabatic (Example 6, Chapter 23).

The most important reason for studying adiabatic processes, however, is that ideal engines use processes that are exactly adiabatic. These ideal engines determine the theoretical limits to the operation and capabilities of real engines. We shall look further into this in Chapter 25.

A process of much theoretical interest is that of *free expansion.* This is an adiabatic process in which no work is performed on or by the system. Something like this can be achieved by connecting one vessel which contains a gas to another evacuated vessel with a stopcock connection, the whole system being enclosed with thermal insulation (Fig. 22-14). If the stopcock is suddenly opened, the gas rushes into the vacuum and expands freely. Because of the heat insulation this process is adiabatic, and because the walls of the vessels are rigid no external work is done on the system. Hence, in the first law we have $Q = 0$ and $W = 0$, so that $U_i = U_f$ for this process. The initial and final internal energies are equal in free expansion.

In free expansion, after we open the stopcock we have no further control over the process. At intermediate states the pressure, volume, and temperature do not have unique values characteristic of the system as a whole, that is, the system passes through nonequilibrium states so that we cannot plot the course of the process by a curve on a *p-V* diagram. We can plot the initial and final states as points on such plots because they are well-defined, equilibrium states. The free expansion is a good example of an irreversible process; see Section 25-2.

Heat–insulating material

Stopcock closed

Gas   Vacuum

Initial equilibrium state

Stopcock open

Gas   Gas

Final equilibrium state

figure 22-14
Free expansion. There is no change of internal energy U since there is no flow of heat Q and no external work W is done.

1. Give examples to distinguish clearly between temperature and heat.
2. (a) Show how heat conduction and calorimetry could be explained by the caloric theory. (b) List some heat phenomena that cannot be explained by the caloric theory.
3. Give an example of a process in which no heat is transferred to or from the system but the temperature of the system changes.
4. Can heat be considered a form of stored (or potential) energy? Would such an interpretation contradict the concept of heat as energy in process of transfer because of a temperature difference?
5. Apply Eq. 22-1 to boiling water.
6. It is difficult to "boil" eggs in water at the top of a high mountain because water boils there at a relatively low temperature. What is a simple, practical way of overcoming this difficulty?
7. Will a three-minute egg cook any faster if the water is boiling furiously than if it is simmering quietly?
8. Can heat be added to a substance without causing the temperature of the substance to rise? If so, does this contradict the concept of heat as energy in process of transfer because of a temperature difference?
9. Why must heat energy be supplied to melt ice—the temperature doesn't change, after all?
10. (a) Can ice be heated to a temperature above 0° C without its melting? Explain. (b) Can water be cooled to a temperature below 0° C without its freezing? Explain. (See "The Undercooling of Liquids" by David Turnbull in *Scientific American*, January 1965.)
11. Does putting sand on it help you to drive on an icy road? Does your answer depend on the temperature? Explain.
12. Explain the fact that the presence of a large body of water nearby, such as a sea or ocean, tends to moderate the temperature extremes of the climate on adjacent land.
13. Theory shows that the coefficient of linear expansion α (see Sec. 21-8) is proportional to the heat capacity C_v. Show that this is to be expected. (*Hint:* Heat capacity measures the rate of change of the vibrational energy with temperature.)
14. If someone told you that a conventional electric fan not only does not cool the air but heats it slightly, how would you reply?
15. Both heat conduction and wave propagation involve the transfer of energy. Is there any difference in principle between these two phenomena?
16. When a hot body warms a cool one are their temperature changes equal in magnitude? Give examples. Can one then say that temperature passes from one to the other?
17. What connection is there between an object's feeling hot or cold and its heat capacity? Between this and its thermal conductivity?
18. A block of wood and a block of metal are at the *same* temperature. When the blocks feel cold the metal feels colder than the wood; when the blocks feel hot the metal feels hotter than the wood. Explain. At what temperature will the blocks feel equally cold or hot?
19. Explain why your finger sticks to a metal ice tray just taken from the refrigerator.
20. On a winter day the temperature of the inside surface of a wall is much lower than room temperature and that of the outside surface is much higher than the outdoor temperature. Explain.

The physiological mechanisms which maintain man's internal temperature operate in a limited range of external temperature. Explain how this range can be extended at each extreme by the use of clothes. (See "Heat, Cold, and Clothing" by James B. Kelley in *Scientific American*, February 1956.)

22. What requirements for thermal conductivity, specific heat capacity, and coefficient of expansion would you want a material to be used as a cooking utensil to satisfy?

23. Consider that heat can be transferred by convection and radiation, as well as by conduction, and explain why a thermos bottle is double-walled, evacuated, and silvered.

24. The system heating the cabin of a rocket ship seems to fail when the rocket ship is far out in free space. Give one possible explanation.

25. In what way is steady-state heat flow analogous to the flow of an incompressible fluid?

26. Is the mechanical equivalent of heat, J, a physical quantity or merely a conversion factor for converting energy from heat units to mechanical units and vice versa?

27. Defend this statement: "In Joule's experiment on the mechanical equivalent of heat, described in Section 22-5, no heat is involved."

28. Is the temperature of an isolated system (no interaction with the environment) conserved?

29. Is heat the same as internal energy? If not, give an example in which a system's internal energy changes without a flow of heat across the system's boundary.

30. Can one distinguish between whether the internal energy of a body was acquired by heat transfer or acquired by performance of work?

31. If the pressure and volume of a system are given, is the temperature always uniquely determined?

32. Does a gas do any work when it expands adiabatically? If so, what is the source of the energy needed to do this work?

33. A quantity of gas occupies an initial volume V_0 at a pressure p_0 and a temperature T_0. It expands to a volume V (a) at constant temperature and (b) at constant pressure. In which case does the gas do more work?

34. Discuss the process of the freezing of water from the point of view of the first law of thermodynamics. Remember that ice occupies a greater volume than an equal mass of water.

35. A thermos bottle contains coffee. The thermos bottle is vigorously shaken. Consider the coffee as the system. (a) Does its temperature rise? (b) Has heat been added to it? (c) Has work been done on it? (d) Has its internal energy changed?

36. We have seen that "energy conservation" is a universal law of nature. At the same time national leaders urge "energy conservation" upon us (driving slower, etc.). Explain the two quite different meanings of these words.

problems

SECTION 22-2

1. Suppose the specific heat of a substance is found to vary with temperature as

$$c = A + BT^2,$$

where A and B are constants and T is Celsius temperature. Compare the mean specific heat of the substance in a temperature range $T = 0$ to $T = T_0$ to the specific heat at the midpoint $T_0/2$.

Answer: Mean specific heat exceeds that at the midpoint by $BT_0^2/12$.

2. By means of a heating coil energy is transferred at a constant rate to a substance in a thermally insulated container. The temperature of the substance is measured as a function of time. (a) Show how we can deduce the way in which the heat capacity of the body depends on the temperature from this information. (b) Suppose that in a certain temperature range it is found that the temperature T is proportional to t^3, where t is the time. How does the heat capacity depend on T in this range?

3. Calculate the specific heat of a metal from the following data. A container

made of the metal weighs 8.0 lb (mass = 3.6 kg) and contains 30 lb (mass = 14 kg) of water. A 4.0 lb (mass = 1.8 kg) piece of the metal initially at a temperature of 350° F (180° C) is dropped into the water. The container and water initially have a temperature of 60° F (16° C) and the final temperature of the entire system is 65° F (18° C).
Answer: 0.14 Btu/lb · F° (0.099 cal/g · C°).

4. Two 50-g ice cubes are dropped into 200 g of water in a glass. If the water was initially at a temperature of 25° C, and if the ice came directly from a freezer operating at a temperature of −15° C, what will be the final temperature of the drink? The specific heat of ice is approximately 0.50 cal/g · C° in this temperature range and the heat required to melt ice to water is approximately 80 cal/g.

5. A thermometer of mass 0.0550 kg and of specific heat 0.200 cal/g · C° reads 15.0° C. It is then completely immersed in 0.300 kg of water and it comes to the same final temperature as the water. If the thermometer reads 44.4° C and is accurate, what was the temperature of the water before insertion of the thermometer, neglecting other heat losses? *Answer:* 45.5° C.

6. A copper ring has a diameter of exactly 1.00000 in. at its temperature of 0° C. An aluminum sphere has a diameter of exactly 1.00200 in. at its temperature of 100° C. The sphere is placed on top of the ring (Fig. 22-15), and the two are allowed to come to thermal equilibrium, no heat being lost to the surroundings. The sphere just passes through the ring at the equilibrium temperature. What is the ratio of the mass of the sphere to the mass of the ring?

figure 22-15
Problem 6

SECTION 22-3

7. Show that the number of atomic mass units per gram of a substance is equal to the number of particles per mole (Avogadro's number).

SECTION 22-4

8. Consider the rod shown in Fig. 22-4. Suppose $L = 25$ cm, $A = 1.0$ cm², and the material is copper. If $T_2 = 125°$ C, $T_1 = 0°$ C, and a steady state is reached, find (a) the temperature gradient, (b) the rate of heat transfer, and (c) the temperature at a point in the rod 10 cm from the high-temperature end.

9. A cylindrical copper rod of length 1.2 m and cross-sectional area 4.8 cm² is insulated to prevent heat loss through its surface. The ends are maintained at a temperature difference of 100 C° by having one end in a water-ice mixture and the other in boiling water and steam. (a) Find the rate at which heat is transferred along the rod. (b) Find the rate at which ice melts at one end. *Answer:* (a) 3.7 cal/s. (b) 0.046 g/s.

10. (a) Calculate the rate at which body heat flows out through the clothing of a skier, given the following data. The body surface area is 1.8 m² and the clothing is 1.0 cm thick; skin surface temperature is 33° C, whereas the outer surface of the clothing is at −5° C; the thermal conductivity of the clothing is 0.04 W/m · K. (b) How would the answer change if, after a fall, the skier's clothes become soaked with water?

11. Two identical square rods of metal are welded end-to-end as shown in Fig. 22-16a. Assume that 10 cal of heat flows through the rods in 2 min. How long would it take for 10 cal to flow through the rods if they are welded as shown in Fig. 22-16b? *Answer:* 0.5 min.

figure 22-16
Problem 11

12. Show that in a compound slab the temperature gradient in each portion is inversely proportional to the thermal conductivity.

13. Assume that the thermal conductivity of copper is twice that of aluminum and four times that of brass. Three metal rods, made of copper, aluminum, and brass, respectively, are each 6.0 in. long and 1.0 in. in diameter. These rods are placed end-to-end, with the aluminum between the other two. The free ends of the copper and brass rods are maintained at 100 and 0° C, respectively. Find the equilibrium temperatures of the copper-aluminum junction and the aluminum-brass junction.
   Answer: Cu-Al, 86° C; Al-Brass, 57° C.

14. (a) What is the rate of heat loss in W/m² through a glass window 3.0 mm thick if the outside temperature is −20° F and the inside temperature is +72° F? (b) If a storm window is installed having the same thickness of glass but with an air gap of 7.5 cm between the two windows, what will be the corresponding rate of heat loss?

15. A tank of water has been outdoors in cold weather until a 5.0-cm thick slab of ice has formed on its surface (Fig. 22-17). The air above the ice is at −10° C. Calculate the rate of formation of ice (in cm/h) on the bottom surface of the ice slab. Take the thermal conductivity, density and heat of fusion of ice to be 0.0040 cal/s · cm · C°, 0.92 g/cm³ and 80 cal/g, respectively. Assume that no heat enters or leaves the water through the walls of the tank.
   Answer: 0.39 cm/h.

16. Assuming k is constant, show that the radial rate of flow of heat in a substance between two concentric spheres is given by

$$H = \frac{(T_1 - T_2)4\pi k r_1 r_2}{r_2 - r_1}$$

   where the inner sphere has a radius r_1 and temperature T_1, and the outer sphere has a radius r_2 and temperature T_2.

figure 22-17
Problem 15

17. Energy released by radioactivity within the earth is conducted outward as heat through the oceans. For purposes of approximate calculation, assume the average temperature gradient within the solid earth beneath the ocean to be 0.07 C°/m and the average thermal conductivity to be 2×10^{-4} kcal/m · s · C°, and (a) determine the rate of heat transfer per square meter. Assume that this is approximately the rate for the entire surface of the earth, and (b) determine how much heat is thereby transferred through the earth's surface each day.
   Answer: (a) 1.4×10^{-5} kcal/m² · s. (b) 6.2×10^{14} kcal/day.

18. Assuming k is constant, show that the radial rate of flow of heat in a substance between two coaxial cylinders is given by

$$H = \frac{(T_1 - T_2)2\pi L k}{\ln (r_2/r_1)}$$

   where the inner cylinder has a radius r_1 and temperature T_1, and the outer cylinder has a radius r_2 and temperature T_2, each cylinder having a length L.

19. A long tungsten heater wire is rated at 3.0 kW/m and is 5.0×10^{-4} m in diameter. It is embedded along the axis of a ceramic cylinder of diameter 0.12 m. When operating at the rated power, the wire is at 1500° C; the outside of the cylinder is at 20° C. Find the thermal conductivity of the ceramic; use the result given in Problem 18.   *Answer:* 1.8 J/m · s · C°.

SECTION 22-5

20. In a Joule experiment, a mass of 6.00 kg falls through a height of 50.0 m and rotates a paddle wheel that stirs 0.600 kg of water. The water is initially at 15.0° C. By how much does its temperature rise?

21. An energetic athlete dissipates all the energy in a diet of 4000 kcal per day. If he were to release this energy at a steady rate, how would this output

compare with the energy output of a 100-W bulb? (*Note:* The calorie of nutrition is really a kilocalorie, as we have defined it.)
Answer: 1.9 times as great.

22. Power is supplied at the rate of 0.40 hp for 2.0 min in drilling a hole in a 1.0-lb copper block. (*a*) How much heat is generated? (*b*) What is the rise in temperature of the copper if only 75% of the power warms the copper? (*c*) What happens to the other 25%?

23. (*a*) Compute the possible increase in temperature for water going over Niagara Falls, 162 ft high. (*b*) What factors would tend to prevent this possible rise?   *Answer:* (*a*) 0.12 C°.

24. A 2.0-g (1.4×10^{-4}-slug) bullet moving at a speed of 200 m/s (660 ft/s) becomes embedded in a 2.0 kg (0.14-slug) wooden block suspended as a pendulum bob (a ballistic pendulum). Calculate the rise in temperature of the bullet, assuming that all the absorbed energy raises the bullet's temperature.

25. A block of ice at 0° C whose mass is initially 50.0 kg slides along a horizontal surface, starting at a speed of 5.38 m/s and finally coming to rest after traveling 28.3 m. Compute the mass of ice melted as a result of the friction between the block and the surface.   *Answer:* 2.16 g.

26. The specific heat of silver, measured at atmospheric pressure, is found to vary with temperature between 50 and 100 K by the empirical equation

$$c_p = 0.076T - 0.00026T^2 - 0.15,$$

where c_p is in cal/mol · K and T is the Kelvin temperature. Calculate the quantity of heat required to raise 216 g of silver from 50 to 100 K.

27. Count Rumford weighed a metal object at low temperature and then at high temperature to see whether its "caloric content" increased. He concluded that (for gold) the "caloric" did not weigh more than 10^{-6} the weight of the sample. (*a*) Should the mass of a sample increase when heated, according to modern theories? (*b*) If so, by what order of magnitude? (*c*) Was Rumford safe in rejecting the caloric theory on this basis, in retrospect?
Answer: (*a*) Yes. (*b*) About 10^{-14} kg. (*c*) No.

28. Take the average specific heat of copper to be 0.090 cal/g · C° in the temperature range 0 to 1000° C. If 1.00 kg of copper is heated from 0 to 1000° C, by how much does its mass increase?

29. A "flow calorimeter" is used to measure the specific heat of a liquid. Heat is added at a known rate to a stream of the liquid as it passes through the calorimeter at a known rate. Then a measurement of the resulting temperature difference between the inflow and the outflow points of the liquid stream enables us to compute the specific heat of the liquid.

   A liquid of density 0.85 g/cm³ flows through a calorimeter at the rate of 8.0 cm³/s. Heat is added by means of a 250-W electric heating coil, and a temperature difference of 15 C° is established in steady-state conditions between the inflow and outflow points. Find the specific heat of the liquid.
Answer: 2500 J/kg · C°.

30. A chef, upon awaking one morning to find his stove out of order, decides to boil the water for his wife's coffee by shaking it in a thermos flask. Suppose that he uses $\frac{1}{2}$ liter of tap water at 59° F, and that the water falls 1.0 ft each shake, the chef making 30 shakes each minute. Neglecting any loss of heat, how long must he shake the flask before the water boils?

SECTION 22-7

31. Determine the value of *J*, the mechanical equivalent of heat, from the following data: 2000 cal (7.936 Btu) of heat are supplied to a system; the system does 3350 J (2471 ft · lb) of external work during that time; the increase of internal energy during the process is 5030 J (3710 ft · lb).
Answer: 4.190 J/cal (779 ft · lb/Btu).

SECTION 22-8

32. A thermodynamic system is taken from an initial state *A* to another *B* and back again to *A*, via state *C*, as shown by the path *A-B-C-A* in the *p-V* dia-

figure 22-18
Problem 32

$$\Delta U$$
$$Q = \Delta U + W$$
$$+ \quad +U$$

(a)

(b)

(a)

(b)

figure 22-19
Problem 33

gram of Fig. 22-18a. (a) Complete the table in Fig. 22-18b by filling in appropriate + or − indications for the signs of the thermodynamic quantities associated with each process. (b) Calculate the numerical value of the work done by the system for the complete cycle A-B-C-A.

33. Figure 22-19a shows a cylinder containing gas and closed by a movable piston. The cylinder is submerged in an ice-water mixture. The piston is *quickly* pushed down from position (1) to position (2). The piston is held at position (2) until the gas is again at 0° C and then is *slowly* raised back to position (1). Figure 22-19b is a p-V diagram for the process. If 100 g of ice are melted during the cycle, how much work has been done *on* the gas?
Answer: 8000 cal.

34. When a system is taken from state i to state f along the path iaf in Fig. 22-20 it is found that $Q = 50$ cal and $W = 20$ cal. Along the path ibf, $Q = 36$ cal. (a) What is W along the path ibf? (b) If $W = -13$ cal for the curved return path fi, what is Q for this path? (c) Take $U_i = 10$ cal. What is U_f? (d) If $U_b = 22$ cal, what is Q for the process ib? For the process bf?

35. An iron ball is dropped onto a concrete floor from a height of 10 m. On the first rebound it rises to a height of 0.50 m. Assume that all the macroscopic mechanical energy lost in the collision with the floor goes into the ball. The specific heat of iron is 0.12 cal/g · C°. During the collision (a) has heat been added to the ball? (b) Has work been done on it? (c) Has its internal energy changed? If so, by how much? (d) How much has the temperature of the ball risen after the first collision?
Answer: (a) No. (b) Yes. (c) Yes, by +93 J/kg. (d) 0.20 C°.

36. A cylinder has a well-fitted 2.0-kg metal piston whose cross-sectional area is 2.0 cm² (Fig. 22-21). The cylinder contains water and steam at 100° C. The piston is observed to fall slowly at a rate of 0.30 cm/s because heat flows out of the cylinder through the cylinder walls. As this happens, some steam condenses in the chamber. The density of the steam inside the chamber is 6.0×10^{-4} g/cm³. (a) Calculate the rate of condensation of steam. (b) What is the rate of change of internal energy of the steam and water inside the chamber? (c) At what rate is heat leaving the chamber?

figure 22-20
Problem 34

figure 22-21
Problem 36

23

kinetic theory
of gases—I

Thermodynamics deals only with macroscopic variables, such as pressure, temperature, and volume. Its basic laws, expressed in terms of such quantities, say nothing at all about the fact that matter is made up of atoms. *Statistical mechanics*, however, which deals with the same areas of science that thermodynamics does, presupposes the existence of atoms. Its basic laws are the laws of mechanics, which are applied to the atoms that make up the system.

No existing electronic computer could solve the problem of applying the laws of mechanics individually to every atom in a gas, say. If there were one, the results of such calculations would be too voluminous to be useful. Fortunately, the detailed life histories of individual atoms in a gas are not important if we want to calculate only the macroscopic behavior of the gas. We apply the laws of mechanics *statistically*, then, and we find that we are able to express all the thermodynamic variables as certain averages of atomic properties. For example, the pressure exerted by a gas on the wall of the containing vessel is the average rate per unit area at which the atoms of the gas transfer momentum to the wall as they collide with it. The number of atoms in a macroscopic system is usually so large that such averages are very sharply defined quantities indeed.

We can apply the laws of mechanics statistically to assemblies of atoms at two different levels. At the level called *kinetic theory* we proceed in a rather physical way, using relatively simple mathematical averaging techniques. In this chapter we will use these methods to enlarge our understanding of pressure, temperature, specific heat, and internal energy at the atomic level. Kinetic theory was developed by Robert Boyle (1627–1691), Daniel Bernoulli (1700–1782), James Joule

23-1
INTRODUCTION

(1818–1889), A. Kronig (1822–1879), Rudolph Clausius (1822–1888), and Clerk Maxwell (1831–1879), among others.* In this book we apply the kinetic theory to gases only, because the interactions between atoms in gases are much weaker than in liquids and solids; this greatly simplifies the mathematical difficulties.

At another level, we can apply the laws of mechanics statistically using techniques that are more formal and abstract than those of kinetic theory. This approach, developed by J. Willard Gibbs (1839–1903) and by Ludwig Boltzmann (1844–1906) among others, is called *statistical mechanics*, a term that includes kinetic theory as a sub-branch. Using these methods one can derive the laws of thermodynamics, thus establishing that science as a branch of mechanics. The fullest flowering of statistical mechanics (*quantum statistics*) involves the statistical application of the laws of quantum mechanics—rather than those of classical mechanics—to many-atom systems.

23-2
IDEAL GAS—A MACROSCOPIC DESCRIPTION

Let a mass nM of a gas be confined in a container of volume V; M is the molecular weight (grams/mole) and n is the number of moles. The density ρ of the gas is nM/V and it is clear that we can reduce ρ either by removing some gas from the container (reducing n) or by putting the gas in a larger container (increasing V). We find from experiment that, at low enough densities, all gases, no matter what their chemical composition, tend to show a certain simple relationship among the thermodynamic variables p, V, and T. This suggests the concept of an *ideal gas*, one that would have the same simple behavior under all conditions. In this section we give a macroscopic or thermodynamic definition of an ideal gas. In Section 23-3 we will define an ideal gas microscopically, from the standpoint of kinetic theory, and we will see what we can learn by comparing these two approaches.

Given a mass nM of any gas in a state of thermal equilibrium we can measure its pressure p, its temperature T, and its volume V. For low enough values of the density experiment show that (1) for a given mass of gas held at a constant temperature, the pressure is inversely proportional to the volume (Boyle's law) and (2) for a given mass of gas held at a constant pressure, the volume is directly proportional to the temperature (law of Charles and Gay-Lussac). We can summarize these two experimental results by the relation

$$\frac{pV}{T} = \text{a constant} \quad \text{(for a fixed mass of gas).} \qquad (23\text{-}1)$$

The volume occupied by a gas at a given pressure and temperature is proportional to its mass. Thus the constant in Eq. 23-1 must also be proportional to the mass of the gas. In Section 22-2 (see Fig. 22-2) we saw the great simplification that occurs in studies of the specific heats of solids if we compare samples of solids that contain the same number of molecules rather than samples which have the same mass in grams. We did this by using the mole as our mass unit. Let us also do that here.

We therefore write the constant in Eq. 23-1 as nR, where n is the number of moles of the gas and R is a constant that must be determined for each gas by experiment. Our expectation that simplicity will emerge if we compare gases on a molar basis is justified because experiment

---

* See "John James Waterston and the Kinetic Theory of Gases," by S. G. Brush, in *American Scientist*, June 1961, for an interesting aspect of the history of kinetic theory.

shows that, at low enough densities, *R has the same value for all gases*, namely

$$R = 8.314 \ \text{J/mol·K} = 1.986 \ \text{cal/mol·K.}$$

R is called the *universal gas constant.* We then write Eq. 23-1 as

$$pV = nRT \qquad (23\text{-}2)$$

and we define an ideal gas as one that obeys this relation *under all conditions.* There is no such thing as a truly ideal gas, but it remains a useful and simple concept connected with reality by the fact that all real gases approach the ideal gas abstraction in their behavior if the density is low enough. Equation 23-2 is called the *equation of state* of an ideal gas.

If we could fill the bulb of an (ideal) constant-volume gas thermometer with an ideal gas, we see from Eq. 23-2 that we could define temperature in terms of its pressure readings, that is,

$$T = 273.16 \ \text{K} \ \frac{p}{p_{tr}} \quad \text{(ideal gas).}$$

Here p_{tr} is the gas pressure at the triple point, at which the temperature T_{tr} is 273.16 K by definition. In practice we must fill our thermometer with a real gas and measure the temperature by extrapolating to zero density using Eq. 21-4,

$$T = 273.16 \ \text{K} \lim_{p_{tr} \to 0} \frac{p}{p_{tr}} \quad \text{(real gas).}$$

If we had an ideal gas available (which we do not), the extrapolation would be unnecessary.

EXAMPLE 1

A cylinder contains oxygen at a temperature of 20° C and a pressure of 15 atm in a volume of 100 liters. A piston is lowered into the cylinder decreasing the volume occupied by the gas to 80 liters and raising the temperature to 25° C. Assuming oxygen to behave like an ideal gas under these conditions, what then is the gas pressure?

From Eq. 23-1, since the mass of gas remains unchanged, we may write

$$\frac{p_i V_i}{T_i} = \frac{p_f V_f}{T_f}.$$

Our initial conditions are

$$p_i = 15 \ \text{atm,} \qquad T_i = 293 \ \text{K,} \qquad V_i = 100 \ \text{liters.}$$

Our final conditions are

$$p_f = ?, \qquad T_f = 298 \ \text{K,} \qquad V_f = 80 \ \text{liters.}$$

Hence,

$$p_f = \left(\frac{T_f}{V_f}\right)\left(\frac{p_i V_i}{T_i}\right) = \left(\frac{298 \ \text{K}}{80 \ \text{liters}}\right)\left(\frac{15 \ \text{atm} \times 100 \ \text{liters}}{293 \ \text{K}}\right) = 19 \ \text{atm.}$$

EXAMPLE 2

Calculate the work per mole done by an ideal gas which expands isothermally, that is, at constant temperature, from an initial volume V_i to a final volume V_f.

The work done may be represented as

$$W = \int_{V_i}^{V_f} p \ dV.$$

From the ideal gas law we have

$$p = \frac{nRT}{V},$$

so that W/n, the work per mole, is

$$\frac{W}{n} = \int_{V_i}^{V_f} \frac{RT}{V} \, dV.$$

The temperature is constant so that

$$\frac{W}{n} = RT \int_{V_i}^{V_f} \frac{dV}{V} = RT \ln \frac{V_f}{V_i}$$

is the work per mole done by an ideal gas in an isothermal expansion at temperature T from an initial volume V_i to a final volume V_f.

Notice that when the gas expands, so that $V_f > V_i$, the work done by the gas is positive; when the gas is compressed, so that $V_f < V_i$, the work done by the gas is negative. This is consistent with the sign convention adopted for W in the first law of thermodynamics. The work done is shown as the shaded area in Fig. 23-1. The solid line is an isotherm, that is, a curve giving the relation of p to V at a constant temperature.

In practice, how can we keep an expanding or contracting gas at constant temperature?

figure 23-1
Example 2. The shaded area represents the work done by n moles of gas in expanding from V_i to V_f with the temperature held constant.

From the microscopic point of view we define an ideal gas by making the following assumptions; it will then be our task to apply the laws of classical mechanics statistically to the gas atoms and to show that our microscopic definition is consistent with the macroscopic definition of the preceding section.

23-3
AN IDEAL GAS–MICROSCOPIC DEFINITION

1. *A gas consists of particles, called molecules.* Depending on the gas, each molecule will consist of one atom or a group of atoms. If the gas is an element or a compound and is in a stable state, we consider all its molecules to be identical.

2. *The molecules are in random motion and obey Newton's laws of motion.* The molecules move in all directions and with various speeds. In computing the properties of the motion, we assume that Newtonian mechanics works at the microscopic level. As for all our assumptions, this one will stand or fall depending on whether or not the experimental results it predicts are correct.

3. *The total number of molecules is large.* The direction and speed of motion of any one molecule may change abruptly on collision with the wall or another molecule. Any particular molecule will follow a zigzag path because of these collisions. However, because there are so many molecules we assume that the resulting large number of collisions maintains the over-all distribution of molecular velocities and the randomness of the motion.

4. *The volume of the molecules is a negligibly small fraction of the volume occupied by the gas.* Even though there are many molecules, they are extremely small. We know that the volume occupied by a gas can be changed through a large range of values with little difficulty, and that when a gas condenses the volume occupied by the liquid may be thousands of times smaller than that of the gas. Hence, our assumption is plausible. Later we shall investigate the actual size of molecules and see whether we need to modify this assumption.

5. *No appreciable forces act on the molecules except during a collision.* To the extent that this is true a molecule moves with uniform velocity between collisions. Because we have assumed the molecules to be so small, the average distance between molecules is large compared to the size of a molecule. Hence, we assume that the range of molecular forces is comparable to the molecular size.

6. *Collisions are elastic and are of negligible duration.* Collisions between molecules and with the walls of the container conserve momentum and (we assume) kinetic energy. Because the collision time is negligible compared to the time spent by a molecule between collisions, the kinetic energy which is converted to potential energy during the collision is available again as kinetic energy after such a brief time that we can ignore this exchange entirely.

Let us now calculate the pressure of an ideal gas from kinetic theory. To simplify matters, we consider a gas in a cubical vessel whose walls are perfectly elastic. Let each edge be of length l. Call the faces normal to the x-axis (Fig. 23-2) A_1 and A_2, each of area l^2. Consider a molecule which has a velocity **v**. We can resolve **v** into components v_x, v_y, and v_z in the directions of the edges. If this particle collides with A_1, it will rebound with its x-component of velocity reversed. There will be no effect on v_y or v_z, so that the change in the particle's momentum will be

final momentum − initial momentum $= -mv_x - (mv_x) = -2mv_x$,

normal to A_1. Hence, the momentum imparted to A_1 will be $2mv_x$, since the total momentum is conserved.

Suppose that this particle reaches A_2 without striking any other particle on the way. The time required to cross the cube will be l/v_x. At A_2 it will again have its x-component of velocity reversed and will return to A_1. Assuming no collisions in between, the round trip will take a time $2l/v_x$. Hence, the number of collisions per unit time this particle makes with A_1 is $v_x/2l$, so that the rate at which it transfers momentum to A_1 is

$$2mv_x \frac{v_x}{2l} = \frac{mv_x^2}{l}.$$

To obtain the total force on A_1, that is, the rate at which momentum is imparted to A_1 by all the gas molecules, we must sum up mv_x^2/l for all the particles. Then, to find the pressure, we divide this force by the area of A_1, namely l^2.

If m is the mass of each molecule, we have

$$p = \frac{m}{l^3}(v_{x1}^2 + v_{x2}^2 + \cdots),$$

where v_{x1} is the x-component of the velocity of particle 1, v_{x2} is that of particle 2, etc. If N is the total number of particles in the container and n_v is the number per unit volume, then $N/l^3 = n_v$ or $l^3 = N/n_v$. Hence,

$$p = mn_v\left(\frac{v_{x1}^2 + v_{x2}^2 + \cdots}{N}\right).$$

But mn_v is simply the mass per unit volume, that is, the density ρ. The quantity $(v_{x1}^2 + v_{x2}^2 + \cdots)/N$ is the average value of v_x^2 for all the particles in the container. Let us call this $\overline{v_x^2}$. Then

23-4
KINETIC CALCULATION OF THE PRESSURE

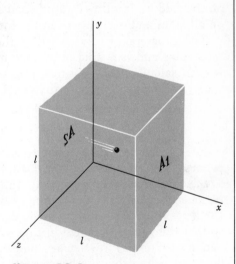

figure 23-2
A cubical box of side l, containing an ideal gas. A molecule is shown moving toward A_1.

$$p = \rho \overline{v_x{}^2}.$$

For any particle $v^2 = v_x{}^2 + v_y{}^2 + v_z{}^2$. Because we have many particles and because they are moving entirely at random, the average values of $v_x{}^2$, $v_y{}^2$, and $v_z{}^2$ are equal and the value of each is exactly one-third the average value of v^2. There is no preference among the molecules for motion along any one of the three axes. Hence, $\overline{v_x{}^2} = \frac{1}{3}\overline{v^2}$, so that

$$p = \rho \overline{v_x{}^2} = \tfrac{1}{3}\rho \overline{v^2}. \qquad (23\text{-}3)$$

Although we derived this result by neglecting collisions between particles, the result is true even when we consider collisions. Because of the exchange of velocities in an elastic collision between identical particles, there will always be some one molecule that will collide with A_2 with momentum mv_x corresponding to the one that left A_1 with this same momentum. Also, the time spent during collisions is negligible compared to the time spent between collisions. Hence, our neglect of collisions is merely a convenient device for calculation. Similarly, we could have chosen a container of any shape—the cube merely simplifies the calculation. Although we have calculated the pressure exerted only on the side A_1, it follows from Pascal's law that the pressure is the same on all sides and everywhere in the interior.*

The square root of $\overline{v^2}$ is called the *root-mean-square* speed of the molecules and is a kind of average molecular speed.† Using Eq. 23-3, we can calculate this root-mean-square speed from measured values of the pressure and density of the gas. Thus,

$$v_{\text{rms}} = \sqrt{\overline{v^2}} = \sqrt{\frac{3p}{\rho}}. \qquad (23\text{-}4a)$$

In Eq. 23-3 we relate a macroscopic quantity (the pressure p) to an average value of a microscopic quantity (that is, to $\overline{v^2}$ or $v_{\text{rms}}{}^2$). However, averages can be taken over short times or over long times, over small regions of space or large regions of space. The average computed in a small region for a short time might depend on the time or region chosen, so that the values obtained in this way may fluctuate. This could happen in a gas of very low density, for example. We can ignore fluctuations, however, when the number of particles in the system is large enough.

Calculate the root-mean-square speed of hydrogen molecules at $0.00°$ C and 1.00-atm pressure, assuming hydrogen to be an ideal gas. Under these conditions hydrogen has a density ρ of 8.99×10^{-2} kg/m³. Then, since $p = 1.00$ atm $= 1.01 \times 10^5$ N/m²,

EXAMPLE 3

$$v_{\text{rms}} = \sqrt{\frac{3p}{\rho}} = 1840 \text{ m/s}.$$

This is of the order of a mile per second, or 3600 mi/h.

Table 23-1 gives the results of similar calculations for some gases at $0°$ C. These molecular speeds are roughly of the same order as the speed

---

* We neglect the weight of the gas, a negligible effect unless the gas is of very large extent, as in the atmosphere. (See Section 17-3 and Problem 26.)

† We will consider this further in Section 24-2 in which we discuss the molecular distribution of speeds.

Table 23-1

Gas	Molecular weight,* g/mol	v_{rms} (at 0° C), m/s	Translational kinetic energy per mole (at 0° C), $\frac{1}{2}Mv_{rms}^2$, J/mol
H_2	2.02	1838	3370
He	4.0	1311	3430
H_2O	18	615	3400
Ne	20.1	584	3420
N_2	28	493	3390
CO	28	493	3390
Air	28.8	485	3280
O_2	32	461	3400
CO_2	44	393	3400

* The molecular weight and the mole are defined on page 479. We will discuss the last column in this table in the next section.

of sound at the same temperature. For example, in air at 0° C, $v_{rms} = 485$ m/s and the speed of sound is 331 m/s and in hydrogen $v_{rms} = 1838$ m/s and sound travels at 1286 m/s. These results are to be expected in terms of our model of a gas; see Prob. 34. We visualize the propagation of sound waves as a directional motion of the molecules as a whole superimposed on their random motion. Hence, the energy of the sound wave is carried as kinetic energy from one molecule to the next one with which it collides. The molecules themselves, in spite of their high speeds, do not move very far during a period of the sound vibration; they are confined to a rather small space by the effects of a large number of collisions.† However, the energy of the sound wave is communicated from one molecule to the next with that high speed, even though we do not expect the speed of sound to be *exactly* equal to v_{rms}, a point that we will clarify in Example 6.

EXAMPLE 4

Assuming that the speed of sound in a gas is the same as the root-mean-square speed of the molecules, show how the speed of sound for an ideal gas would depend on the temperature. (Actually this assumption is only crudely correct. Compare Eq. 23-4a and Eq. 23-15.)

The density of a gas is

$$\rho = \frac{nM}{V}$$

in which M is the molecular weight (grams/mole) and n is the number of moles. Combining this with the ideal gas law

$$pV = nRT$$

yields

$$\frac{p}{\rho} = \frac{RT}{M}.$$

We obtain from Eq. 23-4a

$$v_{rms} = \sqrt{\frac{3p}{\rho}} = \sqrt{\frac{3RT}{M}}, \qquad (23\text{-}4b)$$

so that the speed of sound v_1 at a temperature T_1 is related to the speed of sound

---

† This explains why there is a time lag between opening an ammonia bottle at one end of a room and smelling it at the other end. Although molecular speeds are high, the large number of collisions restrains the advance of the ammonia molecules. They diffuse through the air at speeds that are very much less than molecular speeds.

v_2 in the same gas at a temperature T_2 by

$$\frac{v_1}{v_2} = \sqrt{\frac{T_1}{T_2}}.$$

For example, if the speed of sound at 273 K is 332 m/s in air, its speed in air at 300 K will be

$$\sqrt{\tfrac{300}{273}} \times 332 \text{ m/s} = 348 \text{ m/s}.$$

Would our result change if the speed of sound were proportional to, rather than equal to, the root-mean-square speed of the molecules of a gas?

If we multiply each side of Eq. 23-3 by the volume V, we obtain

$$pV = \tfrac{1}{3}\rho V \overline{v^2},$$

where ρV is simply the total mass of gas, ρ being the density. We can also write the mass of gas as nM, in which n is the number of moles and M is the molecular weight. Making this substitution yields

$$pV = \tfrac{1}{3}nM\overline{v^2}.$$

The quantity $\tfrac{1}{3}nM\overline{v^2}$ is two-thirds the total kinetic energy of translation of the molecules, that is, $\tfrac{2}{3}(\tfrac{1}{2}nM\overline{v^2})$*. We can then write

$$pV = \tfrac{2}{3}(\tfrac{1}{2}nM\overline{v^2}).$$

The equation of state of an ideal gas is

$$pV = nRT.$$

Combining these two expressions, we obtain

$$\tfrac{1}{2}M\overline{v^2} = \tfrac{3}{2}RT. \qquad (23\text{-}5)$$

That is, *the total translational kinetic energy per mole of the molecules of an ideal gas is proportional to the temperature.* We may say that this result, Eq. 23-5, is necessary to fit the kinetic theory to the equation of state of an ideal gas, or we may consider Eq. 23-5 as a definition of gas temperature on a kinetic theory or microscopic basis. In either case, we gain some insight into the meaning of temperature for gases.

The temperature of a gas is related to the total translational kinetic energy measured with respect to the center of mass of the gas. The kinetic energy associated with the motion of the center of mass of the gas has no bearing on the gas temperature. In Section 23-3 we assumed random motion as part of our statistical definition of an ideal gas and in Section 23-4 we calculated $\overline{v^2}$ on this basis. For a random distribution of molecular velocities with direction the center of mass would be at rest, so that we must use a reference frame in which the center of mass of the gas is at rest. For all other frames the molecules will each have velocities greater by **u** (the velocity of the center of mass in that frame) than in the center of mass frame; hence, the motions will no longer be random and we will obtain different values for $\overline{v^2}$. The temperature of a gas in a container does not increase when we put the container on a moving train!

Let us now divide each side of Eq. 23-5 by Avogadro's number, N_0, which (see page 479, footnote) is the number of molecules per mole of a gas. Thus M/N_0 $(= m)$ is the mass of a single molecule and we have

$$\tfrac{1}{2}(M/N_0)\overline{v^2} = \tfrac{1}{2}m\overline{v^2} = \tfrac{3}{2}(R/N_0)T.$$

*If N is the total number of molecules and m is the mass of each molecule, then $\tfrac{1}{2}mv_1{}^2 + \tfrac{1}{2}mv_2{}^2 + \cdots = \tfrac{1}{2}mN\left[\dfrac{v_1{}^2 + v_2{}^2 + \cdots}{N}\right] = \tfrac{1}{2}mN\overline{v^2}$ in which mN $(= nM)$ is the total mass of the gas.

23-5
KINETIC INTERPRETATION OF TEMPERATURE

Now $\frac{1}{2}m\overline{v^2}$ is the average translational kinetic energy per molecule. The ratio R/N_0—which we call k, the *Boltzmann constant*—plays the role of the gas constant per molecule. We have

$$\tfrac{1}{2}m\overline{v^2} = \tfrac{3}{2}kT \qquad (23\text{-}6)$$

in which*

$$k = \frac{R}{N_0} = \frac{8.314 \text{ J/mol·K}}{6.023 \times 10^{23} \text{ molecules/mol}} = 1.380 \times 10^{-23} \text{ J/molecule·K}$$

We shall return to Boltzmann's constant in Chapter 24.

In the last column of Table 23-1 we list calculated values of $\frac{1}{2}Mv_{rms}^2$. As Eq. 23-5 predicts, this quantity (the translational kinetic energy per mole) has (closely) the same value for all gases at the same temperatures, $0°$ C in this case. From Eq. 23-6 we conclude that at the same temperature T the ratio of the root-mean-square speeds of molecules of two different gases is equal to the square root of the inverse ratio of their masses. That is, from

$$T = \frac{2}{3k}\frac{m_1\overline{v_1^2}}{2} = \frac{2}{3k}\frac{m_2\overline{v_2^2}}{2}$$

we obtain

$$\sqrt{\frac{\overline{v_1^2}}{\overline{v_2^2}}} = \frac{v_{1rms}}{v_{2rms}} = \sqrt{\frac{m_2}{m_1}}. \qquad (23\text{-}7)$$

We can apply Eq. 23-7 to the diffusion of two different gases in a container with porous walls placed in an evacuated space. The lighter gas, whose molecules move more rapidly on the average, will escape faster than the heavier one. The ratio of the numbers of molecules of the two gases which find their way through the porous walls for a short time interval is equal to the square root of the inverse ratio of their masses, $\sqrt{m_2/m_1}$. This diffusion process is one method of separating (fissionable) U^{235} (0.7% abundance) from a normal sample of uranium containing mostly (nonfissionable) U^{238} (99.3% abundance). To quote from the Smyth report,†

As long ago as 1896 Lord Rayleigh showed that a mixture of gases of different atomic weight could be partly separated by allowing some of it to diffuse through a porous barrier into an evacuated space. Because of their higher average speed the molecules of the light gas diffuse through the barrier faster so that the gas which has passed through the barrier (i.e., the "diffusate") is enriched in the lighter constituent and the residual gas which has not passed through the barrier is impoverished in the lighter constituent. The gas most highly enriched in the lighter constituent is the so-called "instantaneous diffusate"; it is the part that diffuses before the impoverishment of the residue has become appreciable. . . . On the assumption that the diffusion rates are inversely proportional to the square roots of the molecular weights‡ the separation factor for the instantaneous diffusate, called the "ideal separation factor" α, is given by

$$\alpha = \sqrt{M_2/M_1},$$

where M_1 is the molecular weight of the lighter gas and M_2 that of the heavier. Applying this formula to the case of uranium will illustrate the magnitude of the separation problem. Since uranium itself is not a gas, some gaseous com-

---

* See footnote, p. 479.
† *A General Account of the Development of Methods of Using Atomic Energy for Military Purposes* . . . , H. D. Smyth, U.S. Government Printing Office, 1945.
‡ Note that the ratio m_2/m_1 of the masses of the two molecules of different gases is the same as the ratio M_2/M_1 of their molecular weights, because the molecular weights refer to the same number of molecules. Compare Eq. 23-7.

pound of uranium must be used. The only one obviously suitable is uranium hexafluoride, UF_6. . . . Since fluorine has only one isotope, the two important uranium hexafluorides are $U^{235}F_6$ and $U^{238}F_6$; their molecular weights are 349 [g/mol] and 352 [g/mol]. Thus if a small fraction of a quantity of uranium hexafluoride is allowed to diffuse through a porous barrier, the diffusate will be enriched in $U^{235}F_6$ by a factor

$$\alpha = \sqrt{\tfrac{352}{349}} = 1.0043 \ldots$$

To separate the uranium isotopes, many successive diffusion stages (i.e., a cascade) must be used. . . . Studies by Cohen and others show that the best flow arrangement for the successive stages is that in which half the gas pumped into each stage diffused through the barrier, the other (impoverished) half being returned to the feed of the next lower stage. . . . If one desires to produce 99 per cent pure $U^{235}F_6$, and if one uses a cascade in which each stage has a reasonable overall enrichment factor, then it turns out that roughly 4000 stages are required. . . . Most of the material that eventually emerges from the cascade has been recycled many times. Calculation shows that for an actual uranium-separation plant it may be necessary to force through the barriers of the first stage 100,000 times the volume of gas that comes out the top of the cascade (i.e., as desired product $U^{235}F_6$).

Forces between molecules are of electromagnetic origin. All molecules contain electric charges in motion. These molecules are electrically neutral in the sense that the negative charge of the electrons is equal and opposite to the charge of the nuclei. This does not mean, however, that molecules do not interact electrically. For example, when two molecules approach each other, the charges on each are disturbed and depart slightly from their usual positions in such a way that the average distance between opposite charges in the two molecules is a little smaller than that between like charges. Hence, an attractive intermolecular force results. This internal rearrangement takes place only when molecules are fairly close together, so that these forces act only over short distances; they are short-range forces. If the molecules come very close together, so that their outer charges begin to overlap, the intermolecular force becomes repulsive. The molecules repel each other because there is no way for a molecule to rearrange itself internally to prevent repulsion of the adjacent external electrons. It is this repulsion on contact that accounts for the billiard-ball character of molecular collisions in gases. If it were not for this repulsion, molecules would move right through each other instead of rebounding on collision.

Let us assume that molecules are approximately spherically symmetrical. Then we can describe intermolecular forces graphically by plotting the mutual potential energy of two molecules, U, as a function of distance r between their centers. The force F acting on each molecule is related to the potential energy U by $F = -dU/dr$. In Fig. 23-3a we plot a typical $U(r)$. Here we can imagine one molecule to be fixed at O. Then the other molecule will be repelled from O when the slope of U is negative and will be attracted to O when the slope is positive. At r_0 no force acts between the molecules; the slope is zero there. In Fig. 23-3b we plot the mutual force $F(r)$ corresponding to this potential energy function. The line E in Fig. 23-3a represents the total mechanical energy of the colliding molecules. The intersection of $U(r)$ with this line is a "turning point" of the motion (see Section 8-5). The separation of the centers of two molecules at the turning point is the distance of closest approach. The separation distance at which the mutual potential energy is zero may be taken as the approximate distance of closest approach in a collision and hence as the diameter of the molecule. For simple molecules the diameter is about 2.5×10^{-10} m. The forces between molecules practically cease at about 10^{-9} m or 4 diameters apart, so that molecular forces are very short-range ones. The distance r_0 at which the potential is a minimum (the equilibrium point) is about 3.5×10^{-10} m for simple molecules. Of course, different molecules have different sizes and internal arrangement of charges so that intermolecular forces vary from one molecule to

23-6
INTERMOLECULAR FORCES

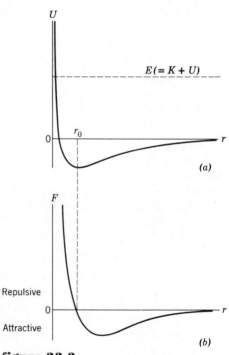

figure 23-3
(a) The mutual potential energy of two molecules versus their separation. E shows their total mechanical energy ($= K + U$). (b) The mutual force, $-dU/dr$, corresponding to this potential energy. U is a minimum at r_0, at which separation $F = 0$.

another. However, they always show the qualitative behavior indicated in the figures.*

In the solid state molecules vibrate about the equilibrium position r_0, their total energy E being negative, that is, lying below the horizontal axis in Fig. 23-3a. The molecules do not have enough energy to escape from the potential valley (that is, from the attractive binding force). The centers of vibration O are more or less fixed in a solid. In a liquid the molecules have greater vibrational energy about centers which are free to move but which remain about the same distance from one another. These molecules have their greatest kinetic energy in the gaseous state. In a gas the average distance between the molecules is considerably greater than the effective range of intermolecular forces, and the molecules move in straight lines between collisions. Clerk Maxwell discusses the relation between the kinetic theory model of a gas and the intermolecular forces as follows: "Instead of saying that the particles are hard, spherical, and elastic, we may if we please say that the particles are centers of force, of which the action is insensible except at a certain small distance, when it suddenly appears as a repulsive force of very great intensity. It is evident that either assumption will lead to the same results."

It is interesting to compare the measured intermolecular forces with the gravitational force of attraction between molecules. If we choose a separation distance of 4×10^{-10} m, for example, the force between two helium atoms is about 6×10^{-13} N. The gravitational force at that separation is about 7×10^{-42} N, smaller than the intermolecular force by a factor of 10^{29}! This is a typical result and shows that gravitational forces are negligible in comparison with intermolecular forces. Although the intermolecular forces appear to be small by ordinary standards, we must remember that the mass of a molecule is so small (about 10^{-26} kg) that these forces can impart instantaneous accelerations of the order of 10^{15} m/s² ($10^{14}g$). These accelerations may last for only a very short time, of course, because one molecule can very quickly move out of the range of influence of the other.

23-7
SPECIFIC HEATS OF AN IDEAL GAS

We picture the molecules in an ideal gas as hard elastic spheres; that is, we assume that there are no forces between the molecules except during collisions and that the molecules are not deformed by collisions. If this is so, there is no internal potential energy and the internal energy of an ideal gas is entirely kinetic. We have already found that the average translational kinetic energy per molecule is $\frac{3}{2}kT$, so that the internal energy U of an ideal gas containing N molecules is†

$$U = \tfrac{3}{2}NkT = \tfrac{3}{2}nRT. \qquad (23\text{-}8)$$

This prediction of kinetic theory says that *the internal energy of an ideal gas is proportional to the Kelvin temperature and depends only on the temperature*, being independent of pressure and volume. With this result we can now obtain information about the specific heats of an ideal gas.

The specific heat of a substance is the heat required per unit mass per unit temperature change. A convenient unit of mass is the mole. The corresponding specific heat is called the molar heat capacity and is represented by C. Only two varieties of molar heat capacity are important for gases, namely, that at constant volume, C_v, and that at constant pressure, C_p.

---

* See "The Force between Molecules," by B. V. Derjaguin, *Scientific American*, July 1960, for a discussion of the measurement of molecular attractions between macroscopic bodies.
† We will see in Section 23-8 that this result applies only to monatomic gases, for which rotational and vibrational energies are not possible. Only in this case can we equate U to the *translational* kinetic energy.

Constant-
pressure
process

$T + \Delta T$
$p, V + \Delta V$

(b)

T
p, V

(a)

Constant-
volume
process

$T + \Delta T$
$p + \Delta p, V$

(c)

(d)

figure 23-4
The temperature of a given mass of
gas is raised by the same amount
by a constant-pressure process
$(a \rightarrow b)$ and by a constant-volume
process $(a \rightarrow c)$.

Let us confine a certain number of moles of an ideal gas in a piston-
cylinder arrangement as in Fig. 23-4a. The cylinder rests on a heat
reservoir whose temperature can be raised or lowered at will, so that
we may add heat to the system or remove it, as we wish. The gas has
a pressure p such that its upward force on the (frictionless) piston just
balances the weight of the piston and its sand load. The state of the
system is represented by point a in the p-V diagram of Fig. 23-4d; this
diagram shows two isothermal lines, all points on one corresponding
to a temperature T and all points on the other to a (higher) temperature
$T + \Delta T$.

Now let us raise the temperature of the system by ΔT, by slowly increasing the reservoir temperature. As we do this let us add sand to
the piston so that the volume V does not change. This *constant-volume
process* carries the system from the initial state of Fig. 23-4a to the final
state of Fig. 23-4c. Equivalently, it goes from point a to point c in Fig.
23-4d. Let us apply the first law of thermodynamics

$$\Delta U = Q - W$$

to this process. By definition of C_v we have $Q = nC_v \, \Delta T$. Also, W
$(= p \, \Delta V) = 0$ because $\Delta V = 0$. Thus

$$\Delta U = nC_v \, \Delta T. \tag{23-9}$$

Let us restore the system to its original state and again raise its temperature by ΔT, this time leaving the sand load undisturbed so that
the pressure p does not change. This *constant-pressure process* carries
the system from the initial state of Fig. 23-4a to the final state of Fig.
23-4b. Equivalently, it goes from point a to point b in Fig. 23-4d. Let
us apply the first law to *this* process. By definition of C_p we have $Q =
nC_p \, \Delta T$. Also, $W = p \, \Delta V$. Now for an ideal gas, U depends only on the

temperature. Since processes $a \rightarrow b$ and $a \rightarrow c$ in Fig. 23-4 involve the same change ΔT in temperature, they must also involve the same change ΔU in internal energy, namely, that given by Eq. 23-9. Thus for the constant-pressure process the first law yields

$$nC_p \,\Delta T = nC_v \,\Delta T + p \,\Delta V.$$

Let us apply the equation of state $pV = nRT$ to the constant-pressure process $a \rightarrow b$. For p constant we have, by taking differences,

$$p \,\Delta V = nR \,\Delta T.$$

Combining these equations yields

$$C_p - C_v = R. \tag{23-10}$$

This shows that the molar heat capacity of an ideal gas at constant pressure is always larger than that at constant volume by an amount equal to the universal gas constant R ($= 8.31$ J/mol·K or 1.99 cal/mol·K). Although Eq. 23-10 is exact only for an ideal gas, it is nearly true for real gases at moderate pressure (see Table 23-2). Notice that in obtaining this result we did not use the specific relation $U = \frac{3}{2}nRT$, but only the fact that U depends on temperature alone.

If we can compute C_v, then Eq. 23-10 will give us C_p and vice versa. We *can* obtain C_v by combining Eq. 23-9 with the kinetic theory result for the internal energy of an ideal gas, $U = \frac{3}{2}nRT$ (Eq. 23-8). Thus, in the limit of differential changes,

$$C_v = \frac{dU}{n \, dT} = \frac{d}{n \, dT} \left[\tfrac{3}{2}nRT \right] = \tfrac{3}{2}R. \tag{23-11}$$

This result (about 3 cal/mol·K) turns out to be rather good for monatomic gases. It is, however, in serious disagreement with values obtained for diatomic and polyatomic gases (see Table 23-2). This suggests that Eq. 23-8 is not generally correct (see footnote on p. 507). Since that relation followed directly from the kinetic theory model, we conclude that we must change the model if kinetic theory is to survive as a useful approximation to the behavior of real gases.

EXAMPLE 5

Show that for an ideal gas undergoing an adiabatic process $pV^\gamma = a$ constant, where $\gamma = C_p/C_v$.

Let us apply the first law of thermodynamics

$$Q = \Delta U + W.$$

For an adiabatic process $Q = 0$. For W we put $p \,\Delta V$. Since the gas is assumed to be ideal, U depends only on temperature and, from Eq. 23-9, $\Delta U = nC_v \,\Delta T$. With these substitutions we have

$$0 = nC_v \,\Delta T + p \,\Delta V$$

or

$$\Delta T = -\frac{p \,\Delta V}{nC_v}.$$

For an ideal gas $pV = nRT$, so that, if p, V, and T are allowed to take on small variations,

$$p \,\Delta V + V \,\Delta p = nR \,\Delta T$$

or

$$\Delta T = \frac{p \,\Delta V + V \,\Delta p}{nR}.$$

Equating these two expressions and using Eq. 23-10 $(C_p - C_v = R)$, we obtain, after some rearrangement,

$$p\,\Delta V C_p + V\,\Delta p C_v = 0$$

Dividing by pVC_v and recalling that, by definition, $C_p/C_v = \gamma$, we get

$$\frac{\Delta p}{p} + \gamma\frac{\Delta V}{V} = 0.$$

In the limiting case of differential changes this reduces to

$$\frac{dp}{p} + \gamma\frac{dV}{V} = 0,$$

which (assuming γ to be constant) we can integrate as

$$\ln p + \gamma\ln V = \text{a constant}$$

or

$$pV^\gamma = \text{a constant.} \qquad (23\text{-}12)$$

The value of the constant is proportional to the quantity of gas. In Fig. 23-5 we compare the isothermal and adiabatic behaviors of an ideal gas.

figure 23-5
T_1, T_2, T_3 and T_4 show how the pressure of one mole of an ideal gas changes as its volume is changed, the temperature being held constant (isothermal process). A_1, A_2, A_3 and A_4 show how the pressure of an ideal gas changes as its volume is changed, no heat being allowed to flow to or from the gas (adiabatic process). An adiabatic *increase* in volume (for example going from a to b along A_3) is always accompanied by a *decrease* in temperature, since at a, $T = 400$ K, whereas at b, $T = 300$ K.

The compressions and rarefactions in a sound wave are practically adiabatic at audio frequencies. Show that in such a case the speed of sound in an ideal gas is given by

EXAMPLE 6

$$v = \sqrt{\frac{\gamma p}{\rho}}.$$

In Chapter 20 we showed the speed of sound to be $v = \sqrt{B/\rho}$, where ρ is the gas density and B is the bulk modulus of the gas, $B = -V(\Delta p/\Delta V)$. However, B will depend on the conditions that prevail as the pressure is changed. If we assume the temperature to remain constant we have, in the limit of differential changes,

$$B_{isothermal} = -V\left(\frac{dp}{dV}\right)_{isothermal} \qquad (23\text{-}13)$$

In an isothermal process for an ideal gas we have

$$pV = \text{a constant}$$

or, by differentiation with respect to V,

$$p + V\left(\frac{dp}{dV}\right)_{isothermal} = 0.$$

Combined with Eq. 23-13 this yields

$$B_{isothermal} = p.$$

In a sound wave, however, the conditions are not isothermal but closely adiabatic. The appropriate bulk modulus is then

$$B_{adiabatic} = -V\left(\frac{dp}{dV}\right)_{adiabatic} \qquad (23\text{-}14)$$

In an adiabatic process for an ideal gas we have

$$pV^{\gamma} = \text{a constant}$$

or, by differentiating with respect to V,

$$p\gamma V^{\gamma-1} + V^{\gamma}\left(\frac{dp}{dV}\right)_{adiabatic} = 0.$$

This, combined with Eq. 23-14, yields

$$B_{adiabatic} = \gamma p$$

and, for the speed of sound,

$$v = \sqrt{\frac{B}{\rho}} = \sqrt{\frac{\gamma p}{\rho}}. \qquad (23\text{-}15)$$

To understand why the compressions and rarefactions are adiabatic rather than isothermal, recall that compression of a gas causes a temperature rise and rarefaction a temperature fall unless heat energy is removed or added. Hence, in a gas through which sound propagates, the compressed regions are warmer than the rarefied ones. In principle, heat will be conducted from compression to rarefaction. The rate of heat conduction per unit area, however, depends (see Section 22-4) on the thermal conductivity of the gas and on the distance between compression and adjacent rarefaction, which is half a wavelength. The wavelength of audible sound is much too large for any significant rate of heat flow even in gases that are the best heat conductors. Hence, the conditions are essentially adiabatic in sound propagation and not isothermal. Actually the condition for breakdown of the adiabatic approximation is that the wavelength of the wave be comparable with the mean free path of molecules in the gas, an extreme situation (see Section 24-1).

Newton derived a formula for the speed of sound in 1710, when the only gas law formulated was Boyle's law. He assumed isothermal rather than adiabatic conditions and obtained $v = \sqrt{p/\rho}$ rather than the (correct) value of $\sqrt{\gamma\,p/\rho}$. Newton was able to get good agreement with experimental values by making (then) reasonable corrections to his basic model.* His error and the correct model were pointed out by Laplace in 1816, more than a century later. We must remember that, at that date, the concept of energy was not yet understood and the science of thermodynamics did not exist.

Does this result modify the result obtained in Example 4? Can you now explain why the speed of sound in a gas is not the same as the root-mean-square speed of the gas molecules?

---

* See "Newton's Derivation of the Velocity of Sound" by Haven Whiteside, *American Journal of Physics,* May 1964.

A modification of the kinetic theory model designed to explain the specific heats of gases was first suggested by Clausius in 1857. Recall that in our model we assumed a molecule to behave like a hard elastic sphere and we treated its kinetic energy as purely translational. The specific heat prediction was satisfactory for monatomic molecules. Further, because of the success of this simple model in other respects in predicting the correct behavior of gases of all kinds over wide temperature ranges, we feel confident that it is the average kinetic energy of translation which determines what we measure as the temperature of a gas.

In the case of specific heats, however, we are concerned with all possible ways of absorbing energy and we must ask whether or not a molecule can store energy internally, that is, in a form other than kinetic energy of translation. This would certainly be so if we pictured a molecule, not as a rigid particle, but as an object with internal structure. For then a molecule could rotate and vibrate as well as move with translational motion. In collisions, the rotational and vibrational modes of motion could be excited, and this would contribute to the internal energy of the gas. Here then is a model which enables us to modify the kinetic theory formula for the internal energy of a gas.

Let us now find the total energy of a system containing a large number of such molecules, where each molecule is thought of as an object having internal structure. The energy will consist of kinetic energy of translation, with terms like $\frac{1}{2}mv_x^2$; of kinetic energy of rotation, with terms like $\frac{1}{2}I\omega_x^2$; of kinetic energy of vibration of the atoms in a molecule, with terms like $\frac{1}{2}\mu v^2$ (where μ is the reduced mass), and of potential energy of vibration of the atoms in a molecule, with terms like $\frac{1}{2}kx^2$. Although other kinds of energy contributions exist, such as magnetic, for gases we can describe the total energy quite accurately by terms such as these. Although these terms have different origins, they all have the same mathematical form, namely, a positive constant times the square of a quantity which can take on negative or positive values. We can show from statistical mechanics that *when the number of particles is large and Newtonian mechanics holds, all these terms have the same average value, and this average value depends only on the temperature.* In other words, the available energy depends only on the temperature and distributes itself in equal shares to each of the independent ways in which the molecules can absorb energy. This theorem, stated here without proof, is called the *equipartition of energy* and was deduced by Clerk Maxwell. Each such independent mode of energy absorption is called a *degree of freedom*.

From Eq. 23-8 we know that the kinetic energy of translation per mole of gaseous molecules is $\frac{3}{2}RT$. The kinetic energy of translation per mole is the sum of three terms, however, namely $\frac{1}{2}M\overline{v_x^2}$, $\frac{1}{2}M\overline{v_y^2}$, and $\frac{1}{2}M\overline{v_z^2}$. The theorem of equipartition requires that each such term contribute the same amount to the total energy per mole, or $\frac{1}{2}RT$ per degree of freedom.

For *monatomic gases* the molecules have only translational motion (no internal structure in kinetic theory), so that $U = \frac{3}{2}nRT$. It follows from Eq. 23-11 that $C_v = \frac{3}{2}R \cong 3$ cal/mol K. Then from Eq. 23-10, $C_p = \frac{5}{2}R$, and the ratio of specific heat is

$$\gamma = \frac{C_p}{C_v} = \frac{5}{3} = 1.67.$$

For a *diatomic gas* we can think of each molecule as having a dumb-

23-8
EQUIPARTITION OF ENERGY

bell shape (two spheres joined by a rigid rod). Such a molecule can rotate about any one of three mutually perpendicular axes. However, the rotational inertia about an axis along the rigid rod should be negligible compared to that about axes perpendicular to the rod, so that the rotational energy should consist of only two terms,* such as $\frac{1}{2}I\omega_y{}^2$ and $\frac{1}{2}I\omega_z{}^2$. Each rotational degree of freedom is required by equipartition to contribute the same energy as each translational degree, so that for a diatomic gas having both rotational and translational motion,

$$U = 3n(\tfrac{1}{2}RT) + 2n(\tfrac{1}{2}RT) = \tfrac{5}{2}nRT,$$

or

$$C_v = \frac{dU}{n\,dT} = \tfrac{5}{2}R \cong 5 \text{ cal/mol·K}$$

and

$$C_p = C_v + R = \tfrac{7}{2}R,$$

or

$$\gamma = \frac{C_p}{C_v} = \frac{7}{5} = 1.40.$$

For *polyatomic gases*, each molecule contains three or more spheres (atoms) joined together by rods in our model, so that the molecule is capable of rotating energetically about each of three mutually perpendicular axes. Hence, for a polyatomic gas having both rotational and translational motion,

$$U = 3n(\tfrac{1}{2}RT) + 3n(\tfrac{1}{2}RT) = 3nRT,$$

or

$$C_v = \frac{dU}{n\,dT} = 3R = 6 \text{ cal/mol·K},$$

and

$$C_p = 4R,$$

or

$$\gamma = \frac{C_p}{C_v} = 1.33.$$

Let us now turn to experiment to test these ideas. In Table 23-2 we list the experimentally determined molar heat capacities for common gases at 20° C and 1.0 atm. Notice that for monatomic and diatomic

Table 23-2

Type of Gas	Gas	C_p, cal/mol·K	C_v, cal/mol·K	$C_p - C_v$	$\gamma = C_p/C_v$
Monatomic	He	4.97	2.98	1.99	1.67
	A	4.97	2.98	1.99	1.67
Diatomic	H_2	6.87	4.88	1.99	1.41
	O_2	7.03	5.03	2.00	1.40
	N_2	6.95	4.96	1.99	1.40
	Cl_2	8.29	6.15	2.14	1.35
Polyatomic	CO_2	8.83	6.80	2.03	1.30
	SO_2	9.65	7.50	2.15	1.29
	NH_3	8.80	6.65	2.15	1.31
	C_2H_6	12.35	10.30	2.05	1.20

* We have already ruled out the possibility that a monatomic molecule could rotate. Actually it could spin about any one of three mutually perpendicular axes if it had any extent, such as a finite sphere. Implicitly, therefore, we have adopted a point mass as our model of the atom. Hence, in a diatomic molecule we are rid of one rotational degree of freedom, for point masses joined by a rigid line have no rotational energy about an axis along that line.

gases the values of C_v, C_p, and γ are close to the ideal gas predictions. In some diatomic gases, like chlorine, and in most polyatomic gases the specific heats are larger than the predicted values. Even γ shows no simple regularity for polyatomic gases. This suggests that our model is not yet close enough to reality.

We have not yet considered energy contributions from the vibrations of the atoms in diatomic and polyatomic molecules. That is, we can modify the dumbbell model and join the spheres instead by springs. This new model will greatly improve our results in some cases. Instead of having a theoretical model for all gases, however, we now require an empirical model which differs from gas to gas. We can obtain a reasonably good picture of molecular behavior this way and the empirical model is therefore useful; however it ceases to be fundamental.

figure 23-6
Variation of the molar heat C_v of hydrogen with temperature. Note that T is drawn on a logarithmic scale. Hydrogen dissociates before 3200 K is reached. The dashed curve is for a diatomic molecule that does not dissociate before 10,000 K is reached.

To see this more clearly, let us consider Fig. 23-6, which shows the variation of the molar heat capacity of hydrogen with temperature. The value of 5 cal/mol·K, which is predicted for diatomic molecules by our model, is characteristic of hydrogen only in the temperature range from about 250 to 750 K. Above 750 K, C_v increases steadily toward 7 cal/mol·K and below 250 K, C_v decreases steadily to 3 cal/mol K. Other gases show similar variations of molar heat with temperature.

Here is a possible explanation. At low temperatures apparently (see Example 7) the hydrogen molecule has translational energy only and, for some reason, cannot rotate. As the temperature rises rotation becomes possible so that at "ordinary" temperatures a hydrogen molecule acts like our dumbbell model. At high temperatures the collisions between molecules cause the atoms in the molecule to vibrate and the molecule ceases to behave as a rigid body. Different gases, because of their different molecular structure, may show these effects at different temperatures. Thus a chlorine molecule appears to vibrate at room temperature.

Although this description is essentially correct, and we have obtained much insight into the behavior of molecules, this behavior contradicts classical kinetic theory. For kinetic theory is based on Newtonian mechanics applied to a large collection of particles, and the equipartition of energy is a necessary consequence of this classical statistical mechanics. But *if equipartition of energy holds, then, no matter what happens to the total internal energy as the temperature* changes, *each part of the energy*—translational, rotational, and vibrational—*must share equally in the change.* There is no classical mechanism for changing one mode of mechanical energy at a time in such a system. Kinetic theory requires that the specific heats of gases be independent of the temperature.

Hence, we have come to the limit of validity of classical mechanics when we try to explain the structure of the atom (or molecule). Just as Newtonian principles break down at very high speeds (near the speed of light), so here in the region of very small dimensions they also break down. Relativity theory modifies Newtonian ideas to account for the behavior of physical systems in the region of high speeds. It is quantum physics that modifies Newtonian ideas to account for the behavior of physical systems in the region of small dimensions. Both relativity theory and quantum mechanics are generalizations of classical theory in the sense that they give the (correct) Newtonian results in the regions in which Newtonian physics has accurately described experimental observations. In the following two chapters we shall confine our attention to the very fruitful application of thermodynamics and the kinetic theory to "classical" systems.

EXAMPLE 7

According to quantum theory the internal energy of an atom (or molecule) is "quantized"; that is, the atom cannot have any of a continuous set of internal energies but *only certain discrete ones.* After being raised from its lowest energy state to some higher one the atom can give up this energy by emitting radiation whose energy equals the difference in energy between the upper and lower internal energy states of the atom.

When two atoms collide, some of their translational kinetic energy may be converted into internal energy of one or both of the atoms. In such a case the collision is inelastic, for translational kinetic energy is not conserved. In a gas, the average translational kinetic energy of an atom is $\frac{3}{2}kT$. When the temperature is raised to a value where $\frac{3}{2}kT$ is about equal to some allowed internal excitation energy of the atom, then an appreciable number of the atoms can absorb enough energy through inelastic collisions to be raised to this higher internal energy state. We can detect this because, after an interval, radiation corresponding to the absorbed energy will be emitted.

(a) Compute the average translational kinetic energy per molecule in a gas at room temperature.

We have, for $T = 300$ K,

$$\tfrac{3}{2}kT = \tfrac{3}{2}(1.38 \times 10^{-23} \text{ J/molecule·K})(300 \text{ K})$$

$$= 6.21 \times 10^{-21} \text{ J/molecule}$$

$$= 3.88 \times 10^{-2} \text{ eV/molecule}.$$

This is about $\frac{1}{25}$ eV per molecule. Some molecules will have larger energies and some smaller energies than this average value.

(b) The first allowed (internal) excited state of a hydrogen atom is 10.2 eV above its lowest (ground) state. What temperature is needed to excite a "large" number of hydrogen atoms to emit radiation of this energy?

We require

$$\tfrac{3}{2}kT = 10.2 \text{ eV}$$

and we have from above

$$\tfrac{3}{2}k(300 \text{ K}) = \tfrac{1}{25} \text{ eV}.$$

Hence

$$T = 300 \text{ K} \times 10.2/(\tfrac{1}{25}) \simeq 7.5 \times 10^4 \text{ K}.$$

Actually, because many molecules have energies much greater than the average value, appreciable excitation may occur at somewhat lower temperatures.

We can now appreciate why the kinetic theory assumption, that molecules can be regarded as having no internal structure and collide elastically with one another, holds true at ordinary temperatures. Only at temperatures high enough to give the molecules an average translational kinetic energy comparable to the energy difference between the lowest and the first allowed excited state of the molecule will the internal structure of the molecule change and the collisions become inelastic. Indeed, in retrospect one may say that early evidence that the internal energy of an atom is quantized existed in experiments with gas collisions and that the seeds of quantum theory lay in the kinetic theory of gases.*

questions

1. In discussing the fact that it is impossible to apply the laws of mechanics individually to atoms in a macroscopic system, Mayer and Mayer state: "The very complexity of the problem [that is, the fact that the number of atoms is large] is the secret of its solution." Discuss this sentence.

2. Is there any such thing as a truly continuous body of matter?

3. In kinetic theory we assume that there are a large number of molecules in a gas. Real gases behave like an ideal gas at low densities. Are these statements contradictory? If not, what conclusion can you draw from them?

4. We have assumed that the walls of the container are elastic for molecular collisions. Actually, the walls may be inelastic. In practice this makes no difference as long as the walls are at the same temperature as the gas. Explain.

5. In large-scale inelastic collisions mechanical energy is lost through internal friction resulting in a rise of temperature owing to increased internal molecular agitation. Is there a loss of mechanical energy to heat in an inelastic collision between molecules?

6. What justification is there in neglecting the change in gravitational potential energy of molecules in a gas?

7. We have assumed that the force exerted by molecules on the wall of a container is steady in time. How is this justified?

8. The average velocity of the molecules in a gas must be zero if the gas as a whole and the container are not in translational motion. Explain how it can be that the average *speed* is not zero.

9. Consider a hot, stationary golf ball sitting on a tee and a cold golf ball just moving off the tee after being hit. Can the numerical value of the kinetic energy of the molecules' motion relative to the tee be the same in each case? If so, what is the difference between the two cases?

10. By considering quantities which must be conserved in an elastic collision, show that in general molecules of a gas cannot have the same speeds after a collision as they had before. Is it possible, then, for a gas to consist of molecules which all have the same speed?

11. Justify the fact that the pressure of a gas depends on the *square* of the speed of its particles by explaining the dependence of pressure on the collision frequency and the momentum transfer of the particles.

12. Why does the boiling temperature of a liquid increase with pressure?

13. Pails of hot and cold water are set out in freezing weather. Explain (a) if the pails have lids, the cold water will freeze first but (b) if the pails do not have lids, it is possible for the hot water to freeze first. (*Hint:* If equal masses of water are taken at two starting temperatures, more rapid evaporation from the hotter one may diminish its mass enough to compensate for the greater temperature range it must cover to reach freezing. See "The Freezing of Hot and Cold Water" by G. S. Kell, *American Journal of Physics*, May 1969.)

---

* See "On Teaching Quantum Phenomena" by Sir N. F. Mott in *Contemporary Physics*, August 1964.

14. How is the speed of sound related to gas variables in the kinetic theory model?

15. Far above the earth's surface the gas kinetic temperature (see Eq. 23-5) is reported to be the order of 1000 K. However, a person placed in such an environment would freeze to death rather than vaporize. Explain.

16. Why must the time allowed for diffusion separation be relatively short?

17. Suppose we want to obtain U^{238} instead of U^{235} as the end product of a diffusion process. Would we use the same process? If not, explain how the separation process would have to be modified.

18. Considering the diffusion of gases into each other (see footnote on page 503), can you draw an analogy to a large jostling crowd with many "collisions" on a large inclined plane with a slope of a few degrees?

19. Can you describe a centrifugal device for gaseous separation? Is a centrifuge better than a diffusion chamber for separation of gases?

20. Would you expect real molecules to be spherically symmetrical? If not, how would the potential energy function of Fig. 23-3 change?

21. Explain how we might keep a gas at a constant temperature during a thermodynamic process.

22. Explain why the temperature of a gas drops in an adiabatic expansion.

23. If hot air rises, why is it cooler at the top of a mountain than near sea level?

24. Comment on this statement: "There are two ways to carry out an adiabatic process. One is to do it quickly and the other is to do it in an insulated box."

25. A sealed rubber balloon contains a very light gas. The balloon is released and it rises high into the atmosphere. Describe and explain the thermal and mechanical behavior of the balloon.

26. Explain why the specific heat at constant pressure is greater than the specific heat at constant volume.

27. It is more common to excite radiation from gaseous atoms by use of electrical discharge than by thermal methods. Why?

28. *Extensive* quantities have values that depend on what the system's boundaries are, whereas *intensive* quantities are independent of the choice of boundaries. That is, extensive quantities are necessarily defined for a whole system, whereas intensive quantities apply uniformly to any small part of the system. Of the following quantities, determine which are extensive and which are intensive: pressure, volume, temperature, density, mass, internal energy.

problems

SECTION 23-2

1. At 0° C and 1.000-atm pressure the densities of air, oxygen, and nitrogen are, respectively, 1.293 kg/m³, 1.429 kg/m³, and 1.251 kg/m³. Calculate the percentage of nitrogen in the air from these data, assuming only these two gases to be present.　　　　　　　　　*Answer:* 76.4%, by mass.

2. (a) What is the volume occupied by one mole of an ideal gas at standard conditions, that is, pressure of one atmosphere and temperature of 0° C? (b) Show that the number of molecules per cubic centimeter (Loschmidt number) at standard conditions is 2.687×10^{19}.

3. The best vacuum that can be attained in the laboratory corresponds to a pressure of about 10^{-14} atm, or about 10^{-10} mm-Hg. How many molecules are there per cubic centimeter in such a "vacuum" at room temperature? *Answer:* 2.7×10^5.

4. An air bubble of 20 cm³ volume is at the bottom of a lake 40 m deep where the temperature is 4° C. The bubble rises to the surface which is at a temperature of 20° C. Take the temperature of the bubble to be the same as that of the surrounding water and find its volume just before it reaches the surface?

5. Oxygen gas having a volume of 1.0 liter at 40° C and a pressure of 76 cm-Hg expands until its volume is 1.5 liters and its pressure is 80 cm-Hg. Find (a) the mass in moles of oxygen in the system and (b) its final temperature. *Answer:* (a) 0.039 mol. (b) 220° C.

6. An automobile tire has a volume of 1000 in.³ and contains air at a gauge pressure of 24 lb/in.² when the temperature is 0° C. What is the gauge pressure of the air in the tires when its temperature rises to 27° C and its volume increases to 1020 in.³?

7. Compute the number of molecules in a gas contained in a volume of 1.00 cm³ at a pressure of 1.00×10^{-3} atm and a temperature of 200 K. *Answer:* 3.67×10^{16}.

8. If the water molecules in 1.0 g of water were distributed uniformly over the surface of the earth, how many such molecules would there be on 1.0 cm² of the earth's surface?

9. Calculate the work done in compressing 1.00 mol of oxygen from a volume of 22.4 l at 0° C and 1.00-atm pressure to 16.8 l at the same temperature. *Answer:* 648 J.

10. Suppose that, as happened historically, we are given Boyle's law

$$pV = a \text{ constant} \quad (\text{constant } T)$$

and Charles' law

$$V/T = a \text{ constant} \quad (\text{constant } p)$$

separately. Show how these two laws may be combined to yield

$$pV/T = a \text{ constant}.$$

figure 23-7
Problem 11

11. A mercury-filled manometer with two unequal arms is sealed off with the same pressure p_0 in the two arms as in Fig. 23-7. The cross-sectional area of the manometer arms is 1.0 cm². With the temperature constant, an additional 10 cm³ of mercury is admitted through the stopcock at the bottom; the level on the left increases 6.0 cm and that on the right increases 4.0 cm. Find the pressure p_0.   *Answer:* 1.5×10^5 Pa.

12. Air that occupies 5.0 ft³ (0.14 m³) at 15 lb/in.² (1.034×10^5 Pa) gauge pressure is expanded isothermally to atmospheric pressure and then cooled at constant pressure until it reaches its initial volume. Compute the work done by the gas.

SECTION 23-4

13. The mass of the H_2 molecule is 3.3×10^{-24} g (2.3×10^{-28} slug). If 10^{23} hydrogen molecules per second strike 2.0 cm² (0.31 in.²) of wall at an angle of 45° with the normal when moving with a speed of 10^5 cm/s (3.3×10^3 ft/s), what pressure do they exert on the wall?   *Answer:* 2300 Pa (0.35 lb/in.²).

SECTION 23-5

14. At 273 K. and 1.00×10^{-2} atm the density of a gas is 1.24×10^{-5} g/cm³. (a) Find v_{rms} for the gas molecules. (b) Find the molecular weight of the gas and identify it.

15. (a) Compute the root-mean-square speed of an argon atom at room temperature (20° C). (b) At what temperature will the root-mean-square speed be half that value? Twice that value? *Answer:* (a) 430 m/s. (b) 73 K; 1200 K.

16. In a gas of uranium hexafluoride there are isotopes $U^{235}F_6$ and $U^{238}F_6$ having molecular weights 349 and 352, respectively. (a) What is the ratio of the rms speeds of these two molecular isotopes? (b) How could this fact be used to separate the isotopes?

17. (a) Determine the average value of the kinetic energy of the particles of an ideal gas at 0.0° C and 100° C. (b) What is the kinetic energy per mole of an ideal gas at these temperatures? *Answer:* (a) 5.65×10^{-21} J; 7.72×10^{-21} J. (b) 3400 J; 4650 J.

18. At what temperature is the average translational kinetic energy of a mole-

cule equal to the kinetic energy of an electron accelerated from rest through a potential difference of one volt (that is, an energy of 1.0 eV)?

19. Oxygen gas at 273 K and 1.00-atm pressure is confined to a cubical container 10 cm on a side. Compare the change in gravitational potential energy of an oxygen molecule falling the height of the box with its mean translational kinetic energy.

    Answer: Ratio of the mean translational kinetic energy to the change in gravitational potential energy is 1.1×10^5.

20. Find the root-mean-square speeds of (a) helium and (b) argon molecules at 40° C from that of oxygen molecules (460 m/s at 0.00° C). The molecular weight of oxygen is 32 g/mol, of argon 40, of helium 4.0.

21. (a) Compute the temperature at which the root-mean-square speed is equal to the speed of escape from the surface of the earth for molecular hydrogen. For molecular oxygen. (b) Do the same for the moon, assuming gravity on its surface to be 0.16 g. (c) The temperature high in the earth's upper atmosphere is about 1000 K. Would you expect to find much hydrogen there? Much oxygen?

    Answer: (a) 1.0×10^4 K; 1.6×10^5 K. (b) 440 K; 7000 K.

22. (a) Consider an ideal gas at 273 K and 1.0-atm pressure. Imagine that the molecules are for the most part evenly spaced at the centers of identical cubes. Using Avogadro's number and taking the diameter of a molecule to be 3.0×10^{-8} cm, find the length of an edge of such a cube and compare this length to the diameter of a molecule. (b) Now consider a mole of water having a volume of 18 cm³. Again imagine the molecules to be evenly spaced at the centers of identical cubes. Find the length of an edge of such a cube and compare this length to the diameter of a molecule.

23. Plot and physically interpret (a) the variation of gas density with temperature for an isobaric (constant-pressure) process and (b) the variation of gas density with pressure for an isothermal process.

24. Water standing in the open at 27° C evaporates due to the escape of some of the surface molecules. The heat of vaporization (540 cal/g) may be found approximately from ϵn, where ϵ is the average energy of the escaping molecules and n is the number of molecules per gram. (a) Find ϵ. (b) How many times greater is ϵ than the average kinetic energy of H_2O molecules, assuming that the kinetic energy is related to temperature in the same way as it is for gases?

25. Consider a given mass of an ideal gas. Compare curves representing constant-pressure, constant-volume, and isothermal processes on (a) a p-V diagram, (b) a p-T diagram, and (c) a V-T diagram. (d) How do these curves depend on the mass of gas chosen?

26. (a) Show that the variation in pressure in the earth's atmosphere, assumed to be isothermal, is given by $p = p_0 e^{-Mgy/RT}$ where M is the molecular weight of the gas. (See Example 1, Chapter 17.) (b) Show also that $n_v = n_{v_0} e^{-Mgy/RT}$ where n_v is the number of molecules per unit volume.

SECTION 23-7

27. (a) What is the internal energy of one mole of an ideal gas at 273 K? (b) Does it depend on volume or pressure? Does it depend on the nature of the gas?

    Answer: (a) 3400 J.

28. One mole of an ideal gas expands adiabatically from an initial temperature T_1 to a final temperature T_2. Prove that the work done by the gas is $C_v(T_1 - T_2)$, where C_v is the molar heat capacity.

29. One mole of an ideal gas undergoes an isothermal expansion. Find the heat flow into the gas in terms of the initial and final volumes and the temperature.   *Answer:* $RT \ln V_f/V_i$.

30. The mass of a gas molecule can be computed from the specific heat at constant volume. Take $C_v = 0.075$ kcal/kg · K for argon and calculate (a) the mass of an argon atom and (b) the atomic weight of argon.

31. Take the mass of a helium atom to be 6.66×10^{-27} kg. Compute the specific heat at constant volume for helium gas. *Answer:* 3.11×10^3 J/kg · K.

32. Air at $0.00°$ C and 1.00-atm pressure has a density of 1.291×10^{-3} g/cm³ and the speed of sound in air is 332 m/s at that temperature. Compute the ratio of specific heats of air.

33. Show that the speed of sound in an ideal gas is independent of the pressure and density.

34. The speed of sound in different gases at the same temperature depends on the molecular weight of the gas. Show that $v_1/v_2 = \sqrt{M_2/M_1}$ (constant T) where v_1 is the speed of sound in the gas of molecular weight M_1 and v_2 is the speed of sound in the gas of molecular weight M_2.

35. Show that the speed of sound in air increases about 0.61 m/s for each Celsius degree rise in temperature near $0°$ C.

36. From the knowledge that c_v, the specific heat at constant volume, for a gas in a container is $5R$, what can you conclude about the ratio of the speed of sound in that gas to the root-mean-square speed of its molecules at a temperature T?

37. The following data are the result of accurate experimental measurements: 1.000 mol of a gas occupies a volume of 2.541×10^{-2} m³ at a pressure of 9.480×10^4 Pa when its temperature is 290.0 K. The same mass of gas requires 125.0 cal to raise its temperature from 290.0 to 315.0 K while its volume is held constant. The ratio (c_p/c_v) of its specific heats is 1.430. (a) Use these data to compute the mechanical equivalent of heat J. (b) Account for the fact that your value of J differs from the accepted three-figure value—namely, 4.19 J/cal. *Answer:* (a) 3.86 J/cal.

38. A mass of gas occupies a volume of 4.0 liters at a pressure of 1.0 atm and a temperature of 300 K. It is compressed adiabatically to a volume of 1.0 liter. Determine (a) the final pressure and (b) the final temperature, assuming it to be an ideal gas for which $\gamma = 1.5$.

39. (a) A liter of gas with $\gamma = 1.3$ is at 273 K and 1.0-atm pressure. It is suddenly compressed to half its original volume. Find its final pressure and temperature. (b) The gas is now cooled back to $0°$ C at constant pressure. What is its final volume? *Answer:* (a) 2.5 atm; 340 K. (b) 0.40 liter.

40. A reversible heat engine carries 1.00 mol of an ideal monatomic gas around the cycle shown in Fig. 23-8. Process 1-2 takes place at constant volume, process 2-3 is adiabatic, and process 3-1 takes place at a constant pressure. (a) Compute the heat Q, the change in internal energy ΔU, and the work done W, for each of the three processes and for the cycle as a whole. (b) If the initial pressure at point 1 is 1.00 atm, find the pressure and the volume at points 2 and 3.

figure 23-8
Problem 40

41. A quantity of ideal gas occupies an initial volume V_0 at a pressure p_0 and a temperature T_0. It expands to a volume V_1 (a) at constant pressure, (b) at constant temperature, (c) adiabatically. Graph each case on a P-V diagram. In which case is Q greatest? Least? In which case is W greatest? Least? In which case is ΔU greatest? Least?

    Answer:

	greatest	least
Q	a	c
W	a	c
ΔU	a	c

42. A thin tube, sealed at both ends, is 1.0 m long. It lies horizontally, the middle 10 cm containing mercury and the two equal ends containing air at standard atmospheric pressure. If the tube is now turned to a vertical position, by what amount will the mercury be displaced? Assume that the process is (a) isothermal and (b) adiabatic. Which assumption is more reasonable?

SECTION 23-8

43. One mole of oxygen is heated at a constant pressure starting at $0.00°$ C. How much heat energy must be added to the gas to double its volume? *Answer:* 8040 J.

44. An ideal diatomic gas (4.0 moles) at high temperature experiences a temperature increase of 60 K under constant pressure conditions. (a) How much heat was added to the gas? (b) By how much did the internal energy of the gas increase? (c) How much work was done by the gas? (d) By how much did the internal translational kinetic energy of the gas increase?

45. Ten grams of oxygen are heated at constant atmospheric pressure from 27.0 to 127° C. (a) How much heat is transferred to the oxygen? (b) What fraction of the heat is used to raise the internal energy of the oxygen?
    Answer: (a) 920 J. (b) 71%.

46. Calculate the mechanical equivalent of heat from the value of R and the values of C_v and γ for oxygen from Table 23-2.

47. *Avogadro's law* states that under the same condition of temperature and pressure equal volumes of gas contain equal numbers of molecules. Derive this law from kinetic theory using Eq. 23-3 and the equipartition of energy assumption.

48. A room of volume V is filled with a diatomic ideal gas (air) at temperature T_1 and pressure p_0. The air is heated to a higher temperature T_2, the pressure remaining constant at p_0 because the walls of the room are not air-tight. Show that the internal energy content of the air remaining in the room is the same at T_1 and T_2, and that the energy supplied by the furnace to heat the air has all gone to heat the air *outside* the room. If we add no energy to the air, why bother to light the furnace? (Ignore the furnace energy used to raise the temperature of the walls, and consider only the energy used to raise the air temperature.)

49. The atomic weight of iodine is 127. A standing wave in a tube filled with iodine gas at 400 K has nodes that are 6.77 cm apart when the frequency is 1000 Hz. Is iodine gas monatomic or diatomic?   *Answer:* Diatomic.

50. How would you explain the observed value of $C_v = 7.50$ cal/mol · K for gaseous SO_2 at 15.0° C and 1.00 atm?

51. *Dalton's law* states that when mixtures of gases having no chemical interaction are present together in a vessel, the pressure exerted by each constituent at a given temperature is the same as it would exert if it alone filled the whole vessel, and that the total pressure is equal to the sum of the partial pressures of each gas. Derive this law from kinetic theory, using Eq. 23-3.

52. A hydrogen atom, in its lowest (ground) state and moving with 13-eV kinetic energy, collides head-on with another hydrogen atom which is *at rest* in its ground state. (a) Use the conservation laws of energy and momentum to show that this collision must be elastic. The first allowed excited state is about 10.2 eV above the ground state. (b) Show that the minimum initial kinetic energy that the incident atom needs to raise one of the atoms to the first excited state is *twice* the energy difference between ground state and first excited state.

53. (a) A monatomic ideal gas initially at 17° C is suddenly compressed to one-tenth its original volume. What is its temperature after compression? (b) Make the same calculation for a diatomic gas.
    Answer: (a) 1350 K. (b) 730 K.

24
kinetic theory
of gases—II

Between successive collisions a molecule in a gas moves with constant speed along a straight line. The average distance between such successive collisions is called the *mean free path* (Fig. 24-1). If molecules were points, they would not collide at all and the mean free path would be infinite. Molecules, however, are not points and hence collisions occur. If they were so numerous that they completely filled the space available to them, leaving no room for translational motion, the mean free path would be zero. Thus the mean free path is related to the size of the molecules and to their number per unit volume.

Consider the molecules of a gas to be spheres of diameter d. The cross section for a collision is then πd^2. That is, a collision will take place when the centers of two molecules approach within a distance d of one another. An equivalent description of collisions made by any one molecule is to regard that molecule as having a diameter $2d$ and all other molecules as point particles (see Fig. 24-2).

Imagine a typical molecule of equivalent diameter $2d$ moving with speed v through a gas of equivalent point particles and let us assume, for the time being, that the molecule and the point particles exert no forces on each other. In time t our molecule will sweep out a cylinder of cross-sectional area πd^2 and length vt. If n_v is the number of molecules per unit volume, the cylinder will contain $(\pi d^2 vt)n_v$ particles (see Fig. 24-3). Since our molecule and the point particles *do* exert forces on each other, this will be the number of collisions experienced by the molecule in time t. The cylinder of Fig. 24-3 will, in fact, be a broken one, changing direction with every collision.

The mean free path $\bar{l}$ is the average distance between successive collisions. Hence, $\bar{l}$ is the total distance vt covered in time t divided by the

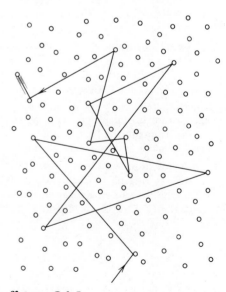

figure 24-1
A molecule traveling through a gas, colliding with other molecules in its path. Of course, all the other molecules are moving in a similar fashion.

number of collisions that take place in this time, or

$$\bar{l} = \frac{vt}{\pi \, d^2 n_v vt} = \frac{1}{\pi n_v d^2}.$$

This equation is based on the picture of a molecule hitting stationary targets. Actually the molecule hits moving targets. The collision frequency is increased as a result (see below) and the mean free path is reduced to

$$\bar{l} = \frac{1}{\sqrt{2} \, \pi \, n_v d^2}. \qquad (24\text{-}1)$$

When the target molecules are moving, the two v's in the first equation above are not the same. The one in the numerator $(= \bar{v})$ is the mean molecular speed measured with respect to the container. The one in the denominator $(= \bar{v}_{rel})$ is the mean *relative* speed with respect to other molecules; it is this relative speed that determines the collision rate.

We can see qualitatively that $\bar{v}_{rel} > \bar{v}$. Thus two molecules of speed v moving toward each other have a relative speed of $2v$ $(> v)$; two molecules with speed v moving at right angles on a collision course have a relative speed of $\sqrt{2} \, v$ (also $> v$); two molecules moving with speed v in the same direction have a relative speed of zero $(< v)$. Thus molecules arriving from *all of the forward hemisphere* and *part of the backward hemisphere* have $\bar{v}_{rel} > \bar{v}$. The molecules arriving from the rest of the backward hemisphere have $\bar{v}_{rel} < \bar{v}$ but, since their numbers are smaller, they do not determine the nature of the average over both hemispheres, which yields $\bar{v}_{rel} > \bar{v}$. A quantitative calculation, taking into account the actual speed distribution of the molecules, gives $\bar{v}_{rel} = \sqrt{2} \, \bar{v}$.

Let us calculate the magnitude of the mean free path and the collision frequency for air molecules at 0° C and 1-atm pressure.

We take 2×10^{-8} cm as an effective molecular diameter d. For the conditions stated, the average speed of air molecules is about 1×10^5 cm/s and there are about 3×10^{19} molecules/cm³. The mean free path is then

$$\bar{l} = \frac{1}{\pi \sqrt{2} \, n_v d^2} = \frac{1}{\pi \sqrt{2} \, (3 \times 10^{19}/\text{cm}^3)(2 \times 10^{-8} \text{ cm})^2}$$

$$= 2 \times 10^{-5} \text{ cm}.$$

This is about a thousand molecular diameters.

The corresponding collision frequency is

$$\frac{v}{\bar{l}} = (1 \times 10^5 \text{ cm/s})/(2 \times 10^{-5} \text{ cm})$$

$$= 5 \times 10^9/\text{s}.$$

Thus, on the average, *each molecule* makes five billion collisions per second!

In the earth's atmosphere we have seen that the mean free path of air molecules at sea level (760 mm-Hg) is 2×10^{-5} cm. At 100 km above the earth (10^{-3} mm-Hg) the mean free path is 2 mm. At 300 km (10^{-6} mm-Hg) it is 15 cm, and yet there are about 10^8 molecules/cm³ in this region. This emphasizes that molecules are indeed small. At great enough heights the mean free path concept fails because the upward-directed molecules follow ballistic paths and may escape from the atmosphere.

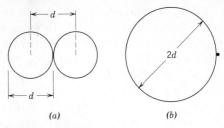

figure 24-2
(a) If a collision occurs when two molecules come within a distance d of each other, the process can be treated equivalently (b) by thinking of one molecule as having an effective diameter $2d$ and the other as being a point mass.

EXAMPLE 1

figure 24-3
A molecule of equivalent diameter $2d$ traveling with speed v sweeps out a cylinder of base πd^2 and length vt in a time t. It suffers a collision with every other molecule whose center lies within this cylinder.

In the laboratory the mean free path concept is useful in situations such as that of Example 1. In even modest laboratory vacuums, however, it loses some of its meaning because nearly all the collisions are with the wall of the containing vessel rather than with other molecules. Consider a box 10 cm on edge containing air at 10^{-7} mm-Hg pressure. The mean free path is 150 cm, so that collisions between molecules are rare indeed. And yet this box contains about 10^{12} molecules!

Even in a finite "box," however, there are some conditions in which particles can travel great distances without striking the walls. In a typical proton synchrotron, used to accelerate protons to the billion-electron-volt range of energies, the protons are constrained by a magnetic field to move in a circular path and may travel *several hundred thousand miles* during the acceleration process. Mean free path considerations are important if the accelerating protons are to have essentially no collisions with residual air molecules. In this case the effective cross section of the proton is so much smaller than that of the air molecules that if we have a vacuum of about 10^{-6} mm-Hg, there is essentially no beam loss by proton scattering from gas molecules inside the vacuum chamber.

24-2
DISTRIBUTION OF MOLECULAR SPEEDS

In Chapter 23 we discussed the root-mean-square speed of the molecules of a gas. However, the speeds of individual molecules vary over a wide range of magnitude; there is a characteristic distribution of molecular speeds for a given gas which depends, as we will see below, on the temperature. If all the molecules of a gas had the same speed v, this situation would not persist for very long because the molecular speeds would be changed by collisions. However, we do not expect many molecules to have speeds $\ll v_{rms}$ (that is, near zero) or $\gg v_{rms}$ because such extreme speeds would require an unlikely sequence of preferential collisions.

Clerk Maxwell first solved the problem of the most probable distribution of speeds in a large number of molecules of a gas. His molecular speed distribution law, for a sample of gas containing N molecules, is*

$$N(v) = 4\pi N (m/2\pi kT)^{3/2} v^2 e^{-mv^2/2kT}. \qquad (24\text{-}2)$$

In this equation $N(v)\,dv$ is the number of molecules in the gas sample having speeds between v and $v + dv$. T is the absolute temperature, k is Boltzmann's constant, and m is the mass of a molecule. Note that for a given gas the speed distribution depends only on the temperature. We find N, the total number of molecules in the sample, by adding up (that is, by integrating) the number present in each differential speed interval from zero to infinity, or

$$N = \int_0^\infty N(v)\,dv. \qquad (24\text{-}3)$$

The unit of $N(v)$ is, say, molecules/(cm/s).

In Fig. 24-4 we plot the Maxwell distribution of speeds for molecules of oxygen at two different temperatures. The number of molecules having a speed between v_1 and v_2 equals the area under the curve between the vertical lines at v_1 and v_2. As Eq. 24-3 shows, the area under the speed distribution curve, which is the integral in that equation, is equal to the total number of molecules in the sample. At any

---

* A derivation of Eq. 24-2 appears in Supplementary Topic IV.

figure 24-4
The Maxwellian distribution of speeds of 10^6 oxygen molecules at two different temperatures. The number of molecules within a certain range of speeds (say, 300 to 600 m/s) is the area under this section of the curve. The complete area under either curve is the total number of molecules (equals 10^6); this area must be the same for each temperature if, as in this case, the curves refer to a given number of molecules. The pressure is lower than atmospheric because oxygen is a liquid at 1.0 atm and 73 K.

temperature the number of molecules in a given speed interval* Δv increases as the speed increases up to a maximum (the most probable speed v_p) and then decreases asymptotically toward zero. The distribution curve is not symmetrical about the most probable speed because the lowest speed must be zero, whereas there is no classical limit to the upper speed a molecule can attain. In this case the *average speed* $\bar{v}$ is somewhat larger than the most probable value. The root-mean-square value, v_{rms}, being the square root of the average of the *squares* of the speeds, is still larger.

As the temperature increases, the root-mean-square speed v_{rms} (and $\bar{v}$ and v_p as well) increases, in accord with our microscopic interpretation of temperature. The range of typical speeds is now greater, so that the distribution broadens. Since the area under the distribution curve (which is the total number of molecules in the sample) remains the same, the distribution must also flatten as the temperature rises. Hence the number of molecules which have speeds greater than some given speed increases as the temperature increases (see Fig. 24-4). This explains many phenomena, such as the increase in the rates of chemical reactions with rising temperature.

The distribution of speeds of molecules in a liquid also resembles the curves of Fig. 24-4. This explains why some molecules in a liquid (the fast ones) can escape through the surface (evaporate) at temperatures well below the normal boiling point. Only these molecules can overcome the attraction of the molecules in the surface and escape by evaporation. The average kinetic energy of the remaining molecules drops correspondingly, leaving the liquid at a lower temperature. This explains why evaporation is a cooling process.

From Eq. 24-2 we see that the distribution of molecular speeds depends on the mass of the molecule as well as on the temperature. The smaller the mass, the larger the proportion of high-speed molecules at

---

* We cannot simply plot the "number of particles having speed v" against v, because there are a finite number of particles and an infinite number of possible speeds. Hence, the probability that a particle has a precisely stated speed, such as $279.343267 \cdots$ m/s, is exactly zero. However, we can divide the range of speeds into intervals and the probability that a particle has a speed somewhere in a given interval (such as 279 m/s to 280 m/s) has a definite nonzero value.

any given temperature. Hence, hydrogen is more likely to escape from the atmosphere at high altitudes than oxygen or nitrogen. The moon has a tenuous atmosphere. For the molecules in this atmosphere not to have a great probability of escaping from the weak gravitational pull of the moon, even at the low temperatures there, we would expect them to be molecules or atoms of the heavier elements. Evidence points to the heavy inert gases, such as krypton and xenon, which were produced largely by radioactive decay early in the moon's history. The atmospheric pressure on the moon is about 10^{-13} of the earth's atmospheric pressure.

EXAMPLE 2

The speeds of ten particles in m/s are 0, 1.0, 2.0, 3.0, 3.0, 3.0, 4.0, 4.0, 5.0, and 6.0. Find (a) the average speed, (b) the root-mean-square speed, and (c) the most probable speed of these particles.

(a) The average speed is

$$\bar{v} = \frac{0 + 1.0 + 2.0 + 3.0 + 3.0 + 3.0 + 4.0 + 4.0 + 5.0 + 6.0}{10} = 3.1 \text{ m/s}.$$

(b) The mean-square speed is

$$\overline{v^2} = \frac{0 + (1.0)^2 + (2.0)^2 + (3.0)^2 + (3.0)^2 + (3.0)^2 + (4.0)^2 + (4.0)^2 + (5.0)^2 + (6.0)^2}{10}$$

$$= 12.5 \text{ m}^2/\text{s}^2$$

and the root-mean-square speed is

$$v_{\text{rms}} = \sqrt{12.5 \text{ m}^2/\text{s}^2} = 3.5 \text{ m/s}.$$

(c) Of the ten particles three have speeds of 3.0 m/s, two have speeds of 4.0 m/s, and the other five each have a different speed. Hence, the most probable speed of a particle v_p is

$$v_p = 3.0 \text{ m/s}.$$

EXAMPLE 3

Use Eq. 24-2 to determine the average speed $\bar{v}$, the root-mean-square speed v_{rms}, and the most probable speed v_p of the molecules in a gas in terms of the gas parameters.

The quantity $N(v)\, dv$ is the number of particles in the sample having a speed between v and $v + dv$, $N(v)$ being given by Eq. 24-2. We find the average speed $\bar{v}$ in the usual way: we multiply the number of particles in each speed interval by a speed v characteristic of that interval; we sum these products over all speed intervals and we divide by the total number of particles. Replacing the summation by an integration, we obtain

$$\bar{v} = \frac{\int_0^\infty N(v)v\, dv}{N}$$

Substituting Eq. 24-2 for $N(v)$ and integrating* we obtain

$$\bar{v} = \sqrt{\frac{8}{\pi}\frac{kT}{m}} = 1.59 \sqrt{\frac{kT}{m}} \quad \text{(average speed)}.$$

The mean-square speed is given by

* Let $\lambda = m/2kT$. From tables of integrals,

$$\int_0^\infty v^2 e^{-\lambda v^2}\, dv = \frac{1}{4}\sqrt{\frac{\pi}{\lambda^3}}; \quad \int_0^\infty v^3 e^{-\lambda v^2}\, dv = \frac{1}{2\lambda^2}; \quad \int_0^\infty v^4 e^{-\lambda v^2}\, dv = \frac{3}{8}\sqrt{\frac{\pi}{\lambda^5}}.$$

$$\overline{v^2} = \frac{\int_0^\infty N(v)v^2 \, dv}{N}$$

which yields

$$v_{\text{rms}} = \sqrt{\overline{v^2}} = \sqrt{\frac{3kT}{m}} = 1.73 \sqrt{\frac{kT}{m}} \quad \text{(root-mean-square speed)}.$$

The most probable speed v_p is the speed at which $N(v)$ has its maximum value. It is given by requiring that

$$\frac{dN(v)}{dv} = 0.$$

By substituting Eq. 24-2 for $N(v)$ we obtain, as you should show,

$$\overline{v}_p = \sqrt{\frac{2kT}{m}} = 1.41 \sqrt{\frac{kT}{m}} \quad \text{(most probable speed)}.$$

In Fig. 24-4 we show v_p, $\overline{v}$, and v_{rms} at 0° C for a molecular speed distribution in oxygen.

24-3
EXPERIMENTAL CONFIRMATION OF THE MAXWELLIAN DISTRIBUTION

Maxwell derived his distribution law for molecular speeds (Eq. 24-2) in 1859. At that early date it was not possible to check this law by direct measurement and, indeed, it was not until 1920 that Stern made the first serious attempt to do so. Techniques improved rapidly in the hands of various workers but it was not until 1955 that a high-precision experimental verification of the law (for gas molecules) was provided, by Miller and Kusch of Columbia University.

Their apparatus is shown in Fig. 24-5. The walls of oven O were heated, in one set of experiments, to a uniform temperature of 870 ± 4K, some thallium having been placed in the oven. At this temperature thallium vapor, at a pressure of 3.2×10^{-3} mm-Hg, fills the oven. Some molecules of thallium vapor escape from slit S into the highly evacuated space outside the oven, falling on the rotating cylinder R. This cylinder, of length l, has a number of helical grooves cut into it, only one of them being shown in Fig. 24-5. For a given angular speed ω of the cylinder, only molecules of a sharply defined speed v can pass along the grooves without striking the walls. The speed v can be found from:

$$\text{time of travel along the groove} = \frac{l}{v} = \frac{\phi}{\omega},$$

or

$$v = l\omega/\phi \tag{24-4}$$

in which ϕ (see Fig. 24-5) is the angular displacement between the entrance and

figure 24-5
The apparatus used by Miller and Kusch to verify the Maxwell speed distribution law. The mechanism for rotating the cylinder is not shown. The whole apparatus is highly evacuated to reduce collisions with the residual gas molecules of the thallium molecules in the beam emerging from slit S.

the exit of a helical groove. Thus the rotating cylinder is a *velocity selector*, the speed selected being proportional to the (controllable) angular speed ω, as Eq. 24-4 shows. One observes the beam intensity recorded by detector D as a function of the selected speed v. Figure 24-6 shows the remarkable agreement between theory (the solid line) and experiment (the triangles and circles) for thallium vapor.

The distribution of speeds in the *beam* (as distinguished from the distribution of speeds in the *oven*) is not proportional to $v^2e^{-mv^2/2kT}$, as in Eq. 24-2, but to $v^3e^{-mv^2/2kT}$. Consider a group of molecules in the oven whose speeds lie within a certain small range v_1 to $v_1 + \Delta v$, where v_1 is less than the most probable speed v_p. We can always find another equal speed interval Δv, extending from v_2 to $v_2 + \Delta v$, where v_2, which will be greater than v_p, is chosen so that the two speed intervals contain the same number of molecules. However, more molecules in the latter interval than in the former will escape from slit S to form the beam, because molecules in the latter interval "bombard" the slit with a greater frequency, by precisely the factor v_2/v_1. Thus, other things being equal, fast molecules are favored in escaping from the oven, just in proportion to their speeds, and the molecules in the beam have a "v^3" rather than a "v^2" distribution. This effect is allowed for in computing the theoretical curve of Fig. 24-6.

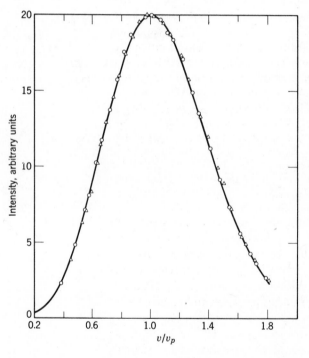

figure 24-6
The solid line shows Maxwell's molecular speed distribution. The circles (O) are experimental points for thallium atoms emerging from an oven at 870 K; the triangles ($\triangle$) correspond to 944 K. The horizontal scale is a plot of v/v_p where v_p is the most probable speed. When speeds are plotted in this way the distributions for different temperatures should fall on the same curve. At 870 K, $v_p = 376$ m/s and at 944 K, it is 395 m/s. From R. C. Miller and P. Kusch, *Physical Review*, 99, 1314 (1955).

Rainwater and Havens (1946), also of Columbia University, provided a convincing experimental check of the Maxwell speed distribution law by using a "gas" of neutrons. The neutrons were produced (as fast neutrons) in continuous series of short bursts in a cyclotron and allowed to fall on a block of paraffin. By repeated collisions with the nuclei of the block, the neutrons rapidly slowed down and came into thermal equilibrium with the block, behaving like a "neutron gas" in a container. The container, however, is a leaky one because neutrons diffuse out through the walls of the block and move across the laboratory. It is possible, by electronic means, to measure the time between the production of the neutrons in the cyclotron and their arrival at a distant detector after escaping from the paraffin block. Thus one can measure the speed distribution in a collimated beam of escaping neutrons and can compare it to the prediction of Maxwell; the agreement of theory and experiment is excellent.

Although the Maxwell speed distribution for gases agrees remarkably well with observations under ordinary conditions, it fails at high densities, where the basic assumptions of the classical kinetic theory fail. In these regions we must use speed distributions founded on the principles of quantum physics, the Fermi-Dirac and the Bose-Einstein distributions. These quantum distributions closely agree with the Maxwell distribution in the classical region (low density) and agree with experiment where the classical distribution fails. Hence, there are limits to the applicability of the Maxwell distribution, as in fact there are to any theory.

24-4
BROWNIAN MOTION

The prominence given to atomic and molecular theory during the last quarter of the nineteenth century was deplored by many able scientists. In spite of the many quantitative agreements between kinetic theory and the behavior of gases, no proof of the separate existence of atoms and molecules had been obtained, nor had any observation been made that could really demonstrate the continuous motions of the molecules. Ernst Mach (1838–1916) saw no point to "thinking of the world as a mosaic, since we cannot examine its individual pieces of stone." It had been established rather early in the development of kinetic theory that an atom should be about 10^{-7} cm or 10^{-8} cm in diameter. No one actually expected to see an atom or detect the effect of a single atom.

The leader of the opposition to the atomic theory was Wilhelm Ostwald, rightly regarded as "the father of physical chemistry." He was a champion of the principle of the conservation of energy and regarded energy as the ultimate reality. Ostwald argued that with a thermodynamical treatment of a process we know all that is essential about the process and that further mechanical assumptions about the mechanism of the reactions are unproved hypotheses. He abandoned the atomic and molecular theories and fought to free science "from hypothetical conception which lead to no immediate experimentally verifiable conclusions." Other prominent scientists were reluctant to admit the atom as an established scientific fact.

Ludwig Boltzmann felt compelled to protest this attitude in an article in 1897, stressing the indispensability of atomism in natural science. The progress of science is often guided by the analogies of nature's processes which occur in the minds of investigators. Kinetic theory was such a mechanical analogy. As with most analogies it suggests experiments to test the validity of our mental pictures and leads to further investigations and clearer knowledge.

As is always true in such controversies in science, the decision rests with experiment. The earliest and most direct experimental evidence for the reality of atoms was the proof of the atomic kinetic theory provided by the quantitative studies of Brownian motion. These observations convinced both Mach and Ostwald of the validity of the kinetic theory and the atomic description of matter on which it rests. The atomic theory gained unquestioned acceptance in later years when a wide variety of experiments all led to the same values of the fundamental atomic constants.

Brownian motion is named after the English botanist Robert Brown who discovered in 1827 that pollen suspended in water shows a continuous random motion when viewed under a microscope. At first these motions were considered a form of life, but it was soon found that small inorganic particles behave similarly. There was no quantitative explanation of this phenomenon until the development of kinetic theory. Then, in 1905, Albert Einstein developed a theory of Brownian motion.* In his *Autobiographical Notes*, Einstein writes, "My major aim in this was to find facts which would guarantee as much as possible the existence of atoms of definite size. In the midst of this I discovered that, according to atomistic theory, there would have to be a movement of suspended microscopic particles open to observation, without knowing that observations concerning the Brownian motion were already long familiar."

The basic assumption made by Einstein was that particles suspended in a liquid or a gas share in the thermal motions of the medium and that on the average the translational kinetic energy of each particle is $\frac{3}{2} kT$, in accordance with the principle of equipartition of energy. In this view the Brownian motions result from impacts by molecules of the fluid, and the suspended particles acquire the same mean kinetic energy as the molecules of the fluid.

The suspended particles are extremely large compared to the molecules of the fluid and are being continually bombarded on all sides by them. If the particles are sufficiently large and the number of molecules is sufficiently great, equal numbers of molecules strike the particles on all sides at each instant. For smaller particles and fewer molecules the number of molecules striking various sides of the particle at any instant, being merely a matter of chance, may not be equal; that is, fluctuations occur. Hence the particle at each instant suffers an unbalanced force causing it to move this way or that. The particles therefore act just like very large molecules in the fluid, and their motions should be qualitatively the same as the motions of the fluid molecules. If Avogadro's number were infinite, there would be no statistical unbalance (fluctuations) and no Brownian motion. If Avogadro's number were very small, the Brownian motion would be very large. Hence we should be able to deduce the value of Avogadro's number from observations of the Brownian motion. Deeply ingrained in this picture is the idea of molecular motion and the smallness of molecules. The Brownian motion therefore offers a striking experimental test of the kinetic theory hypotheses.

The suspended particles are under the influence of gravity and would settle to the bottom of the fluid were it not for the molecular bombardment opposing this tendency. Since the suspended particles behave like gas molecules we are not surprised to learn that, as for molecules in the atmosphere, their density drops off exponentially with respect to height in the fluid; they form a "miniature atmosphere"; see Example 1, Chapter 17; Problem 26, Chapter 23; and Problem 21, this chapter. Jean Perrin, a French physical chemist, confirmed this prediction in 1908 by determining the numbers of small particles of gum resin suspended at different heights in a liquid drop (Fig. 24-7, left). From his data he deduced a value of Avogadro's number $N_0 = 6 \times 10^{23}$ particles/mol. Perrin also made measurements of the displacements of Brownian particles during many equal time intervals and found that they have the statistical distribution required by kinetic theory and the root-mean-square displacement predicted by Einstein (Fig. 24-7, right).

Among the many subsequent experiments was that of Kappler, in 1931, who observed the Brownian motion of a rather gross object, a small mirror (area 0.77 mm²), mounted on a fine torsion fiber with light reflected from the mirror

---

*Einstein's theory appeared as an article in the same volume of the *Annalen der Physik* which contained his famous paper on the theory of relativity and also his paper on the theory of the photoelectric effect. It was for his work on the photoelectric effect that he won the Nobel prize in 1921.

(a)   (b)

figure 24-7

(a) A gum resin suspension contained in a glass vessel viewed in a
microscope by Perrin in 1909. At first the distribution of particles was
uniform, but in time they settled to the steady state distribution shown.
The particles have a diameter of 0.6×10^{-3} cm and the horizontal lines are
10×10^{-3} cm apart. *(b)* Sketch by V. Henri in 1908 from his cinematographic
study of Brownian movement. Henri used a microscope with a motion-
picture camera which ran 20 frames/s, each exposure being $\frac{1}{320}$ s. The zigzag
lines show the position of five rubber particles as recorded by successive
frames. The lines do not represent the actual paths of the particles for
between exposures the particles may have traveled a similar erratic path.
The scale at the bottom is divided into micrometers (abbr. μm; value 10^{-6} m).

to a moving photographic film. The mirror is mounted in a chamber with gas
at low pressure (10^{-2} mm-Hg); the record on the moving film yields the function $\theta(t)$ (angular displacement as a function of time). This shows clearly the
rotational Brownian motion of the mirror which consists of a series of angular
displacements produced by unbalanced impacts from the molecules. As the
gas pressure is lowered, there is a gradual decrease in the motion. From the
photographic record we can find the angular displacement θ and the angular
velocity ω. The equipartition of energy principle requires that

$$\tfrac{1}{2}I\overline{\omega^2} = \tfrac{1}{2}\kappa\overline{\theta^2} = \tfrac{1}{2}kT,$$

for $\tfrac{1}{2}I\overline{\omega^2}$ is the average rotational kinetic energy of the system and $\tfrac{1}{2}\kappa\overline{\theta^2}$ is the
average potential energy of the system. Here I is the rotational inertia of the
system and κ the torsion constant of the fiber. From his observations Kappler
could calculate Boltzmann's constant k and from the relation $N_0 = R/k$ he
could obtain Avogadro's number. His values were $k = 1.36 \times 10^{-23}$ J/molecule
K $\pm 3\%$ (the accepted value today of 1.380×10^{-23} J/molecule·K being within the
limits of error) and $N_0 = 6.1 \times 10^{23}$ particles/mole.

In the preceding chapter we discussed the behavior of an ideal gas. On
the macroscopic scale its fundamental relationship is the equation of
state

$$pV = nRT.$$

24-5
THE VAN DER WAALS
EQUATION OF STATE

From this equation and the principles of thermodynamics we can show that the internal energy U of a gas depends only on the temperature. Real gases obey these relations fairly well at low densities, but their behavior may become markedly different as the density increases. We cannot neglect these deviations from ideal behavior in accurate scientific work. For example, to establish the Kelvin thermodynamic scale in the laboratory we must know how to make the necessary corrections to the scale of a constant-volume gas thermometer. We must therefore know the behavior of real gases rather accurately. Even more important, perhaps, is the fact that the behavior of real gases gives us information on the nature of intermolecular forces and the structure of molecules.

Kinetic theory provides the microscopic description of the behavior of an ideal gas. We have already suggested how the assumptions of kinetic theory could become invalid if applied to a real gas. Under some conditions we may not be justified in neglecting the facts that the molecules occupy a fraction of the volume available to the gas and that the range of molecular forces is greater than the size of the molecule. At high densities we cannot ignore these effects.

J. D. van der Waals (1837–1923) deduced a modified equation of state which takes these factors into account in a simple way. Let us imagine the molecules to be hard spheres of diameter d. The diameter of such a sphere would correspond to the distance between the centers of molecules at which strong collision forces come into play. During its motion the center of a molecule cannot approach within a distance $d/2$ from a wall or a distance d from the center of another molecule. Hence the actual volume available to a molecule is smaller than the volume of the containing vessel. Just how much smaller depends on how many molecules there are. Let us represent the volume per mole, V/n, by v. Then the "free volume" per mole would be less than this by the "covolume" b. Hence we modify the equation of state from the ideal relation $pv = RT$ to

$$p(v - b) = RT$$

to allow for this. Because of the reduced volume, the number of impacts on the wall increases, thereby increasing the pressure; this relationship was first derived by Clausius.

We can also allow for the effect of attractive forces between molecules in a simple way. Imagine a plane passed through a gas and consider, at any instant, the intermolecular forces which act across it. Each molecule on the left, say, will attract and be attracted by some small number n of those on the right. Now compare this situation with another similar in every way except that the number of molecules per unit volume is doubled. Here any particular molecule on the left will interact on the average with $2n$ of those on the right, for the range of the molecular force is the same, and twice as many molecules now fall into this range. Since there also are twice as many molecules on the left as before which attract in this way, it is clear that the number of attractive pairs across the plane has increased fourfold. Therefore, the effect of these forces varies as the *square* of the number of particles per unit volume or inversely as the square of the volume per mole, that is, as $(1/v)^2$. Because of these intermolecular force bonds, the gas should, for a given external pressure, occupy a volume less than the volume it would occupy as an ideal gas, in which there are no such attractive forces. Or, equivalently, the gas acts as though it is subject to a pressure in excess of the externally applied pressure. This excess pressure is proportional to $(1/v)^2$, or equal to a/v^2 where a is a constant. Hence, we obtain the *van der Waals equation of state* of a gas,

$$\left(p + \frac{a}{v^2}\right)(v - b) = RT. \tag{24-5}$$

The values of a and b are to be found from experiment, and in this respect the equation is empirical. We must realize that these corrections to the ideal gas equation of state are of the simplest kind, and that failure of the van der Waals equation in any particular case is evidence that our assumptions are oversimplified for that case. No one simple formula is known which applies to all gases under all conditions.*

We have seen that real gases do not follow the ideal gas law exactly. Our discussion suggests also that for real gases the internal energy U depends on the volume as well as on the temperature. For if there are (long range) attractive forces between molecules, the potential energy increases as the average distance between molecules increases. Hence, we would expect the internal energy of most real gases to increase slightly with the volume at ordinary temperatures, and this is found to be the case. Of course, collisions can be regarded as arising from repulsive forces. If the molecules move rapidly so as to make many collisions, the potential energy of the (short range) repulsive forces may be more important than that of the attractive forces and the internal energy could decrease as the volume increases. This is true for hydrogen and helium at ordinary temperatures. In either case, however, the internal energy U is not a function of temperature alone but depends also on the volume. The dependence of the internal energy of a gas on the volume can be deduced readily from the observed results of the free expansion experiment, discussed in Chapter 22.

EXAMPLE 4

On a pressure-volume diagram compare the behavior of an ideal gas at constant temperature to that of a van der Waals gas.

In Fig. 24-8a we draw the isotherms (curves of constant T) according to the law $pv = RT$. Figure 24-8b shows the isotherms according to the law

$$(p + a/v^2)(v - b) = RT.$$

(a)   (b)

figure 24-8
(a) Isotherms for an ideal gas. (b) Isotherms for a van der Waals gas. We have assumed $a = 3.59$ liter²atm/mole² and $b = 0.0427$ liter/mole in Eq. 24.4. These values give the best fit of this equation to p-V-T data for the real gas CO_2. T_{cr} ($= 304$ K) is the critical temperature.

---

* For an interesting discussion of these and related matters see 'Liquids — The Awkward In-between' by J. G. Powles, in *Contemporary Physics*, September 1974.

The ideal gas isotherms are each one branch of a rectangular hyperbola, $pv =$ constant. For the van der Waals gas the pressure varies with volume as

$$p = \frac{RT}{(v - b)} - \frac{a}{v^2}. \qquad (24\text{-}6)$$

As the volume per mole v decreases from large values, the pressure rises, but the a/v^2 term, which diminishes the pressure, climbs rapidly so that for sufficiently low T the pressure passes through a maximum at A. As v is further decreased, the $RT/(v - b)$ term climbs more rapidly so that the pressure goes through a minimum at B and then rises rapidly without bound as v tends to the value b. At neighboring higher temperatures, the maxima and minima are less pronounced and are closer to the inflection point that lies between them. At the so-called critical temperature $(T = T_{cr})$, they coincide in a horizontal inflection point called the critical point. For temperatures sufficiently higher than the critical temperature T_{cr} the van der Waals isotherms have no inflection point and approach the rectangular-hyperbola behavior of the ideal-gas isotherms. For carbon dioxide the critical temperature is 304 K and the pressure at the critical point is 72.9 atm.

We can obtain the pressure p_{cr}, the molar volume v_{cr}, and the temperature T_{cr} of the critical point quite generally from the conditions that the tangent to the isotherm is horizontal, $dp/dv = 0$ when $T = $ constant, and that the point is an inflection point, $d^2p/dv^2 = 0$ when $T = $ constant. We obtain

$$\frac{dp}{dv} = -\frac{RT}{(v - b)^2} + \frac{2a}{v^3} = 0 \quad (T = \text{constant})$$

and

$$\frac{d^2p}{dv^2} = \frac{2RT}{(v - b)^3} - \frac{6a}{v^4} = 0 \quad (T = \text{constant}).$$

This gives us

$$v_{cr} = 3b$$

and

$$T_{cr} = \frac{8a}{27bR}.$$

Putting these in Eq. 24-6, we obtain

$$p_{cr} = \frac{a}{27b^2}.$$

The isotherms suggest the actual experimental behavior of liquids and gases. The maxima and minima of the isotherms below the critical temperature are not usually observed experimentally. At some point x the gas begins to condense. As the volume is decreased, the pressure remains constant (dotted line) until at y all the gas has been transformed into liquid. Beyond y, as we decrease the volume, we are compressing a liquid, with the consequent sharp rise in pressure needed to make even small volume changes. Actually the portions xA and By of the isotherms can be obtained experimentally by using very pure gases and liquids. We call these supersaturated vapors and supercooled liquids,* and they are in metastable states. The portion AB cannot be reproduced experimentally and is unstable.

The constants a and b in van der Waals equation can be calculated from the experimental values of the critical quantities. The term a/v^2 is called an *internal pressure*. Some values for air are of interest. For air at $0°$ C and external pressure p of 1.00 atm, the internal pressure is 0.0028 atm; at $0°$ C and external pressure p of 100 atm, the internal pressure is 26 atm. For air at $-75°$ C the corresponding values of the internal pressure are 0.0056 atm and 84.5 atm. When a gas expands under pressure

---

* See "The Undercooling of Liquids" by David Turnbull in *Scientific American*, January 1965.

and does work against outside compressing forces, it must also do work against these internal forces. For air at $-75°$ C and 100 atm, the work done against internal forces is nearly as great as that done against external forces. There is an important distinction between internal and external work, however. In the case of external work, energy is transferred from the body to an outside body; in the case of internal work, there is merely a transfer from one kind of energy to another within the body, as from potential to kinetic. The constant b varies from gas to gas, but is usually of the order of 30 cm³/mol. Hence the covolume is about 0.15% of the free volume available to a gas at standard conditions.

Although the van der Waals formula is a good qualitative guide, the quantitative experimental data cannot be matched everywhere with constant values for a and b. The reason is that the model on which the formula is based is still an oversimplification. Instead of assuming that the molecules always have a well-defined diameter, for example, we must use the actual intermolecular force (Fig. 23-3). In this way a more accurate correction to the ideal gas law can be made. Van der Waals knew this would be necessary for accurate quantitative work.

questions

1. Consider the case in which the mean free path is greater than the longest straight line in a vessel. Is this a perfect vacuum for a molecule in this vessel?

2. List effective ways of increasing the number of molecular collisions per unit time in a gas.

3. Give a qualitative explanation of the connection between the mean free path of ammonia molecules in air and the time it takes to smell the ammonia when a bottle is opened across the room.

4. Consider Archimedes' principle applied to a gas. Isn't it true that once we accept a kinetic theory model of a gas, we need a new explanation for this principle? For example, suppose the mean free path of a gas molecule is comparable to the depth of the body immersed in the gas, or greater; what is the origin of the buoyant force then? (See "Archimedes' Principle in Gases" by Alan J. Walton in *Contemporary Physics*, March 1969.)

5. The two opposite walls of a container of gas are kept at different temperatures. Describe the mechanism of heat conduction through the gas.

6. A gas can transmit only those sound waves whose wavelength is long compared with the mean free path. Can you explain this? Where might this limitation arise?

7. If molecules are not spherical, what meaning can we give to d in Eq. 24-1 for the mean free path? In which gases would the molecules act the most nearly as rigid spheres?

8. Suppose we dispense with the hypothesis of elastic collisions in kinetic theory and consider the molecules as centers of force acting at a distance. Does the concept of mean free path have any meaning under these circumstances?

9. Since the actual force between molecules depends on the distance between them, forces can cause deflections even when molecules are far from "contact" with one another. Furthermore, the deflection caused should depend on how long a time these forces act and hence on the relative speed of the molecules. (a) Would you then expect the measured mean free path to depend on temperature, even though the density remains constant? (b) If so, would you expect $\bar{l}$ to increase or decrease with temperature? (c) How does this dependence enter into Eq. 24-1?

10. Justify qualitatively the statement that, in a mixture of molecules of different kinds in complete equilibrium, each kind of molecule has the same

Maxwellian distribution in speed that it would have if the other kinds were not present.

11. What observation is good evidence that not all molecules of a body are moving with the same speed at a given temperature?

12. The Maxwellian distribution of speeds among molecules in a gas is shown in Fig. 24-4. How would you expect the Maxwellian distribution of *velocities* to look? What would the average velocity be?

13. The fraction of molecules within a given range Δv of the root-mean-square speed decreases as the temperature of a gas rises. Explain why.

14. (a) Do half the molecules in a gas in thermal equilibrium have speeds greater than v_p? Than v? Than v_{rms}?
    (b) Which speed, v_p, $\bar{v}$, or v_{rms}, corresponds to a molecule having average kinetic energy?

15. The slit system in Fig. 24-5 selects only those molecules moving in the $+x$-direction. Does this destroy the validity of the experiment as a measure of the distribution of speeds of molecules moving in all directions?

16. Why did Rainwater and Havens, in their investigation of the speed distribution of neutrons (page 528), select paraffin as a material to bring fast neutrons rather quickly into thermal equilibrium?

17. List examples of the Brownian motion in physical phenomena.

18. Would Brownian motion occur in gravity-free space?

19. A golf ball is suspended from the ceiling by a long thread. Explain in detail why its Brownian motion is not readily apparent.

20. We have defined n_v to be the number of molecules per unit volume in a gas. If we define n_v for a very small volume in a gas, say one equal to ten times the volume of an atom, then n_v fluctuates with time through the range of values zero to some maximum value. How then can we justify a statement that n_v has a definite value at every point in the gas?

21. Show that as the volume per mole of a gas increases, the van der Waals equation tends to the equation of state of an ideal gas.

22. The covolume b in van der Waals equation is often taken to be four times the actual volume of the gas molecules themselves. What factors would have to be taken into account to obtain such a result?

23. Keeping in mind that internal energy of a body consists of kinetic energy and potential energy of its particles, how would you distinguish between the internal energy of a body and its temperature?

problems

SECTION 24-1

1. The mean free path of nitrogen molecules at $0°$ C and 1 atm is 0.80×10^{-5} cm. At this temperature and pressure there are 2.7×10^{19} molecules/cm³. What is the molecular diameter? *Answer:* 3.2×10^{-8} cm.

2. In a certain particle accelerator the protons travel around a circular path of diameter 75 ft in a chamber of 10^{-6} mm-Hg pressure and 273 K temperature. (a) Estimate the number of gas molecules per cubic centimeter at this pressure. (b) What is the mean free path of the gas molecules under these conditions if the molecular diameter is 2.0×10^{-8} cm?

3. At what frequency would the wavelength of sound be of the order of the mean free path in oxygen at 1-atm pressure and $0°$ C? Take the diameter of the oxygen molecule to be 3.00×10^{-8} cm. *Answer:* 3.5×10^{9} Hz.

4. What is the mean free path for 15 spherical jelly beans in a bag that is vigorously shaken? Take the volume of the bag to be 1.0 l and the diameter of a jelly bean to be 1.0 cm.

5. At 2500 km above the earth's surface the density is about one molecule/cm³. (a) What mean free path is predicted by Eq. 24-1 and (b) what is its significance under these conditions?
   Answer: (a) 7×10^{9} km. (b) The answer to (a) has little significance because,

at this altitude, nearly all molecules would follow collisionless ballistic paths in the earth's gravitational field, and many would escape from the atmosphere.

6. The mean free path $\bar{l}$ of the molecules of a gas may be determined from measurements (e.g., from measurement of the viscosity of the gas). At 20° C and 75 cm-Hg pressure such measurements yield values of $\bar{l}_A$ (argon) = 9.9×10^{-6} cm and $\bar{l}_{N_2}$ (nitrogen) = 27.5×10^{-6} cm. (a) Find the ratio of the effective cross-section diameters of argon and nitrogen. (b) What would the value be of the mean free path of argon at 20° C and 15 cm-Hg? (c) What would the value be of the mean free path of argon at −40° C and 75 cm-Hg?

7. A molecule of hydrogen (diameter 1.0×10^{-8} cm) escapes from a furnace $(T = 4000$ K) with the root-mean-square speed into a chamber containing atoms of cold argon (diameter 3.0×10^{-8} cm) at a density of 4.0×10^{19} atoms/cm³. (a) What is the speed of the hydrogen molecule? (b) If the molecule and an argon atom collide, what is the closest distance between their centers, considering each as spherical? (c) What is the initial number of collisions per unit time experienced by the hydrogen molecule?
   Answer: (a) 7.1 km/s. (b) 2.0×10^{-8} cm. (c) 5.0×10^{10} collisions/s.

8. The mean free path of a molecule is $\bar{l}$. Prove that the probability that a molecule will go *at least* a distance x before having its next collision is $e^{-x/\bar{l}}$.

9. For a gas in which all molecules travel with the same speed $\bar{v}$, show that $\bar{v}_{rel} = \frac{4}{3}v$ rather than $\sqrt{2}\,\bar{v}$ (which is the result obtained when we consider the actual distribution of molecular speeds). See p. 523.

SECTION 24-2

10. It is found that the most probable speed of molecules in a gas at an equilibrium temperature T_2 is the same as the root-mean-square speed of the molecules in this gas when its equilibrium temperature is T_1. Find T_2/T_1.

11. You are given the following group of particles (N_i represents the number of particles which have a speed v_i).

N_i	v_i(cm/s)
2	1.00
4	2.00
6	3.00
8	4.00
2	5.00

   (a) Compute the average speed $\bar{v}$. (b) Compute the root-mean-square speed v_{rms}. (c) Among the five speeds shown, which is the most probable speed v_p for the entire group?   *Answer:* (a) 3.2 cm/s. (b) 3.4 cm/s. (c) 4.0 cm/s.

12. Consider the distribution of speeds shown in Fig. 24-9. (a) List v_{rms}, $\bar{v}$, and v_p in the order of increasing speed. (b) How does this compare with the Maxwellian distribution?

figure 24-9
Problem 12

13. A gas consists of N particles. (a) Show that $v_{rms} \geq \bar{v}$ regardless of the form of the distribution of speeds. (b) When does the equality hold?
   Answer: (b) When all the speeds are the same.

14. A hypothetical gas of N particles has the speed distribution shown in Fig. 24-10. ($N_v = 0$ for $v > 2v_0$.) (a) Evaluate a in terms of N and v_0. (b) Find the number of particles with speeds between $1.5v_0$ and $2.0v_0$. (c) Find the average speed of the particles.

15. A container of volume 1000 cm³ contains argon at a pressure of 3.0×10^5 Pa and a temperature of 300 K. The atomic weight of argon is 40. (a) How many argon atoms are in the container? (b) What is the average speed of these atoms? (c) How many atoms strike an area of 1.0×10^{-3} cm² on one of the container walls in one second? (d) If this area is a hole, and all the atoms striking the hole leave the container, how long will it take for the number of atoms in the container to fall to $1/e$ of its initial value?
   Answer: (a) 7.2×10^{22}. (b) 400 m/s. (c) 7.2×10^{20}. (d) 100 s.

figure 24-10
Problem 14

SECTION 24-3

16. In the apparatus of Miller and Kusch (Fig. 24-5) the length l of the rotating cylinder is 20.4 cm and the angle ϕ is $(2\pi/74.7)$ rad. What rotational speed corresponds to a selected speed v of 200 m/s?

SECTION 24-4

17. Calculate the root-mean-square speed of smoke particles of mass 5.0×10^{-14} g in air at $0°$ C and 1.0-atm pressure.   *Answer:* 1.5 cm/s.

18. Particles of mass 6.2×10^{-14} g are suspended in a liquid at $27°$ C and are observed to have a root-mean-square speed of 1.4 cm/s. Calculate Avogadro's number from the equipartition theorem and these data.

19. The average speed of hydrogen molecules at $0°$ C is 1694 m/s. Compute the average speed of colloidal particles of "molecular weight" 3.2×10^{6} g/mol. *Answer:* 1.3 m/s.

20. Very small solid particles, called grains, exist in interstellar space. They are continually bombarded by hydrogen atoms of the surrounding interstellar gas. As a result of these collisions, the grains execute Brownian movement in both translation and rotation. Assume the grains are uniform spheres of diameter 4.0×10^{-6} cm and density 1.0 g/cm³, and that the temperature of the gas is 100 K. Find (a) the root-mean-square speed of the grains between collisions and (b) the approximate rate (rev/s) at which the grains are spinning.

21. Colloidal particles in solution are buoyed up by the liquid in which they are suspended. Let ρ' be the density of liquid and ρ the density of the particles. If V is the volume of a particle, show that the number of particles per unit volume in the liquid varies with height as

$$n_v = n_{v0} \exp\left[-\frac{N_0}{RT} V(\rho - \rho')gh\right].$$

This equation was tested by Perrin in his Brownian motion studies.

SECTION 24-5

22. The constant a in van der Waals equation is (a) 0.37 N · m⁴/mol² for CO_2 and (b) 0.025 N · m⁴/mol² for hydrogen. Compute the internal pressures for these gases for values of v/v_0 (where $v_0 = 22.4$ l/mol) of 1, 0.01, and 0.001.

23. (a) The constant b in van der Waals equation is 43 cm³/mol for CO_2. Using the value for a in the previous problem, compute the pressure at $0°$ C for a specific volume of 0.55 l/mol, assuming van der Waals equation to be strictly true. (b) What is the pressure under these same conditions, assuming CO_2 behaves as an ideal gas?
    Answer: (a) 3.3×10^{6} Pa. (b) 4.1×10^{6} Pa.

24. Van der Waals b for oxygen is 32 cm³/mol. Assume b is four times the actual volume of a mole of "billiard-ball" O_2 molecules and compute the diameter of an O_2 molecule.

25. Calculate the work done in an isothermal expansion of one mole of a van der Waals gas from specific volume v_i to v_f.

    Answer: $RT \ln \dfrac{v_f - b}{v_i - b} + a(1/v_f - 1/v_i)$.

26. The constants a and b in the van der Waals equation are different for different substances. Show, however, that if we take v_{cr}, p_{cr}, and T_{cr} as the units of specific volume, pressure, and temperature, the van der Waals equation becomes identical for all substances.

25
entropy and the second law of thermodynamics

The first law of thermodynamics states that energy is conserved. However, we can think of many thermodynamic processes which conserve energy but which actually never occur. For example, when a hot body and a cold body are put into contact, it simply does not happen that the hot body gets hotter and the cold body colder. Or again, a pond does not suddenly freeze on a hot summer day by giving up heat to its environment. *And yet neither of these processes violates the first law of thermodynamics.* Similarly, the first law does not restrict our ability to convert work into heat or heat into work, except that energy must be conserved in the process. And yet in practice, although we can convert a given quantity of work completely into heat, we have never been able to find a scheme that converts a given amount of heat completely into work. The second law of thermodynamics deals with this question of whether processes, assumed to be consistent with the first law, do or do not occur in nature. Although the ideas contained in the second law may seem subtle or abstract, in application they prove to be extremely practical.

Consider a typical system in thermodynamic equilibrium, say a mass M of a (real) gas confined in a cylinder-piston arrangement of volume V, the gas having a pressure p and a temperature T. In an equilibrium state these thermodynamic variables remain constant with time. Suppose that the cylinder, whose walls are an (ideal) heat insulator but whose base is an (ideal) heat conductor is placed on a large heat reservoir maintained at this same temperature T, as in Fig. 22-9. Now let us change the system to another equilibrium state in which the temperature T is the

539

same but the volume V is reduced by one-half. Of the many ways in which we could do this we discuss two extreme cases.

I. We depress the piston very rapidly; we then wait for equilibrium with the reservoir to be re-established. During this process the gas is turbulent and its pressure and temperature are not well defined; we cannot plot the process as a continuous line on a p-V diagram because we would not know what value of pressure (or temperature) to associate with a given volume. The system passes from one equilibrium state i to another f through a series of nonequilibrium states (Fig. 25-1a).

II. We depress the piston (assumed to be frictionless) exceedingly slowly—perhaps by adding sand to the top of the piston—so that the pressure, volume, and temperature of the gas are, at all times, well-defined quantities. We first drop a few grains of sand on the piston. This will reduce the volume of the system a little and the temperature will tend to rise; the system will depart from equilibrium, but only slightly. A small amount of heat will be transferred to the reservoir and in a short time the system will reach a new equilibrium state, its temperature again being that of the reservoir. Then we drop a few more grains of sand on the piston, reducing the volume further. Again we wait for a new equilibrium state to be established, and so forth. By many repetitions of this procedure we finally reduce the volume by one-half. During this entire process the system is never in a state differing much from an equilibrium state. If we imagine carrying out this procedure with still smaller successive increases in pressure, the intermediate states will depart from equilibrium even less. By indefinitely increasing the number of changes and correspondingly decreasing the size of each change, we arrive at an ideal process in which the system passes through a continuous succession of equilibrium states, which we can plot as a continuous line on a p-V diagram (Fig. 25-1b). During this process a certain amount of heat Q is transferred from the system to the reservoir.

Processes of type I are called *irreversible* and those of type II are called *reversible*. *A reversible process is one that, by a differential change in the environment, can be made to retrace its path.* Thus as we cause the piston to move slowly downward, in II, the external pressure on the piston exceeds the pressure exerted on it by the gas by only a differential amount dp. If at any instant we reduce the external pressure ever so slightly (by removing a few sand grains), so that it is *less than* the internal gas pressure by dp, the gas will expand instead of contracting and the system will retrace the equilibrium states through which it has just passed.* In practice all processes are irreversible, but we can approach reversibility arbitrarily closely by making appropriate experimental refinements. The strictly reversible process is a simple and useful abstraction that bears a similar relation to real processes that the ideal gas abstraction does to real gases.

The process described in II is not only reversible but *isothermal*, be-

---

* Not all processes carried out very slowly are reversible. For example, if the piston in our example exerted a frictional force against the cylinder walls, it would not reverse its motion if we made only a differential change dp in the external pressure. We would have to make a change Δp that might be an appreciable fraction of p. Thus our criterion for reversibility, which involves a response of the system to a *differential* change in the environment, is not met. The word *quasi-static* is used to describe processes that are carried out slowly enough so that the system passes through a continuous sequence of equilibrium states; a quasi-static process may or may not be reversible. See "Thermodynamics of an Irreversible Quasi-Static Process" by John S. Thomsen, *American Journal of Physics,* **28,** 119, 1960.

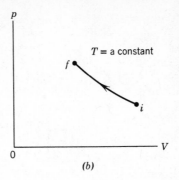

figure 25-1
We cause a real gas to go from an initial equilibrium state i described by p_i, V_i, T_i to a final equilibrium state f described by p_f, V_f $(= \frac{1}{2} V_i)$, and T_f $(= T_i)$. We carry out the process (a) irreversibly, and (b) reversibly.

cause we have assumed that the temperature of the gas differs at all times by only a differential amount dT from the (constant) temperature of the reservoir on which the cylinder rests.

We could also reduce the volume *adiabatically* by removing the cylinder from the heat reservoir and putting it on a nonconducting stand. In an adiabatic process no heat is allowed to enter or to leave the system. An adiabatic process can be either reversible or irreversible— the definition does not exclude either. In a reversible adiabatic process we move the piston exceedingly slowly—perhaps using the sand-loading technique; in an irreversible adiabatic process we shove the piston down quickly.

The temperature of the gas will rise during an adiabatic compression because, from the first law, with $Q = 0$, the work W done in pushing down the piston must appear as an increase ΔU in the internal energy of the system. W will have different values for different rates of pushing down the piston, being given by $\int p \, dV$—that is, by the area under a curve on a p-V diagram—only for reversible processes (for which p has a well-defined value). Thus ΔU and the corresponding temperature change ΔT will not be the same for reversible and irreversible adiabatic processes.

Suppose that we have a system (a real gas, say) in an equilibrium state in a cylinder-piston arrangement. By using our ability to make changes in the environment of the system we can carry out, at our pleasure, a wide variety of processes. We can let the gas expand or we can compress it; we can add or subtract energy in the form of heat; we can do these things and others irreversibly or reversibly. We can also choose to carry out a sequence of processes such that the system returns to its original equilibrium state; we call this a *cycle*. If the processes involved are all reversible, we call it a *reversible cycle*.

Figure 25-2 shows a reversible cycle on a p-V diagram. Along the curve *abc* we allow the system to expand, and the area under this curve represents the work done by the system during the expansion. Along the curve *cda*, which returns the system to its original state, we compress the system, and the area under this curve represents the work we must do on the system during the compression. Hence, the *net* work done by the system is represented by the area enclosed by the curve and is positive. If we decided to traverse the cycle in the opposite sense, that is, expanding along *adc* and compressing along *cba*, the net work done by the system would be the negative of that of the previous case.

An important reversible cycle is the *Carnot cycle*, introduced by Sadi Carnot in 1824. We shall see later that this cycle will determine the

25-3
THE CARNOT CYCLE

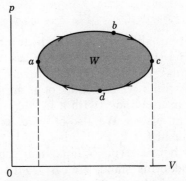

figure 25-2
A p-V diagram of a gas undergoing a reversible cycle. The shaded area W represents the net work done by the gas in the cycle.

figure 25-3
A Carnot cycle. The points *a, b, c,*
and *d* correspond to the points so
labeled in Fig. 25-4. The
cylinder-piston arrangements show
intermediate steps in the processes
that connect adjacent points. The
arrows on the pistons suggest
expansions (caused by removing
sand) and compressions (caused by
adding sand).

limit of our ability to convert heat into work. The system consists of a
"working substance," such as a gas, and the cycle is made up of two
isothermal and two adiabatic reversible processes. The working substance, which we can think of as an ideal gas for concreteness, is contained in a cylinder with a heat-conducting base and nonconducting
walls and piston. We also provide, as part of the environment, a heat
reservoir in the form of a body of large heat capacity at a temperature
T_1, another reservoir of large heat capacity at a temperature T_2, and a
nonconducting stand. We carry out the Carnot cycle in four steps, as
shown in Fig. 25-3. The cycle is shown on the *p-V* diagram of Fig. 25-4.

Step 1. The gas is in an initial equilibrium state represented by p_1, V_1,
T_1 (*a*, Fig. 25-4). We put the cylinder on the heat reservoir at
temperature T_1, and allow the gas to expand slowly to p_2, V_2, T_1

(*b*, Fig. 25-4). During the process heat energy Q_1 is absorbed by the gas by conduction through the base. The expansion is isothermal at T_1 and the gas does work in raising the piston and its load.

Step 2. We put the cylinder on the nonconducting stand and allow the gas to expand slowly further (by reducing the piston load) to p_3, V_3, T_2 (*c*, Fig. 25-4). The expansion is adiabatic because no heat can enter or leave the system. The gas does work in raising the piston and its temperature falls to T_2.

Step 3. We put the cylinder on the (colder) heat reservoir T_2 and compress the gas slowly to p_4, V_4, T_2 (*d*, Fig. 25-4). During the process heat energy Q_2 is transferred from the gas to the reservoir by conduction through the base. The compression is isothermal at T_2 and work is done *on* the gas by the piston and its load.

Step 4. We put the cylinder on the nonconducting stand and compress the gas slowly to the initial condition p_1, V_1, T_1. The compression is adiabatic because no heat can enter or leave the system. Work is done on the gas and its temperature rises to T_1.

The net work W done by the system during the cycle is represented by the area enclosed by path *abcd* of Fig. 25-4. The net amount of heat energy received by the system in the cycle is $Q_1 - Q_2$, where Q_1 is the heat absorbed in Step 1 and Q_2 is that given up in Step 3. The initial and final states are the same so that there is no net change in the internal energy U of the system. Hence, from the first law of thermodynamics,

$$W = Q_1 - Q_2 \tag{25-1}$$

for the cycle, in which Q_1 and Q_2 are taken as positive quantities. The result of the cycle is that heat has been converted into work by the system. Any required amount of work can be obtained by simply repeating the cycle. Hence, the system acts like a *heat engine*.

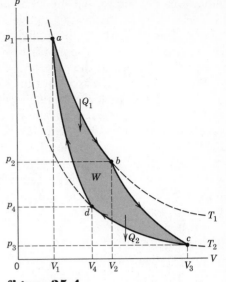

figure 25-4
The Carnot cycle illustrated in the previous figure, plotted on a *p-V* diagram for an ideal gas as the working substance.

We have used an ideal gas as an example of a working substance. The working substance can be anything at all, although the *p-V* diagrams for other substances would be different. Common heat engines use steam or a mixture of fuel and air, or fuel and oxygen as their working substance. Heat may be obtained from the combustion of a fuel such as gasoline or coal, or from the annihilation of mass in nuclear fission processes in nuclear reactors. Heat may be discharged at the exhaust or to a condenser. Although real heat engines do not operate on a reversible cycle, the Carnot cycle, which is reversible, gives useful information about the behavior of any heat engine. It is especially important because, as we shall see, it sets an upper limit to the performance of real engines and thereby gives us a goal to work toward.

The efficiency e of a heat engine is the ratio of the net work done by the engine during one cycle to the heat taken in from the high temperature source in one cycle.* Hence,

$$e = \frac{W}{Q_1} = \frac{Q_1 - Q_2}{Q_1} = 1 - \frac{Q_2}{Q_1}. \tag{25-2}$$

---

* The definition reflects the economic importance of engines. Work W is the desirable output; the heat Q_1, is the input that must be paid for in the form, say, of a fuel bill. An efficient engine has a large ratio of W to Q_1.

Equation 25-2 shows that the efficiency of a heat engine is less than one (100%) so long as the heat Q_2 delivered to the exhaust is not zero. Experience shows that every heat engine rejects some heat during the exhaust stroke. This represents the heat absorbed by the engine that is not converted to work in the process.

We may choose to carry out the Carnot cycle by starting at any point, such as a in Fig. 25-4, and traversing each process in a direction opposite to that of the arrowheads in that figure. Then an amount of heat Q_2 is *removed* from the lower temperature reservoir at T_2, and an amount of heat Q_1 is *delivered* to the higher temperature reservoir at T_1; work must be done *on* the system by an outside agency. In this reversed cycle work must be done *on* the system which extracts heat from the lower temperature reservoir. Any amount of heat can be removed from this reservoir by simply repeating the reverse cycle. Hence, the system acts like a *refrigerator*, transferring heat from a body at a lower temperature (the freezing compartment) to one at a higher temperature (the room) by means of work supplied to it (the electric power input).

EXAMPLE 1

Show that the efficiency of a Carnot engine using an ideal gas as the working substance is $e = (T_1 - T_2)/T_1$.

Along the isothermal path ab, the temperature, and hence the internal energy of the ideal gas, remains constant. From the first law, the heat Q_1 absorbed by the gas in its expansion must be equal to the work W_1 done in this expansion. From Example 2, Chapter 23, we have,

$$Q_1 = W_1 = nRT_1 \ln (V_2/V_1).$$

Likewise, in the isothermal compression along the path cd, we have

$$Q_2 = W_2 = nRT_2 \ln (V_3/V_4).$$

On dividing the first equation by the second, we obtain

$$\frac{Q_1}{Q_2} = \frac{T_1 \ln (V_2/V_1)}{T_2 \ln (V_3/V_4)}.$$

From the equation describing an isothermal process for an ideal gas we obtain for the paths ab and cd

$$p_1V_1 = p_2V_2,$$

$$p_3V_3 = p_4V_4.$$

From the equation describing an adiabatic process for an ideal gas we have for paths bc and da

$$p_2V_2^\gamma = p_3V_3^\gamma,$$

$$p_4V_4^\gamma = p_1V_1^\gamma.$$

Multiplying these four equations together and canceling the factor $p_1p_2p_3p_4$ appearing on both sides, we obtain

$$V_1V_2^\gamma V_3V_4^\gamma = V_2V_3^\gamma V_4V_1^\gamma,$$

from which

$$(V_2V_4)^{\gamma-1} = (V_3V_1)^{\gamma-1}$$

and

$$V_2/V_1 = V_3/V_4.$$

Using this result in our expression for Q_1/Q_2, we see that

$$Q_1/Q_2 = T_1/T_2, \qquad (25\text{-}3)$$

so that

$$e = 1 - Q_2/Q_1 = 1 - T_2/T_1$$

or

$$e = \frac{Q_1 - Q_2}{Q_1} = \frac{T_1 - T_2}{T_1}.$$

The temperatures T_1 and T_2 are those measured on the ideal gas scale described in Chapter 21.

The first heat engines constructed were very inefficient devices. Only a small fraction of the heat absorbed at the high-temperature source could be converted to useful work. Even as engineering design improved, a sizable fraction of the absorbed heat was still discharged at the lower-temperature exhaust of the engine, remaining unconverted to mechanical energy. It remained a hope to devise an engine that could take heat from an abundant reservoir, like the ocean, and convert it completely into useful work. Then it would not be necessary to provide a source of heat at a higher temperature than the outside environment by burning fuels (Fig. 25-5). Likewise, we might hope to be able to devise a refrigerator that simply transfers heat from a cold body to a hot body, without requiring the expense of outside work (Fig. 25-6). *Neither of these hopeful ambitions violates the first law of thermodynamics.* The heat engine would simply convert heat energy completely into mechanical energy, the total energy being conserved in the process. In the refrig-

25-4
THE SECOND LAW OF THERMODYNAMICS

Actual heat engine   "Perfect" heat engine

figure 25-5
In an actual heat engine, some of the heat Q_1 taken in by the engine is converted into work W, but the rest is rejected as heat Q_2. In a "perfect" heat engine all the heat input would be converted into work output.

Actual refrigerator   "Perfect" refrigerator

figure 25-6
In an actual refrigerator, work W is needed to transfer heat from a low-temperature to a high-temperature reservoir. In a "perfect" refrigerator, heat would flow from the low-temperature to the high-temperature reservoir without any work being done on the engine.

erator, the heat energy would simply be transferred from cold body to hot body without any loss of energy in the process. Nevertheless *neither of these ambitions has ever been achieved*, and there is reason to believe they never will be.

The *second law of thermodynamics*, which is a generalization of experience, is an assertion that such devices do not exist. There have been many statements of the second law, each emphasizing another facet of the law, but all can be shown to be equivalent to one another. Clausius stated it as follows: *It is impossible for any cyclical machine to produce no other effect than to convey heat continuously from one body to another at a higher temperature.* This statement rules out our ambitious refrigerator, for it implies that to convey heat continuously from a cold to a hot object it is necessary to supply work by an outside agent. We know from experience that when two bodies are in contact, heat energy flows from the hot body to the cold body. The second law rules out the possibility of heat energy flowing from cold to hot body in such a case and so determines the direction of transfer of heat. The direction can be reversed only by an expenditure of work.

Kelvin (with Planck) stated the second law in words equivalent to these: *A transformation whose only final result is to transform into work heat extracted from a source which is at the same temperature throughout is impossible.** This statement rules out our ambitious heat engine, for it implies that we cannot produce mechanical work by extracting heat from a single reservoir without returning any heat to a reservoir at a lower temperature.

To show that the two statements are equivalent we need to show that, if either statement is false, the other statement must be false also. Suppose Clausius' statement were false so that we could have a refrigerator operating without needing a work input. We could use an ordinary engine to remove heat from a hot body, to do work and to return part of the heat to a cold body. But by connecting our "perfect" refrigerator into the system, this heat would be returned to the hot body without expenditure of work and would become available again for use by the heat engine. Hence, the combination of an ordinary engine and the "perfect" refrigerator would constitute a heat engine which violates the Kelvin-Planck statement. Or we can reverse the argument. If the Kelvin-Planck statement were incorrect, we could have a heat engine which simply takes heat from a source and converts it completely into work. By connecting this "perfect" heat engine to an ordinary refrigerator, we could extract heat from the hot body, convert it completely to work, use this work to run the ordinary refrigerator, extract heat from the cold body, and deliver it plus the work converted to heat by the refrigerator to the hot body. The net result is a transfer of heat from cold to hot body without expenditure of work and this violates Clausius' statement.

The second law tells us that many processes are irreversible. For example, Clausius' statement specifically rules out a simple reversal

---

* This statement needs to be supplemented if we extend thermodynamics to the region of negative Kelvin temperatures. All other formulations of the second law, and indeed, all other laws of thermodynamics apply to negative temperatures without revision. See an article, "Thermodynamics and Statistical Mechanics at Negative Absolute Temperatures" by N. F. Ramsey, in *Temperature, Its Measurement and Control in Science and Industry*, Vol. 3, Part 1, Reinhold Publishing Co., New York, 1962, or "Negative Temperatures and Negative Dissipation" by Stefan Machlup, in *American Journal of Physics*, November 1975.

of the process of heat transfer from hot body to cold body. Not only will some processes not run backward by themselves, but no combination of processes can undo the effect of an irreversible process without causing another corresponding change elsewhere. In later sections we shall develop these ideas more fully and formulate the second law quantitatively.

Carnot first wrote scientifically on the theory of heat engines. In 1824 he published *Reflections on the Motive Power of Heat.* By then the steam engine was commonly used in industry. Carnot wrote:

25-5
THE EFFICIENCY OF ENGINES

In spite of labor of all sorts expended on the steam engine, and in spite of the perfection to which it has been brought, its theory is very little advanced. . . .

The production of motion in the steam engine is always accompanied by a circumstance which we should particularly notice. This circumstance is the passage of caloric from one body where the temperature is more or less elevated to another where it is lower. . . .

The motive power of heat is independent of the agents employed to develop it; its quantity is determined solely by the temperature of the bodies between which, in the final result, the transfer of the caloric occurs.

Hence, Carnot directed attention to the facts that the difference in temperature was the real source of "motive power," that the transfer of heat played a significant role, and that the choice of working substance was of no theoretical importance.

Carnot's achievement was remarkable when we recall that the mechanical equivalence of heat and the conservation of energy principle were not known in 1824. In his later papers, published posthumously in 1872, it became clear that Carnot had foreseen the principle of the conservation of energy and had made an accurate determination of the mechanical equivalent of heat. He had planned a program of research which included all the important developments in the field made by other investigators during the following several decades. However, he died during a cholera epidemic in 1832 at the age of 36, leaving it to others to extend his work. It was William Thomson (later Lord Kelvin) who modified Carnot's reasoning to bring it into accord with the mechanical theory of heat, and who, together with Clausius, successfully developed the science of thermodynamics.

Carnot developed the concept of a reversible engine and the reversible cycle named after him. He stated a theorem of great practical importance: *The efficiency of all reversible engines operating between the same two temperatures is the same, and no irreversible engine working between the same two temperatures can have a greater efficiency than this.* Clausius and Kelvin showed that this theorem was a necessary consequence of the second law of thermodynamics. Notice that nothing is said about the working substance, so that the efficiency of a reversible engine is independent of the working substance and depends only on the temperatures. Furthermore, a reversible engine operates at the maximum efficiency possible for any engine working between the same two temperature limits. The proof of this theorem follows.

Let us call the two reversible engines H and H'. They operate between the temperatures T_1 and T_2 where $T_1 > T_2$. They may differ, say, in their working substance or in their initial pressures and lengths of stroke. We choose H to run forward and H' to run backward (as a refrigerator). The forward-running engine

H takes in heat energy Q_1 at T_1 and gives out heat energy Q_2 at T_2. The backward-running engine (refrigerator) H' takes in heat Q_2' at T_2 and gives out heat Q_1' at T_1. We now connect the engines mechanically and adjust the stroke lengths so that the work done per cycle by H is just sufficient to operate H' (Fig. 25-7). Suppose the efficiency e of H were greater than the efficiency e' of H'. Then

$$e > e', \quad \text{(assumption)}$$

or

$$\frac{Q_1 - Q_2}{Q_1} > \frac{Q_1' - Q_2'}{Q_1'}.$$

Since the work per cycle done by one engine equals the work per cycle done on the other engine,

$$W = W',$$

or

$$Q_1 - Q_2 = Q_1' - Q_2'.$$

Comparing these relations, we see that (since $Q_1 - Q_2 > 0$)

$$\frac{1}{Q_1} > \frac{1}{Q_1'}$$

or

$$Q_1 < Q_1'.$$

Hence (from the work equality),

$$Q_2 < Q_2'.$$

Thus, the hot source gains heat $Q_1' - Q_1$ (positive) and the cool source loses heat $Q_2' - Q_2$ (positive). But no work is done in the process by the combined system $H + H'$ so that we have transferred heat from a body at one temperature to a body at a higher temperature without performing work—in direct contradiction to Clausius' statement of the second law. Hence, we conclude that e cannot be greater than e'. Likewise, by reversing the engines we can use the same reasoning to prove that e' cannot be greater than e. Hence,

$$e = e',$$

proving the first part of Carnot's theorem.

Now suppose that H is an *irreversible* engine. Then by the exact same procedure we can prove that e_{ir} cannot be greater than e'. But H cannot be reversed, so we cannot prove that e' cannot be greater than e_{ir}. Therefore, e_{ir} is either equal to or less than e'. Since $e' = e = e_{reversible}$, we have

$$e_{irreversible} \leq e_{reversible},$$

thus proving the second part of Carnot's theorem.

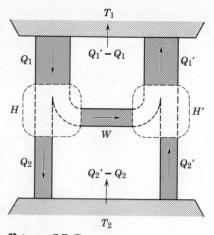

figure 25-7
Proof of Carnot's theorem.

A steam engine takes steam from the boiler at 200° C (225 lb/in.² pressure) and exhausts directly into the air (14 lb/in.² pressure) at 100° C. What is its maximum possible efficiency?

Using the result of Example 1 (which applies to this case by virtue of Carnot's theorem, which we have just proved) we have

$$e = \frac{T_1 - T_2}{T_1} = \frac{473 \text{ K} - 373 \text{ K}}{473 \text{ K}} \times 100\% = 21.1\%.$$

Actual efficiencies of about 15% are usually realized. Energy is lost by friction, turbulence, and heat conduction. Lower exhaust temperatures on more complicated steam engines may raise the maximum possible efficiency to 35% and the actual efficiency to 20%. The efficiency of an ordinary automobile engine is about 22% and that of a large Diesel oil engine about 40%.

EXAMPLE 2

The efficiency of a reversible engine is independent of the working substance and depends only on the two temperatures between which the engine works. Since $e = 1 - Q_2/Q_1$, then Q_2/Q_1 can depend only on the temperatures. This led Kelvin to suggest a new scale of temperature. If we let θ_1 and θ_2 represent these two temperatures, his defining equation is

$$\theta_1/\theta_2 = Q_1/Q_2.$$

That is, two temperatures on this scale are to each other as the heats absorbed and rejected, respectively, by a Carnot engine operating between these temperatures. Such a temperature scale is called the *thermodynamic* (or *Kelvin*) *temperature scale.*

To complete the definition of the thermodynamic scale, we assign the arbitrary value of 273.16 to the temperature of the triple point of water. Hence, $\theta_{tr} = 273.16$ K. Then for a Carnot engine operating between reservoirs at the temperatures θ and θ_{tr}, we have

$$\frac{\theta}{\theta_{tr}} = \frac{Q}{Q_{tr}}$$

or

$$\theta = 273.16 \text{ K } \frac{Q}{Q_{tr}}. \qquad (25\text{-}4)$$

If we compare this with the corresponding equation for the ideal gas temperature T, namely

$$T = 273.16 \text{ K } \lim_{p_{tr} \to 0} \frac{p}{p_{tr}}, \qquad (25\text{-}5)$$

we see that on the thermodynamic scale Q plays the role of a thermometric property. However, Q does not depend on the characteristics of any substance because a Carnot engine is independent of the nature of the working substance. Therefore, we obtain a scale of temperature which is free of the objection we can raise to the ideal gas scale of Chapter 21, and in fact we arrive at a fundamental definition of temperature.

The definition of thermodynamic temperature enables us to rewrite the equation for the efficiency of a reversible engine as

$$e = \frac{Q_1 - Q_2}{Q_1} = \frac{\theta_1 - \theta_2}{\theta_1}. \qquad (25\text{-}6)$$

But we have shown (Example 1) that the efficiency of a Carnot engine using an ideal gas as working substance is

$$e = \frac{Q_1 - Q_2}{Q_1} = \frac{T_1 - T_2}{T_1} \qquad (25\text{-}7)$$

where T is the temperature given by the constant-volume thermometer containing the ideal gas. Hence, $Q_1/Q_2 = T_1/T_2$ and $Q_1/Q_2 = \theta_1/\theta_2$. Since $\theta_{tr} = T_{tr} = 273.16$ and $\theta/\theta_{tr} = T/T_{tr}$, it follows that $\theta = T$. Hence, *if an ideal gas were available for use in a constant-volume thermometer, the thermometer would yield the thermodynamic* (or *Kelvin*) *temperature.* We have seen that, although an ideal gas is not available, measurements made using the limiting process of Eq. 25-5 with real gases correspond to ideal gas behavior. We shall treat the ideal gas scale and the thermodynamic scale as identical and we shall use the designation K interchangeably for each, as in fact we have already done.

In practice, we cannot have a gas below 1 K. One of the methods used in measuring temperature below 1 K employs the thermodynamic scale directly. The ratio of two thermodynamic temperatures is the ratio of two heats transferred during two isothermal processes bounded by the same two adiabatics (Fig. 25-8). The location of the adiabatic boundaries (on the *p-V* diagram) can be found experimentally, and the heats transferred during two nearly reversible isothermal processes can be measured with great precision.

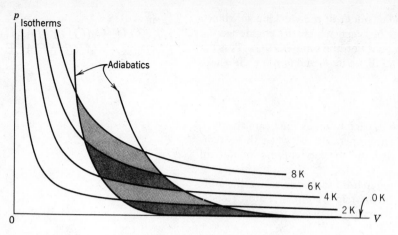

figure 25-8
A series of Carnot cycles tending toward absolute-zero temperature, as used in establishing the thermodynamic scale of temperature. The difference in slope between isothermals and adiabatics has here been exaggerated for clarity.

From the equations

$$T = 273.16 \text{ K} \frac{Q}{Q_{tr}} \qquad \text{or} \qquad \frac{T}{T_{tr}} = \frac{Q}{Q_{tr}}$$

it is clear that the heat Q transferred in an isothermal process between two given adiabatics decreases as the temperature T decreases. Conversely, the smaller Q is the lower the corresponding temperature T is. Now the smallest possible value of Q is zero and the corresponding T is absolute zero. That is, *if a system undergoes a reversible isothermal process with no transfer of heat, the temperature at which this process takes place is the absolute zero.* Hence, at absolute zero, an isothermal and an adiabatic process are identical (Fig. 25-8).

This definition of absolute zero applies to all substances and is independent of the properties of any one of them. Notice that no reference is made to molecules or molecular energy and that we have obtained a purely macroscopic definition of absolute zero.

The efficiency of a Carnot engine is

$$e = 1 - \frac{T_2}{T_1},$$

which is the maximum possible efficiency any engine can have operating between temperatures T_1 and T_2. To obtain 100% efficiency, T_2 must be zero. Only when the low-temperature reservoir is at absolute zero will all the heat absorbed at the high-temperature reservoir be converted to work.

The fundamental feature of all cooling processes is that the lower the temperature, the more difficult it is to go still lower. This experience has led to the formulation of the *third law of thermodynamics,* which can be stated in one form as follows: *It is impossible by any procedure, no matter how idealized, to reduce any system to the absolute zero of temperature in a finite number of operations.* Hence, because we cannot obtain a reservoir at absolute zero, a heat engine with 100% efficiency is a practical impossibility.

25-7 ENTROPY—REVERSIBLE PROCESSES

The zeroth law of thermodynamics is related to the concept of *temperature T* and the first law is related to the concept of *internal energy U.* In this and the following sections we show that the second law of thermodynamics is related to a thermodynamic variable called *entropy, S,* and that we can express the second law quantitatively in terms of this variable. We start by considering a Carnot cycle. For such a cycle we have seen (Eq. 25-3) that

$$\frac{Q_1}{T_1} = \frac{Q_2}{T_2},$$

in which the Q's were taken as positive quantities, that is, we dealt with the magnitudes, or absolute values, only of the Q's. *If we now interpret them again as algebraic quantities, Q being positive when heat enters the system and negative when heat leaves the system,* we can write this relation as

$$\frac{Q_1}{T_1} + \frac{Q_2}{T_2} = 0.$$

This equation states that the sum of the algebraic quantities Q/T is zero for a Carnot cycle.

As a next step, we assert that *any* reversible cycle is equivalent, to as close an approximation as we wish, to an assembly of Carnot cycles. Figure 25-9a shows an arbitrary reversible cycle superimposed on a family of isotherms. We can approximate the actual cycle by connecting the isotherms by suitably chosen adiabatic lines (Fig. 25-9b), thus forming an assembly of Carnot cycles. You should convince yourself that traversing the individual Carnot cycles in Fig. 25-9b is exactly equivalent, in terms of heat transferred and work done, to traversing the jagged sequence of isotherms and adiabatic lines that approximates the actual cycle. This is so because adjacent Carnot cycles have a common isotherm and the two traversals, in opposite directions, cancel each other in the region of overlap as far as heat transfer and work done are concerned. By making the temperature interval between the isotherms in Fig. 25-9b small enough we can approximate the actual cycle as closely as we wish by an alternating sequence of isotherms and adiabatic lines.

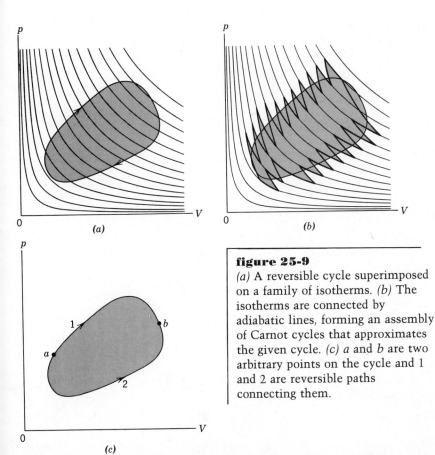

figure 25-9
(a) A reversible cycle superimposed on a family of isotherms. (b) The isotherms are connected by adiabatic lines, forming an assembly of Carnot cycles that approximates the given cycle. (c) a and b are two arbitrary points on the cycle and 1 and 2 are reversible paths connecting them.

We can write, then, for the isothermal-adiabatic sequence of lines in Fig. 25-9*b*,

$$\sum \frac{Q}{T} = 0,$$

or, in the limit of infinitesimal temperature differences between the isotherms of Fig. 25-9*b*,*

$$\oint \frac{dQ}{T} = 0, \tag{25-8}$$

in which $\oint$ indicates that the integral is evaluated for a complete traversal of the cycle, starting (and ending) at any arbitrary point of the cycle.

If the integral of a quantity around any closed path is zero, that quantity is called a state variable, that is, it has a value that is characteristic only of the state of the system, regardless of how that state was arrived at. We call the variable in this case the *entropy S* and we have, from Eq. 25-8,

$$dS = \frac{dQ}{T} \quad \text{and} \quad \oint dS = 0. \tag{25-9}$$

Common units for entropy are J/K or cal/K.

Gravitational potential energy U_g, internal energy U, pressure p, and temperature T are other state variables and equations of the form $\oint dX = 0$ hold for each of them, where for X we substitute the appropriate symbol. Heat Q and work W are *not* state variables and we know that, in general, $\oint dQ \neq 0$ and $\oint dW \neq 0$, as the student can easily show for the special case of a Carnot cycle.

The property of a state variable expressed by $\oint dX = 0$ can also be expressed by saying that $\int dX$ between any two equilibrium states has the same value for all (reversible) paths connecting those states. Let us prove this for the state variable called entropy. We can write Eq. 25-9 (see Fig. 25-9*c*) as

$$_1\int_a^b dS + {}_2\int_b^a dS = 0 \tag{25-10}$$

where a and b are arbitrary points and 1 and 2 describe the paths connecting these points. Since the cycle is reversible, we can write Eq. 25-10 as

$$_1\int_a^b dS - {}_2\int_a^b dS = 0$$

or

$$_1\int_a^b dS = {}_2\int_a^b dS \tag{25-11}$$

In Eq. 25-11 we have simply decided to traverse path 2 in the opposite direction, that is, from a to b rather than from b to a. We do this by changing the order of the limits in the second integral of Eq. 25-10, which requires that we also change the sign of the integral, thus yielding

---

* See footnote on page 487. dQ represents an inexact differential because Q is not a function of the state of the system. The central point of this section is that although dQ is an inexact differential $dQ/T (= dS)$ *is* exact, so that S, like p, V, T, etc. (but not like Q or W), is a state variable.

Eq. 25-11. This latter equation tells us that the quantity $\int_a^b dS$ between any two equilibrium states of the system, such as a and b, is independent of the path connecting those states, for 1 and 2 are quite arbitrary paths. Recall our almost identical discussion in Section 8-2, where we introduced the concept of a conservative force.

The change in entropy between a and b in Fig. 25-9c is, then

$$S_b - S_a = \int_a^b dS = \int_a^b \frac{dQ}{T} \qquad \text{(reversible process),} \qquad (25\text{-}12)$$

where the integral is evaluated over *any reversible path* connecting these two states.

In Section 25-7 we spoke only of reversible processes. However, entropy, like all state variables, depends only on the state of the system and we must be able to calculate the change in entropy for irreversible processes, provided only that they begin and end in equilibrium states. Let us consider two examples.

1. *Free Expansion.* As in Section 22-7 (see Fig. 22-14) let a gas double its volume by expanding into an evacuated enclosure. Since no work is done against the vacuum, $W = 0$ and, since the gas is enclosed by nonconducting walls, $Q = 0$. From the first law, then $\Delta U = 0$ or

$$U_i = U_f \qquad (25\text{-}13)$$

where i and f refer to the initial and final (equilibrium) states. If the gas is an ideal gas, then U depends on temperature alone and not on the pressure or the volume so that Eq. 25-13 implies $T_i = T_f$.

The free expansion is certainly irreversible because we lose control of the environment once we turn the stopcock in Fig. 22-14. There is, however, an entropy difference $S_f - S_i$ between the initial and final equilibrium states, but we cannot calculate it from Eq. 25-12 because that relation applies only to reversible paths; if we tried to use that equation, we would have the immediate difficulty that $Q = 0$ for the free expansion and—further—we would not know how to assign meaningful values of T to the intermediate, nonequilibrium states.

How, then, do we calculate the difference $S_f - S_i$ between these two states? We do so by finding a *reversible* path (*any* reversible path) that connects the states i and f and we calculate the entropy change for that path. In the free expansion a convenient reversible path (assuming an ideal gas) is an isothermal expansion from V_i to $V_f (= 2V_i)$. This corresponds to the isothermal expansion carried out between the points a and b of the Carnot cycle of Fig. 25-4. It represents quite a different set of operations from the free expansion and has in common with it *only* the fact it connects the same set of equilibrium states, i and f. From Eq. 25-12 and Example 1 we have

$$S_f - S_i = \int_i^f \frac{dQ}{T} = nR \ln (V_f/V_i)$$

$$= nR \ln 2.$$

This is positive so that the entropy of the system *increases* in this irreversible, adiabatic process.

2. *Heat Conduction.* For another example consider two bodies that are similar in every respect except that one is at a temperature T_1 and the other at temperature T_2, where $T_1 > T_2$. If we put both objects in contact inside a box with nonconducting walls, they will eventually reach a common temperature T_m, somewhere between T_1 and T_2. Like the free expansion, the process is irreversible because we lose control of the environment once we put the two bodies in the box. Like the free expansion this process is also (irreversibly) adiabatic because no heat enters or leaves the system during the process.

To calculate the entropy change for the system during this process we must again find a *reversible* process connecting the same initial and final states and calculate the system entropy change by applying Eq. 25-12 to that process. We can do so if we imagine that we have at our disposal a heat reservoir of large heat capacity whose temperature T is at our control, by turning a knob, say. We first adjust the reservoir temperature to T_1 and put the first (hotter) object in contact with the reservoir. We then *slowly* (reversibly) lower the reservoir temperature from T_1 to T_m, extracting heat from the hot body as we do so. The hot body *loses* entropy in this process, the amount being approximately

$$\Delta S_1 = -\frac{Q}{T_{1,m}}$$

where $T_{1,m}$ is the average of T_1 and T_m and Q is the heat extracted.

We then adjust our reservoir temperature to T_2 and place it in contact with the second (cooler) object. We then *slowly* (reversibly) raise the reservoir temperature from T_2 to T_m, adding heat to the cool body as we do so. The cool body *gains* entropy in this process, the amount being approximately

$$\Delta S_2 = +\frac{Q}{T_{2,m}},$$

where $T_{2,m}$ is the average of T_2 and T_m and Q is the heat added. Note that the two Q's are identical.

The two bodies are now at the same temperature T_m and the system, which consists of these two bodies, is now in its final equilibrium state. The change in entropy for the complete system is

$$S_f - S_i = \Delta S_1 + \Delta S_2$$

$$= -\frac{Q}{T_{1,m}} + \frac{Q}{T_{2,m}}.$$

Since $T_{1,m} > T_{2,m}$ we have $S_f > S_i$. Again, as for the free expansion, the entropy of the system has *increased* in this irreversible, adiabatic process.

In each of these examples we must distinguish carefully between the actual (irreversible) process (free expansion or heat conduction) and the reversible process that we introduce just so that we can calculate the entropy change in the actual process. We can choose *any* reversible process, as long as it connects the same initial and final state as the actual process; all such reversible processes will yield the same entropy change because this depends only on the initial and final states and not on the process connecting them — be it reversible or irreversible.

We are now ready to formulate the second law of thermodynamics in terms of entropy. Since this law is a generalization from experience we cannot *prove* it but can write it down and show that our statement is in agreement with experiment and is equivalent to other formulations of the second law that we have given earlier. In this spirit we assert that the second law is: *A natural process that starts in one equilibrium state and ends in another will go in the direction that causes the entropy of the system plus environment to increase.*

25-9
ENTROPY AND THE SECOND LAW

Following our pattern for the zeroth law and the first law of thermodynamics (see page 487) the essence of the second law, speaking loosely, is: *There exists a useful thermodynamic variable called entropy.* The second law also tells us how to use this variable to predict whether a particular process will occur in nature.

The two experiments of Section 25-8 (free expansion and heat conduction) are consistent with the second law. The entropy of the system *increased* in each of these irreversible processes. Note that the entropy of the environment in these two cases remains unchanged because, both being carried out in adiabatic enclosures, there was no interchange of heat with the environment. Thus, as required by our statement of the second law, the entropy of the system plus environment increased for each of these (natural) processes.

In the form that we have written it the second law applies only to irreversible processes because only such processes have a "natural direction." Indeed (see Section 25-1) the understanding of the natural directions of such processes is the main concern of the second law. Reversible processes can go equally well in either direction, however, and *for reversible processes the entropy of the system plus environment remains unchanged.* This is so because if heat dQ is transferred from the environment to the system, the entropy of the environment *decreases* by dQ/T whereas that of the system *increases* by dQ/T, the net change for the system plus environment being zero. The fact that the process is reversible means that the environment and the system can differ in temperature by only a differential amount dT when the heat transfer takes place; this is in sharp contrast to our (irreversible) heat conduction problem of the previous section, in which the temperature difference of the two bodies placed in contact was large.

Another class of processes of particular interest are adiabatic processes (reversible or irreversible); they involve no transfer of heat with the environment so that the only entropy change possible is that of the system. From our statement of the second law and from our remarks about reversible processes in the paragraph above, we conclude that

$$S_f = S_i \qquad \text{(reversible adiabatic process)}$$

and

$$S_f > S_i \qquad \text{(irreversible adiabatic process)},$$

where S_f and S_i are the final and initial entropies of the system.

Our statement of the second law is consistent with the Clausius statement (page 546) which declares that there is no such thing as a "perfect" refrigerator (see Fig. 25-6). If there were, the entropy of the lower temperature reservoir would *decrease* by Q/T_2; that of the upper temperature reservoir would *increase* by Q/T_1; that of the system would remain unchanged because the system traverses a cycle, returning to its starting point. Thus the net change in the entropy of the system plus environment is a *decrease*, because $T_2 < T_1$. This violates the statement of the second law that we have just given and, if we wish to retain the statement, we must conclude (with Clausius) that there is no such thing as a "perfect" refrigerator.

Our statement of the second law is also consistent with the Kelvin-Planck statement (page 546) which declares that there is no such thing as a "perfect" heat engine (see Fig. 25-5). If there were, the entropy of the reservoir at temperature T would *decrease* by Q/T; that of the system would remain unchanged because the system traverses a cycle, returning to its starting point. Thus the net change of entropy of the system plus environment is a *decrease*. This violates the statement of the second law that we have just given and, if we wish to retain the statement, we must conclude (with Kelvin) that there is no such thing as a "perfect" heat engine.

EXAMPLE 3

Compute the entropy change of a system consisting of 1.00 kg of ice at 0° C which melts (reversibly) to water at that same temperature. The latent heat of melting is 79.6 cal/g.

The requirement that we melt the ice *reversibly* means that we must put it in contact with a heat reservoir whose temperature exceeds 0° C by only a differential amount; if we lower the reservoir temperature until it is a differential amount below 0° C, the melted ice will begin to freeze. Since the process is reversible, we can use Eq. 25-12 to compute the entropy change of the system. The temperature remains constant at 273 K. Therefore,

$$S_{\text{water}} - S_{\text{ice}} = \int_0^Q \frac{dQ}{T} = \frac{1}{T}\int_0^Q dQ = \frac{Q}{T}.$$

But

$$Q = 10^3 \text{ g} \times 79.6 \text{ cal/g} = 7.96 \times 10^4 \text{ cal}$$

or

$$S_{\text{water}} - S_{\text{ice}} = \frac{7.96 \times 10^4}{273} \text{ cal/K} = 292 \text{ cal/K}$$
$$= 1220 \text{ J/K}.$$

In this example of reversible melting the entropy change of the *system plus environment* is zero, as it must be for all reversible processes. The entropy change calculated above is the increase in entropy of the *system*; there is an exactly equal decrease in entropy of the environment (-1220 J/K) associated with the heat that leaves the reservoir (environment), at 273 K, to melt the ice.

In practice, melting is likely to be irreversible, as when we put an ice cube in a glass of water at room temperature. This process has only one natural direction—the ice will melt. The entropy of the system plus environment will *increase* in this process as required by the second law. The (irreversible) heat conduction example of the previous section should make this understandable.

EXAMPLE 4

Calculate the entropy change that an ideal gas undergoes in a reversible isothermal expansion from a volume V_i to a volume V_f.

From the first law

$$dU = dQ - p\, dV.$$

But $dU = 0$, since U depends only on temperature for an ideal gas and the temperature is constant. Hence,

$$dQ = p\, dV$$

and

$$dS = \frac{dQ}{T} = \frac{p\, dV}{T}.$$

But

$$pV = nRT,$$

so that

$$dS = nR \frac{dV}{V}$$

and

$$S_f - S_i = \int_{V_i}^{V_f} nR \frac{dV}{V} = nR \ln \frac{V_f}{V_i}. \qquad (25\text{-}14)$$

Since $V_f > V_i$, $S_f > S_i$ and the *entropy of the gas increases*.

In order to carry out this process we must have a reservoir at temperature T which is in contact with the system and supplies the heat to the gas. Hence, the *entropy of the reservoir decreases* by $\int dQ/T[= nR \ln (V_f/V_i)]$, so that in this process the entropy of system plus environment does not change. As in the previous example, this is characteristic of a reversible process.

25-10
ENTROPY AND DISORDER

Entropy is associated with disorder and the second law statement that in natural processes the entropy of the (system + environment) tends to increase is equivalent to saying that the disorder of the (system + environment) tends to increase.

There are two approaches to this point of view and we discuss each in turn. The first approach is qualitative and provides an intuitive sense of the equivalence of entropy and disorder. The second is quite formal and provides the solid quantitative base for this equivalence.

From the qualitative point of view let us consider three examples, the first two of which we have discussed in Section 25-8. All are "natural processes" in that there is no doubt whatever as to the direction in which, left to themselves, they will go. Let us now see *qualitatively* in what sense the final (equilibrium) state is more disordered than the initial state.

1. *Free Expansion.* In a free expansion (Section 22-7) the gas molecules confined to one-half of a box are permitted to fill the entire box. By any reasonable definition of the word disorder the system has become more disordered, in the same sense that disorder increases if the litter on one vacant lot is spread over two lots. More precisely, the disorder has increased because we have lost some of our ability to classify molecules. The statement: "The molecules are in the box" is weaker from this point of view than the statement: "The molecules are in the left half of the box."

2. *Heat Conduction.* In this example two bodies of different temperatures T_1 and T_2 come to a uniform intermediate temperature T when they are placed in contact. Here again the system has become more disordered in this natural process because we have lost some of our ability to classify molecules. The statement: "All molecules in the system correspond, by way of Eq. 23-6, to temperature T" is weaker from this point of view than the statement: "All molecules in body A correspond to temperature T_1 and all molecules in body B correspond to temperature T_2." It is clear that some order has been lost in this process.

3. *A Stirred Coffee Cup.* Suppose that you stir a cup of coffee and then remove the spoon. In the initial state there is an ordered motion of the swirling coffee. In the final equilibrium state there is random molecular motion. Surely disorder is increased in this natural and irreversible process.

Let us now discuss the *quantitative* relationship between entropy and disorder. In statistical mechanics we give a precise meaning to disorder and we express its connection with entropy by the relation

$$S = k \ln w. \qquad (25\text{-}15)$$

Here, k is Boltzmann's constant, S is the entropy of the system, and w, which we may call the *disorder parameter*, is the probability that the system will exist in the state it is in relative to all the possible states it could be in. This equation connects a thermodynamic or macroscopic quantity, the entropy, with a statistical or microscopic quantity, the probability.

Let us illustrate by computing the change in entropy of an ideal gas in an isothermal expansion. Here the number of molecules and the temperature do not change, but the volume does. The probability that a given molecule may be found in a region having a volume V is proportional to V; that is, the greater V is, the greater the chance of finding it in V. Hence, the probability of finding a *single* molecule in V is

$$w_1 = cV$$

where c is a constant. The probability of finding N molecules simultaneously in the volume V is the N-fold product of w_1. That is, the probability of a state consisting of N molecules in a volume V is

$$w = w_1{}^N = (cV)^N. \tag{25-16}$$

For example, if the probability of finding a single molecule in V is $\frac{1}{2}$ (that is, there is a 50% chance of its being in V and a 50% chance of its being outside V), the probability of finding two molecules in V is $\frac{1}{4}$. There are four equally probable states here (both in; both out; one in, the other out; one out, the other in), and only one of them is a state with both molecules in V.

If we now combine Eq. 25-15 and Eq. 25-16, we obtain

$$S = kN \left(\ln c + \ln V\right).$$

Hence, the difference in entropy between a state of volume V_f and a state of volume V_i (temperature and number of molecules remaining constant) is

$$S_f - S_i = kN \left(\ln c + \ln V_f\right) - kN \left(\ln c + \ln V_i\right)$$

$$= kN \ln \frac{V_f}{V_i} = \frac{RN}{N_0} \ln \frac{V_f}{V_i} = nR \ln \frac{V_f}{V_i}$$

in exact agreement with the strictly thermodynamic result of Eq. 25-14.

It is on the basis of Eq. 25-16 that we stated above that disorder increases during a free expansion; that equation yields $(cV)^N$ for the disorder parameter before expansion and $(c2V)^N$ for that parameter when the volume is doubled by the expansion.

One must use care not to identify intuitive qualitative ideas of "disorder" as mixed-up-ness with the quantitative meaning we have given the term here. There is a *correlation*, of course, between the qualitative idea of "disorder" and entropy defined either on the macroscopic or microscopic level, but the *identity* exists only for the precise meaning we have given to disorder.*

The statistical definition of entropy, Eq. 25-15, connects the thermodynamic and the statistical mechanical pictures and enables us to put the second law of thermodynamics on a statistical basis. The direction in which natural processes take place (toward higher entropy) is determined by the laws of probability (toward a more probable state). The equilibrium state is the state of maximum entropy thermodynamically and the most probable state statistically. We have seen, however, that fluctuations may occur about an equilibrium distribution (for example, the Brownian motion). From this point of view, then, it is not absolutely certain that the entropy increases in every spontaneous process. The entropy may sometimes decrease. If we waited long enough, even the most improbable states might occur: the water in a pond suddenly freezing on a hot summer day or a local vacuum occurring suddenly in a room. Although such occurrences are possible, the probability of their happening, when computed, turns out to be incredibly small. Hence, the second law of thermodynamics shows us the most probable course of events, not the only possible ones. But its area of application is so broad and the chance of nature's contradicting it so small that it occupies the distinction of being one of the most useful and general laws in all sciences.

---

* For specific examples, see "Entropy and Disorder" by P. G. Wright, in *Contemporary Physics*, November 1970.

questions

1. What requirements should a system meet in order to be in thermodynamic equilibrium?

2. Are any of these phenomena reversible? (a) breaking an empty soda bottle; (b) mixing a cocktail; (c) melting an ice cube in a glass of iced tea; (d) burning a log of firewood; (e) puncturing an automobile tire; (f) finishing the "Unfinished Symphony"; (g) writing this book.

3. Give some examples of irreversible processes in nature.

4. In the irreversible process of Fig. 25-1a can we calculate the work done in terms of an area on a p-V diagram? Is any work done?

5. Can a given amount of mechanical energy be converted completely into heat energy? If so, give an example.

6. Can you suggest a reversible process whereby heat can be added to a system? Would adding heat by means of a Bunsen burner be a reversible process?

7. Give a qualitative explanation of how frictional forces between moving surfaces produce heat energy. Does the reverse process (heat energy producing relative motion of these surfaces) occur? Can you give a plausible explanation?

8. A block returns to its initial position after dissipating mechanical energy to heat through friction. Is this process thermodynamically reversible?

9. To carry out a Carnot cycle we need not start at point a in Fig. 25-4, but may equally well start at points b, c, or d, or indeed any intermediate point. Explain.

10. If a Carnot engine is independent of the working substance, then perhaps real engines should be similarly independent, to a certain extent. Why then, for real engines, are we so concerned to find suitable fuels such as coal, gasoline, or fissionable material? Why not use stones as a fuel?

11. Couldn't we just as well define the efficiency of an engine as $e = W/Q_2$ rather than as $e = W/Q_1$? Why don't we?

12. Under what conditions would an ideal heat engine be 100% efficient?

13. What factors reduce the efficiency of a heat engine from its ideal value?

14. In order to increase the efficiency of a Carnot engine most effectively, would you increase T_1, keeping T_2 constant, or would you decrease T_2, keeping T_1 constant?

15. Can a kitchen be cooled by leaving the door of an electric refrigerator open? Explain.

16. Is a heat engine operating between the warm surface water of a tropical ocean and the cooler water beneath the surface a possible concept? Is the idea practical? (See "Solar Sea Power" by Clarence Zener, *Physics Today*, January 1973.)

17. Is there a change in entropy in purely mechanical motions?

18. Two samples of a gas initially at the same temperature and pressure are compressed from a volume V to a volume $(V/2)$, one isothermally, the other adiabatically. In which sample is the final pressure greater? Does the entropy of the gas change in either process?

19. Suppose we had chosen to represent the state of a system by its entropy and its absolute temperature rather than by its pressure and volume. (a) What would a Carnot cycle look like on a T-S diagram? (b) What physical significance, if any, can be attached to the area under a curve on a T-S diagram?

20. Consider a box containing a very small number of molecules, say five. It must sometimes happen by chance that all of these molecules find themselves in the left half of the box, the right half being completely empty. This is just the reverse of a free expansion, a process that we have declared to be *irreversible*. What is your explanation?

21. Show that the total entropy increases when work is converted into heat by friction between sliding surfaces. Describe the increase in disorder.

22. Comment on the statement "A heat engine converts disordered mechanical motion into organized mechanical motion."

23. When we put cards together in a deck or put bricks together to build a house, for example, we increase the order in the physical world. Does this violate the second law of thermodynamics? Explain.

24. The process of human birth seems to involve an increase in order. Does this process then violate the rule governing the entropy of a system? (See "Thermodynamics of Evolution" by Prigogine, Nicolis, and Babloyantz in *Physics Today*, November 1972.)

25. A rubber band feels warmer than its surroundings immediately after it is quickly stretched; it becomes noticeably cooler when it is allowed to contract rapidly; and a rubber band supporting a load contracts on being heated. Explain these observations using the fact that the molecules of rubber consist of intertwined and cross-linked long chains of atoms in roughly random orientation.

26. Explain the statement "Cosmic rays continually *decrease the entropy of the earth* on which they fall." Does this contradict the second law of thermodynamics?

27. Heat energy flows from the sun (surface temperature 6000 K) to the earth (surface temperature 300 K). Show that the entropy of the earth-sun system increases during this process.

28. Is it true that the heat energy of the universe is steadily growing less available? Is so, why?

29. Can one use terrestrial thermodynamics, which is known to apply to bounded and isolated bodies, for the whole universe? If so, is the universe bounded and from what is the universe isolated?

30. The first, second and third laws of thermodynamics may be paraphrased respectively as follows: (1) You can't win; (2) You can't even break even; (3) You can't get out of the game. Explain in what sense these are permissible restatements.

31. Discuss the following comment of Panofsky and Phillips: "From the standpoint of formal physics there is only one concept which is asymmetric in the time, namely entropy. But this makes it reasonable to assume that the second law of thermodynamics can be used to ascertain the sense of time independently in any frame of reference; that is, we shall take the positive direction of time to be that of statistically increasing disorder, or increasing entropy. . . ." (See, in this connection, "The Arrow of Time" by David Layzer, in *Scientific American*, December 1975.)

SECTION 25-3

1. An ideal gas heat engine operates in a Carnot cycle between 227 and 127° C. It absorbs 6.0×10^4 cal at the higher temperature. (a) How much work per cycle is this engine capable of performing? (b) What is the efficiency of the engine?   *Answer:* (a) 1.2×10^4 cal. (b) 20%.

2. In a Carnot cycle, the isothermal expansion of an ideal gas takes place at 400 K and the isothermal compression at 300 K. During the expansion 500 cal of heat energy are transferred to the gas. Determine (a) the work performed by the gas during the isothermal expansion, (b) the heat rejected from the gas during the isothermal compression, (c) the work done on the gas during the isothermal compression.

3. If the Carnot cycle is run backward, we have an ideal refrigerator. A quantity of heat Q_2 is taken in at the lower temperature T_2 and a quantity of heat Q_1 is given out at the higher temperature T_1. The difference is the work W that must be supplied to run the refrigerator. (a) Show that

$$W = Q_2 \frac{T_1 - T_2}{T_2}.$$

problems

(b) The *coefficient of performance K* of a refrigerator is defined as the ratio of the heat extracted from the cold source to the work needed to run the cycle. Show that ideally

$$K = \frac{T_2}{T_1 - T_2}.$$

In actual refrigerators K has a value of 5 or 6.

(c) In a mechanical refrigerator the low-temperature coils are at a temperature of $-13°$ C, and the compressed gas in the condenser has a temperature of $27°$ C. What is the theoretical coefficient of performance? *Answer:* (c) 6.5.

4. How is the efficiency of a reversible ideal heat engine related to the coefficient of performance of the reversible refrigerator obtained by running the engine backward?

5. (a) A Carnot engine operates between a hot reservoir at 320 K and a cold reservoir at 260 K. If it absorbs 500 joules of heat at the hot reservoir, how much work does it deliver? (b) If the same engine, working in reverse, functions as a refrigerator between the same two reservoirs, how much work must be supplied to remove 1000 J of heat from the cold reservoir? *Answer:* (a) 94 J. (b) 230 J.

6. (a) How much work must be done to extract 1.0 J of heat from a reservoir at $7°$ C and transfer it to one at $27°$ C by means of a refrigerator using a Carnot cycle? (b) From one at $-73°$ C to one at $27°$ C? (c) From one at $-173°$ C to one at $27°$ C? (d) From one at $-223°$ C to one at $27°$ C?

7. (a) Plot accurately a Carnot cycle on a p-V diagram for 1.0 mol of an ideal gas. Let point a correspond to $p = 1.0$ atm, $T = 300$ K, and let b correspond to $p = 0.50$ atm, $T = 300$ K; take the low temperature reservoir to be at 100 K. Let $\gamma = 1.5$. (b) Compute graphically the work done in this cycle. (c) Compute the work analytically.   *Answer:* (c) 1150 J.

8. In a two-stage Carnot heat engine a quantity of heat Q_1 is absorbed at a temperature T_1, work W_1 is done, and a quantity of heat Q_2 is expelled at a lower temperature T_2 by the first stage. The second stage absorbs the heat expelled by the first, does work W_2, and expels a quantity of heat Q_3 at a lower temperature T_3. Prove that the efficiency of the combination engine is $(T_1 - T_3)/T_1$.

SECTION 25-5

9. A combination mercury-steam turbine takes saturated mercury vapor from a boiler at $876°$ F and exhausts it to heat a steam boiler at $460°$ F. The steam turbine receives steam at this temperature and exhausts it to a condenser at $100°$ F. What is the maximum efficiency of the combination? *Answer:* 58%.

10. Apparatus that liquefies helium is in a room at 300 K. If the helium in the apparatus is at 5 K, what is the minimum ratio of heat energy delivered to the room to the heat energy removed from the helium?

11. Suppose a deep shaft were drilled in the earth's crust near one of the poles where the surface temperature is $-40°$ C to a depth where the temperature is $800°$ C. (a) What is the theoretical limit to the efficiency of an engine operating between these temperatures? (b) If all of the heat released into the low temperature reservoir were used to melt ice that was initially at $-40°$ C, at what rate could water at $0°$ C be produced by a power plant having an output of 100 MW? The specific heat of ice is 0.50 cal/g · C°; its heat of fusion is 80 cal/g.   *Answer:* (a) 78%. (b) 239 kg/s.

12. The motor in a refrigerator has a power output of 200 W. If the freezing compartment is at 270 K and the outside air is at 300 K, assuming ideal efficiency, what is the maximum amount of heat that can be extracted from the freezing compartment in 10 min?

13. In a heat pump, heat Q_2 is extracted from the outside atmosphere at T_2 and a larger quantity of heat Q_1 is delivered to the inside of the house at T_1,

with the performance of work W. (a) Draw a schematic diagram of a heat pump. (b) How does it differ in principle from a refrigerator? In practical use? (c) How are Q_1, Q_2, and W related to one another? (d) Can a heat pump be reversed for use in summer? Explain. (e) What advantages does such a pump have over other heating devices?

14. In a heat pump, heat from the outdoors at $-5°$ C is transferred to a room at $17°$ C, energy being supplied by an electric motor. How many joules of heat will be delivered to the room for each joule of electric energy consumed, ideally?

15. A gasoline internal combustion engine can be approximated by the cycle shown in Figure 25-10. Assume an ideal gas and use a compression ratio of $4 : 1$ $(V_4 = 4V_1)$. Assume $p_2 = 3p_1$. (a) Determine the pressure and temperature of each of the vertex points of the p-V diagram in terms of p_1, T_1 and the ratio of specific heats of the gas. (b) What is the efficiency of this cycle?

$Answer$: (a) $T_2 = 3T_1$
$T_3 = 3(4)^{1-\gamma}T_1$
$T_4 = (4)^{1-\gamma}T_1$
$p_2 = 3p_1$
$p_3 = (3)(4)^{-\gamma}p_1$
$p_4 = (4)^{-\gamma}p_1$.
(b) $1 - (4)^{1-\gamma}$.

figure 25-10
Problem 15

SECTION 25-6

16. Using the equation of state of an ideal gas and the equation describing an adiabatic process for an ideal gas, show that the slope, dp/dV, on a p-V diagram of an adiabatic can be written as $-\gamma p/V$ and of an isothermal can be written as $-p/V$. From these results prove that adiabatics are steeper curves than isothermals.

17. Suppose that we were to take as our measure of temperature $-1/T$ rather than T. The unit of this new measure might be the Nivlek (Kelvin spelled backwards) degree (°N). Write a sequence of temperatures in °N extending from positive to negative values of T. (See footnote, page 464.)

SECTION 25-7

18. A mole of a monatomic ideal gas is taken from an initial state of pressure p and volume V to a final state of pressure $2p$ and volume $2V$ by two different processes. (I) It expands isothermally until its volume is doubled, and then its pressure is increased at constant volume to the final state. (II) It is compressed isothermally until its pressure is doubled, and then its volume is increased at constant pressure to the final state.

Show the path of each process on a p-V diagram. For each process calculate in terms of p and V (a) the heat absorbed by the gas in each part of the process; (b) the work done by the gas in each part of the process; (c) the change in internal energy of the gas $U_f - U_i$; (d) the change in entropy of the gas $S_f - S_i$.

SECTION 25-8

19. Heat can be removed from water at $0°$ C and atmospheric pressure without causing the water to freeze, if done with little disturbance of the water. Suppose the water is cooled to $-5.0°$ C before ice begins to form. What is the change in entropy per unit mass occurring during the sudden freezing that then takes place? $Answer$: -0.30 cal/g · K.

20. In a specific heat experiment, 200 g of aluminum ($c_p = 0.215$ cal/g · C°) at $100°$ C is mixed with 50 g of water at $20°$ C. Find the difference in entropy of the system at the end from its value before mixing.

21. An 8.00 g ice cube at $-10.0°$ C is dropped into a thermos flask containing 100 cm³ of water at $20.0°$ C. What is the change in entropy of the system when a final equilibrium state is reached? The specific heat of ice is 0.52 cal/g · C°. $Answer$: $+0.15$ cal/K.

22. A 10-g ice cube at $-10°$ C is placed in a lake whose temperature is $+15°$ C. Calculate the change in entropy of the system as the ice cube comes to thermal equilibrium with the lake.

SECTION 25-9

23. (a) Show that when a substance of mass m having a constant specific heat c is heated from T_1 to T_2 the entropy change is

$$S_2 - S_1 = mc \ln \frac{T_2}{T_1}.$$

(b) Does the entropy of the substance decrease on cooling? (c) If so, does the total entropy of the universe decrease in such a process?
Answer: (b) Yes. (c) No.

24. Four moles of an ideal gas are caused to expand from a volume V_1 to a volume V_2 $(= 2V_1)$. (a) If the expansion is isothermal at the temperature $T = 400$ K, find the work done by the expanding gas. (b) Find the change in entropy, if any. (c) If the expansion were reversibly adiabatic instead of isothermal, would the change in entropy be positive, negative, or zero?

25. A brass rod is in contact thermally with a heat reservoir at $127°$ C at one end and a heat reservoir at $27°$ C at the other end. (a) Compute the total change in the entropy arising from the process of conduction of 1200 cal of heat through the rod. (b) Does the entropy of the rod change in the process?   *Answer:* (a) $+1.0$ cal/K. (b) No.

26. One mole of hydrogen gas and 1.0 mole of nitrogen gas are in adjacent containers at the same pressure p and temperature T. The pressure and temperature are such that both gases behave virtually ideally. (a) If the rms speed of the H_2 molecules is 1850 m/s at temperature T, what will the rms speed be of the N_2 molecules? (b) For which gas will a larger percentage or fraction of the molecules have speeds within ± 50 m/s of the rms speed? (c) If the containers are connected so that the H_2 and N_2 mix, will the change in entropy be positive, negative, or zero?

27. An object of constant heat capacity C is heated from an initial temperature T_i to a final temperature T_f, by being placed in contact with a heat reservoir at T_f. Represent the process on a graph of C/T versus T and (a) show graphically that the total change in entropy ΔS (object plus reservoir) is positive, and (b) show how the use of heat reservoirs at intermediate temperatures would allow the process to be carried out in a way that makes ΔS as small as desired.

28. (a) A body of finite mass is originally at temperature T_2, higher than that of a heat reservoir at a temperature T_1. An engine operates in infinitesimal cycles between the body and the reservoir until it lowers the temperature of the body from T_2 to T_1. Prove that the maximum work obtainable from the engine is $W_{max} = Q - T_1(S_2 - S_1)$, where $S_1 - S_2$ is the entropy change in the body and Q is the heat extracted from the body by the engine. (b) A body of finite mass is originally at temperature T_1, the same as that of a heat reservoir. A refrigerator operates in infinitesimal cycles between the body and reservoir until it lowers the temperature of the body from T_1 to T_0. Prove that the minimum amount of work which must be supplied to the refrigerator is $W_{min} = T_1(S_1 - S_0) - Q$, where $S_0 - S_1$ is the entropy change in the body and Q is the heat extracted from the body by the refrigerator.

SECTION 25-10

29. In general, the probability w_{12} of a complex event, which consists of two unrelated simple events, is equal to the product of their respective probabilities w_1, w_2. The entropy S_{12} of a complex system which consists of two simple systems is just the sum of their respective entropies, S_1, S_2. Show that Eq. 25-15, which relates probability and entropy, is consistent with the additive property of entropy and the multiplicative property of probability for a complex system.

30. (a) Compute the change in entropy of a deck of cards caused by taking the 52 cards of a particular randomly dealt bridge hand and stacking them into a pile, the cards arranged in a specific, predetermined order. (b) Compare, in order of magnitude, this entropy change with thermodynamic entropy changes.

supplementary topics

In Section 11-6 we discussed the relations between the linear and angular kinematic variables for a particle moving in a plane but confined to move in a circle about an axis at right angles to the plane. Such a particle might be any particle in a rigid body rotating about a fixed axis. Here we relax the restriction and allow the particle to move freely in the plane. A planet moving in an elliptical orbit about the sun is an example.

We start from Eq. 11-11, $\mathbf{r} = \mathbf{u}_r r$, in which, however, we now take *both* $\mathbf{u}_r$ and r to be variables; the particle is no longer confined to a circle of constant radius. We find the velocity by differentiation, or

$$\mathbf{v} = \frac{d\mathbf{r}}{dt} = \mathbf{u}_r \frac{dr}{dt} + r \frac{d\mathbf{u}_r}{dt}.$$

Equation 11-13 shows us that $d\mathbf{u}_r/dt = \mathbf{u}_\theta \omega$. Thus we can write

$$\mathbf{v} = \mathbf{u}_r \frac{dr}{dt} + \mathbf{u}_\theta \omega r, \tag{I-1}$$

which shows that $\mathbf{v}$ has two components, a radial component $v_r = dr/dt$ and a component at right angles, $v_\theta = \omega r$. If we hold r constant, then $dr/dt = 0$ and Eq. I-1 reduces to Eq. 11.14a, as it must.

To find the acceleration we differentiate Eq. I-1, remembering that *all five* quantities on the right are variables. We obtain

$$\mathbf{a} = \frac{d\mathbf{v}}{dt} = \mathbf{u}_r \frac{d^2r}{dt^2} + \frac{dr}{dt} \frac{d\mathbf{u}_r}{dt} + (\mathbf{u}_\theta)\left(\omega \frac{dr}{dt} + r \frac{d\omega}{dt} \right) + (\omega r)\left(\frac{d\mathbf{u}_\theta}{dt} \right).$$

Now $d\mathbf{u}_r/dt = \mathbf{u}_\theta \omega$, $d\mathbf{u}_\theta/dt = -\mathbf{u}_r \omega$ (see Eq. 11-16), and $d\omega/dt = \alpha$. Substituting and rearranging leads us finally to

$$\mathbf{a} = \mathbf{u}_r\left(\frac{d^2r}{dt^2} - \omega^2 r \right) + \mathbf{u}_\theta\left(\alpha r + 2\omega \frac{dr}{dt} \right). \tag{I-2}$$

$$= \mathbf{u}_r(a_r - \omega^2 r) + \mathbf{u}_\theta(\alpha r + 2\omega v_r).$$

SUPPLEMENTARY TOPIC I
RELATION BETWEEN LINEAR AND ANGULAR KINEMATICS FOR A PARTICLE MOVING IN A PLANE

Once again, if r = a constant, then $dr/dt = d^2r/dt^2 = 0$ and Eq. I-2 reduces to Eq. 11-17, which we derived especially for this case.

The two new terms in Eq. I-2, $\mathbf{u}_r d^2r/dt^2$ and $\mathbf{u}_\theta 2\omega\, dr/dt$, need a little explanation. We can understand the first of these terms by imagining that the particle moving in the plane is *not* rotating about the axis. If we put $\omega = \alpha = 0$ in Eq. I-2 this equation reduces to

$$\mathbf{a} = \mathbf{u}_r \frac{d^2r}{dt^2} = \mathbf{u}_r a_r,$$

which is just the familiar acceleration of a particle moving along a straight line. Hence this term in Eq. I-2 gives the radial acceleration due to the change in the *magnitude* of $\mathbf{r}$, the other radial acceleration term arising from the changing *direction* of $\mathbf{r}$ as the particle rotates.

There are also two θ-directed acceleration terms. The first one, $\mathbf{u}_\theta \alpha r$, arises simply from the angular acceleration α of a particle in circular motion (r = constant) and is the tangential acceleration of Section 11-5. To understand the second term, $\mathbf{u}_\theta 2\omega\, dr/dt$, consider a man walking outward along a radial line painted on the floor of a merry-go-round. The merry-go-round is rotating with constant angular velocity ω so that its angular acceleration α is now zero. If the man were simply to stand still on the merry-go-round, ($d^2r/dt^2 = dr/dt = 0$, and r = constant) his acceleration, as seen by an observer in a reference frame on the ground (see Eq. I-2), would be simply the familiar centripetal acceleration $-\mathbf{u}_r \omega^2 r$, directed radially inward. If he walks outward, however, $dr/dt \neq 0$ and then Eq. I-2 predicts that the ground observer would also measure a θ-directed acceleration given by $\mathbf{u}_\theta 2\omega v_r$, where $v_r = dr/dt$. This is called a *Coriolis acceleration*. It arises from the fact that even though the angular velocity of the man is constant his speed increases as r increases. Let us convince ourselves that this effect really exists.*

Figure I-1a shows the walking man (point P) as he appears to the ground observer at times t and $t + \Delta t$. We show at time t his radially directed velocity $\mathbf{v}_r (= \mathbf{u}_r\, dr/dt)$ and also a θ-directed velocity caused by the rotation of the merry-go-round and given by $\mathbf{v}_\theta (= \mathbf{u}_\theta \omega r)$. At a time Δt later each of these velocities has changed. The radial velocity has changed in direction, although its magnitude remains dr/dt. The θ-directed velocity has not only changed direction (we have learned to account for this as a centripetal acceleration), but, because the man has moved outward to a point at which the floor is moving faster, its *magnitude* has also changed, from ωr to $\omega(r + \Delta r)$.

Figure I-1b shows the change in velocity caused by the change in direction of the radial line along which the man is walking. If $\Delta\theta$ in the triangle shown is small enough, we have

$$\Delta v_r = v_r\, \Delta\theta.$$

Dividing by Δt and letting Δt approach zero yields

$$a' = \frac{dv_r}{dt} = v_r \frac{d\theta}{dt} = v_r \omega.$$

This is just half the term $2\omega v_r$ in Eq. I-2. However, we have considered only the change in the *radial* velocity; there is also a change in the *tangential* velocity.

The change in tangential velocity, caused by the fact that the man is moving radially outward, is

$$\Delta v_\theta = \omega(r + \Delta r) - \omega r = \omega \Delta r.$$

Dividing by Δt and letting Δt approach zero yields

$$a'' = \frac{dv_\theta}{dt} = \omega \frac{dr}{dt} = \omega v_r.$$

Now both a' and a'' are magnitudes of vectors that point in the same direction,

---

* See "The Coriolis Effect," James E. McDonald, *Scientific American*, May 1952 and also "The Case of the Coriolis Force," Malcolm Correll, *The Physics Teacher*, January 1976.

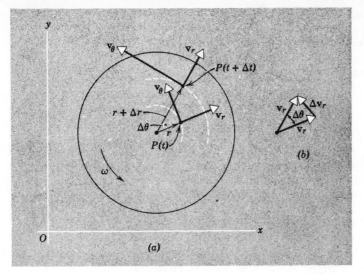

figure 1-1

(a) A merry-go-round, rotating about a fixed axis, is observed by an observer in inertial reference frame *x, y*. A man walks along a radial line at constant speed *v*. In a time interval Δt this line, as seen by the ground observer, sweeps through an angle $\Delta\theta$ and the man moves between the positions shown. His *r*- and θ-directed velocities are shown for each position. *(b)* Showing the change Δv_r in the walking man's *r*-directed velocity. Note that, as $\Delta t \to 0$, Δv_r points in the θ-direction at *P*.

namely the direction of increasing θ at point $P(t)$. The total acceleration in this direction is then

$$a' + a'' = v_r\omega + \omega v_r = 2\omega v_r,$$

which is just what we set out to prove.

If there is indeed an acceleration in the θ-direction in Fig. I-1, there must be a force in this direction. For a man walking outward along a radial line on a rotating merry-go-round this force can only be provided by the friction between his feet and the floor.

We remember that we can interpret classical mechanics most simply if we always view events from an inertial frame. If we do so, we can always associate accelerations with forces exerted by bodies that we can point to in the environment. We can still apply classical mechanics, however, if we select a noninertial reference frame, such as a rotating frame. The small penalty that we must pay is that we must introduce inertial forces, that is, forces that we cannot associate with objects in the environment and which cannot be detected by an observer in an inertial frame. In Section 6-4 we saw that centrifugal force is such an inertial force.

Consider an observer on the rotating merry-go-round watching a man walk along a radial line at a constant speed $v_r = dr/dt$. He would say that the man is in equilibrium because he has no acceleration. Yet the floor is exerting a (very real) frictional force on the soles of the man's feet. This force has one component $(-\mathbf{u}_r F_r)$ that points radially inward and one $(\mathbf{u}_\theta F_\theta)$ that points in the θ-direction, that is, in the direction of rotation.

From the point of view of the ground observer these forces are understandable and, indeed, quite necessary. F_r is associated with the centripetal acceleration $\omega^2 r$ and F_θ with the Coriolis acceleration $2\omega v_r$. The observer on the merry-go-round does not see either of these accelerations however; to him the walking man is in equilibrium. How can this be, in view of the frictional forces that act on the soles of the walking man's shoes? The man himself is well aware of these forces; if he did not lean to compensate for their turning effect, they would knock him off his feet!

The observer on the merry-go-round saves the situation by declaring that two inertial forces act on the walking man, just canceling the (real) frictional forces. One of these inertial forces, called the *centrifugal force*, has magnitude F_r and acts radially *outward*. The other, called the *Coriolis force*, has magnitude F_θ and acts in the negative θ-direction, that is, *opposite* to the direction of rotation. By introducing these forces, which seem quite "real" to him although he cannot point to any body in the environment that is causing them, the observer in the rotating (noninertial) reference frame can apply classical mechanics in the usual way. The ground observer, who is in an inertial frame, cannot detect these inertial forces. Indeed there is no need for them—and no room for them—in his applications of classical mechanics.

Equations I-1 and I-2 are general kinematical descriptions for the motion of a particle in two dimensions. An obvious extension, which we will not attempt here, is to derive corresponding descriptions for motion in three dimensions; this will require us to introduce a third unit vector to define the third dimension.*

SUPPLEMENTARY TOPIC II
POLAR VECTORS AND AXIAL VECTORS

Some vectors called *axial vectors*, such as ω, α, τ, and l, differ in a rather important way from other vectors called *polar vectors*, of which r, v, a, F, and p are examples. Although we shall not need to take this difference into account in this book, it may prove to be instructive and interesting to examine briefly what the difference is.

Consider a typical polar vector such as r. If a student leaves his dormitory and goes to a classroom, his displacement vector r points *from* the dormitory *to* the classroom; there is no question as to our choice of direction. This direction is both "physical" and "natural." Similar remarks apply to the other typical polar vectors listed, namely, v, a, F, and p.

If a student sees a wheel rotating about a fixed axis, he can assign an angular velocity ω to the wheel and can give direction to ω by the right-hand rule (see Section 11-4). This direction, however, *is a convention only*, based on this arbitrary rule. A left-hand rule would have given the opposite direction. The things that are "physical" and "natural" about the wheel are the axis of rotation and the sense of rotation, that is, is it going clockwise or counterclockwise as the student looks at it from a particular end of the axis? Whether ω is chosen to point in one way or the other along the axis does not really matter as long as we are consistent. The same remarks apply to the angular acceleration α and to the other axial vectors listed, namely, $\tau (= r \times F)$ and $l (= r \times p)$. It is for this reason that we sometimes find it more comfortable to say "torque *around* an axis" than "torque *along* an axis" although they mean the same thing. All vectors defined as the vector product of two *polar* vectors are axial vectors because they all depend for their direction assignment on the (arbitrary) right-hand rule.

We have stressed that the laws of physics remain the same no matter how we change the inertial reference frame in which they are expressed. In Section 2-5 we discussed this for translations and rotations of the reference frame and noted that laws expressed in vector form remained unchanged (that is, *invariant*) under such transformations. We also noted that something special may occur when we change the reference frame in another way, namely, by substituting a left-handed frame for a right-handed one. There is an easy way to make such a transformation: Build a right-handed frame and look at its image in a mirror; it will be converted to a left-handed frame (see Fig. II-1) because of the well-known property of a mirror to reverse right and left.

Figure II-1a shows the vector displacement of a student from his dormitory to each of three classrooms. In the mirror each displacement is *still* from the dormitory D to a classroom C. In Fig. II-1b, however, we show a rotating wheel in three orientations. If we establish the directions of ω for both the wheels and their mirror images by the right-hand rule, we see that the image vectors are reversed in comparison to the corresponding image vectors in Fig. II-1a (toward the origin rather than away from the origin). Polar vectors and axial vectors behave differently when we transform reference frames by mirror reflection! This behavior of axial vectors under mirror reflection is not hard to understand. If we imagine ourselves physically applying the right-hand rule to a real rotating wheel, in the mirror, we shall *seem* to be applying a left-hand rule because the image of our right hand is our left hand. A left-hand rule, of course, will give us the opposite direction for ω.

Hence an axial vector is a vector whose sense of direction depends on the

---

* See, for example, *Mechanics*, Section 3-5, by Keith R. Symon, Addison-Wesley Publishing Co., 3d ed., 1971.

figure II-1
(a) Polar vectors, showing, on the right, the displacements r_1, r_2, and r_3 between a dormitory D and three classrooms C_1, C_2, and C_3. On the left we have the mirror images of D, C_1, C_2, and C_3, along with the corresponding displacements. *(b)* *Axial vectors*, showing, on the right, the angular velocities ω_1, ω_2, and ω_3 of three wheels rotating as shown. On the left we have the mirror images of these wheels, along with the angular velocities assigned using the usual right hand rule.

handedness of the reference frame. It is sometimes called a *pseudovector*. A polar vector is a vector that has a direction independent of the reference frame. We mention these facts (1) to stress the arbitrary character of the direction assigned to axial vectors and (2) to stress the importance of testing experiments and physical laws for invariance under translation, rotation, and mirror reflection of the inertial reference frame. In Section 2-5 we referred briefly to some experiments that were *not* invariant under a reflection transformation. This fact, which constituted a violation under certain circumstances of a law of physics previously thought to be well founded (the law of *conservation of parity*), has posed some challenging problems and is leading us to an understanding of the physical world at a deeper level.*

SUPPLEMENTARY TOPIC III
THE WAVE EQUATION FOR A STRETCHED STRING

Figure III-1 shows a section of a long string which is under tension F. The string has been pulled transversely in the y-direction so that a displacement wave travels along the string in the x-direction. We consider a differential element of the string dx and apply Newton's second law of motion to it in order to find how the wave moves along the string.

Let μ be the mass per unit length of the string, so that the mass of element dx is $\mu \, dx$. The net force in the y-direction acting on this element is

$$F \sin \theta_{x+dx} - F \sin \theta_x.$$

We consider only small transverse displacements of the string, so that the restoring force will vary linearly with displacement and the principle of superposition will hold (see Section 19-4). This means that θ in Fig. III-1 will be small, so that we may replace $\sin \theta$ by $\tan \theta$. Now $\tan \theta$ is simply the slope of the string, that is, it equals $\partial y/\partial x$. We must use partial derivatives because the transverse displacement y depends not only on x but also on t. The net force in the y-direction is then

$$F\left(\frac{\partial y}{\partial x}\right)_{x+dx} - F\left(\frac{\partial y}{\partial x}\right)_x,$$

figure III-1

* See "The Overthrow of Parity," by Philip Morrison, *Scientific American*, April 1957.

which we may write as

$$F \frac{\partial}{\partial x}\left(\frac{\partial y}{\partial x}\right) dx$$

or

$$F \frac{\partial^2 y}{\partial x^2} dx.$$

The mass of the element of the string is $\mu\, dx$ and its transverse acceleration is simply $\partial^2 y/\partial t^2$. Hence, Newton's second law, applied to the transverse motion of the string, is

$$F \frac{\partial^2 y}{\partial x^2} dx = (\mu\, dx) \frac{\partial^2 y}{\partial t^2}$$

or

$$\frac{\partial^2 y}{\partial x^2} = \frac{\mu}{F} \frac{\partial^2 y}{\partial t^2}. \tag{III-1}$$

Equation III-1, called the *wave equation*, is the differential equation that describes wave propagation in a string of mass per unit length μ and tension F.

To prove this we show that Eqs. 19-2 and 19-3

$$y = f(x \pm vt), \tag{III-2}$$

which is the general equation representing a wave of any shape traveling along x, is a solution of Eq. III-1. Recall that v in Eq. III-2 is the speed of the wave disturbance and f is any reasonable function of $(x \pm vt)$.

Let us see whether Eq. III-2 is indeed a solution of Eq. III-1 by substituting the former equation into the latter. To do so we note that the two second partial derivatives of y are

$$\frac{\partial^2 y}{\partial x^2} = f'' \quad \text{and} \quad \frac{\partial^2 y}{\partial t^2} = v^2 f''$$

in which f'' is the second derivative of the function f of Eq. III-2 with respect to $(x \pm vt)$. Substitution of these derivatives into Eq. III-1 yields

$$f'' = \frac{\mu}{F} v^2 f'',$$

which we may write as (see Eq. 19-12)

$$v = \sqrt{\frac{F}{\mu}}. \tag{III-3}$$

Thus we conclude that Eq. III-2 is indeed a solution of the partial differential equation Eq. III-1 if the speed of the wave disturbance described by this equation is given by Eq. III-3.

In particular, let us check that Eq. 19-10

$$y = y_m \sin (kx \pm \omega t) \tag{19-10}$$

is a solution of Eq. III-1. We know that it must be because Eq. 19-10 is simply a special case of the general relation Eq. III-2, which we have just shown to be a solution. Even so it is instructive to test this important specific function of $(x \pm vt)$ by substitution into Eq. III-1.

The second derivatives of Eq. 19-10 are

$$\frac{\partial^2 y}{\partial x^2} = -k^2 y_m \sin (kx \pm \omega t)$$

and

$$\frac{\partial^2 y}{\partial t^2} = -\omega^2 y_m \sin (kx \pm \omega t).$$

Substitution into Eq. III-1 yields

$$-k^2 y_m \sin (kx \pm \omega t) = \left(\frac{\mu}{F}\right)[-\omega^2 y_m \sin (kx \pm \omega t)]$$

or

$$\frac{\omega}{k} = \sqrt{\frac{F}{\mu}}.$$

Since $\omega/k = v$ (see Eq. 19-11), this relation is identical with Eq. III-3, and Eq. 19-10, as we expect, is indeed a solution of Eq. III-1.

SUPPLEMENTARY TOPIC IV
DERIVATION OF MAXWELL'S SPEED DISTRIBUTION LAW

Boltzmann, in 1876, derived the Maxwell speed distribution law (see Eq. 24-2) from this line of argument: Let a uniform gravitational field **g** act on an ideal gas maintained at a fixed temperature T. The number of molecules per unit volume n_v will then decrease with altitude z according to the law of atmospheres (see Example 1, Chapter 17). From what we know about the statistical-mechanical interpretation of temperature, however, the speed distribution law — whose form we assume that we do not yet know — must remain the same at all altitudes because it depends only on the temperature. However, this law determines the rate at which molecules move vertically in the atmosphere at any altitude and must thus be intimately related to the decrease of n_v with z. By exploring this relationship in detail we can, in fact, deduce the speed distribution law.

The weight of gas per unit area between the levels z and $z + dz$ in Fig. IV-1 is $n_v mg\, dz$ in which m is the mass of a single molecule. For equilibrium, this weight per unit area must equal the difference in pressure between z and $z + dz$, or

$$n_v mg\, dz = -dp \tag{IV-1}$$

in which we have inserted a minus sign because p decreases as z increases.

We can write the equation of state of an ideal gas, $pV = nRT$, as

$$p = n_v kT \tag{IV-2}$$

because $n = n_v V/N_0$, where $N_0\ (= R/k)$ is Avogadro's number, the number of molecules per mole, and k is Boltzmann's constant. Combining Eqs. IV-1 and IV-2 yields

$$\frac{dp}{p} = \frac{dn_v}{n_v} = -\frac{mg}{kT} dz.$$

figure IV-1

For a constant temperature, we can integrate this relation to yield

$$n_v = \text{constant } e^{-mgz/kT} \tag{IV-3}$$

which, in view of Eq. IV-2, agrees with the result of Example 1, Chapter 17.

We can find the change in n_v as we go from z to $z + dz$ by differentiating Eq. IV-3, or

$$dn_v = -\text{constant } e^{-mgz/kT}\, dz. \tag{IV-4}$$

We associate this decrease in n_v over the interval dz with the fact that, at $z = 0$ (which can be any level we choose) there are some upward-directed molecules — we call them "special molecules" temporarily for convenience — whose vertical velocity components lie in a particular range v_z to $v_z + dv_z$ such that (neglecting collisions; see below) they can rise as high as z but not as high as $z + dz$. Such molecules pass upward through the level z, reverse their direction and pass downward again, as Fig. IV-1 shows. At this point we see more clearly the relationship between Eq. IV-3 and the speed distribution law. Molecules that pass through the interval dz (from above or below) or molecules that never reach the interval cannot contribute to the decrease dn_v of Eq. IV-4.

The rate per unit area at which "special molecules" leave level $z = 0$ (and arrive at level z) is $v_z n_v(v_z)\, dv_z$. Here $n_v(v_z)\, dv_z$ is the number of molecules per unit volume whose vertical velocity components lie between v_z and $v_z + dv_z$.

Now the rate per unit area at which the "special molecules" arrive at level z, but not as high as level $z + dz$, is proportional to the magnitude of the density difference dn_v between z and $z + dz$, or, from Eq. IV-4,

$$v_z n_r(v_z) \, dv_z = \text{constant } e^{-mgz/kT} \, dz, \qquad \text{(IV-5)}$$

in which the constant is independent of z. Equation IV-5, which requires that the change dn_v be accounted for by the "special molecules" is, in fact, the defining equation for $n_v(v_z)$.

From conservation of energy the special molecules have the property that*

$$\tfrac{1}{2}mv_z^2 = mgz$$

or
$$mv_z \, dv_z = mg \, dz.$$

We use these two relations to eliminate z and dz from Eq. IV-5, obtaining, as you should verify,

$$n_v(v_z) \, dv_z = \text{constant } e^{-mv_z^2/2kT} \, dv_z \qquad \text{(IV-6a)}$$

in which $n_v(v_z) \, dv_z$ is the number of molecules per unit volume whose vertical velocity components lie between v_z and $v_z + dv_z$. Note that Eq. IV-6a does not contain g or z. The gravitational field of Fig. IV-1, introduced to allow us to calculate the speed distribution, has served its purpose. We may apply Eq. IV-6a to a gas for which $g = 0$ or in which gravitational effects are negligible. In such a case the vertical direction, which we have identified as the z-direction, no longer has any special meaning. That is, the speed distribution for one component of velocity should be the same for another component of velocity since there is no special or preferred direction in a gas in equilibrium free of external forces. Thus we can write

$$n_v(v_x) \, dv_x = \text{constant } e^{-mv_x^2/2kT} \, dv_x \qquad \text{(IV-6b)}$$

and
$$n_v(v_y) \, dv_y = \text{constant } e^{-mv_y^2/2kT} \, dv_y, \qquad \text{(IV-6c)}$$

for the other two velocity components.

We now seek to find Maxwell's speed distribution (Eq. 24-2); it is expressed in terms of the speed v, rather than in terms of the separate components v_x, v_y, and v_z. We are not concerned here with the direction of **v**, because we assume it to be completely random. We can represent any velocity **v** as a vector extending from the origin in Fig. IV-2; the projections of the vector in the x-y- and z-directions are v_x, v_y, and v_z, respectively. We commonly say that the axes of Fig. IV-2 define a "velocity space," which has many formal similarities to ordinary (or coordinate) space, in which the axes are x, y, and z.

We also show in Fig. IV-2 a small "volume" element, whose sides are dv_x, dv_y, and dv_z; we say that this element has a volume $dv_x \, dv_y \, dv_z$ in velocity space. A point in this element corresponds to a particle whose velocity components lie between v_x and $v_x + dv_x$; v_y and $v_y + dv_y$; and v_z and $v_z + dv_z$. We can regard $n_v(v_z)$ in Eq. IV-6a as giving the *probability* that a given molecule will have a velocity component in the specified range v_z to $v_z + dv_z$, with similar interpretations for $n_v(v_x)$ and $n_v(v_y)$. The probability that a given molecule will have *all three* of its velocity components in the specified ranges, that is, the probability that the tip of the velocity vector **v** will lie inside the volume element of Fig. IV-2, is the *product* of the three (independent) probabilities given in Eq. IV-6, or

$$\text{constant } e^{-mv_x^2/2kT} e^{-mv_y^2/2kT} e^{-mv_z^2/2kT} \, dv_x \, dv_y \, dv_z$$

which, since

$$v^2 = v_x^2 + v_y^2 + v_z^2,$$

we may write as

$$\text{constant } e^{-mv^2/2kT}(dv_x \, dv_y \, dv_z). \qquad \text{(IV-7)}$$

figure IV-2

* If we consider collisions, this result is still true *on the average* for the many molecules that start at $z = 0$ with a given value of v_z and move to the interval z to $z + dz$, having $v_z = 0$ there, even though such molecules would follow very erratic paths because of the collisions.

The quantity in parentheses above is a volume element in velocity space. Since in Maxwell's speed distribution law we are not concerned with the direction of molecular velocities but only with their speeds, it is more convenient to substitute a different volume element for the above, namely, one corresponding to all molecules whose speeds lie between v and $v + dv$, *regardless of direction*. This volume element is not a "cube" but is the space between two concentric spheres, one of radius v and one of radius $v + dv$. The volume of this element in velocity space is $(4\pi v^2)(dv)$. Substituting this for the quantity enclosed in parentheses in Eq. IV-7 yields for the number of molecules per unit volume whose speeds lie between v and $v + dv$,

$$n_v(v)\, dv = \text{constant } e^{-mv^2/2kT}(4\pi v^2\, dv)$$

or

$$n_v(v) = Cv^2 e^{-mv^2/2kT}$$

in which C is a constant. If we sum up over all possible speeds, we simply obtain the total number of molecules per unit volume, regardless of speed. Hence, we can find C by requiring that

$$\int_0^\infty n_v(v)\, dv = n_v,$$

where n_v is the total number of particles per unit volume, regardless of speed. Guided by the methods of Example 3 (Chapter 24), you should show that

$$C = 4\pi n_v (m/2\pi kT)^{3/2}$$

so that

$$n_v(v) = 4\pi n_v (m/2\pi kT)^{3/2} v^2 e^{-mv^2/2kT}. \tag{IV-8}$$

Let us consider a finite number N of molecules contained in a box of volume V. If we multiply each side of the above equation by V, we can replace $n_v V$ on the right by N and $n_v(v)V$ on the left by $N(v)$, which gives us Eq. 24-2.

Here we simply display in one place some conclusions drawn from the special theory of relativity (hereafter, SR) proposed by Einstein in 1905. We omit all proofs and make only a modest attempt to make the conclusions acceptable in terms of "common sense."

SUPPLEMENTARY TOPIC V
SPECIAL RELATIVITY— A SUMMARY OF CONCLUSIONS*

V-1
Introduction

V-2
The Postulates (RR, Section 1.9)

Einstein based his theory on two postulates and *all* of the conclusions of SR derive from them.

a. The First Postulate. From the time of Galileo it was known that the laws of mechanics were the same in all inertial frames (see Fig. V-1 and p. 66). This means that all inertial observers having relative motion, even though they may measure different values for the velocities, momenta, etc., of the particles involved in a given experiment (a game of pool, perhaps), would nevertheless agree on the laws of mechanics involved (conservation of linear momentum, etc.) and on the outcome of the experiment (who won).

Einstein took the bold step of extending this invariance principle to *all* of physics, not only mechanics, including especially electromagnetism. Einstein's first postulate is:

*For a fuller treatment, geared to the level of this book, see *Introduction to Special Relativity*, Robert Resnick, John Wiley and Sons, Inc., New York, 1968. References to this work will be in the style RR, p. 187; RR, Section 1.9, etc.

The laws of physics are the same in all inertial frames. No preferred inertial frame exists.

b. The Second Postulate. In pre-SR days a vexing question was this: Given that the speed of light c is 2.988×10^8 m/s, with respect to what is this speed measured? For sound waves in air the answer is simple—it is with respect to the medium (air) through which the sound wave travels. Light, however, travels through a vacuum. Even so, is there a tenuous medium (the luminiferous, or light carrying, ether) that plays the same role for light that air does for sound? If such an ether exists, can we detect it? Alternatively, should c be measured with respect to the source that emits the light?

All attempts to make experimental verifications along these lines failed completely (see Section 45-8* and RR, Sections 1.5 through 1.8). Einstein made a second bold postulate.

The speed of light is the same in all inertial frames.

Note that no ether is needed or involved. This second postulate means, for example, that if you consider three light sources (a) one at rest with respect to you, (b) one moving toward you at speed $0.9\ c$, say, and (c) one moving away from you at speed $0.9\ c$, you would measure the *same* speed of light from all three sources.

This second postulate has been tested directly (see RR, p. 34) using as a moving "light" source π^0 mesons generated in a proton synchrotron at speeds of $0.99975\ c$. These mesons disintegrate by emitting γ-rays which, like light, are electromagnetic in character and travel with the same speed. The speed measured for the radiation emitted by these fast moving sources was, within experimental error, just c, as Einstein's second postulate predicts.

Many of the conclusions of SR simply don't seem reasonable on the basis of everyday experience. Even Einstein's second postulate seems to violate common sense. If you catch a pitched baseball thrown by a pitcher (a) at rest with respect to you, (b) moving toward you (in an automobile, say) at 30 mi/h and, (c) moving away from you at this same speed, you expect a different baseball speed in each case with respect to you. But if you extend this experience to a source (the pitcher) emitting light (photons), you would contradict Einstein's second postulate. And yet experiment shows that light does have the same speed in each case, in support of Einstein's postulate.

The solution to this dilemma comes about when we realize that the basis of our "common sense" experience is very limited indeed. It is restricted to speeds v such that $v \ll c$, where c is the speed of light. For example, the speed of a satellite in earth orbit may be about 8000 m/s, which seems fast to us, but in terms of the speed of light $(3.0 \times 10^8$ m/s) it is only $0.000027\ c$. We simply have no personal experience in regions of high relative velocity.

As an example, to accelerate an average person (to say nothing of a spaceship) to $0.90\ c$ would require no less than 13 percent of this country's 1971 total energy consumption. However, the particles of physics (electrons, mesons, protons, etc.) can readily be accelerated to high speeds. Electrons emerging from the two-mile long linear accelerator at Stanford University have speeds of $0.999\ c$, for example. In the arena of particle physics SR is absolutely necessary for the solution of mechanical problems.

It turns out that in nature there is a certain finite speed that cannot be exceeded and which we call the limiting speed. This limiting speed is the speed

V-3

Special Relativity and Newtonian Mechanics (RR, Section 2.8)

---

*This book is published in a combined volume (Chapters 1–50) and separate volumes (Part I, Chapters 1–25; Part II, Chapters 26–50). Whether cited references are accessible depends on which of these three volumes you are reading.

of light, c, the greatest speed with which signals can be transmitted. Classical physics assumes that signals can be transmitted with infinite speed, but nature contradicts that assumption, and it really does seem fanciful that such a signal could exist. Experiment confirms c as the limiting speed, so that in a sense the speed of light plays the role in relativity that infinity does in classical physics. It is then not difficult to understand — in fact, it becomes very plausible — that the finite speed of the source of light cannot affect the measured value of the speed of an emitted signal already having the limiting value.

The world in which we live and develop our sense perceptions is a world of Newtonian mechanics, in which $v \ll c$. Newtonian mechanics is revealed as a special case of SR for the limit of low speeds. Indeed, a test of SR is to allow $c \rightarrow \infty$ (in which case $v \ll c$ always holds true) and see that the corresponding formulas of Newtonian mechanics emerge.

Newtonian mechanics, although a special case, is an all-important one. It describes the essential motions of our solar system, the tides, our space ventures, the behavior of baseballs and pinball machines, etc. It works beautifully in the vastly important realm $v \ll c$. But it breaks down at speeds approaching that of light.

Few theories have been subject to more rigorous experimental tests than SR. Not the least among them is the fact that particle accelerators work. They are designed using SR at the level of engineering and technology. An accelerator designed on the basis of Newtonian mechanics simply would not work. Nuclear reactors and, alas, nuclear bombs, are further proof that SR really works.

Einstein once said that no number of experiments could prove him right but a single experiment could prove him wrong. To date this single experiment has not been found.

The basic observation made in SR (or in Newtonian mechanics for that matter) is this. Consider observers to be in different inertial frames, S and S' (Fig. V-1). The corresponding axes of S and S' are parallel, the x-x' axes being common, and S' moves to the right with speed v as seen by S; the two origins coincide at $t = t' = 0$. Each observer, S and S', records the same event, which might be the detonation of a tiny flashbulb, and assigns space and time coordinates to the event, namely, x, y, z, t and x', y', z', t'. What are the relations between these two sets of numbers written down in the observers' notebooks?

Before SR the accepted relations were

$$x' = x - vt \qquad y' = y$$
$$t' = t \qquad z' = z, \tag{V-1}$$

called the *Galilean transformation equations* (RR, Section 1.2). Though impressively correct in the important region $v \ll c$ they nevertheless fail as $v \rightarrow c$.

The corresponding equations used in SR, called the *Lorentz transformation equations*, are (RR, Table 2-1)

$$x' = \frac{x - vt}{\sqrt{1 - (v/c)^2}} \qquad y' = y$$
$$t' = \frac{t - (v/c^2)x}{\sqrt{1 - (v/c)^2}} \qquad z' = z \tag{V-2}$$

Note certain things about these equations. (a) Space and time coordinates are thoroughly intertwined. In particular, time is not the same for each observer; t' depends on x as well as on t. (b) If we let $c \rightarrow \infty$, the Lorentz equations reduce to the Galilean equations, as promised! Finally, (c) We must have $v < c$ or else the quantities x' and t' become indeterminate ($v = c$) or imaginary ($v > c$). The speed of light is an upper limit for the speeds of material objects.

The Lorentz equations, like everything else in SR, can be derived from Einstein's two postulates (RR, Section 2.2).

V-4
The Transformation Equations
(RR, Section 2.2)

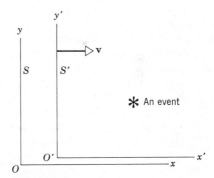

figure V-1
Two inertial frames with parallel axes, the $x - x'$ axis being common. S' moves to the right with speed v as seen by S. At $t = t' = O$ the two origins, O and O', coincide.

Let S' observe two events that occur at the same place in his reference frame. They might be two successive positions of the hand of a clock located at a fixed position, x'. Let S' measure the time interval $\Delta t'$ between these events. S, for whom the clock appears to be moving, observes the same two events and measures a different time interval Δt, which is given by

$$\Delta t = \frac{\Delta t'}{\sqrt{1 - (v/c)^2}}. \tag{V-3}$$

This fact, that $\Delta t > \Delta t'$, is called *time dilation*, and we often verbalize it as "moving clocks run slow." Observer S records a longer time interval than shown to have transpired on the moving clock.

Equation V-3 has been tested experimentally and found to be correct. In one test the "moving clocks" were fast particles called pions $(\pi \pm)$. Pions are radioactive, and their rate of radioactive decay is a measure of their time-keeping ability. See RR, Example 3, p. 75.

Now let us consider a rod, parallel to the $x - x'$ axes, to be at rest in the S' frame. S' will measure a length $\Delta x'$ for it. S, however, measuring the same rod, which is moving with respect to him, would find a length Δx, which is given by

$$\Delta x = \sqrt{1 - (v/c)^2}\, \Delta x' \tag{V-4}$$

This fact, that $\Delta x < \Delta x'$, is called *length contraction*.

The length contraction has been verified in the design of, say, the linear electron accelerator at Stanford University. At an exit speed of $v = 0.999975c$, each meter of the accelerating tube seems like 7.1 mm to an observer moving with the electron. If these length contraction considerations had not been taken into account the machine simply would not work.

The simplest way to understand these results—the time dilation and the length contraction—is to note that one observer, S', is at rest with respect to what he is measuring (clock or rod) whereas for the other observer, S, the objects are in motion. Relativity therefore asserts that *motion affects measurement*. If we had interchanged frames, letting the clock and rod be at rest in S, for example, we would have found the observers again disagreeing on the measured values, but now we would have $\Delta x' < \Delta x$ and $\Delta t' > \Delta t$. So the results are reciprocal, neither observer being "absolutely" right or wrong.

What both observers *will* agree on however, is the *rest length* of a given rod (they will both measure the rod to have the same length when the rod is at rest with respect to their measuring instruments) and the *proper time interval* of a given clock (they will both measure the successive positions of the hand of the clock to have taken the same elapsed time when the clock is at rest with respect to their measuring instruments).

That motion should affect measurement is not a strange idea, even in classical physics. For example, the measured frequency of sound or of light depends on the motion of the source with respect to the observer; we call this the Doppler effect and everyone is familiar with it. And, in mechanics the measured values of the speed, the momentum, or kinetic energy of moving particles are different for observers on the ground than those on a moving train. However, in classical physics measurements of space intervals and time intervals *are* absolute whereas in SR such measurements are relative to the observer. Not only does experiment contradict classical physics but only by adopting the relativity of space and time do we arrive at the invariance (the absoluteness) of all of the laws of physics for all observers. Surely, giving up the absoluteness of the laws of physics (would they then be laws?), as classical notions of time and length require, would leave us with an arbitrary and complex world. By comparison, relativity is absolute and simple.

Let S observe a particle moving with speed u' parallel to the x'-axis. What speed u would S measure? From the Galilean transformation equations (Eq. V-1) we can easily show that

$$u = u' + v \tag{V-5}$$

This relation, which seems to most of us to be intuitively obvious, is, alas, not

V-5
Time Dilation and Length Contraction (RR, Sections 2.3 and 2.4)

V-6
Relativistic Addition of Velocities and the Doppler Effect (Sections 4.6, 6.5, 42.4, and 42.5; RR, Sections 2.6 and 2.7)

true (except for the very important special case of $v \ll c$). The Lorentz transformation equations lead us to

$$u = \frac{u' + v}{1 + (u'v/c^2)}. \tag{V-6}$$

As we expect, for $c \to \infty$, Eq. V-6 reduces to Eq. V-5. Prove that if $u' < c$ and $v < c$, then it must always be true that $u < c$. There is no way to generate speeds $\geqslant c$ by compounding velocities.

Using the relativistic velocity addition result (Eq. V-6), we can deduce the Doppler effect for light. In relativity theory there is no difference between the two cases, which are different in classical theory (namely, source at rest — observer moving and observer at rest — source moving); only the relative motion v of source and observer counts. This fact and the result

$$\nu = \nu' \sqrt{\frac{c \pm v}{c \mp v}} \tag{V-7}$$

are in agreement with experiment. Here, ν' is the frequency of the source at rest in S' and ν is the frequency observed in frame S with respect to which the source moves at speed v; the upper signs refer to source and observer moving *toward* one another and the lower signs refer to source and observer moving *away* from one another. Equation V-7 is called the *longitudinal* Doppler effect, and v refers to the relative velocity of source and observer along the line connecting them.

There is in relativity, however, an effect not predicted by classical physics, namely a *transverse* Doppler effect; that is, if the relative velocity v is at right angles to the line connecting source and observer, we find

$$\nu = \nu' \sqrt{1 - v^2/c^2}. \tag{V-8}$$

This result, confirmed by experiment, can be interpreted simply as a time dilation, moving clocks appearing to run slow.

We have seen that time and length measurements are functions of velocity v. Should mass be any different? SR tells us that the *relativistic mass* m of a particle moving at speed v with respect to the observer is

$$m = \frac{m_0}{\sqrt{1 - (v/c)^2}} \tag{V-9}$$

in which m_0 is the rest mass, that is, the mass measured when the particle is at rest ($v = 0$) with respect to the observer.

It is m and not m_0 that must be taken into account when designing magnets to bend charged particles in arcs of circles. By these techniques, Eq. V-9 has been thoroughly tested. Incidentally, the ratio m/m_0 for electrons emerging from the Stanford University linear accelerator at $K = 30$ GeV is the order of 60,000.

To preserve the law of conservation of linear momentum in SR, we redefine the momentum of a particle of rest mass m_0 and speed v as,

$$p = mv = \frac{m_0 v}{\sqrt{1 - (v/c)^2}}.$$

As a result of the considerations above, in SR the kinetic energy of a particle is no longer given by $\frac{1}{2} m_0 v^2$ but by

$$K = mc^2 - m_0 c^2$$

$$= m_0 c^2 \left(\frac{1}{\sqrt{1 - (v/c)^2}} - 1 \right). \tag{V-10}$$

Can you show that $K \to \frac{1}{2} m_0 v^2$ as $c \to \infty$?

V-7
Mass, Momentum, and Kinetic Energy (Sections 8.9 and 9.3; RR, Sections 3.3 and 3.5)

The best known result of SR is the so-called mass-energy equivalence. That is, the conservation of total energy is equivalent to the conservation of relativistic mass. Mass and energy are equivalent; they form a single invariant that we can call mass-energy. The relation

$$E = mc^2 \qquad\qquad \text{(V-11)}$$

expresses the fact that mass-energy can be expressed in energy (E) units or equivalently in mass $(m = E/c^2)$ units. In fact, it has become common practice to refer to masses in terms of electron volts, such as saying that the rest mass of an electron is 0.51 MeV, for convenience in energy calculations. Likewise entities of zero rest mass, such as photons, may be assigned an effective mass equivalent to their energy. We associate mass with each of the various forms of energy.

In classical physics we had two separate conservation principles: (1) the conservation of (classical) mass, as in chemical reactions, and (2) the conservation of energy. In relativity, these merge into one conservation principle, that of the conservation of mass-energy. The two classical laws may be viewed as special cases that would be expected to agree with experiment only if energy transfers into or out of the system are so small compared with the system's rest mass that the corresponding fractional change in rest mass of the system is too small to be measured.

For example, the rest mass of a hydrogen atom is 1.00797 u (= 938.8 MeV). If enough energy (13.58 eV) is added to ionize hydrogen, that is, to break it up into its constituent parts, a proton and an electron, the fractional change in rest mass of the system is

$$\frac{13.58 \text{ eV}}{938.8 \times 10^6 \text{ eV}} = 1.45 \times 10^{-8}$$

or 1.45×10^{-6} percent, too small to measure. However, for a nucleus such as the deuteron, whose rest mass is 2.01360 u (= 1876.4 MeV), one must add an energy of 2.22 MeV to break it up into its constituent parts, a proton and a neutron. The fractional change in rest mass of the system is

$$\frac{2.22 \text{ MeV}}{1876.4 \text{ MeV}} = 1.18 \times 10^{-3}$$

or 0.12 percent, which is readily measureable. This is characteristic of the fractional rest-mass changes in nuclear reactions, so that one must use the relativistic law of conservation of mass-energy to get agreement between theory and experiment in nuclear reactions. The classical (rest) mass is *not* conserved, but total energy (mass-energy) is.

V-8

The Equivalence of Mass and Energy (Section 8.9; RR, Section 3.6)

appendices

SI base units[a]

Quantity	Name	Symbol	Definition
length	meter	m	" . . . the length equal to 1,650,763.73 wavelengths in vacuum of the radiation corresponding to the transition between the levels $2p_{10}$ and $5d_5$ of the krypton-86 atom." (1960)
mass	kilogram	kg	" . . . this prototype [a certain platinum-iridium cylinder] shall henceforth be considered to be the unit of mass." (1889)
time	second	s	" . . . the duration of 9,192,-631,770 periods of the radiation corresponding to the transition between the two hyperfine levels of the ground state of the cesium-133 atom." (1967)

APPENDIX A
THE INTERNATIONAL SYSTEM OF UNITS (SI)*

* Adapted from "The International System of Units (SI)," National Bureau of Standards Special Publication 330, 1972 edition.

[a] The definitions of these base units were adopted by the General Conference of Weights and Measures, an international body, on the dates shown. In this book we will not use the candela.

SI base units (Continued)

Quantity	Name	Symbol	Definition
electric current	ampere	A	" . . . that constant current which, if maintained in two straight parallel conductors of infinite length, of negligible circular cross section, and placed 1 meter apart in vacuum, would produce between these conductors a force equal to 2×10^{-7} newton per meter of length." (1946)
thermodynamic temperature	kelvin	K	" . . . the fraction 1/273.16 of the thermodynamic temperature of the triple point of water." (1967)
amount of substance	mole	mol	" . . . the amount of substance of a system which contains as many elementary entities as there are atoms in 0.012 kilogram of carbon-12." (1971)
luminous intensity	candela	cd	" . . . the luminous intensity, in the perpendicular direction, of a surface of 1/600,000 square meter of a blackbody at the temperature of freezing platinum under a pressure of 101,325 newton per square meter." (1967)

Some SI derived units with special names

Quantity	SI unit			
	Name	Symbol	Expression in terms of other units	Expression in terms of SI base units
frequency	hertz	Hz		s^{-1}
force	newton	N		$m \cdot kg/s^2$
pressure	pascal	Pa	N/m^2	$kg/m \cdot s^2$
energy, work, quantity of heat	joule	J	$N \cdot m$	$kg \cdot m^2/s^2$
power, radiant flux	watt	W	J/s	$kg \cdot m^2/s^3$
quantity of electricity, electric charge	coulomb	C		$A \cdot s$
electric potential, potential difference, electromotive force	volt	V	W/A	$kg \cdot m^2/A \cdot s^3$
capacitance	farad	F	C/V	$A^2 \cdot s^4/kg \cdot m^2$
electric resistance	ohm	Ω	V/A	$kg \cdot m^2/A^2 \cdot s^3$
conductance	siemens	S	A/V	$A^2 \cdot s^3/kg \cdot m^2$
magnetic flux	weber	Wb	$V \cdot s$	$kg \cdot m^2/A \cdot s^2$
magnetic field	tesla	T	Wb/m^2	$kg/A \cdot s^2$
inductance	henry	H	Wb/A	$kg \cdot m^2/A^2 \cdot s^2$

Some symbols for units of physical quantities

SI Symbols		Symbols other than SI that are Commonly Used	
Name	Abbreviation	Name	Abbreviation
ampere	A	angstrom	Å
candela	cd	British thermal unit	Btu
coulomb	C	calorie	cal
farad	F	day	d
henry	H	degree	°
hertz	Hz	dyne	dyn
joule	J	electron volt	eV
kelvin	K	foot	ft
kilogram	kg	gauss	G
meter	m	gram	g
mole	mol	horsepower	hp
newton	N	hour	h
ohm	Ω	inch	in.
pascal	Pa	mile	mi
radian	rad	minute (of arc)	'
second	s	minute (of time)	min
siemens	S	pound	lb
steradian	sr	revolution	rev
tesla	T	second (of arc)	"
volt	V	standard atmosphere	atm
watt	W	unified atomic mass unit	u
weber	Wb	year	yr

Over the years many hundreds of measurements of fundamental physical quantities, alone and in combination, have been made by hundreds of scientists in many countries. These measurements have different precisions and, most important, they are interdependent. For example, the direct measurements of e, e/m, h/e, etc., are obviously interrelated. Sorting out the best values of e, m, h, etc., from a large mass of overlapping data is not simple.†

For most problems in this book three significant figures will do, and the computational (rounded) values may be used.

APPENDIX B
SOME FUNDAMENTAL CONSTANTS OF PHYSICS*

Constant	Symbol	Computational value	Best (1973) Value Value[a]	Best (1973) Value Uncertainty[b]
Speed of light in a vacuum	c	3.00×10^8 m/s	2.99792458	0.004
Elementary charge	e	1.60×10^{-19} C	1.6021892	2.9
Electron rest mass	m_e	9.11×10^{-31} kg	9.109534	5.1
Permitivity constant	ϵ_0	8.85×10^{-12} F/m	8.854187818	0.008
Permeability constant	μ_0	12.6×10^{-7} H/m	4π (exactly)	—
Electron charge to mass ratio	e/m_e	1.76×10^{11} C/kg	1.7588047	2.8
Proton rest mass	m_p	1.67×10^{-27} kg	1.6726485	5.1
Ratio of proton mass to electron mass	m_p/m_e	1840	1836.15152	0.38
Neutron rest mass	m_n	1.68×10^{-27} kg	1.6749543	5.1
Muon rest mass	m_μ	1.88×10^{-28} kg	1.883566	5.6
Planck constant	h	6.63×10^{-34} J·s	6.626176	5.4
Electron Compton wavelength	λ_C	2.43×10^{-12} m	2.4263089	1.6
Molar gas constant	R	8.31 J/mol·K	8.31441	31
Avogadro constant	N_A	6.02×10^{23}/mol	6.022045	5.1
Boltzmann constant	k	1.38×10^{-23} J/K	1.380662	32
Molar volume of ideal gas at STP[c]	V_m	2.24×10^{-2} m³/mol	2.241383	31
Faraday constant	F	9.65×10^4 C/mol	9.648456	2.8
Stefan–Boltzmann constant	σ	5.67×10^{-8} W/m²·K⁴	5.67032	125
Rydberg constant	R	1.10×10^7/m	1.097373177	0.075
Gravitational constant	G	6.67×10^{-11} m³/s²·kg	6.6720	615
Bohr radius	a_0	5.29×10^{-11} m	5.2917706	0.82
Electron magnetic moment	μ_e	9.28×10^{-24} J/T	9.284832	3.9
Proton magnetic moment	μ_p	1.41×10^{-26} J/T	1.4106171	3.9
Bohr magneton	μ_B	9.27×10^{-24} J/T	9.274078	3.9
Nuclear magneton	μ_N	5.05×10^{-27} J/T	5.050824	3.9

[a] Same unit and power of ten as the computational value.
[b] Parts per million.
[c] STP-standard temperature and pressure = 0° C and 1.0 atm.

* The values in this table were selected from a longer list developed by E. Richard Cohen and B. N. Taylor, *Journal of Physical and Chemical Reference Data*, vol. 2, no. 4 (1973).
† See "A Pilgrim's Progress in Search of the Fundamental Constants," by J. W. M. Du Mond, *Physics Today*, October 1965, and "The Fundamental Physical Constants" by Taylor, Langenberg, and Parker, *Scientific American*, October, 1970.

The sun

Mass	1.99×10^{30} kg
Radius	6.96×10^5 km
Mean density	1,410 kg/m^3
Surface gravity	274 m/s^2
Surface temperature	6000 K
Total radiation rate	3.92×10^{26} W

The earth

Mass	5.98×10^{24} kg
Equatorial radius	6.378×10^6 m
	3963 mi
Polar radius	6.357×10^6 m
	3950 mi
Radius of a sphere of the same volume	6.37×10^6 m
	3600 mi
Mean density	5522 kg/m^3
Acceleration of gravity[a]	9.80665 m/s^2
	32.1740 ft/s^2
Mean orbital speed	29,770 m/s
	18.50 mi/s
Angular speed	7.29×10^{-5} rad/s
Solar constant[b]	1340 W/m^2
Magnetic field (at Washington, D.C.)	5.7×10^{-5} T
Magnetic dipole moment	8.1×10^{22} A·m^2
Standard atmosphere	1.013×10^5 Pa
	14.70 lb/in.^{2}
	760.0 mm-Hg
Density of dry air at STP[c]	1.29 kg/m^3
Speed of sound in dry air at STP	331.4 m/s
	1089 ft/s
	742.5 mi/h

[a] This value, adopted by the General Committee on Weights and Measures in 1901, approximates the value at 45° latitude at sea level.
[b] This is the rate per unit area at which solar energy falls, at normal incidence, just outside the earth's atmosphere.
[c] STP = standard temperature and pressure = 0° C and 1 atm.

The moon

Mass	7.36×10^{22} kg
Radius	1738 km
Mean density	3340 kg/m^3
Surface gravity	1.67 m/s^2
Mean earth-moon distance	3.80×10^5 km

APPENDIX D
THE SOLAR SYSTEM*

	MERCURY	VENUS	EARTH	MARS	JUPITER	SATURN	URANUS	NEPTUNE	PLUTO
Maximum distance from sun (10⁶ km)	69.7	109	152.1	249.1	815.7	1,507	3,004	4,537	7,375
Minimum distance from sun (10⁶ km)	45.9	107.4	147.1	206.7	740.9	1,347	2,735	4,456	4,425
Mean distance from sun (10⁶ km)	57.9	108.2	149.6	227.9	778.3	1,427	2,869.6	4,496.6	5,900
Mean distance from sun (astronomical units)	.387	.723	1	1.524	5.203	9.539	19.18	30.06	39.44
Period of revolution	88 d	224.7 d	365.26 d	687 d	11.86 y	29.46 y	84.01 y	164.8 y	247.7 y
Rotation period	59 d	−243 d retrograde	23 h 56 min 4 s	24 h 37 min 23 s	9 h 50 min 30 s	10 h 14 min	−11 h retrograde	16 h	6 d 9 h
Orbital velocity (km/s)	47.9	35	29.8	24.1	13.1	9.6	6.8	5.4	4.7
Inclination of axis	<28°	3°	23°27′	23°59′	3°05′	26°44′	82°5′	28°48′	?
Inclination of orbit to ecliptic	7°	3.4°	0°	1.9°	1.3°	2.5°	.8°	1.8°	17.2°
Eccentricity of orbit	.206	.007	.017	.093	.048	.056	.047	.009	.25
Equatorial diameter (km)	4,880	12,104	12,756	6,787	142,800	120,000	51,800	49,500	6,000 (?)
Mass (earth = 1)	.055	.815	1	.108	317.9	95.2	14.6	17.2	.1 (?)
Volume (earth = 1)	.06	.88	1	.15	1,316	755	67	57	.1 (?)
Density (water = 1)	5.4	5.2	5.5	4.0	1.3	.7	1.2	1.7	?
Oblateness	0	0	.003	.009	.06	.1	.06	.02	?
Atmosphere (main components)	none	carbon dioxide	nitrogen, oxygen	carbon dioxide, argon	hydrogen, helium	hydrogen, helium	hydrogen, helium, methane	hydrogen, helium, methane	none detected
Mean temperature at visible surface (degrees Celsius) S = solid, C = clouds	350(S) d −170(S) night	−33 (C) 480 (S)	22 (S)	−23 (S)	−150 (C)	−180 (C)	−210 (C)	−220 (C)	−230(?)
Atmospheric pressure at surface (millibars)	10⁻⁹	90,000	1,000	6	?	?	?	?	?
Surface gravity (earth = 1)	.37	.88	1	.38	2.64	1.15	1.17	1.18	?
Mean apparent diameter of sun as seen from planet	1°22′40″	44′15″	31′59″	21′	6′09″	3′22″	1′41″	1′04″	49″
Known satellites	0	0	1	2	14	10 (plus rings)	5 (plus rings)	2	0

* Reprinted by permission (with addition of later discoveries) from "The Solar System," Carl Sagan, *Scientific American*, September 1975.

APPENDIX E
PERIODIC TABLE OF THE ELEMENTS

Key:

| atomic number |
| H |
| atomic mass |

1
H
1.0079

Main Table

IA	IIA	IIIB	IVB	VB	VIB	VIIB	VIII	VIII	VIII	IB	IIB	IIIA	IVA	VA	VIA	VIIA	NOBLE GASES O
1 H 1.00797																	2 He 4.00260
3 Li 6.941	4 Be 9.01218											5 B 10.81	6 C 12.01115	7 N 14.0067	8 O 15.9994	9 F 18.99840	10 Ne 20.179
11 Na 22.98977	12 Mg 24.305											13 Al 26.98154	14 Si 28.086†	15 P 30.97376	16 S 32.06	17 Cl 35.453	18 Ar 39.948
19 K 39.098	20 Ca 40.08	21 Sc 44.9559	22 Ti 47.90	23 V 50.9414	24 Cr 51.996	25 Mn 54.9380	26 Fe 55.847	27 Co 58.9332	28 Ni 58.71	29 Cu 63.546	30 Zn 65.38	31 Ga 69.72	32 Ge 72.59	33 As 74.9216	34 Se 78.96	35 Br 79.904	36 Kr 83.80
37 Rb 85.4678	38 Sr 87.62	39 Y 88.9059	40 Zr 91.22	41 Nb 92.9064	42 Mo 95.94	43 Tc 98.9062	44 Ru 101.07	45 Rh 102.9055	46 Pd 106.4	47 Ag 107.868	48 Cd 112.40	49 In 114.82	50 Sn 118.69	51 Sb 121.75	52 Te 127.60	53 I 126.9045	54 Xe 131.30
55 Cs 132.9054	56 Ba 137.34	57 *La 138.9055	72 Hf 178.49	73 Ta 180.9479	74 W 183.85	75 Re 186.2	76 Os 190.2	77 Ir 192.22	78 Pt 195.09	79 Au 196.9665	80 Hg 200.59	81 Tl 204.37	82 Pb 207.19	83 Bi 208.9804	84 Po (210)	85 At (210)	86 Rn (222)
87 Fr (223)	88 Ra 226.0254	89 †Ac (227)	104 (261)	105 (260)	106 (263)												

PERIODS: 1, 2, 3, 4, 5, 6, 7

* Lanthanides

58 Ce 140.12	59 Pr 140.9077	60 Nd 144.24	61 Pm (147)	62 Sm 150.4	63 Eu 151.96	64 Gd 157.25	65 Tb 158.9254	66 Dy 162.50	67 Ho 164.9304	68 Er 167.26	69 Tm 168.9342	70 Yb 173.04	71 Lu 174.97

† Actinides

90 Th 232.0381	91 Pa 231.0359	92 U 238.029	93 Np 237.0482	94 Pu (244)	95 Am (243)	96 Cm (247)	97 Bk (247)	98 Cf (251)	99 Es (254)	100 Fm (257)	101 Md (258)	102 No (255)	103 Lr (256)

APPENDIX F
THE PARTICLES OF PHYSICS*

Family name	Particle name	Symbol		Spin	Charge, e	Strangeness	Rest mass, MeV	Mean life, seconds	Typical decay mode
		Particle	Antiparticle						
–	Photon	γ	γ	1	0	0	0	Stable	–
L E P T O N S	Electron	e^-	$\overline{e^-}$	$\frac{1}{2}$	∓ 1	0	0.5110	Stable	–
	Muon	μ^+	$\overline{\mu^+}$	$\frac{1}{2}$	± 1	0	105.7	2.197×10^{-6}	$e + \nu + \overline{\nu}$
	Electron's neutrino	ν_e	$\overline{\nu_e}$	$\frac{1}{2}$	0	0	0	Stable	–
	Muon's neutrino	ν_μ	$\overline{\nu_\mu}$	$\frac{1}{2}$	0	0	0	Stable	–
H A D R O N S — M E S O N S	Pion	π^+	$\overline{\pi^+}$	0	± 1	0	139.6	2.603×10^{-8}	$\mu + \nu$
		π^0	π^0	0	0	0	135.0	8.28×10^{-17}	$\gamma + \gamma$
	K-meson	K^+	$\overline{K^+}$	0	± 1	± 1	493.7	1.237×10^{-8}	$\mu + \nu$
		K^0	$\overline{K^0}$	0	0	± 1	497.7	$\begin{cases} 8.930 \times 10^{-11} \\ 5.181 \times 10^{-8} \end{cases}$	$\pi^+ + \pi^-$ $\pi^0 + \pi^0 + \pi^0$
	Eta-meson	η^0	η^0	0	0	0	548.8	?	$\gamma + \gamma$
H A D R O N S — B A R Y O N S (NUCLEON)	Proton	p	$\overline{p}$	$\frac{1}{2}$	± 1	0	938.3	Stable	–
	Neutron	n	$\overline{n}$	$\frac{1}{2}$	0	0	939.6	918	$p + e^- + \nu$
	Lambda particle	Λ^0	$\overline{\Lambda^0}$	$\frac{1}{2}$	0	∓ 1	1116	2.578×10^{-10}	$p + \pi^-$
	Sigma particle	Σ^+	$\overline{\Sigma^+}$	$\frac{1}{2}$	$+1$	∓ 1	1189	8.00×10^{-11}	$p + \pi^0$
		Σ^0	$\overline{\Sigma^0}$	$\frac{1}{2}$	0	∓ 1	1192	$< 1.0 \times 10^{-14}$	$\Lambda^0 + \gamma$
		Σ^-	$\overline{\Sigma^-}$	$\frac{1}{2}$	-1	∓ 1	1197	1.482×10^{-10}	$n + \pi^-$
	Xi particle	Ξ^0	$\overline{\Xi^0}$	$\frac{1}{2}$	0	∓ 2	1315	2.96×10^{-10}	$\Lambda^0 + \pi^0$
		Ξ^-	$\overline{\Xi^-}$	$\frac{1}{2}$	∓ 1	∓ 2	1321	1.652×10^{-10}	$\Lambda^0 + \pi^-$
	Omega particle	Ω^-	$\overline{\Omega^-}$	$\frac{3}{2}$	∓ 1	∓ 3	1672	1.3×10^{-10}	$\Xi^0 + \pi^-$

*See (1) "Review of Particle Properties," *Reviews of Modern Physics*, vol. 48, no. 2, Part II, April (1976).
(2) "Quarks with Color and Flavor," by Sheldon Lee Glashow, *Scientific American*, October (1975).
(3) "The New Elementary Particles and Charm," by Lewis Ryder, *Physics Education*, January (1976) for fuller information. Particle physics is one of the sharp cutting edges of contemporary physics.

Conversion factors may be read off directly from the tables. For example, 1 degree $= 2.778 \times 10^{-3}$ revolutions, so $16.7° = 16.7 \times 2.778 \times 10^{-3}$ rev. The SI quantities are capitalized. The prefix "ab" refers to electromagnetic units (emu); "stat" refers to electrostatic units (esu). Adapted in part from G. Shortley and D. Williams, *Elements of Physics*, Prentice-Hall, Englewood Cliffs, N.J., 1965.

Plane angle

	°	′	″	RADIAN	rev
1 degree =	1	60	3600	1.745×10^{-2}	2.778×10^{-3}
1 minute =	1.667×10^{-2}	1	60	2.909×10^{-4}	4.630×10^{-5}
1 second =	2.778×10^{-4}	1.667×10^{-2}	1	4.848×10^{-6}	7.716×10^{-7}
1 RADIAN =	57.30	3438	2.063×10^{5}	1	0.1592
1 revolution =	360	2.16×10^{4}	1.296×10^{6}	6.283	1

Solid angle

1 sphere $= 4\pi$ steradians $= 12.57$ steradians

Length

	cm	METER	km	in.	ft	mi
1 centimeter =	1	10^{-2}	10^{-5}	0.3937	3.281×10^{-2}	6.214×10^{-6}
1 METER =	100	1	10^{-3}	39.3	3.281	6.214×10^{-4}
1 kilometer =	10^{5}	1000	1	3.937×10^{4}	3281	0.6214
1 inch =	2.540	2.540×10^{-2}	2.540×10^{-5}	1	8.333×10^{-2}	1.578×10^{-5}
1 foot =	30.48	0.3048	3.048×10^{-4}	12	1	1.894×10^{-4}
1 mile =	1.609×10^{5}	1609	1.609	6.336×10^{4}	5280	1

1 angstrom $= 10^{-10}$ m   1 light-year $= 9.4600 \times 10^{12}$ km   1 yard $= 3$ ft
1 nautical mile $= 1852$ m   1 parsec $= 3.084 \times 10^{13}$ km   1 rod $= 16.5$ ft
  $= 1.151$ miles $= 6076$ ft   1 fathom $= 6$ ft   1 mil $= 10^{-3}$ in.

Area

	METER²	cm²	ft²	in.²	circ mil
1 SQUARE METER =	1	10^{4}	10.76	1550	1.974×10^{9}
1 square centimeter =	10^{-4}	1	1.076×10^{-3}	0.1550	1.974×10^{5}
1 square foot =	9.290×10^{-2}	929.0	1	144	1.833×10^{8}
1 square inch =	6.452×10^{-4}	6.452	6.944×10^{-3}	1	1.273×10^{6}
1 circular mil =	5.067×10^{-10}	5.067×10^{-6}	5.454×10^{-9}	7.854×10^{-7}	1

1 square mile $= 2.788 \times 10^{8}$ ft² $= 640$ acres   1 acre $= 43,600$ ft²
1 barn $= 10^{-28}$ m²

Volume

	METER³	cm³	li	ft³	in.³
1 CUBIC METER =	1	10^6	1000	35.31	6.102×10^4
1 cubic centimeter =	10^{-6}	1	1.000×10^{-3}	3.531×10^{-5}	6.102×10^{-2}
1 liter =	1.000×10^{-3}	1000	1	3.531×10^{-2}	61.02
1 cubic foot =	2.832×10^{-2}	2.832×10^4	28.32	1	1728
1 cubic inch =	1.639×10^{-5}	16.39	1.639×10^{-2}	5.787×10^{-4}	1

1 U. S. fluid gallon = 4 U. S. fluid quarts = 8 U. S. pints = 128 U. S. fluid ounces = 231 in.³
1 British imperial gallon = 277.4 in.³   1 liter = 10^{-3} m³.

Mass

Quantities in the shaded areas are not mass units but are often used as such. When we write, for example, 1 kg "=" 2.205 lb this means that a kilogram is a *mass* that *weighs* 2.205 pounds under standard condition of gravity (g = 9.80665 m/s²).

	gm	KG	slug	u	oz	lb	ton
1 gram =	1	0.001	6.852×10^{-5}	6.024×10^{23}	3.527×10^{-2}	2.205×10^{-3}	1.102×10^{-6}
1 KILOGRAM =	1000	1	6.852×10^{-2}	6.024×10^{26}	35.27	2.205	1.102×10^{-3}
1 slug =	1.459×10^4	14.59	1	8.789×10^{27}	514.8	32.17	1.609×10^{-2}
1 u =	1.660×10^{-24}	1.660×10^{-27}	1.137×10^{-28}	1	5.855×10^{-26}	3.660×10^{-27}	1.829×10^{-30}
1 ounce =	28.35	2.835×10^{-2}	1.943×10^{-3}	1.708×10^{25}	1	6.250×10^{-2}	3.125×10^{-5}
1 pound =	453.6	0.4536	3.108×10^{-2}	2.732×10^{26}	16	1	0.0005
1 ton =	9.072×10^5	907.2	62.16	5.465×10^{29}	3.2×10^4	2000	1

Density

Quantities in the shaded areas are weight densities and, as such, are dimensionally different from mass densities. See note for mass table.

	slug/ft³	KG/METER³	g/cm³	lb/ft³	lb/in.³
1 slug per ft³ =	1	515.4	0.5154	32.17	1.862×10^{-2}
1 KILOGRAM per METER³ =	1.940×10^{-3}	1	0.001	6.243×10^{-2}	3.613×10^{-5}
1 gram per cm³ =	1.940	1000	1	62.43	3.613×10^{-2}
1 pound per ft³ =	3.108×10^{-2}	16.02	1.602×10^{-2}	1	5.787×10^{-4}
1 pound per in.³ =	53.71	2.768×10^4	27.68	1728	1

Time

	yr	d	h	min	SECOND
1 year =	1	365.2	8.766×10^3	5.259×10^5	3.156×10^7
1 day =	2.738×10^{-3}	1	24	1440	8.640×10^4
1 hour =	1.141×10^{-4}	4.167×10^{-2}	1	60	3600
1 minute =	1.901×10^{-6}	6.944×10^{-4}	1.667×10^{-2}	1	60
1 SECOND =	3.169×10^{-8}	1.157×10^{-5}	2.778×10^{-4}	1.667×10^{-2}	1

Speed

	ft/s	km/h	METER/ SECOND	mi/h	cm/s	knot
1 foot per second =	1	1.097	0.3048	0.6818	30.48	0.5925
1 kilometer per hour =	0.9113	1	0.2778	0.6214	27.78	0.5400
1 METER per SECOND =	3.281	3.6	1	2.237	100	1.944
1 mile per hour =	1.467	1.609	0.4470	1	44.70	0.8689
1 centimeter per second =	3.281×10^{-2}	3.6×10^{-2}	0.01	2.237×10^{-2}	1	1.944×10^{-2}
1 knot =	1.688	1.852	0.5144	1.151	51.44	1

1 knot = 1 nautical mi/h   1 mi/min = 88.00 ft/s = 60.00 mi/h

Force

Quantities in the shaded areas are not force units but are often used as such. For instance, if we write 1 gram-force "=" 980.7 dynes, we mean that a gram-*mass* experiences a *force* of 980.7 dynes under standard conditions of gravity ($g = 9.80665$ m/s²).

	dyne	NEWTON	lb	pdl	gf	kgf
1 dyne =	1	10^{-5}	2.248×10^{-6}	7.233×10^{-5}	1.020×10^{-3}	1.020×10^{-6}
1 NEWTON =	10^5	1	0.2248	7.233	102.0	0.1020
1 pound =	4.448×10^5	4.448	1	32.17	453.6	0.4536
1 poundal =	1.383×10^4	0.1383	3.108×10^{-2}	1	14.10	1.410×10^{-2}
1 gram-force =	980.7	9.807×10^{-3}	2.205×10^{-3}	7.093×10^{-2}	1	0.001
1 kilogram-force =	9.807×10^5	9.807	2.205	70.93	1000	1

Pressure

	atm	dyne/cm²	inch of water	cm-Hg	PASCAL	lb/in.²	lb/ft²
1 atmosphere =	1	1.013×10^6	406.8	76	1.013×10^5	14.70	2116
1 dyne per cm² =	9.869×10^{-7}	1	4.015×10^{-4}	7.501×10^{-5}	0.1	1.450×10^{-5}	2.089×10^{-3}
1 inch of water[a] at 4° C =	2.458×10^{-3}	2491	1	0.1868	249.1	3.613×10^{-2}	5.202
1 centimeter of mercury[a] at 0° C =	1.316×10^{-2}	1.333×10^4	5.353	1	1333	0.1934	27.85
1 PASCAL =	9.869×10^{-6}	10	4.015×10^{-3}	7.501×10^{-4}	1	1.450×10^{-4}	2.089×10^{-2}
1 pound per in.² =	6.805×10^{-2}	6.895×10^4	27.68	5.171	6.895×10^3	1	144
1 pound per ft² =	4.725×10^{-4}	478.8	0.1922	3.591×10^{-2}	47.88	6.944×10^{-3}	1

[a] Where the acceleration of gravity has the standard value 9.80665 m/s².

1 bar = 10^6 dyne/cm² = 0.1 MPa   1 millibar = 10^2 dyne/cm² = 10^2 Pa

Energy, work, heat

Quantities in the shaded areas are not properly energy units but are included for convenience. They arise from the relativistic mass-energy equivalence formula $E = mc^2$ and represent the energy released if a kilogram or unified atomic mass unit (u) is completely converted to energy.

	Btu	erg	ft·lb	hp·h	JOULE	cal	kW·h	eV	MeV	kg	u
1 British thermal unit =	1	1.055×10^{10}	777.9	3.929×10^{-4}	1055	252.0	2.930×10^{-4}	6.585×10^{21}	6.585×10^{15}	1.174×10^{-14}	7.074×10^{12}
1 erg =	9.481×10^{-11}	1	7.376×10^{-8}	3.725×10^{-14}	10^{-7}	2.389×10^{-8}	2.778×10^{-14}	6.242×10^{11}	6.242×10^{5}	1.113×10^{-24}	670.5
1 foot-pound =	1.285×10^{-3}	1.356×10^{7}	1	5.051×10^{-7}	1.356	0.3239	3.766×10^{-7}	8.464×10^{18}	8.464×10^{12}	1.509×10^{-17}	9.092×10^{9}
1 horsepower-hour =	2545	2.685×10^{13}	1.980×10^{6}	1	2.685×10^{6}	6.414×10^{5}	0.7457	1.676×10^{25}	1.676×10^{19}	2.988×10^{-11}	1.800×10^{16}
1 JOULE =	9.481×10^{-4}	10^{7}	0.7376	3.725×10^{-7}	1	0.2389	2.778×10^{-7}	6.242×10^{18}	6.242×10^{12}	1.113×10^{-17}	6.705×10^{9}
1 calorie =	3.968×10^{-3}	4.186×10^{7}	3.087	1.559×10^{-6}	4.186	1	1.163×10^{-6}	2.613×10^{19}	2.613×10^{13}	4.659×10^{-17}	2.807×10^{10}
1 kilowatt-hour =	3413	3.6×10^{13}	2.655×10^{6}	1.341	3.6×10^{6}	8.601×10^{5}	1	2.247×10^{25}	2.247×10^{19}	4.007×10^{-11}	2.414×10^{16}
1 electron volt =	1.519×10^{-22}	1.602×10^{-12}	1.182×10^{-19}	5.967×10^{-26}	1.602×10^{-19}	3.827×10^{-20}	4.450×10^{-26}	1	10^{-6}	1.783×10^{-36}	1.074×10^{-9}
1 million electron volts =	1.519×10^{-16}	1.602×10^{-6}	1.182×10^{-13}	5.967×10^{-20}	1.602×10^{-13}	3.827×10^{-14}	4.450×10^{-20}	10^{6}	1	1.783×10^{-30}	1.074×10^{-3}
1 kilogram =	8.521×10^{13}	8.987×10^{23}	6.629×10^{16}	3.348×10^{10}	8.987×10^{16}	2.147×10^{16}	2.497×10^{10}	5.610×10^{35}	5.610×10^{29}	1	6.025×10^{26}
1 unified atomic mass unit =	1.415×10^{-13}	1.492×10^{-3}	1.100×10^{-10}	5.558×10^{-17}	1.492×10^{-10}	3.564×10^{-11}	4.145×10^{-17}	9.31×10^{8}	931.0	1.660×10^{-27}	1

Power

	Btu/h	ft·lb/s	hp	cal/s	kW	WATT
1 British thermal unit per hour =	1	0.2161	3.929×10^{-4}	7.000×10^{-2}	2.930×10^{-4}	0.2930
1 foot-pound per second =	4.628	1	1.818×10^{-3}	0.3239	1.356×10^{-3}	1.356
1 horsepower =	2545	550	1	178.2	0.7457	745.7
1 calorie per second =	14.29	3.087	5.613×10^{-3}	1	4.186×10^{-3}	4.186
1 kilowatt =	3413	737.6	1.341	238.9	1	1000
1 WATT =	3.413	0.7376	1.341×10^{-3}	0.2389	0.001	1

Charge

	abcoul	A·h	COULOMB	statcoul
1 abcoulomb =	1	2.778×10^{-3}	10	2.998×10^{10}
1 ampere-hour =	360	1	3600	1.079×10^{13}
1 COULOMB =	0.1	2.778×10^{-4}	1	2.998×10^{9}
1 statcoulomb =	3.336×10^{-11}	9.266×10^{-14}	3.336×10^{-10}	1

1 electronic charge = 1.602×10^{-19} coulomb

Current

	abamp	AMPERE	statamp
1 abampere =	1	10	2.998×10^{10}
1 AMPERE =	0.1	1	2.998×10^{9}
1 statampere =	3.336×10^{-11}	3.336×10^{-10}	1

Potential, electromotive force

	abvolt	VOLT	statvolt
1 abvolt =	1	10^{-8}	3.336×10^{-11}
1 VOLT =	10^{8}	1	3.336×10^{-3}
1 statvolt =	2.998×10^{10}	299.8	1

Resistance

	abohm	OHM	statohm
1 abohm =	1	10^{-9}	1.113×10^{-21}
1 OHM =	10^{9}	1	1.113×10^{-12}
1 statohm =	8.987×10^{20}	8.987×10^{11}	1

Capacitance

	abf	FARAD	μF	statf
1 abfarad =	1	10^{9}	10^{15}	8.987×10^{20}
1 FARAD =	10^{-9}	1	10^{6}	8.987×10^{11}
1 microfarad =	10^{-15}	10^{-6}	1	8.987×10^{5}
1 statfarad =	1.113×10^{-21}	1.113×10^{-12}	1.113×10^{-6}	1

Inductance

	abhenry	HENRY	μH	mH	stathenry
1 abhenry =	1	10^{-9}	0.001	10^{-6}	1.113×10^{-21}
1 HENRY =	10^{9}	1	10^{6}	1000	1.113×10^{-12}
1 microhenry =	1000	10^{-6}	1	0.001	1.113×10^{-18}
1 millihenry =	10^{6}	0.001	1000	1	1.113×10^{-15}
1 stathenry =	8.987×10^{20}	8.987×10^{11}	8.987×10^{17}	8.987×10^{14}	1

Magnetic flux

	maxwell	WEBER
1 maxwell =	1	10^{-8}
1 WEBER =	10^{8}	1

Magnetic field

	gauss	TESLA	milligauss
1 gauss =	1	10^{-4}	1000
1 TESLA =	10^{4}	1	10^{7}
1 milligauss =	0.001	10^{-7}	1

1 tesla = 1 weber/meter^2

Mathematical Signs and Symbols

= equals
$\cong$ equals approximately
$\neq$ is not equal to
$\equiv$ is identical to, is defined as
$>$ is greater than ($\gg$ is much greater than)
$<$ is less than ($\ll$ is much less than)
$\geq$ is more than or equal to (or, is no less than)
$\leq$ is less than or equal to (or, is no more than)
$\pm$ plus or minus ($\sqrt{4} = \pm 2$)
$\propto$ is proportional to (Hooke's law: $F \propto x$, or $F = -kx$)
Σ the sum of
$\bar{x}$ the average value of x

The Greek Alphabet

Alpha	A	α	Nu	N	ν
Beta	B	β	Xi	Ξ	ξ
Gamma	Γ	γ	Omicron	O	o
Delta	Δ	δ	Pi	Π	π
Epsilon	E	ϵ	Rho	P	ρ
Zeta	Z	ζ	Sigma	Σ	σ
Eta	H	η	Tau	T	τ
Theta	Θ	θ	Upsilon	Υ	υ
Iota	I	ι	Phi	Φ	ϕ, φ
Kappa	K	κ	Chi	X	χ
Lambda	Λ	λ	Psi	Ψ	ψ
Mu	M	μ	Omega	Ω	ω

Geometry

Circle of radius r: circumference $= 2\pi r$; area $= \pi r^2$.
Sphere of radius r: area $= 4\pi r^2$; volume $= \frac{4}{3}\pi r^3$.
Right circular cylinder of radius r and height h: area $= 2\pi r^2 + 2\pi rh$; volume $= \pi r^2 h$.

Quadratic Formula

If $ax^2 + bx + c = 0$, then $x = \dfrac{-b \pm \sqrt{b^2 - 4ac}}{2a}$.

Trigonometric Functions of Angle θ

$$\sin \theta = \frac{y}{r} \qquad \cos \theta = \frac{x}{r}$$

$$\tan \theta = \frac{y}{x} \qquad \cot \theta = \frac{x}{y}$$

$$\sec \theta = \frac{r}{x} \qquad \csc \theta = \frac{r}{y}$$

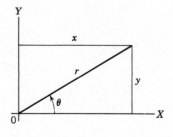

Pythagorean Theorem

$$x^2 + y^2 = r^2$$

Trigonometric Identities

$$\sin^2 \theta + \cos^2 \theta = 1 \qquad \sec^2 \theta - \tan^2 \theta = 1 \qquad \csc^2 \theta - \cot^2 \theta = 1$$

$$\sin 2\theta = 2 \sin \theta \cos \theta$$

$$\cos 2\theta = \cos^2 \theta - \sin^2 \theta = 2 \cos^2 \theta - 1 = 1 - 2 \sin^2 \theta$$

$$\sin \theta = \frac{e^{i\theta} - e^{-i\theta}}{2i} \qquad \cos \theta = \frac{e^{i\theta} + e^{-i\theta}}{2}$$

$$e^{\pm i\theta} = \cos \theta \pm i \sin \theta$$

$$\sin (\alpha \pm \beta) = \sin \alpha \cos \beta \pm \cos \alpha \sin \beta$$

$$\cos (\alpha \pm \beta) = \cos \alpha \cos \beta \mp \sin \alpha \sin \beta$$

$$\tan (\alpha \pm \beta) = \frac{\tan \alpha \pm \tan \beta}{1 \mp \tan \alpha \tan \beta}$$

$$\sin \alpha \pm \sin \beta = 2 \sin \tfrac{1}{2}(\alpha \pm \beta) \cos \tfrac{1}{2}(\alpha \mp \beta)$$

Taylor's Series

$$f(x_0 + x) = f(x_0) + f'(x_0)x + f''(x_0) \frac{x^2}{2!} + f'''(x_0) \frac{x^3}{3!} + \cdots$$

Binomial Expansion

$$(1 + x)^n = 1 + \frac{nx}{1!} + \frac{n(n-1)}{2!} x^2 + \cdots$$

Exponential Expansion

$$e^x = 1 + x + \frac{x^2}{2!} + \frac{x^3}{3!} + \cdots$$

Logarithmic Expansion

$$\ln(1 + x) = x - \tfrac{1}{2}x^2 + \tfrac{1}{3}x^3 - \cdots$$

Trigonometric Expansions (θ in radians)

$$\sin \theta = \theta - \frac{\theta^3}{3!} + \frac{\theta^5}{5!} - \cdots$$

$$\cos \theta = 1 - \frac{\theta^2}{2!} + \frac{\theta^4}{4!} - \cdots$$

Derivatives and Indefinite Integrals

In what follows, the letters u and v stand for any functions of x, and a and m are constants. To each of the integrals should be added an arbitrary constant of integration. The *Handbook of Chemistry and Physics* (Chemical Rubber Publishing Co.) gives a more extensive tabulation.

1. $\dfrac{dx}{dx} = 1$

1. $\int dx = x$

2. $\dfrac{d}{dx} (au) = a \dfrac{du}{dx}$

2. $\int au \, dx = a \int u \, dx$

3. $\dfrac{d}{dx} (u + v) = \dfrac{du}{dx} + \dfrac{dv}{dx}$

3. $\int (u + v) \, dx = \int u \, dx + \int v \, dx$

4. $\dfrac{d}{dx} x^m = mx^{m-1}$

4. $\int x^m \, dx = \dfrac{x^{m+1}}{m + 1} \qquad (m \neq -1)$

5. $\dfrac{d}{dx} \ln x = \dfrac{1}{x}$

5. $\displaystyle\int \dfrac{dx}{x} = \ln |x|$

6. $\dfrac{d}{dx} (uv) = u\dfrac{dv}{dx} + v\dfrac{du}{dx}$

6. $\displaystyle\int u\dfrac{dv}{dx}\, dx = uv - \int v\dfrac{du}{dx}\, dx$

7. $\dfrac{d}{dx} e^x = e^x$

7. $\displaystyle\int e^x\, dx = e^x$

8. $\dfrac{d}{dx} \sin x = \cos x$

8. $\int \sin x\, dx = -\cos x$

9. $\dfrac{d}{dx} \cos x = -\sin x$

9. $\int \cos x\, dx = \sin x$

10. $\dfrac{d}{dx} \tan x = \sec^2 x$

10. $\int \tan x\, dx = \ln |\sec x|$

11. $\dfrac{d}{dx} \cot x = -\csc^2 x$

11. $\int \cot x\, dx = \ln |\sin x|$

12. $\dfrac{d}{dx} \sec x = \tan x \sec x$

12. $\int \sec x\, dx = \ln |\sec x + \tan x|$

13. $\dfrac{d}{dx} \csc x = -\cot x \csc x$

13. $\int \csc x\, dx = \ln |\csc x - \cot x|$

14. $\dfrac{d}{dx} \arctan x = \dfrac{1}{1 + x^2}$

14. $\displaystyle\int \dfrac{dx}{1 + x^2} = \arctan x$

15. $\dfrac{d}{dx} \arcsin x = \dfrac{1}{\sqrt{1 - x^2}}$

15. $\displaystyle\int \dfrac{dx}{\sqrt{1 - x^2}} = \arcsin x$

16. $\dfrac{d}{dx} \operatorname{arcsec} x = \dfrac{1}{x\sqrt{x^2 - 1}}$

16. $\displaystyle\int \dfrac{dx}{x\sqrt{x^2 - 1}} = \operatorname{arcsec} x$

Vector Products

Let $\mathbf{i}, \mathbf{j}, \mathbf{k}$ be unit vectors in the x, y, z directions. Then

$$\mathbf{i} \cdot \mathbf{i} = \mathbf{j} \cdot \mathbf{j} = \mathbf{k} \cdot \mathbf{k} = 1, \quad \mathbf{i} \cdot \mathbf{j} = \mathbf{j} \cdot \mathbf{k} = \mathbf{k} \cdot \mathbf{i} = 0,$$

$$\mathbf{i} \times \mathbf{i} = \mathbf{j} \times \mathbf{j} = \mathbf{k} \times \mathbf{k} = 0,$$

$$\mathbf{i} \times \mathbf{j} = \mathbf{k}, \quad \mathbf{j} \times \mathbf{k} = \mathbf{i}, \quad \mathbf{k} \times \mathbf{i} = \mathbf{j}.$$

Any vector $\mathbf{a}$ with components a_x, a_y, a_z along the x, y, z axes can be written

$$\mathbf{a} = a_x\mathbf{i} + a_y\mathbf{j} + a_z\mathbf{k}.$$

Let $\mathbf{a}, \mathbf{b}, \mathbf{c}$ be arbitrary vectors with magnitudes a, b, c. Then

$$\mathbf{a} \times (\mathbf{b} + \mathbf{c}) = \mathbf{a} \times \mathbf{b} + \mathbf{a} \times \mathbf{c}$$

$$(s\mathbf{a}) \times \mathbf{b} = \mathbf{a} \times (s\mathbf{b}) = s(\mathbf{a} \times \mathbf{b}) \qquad (s = \text{a scalar}).$$

Let θ be the smaller of the two angles between $\mathbf{a}$ and $\mathbf{b}$. Then

$$\mathbf{a} \cdot \mathbf{b} = \mathbf{b} \cdot \mathbf{a} = a_x b_x + a_y b_y + a_z b_z = ab \cos \theta$$

$$\mathbf{a} \times \mathbf{b} = -\mathbf{b} \times \mathbf{a} = \begin{vmatrix} \mathbf{i} & \mathbf{j} & \mathbf{k} \\ a_x & a_y & a_z \\ b_x & b_y & b_z \end{vmatrix} = (a_y b_z - b_y a_z)\mathbf{i} + (a_z b_x - b_z a_x)\mathbf{j} + (a_x b_y - b_x a_y)\mathbf{k}$$

$$|\mathbf{a} \times \mathbf{b}| = ab \sin \theta$$

$$\mathbf{a} \cdot (\mathbf{b} \times \mathbf{c}) = \mathbf{b} \cdot (\mathbf{c} \times \mathbf{a}) = \mathbf{c} \cdot (\mathbf{a} \times \mathbf{b})$$

$$\mathbf{a} \times (\mathbf{b} \times \mathbf{c}) = (\mathbf{a} \cdot \mathbf{c})\mathbf{b} - (\mathbf{a} \cdot \mathbf{b})\mathbf{c}$$

Degrees	Radians	Sine	Tangent	Cotangent	Cosine		
0	0	0	0	∞	1.0000	1.5708	**90**
1	.0175	.0175	.0175	57.290	.9998	1.5533	89
2	.0349	.0349	.0349	28.636	.9994	1.5359	88
3	.0524	.0523	.0524	19.081	.9986	1.5184	87
4	.0698	.0698	.0699	14.301	.9976	1.5010	86
5	.0873	.0872	.0875	11.430	.9962	1.4835	**85**
6	.1047	.1045	.1051	9.5144	.9945	1.4661	84
7	.1222	.1219	.1228	8.1443	.9925	1.4486	83
8	.1396	.1392	.1405	7.1154	.9903	1.4312	82
9	.1571	.1564	.1584	6.3138	.9877	1.4137	81
10	.1745	.1736	.1763	5.6713	.9848	1.3963	**80**
11	.1920	.1908	.1944	5.1446	.9816	1.3788	79
12	.2094	.2079	.2126	4.7046	.9781	1.3614	78
13	.2269	.2250	.2309	4.3315	.9744	1.3439	77
14	.2443	.2419	.2493	4.0108	.9703	1.3265	76
15	.2618	.2588	.2679	3.7321	.9659	1.3090	**75**
16	.2793	.2756	.2867	3.4874	.9613	1.2915	74
17	.2967	.2924	.3057	3.2709	.9563	1.2741	73
18	.3142	.3090	.3249	3.0777	.9511	1.2566	72
19	.3316	.3256	.3443	2.9042	.9455	1.2392	71
20	.3491	.3420	.3640	2.7475	.9397	1.2217	**70**
21	.3665	.3584	.3839	2.6051	.9336	1.2043	69
22	.3840	.3746	.4040	2.4751	.9272	1.1868	68
23	.4014	.3907	.4245	2.3559	.9205	1.1694	67
24	.4189	.4067	.4452	2.2460	.9135	1.1519	66
25	.4363	.4226	.4663	2.1445	.9063	1.1345	**65**
26	.4538	.4384	.4877	2.0503	.8988	1.1170	64
27	.4712	.4540	.5095	1.9626	.8910	1.0996	63
28	.4887	.4695	.5317	1.8807	.8829	1.0821	62
29	.5061	.4848	.5543	1.8040	.8746	1.0647	61
30	.5236	.5000	.5774	1.7321	.8660	1.0472	**60**
31	.5411	.5150	.6009	1.6643	.8572	1.0297	59
32	.5585	.5299	.6249	1.6003	.8480	1.0123	58
33	.5760	.5446	.6494	1.5399	.8387	.9948	57
34	.5934	.5592	.6745	1.4826	.8290	.9774	56
35	.6109	.5736	.7002	1.4281	.8192	.9599	**55**
36	.6283	.5878	.7265	1.3764	.8090	.9425	54
37	.6458	.6018	.7536	1.3270	.7986	.9250	53
38	.6632	.6157	.7813	1.2799	.7880	.9076	52
39	.6807	.6293	.8098	1.2349	.7771	.8901	51
40	.6981	.6428	.8391	1.1918	.7660	.8727	**50**
41	.7156	.6561	.8693	1.1504	.7547	.8552	49
42	.7330	.6691	.9004	1.1106	.7431	.8378	48
43	.7505	.6820	.9325	1.0724	.7314	.8203	47
44	.7679	.6947	.9657	1.0355	.7193	.8029	46
45	.7854	.7071	1.0000	1.0000	.7071	.7854	**45**
	Cosine	Cotangent	Tangent	Sine		Radians	Degrees

1901	Wilhelm Konrad Röntgen	1845–1923	for the discovery of the remarkable rays subsequently named after him
1902	Hendrik Antoon Lorentz Pieter Zeeman	1853–1928 1865–1943	for their researches into the influence of magnetism upon radiation phenomena
1903	Antoine Henri Becquerel	1852–1908	for his discovery of spontaneous radioactivity

* See *Nobel Lectures, Physics,* 1901–1970, Elsevier Publishing Company, for the Nobel presentations, lectures and biographies. The attributions are, almost without exception, quotations from the Nobel Prize citations.

	Pierre Curie Marie Sklowdowska-Curie	1859–1906 1867–1934	for their joint researches on the radiation phenomena discovered by Professor Henri Becquerel
1904	Lord Rayleigh (John William Strutt)	1842–1919	for his investigations of the densities of the most important gases and for his discovery of argon
1905	Philipp Eduard Anton von Lenard	1862–1947	for his work on cathode rays
1906	Joseph John Thomson	1856–1940	for his theoretical and experimental investigations on the conduction of electricity by gases
1907	Albert Abraham Michelson	1852–1931	for his optical precision instruments and metrological investigations carried out with their aid
1908	Gabriel Lippmann	1845–1921	for his method of reproducing colors photographically based on the phenomena of interference
1909	Guglielmo Marconi Carl Ferdinand Braun	1874–1937 1850–1918	for their contributions to the development of wireless telegraphy
1910	Johannes Diderik van der Waals	1837–1923	for his work on the equation of state for gases and liquids
1911	Wilhelm Wien	1864–1928	for his discoveries regarding the laws governing the radiation of heat
1912	Nils Gustaf Dalén	1869–1937	for his invention of automatic regulators for use in conjunction with gas accumulators for illuminating lighthouses and buoys
1913	Heike Kamerlingh Onnes	1853–1926	for his investigations of the properties of matter at low temperatures which led, *inter alia*, to the production of liquid helium
1914	Max von Laue	1879–1960	for his discovery of the diffraction of Röntgen rays by crystals
1915	William Henry Bragg William Lawrence Bragg	1862–1942 1890–1971	for their services in the analysis of crystal structure by means of Röntgen rays
1917	Charles Glover Barkla	1877–1944	for his discovery of the characteristic Röntgen radiation of the elements
1918	Max Planck	1858–1947	for his discovery of energy quanta
1919	Johannes Stark	1874–1957	for his discovery of the Doppler effect in canal rays and the splitting of spectral lines in electric fields
1920	Charles-Édouard Guillaume	1861–1938	for the service he has rendered to precision measurements in Physics by his discovery of anomalies in nickel steel alloys
1921	Albert Einstein	1879–1955	for his services to Theoretical Physics, and especially for his discovery of the law of the photoelectric effect
1922	Niels Bohr	1885–1962	for the investigation of the structure of atoms, and of the radiation emanating from them
1923	Robert Andrews Millikan	1868–1953	for his work on the elementary charge of electricity and on the photoelectric effect
1924	Karl Manne Georg Siegbahn	1886–1954	for his discoveries and research in the field of x-ray spectroscopy
1925	James Franck Gustav Hertz	1882–1964 1887–1975	for their discovery of the laws governing the impact of an electron upon an atom

1926	Jean Baptiste Perrin	1870–1942	for his work on the discontinuous structure of matter, and especially for his discovery of sedimentation equillibrium
1927	Arthur Holly Compton	1892–1962	for his discovery of the effect named after him
	Charles Thomson Rees Wilson	1869–1959	for his method of making the paths of electrically charged particles visible by condensation of vapor
1928	Owen Willans Richardson	1879–1959	for his work on the thermionic phenomenon and especially for the discovery of the law named after him
1929	Prince Louis-Victor de Broglie	1892–	for his discovery of the wave nature of electrons
1930	Sir Chandrasekhara Venkata Raman	1888–1970	for his work on the scattering of light and for the discovery of the effect named after him
1932	Werner Heisenberg	1901–1976	for the creation of quantum mechanics, the application of which has, among other things, led to the discovery of the allotropic forms of hydrogen
1933	Erwin Schrödinger Paul Adrien Maurice Dirac	1887–1961 1902–	for the discovery of new productive forms of atomic theory
1935	James Chadwick	1891–1974	for his discovery of the neutron
1936	Victor Franz Hess	1883–1964	for his discovery of cosmic radiation
	Carl David Anderson	1905–	for his discovery of the positron
1937	Clinton Joseph Davisson George Paget Thomson	1881–1958 1892–1975	for their experimental discovery of the diffraction of electrons by crystals
1938	Enrico Fermi	1901–1954	for his demonstrations of the existence of new radioactive elements produced by neutron irradiation, and for his related discovery of nuclear reactions brought about by slow neutrons
1939	Ernest Orlando Lawrence	1901–1958	for the invention and development of the cyclotron and for results obtained with it, especially with regard to artificial radioactive elements
1943	Otto Stern	1888–1969	for his contribution to the development of the molecular ray method and his discovery of the magnetic moment of the proton
1944	Isidor Isaac Rabi	1898–	for his resonance method for recording the magnetic properties of atomic nuclei
1945	Wolfgang Pauli	1900–1958	for the discovery of the Exclusion Principle, also called the Pauli Principle
1946	Percy Williams Bridgman	1882–1961	for the invention of an apparatus to produce extremely high pressures, and for the discoveries he made therewith in the field of high-pressure physics
1947	Sir Edward Victor Appleton	1892–1965	for his investigations of the physics of the upper atmosphere, especially for the discovery of the so-called Appleton layer
1948	Patrick Maynard Stuart Blackett	1897–1974	for his development of the Wilson cloud chamber method, and his discoveries therewith in the fields of nuclear physics and cosmic radiation

1949	Hideki Yukawa	1907–	for his prediction of the existence of mesons on the basis of theoretical work on nuclear forces
1950	Cecil Frank Powell	1903–1969	for his development of the photographic method of studying nuclear processes and his discoveries regarding mesons made with this method
1951	Sir John Douglas Cockcroft	1897–1967	for their pioneer work on the transmutation of atomic nuclei by artificially accelerated atomic particles
	Ernest Thomas Sinton Walton	1903–	
1952	Felix Bloch	1905–	for their development of new methods for nuclear magnetic precision methods and discoveries in connection therewith
	Edward Mills Purcell	1912–	
1953	Frits Zernike	1888–1966	for his demonstration of the phase-contrast method, especially for his invention of the phase-contrast microscope
1954	Max Born	1882–1970	for his fundamental research in quantum mechanics, especially for his statistical interpretation of the wave function
	Walther Bothe	1891–1957	for the coincidence method and his discoveries made therewith
1955	Willis Eugene Lamb	1913–	for his discoveries concerning the fine structure of the hydrogen spectrum
	Polykarp Kusch	1911–	for his precision determination of the magnetic moment of the electron
1956	William Shockley	1910–	for their researches on semiconductors and their discovery of the transistor effect
	John Bardeen	1908–	
	Walter Houser Brattain	1902–	
1957	Chen Ning Yang	1922–	for their penetrating investigation of the so-called parity laws which has led to important discoveries regarding the elementary particles
	Tsung Dao Lee	1926–	
1958	Pavel Aleksejevič Čerenkov	1904–	for the discovery and the interpretation of the Čerenkov effect
	Il' ja Michajlovič Frank	1908–	
	Igor' Evgen'evič Tamm	1895–1971	
1959	Emilio Gino Segrè	1905–	for their discovery of the antiproton
	Owen Chamberlain	1920–	
1960	Donald Arthur Glaser	1926–	for the invention of the bubble chamber
1961	Robert Hofstadter	1915–	for his pioneering studies of electron scattering in atomic nuclei and for his thereby achieved discoveries concerning the structure of the nucleons
	Rudolf Ludwig Mössbauer	1929–	for his researches concerning the resonance absorption of γ-radiation and his discovery in this connection of the effect which bears his name
1962	Lev Davidovič Landau	1908–	for his pioneering theories of condensed matter, especially liquid helium
1963	Eugene P. Wigner	1902–	for his contributions to the theory of the atomic nucleus and the elementary particles, particularly through the discovery and application of fundamental symmetry principles
	Maria Goeppert Mayer	1906–1972	for their discoveries concerning nuclear shell structure
	J. Hans D. Jensen	1907–1973	

1964	Charles H. Townes	1915–	for fundamental work in the field of quantum electronics which has led to the construction of oscillators and amplifiers based on the maser-laser principle
	Nikolai G. Basov	1922–	
	Alexander M. Prochorov	1916–	
1965	Sin-Itiro Tomonaga	1906–	for their fundamental work in quantum electrodynamics, with deep-ploughing consequences for the physics of elementary particles
	Julian Schwinger	1918–	
	Richard P. Feynman	1918–	
1966	Alfred Kastler	1902–	for the discovery and development of optical methods for studying Hertzian resonance in atoms
1967	Hans Albrecht Bethe	1906–	for his contributions to the theory of nuclear reactions, especially his discoveries concerning the energy production in stars
1968	Luis W. Alvarez	1911–	for his decisive contribution to elementary particle physics, in particular the discovery of a large number of resonance states, made possible through his development of the technique of using hydrogen bubble chamber and data analysis
1969	Murray Gell-Mann	1929–	for his contributions and discoveries concerning the classification of elementary particles and their interactions
1970	Hannes Alvén	1908–	for fundamental work and discoveries in magneto-hydrodynamics with fruitful applications in different parts of plasma physics
	Louis Néel	1904–	for fundamental work and discoveries concerning antiferromagnetism and ferrimagnetism which have led to important applications in solid state physics
1971	Dennis Gabor	1900–	for his discovery of the principles of holography
1972	John Bardeen	1908–	for their development of a theory of superconductivity
	Leon N. Cooper	1930–	
	J. Robert Schrieffer	1931–	
1973	Leo Esaki	1925–	for his discovery of tunneling in semiconductors
	Ivar Giaever	1929–	for his discovery of tunneling in superconductors
	Brian D. Josephson	1940–	for his theoretical prediction of the properties of a super-current through a tunnel barrier
1974	Antony Hewish	1924–	for the discovery of pulsars
	Sir Martin Ryle	1918–	for his pioneering work in radioastronomy
1975	Aage Bohr	1922–	for the discovery of the connection between collective motion and particle motion and the development of the theory of the structure of the atomic nucleus based on this connection
	Ben Mottelson	1926–	
	James Rainwater	1917–	
1976	Burton Richter	1931–	for their (independent) discovery of an important fundamental particle.
	Samuel Chao Chung Ting	1936–	
1977	Philip Warren Anderson	1923-	for their fundamental theoretical investigations of the electronic structure of magnetic and disordered solids
	Nevill Francis Mott	1905-	
	John Hasbrouck Van Vleck	1899-	

index